ENTOMOLOGY IN HUMAN AND ANIMAL HEALTH

ENTOMOLOGY IN HUMAN AND ANIMAL HEALTH

Seventh Edition

Robert F. Harwood, Ph. D.

Professor of Entomology and Chairman,
Department of Entomology,
Washington State University,
Pullman

Maurice T. James, Ph. D.

Professor Emeritus of Entomology,
Washington State University,
Pullman

Macmillan Publishing Co., Inc.
NEW YORK

Collier Macmillan Canada, Ltd.
TORONTO

Baillière Tindall
LONDON

Earlier editions entitled *Medical and Veterinary Entomology,* by W. B. Herms,
copyright 1915 and 1923 by Macmillan Publishing Co., Inc.,
copyright renewed 1943 by William B. Herms and 1951 by Lillie M. Herms.
Earlier editions entitled *Medical Entomology* by W. B. Herms,
copyright 1939 and 1950 by Macmillan Publishing Co., Inc.,
copyright renewed 1967 by William M. Herms.
Earlier edition entitled *Medical Entomology* by William B. Herms
and Maurice T. James, © Macmillan Publishing Co., Inc., 1961.
Earlier edition entitled *Herms's Medical Entomology* by Maurice T. James
and Robert F. Harwood, © Copyright Macmillan Publishing Co., Inc., 1969.

MACMILLAN PUBLISHING CO., INC.
866 THIRD AVENUE, NEW YORK, NEW YORK 10022

COLLIER MACMILLAN CANADA, LTD.
BAILLIÈRE TINDALL • LONDON

Library of Congress Cataloging in Publication Data

Harwood, Robert Frederick (date)

Entomology in human and animal health.

Bibliography: p.
Includes index.
1. Arthropod vectors.   I.   James, Maurice Theodore
(date) joint author.   II.   Herms, William Brodbeck,
1876–1949.   Herms's medical entomology.   II.   Title.
RA641.A7H37   1979   614.4′32   78-21685
ISBN 0-02-351600-3
Printing: 1 2 3 4 5 6 7 8   Year: 9 0 1 2 3 4 5

PREFACE

The fields of medical and veterinary entomology continue to amass more information and to encompass innovations, particularly regarding procedures for control. Perhaps most important among recent trends is the amalgamation of arthropod life history and behavior information, as well as awareness of problems with environmental disruptions, into the development of more holistic control schemes. It is this more-inclusive approach that has motivated the change in title and orientation in this seventh edition. Although emphasizing entomological aspects of arthropod-associated diseases and discomfort, this revision attempts to draw attention to such techniques as immunization and chemotherapy as effective and practical means of disease control. As in previous editions, the book is designed not only as a text for introductory and advanced instruction in the fields of medical-veterinary entomology and certain aspects of public health but also as a general reference for established professionals. To meet the requirements of a broader approach we have increased the information on animal health problems and included additional emphasis on outdoor recreational concerns. We have also tried to be more worldwide in viewpoint, realizing that arthropod-associated health problems are readily introduced from one region to another, indeed even between hemispheres.

To provide a rapid summary and overview of important and often complex diseases, we have included a standardized introductory statement for what we consider to be the more important examples. In most instances a distribution map is included for these examples. Such maps help to visualize at a glance the extent of disease problem areas, but they can also be somewhat misleading by not suggesting the relative abundance of clinical cases within the overall distribution.

Several changes have been necessary in order to conserve space. The general task of literature review has been eased through the increasing availability of excellent review articles. We have concentrated on such reviews in the belief that they also provide references to original papers. Unfortunately, it has also been necessary to reduce the number of citations to the older and more basic literature, but in many instances the reader will find these works listed in the bibliography of the previous edition. Our basic means of literature review was through a systematic survey of entries in the *Review of Applied Entomology, Series B,* with additional citations gleaned through correspondence, chapter reviews by specialists, an examination of new issues of journals, and other sources. In the control chapter we have attempted to confine our discussion of the subject to basic principles, with specific procedures for a particular group of arthropods stated in the relevant chapters. As to scientific names, authors of species are listed in the index only.

The source of illustrations is credited in the accompanying legend. However, special acknowledgment is made to Ken Gray and Pacific Products Corporation for making available a number of excellent color slides. Roger D. Akre, Washington State University, and certain students under his direction, converted these and other slides into black-and-white prints suitable for publication.

Our appreciation is offered to the many teachers and research professionals who responded to a questionnaire that we distributed in order to obtain ideas for the present edition. Discussions were also held with several other professionals and students. Improvements can in large part be attributed to these and specifically credited contributions; errors and shortcomings are our responsibility. Specific assistance for chapter reviews is gratefully acknowledged to R. D. Akre, Washington State University, Pullman, Washington; A. R. Barr, University of California, Los Angeles, California; K. J. Capelle, Brigham City, Utah; D. G. Cochran, Virginia Polytechnic Institute and State University, Blacksburg, Virginia; K. R. Depner, Canada Agriculture, Lethbridge, Alberta; B. F. Eldridge, Oregon State University, Corvallis, Oregon; D. P. Furman, University of California, Berkeley, California; B. Greenberg, University of Illinois, Chicago Circle, Chicago, Illinois; E. J. Hansens, Rutgers University, New Brunswick, New Jersey; H. Hoogstraal, U.S. Naval Medical Research Unit 3, Cairo, Egypt; H. L. Keegan, University of Mississippi Medical Center, Jackson, Mississippi; K. C. Kim, Pennsylvania State University, University Park, Pennsylvania; W. L. Krinsky, Yale University, New Haven, Connecticut; S. B. McIver, Toronto University, Toronto, Ontario; L. L. Pechuman, Cornell University, Ithaca, New York; R. E. Ryckman, Loma Linda University, Loma Linda, California; J. G. Stoffolano, Jr., University of Massachusetts, Amherst, Massachusetts; C. H. Schaefer, University of California, Berkeley, California; T. J. Spilman, USNM, Washington, D.C.; R. Traub, Bethesda, Maryland; W. J. Turner, Washington State University, Pullman, Washington; R. A. Ward, Smithsonian Institution, Washington, D.C.; D. G. Young, University of Florida, Gainesville, Florida; and R. Zeledón, University of Costa Rica, San Jose, Costa Rica. Other substantial help was received from C. M. Clifford, Jr., Rocky Mountain Laboratory, U.S. Public Health Service, Hamilton, Montana; N. Gratz, World Health Organization, Geneva, Switzerland; S. McKeever, Georgia Southern College, Statesboro, Georgia; and R. Pal, World Health Organization, Geneva, Switzerland.

One of us (R.F.H.) was given the opportunity to write in a creative atmosphere during a month of residence at the Villa Serbelloni, Bellagio, Italy, a study and conference center of the Rockefeller Foundation.

ROBERT F. HARWOOD
MAURICE T. JAMES

CONTENTS

ENTOMOLOGY IN HUMAN AND ANIMAL HEALTH

1

INTRODUCTION

Many insects and other arthropods are of medical and veterinary importance in that they either cause pathological conditions or transmit pathogenic organisms to man or animals. The arthropods involved may be (1) causal agents themselves (e.g., the scabies mite); (2) developmental transfer hosts (e.g., certain beetles that serve in this capacity for helminth parasites); or (3) vectors of pathogens (e.g., *Anopheles* mosquitoes in relation to malaria). The study of these conditions includes broad aspects of the biology and control of the offending arthropods and recognition of the damage they do and of the way that they do it. The bearing on public and individual health and on the health of domestic and wild animals is obvious. Both mental and physical health, plus general comfort and well-being, are concerned.

GENERAL CONSIDERATIONS

Relation to Other Sciences. The study of insects and their relatives in relation to human health was termed *medical entomology* in about 1909. Today, fields of medical and veterinary entomology are complex and specialized, and it is impossible to be wholly proficient in all the other relevant sciences and disciplines. For investigation of the broader aspects of the subject, teams of researchers rather than individuals, or cooperation with specialists in other areas, is often necessary.

Entomology, along with some arachnology and general arthropodology, furnishes the biological and taxonomic foundation of this field, but a basic knowledge of mammals and birds is required, and the lower vertebrates are not without importance. In respect to invertebrates other than arthropods, a knowledge of the several helminth phyla and of the Protozoa, including some understanding of their life cycles, is of great importance. Aquatic invertebrates of other phyla may be involved in the study of life histories or in determining public health implications, for example, when these animals are found in domestic water supplies or in water with which humans or domestic animals have contact. Finally, it is impossible to discuss arthropods in relation to disease without assuming that the student has some background in microbiology, including bacteriology, virology, protozoology, epidemiology, and virology.

Ecological relationships involving medically important arthropods are often very complex. Many illustrations of this are indicated in this text; the specific complexities, however, are often so great that only a drastic abbreviation of them can be presented here. As examples one might consider the problems involved in understanding the relationships of mosquitoes and plasmodia to malaria and of tsetse and trypanosomes to human sleeping sickness and nagana. Many factors are involved, including the species of vectors and mammals concerned and the biologies and ecology of all the interacting organisms. Human factors affecting the dynamics of these diseases involve human ecology and nutrition, economics, and sociological problems such as superstitions, taboos, and established life patterns (cf. McKelvey, 1973, regarding African trypanosomiases). Above all, current emphasis on the combination of tactics for developing pest management strategies to control arthropods requires a sound understanding of ecological principles.

Engineering problems may be involved in such matters as controlling mosquitoes in areas where drainage, filling, and other management of waters are needed, or controlling filth-breeding or filth-frequenting flies in areas where sewage, garbage, or excrement is disposed of. Training in the physical sciences is useful in insecticidal investigations, physiological studies, and radiological sterilization of pests. Mathematical models and statistical methods may be needed. An understanding of anatomy, pathology, histology, hematology, toxicology, and dermatology is helpful. To aid in the control decision process, it is well to understand economic principles, as they affect human productivity, recreational developments, and the production of meat and dairy products.

Objectives. The aim of medical and veterinary entomology is to control, prevent, and, if possible, eventually eradicate arthropod-related human and animal diseases. To attain this objective, and, consequently, within the general scope of objectives, knowledge of all kinds is pursued concerning the arthropods involved, their bi-

ology and control, and the pertinent human and animal factors.

Development of Arthropod-Vertebrate-Pathogen Relationships. For millennia prior to the appearance of any human records of arthropod-associated annoyance and disease, systems of casual and purposive relationships evolved between arthropods and vertebrate hosts. Microorganisms and metazoan parasites exploited these developing relationships, probably through several pathways. The subject is of great interest and of practical value in understanding present-day problems. Most of our understanding is based on studying present-day circumstances and speculating how these arose, though the fossil record leaves some clues about arthropods. It is not possible to devote much space in this text for a discussion of the subject, so the interested student is referred to the previous text edition (James and Harwood, 1969) for some additional details. A few selected comments are entered here, and certain specifics of evolutionary development are mentioned in chapter accounts of groups of arthropods.

Insects and related arthropods preceded man on earth by at least 400 million years, and microorganisms were present long before that. It is natural to surmise that arthropods and microorganisms established a variety of relationships, and that the progenitors of man were next involved in these systems. But this is not necessarily the case, for there has been ample time for microorganisms and metazoan parasites to become associated with man and other vertebrates, and for arthropods to become secondarily involved. The key connections lie perhaps in how arthropods developed the blood- and tissue-feeding habit, and the kinds of parasitic relationships that evolved between arthropods and vertebrates.

Parasitism of vertebrates by terrestrial arthropods seems to have begun chiefly and most successfully in lairs, nests, and other host habitations. Parasitism may have developed through arthropods scavenging on lair detritus (mites and lice) or their predation on other lair invertebrates (triatomine bugs), coupled with a general tendency of many arthropods to probe and attempt to feed on a variety of substrates. From these circumstances certain arthropods could have thrived on vertebrate hosts that were a previously unexploited niche providing dependable and nutritious food. As humans, who are essentially tropical creatures, pushed into temperate regions, they modified their surrounding environment for habitation. Associated arthropods were also favored by the food, warmth, and shelter provided, and some developed close associations without feeding on humans, a loose form of dependence termed SYNANTHROPY. Synanthropy is noted in house-infesting cockroaches, some flies, and various ants. Several structural adaptations of arthropods were present, or developed to improve the parasitic existence. These include piercing-sucking mouth parts for partaking of blood or other tissue fluids, winglessness for ectoparasitism, and diminutive size to permit invasion of the vertebrate body. Nutritional dependency included the need for a high protein intake, such as that available from blood meals, for eggs to develop properly in female arthropods.

HISTORICAL REVIEW

Man's awareness of medically important insects undoubtedly goes back to prehistoric times, although then he was concerned primarily with pests that directly annoyed him. Some records have come down to us through legends, art, and even speech. Some variations of the Pandora's box story have substituted lice and fleas for the troubles released by the unwise girl, thus attempting to explain how man became the victim of these annoyers. Hoeppli (1969) (Figs. 1-1, 1-2), Greenberg (1973), and Busvine (1976) pro-

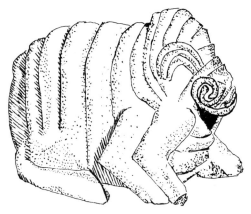

Fig. 1-1 Stone representation of flea, Mexico, ca. A.D. 1200–1500. (Drawn from a photograph in Hoeppli, 1969.)

vide a number of accounts and illustrations of man's awareness of insect attackers from the second millennium B.C. to more recent times. These problems included ectoparasites, filth-associated and biting flies, and diseases caused by arthropod-transmitted

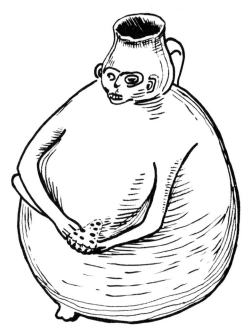

Fig. 1-2 Pottery vessel, Peru, ca. A.D. 400–900 showing holes where chigoe fleas have been removed from foot. (Drawn from a photograph in Hoeppli, 1969.)

Fig. 1-3 Mimbres pottery plate, ca. A.D. 1200. (Drawn from a photograph in Rodeck, 1932.)

pathogens. In approximately A.D. 1200 an American Indian artist unmistakably depicted on a piece of Mimbres, New Mexico, pottery a swarm of mosquitoes poised for attack (Fig. 1-3) (Rodeck, 1932). The Flathead Indians of Montana were obviously aware that maggots were the young of flies, since they used the same term, *xelmalten,* to designate both (Weisel, 1952).

Scant and vague biblical accounts refer to "a grievous swarm of flies" that entered the house of Pharaoh and "corrupted the land" (Exodus 8:24); possibly to lice, fleas, and scabies mite (scab); to vermin (possibly lice) that the Lord created from dust struck by the rod of Aaron (Exodus 8:16–17); and to an epidemic that killed more than fifty thousand Philistines in the city of Bethsames alone, believed to be plague because of "emerods" (buboes) and the abundance of "mice" (rats).

According to Busvine, we first find numerous references to ectoparasitic insects in classical Greece, because of their advanced literacy and their keen interest in the world about them. Legends involving lice date back to Homer (ca. 900 B.C.) and to Phalanthus, founder of Tarentum (eighth century B.C.). Aristotle knew about mites and was aware that one kind of them could be re-

moved from dry boils in the human skin; he did not associate them, however, with the causation of scabies.

The development of medical entomology begins with modern times, when it was made possible by the invention of the microscope, the rejection of spontaneous generation, and the germ theory of disease. However, by the sixteenth century it was known that mites not only lived in but caused "scabs" in the skin. As early as 1577 Mercurialis expressed the belief that flies carried the "virus" of plague from the bodies of those who were ill or had died of the disease to the food of other persons. Although his theory was erroneous in its specific aspects, he did recognize the principal role that flies play as vectors, that is, that they are food contaminators. In 1587 the Brazilian Gabriel Soares de Souza stated, in reference to yaws (*Framboesia tropica*) that flies suck the poisons from sores and leave them in skin abrasions on healthy individuals, thus infecting many persons. Edward Bancroft in 1769 advanced a similar theory, and in 1907 Castellani proved the essential correctness of this by demonstrating experimentally that flies do transmit *Treponema pertenue,* the causal agent of yaws.

An early record of a relationship between insects and disease in America dates back to 1764 (Herrer and Christensen, 1975), when the noted Spanish-born physician Cosme Bueno described succinctly but accurately Carrión's disease and cutaneous leishmaniasis in Peru and attributed both diseases to the bite of "a small insect called uta." The name "uta" is still used in parts of the Peruvian highlands both for cutaneous leishmaniasis and for the sand fly vector of its causative agent. The surprising fact about Bueno's record is that he attributed transmission to the bite of the correct insect, an inconspicuous and easily overlooked fly, though he knew nothing about microorganisms.

Although popular views in many parts of the world associated mosquitoes with various tropical fevers, no well-formulated ideas were presented until Josiah Nott, in 1848, published his belief that mosquitoes "gave rise to" both malaria and yellow fever. Beauperthuy, a French physician in the West Indies, expressed in 1854 the idea that in yellow fever an unknown disease factor was carried by the mosquito from certain decomposing matter and introduced into the human body.

Although, according to Howard (1921), no standard medical treatise prior to 1871 mentioned any specific disease as related to insect transmission of its pathogen, Raimbert (1869) showed that by inoculation of proboscids, wings, and other parts of nonbiting flies into guinea pigs anthrax could be disseminated; this discovery had been foreshadowed by a theory proposed by Montfils, almost a century earlier, in 1776. The formulation of the germ theory of disease by Pasteur, in 1877, led to the extensive activity of the late nineteenth century and the first decade of the twentieth, which saw the birth of medical entomology as a science.

Late Nineteenth Century. Early naturalists and physicians were fairly well informed concerning the larger intestinal parasites, such as roundworms and tapeworms, and these helminths were not unfamiliar objects by the late nineteenth century. The first discovery of primary importance in medical entomology was the observation by Patrick Manson, in 1878, of the development of the nematode, *Wuchereria bancrofti,* in the body of the mosquito, *Culex pipiens quinquefasciatus.* Eventually Manson, in cooperation with Bancroft, Low, and others proved the mosquito to be the intermediate host and vector of the causative organism of human filariasis.

Three important discoveries within the next fifteen years established a new basis for control and prevention of disease in both man and domestic animals. Laveran, in 1880, found a causal organism of human malaria, *Plasmodium malariae,* living parasi-

tically in the red blood cells of man. In 1889 Theobald Smith discovered the causal protozoon of Texas cattle fever, *Babesia bigemina,* also within the red blood cells of the host. In 1893 Smith and F. L. Kilbourne made the third great fundamental discovery in this field, namely that the cattle tick, *Boophilus annulatus,* is the necessary developmental host of the causal agent of Texas cattle fever.

In quick succession there followed a series of famous discoveries. Bruce in 1895 investigated nagana and established that the pathogen is conveyed from animal to animal through the bite of the tsetse, *Glossina morsitans.*

Ronald Ross, in 1897, reported finding the zygotes of the malaria parasite in two "dapple-winged mosquitoes" that he had bred from larvae and fed on a patient whose blood contained crescents. His chief contribution, however, was the discovery of the complete host cycle of the causal organism of bird malaria, with the establishment of the bird-to-mosquito-to-bird cycle. Grassi, Bignami, and Bastianelli in 1899 proved that the human malaria parasites were transmitted by a particular genus of mosquitoes, *Anopheles,* and Sambon and Low, in 1900, demonstrated the fact of transmission without a doubt.

Carlos Finlay, a Cuban physician, as early as 1880 propounded the theory that mosquitoes transmit the yellow fever pathogen, and he conducted experiments in an attempt to prove it. His efforts were not taken seriously at the time. In 1900 the United States Yellow Fever Commission, consisting of Walter Reed, James Carroll, Jesse W. Lazear, and A. Agramonte, in one of the outstanding achievements in the field of experimental medicine, proved conclusively, in Cuba, that the yellow fever pathogen is carried by the mosquito *Aedes aegypti.* In the course of these experiments, Doctor Lazear lost his life.

The discoveries concerning malaria and

yellow fever gave great impetus to the consideration of mosquito control, although Howard had already demonstrated the value of kerosene as a larvicide in the Catskill Mountains in 1882. Howard's pioneer book, *Mosquitoes: How They Live; How They Carry Disease; How They Are Classified; How They May Be Destroyed,* appeared in 1901. The first comprehensive treatise dealing with arthropods in relation to disease belongs to this historical period: it is that of Nuttall (1889), now generally considered the founder of medical entomology.

Mosquitoes in the Twentieth Century. Little progress was made in the knowledge of malaria and yellow fever during the first third of the twentieth century. The complete solution of the problem of control of both diseases seemed to be within reach—that is, simply mosquito control. However, what had originally appeared as simple was complicated, in the case of malaria, by the discovery of the *Anopheles maculipennis* complex of mosquitoes in Europe. Members of this and other complexes are found to differ behaviorally in their importance as vectors of human malarial parasites. To these aspects of the problem one must add other significant advances, notably the discovery by Huff and his associates of the exoerythrocytic cycle of the malaria parasite and the advances in our knowledge of the epidemiology of the malarias.

Jungle yellow fever was first observed in the Valle de Chanaan, Espirito Santo, Brazil, in 1932. It differed from the previously known yellow fever type only in that it occurred under conditions that suggested infection took place away from human habitats and that man might not be essential in the continuity of the infection. Indeed, as Soper (1936) remarked, "man may be but an accident in the course of an epizootic in the lower animals." With the decreasing importance of classical yellow fever that resulted from its successful control and the ever-im-minent threat of the jungle type, increasing attention was paid to the latter.

Graham (1902), working in Syria, found that dengue, or breakbone fever, a widely distributed essentially tropical disease, required transmission by mosquitoes. He, and later Ashbury and Craig (1907), reported that possibly several species of mosquitoes, notably *Aedes aegypti,* can transmit the dengue pathogen. Incrimination of mosquitoes in the transmission of viruses to man and animals has since that time become amazingly extensive and complex (see Chapter 10). Perhaps the most significant more-recent studies of mosquitoes in relation to viruses were made in the United States in the 1930s and early 1940s. The infectious agent of western encephalitis was isolated from sick horses by Meyer in 1930. In 1933 Kelser transmitted the virus of equine "encephalomyelitis" from inoculated guinea pigs to a horse by the bite of *Aedes aegypti.* The isolation of the virus of western equine encephalitis from wild *Culex tarsalis* by Hammon and associates (1941) and the subsequent establishment of the role of this mosquito in the transmission of that virus, constitute landmarks in the study of the epidemiology and epizootiology of the disease.

Mosquitoes have since the 1940s become the most widely and intensely studied arthropods of medical importance, and rightly so. The literature dealing with their taxonomy, biology, and relation to disease and pathogen transmission is voluminous.

Plague. A historical account of plague is included in Wu, Chun, Pollitzer, and Wu (1936). Epidemics and pandemics of this disease have been known since ancient times. The great pandemic in Europe in the fourteenth century claimed 25 million victims, one-fourth the population of the continent, and the London epidemic of 1666 killed 70,000 of the city's population of 450,000.

Modern progress in the study of plague epidemiology dates from the success of Simond (1898) in transmitting the plague pathogen from a sick rat to a healthy one through the agency of fleas. Though at first discredited, the validity of this discovery was confirmed early in the twentieth century. The designation SYLVATIC (selvatic), for the wild rodent or campestral form was proposed by Jorge (1928) to specify plague of wild rodents and their fleas.

Trypanosomiases. African sleeping sickness in man and nagana in cattle were recognized disease entities in Africa well before the turn of the century (see Chapter 12). In 1902 R. M. Forde observed certain parasites in the blood of persons suffering from Gambian sleeping sickness, which J. E. Dutton recognized as trypanosomes (protozoa) named *Trypanosoma gambiense*. Bruce and Nabarro (1903) showed that *Glossina palpalis* was the carrier, thus adding tsetse-transmitted pathogens to the list of arthropod-borne parasites. Stephens and Fantham (1910) described *Trypanosoma rhodesiense* as the causal organism of Rhodesian sleeping sickness, and Kinghorn and York (1912) proved *Glossina morsitans* to be a responsible vector.

In America, Carlos Chagas, in 1909, demonstrated that the conenose bug, *Panstrongylus megistus,* was a vector of *Trypanosoma cruzi,* the causative organism of what was to be known as Chagas' disease. Chagas had already demonstrated the presence of the trypanosome in the human host.

Lice. Although lice have for centuries been associated in the mind of man with filth and disease, apparently little thought was given to these insects as possible vectors, even though Melnikoff in 1869 had shown that the biting louse, *Trichodectes canis,* was a developmental host of the double-pored tapeworm, *Dipylidium caninum.* Mackie (1907), working in India, found that the relapsing fever organism was transmitted by, and multiplied in the body of, the body louse, *Pediculus humanus humanus.* Nicolle and associates (1909), working in Tunis, and Ricketts and Wilder (1910a), working in Mexico, proved experimentally that the louse is a carrier of the typhus organism, *Rickettsia prowazekii.*

Horse Flies and Gnats. Tabanids (horse flies, deer flies, gad flies) were viewed with suspicion as early as 1776, but apparently no satisfactory evidence against them was forthcoming until 1913 when Mitzmain, working in the Philippine Islands, demonstrated transmission of the surra pathogen of carabao by *Tabanus striatus.* Strong evidence that tabanids of the genus *Chrysops* are intermediate hosts of the nematode *Loa loa* was advanced in the same year by Leiper.

Tularemia, under the name of deer fly fever or Pahvant Valley plague, was investigated by Francis in Utah and was shown by Francis and Mayne (1921) to be carried from rodent to rodent by the tabanid *Chrysops discalis;* these authors presumed that transmission from rodent to man took place by the same means. The causal organism, *Francisella tularensis,* had been known since 1911, when it was identified as the cause of a plaguelike disease of rodents in California. The discoveries by Francis and his coworkers are significant even though involvement of arthropods in the transmission is not the chief factor in the epidemiology of tularemia.

Human onchocerciasis, or blinding filarial disease, has been of great concern in parts of Africa and tropical America. Robles, in Guatemala, was the first to suggest, in 1915, that its causal organism, the nematode *Onchocerca volvulus* was transmitted by black flies. In 1926 Blacklock, in Africa, reported *Simulium damnosum* as its vector.

Control. Control measures directed against arthropods of medical and veterinary importance are complex. Some few of the more significant developments are described in Chapter 5; a more extended discussion may be found in Philip and Rozeboom (1973).

Changing Attitudes. A historical review would not be complete without a consideration of the change of attitudes that has accompanied the development of medical and veterinary entomology as a science. Through the course of history, as well as among persons of different levels of civilization today, attitudes have varied from complete indifference to a conviction that all insects are evil and should be destroyed. Many insects and other arthropods are grossly misunderstood by the average person, and it is important to know which ones are harmless and which threaten the health and well-being of man and animals.

Spiders have generally inspired fear in man. In Italy, in the vicinity of Taranto, there occurred a spider scare during the seventeenth century that gave rise to the condition known as "tarantism." The bite of the European tarantula (not to be confused with the American species; see Chapter 17) was supposed to induce a kind of madness that expressed itself in frantic and extravagant contortions of the body that, if not violent enough, would lead eventually to death. The cure consisted of increasing the violence of these dances to the point where it became self-limiting. To aid in inciting the patient to activity that would bring relief, a musical dance, known as the tarantella, arose; under the pen of such classical composers as Chopin and Liszt, this was later to become an outstanding musical form.

Notable changes in attitude involve such household pests as the house fly and such intimate companions of man as the lice that infest him. West (1951) notes that our early attitude toward the house fly was one of "friendly tolerance"; a few flies were "nice things to have around, to make things seem homelike. . . . Those that were knocked into the coffee or the cream could be fished

out; those that went into the soup or hash were never missed" (Doane, quoted by West). This attitude prevails today to an extent among primitive peoples, who may look upon head lice, for example, as something akin to household pets.

During the last quarter of the nineteenth century there came a period of incrimination, during which many signal successes in research connected arthropods with the transmission of several pathogens. During the period of "popular education" preceding World War I the American public became aware of the true situation through the writings of L. O. Howard and others. The stimulation of the great successes led to some blind alleys, however, such as attempts to link the "transmission" of poliomyelitis with the stable fly, that of pellagra with the buffalo gnat, and that of cancer with cockroaches.

Between World War I and World War II activity decreased as a result of "false security," followed by a period of "stern necessity" during World War II, when the control of typhus, scrub typhus, dengue, malaria, and other arthropod-associated diseases became essential to the war effort. Unlike the aftermath of World War I, the importance of medical entomology continues to be recognized to an intense degree and on a worldwide scale, and veterinary entomology has developed as a field vital to the production of food and leather. Three aspects of this recognition are particularly important: first, the idea of "one world," in which rapid transit makes problems of one continent or geographical area of potential importance to others; second, the idea of eradication, rather than control, as applying either to arthropods themselves (e.g., *Anopheles gambiae* in America and the primary screwworms) or to arthropod-transmitted pathogens (e.g., those of malaria and filariasis); and third, the current thought in the developed world that pest control is important for improving standards of living, comfort, and mental health. Thus, the trend has completely reversed itself, from the idea that a few flies might even be desirable, through the belief that they were dangerous, finally to the idea that annoyance by arthropods does not have to be tolerated.

Impact Upon History. The foregoing has been but a brief sketch of the development of medical entomology up to the present time. Historical information is scattered among many publications; some notable sources are Pollitzer (1954), West (1951), Cushing (1957), and Schultz (1968). The classical works of Zinsser (1935) and Haggard (1929) are valuable references. Philip and Rozeboom (1973) give a recent, condensed treatment, including a very useful chronology of pioneer events in medico-veterinary entomology.

Many aspects of modern civilization have been influenced by man's past relations to insects, the pathogens they transmit, and the diseases that result from this transmission. How much such things have contributed to the present politicoeconomic status of the world is, of course, a matter of conjecture. It has been suggested that insect-transmitted pathogens, for example, trypanosomes transmitted by tsetse in America, may have aided in the extinction of segments of the late Tertiary and Quaternary faunas. Undoubtedly social and political development of prehistoric man was affected by epidemics and pandemics associated with arthropods, at least as soon as living aggregations and migratory bands of suitable size were formed. Records of the ravages of yellow fever, typhus, and plague go far back into history, and there is evidence of malaria from the Neolithic period (Mattingly, 1973). In historic times these diseases and others have been instrumental in the development and the decline and fall of empires, as well as in the settlement (or lack of such) of new areas and in the construction of engineering projects.

There is clear evidence that arthropod-associated diseases influenced the course of empire in Greece and Rome. The great plague of Athens in 430 B.C., thought by some to be typhus (Zinsser, 1935), demoralized the Athenians and terrorized the invading Peloponnesians, accounted for the death of Pericles, and probably extended the duration of the Great Peloponnesian War as much as did strategy. In 396 B.C., the Carthaginian siege of Syracuse failed because of an epidemic of what was probably the same disease. If that siege had succeeded, Carthage instead of Rome might have become the dominant power of the Mediterranean.

Malaria also affected Greece, ending the eastward extension of the empire of Alexander the Great and finally, complicated by alcoholism, leading to the death of the conquerer (Haggard, 1929). It is interesting speculation that malaria may have killed off a large part of the vigorous Grecian stock responsible for the country's Golden Age, which led to their replacement largely from the slaves brought in to serve them.

Zinsser describes the various epidemics that occurred within the Roman Empire from the first century A.D. to its fall in A.D. 476 and finally to the time of Justinian in the sixth century. These epidemics, probably of mixed arthropod- and nonarthropod-associated causes, were spread by the Roman armies and the invading Huns, Goths, and others. No doubt their influence on the fall of Rome was great. Zinsser believes the Plague of Justinian gave the *coup de grâce* to the ancient empire.

Though the Incan empire of South America was terminated by the Spanish conquistadores, there is evidence that epidemics, possibly Carrión's disease, may have reduced the Incas to a state of vulnerability at the time of the conquest (Schultz, 1968).

There is no question as to the effect of arthropod-transmitted pathogens on the progress of man of later times. During the Middle Ages and early modern times Europe experienced repeated pandemics of plague.

Malaria was rampant over most of the continent; in England Oliver Cromwell and, presumably, James I died of the disease, and in France Louis XIV contracted it but was cured with quinine. The Roman marshes were notorious for "mal' aria" (bad air). The trypanosomiases prevented development of Africa until the present century, and yellow fever, introduced to the New World via slave trade from Africa, prevented the development of much of the Americas, including earlier attempts to construct the Panama canal. Typhus was responsible, to a large extent, for the collapse of the Russian and Balkan fronts in World War I; except for DDT and louse control, serious epidemics, such as those that were brought under control at Naples, Italy, and Cologne, Germany, would have taken much greater tolls of human life in World War II.

Human attitudes and nutritional levels have contributed greatly to the serious effects of arthropod-associated diseases. Human prejudices and superstitions have stood and still stand in the way of control and eradication programs. As examples of nutritional relationships the American leishmaniases are of much greater clinical importance wherever the level of nutrition is low; malaria may sap the strength of rural populations, which then produce less food, with a consequent lowering of the nutritional level and increase of the susceptibility of the population to the disease. The effects of unsanitary conditions are obvious; where pestholes exist, the levels of sanitation and nutrition are low and those of parasitic and infectious diseases are high.

War and natural disasters, such as widescale floods and serious earthquakes, are important influences. Along with a decrease in sanitation resulting from mass migration, crowding, and a general interference with the ways of life, the usual control programs against medically important arthropods may be interrupted. For example, it was pointed out at the U.N. Food and Agricultural Organization seminar in Nairobi, Kenya, in 1977, that clashing armies and guerilla operations in many parts of Africa had interfered with or forced suspension of tsetse control operations, with the result that the flies were returning to many areas where they had been eradicated and were even invading previously unoccupied territories. Thus, full-scale epidemics of sleeping sickness and epizootics of nagana were threatened.

The present outlook must involve some thought of the future. Human lives saved by the conquest of arthropod-associated diseases add to our population problems. Through socioeconomic and environmental studies and education we must devise ways in which the desirable gains can be made on all sides. Thus some disciplines that may not seem related to the study of arthropods affecting human and animal health are, in actuality, definitely adjuncts to our science.

MEDICAL IMPORTANCE OF ARTHROPODS

The ways in which arthropods relate to human health and well-being can be classified in three divisions, as follows:

A. Arthropods as direct agents of disease or discomfort.
   1. Entomophobia (including delusory parasitosis)
   2. Annoyance and blood loss
   3. Accidental injury to sense organs
   4. Envenomization
   5. Dermatosis
   6. Myiasis and related infestations
   7. Allergy and related conditions
B. Arthropods as vectors of developmental hosts.
   1. Mechanical carriers (transmission more or less incidental)
   2. Obligatory vectors (involving some degree of development, within the arthropod)
   3. Intermediate hosts (in a passive ca-

pacity; if an intermediate host bites or otherwise seeks out a vertebrate host of a pathogen, it is considered a vector under category 2)
4. Phoretic carriers of offending arthropods
C. Arthropods as natural enemies of medically harmful insects.
1. Competitors
2. Parasites or predators

Entomophobia and Delusory Parasitosis. Insects, spiders, and various other arthropods, even wholly innocuous ones, frequently induce annoyance and worry that may lead to a nervous disorder, sometimes with sensory hallucinations, The milder manifestation, common among humans, is the urge to ''kill that bug,'' without consideration that it might even be beneficial. This should not be confused with the justifiable intolerance of insects that are a nuisance or a threat to man's welfare. Abnormal reactions occur in some persons in the presence of stinging insects or insects that look as though they might sting: the individual will flee wildly and beat madly at the insect, which may as a result become a pursuer.

In true entomophobia, not to be confused with mere squeamishness, the arthropod may evoke hysterical reactions or assume the form of ''bugs'' that jump and cause itching and irritation on the skin. Such disturbances may be of long duration. In some cases no actual arthropod is involved; the victim may produce evidence in the form of epidermal scales, cloth fragments, and other debris. This is called delusory parasitosis. These conditions should be viewed sympathetically and, if necessary, be referred to professional help.

Disturbances of this type are not unusual. The subject has been ably discussed by Pomerantz (1959) and Waldron (1962, 1972). Olkowski and Olkowski (1976) have reviewed the matter, particularly from the standpoint of the pest control specialist; they stress the importance of educating the public to put the matter into proper perspective.

Annoyance and Blood Loss. It is difficult to estimate the importance of annoyance by insects, yet everyone is aware of this source of nuisance and discomfort. A blow fly, noisily buzzing through the house, may be very irritating, and flies indoors or at the picnic table can annoy almost anyone. ''Fly worry'' or ''tick worry'' are recognized entities that result in reduced production of animals because of interference with their normal feeding and rest. Ants may spoil a picnic; so may the threat of yellow jackets and hornets, even though one might not actually be stung by them. There are many instances of serious vehicular accidents due to driver reaction to the entry of bees or wasps. Insect bites may cause economic loss, even death, to livestock; they are not as important to humans because of their better ability to protect themselves, though allergic responses to bites may be important.

Even Collembola (springtails) may be a source of annoyance to man. Scott and associates (1962) list 19 Nearctic species that have been reported as being intimately associated with man. One of these, *Orchesella albosa,* was reported to have infested the head and pubic areas of a Texas family but without causing dermatitis; however, these authors cite records involving the Australasian *Entomobrya tenuicauda* and the cosmopolitan *E. nivalis* in which dermatitis did occur.

Accidental Injury to Sense Organs. Insects of various species, especially small flying ones, may accidentally enter the eye or ear. Some of these, notably certain rove beetles, Staphylinidae, cause extreme pain because of irritating secretions. Many insects discharge odoriferous fluids or vapors that, in some instances, are so forcibly ejected that they may be thrown some distance. Roth and Eisner (1962) have reviewed the subject of the use of squirting mechanisms by insects and have added some original observations.

Injury to the eye may also result when persons are accidentally "struck" by the spiny larvae of the sheep bot fly, *Oestrus ovis* (see Chapter 13) or come in contact with urticating hairs of caterpillars (see Chapter 17), which may be carried by the wind.

Envenomization. This is discussed in Chapter 17.

Dermatosis. Various skin irritations are caused by arthropods, either by bites, simple contacts, secretions, or skin invasions. These are discussed in Chapter 17 and the other chapters dealing with the arthropods involved.

Myiasis and Related Infestations. An invasion of organs and tissues of man and vertebrate animals by larvae of Diptera is called MYIASIS. This subject is discussed in Chapter 13. Similar invasions by beetle larvae (CANTHARIASIS), moth larvae (SCOLECIASIS), and others are known to occur but are rare.

Allergy and Related Conditions. These are discussed in Chapter 17.

Arthropods as Vectors. The general role of arthropods as vectors of pathogens and as developmental hosts of parasites is discussed in Chapter 4.

Competitors. Medically important arthropods are no exception to the rule that competition between species for the same ecological niche is disadvantageous to those least able to withstand it. Predominance may depend in part on such competition; for example, the highly successful and medically important blow fly *Chrysomya megacephala* is a very able competitor. On the other hand, an ordinarily very successful species may suffer from it. The black soldier fly, *Hermetia illucens,* has been shown to compete in parts of California with the house fly, *Musca domestica,* to the detriment of the latter (Furman *et al.,* 1969). As these authors remark, biological control of the house fly by this competitor may be considered as trading one pest for another, but of the two the house fly is by far the more undesirable. An example of competition as a means of control involves the introduction of exotic scarabaeid beetles into Australia and elsewhere to compete for the dung in which the bush fly, horn fly, and other dung-associated pest flies breed.

Parasites or Predators. Many arthropods are parasites or predators of insects of medical and veterinary importance. The host or victim is subject to attack in any of its developmental stages. This subject is dealt with under the heading of biological control in Chapter 5, and several examples are provided in the control section of each chapter dealing with the specific arthropod groups.

TAXONOMIC SCOPE

The text follows the customary inclusion of medically important arachnids, centipedes, and millipedes within the scope of medical entomology. Consideration will also be given to certain other groups, especially the crustaceans, that may affect matters of human health and well-being. Crabs and other Malacostraca may serve as developmental hosts. A number of other Crustacea are important parasites of game and food fishes, and their effect on the human food supply, especially as the utilization of marine and fresh water food sources assumes more significance, is definitely of interest to us. A very good chapter in Schmidt and Roberts (1977) deals with this subject.

The tongue-worms, or Pentastomida, should also be mentioned. The taxonomic status of this group has been, and still is, a matter of controversy. The Pentastomida have been considered a separate phylum or a separate class of Arthropoda, but some authors consider them to be closely related to the crustacean subclass Branchiura. Parasitism of man may take one of two forms. Ingestion of eggs may lead to VISCERAL PENTASTOMIASIS, resulting in development of nymphs in the liver, spleen, lungs, and other organs. This is usually asymptomatic, but it may be more serious than is usually thought, and invasion of the eyes may lead to visual damage. Invasion of the region of the throat, nasal passages, larynx, eustachian tubes, and associated areas by infective nymphs of *Linguatula serrata* may lead to NASO-PHARYNGEAL PENTASTOMIASIS of cats, dogs, foxes, and other mammals, including man. The human disease, known as *halzoun* in the Near East, occurs also in Africa, the Indian subcontinent, China, and the islands of the East Indies. It causes severe disturbances of the region involved and sometimes results in death through asphyxiation. A more extended but concise discussion of these parasites and of parasitism by them is given by Schmidt and Roberts (1977). Cockroaches have been recorded by Lavoipierre and Rajamanickam (1973) as developmental hosts of pentastomids in lizards.

LITERATURE

The literature of medical and veterinary entomology has become so voluminous that it is difficult for the worker in one specialty area to keep abreast of more than the major developments in other areas. It has been our aim to supply the student with important citations under the individual topics discussed, but such citations have had to be selective, and some become obsolete with the passage of time.

The *Annual Reviews* series, published by Annual Reviews, Inc., Palo Alto, California, provides much valuable information for the medical entomologist. The *Annual Review of Entomology,* 1956 to date, includes articles that review recent developments or summarizations in some field of medical entomology in each of its issues; the *History of Entomology* (Smith *et al.,* 1973), distributed by these same publishers, is a standard reference. *The Annual Review of Microbiology,*

1947 to date, as well as some of the other *Reviews,* are of interest. In two other series, published by Academic Press, Inc., New York, the *Advances in Parasitology,* 1963 to date, and *Advances in Virus Research,* 1953 to date, there are many articles of special interest; the serious student will need to consult others in the series.

A number of books are available on restricted groups of arthropods affecting man and animals, or on aspects of the biology, behavior, and control of such groups, Many of these are mentioned in relevant succeeding chapters of this text. General books in English on medical entomology include titles by Leclercq (1969), K. G. V. Smith (1973), and Snow (1974). Cavallo-Serra (1973) has produced in French a heavily illustrated manual of arthropods of medical and veterinary interest. There is currently a lack of an overall manual concerned with arthropods affecting livestock, but several publications cover the parasites of laboratory animals (cf. Griffiths, 1971; D. Owen, 1972). Studies on arthropods affecting man and animals relate to other fields, such as urban entomology (Ebeling, 1975) and forensic medicine (Leclerq and Tinant-Dubois, 1973).

Periodical literature as one can easily see by consulting the "Cited References" section in this book, is widely scattered through many publications. The *Journal of Medical Entomology,* 1964 to date, applies directly to our subject, and *Mosquito News,* 1941 to date, has a wider scope than its title would indicate. Several abstracting and indexing media are available; one of these, an abstracting journal, the *Review of Applied Entomology, Series B,* 1913 to date, deals specifically with medical and veterinary entomology, and additional categorized titles or abstracts may be found in the *Cumulated Index Medicus,* 1960 to date, and *Tropical Disease Bulletin,* 1912 to date.

The taxonomic literature is vast, and necessary additions, revisions, and refinements occur constantly. Some of the more important taxonomic works are cited in the proper places in this text. For general treatments, see K. G. V. Smith (1973) and Beklemishev (1958, in Russian).

We have attempted to follow, as nearly as possible, standard nomenclature both in respect to arthropods of medical importance and the pathogens, hosts, reservoirs, and other organisms associated with them. Scientific names of arthropods in the text do not include describers; consult the index for complete nomenclature. Names of bacteria have been used in accordance with *Bergey's Manual of Determinative Bacteriology* (Buchanan and Gibbs, 1974). For viruses, we follow recent systems used in Berge (1975) and Fenner (1976). For English common names of North American insects we follow the list prepared by the Committee on Common Names of Insects of the Entomological Society of America (D. M. Anderson, 1975; Sutherland, 1977); for those not included in that list, we follow either what seems to be the best standard usage or what seems taxonomically most acceptable. No attempt has been made to standardize names other than English ones, although some commonly used names from other languages have been cited.

2

STRUCTURE, DEVELOPMENT, AND CLASSIFICATION OF INSECTS AND ARACHNIDS

THE INSECTA (HEXAPODA)

The Insecta (Hexapoda) constitute the largest class in numbers of species in the phylum Arthropoda, which in turn comprises a greater number of species than all other phyla of the animal kingdom combined. Various estimates as to the number of described species of insects in the world range from 625,000 to 1,500,000, and the number ultimately known will probably be much greater. As members of the phylum Arthropoda, insects share the following arthropod characteristics: segmented body with paired, segmented appendages; bilateral symmetry; dorsal heart; ventral nerve cord; and exoskeleton. The body of insects is divided into three more or less distinct parts, the HEAD, the THORAX, and the ABDOMEN. There are eighteen to twenty-two segments in the insect body, the variable number resulting from differences in the interpretation of the embryological evidence, especially concerning the number of segments that compose the head. However, because of the specialization of the head and posterior terminal segments, the number of clearly recognizable ones is usually much smaller. The HEAD of the adult insect bears a pair of antennae, the mouth parts, and the eyes. The THORAX bears the locomotor appendages, namely, three pairs of segmented legs, and in addition usually two pairs of wings (which are morphologically not appendages), one or both of which may be absent or non-functional as locomotor structures. The ABDOMEN bears no appendages except the terminalia, sometimes cerci, and, in the Collembola, specialized locomotor structures. Respiration is accomplished by means of a complex system of microscopic tracheal tubules that open through the body wall and carry air directly to all parts of the body of the insect. Immature insects belonging to many of the orders, for example, maggots of flies, differ markedly from the mature forms, but almost all possess tracheae.

The discussion of external insect morphology presented here concerns especially the structures that are important in insect classification (Fig. 2-1). Greater details may be obtained from textbooks of general entomology such as Romoser (1973) or Imms (1957).

15

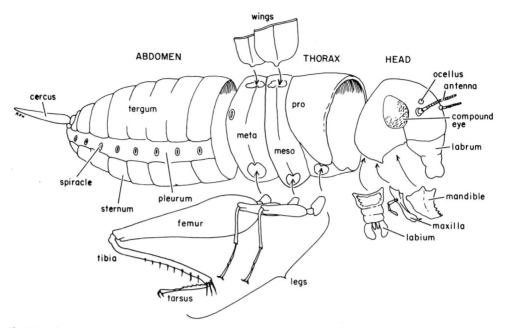

Fig. 2-1 General external structure of an insect.

Wings. The earliest systems of insect classification were based on wing characters; today these, together with mouth parts and metamorphosis, still afford the major elements in the bases for modern classification. The venation of the insect wings is so markedly characteristic for most species that even a part of a wing is sometimes all that is necessary for identification. There are typically two pairs of wings present, situated on the mesothorax and metathorax, although in many parasitic insects, such as the bed bugs, lice, fleas, certain louse flies, and so forth, the wings are vestigial or absent. The wingless insects just mentioned should, of course, not be included with the Apterygota, which are a group of primitively wingless (APTEROUS) forms, of which only one order, the Collembola (considered by some to be distinct from the insects), will be given any consideration here.

In form the wing presents a more or less triangular appearance. The three sides are called margins: the COSTAL margin is anterior; the APICAL and ANAL margins are posterior, the former extending from the wing apex to the area of the anal veins, the latter extending from that area to the wing base. Generally the fore and hind wings differ considerably in size; the fore wing in some groups, such as the mayflies, many butterflies and moths, and the bees and wasps, is larger than the hind wing; in the grasshoppers, cockroaches, beetles, and some others, however, the fore wing is narrow and serves largely as a cover to the hind wing, which folds fanlike. In the dragonflies, termites, and ant lions the fore and hind wings are nearly equal. In the flies, the hind pair of wings is replaced by club-shaped structures known as HALTERES, leaving only one pair of wings, hence the name Diptera (two-winged). In the calyptrate Diptera and some other flies there are present two pairs of lobes (SQUAMAE, also called ALULAE or CALYPTERS) at the junction of the wings and the thorax. One of these squamae, the THORACIC, is more closely associated with the wall of the thorax; the other, the ALAR, is more closely associated with the base of the wing.

There are some differences in the structure of the wings within an order, although for each order a certain general pattern prevails; for example, the Neuroptera have thin membranous wings, often quite filmy, and containing numerous veins and cells; however, the wings of Diptera and many Hemiptera have the same texture throughout, but possess fewer and differently arranged veins. Most Diptera can, of course, be readily distinguished from all but a few insects (such as the male coccids, the Strepsiptera, and the two-winged mayflies) by the presence of but a single pair of wings. In winged Hemiptera the front wings are thickened at the base and the apical portion is membranous. In the Homoptera, the two pairs of wings are of more or less even texture throughout.

The VENATION of the insect wing is an important element in classification, because of the great variety of arrangements and the reliability of this character for the identification of the family, genus, and sometimes even the species. The VEINS are hollow, riblike structures that give strength to the wing. The areas of membrane between the

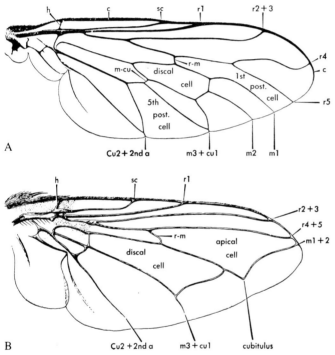

Fig. 2-2 Application of the Comstock-Needham system of nomenclature to two types of dipterous wing (see also Fig. 9-1). The longitudinal veins (with the nomenclature used by those dipterists who follow the old system of numbered veins given in parentheses) are as follows: *c*, costa (costal); *sc*, subcosta (auxillary); *r*, radius (*r*1 = 1st longitudinal, *r*2 + 3 = 2nd longitudinal, *r*4 and *r*5 = 3rd longitudinal); *m*, media (*m*1 and *m*2 = 4th longitudinal, *m*3 = part of 5th longitudinal); *cu*, cubitus (*cu*1 = part of 5th longitudinal, *cu*2 = part of 6th longitudinal; *2nd a*, 2nd anal (part of 6th longitudinal). The 1st anal vein occurs only as a fold in the membrane. The cross veins are: *h*, humeral; *r–m*, radiomedial (anterior or small cross vein); *m–cu*, mediocubital.

A. Wing of Thereva, illustrative of the more primitive type of venation. (USDA drawing.)

B. Wing of *Cochliomyia hominivorax*, illustrative of the muscoid type. (USDA drawing.)

veins are called CELLS; they are said to be OPEN if the membranous area extends to the wing margin, and CLOSED if the cell is surrounded on all sides by veins. By a careful study of the evidence, a fundamental type of wing venation has been constructed by Comstock and Needham and revised by Tillyard. Figure 2-2 illustrates this nomenclature as applied to a primitive type of dipterous wing (Fig. 2-2*A*) and as modified in the muscoid type (Fig. 2-2*B*).

Metamorphosis. To achieve the size and development of the parent, the young insect undergoes greater or less change in size, form, and structure. This series of changes is termed METAMORPHOSIS. The least change is found in the primitively wingless Apterygota (e.g., silverfish, springtail), hence, the newly emerged young individual is externally unlike the parent mainly in size; this type of development is WITHOUT METAMORPHOSIS (Fig. 2-3).

A slightly more obvious metamorphosis occurs in such insects as the true bugs. Not only is there a great difference in size, but also the absence of wings in the young is at once apparent. To reach the winged condition, the young individual casts its skin at intervals and with each ECDYSIS achieves longer wing pads until, after a certain number of molts, the fully developed wings appear. The following stages may be recognized: (1) EGG; (2) NYMPH; and (3) IMAGO, or sexually mature adult. There is, of course, a series of nymphal forms, or INSTARS, one after each molt. This type of metamorphosis is called SIMPLE (Fig. 2-4), and the orders in which this type occurs are known as the Heterometabola.

The greatest difference between the newly hatched young and the parents occurs in such higher insects as flies (Fig. 2-5) and fleas (Fig. 14-1). In these forms the newly hatched insect has no resemblance whatsoever to the adult, but in many cases bears a superficial resemblance to a segmented worm. However, the internal morphology and certain other features are distinctly insectan. The fact that the young are mandibulate and the adults haustellate in Diptera and Siphonaptera indicates an adaptation to different ways of life in the immature and adult forms. To attain the winged condition of the adult, the wingless, wormlike form must undergo many profound changes, and a new stage is interjected, the PUPA, or resting stage, in which this transformation is accomplished. The newly hatched young insect emerging from the egg is called a LARVA, hence the following stages: (1) egg, (2) larva, (3) pupa, and (4) imago. As in insects with simple metamorphosis, there is here again a series of larval instars. This type of metamorphosis is termed COMPLEX metamorphosis, and the orders in which it occurs are known as Holometabola.

Functional wings, and consequently the ability to fly, occur only in adult insects, with one exception, the mayflies, order Ephemeroptera. In this order, the preadult, fully winged SUBIMAGO molts once more to form the adult. The subimaginal skins,

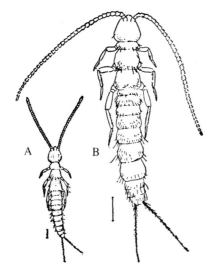

Fig. 2-3 Illustrating ametabolous development (without metamorphosis). *A.* Young of *Campodea* (Order Entotrophi). *B.* Adult of same. (After Kellogg.)

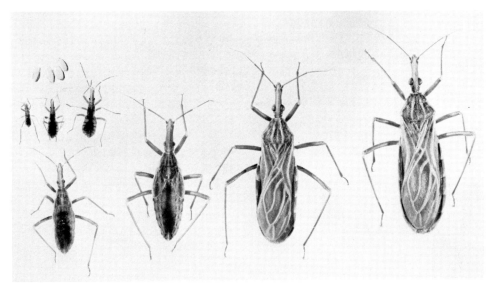

Fig. 2-4 Simple metamorphosis of a bug, *Rhodnius prolixus*. Eggs, nymphs, and adult male and female. (Photograph by R. D. Akre, Washington State University.)

which may be carried in large numbers by the wind, cause allergic responses in susceptible persons.

Basic Internal Morphology. The importance of knowledge of basic internal morphology of insects, especially that of the digestive system including accessory struc-

tures such as the salivary glands, must be considered. In a simple condition, as illustrated in the house fly, pathogenic organisms are sucked up with infectious dejecta, for example, from those ill with cholera or typhoid fever. These organisms then pass out with the feces of the fly, which may be deposited

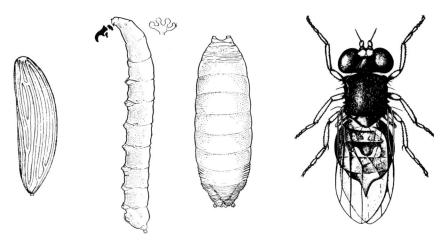

Fig. 2-5 Complex metamorphosis of a fly, *Hippelates collusor*. From *left to right*: egg; larva, showing cephalopharyngeal skeleton and anterior spiracle; pupa; adult. (After Herms and Burgess, except adult fly, which is redrawn after D. G. Hall.)

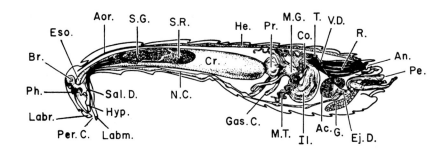

A

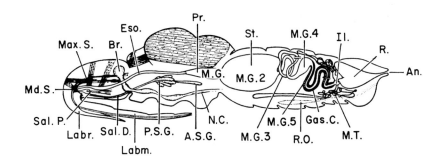

B

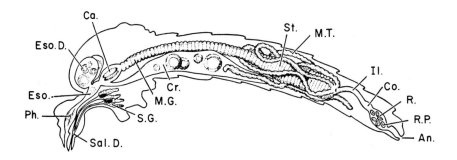

C

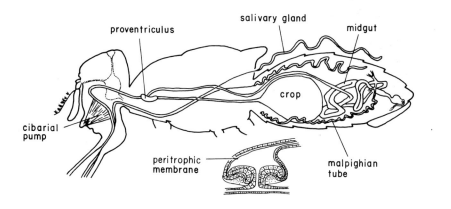

D

20

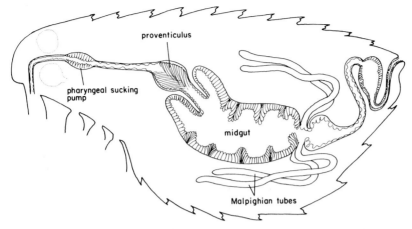

E

Fig. 2-6 Showing digestive tract of the cockroach, Order Dictyoptera (*A*); of assassin bug, Order Hemiptera (*B*); of anopheline mosquito, Order Diptera (*C*); of tsetse, Order Diptera (*D*); and of flea, Order Siphonaptera (*E*). Explanation of abbreviations: *Ac.G.* accessory gland; *An.,* anus; *Aor.,* aorta; *A.S.G.,* accessory salivary gland; *Br.,* brain; *Ca.,* cardia; *Co.,* colon; *Cr.,* crop; *Ej.D.,* ejaculatory duct; *Eso.,* esophagus; *Eso. D.,* esophageal diverticula; *Gas.C.,* gastric caeca; *He.,* heart; *Hyp.,* hypopharynx; *Il.,* ileum; *Labm.,* labium; *Labr.,* labrum; *Max.S.,* maxillary stylet; *Md.S.,* mandibular stylet; *M.G.,* midgut; *M.T.,* malpighian tubules; N.C., nerve cord; *Pe.,* penis; *Ph.,* pharynx; *Pr.,* proventriculus; *Pre.C.,* preoral cavity; *P.S.G.,* principal salivary gland; *R.,* rectum; *R.O.,* reproductive organ; *R.P.,* rectal papilla; *Sal.D.,* salivary duct; *S.G.,* salivary gland; *Sal.P.,* salivary pump; *S.R.,* salivary reservoir; *St.,* stomach; *T.,* testis; *V.D.,* vas deferens. (*A* after Miall; *B* after Elson; *C* after Herms; *D* after Glascow; and *E* after Faarch.)

on human food, either in their original virulent condition or more or less attenuated. Regurgitation on the part of the insect may be equally effective. In this example simple passage through the digestive tract is involved, and not much knowledge of its special anatomy is required.

A more complicated situation exists in the case of the *Anopheles* mosquito, which sucks up pathogenic organisms (plasmodia) with its meal of human blood, whereupon these parasites undergo vital developmental changes within the body of the insect, eventually entering and accumulating in the salivary glands before final introduction by the "bite" into the next victim. The insect in this case is the essential natural vector. An understanding of the part of the cycle of the plasmodium within the mosquito requires at least a basic knowledge of the digestive system of the insect and its spatial relations to other parts of the body.

DIGESTIVE SYSTEM. There are three distinct regions to the insect gut (Fig. 2-6), namely, (1) the FOREGUT, consisting of the mouth, pharynx, esophagus, crop, and proventriculus; (2) the MIDGUT, comprising the stomach; and (3) the HINDGUT, consisting of the ileum, colon, rectum, and anus. In bloodsucking insects, such as the conenose bug and the mosquito, the pharynx, including its pumping organs, becomes associated with the sucking tube of the proboscis. In the mosquito the esophagus has three diverticula that serve as reservoirs for sugars (Fig. 2-6C). The crop is merely a widened portion of the esophagus in the more generalized forms and serves as a food receptacle (Fig. 2-6A). In the more specialized groups, such as adult Diptera and Lepidoptera, the crop is expanded into a capacious pocket or pouch (Fig. 2-6D). In such forms as the cockroach and grasshopper the PROVENTRICULUS consists of a highly muscular dilation provided internally with sclerotized teeth for grinding or straining food (Fig. 2-64). The stomach is a simple sac into which open GASTRIC CAECA, generally few in number, which are important sources of digestive enzymes. At both ends of the stomach are located valves that control the flow of the food. Much variation is seen in the length and the degree of convolution of the hindgut, but usually the three regions—ileum, colon, and rectum —can be located. The MALPIGHIAN TUBULES, which vary in number and length in the several groups of insects, empty into the ileum. These are excretory, not digestive organs, but they are often discussed along with the digestive system because of their location.

The SALIVARY APPARATUS consists of a pair of salivary glands, which may be lobed; they are usually situated within the thorax. Generally each gland empties into a SALIVARY DUCT, and the two ducts discharge into a common duct that opens into the mouth at the base of the labium. In many species of insects a pair of SALIVARY RESERVOIRS is present; these may be located near the opening of the common duct and then present a compound condition, or more often they are situated on either side of the esophagus or within the thorax at the end of a long slender duct.

Insect Larvae and Pupae. When insect larvae are encountered, as parasites or through accidental entry, in the body of man or beast, confuscon may arise because of the wormlike appearance of the invaders; for example, muscoid fly larvae may be incorrectly mistaken for worms. These larvae are short and plump and have eleven or twelve well-marked segments. Microscopic examination of fragments or of the entire larva will reveal tracheae (air tubules), which are not present in worms (Fig. 13-19).

Although the larvae of Diptera (flies, mosquitoes, midges, and such) are characteristically legless and frequently have an un-

developed or poorly developed head, the variations within the order are considerable. MAGGOTS of muscoid flies, for example, are usually smooth, with the body tapering to the apparently headless anterior end (Fig. 12-5), which bears hooklike mouth parts; in contrast, the mosquito larvae (WIGGLERS) have a well-sclerotized, freely moving, conspicuous head with faceted eyes, and both the head and body bear many hairs and setae (Fig. 10-1). The larvae of fleas (Siphonaptera) are also legless; the sclerotized head is well developed; and each of the thoracic and abdominal segments is armed with a band of bristles (Fig. 14-1B). The larvae of beetles (order Coleoptera) commonly have three pairs of legs on the thorax only; the head is well developed; the body may be hairy, spiny, or naked. Some beetle larvae are legless. The larvae of moths and butterflies (order Lepidoptera) have three pairs of thoracic legs and two to five pairs of abdominal prolegs (Fig. 17-6B,C); the head is prominent, and the mouth parts are usually well developed and mandibulate; they are called CATERPILLARS and are hairy, spiny, or naked. The larvae of bees, ants, and wasps (order Hymenoptera) are without legs (APODOUS); the head is more or less well developed; the body is usually fairly smooth and gourd-shaped.

Insect pupae are classified as follows: OBTECT, in which the appendages are closely appressed to the body and held in place by a tightly fitting envelope (Fig. 10-3), as for example in moths; EXARATE, in which the appendages are free from the body (Fig. 9-17), as for example in many beetles; and COARCTATE, in which the appendages are concealed by an enveloping pupal case, or PUPARIUM, as for example in the higher flies (Fig. 2-5). The obtect pupa is sometimes, as in many of the Lepidoptera and Hymenoptera, covered by a silky case, or COCOON (Fig. 14-1c). The puparium of the higher Diptera is formed from the hardened integument of the third stage larva and encloses the developing pupa; the terms *puparium* and *pupa* should not be used interchangeably, as is frequently done. In fact, there is a fourth molt and a fourth instar within the puparium before the formation of the true pupa. The puparium bears the most characteristic features of the third-stage larva, such as the feeding apparatus (cephalopharyngeal skeleton and mouth hooks), the spinous bands, and the anterior and posterior spiracles, so identification of the puparium, and consequently the pupa, can be made from these on the same basis as identification of the larva.

Insect Classification

The medical entomologist must be able to place the insect at hand correctly in at least its proper order and family. To determine the order to which an insect belongs one need usually know only the venation and structure of the wings (if present), the type of mouth parts, the type of metamorphosis, and sometimes one or two other structural characters. Unfortunately, the parasitic forms have undergone many modifications, such as reduction or loss of wings and great alteration in form, but generally the mouth parts, coupled with a salient character or two, will serve as a ready means of crude identification. A list of the orders of insects of medical importance is given below; the usual bases for classification may be tabulated as follows:

1. Wings: (a) presence or absence, (b) form, (c) texture and vestiture, (d) venation.
2. Mouth parts: (a) chewing (mandibulate), (b) sucking (haustellate, of several subtypes).
3. Metamorphosis: (a) none, (b) simple, (c) complex.
4. Special characteristics, such as the modification of the ovipositor as a stinging apparatus in the Hymenoptera.

ORDERS OF INSECTS

The following list includes only those insects that are of some known medical or veterinary importance. For a more complete treatment of the insect orders and keys to separate them the student is referred to such works as Borror and associates (1976), Borror and White (1970), Imms (1957), and Waterhouse (1970, 1974). The classification used here is essentially that of Borror, De Long, and Triplehorn except that the Dictyoptera is considered an order rather than a suborder of Orthoptera.

1. Order Collembola: Springtails, Snowfleas. Wingless; chewing mouth parts, withdrawn into the head; no metamorphosis. These animals have a number of peculiarities: the abdomen is only six-segmented, without external genitalia; appendages of a peculiar nature are present on the first, third, and fourth abdominal segments in the form of a COLLOPHORE, a CATCH, and a SPRING, respectively.

2. Order Dictyoptera: Cockroaches, Mantids. Fore wings modified into leathery covers or TEGMINA, sometimes shortened or absent; chewing mouth parts; simple metamorphosis. Legs fitted for walking or running. Pronotum in cockroaches extending over and concealing head from above, in mantids greatly elongated; fore legs of mantids fitted for grasping and holding prey.

3. Order Mallophaga: Chewing Lice. Wingless; chewing mouth parts; simple metamorphosis.

4. Order Ephemeroptera: Mayflies, Dayflies. Two, sometimes one, pair of triangularly shaped, net-veined, membranous wings; mouth parts vestigial; metamorphosis simple but involving a preadult, fully winged form (SUBIMAGO), found only in insects of this order.

5. Order Thysanoptera: Thrips. Wingless or with very narrow, elongated wings, fringed posteriorly with long hairs and almost without veins; rasping-sucking mouth parts; simple metamorphosis.

6. Order Anoplura: Sucking Lice. Wingless; piercing-sucking mouth parts; simple metamorphosis.

7. Order Hemiptera: The True Bugs. Wings two pairs (rarely one) or none, the fore pair thickened on the basal half, membranous on the apical half, sometimes reduced to small pads; piercing-sucking mouth parts; simple metamorphosis.

8. Order Homoptera: Cicadas, Treehoppers, Leafhoppers, Aphids, Scale Insects, and Others. Wings, when present, membranous or evenly thickened; piercing-sucking mouth parts; simple metamorphosis.

9. Order Trichoptera: Caddisflies. Two pairs of mothlike wings, clothed with hairs (not with scales, as in the moths); chewing, though often vestigial, mouth parts; complex metamorphosis. Included here because of occasional human allergic reactions.

10. Order Coleoptera: Beetles, Weevils. Fore wings thickened into hardened wing covers, or ELYTRA concealing the hind wings when at rest; chewing mouth parts; complex metamorphosis.

11. Order Diptera: Flies, Gnats, Mosquitoes. One pair of wings, the hind pair being replaced by knoblike structures known as halteres; sucking mouth parts; complex metamorphosis.

12. Order Hymenoptera: Bees, Wasps, Ants, Sawflies, Horntails, and Others. Two pairs of membranous wings (sometimes lacking), the anterior pair the larger; chewing or lapping-sucking mouth parts; complex metamorphosis.

13. Order Siphonaptera: Fleas. Wingless; piercing-sucking mouth parts; complex metamorphosis. Body flattened laterally; hind legs usually enlarged, fitted for jumping.

14. Order Lepidoptera: Moths, Butterflies. Two pairs of wings (rarely absent), clothed with scales; sucking (siphoning) mouth parts; complex metamorphosis.

THE ARACHNIDA

The class Arachnida includes the ticks, mites, spiders, scorpions, and related forms. Among the species of arachnids are some of the most important parasites and vectors of pathogens to man and beast; for example, the ticks, which carry the causative organisms of spotted fever and relapsing fever of man, Texas cattle fever, and bovine anaplasmosis. Parasitic mites cause acariasis, often serious, such as mange, scabies, and various forms of itch, and may, like the ticks, serve as vectors, particularly of the scrub typhus rickettsia.

The most important arachnids, for example, ticks, mites, and spiders, lack distinct segmentation of the body; scorpions, pseudoscorpions, and a few others are clearly segmented. The body is divided into two parts (Fig. 2-7): first the CEPHALOTHORAX (prosoma), composed of combined head and thorax, and second the ABDOMEN (opisthosoma). In the ticks and mites there is a strong fusion of the cephalothorax and the abdomen, so that the body becomes saclike in form.

Adult arachnids with few exceptions (such as the eriophyid mites, in which only the first two pairs of legs are developed) have four pairs of legs, though the larvae of ticks and most mites have but three pairs. In spiders there is a pair of PEDIPALPI, which may resemble an additional pair of legs; in the scorpions, whipscorpions, and pseudoscorpions these are CHELATE (that is, the terminal segment of the limb is opposed to the preceding segment, an adaptation for grasping). All arachnids are devoid of wings and antennae. Eyes, when present, are simple. The mouth parts usually consist of a pair of piercing CHELICERAE and the PEDIPALPI, and in ticks and some mites, a HYPOSTOME. The respiratory system of many arachnids, particularly ticks and mites, is tracheal as in insects, except that there is usually but one pair of spiracles. In spiders the respiratory organ is a

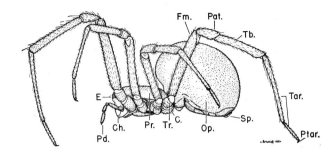

Fig.2-7 Showing external anatomy of a spider. *C.,* coxa; *Ch.,* chelicera; *E.,* eyes; *Fm.,* femur; *Op.,* opisthosoma; *Pat.,* patella; *Pd.,* pedipalp; *Pr.,* prosoma; *Ptar.,* pretarsus; *Sp.,* spinnerets; *Tar.,* tarsus; *Tb.,* tibia; *Tr.,* trochanter.

combination of booklungs and tracheae. There is frequently a strong sexual dimorphism in the arachnids; the males are commonly smaller than the females.

In general, arachnids are predatory or parasitic, although many mites are plant-feeders or scavengers. Most of them are terrestrial, although aquatic mites are of common occurrence.

Arachnid Development. All orders of Arachnida deposit eggs except the scorpions and some mites (e.g., *Pyemotes,* Fig. 15-5), which are viviparous. Eggs are usually numerous, particularly in the ticks, which may deposit as many as 18,000 per female. The newly hatched individuals have the general form of adults, although the number of legs may vary; for example, newly hatched ticks and mites usually have three pairs of legs. Metamorphosis is simple, as in cockroaches and grasshoppers. Molting takes place as in insects, the various stages being termed instars as in the Insecta. The longevity of many arachnids is remarkable: certain ticks have been known to live for as many as sixteen years, and some species are able to endure starvation for several years.

Internal Morphology. The digestive tract of arachnids (Figs. 2-8, 2-9) is characterized by various types of diverticula and branched tubules. The diverticula, which diverge from the tract between the sucking organ of the pharynx and the mesenteron, range, according to Savory (1935), from two short simple sacs directed forward in the cephalothorax to a condition of five pairs, four of which extend laterally, reaching the

bases of the legs and entering the coxae for a short distance; also, there is a very complex type that branches and divides and becomes very large. Leading from the mesenteron is a complex system of branched tubules that occupy most of the abdomen and function partly as a digestive gland and partly as a reservoir. The Arachnida are thus enabled to store large quantities of food and to undergo long periods of fasting.

The excretory organs of the arachnids are MALPIGHIAN TUBULES, which empty into the gut, and COXAL GLANDS, which empty excretory products into tubules and discharge to the exterior from openings that

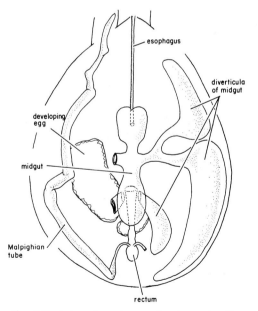

Fig. 2-8 Internal anatomy of spiny rat mite, *Echinolaelaps echidnius.* (After Jakeman.)

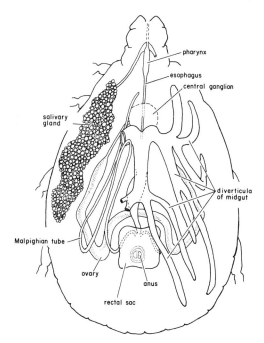

Fig. 2-9 Internal anatomy of a tick, *Dermacentor andersoni*. (After Douglas.)

vary in relation to the coxae among the several orders.

Arachnid Classification. Some authors include among the Arachnida the Pycnogonida (Pantopoda), sea-spiders, and the Tardigrada, waterbears. These and three small terrestrial orders, namely, the Palpigradi, Schizomida, and Ricinulei, as well as the marine Xiphosura, kingcrabs or horseshoecrabs, are omitted from this discussion. For a good discussion of these and other arachnids, the student is referred to Cloudsley-Thompson (1958).

The following orders, all terrestrial, are at least of some medical or veterinary importance: (1) Scorpionida, scorpions; (2) Araneida (Araneae), spiders; (3) Pedipalpida, whipscorpions; (4) Phrynichida (Amblypygi), tailless whipscorpions; (5) Chelonethida (Pseudoscorpionida), pseudoscorpions; (6) Solpugida (Solifugae), sunspiders; (7) Phalangida (Opiliones), harvestmen or harvestspiders; (8) Acarina (Acari), mites and ticks.

Among the more important characters used to separate the terrestrial arachnids into orders are the following: SEGMENTATION of the body; presence or absence of the PEDICEL; presence or absence of TELSON; CHELICERAE, large or small; PEDIPALPI, chelate or unchelate; location and form of SPIRACLES.

CLASSIFICATION, IDENTIFICATION, AND TAXONOMY

It is incorrect to use the three above terms synonymously. CLASSIFICATION is the arrangement of taxonomic units (taxa) in what is hopefully a natural order; IDENTIFICATION is the determination of the position of a given specimen or individual within that order; TAXONOMY is the science governing classification and making possible the art of identification.

The basic importance of taxonomy in medical entomology must be recognized. Correct identification of medically important arthropods is essential. To incriminate mosquitoes as vectors of the plasmodia of malaria is not sufficient; the questions as to which mosquitoes, which plasmodia, and which malaria are involved must be answered. The success of treating the disease and controlling the vector and pathogen, as well as an understanding of the general situation involved, depends upon getting the correct answers.

Identification is often the task of a specialist, although in some cases, where well-defined species are involved or where only approximate results are needed, the general worker can determine the needed answers. Even a specialist, however, cannot extend his work to all phases of arthropod taxonomy. The numbers of species is so vast, and their correlated biologies and taxonomy in many cases so poorly known, that no one person can cover the field. Without the more exact help that a specialist can give, information is often meaningless. In one published account, for example, a particular virus was isolated from a mixed lot of about ten species of flies in two families, without any indication as to which species harbored it. Such information is of very little value.

Often closely similar, even apparently identical, species of arthropods differ greatly in their biologies and medicoveterinary importance. These are called CRYPTIC or SIBLING species. A classical example of sibling species of medical importance is the *Anopheles maculipennis* complex of mosquitoes, but other such complexes, for example, the *Argas persicus* complex of soft ticks, are discussed in the text. In other instances there are pairs or members of a complex that are so similar that they may remain unrecognizable as such over considerable periods of time. Failure to recognize the distinction between the pri-

mary and secondary screwworms of America until the 1930s led to misdirected control operations that cost many human lives and millions of dollars in loss to livestock. Careful taxonomic work could have enabled even the general worker to distinguish these species from each other, in spite of their close similarity.

In the present treatment, descriptive information is mostly reduced to diagnostic characters, and keys are omitted. This course is feasible because laboratory manuals containing such keys and aids for identification are available. Researchers and professional medical entomologists need more help than could be given here. The manual of K. G. V. Smith (1973) will serve the general needs of such workers; for the more specialized areas comprehensive manuals and aids are cited under the appropriate chapters and sections of this text.

3

THE FEEDING APPARATUS (AND ASSOCIATED ASPECTS OF THE DIGESTIVE TRACT)

The feeding apparatus at the anterior end of the digestive tract, the tract itself, and factors causing engorgement are of importance in understanding the most common route of acquisition and transmission of pathogens by medically important arthropods. This same system introduces salivary secretions that may cause toxic and allergic reactions in vertebrates. For a working knowledge one needs to know: (1) the structure of mouth parts and their means of penetrating tissues; (2) the pharyngeal or esophageal pumping apparatus that draws up blood or other fluid; (3) the relationship of salivary glands to the mouth parts, their functions, and the ways that these structures may be involved in development, storage and introduction of pathogens; (4) the basic structure of the digestive tube, its role in the development of pathogens and in the passage of pathogens through the wall of the tract to develop in other host tissues; (5) and factors controlling blood feeding. Some information on the mechanics of body expansion for blood engorgement, stimuli causing feeding, size of blood meals, the process of digestion, and use of artificial feeding techniques is also germane.

IMPORTANCE OF MOUTH PARTS

No doubt all arthropods possessing mouth parts capable of piercing the skin may be regarded as potential vectors of human and animal pathogens, even in cases where blood sucking is not a normal habit. Arthropods with nonpiercing mouth parts obviously cannot introduce pathogens directly into the circulation, but they can do so through natural orifices of the body or through previously injured surfaces. Thus the house fly, which possesses nonpiercing mouth parts, is a vector of the yaws spirochete, *Treponema pertenue,* which it can acquire from open sores and then mechanically infect wounds of man in the act of feeding.

Piercing-sucking mouth parts of insects have the functional portion that actually enters the host developed into a single, often interlocking unit termed the FASCICLE. The units of the fascicle are made up of parts that are called STYLETS. Stylets may interlock and slide on one another or they may function as separate bladelike structures. If action of the fascicle is interpreted on the basis of direct microscopic examination of dissected mouth parts, or on histological preparations of mouth parts fixed in tissue, the impression is gained that this structure is relatively rigid and incapable of much directional control. Techniques developed to observe mouth parts within living tissue have revealed that the fascicle may be extremely flexible and capable of changing the path of its thrust. These observations permit interpretation of the function of individual stylets (Gordon and Lumsden, 1939, and similar more recent studies). It has thus also become possible to determine the specific site of blood uptake. The development of an electric "bitometer" technique (Kashin, 1966) coupled with direct observation during feeding, has provided detailed information on probing, sucking, and salivation.

With respect to the specific site of blood acquisition, Lavoipierre (1965) has coined the terms SOLENOPHAGE for a *vessel feeder* that obtains blood directly from venules or small veins, and TELMOPHAGE for a *pool feeder* that obtains blood directly from a pool of blood formed in tissue after the mouth parts have lacerated blood vessels. The method of blood feeding must surely be of consequence in both acquisition and introduction of pathogens by vectors. In lymphatic filariasis there is evidence that the parasites are found at a higher rate in the blood within a mosquito's stomach than in blood withdrawn from animals by puncture in sampling for parasites. Likewise the introduction of pathogens directly into a blood vessel should ensure their distribution throughout a vertebrate much more rapidly than is possible if release occurs into tissues surrounding vessels.

It is important to know the feeding habits of an arthropod in all stages of its life cycle. Many insects with complete metamorphosis that have sucking mouth parts as adults possess mandibulate (biting-chewing) mouth parts as larvae. The biting mouth parts of the flea larva enable it to ingest particles of excrement or other matter in which eggs of the double-pored dog tapeworm occur, and it

thus becomes an intermediate host of this worm, retaining the infection as an adult flea; if it is ingested by a suitable host, it becomes the agent of infection. Insects with simple metamorphosis, such as the cockroaches and bugs, have the same general type of mouth parts and feeding habits in all active stages of development, although in aquatic groups the mouth parts may change considerably from the larval type (dragonflies and damselflies), or become vestigial in the adult (mayflies). Snodgrass (1944) describes in great detail the feeding apparatus of biting and sucking insects affecting man and animals.

CLASSIFICATION OF MOUTH PARTS

All adult insect mouth parts, however highly specialized, have been derived from a simple primitive chewing type such as exists with some modification in the cockroach. Insect mouth parts are commonly divided into two broad classes: (1) MANDIBULATE (biting and chewing) as in cockroaches, grasshoppers, and beetles; and (2) HAUSTELLATE (sucking) as in bugs, flies, butterflies, and moths. This classification is far too general for a real understanding of functions. For example, the house fly, *Musca domestica,* and the stable fly, *Stomoxys calcitrans,* both possess haustellate mouth parts and are both in the family Muscidae; yet by virtue of its efficient piercing proboscis the stable fly pierces the skin and sucks blood, becoming a direct infector, whereas the house fly cannot pierce the skin because of the structure of its sponging proboscis, and it is therefore more particularly a food contaminator.

Obviously insects could be grouped on the basis of mouth parts and feeding habits, into (1) piercing, as in mosquitoes, and (2) nonpiercing, as in cockroaches. This, however, is too great an oversimplification and provides no information on functional details. For an inclusive classification of arthropod mouth parts Metcalf and coauthors (1962) may be consulted. A more limited scheme, with examples restricted to medically important arthropods that actually feed on vertebrates or serve as intermediate hosts of parasites of vertebrates, is provided here.

1. **Orthopteran Type** (Biting and Chewing). Generalized mouth parts with opposable mandibles are used in biting and chewing; upper and lower lips are easily recognized. Orders Dictyoptera, cockroaches; Coleoptera, beetles; Mallophaga, chewing lice; and others.

2. **Thysanopteran Type** (Puncturing-Sucking). Mouth parts are minute; approach the biting form, but lacerate tissues by puncturing and function as suctorial organs; the right mandible is greatly reduced or possibly even absent, causing a peculiar asymmetry. Order Thysanoptera, thrips.

3. **Hemipteran Type** (Piercing-Sucking). Mouth parts comprise four stylets closely ensheathed within the elongated labium, forming a three- or four-segmented proboscis. Order Hemiptera, true bugs.

4. **Anopluran Type** (Piercing-Sucking). Stylets are in a sac concealed within the head but everted when functioning; three units consist of united maxillae, hypopharynx, and labium; mandibles are vestigial. Order Anoplura, sucking lice.

5. **Dipteran Type** (Suctorial and Piercing or Nonpiercing). No single representative is available to illustrate the entire order Diptera, hence the following subtypes:

a. Mosquito Subtype. Mouth parts consist of six stylets, loosely ensheathed within the labium; Culicidae.

b. Horse Fly Subtype. Mouth parts consist of six short bladelike structures, four of which are used for piercing and cutting, all loosely ensheathed within the labium; Tabanidae. Functionally similar mouth parts are found in black flies (Simuliidae), no-see-ums (Ceratopogonidae), biting snipe flies (Rhagionidae, *Symphoromyia*), and phlebotomine sand flies (Psychodidae, Phlebotominae).

c. Sponging Muscoid Fly Subtype. Mouth parts consist of a muscular proboscis, not suited for piercing; stylets are rudimentary. Higher Diptera, including Calliphoridae, Sarcophagidae, some Muscidae, and others.

d. Biting Muscoid Fly Subtype. Stylets are two in number, forming a tube closely ensheathed within the labium, as in the stable fly and tsetse.

e. Louse Fly Subtype: Mouth parts are closely related to those of the biting muscoid subtype, but the haustellum is fitted for piercing the skin, as in the sheep ked, Hippoboscidae.

6. **Siphonapteran Type** (Piercing-Sucking). This type has a pair of broad maxillary lobes bearing long palpi, a pair of broad maxillary lacinial stylets, a slender labium with parallel palpi, and a median unpaired stylet, the "epipharynx" (the labrum is difficult to interpret, according to Snodgrass). Order Siphonaptera, fleas.

7. **Hymenopteran Type** (Sucking-Chewing). Mouth parts consist of suctorial, lapping organs; mandibles are specialized for portage, combat, and other nonfeeding purposes. Order Hymenoptera in part; bees, wasps, and ants.

8. **Lepidopteran Type** (Sucking). Mouth parts consist of a suctorial coiled tube, the maxillary galeae. Order Lepidoptera; butterflies and moths.

9. **Acarine Type** (Piercing-Sucking). This type has entirely different derivations than insect mouth parts. The cutting or piercing structures are the chelicerae, with an anchoring hypostome present in ticks. Class Arachnida; mites and ticks.

STRUCTURE AND FUNCTION OF THE FEEDING APPARATUS

Orthopteran Type. The cockroach may illustrate the orthopteran type of mouth structure. This type—the mandibulate, or

THE FEEDING APPARATUS / 31

chewing-biting—is the generalized or primitive form and will serve as a basis for later comparisons and derivations. Its main importance in medical and veterinary entomology is to furnish a model for understanding the haustellate or sucking types.

If the head of a cockroach (Fig. 3-1) is viewed from the side and again from the front, the relative position of the separate mouth parts will be better understood. Separating the individual parts, the following structures will be observed. In front, low down on the head, hangs the LABRUM or anterior lip, easily lifted as one would raise a

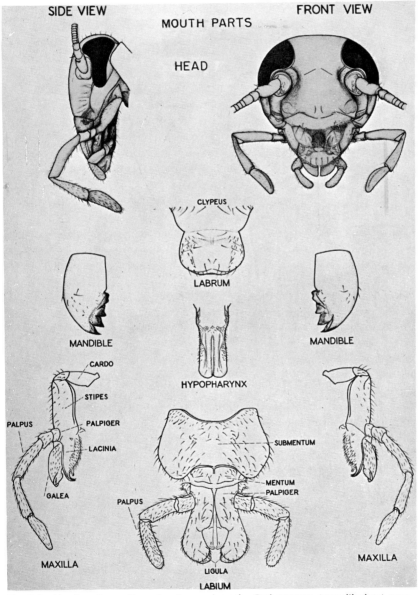

Fig. 3-1 Head and mouth parts of a cockroach. Orthopteran (mandibulate) type mouth parts.

hinged lid, the hinge line being at the lower edge of the sclerite or plate known as the CLYPEUS.

The labrum functions mainly to cover the anterior mouth opening. The inner wall of the labrum is ornamented with ridges and setae and is called the EPIPHARYNX. Because of the close association of these two surfaces, they are often referred to as a double organ, the LABRUM-EPIPHARYNX. When the labrum is removed, a pair of heavy, toothed, laterally opposed jaws, the MANDIBLES, is exposed. Dislodging the mandibles brings into view the pair of MAXILLAE, or accessory jaws. Composite structures separate into *cardo, stipes, lacinia, galea,* and *palpus;* these should be carefully observed, inasmuch as they undergo great modification in the remaining types of insect mouth parts. The two supporting sclerites of the maxillae are *cardo* (basal) and *stipes* (the second); the distal lobes are (1) the *maxillary palpus* (a five-segmented structure) with sensory functions; (2) the *galea* (median and fleshy); (3) the *lacinia* (inner and toothed), capable of aiding in comminuting food.

Underneath the maxillae and forming the floor of the mouth lies the posterior lip or LABIUM, a paired structure similar to the maxillae but fused at the base. On the same structural plan as the maxillae, the labium has a basal sclerite, the *submentum,* followed by the *mentum,* upon which rest the *labial palpi* (a pair of outer three-segmented structures), and the *prementum* bearing the *ligula* (a pair of straplike plates that together correspond to the lower lip). The labium functions as the back wall of the mouth opening, and is also subject to much modification in insects.

The fleshy organ remaining centrally in the mouth cavity after the parts just described have been removed is the HYPO-PHARYNX, an organ comparable in a measure to the tongue of higher animals. The salivary duct enters near the base of the hypopharynx, and in piercing-sucking mouth parts the salivary duct is often located within the hypopharyngeal stylet.

The mandibles are most useful landmarks because they are almost universally present in adult insects, from strong structures in chewing insects to the vestigial structures of fleas. In the Hymenoptera, even though the order is largely haustellate, the mandibles are nevertheless important structures, serving, however, in the honey bee as wax-shaping implements and organs of defense, and in the ants as organs of portage, cutting, and combat. In Hemiptera and many Diptera the mandibles are converted into piercing organs, and the maxillae are also greatly changed in form.

Thysanopteran Type. The thysanopteran type is interesting, as it combines some features of biting-chewing and piercing-sucking mouth parts. The very minute thrips, order Thysanoptera, possess this transitional type of mouth parts. Authors disagree as to the identity of the parts; some believe that the right mandible is reduced, others consider it to be entirely wanting, making the head and mouth parts asymmetrical. The left mandible, both maxillae, and the hypopharynx are elongate, suggesting stylets of the piercing type adapted to move in and out through a circular opening at the apex of the head. No food channel is formed, but the sap from plants is lapped up as it exudes from the abraded surface. Thrips are mentioned because they are on occasion troublesome, biting man when they accidentally land on the skin.

Hemipteran Type. A very different sort of feeding apparatus from those described previously is found in the order Hemiptera (Fig. 3-2). Here the cylindrical labium forms a prominent beaklike proboscis that is usually three- or four-segmented. It is devoid of palpi. The proboscis encloses a fascicle comprised of a pair of mandibles, often terminally barbed, and a pair of maxillae; all four are efficient piercing stylets, the maxillae operating as a unit and the mandibles

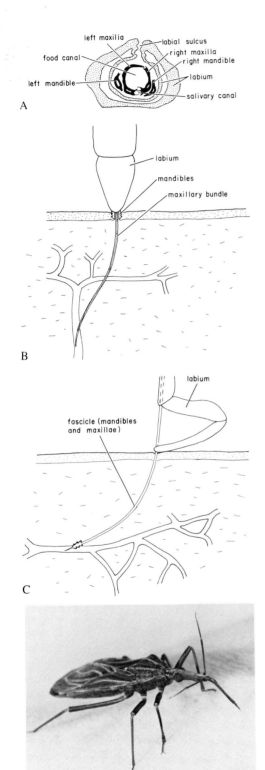

functioning separately. The maxillae are closely apposed, forming the food and salivary tubes (Fig. 3-2*A*); the mandibles may aid rigidity. The labrum is quite short and inconspicuous. The hypopharynx consists of a small complicated lobe at the base of the other mouth parts. Much the same basic plan is found in the Homoptera, which are plant feeders that occasionally bite man when they inadvertently land on the skin.

Hemiptera, as represented in studies of the triatomine bug *Rhodnius* (Lavoipierre, *et al.*, 1959) and the bed bug *Cimex* (Dickerson and Lavoipierre, 1959), are solenophages. However there is a basic difference in the manner in which the fascicle operates. In *Rhodnius* the barbed mandibles anchor into the superficial skin tissue, and it is the maxillary bundle that penetrates to enter a blood vessel (Fig. 3-2*B*). The tips of the maxillae also differ, one being hooked and the other spiny, so that in sliding on one another a very curved path may be followed through the tissue. In *Cimex* the basic details of mandibular and maxillary construction are quite similar, but the mandibles penetrate deeply into the region of the blood vessels (Fig. 3-2*C*). In triatomines the labium is swung forward from its resting position under the body, but in the act of feeding it does not bend (Fig. 3-2*D*). In *Cimex* the labium folds back at the basal segments.

Bennet-Clark (1963) calculated that *Rhodnius* must exert a suction equivalent to at least two atmospheres, possibly as much as nine, to ingest blood at the rate observed. Pinet and colleagues (1969), studying *Triatoma infestans*, demonstrated that the hollow

Fig. 3-2 *A.* Cross section of fascicle sheathed in labium of *Rhodnius*. *B.* Maxillary bundle of *Rhodnius* in blood vessel; note mandibles stay at skin level. *C.* Fascicle of a bedbug, *Cimex*, penetrated deep into host's tissues. *D. Rhodnius* feeding; note that labium is not bent back. (*A-C* after Lavoipierre and Dickerson; photograph *D* by R. D. Akre, Washington State University.)

mandibular and maxillary stylets contain nerves running their length to sensilla that no doubt function as taste receptors. Using split-screen television recording and electrical resistance measurements (bitometer), Smith and Friend (1971) were able to feed *Rhodnius* on artificial diet, and accurately analyze stylet movements, salivation, and ingestion.

Anopluran Type. The mouth parts of the sucking lice are distinctly piercing-sucking in function, but the stylets lie in a sac concealed within the head (Fig. 3-3). The prestomal opening is situated at the extreme anterior portion of a tiny snoutlike proboscis. The proboscis is eversible and thought to consist of the labrum, armed internally with small recurved hooklets that serve as anchorage when everted. The piercing fascicle (three stylets) lies within a long sac and consists of the united maxillae situated dorsally; the hypopharynx and the labium are attached posteriorly to the walls of the enclosing sac. The mandibles are vestigial. The apposed maxillae form the food duct, and the hypopharynx forms the salivary channel. In the act of biting, these parts are pushed forward into the skin by muscular action when firm attachment has been made by means of the circlet of oral eversible teeth. Salivary secretion is poured into the wound, and the cibarial and pharyngeal pumps draw blood. In studying the action of the fascicle of the hog louse, *Haematopinus suis,* Lavoipierre (1967) notes that feeding is from a small blood vessel (solenophagy).

Dipteran Type. Dipteran mouth parts are divided into five subtypes.

MOSQUITO SUBTYPE. A generalized type of dipteran mouth parts is found in the mosquito (Fig. 3-4), with the maximum number of stylets present and loosely ensheathed within the elongated labium, the whole forming a prominent beak or proboscis. The six stylets consist of two mandibles, two maxillae, the hypopharynx, and the labrum-epipharynx. The maxillary palpi are

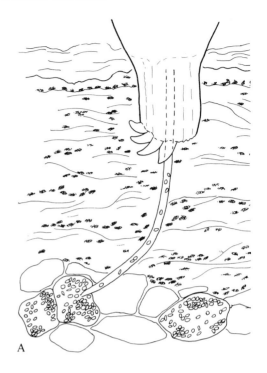

A

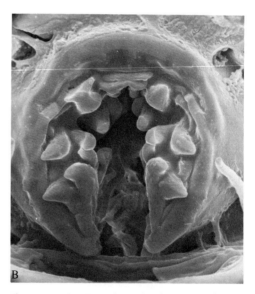

B

Fig. 3-3 *A.* Fascicle of the hog louse, *Haematopinus suis,* in host's tissue. (After Lavoipierre.) *B.* Mouth hooklets of the crab louse, *Pthirus pubis,* representative of structures in Anoplura that anchor the head region to the host's skin in the act of feeding. (Scanning electron micrograph courtesy of D. Corwin and C. Clifford, Rocky Mountain Laboratory, NIH.)

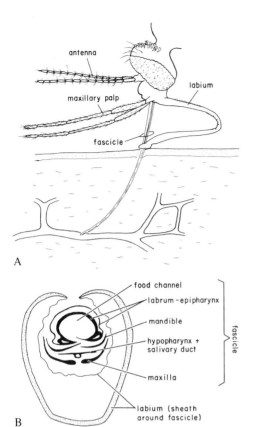

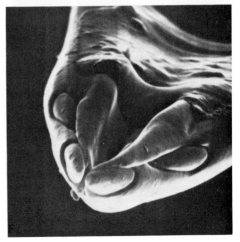

Fig. 3-4 Mosquito mouth parts. *A.* Fascicle thrust deep into host's tissue with labium bent back. *B.* Cross section of mouth parts. *C.* Tip of proboscis, showing fascicle protruding above the labellum. *D.* Cross section of mouth parts from actual specimen, showing general relationships of diagram in *B. E.* Tip of maxillary stylet with toothlike projections. *F.* Tip of labrum-epipharynx with four buttonlike sensilla evident. (*A-B* after Waldbauer. *C-F* are scanning electron micrographs of *Aedes atropalpus;* from Hudson, 1970, *Can. Entomol.,* **102**:501–509; courtesy of A. Hudson, Agriculture Canada, and Entomological Society of Canada.)

conspicuous structures in all mosquitoes; the pair of flattened lobelike organs forming the distal portion of the labium are said to represent the PARAGLOSSAE and are called the LABELLA.

Gordon and Lumsden (1939) observed that the fascicle is highly flexible, and though vessels may be lacerated to form a blood pool, prolonged feeding takes place from the lumen of a vessel. Earlier authors showed that the food channel is formed by the labrum-epipharynx and hypopharynx, but studying the large mosquito *Psorophora ciliata,* Waldbauer (1962) found the mandibles to be interposed between these two structures, citing other authors who had observed the same stylet relationship in other mosquitoes. Further studies by Hudson (1970), using the scanning electron microscope on three species of mosquitoes, show the mandibular stylets positioned outside the maxillary stylets. Apparently the relationship of mandibular and maxillary stylets is variable in mosquitoes. Males lack mandibular stylets and they do not feed on blood. Lee (1974) provides additional interpretations of the structure and function of culicine stylets, based on scanning and transmission electron microscope studies.

Horse Fly Subtype. Though retaining the same number of mouth parts as the mosquito, the horse fly subtype is characterized by the flattened bladelike condition of the stylets (Fig. 3-5). The labium is the conspicuous median portion loosely ensheathing the blades and terminating in a pair of large lobes, the labella. The mandibles, movable transversely, are distinctly flattened and saberlike, and the maxillae are narrower and provided with conspicuous palpi. Both the hypopharynx and labrum-epipharynx are lancetlike. In the males, which do not feed on blood, these piercing parts are very weakly developed.

Studying the fascicle of the tabanid *Haematopota pluvialis* in action in a live host, Dickerson and Lavoipierre (1959) showed that penetration is by means of a thrusting action, with mandibles and maxillae lacerating the tissues. The mandibles move with a scissorlike motion, and the maxillae thrust and retract. The result is a rupture of small and large blood vessels, with the fly feeding on the pool of blood formed in the tissues (telmophagy). The stylets and associated structures of both sexes in four genera have been examined with the aid of the scanning electron microscope (Faucheux, 1975). Toothlike serrations indicate that the mandibles abrade tissue as files do, the maxillae function as saws. The epi-

LABIUM REMOVED

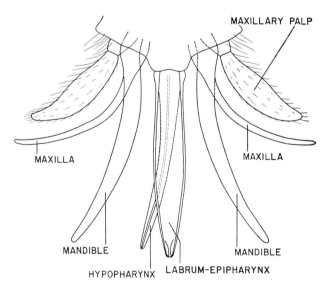

Fig. 3-5 Mouth parts of a horse fly spread apart. (After Snodgrass.)

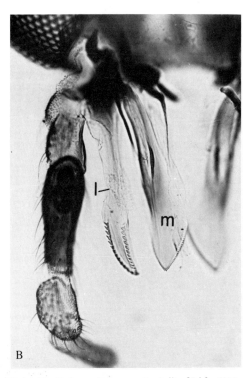

Fig. 3-6 *A.* Scanning electron micrograph of mouth parts of *Simulium venustum* frozen in act of penetrating a membrane. The labellar lobes are particularly evident, the labrum-epipharynx lies between these, and the crossed mandibles with scissorlike cutting action are behind the labrum-epipharynx. *B.* Stylets of a black fly, *Cnephia.* Outer margin of maxillary lacinia (*l*) has recurved anchoring or retrorse teeth; teeth on mandible (*m*) are cutting in nature. (*A* courtesy of J. F. Sutcliffe and S. B. McIver, University of Toronto. *B* from Downes, in Fallis [Ed.], 1971, *Ecology and Physiology of Parasites,* University of Toronto Press; courtesy of J. A. Downes, Agriculture Canada, and University of Toronto Press.)

pharynx has two sensory papillae distally, differing in structure between sexes. The food canal contains seven pairs of sensilla that are probably gustatory in function.

Mouth parts of simuliids and ceratopoganids are similar to those of tabanids, though more delicate. Downes (in Fallis, 1971) illustrates the mandibular and maxillary stylets in representatives of these two families, indicating that the serrated teeth of both edges of the mandibles are cutting in nature, and teeth on the maxillary lacinia curve at the tip toward the base of the structure and are termed *retrorse teeth,* serving to anchor the mouth parts during feeding (Figs. 3-6, 3-7).

In phlebotomine sand flies (Psychodidae) the mouth parts resemble those of ceratopogonids. Lewis (1975) made a functional study of over a hundred New World species,

noting that fascicle structure made it possible to determine that a given species feeds on animals of a particular group, being related to host skin structure.

HOUSE FLY SUBTYPE. In the house fly (Fig. 3-8) the prominent fleshy proboscis consists mainly of the labium, which terminates in a pair of corrugated sponging organs, the labella, and is attached in elbowlike form to the head. The entire structure is highly muscular, and may be either protruded in feeding or partially withdrawn while at rest. Lying on top of the grooved labium is the inconspicuous spadelike labrum-epipharynx that forms, with the hypopharynx, a sucking tube supported by the labium. The maxillae have evidently become fused with the fleshy elbow of the proboscis, and only prominent maxillary palpi remain.

On the bottom surface of the labella are

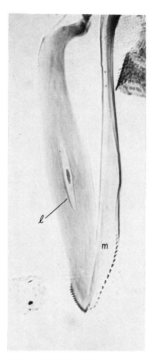

Fig. 3-7 Stylets of *Culicoides variipennis*, showing retrorse teeth on maxillary lacinia (*l*) and cutting teeth on mandibles (*m*). (From Downes, in Fallis [Ed.], 1971, *Ecology and Physiology of Parasites*, University of Toronto Press; courtesy of J. A. Downes, Agriculture Canada, and University of Toronto Press.)

lines of open tubes formed by sclerous partial rings termed PSEUDOTRACHEAE, serving as a sponge through capillary action. The free ends of the pseudotracheal rings may be somewhat abrasive in action as has been noted in the feeding of some species of *Hippelates*.

STABLE FLY SUBTYPE. This subtype (Fig. 3-9) is represented by a group of flies in which the mouth parts are distinctly specialized for piercing.

The proboscis at rest is oriented in front of, and parallel with, the body and is provided with a prominent muscular elbow or knee. This conspicuous proboscis is the labium terminating in the labella, which are provided with a complex series of rasping denticles. The labella are doubtless derived from the same sponging structures in the house fly, but they are proportionally much smaller and more heavily sclerotized. A similar basic structure is found in the horn fly, *Haematobia,* and tsetse, *Glossina.* The proboscis is strongly thrust into the flesh of a victim, and entry is aided by rasping action of the labellar denticles. Lying in the upper groove of the labium are two stylets: the labrum-epipharynx, the uppermost and heavier, and the lower and weaker hypopharynx, the two forming a sucking tube supported within the labium. Maxillary palpi are located at the proximal end of the proboscis. Lavoipierre (1965) states that *Stomoxys calcitrans* is a telmophage.

LOUSE FLY SUBTYPE. The sheep ked and other hippoboscids have mouth parts closely related to those of the stable fly subtype; the characteristic tubular haustellum is adapted for penetration into the skin of the host (Fig. 3-10). The labrum epipharynx is stylet-shaped, its proximal portion strongly sclerotized and rigid and the distal end membranous and very flexible. Nelson and Petrunia (1969) studied the feeding mechanics of the sheep ked, noting it is a solenophage. After the prestomal teeth have abraded through tissue into a venule, they evert and anchor the labella to the inside of the vessel wall.

LARVAL MOUTH PARTS OF DIPTERA. Although not directly involved in transmission of pathogens, two types of larval mouth parts in Diptera are of interest. The more primitive feeding apparatus of mosquito and black fly larvae is involved in removing particulate material from water, a fact that can be used in selectively dispensing pesticides for their control; the mouth hooks of cyclorraphous Diptera are most frequently involved in myiasis.

Feeding Apparatus of Mosquito and Black Fly Larvae. In the larvae of many aquatic Nematocera the external mouth parts form brushlike structures that bring a current of water and suspended particles into the

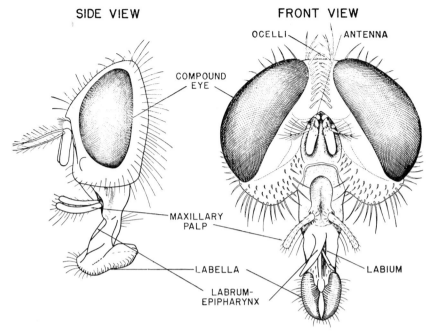

SIDE VIEW   FRONT VIEW

OCELLI   ANTENNA

COMPOUND
EYE

MAXILLARY
PALP

LABELLA   LABIUM

LABRUM-
EPIPHARYNX

Fig. 3-8 Head and mouth parts of the house fly.

pharynx. The significant feature is the development of the pharynx into a pump with internal screens (Fig. 3-11). These comblike screens have closely arranged spacings that permit removal of very fine suspended material (Harwood, 1952). Dadd (1970) noted that the rate of ingestion of suspended particulates by mosquito larvae is proportionate to the time spent actually filtering, which may be increased by the presence of taste stimulants such as yeast extract or adenylic acid. The functional morphology of black fly larval mouth parts has been described by Chance (1970), with a view to more selective insecticide applications through the use of appropriate-sized particulate formulations.

Mouth Hooks of Cyclorrapha. The mouth hooks of maggots are attached to a prominent pharyngeal sclerite. Unlike opposed mandibles that have a biting action, mouth hooks move parallel to each other in a more or less tearing motion. There is some question as to their derivation, but Snodgrass

(1953) states that, if not mandibles, the mouth hooks cannot be homologized with any other insect structures. Some muscoid larvae, such as *Musca domestica* and *Stomoxys calcitrans,* have only a single mouth hook.

Siphonapteran Type. The mouth parts of fleas (Fig. 3-12), though typically of a piercing-sucking type, are peculiar to this order of insects. The broad maxillary lobes bearing long palpi are conspicuous landmarks; the other organ (slender) bearing long parallel palpi is the labium. The principal bladelike piercing organs are a pair of independently movable structures commonly referred to as mandibles, but they are said by Snodgrass and others to be maxillary laciniae. The mandibles are believed to be rudimentary. The median unpaired stylet is said to be the epipharynx, not the labrum of many authors, the labrum being difficult to demonstrate. The epipharynx is closely embraced by the lacinial blades. The paired maxillary stylets and the epipharyngeal stylet form the

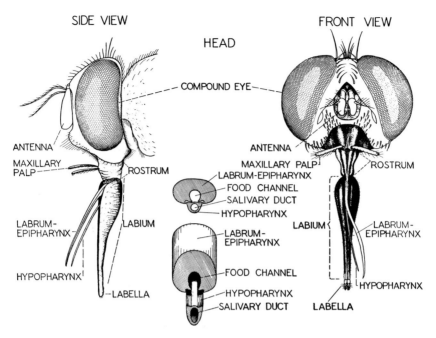

Fig. 3-9 Head and mouth parts of a stable fly.

fascicle held in a channel formed by grooves on the inner margin of the labial palps. The labium is rudimentary, the existence of a hypopharynx not demonstrable. The wound

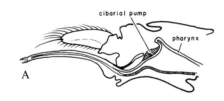

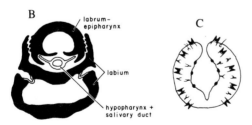

Fig. 3-10 Mouth parts of a hippoboscid, *Pseudolynchia canariensis. A.* Mouth parts protracted in the act of feeding. *B.* Cross section of haustellum. *C.* Tip of labellum with prestomal teeth. (After Bequaert.)

is made by the protraction and retraction of the laciniae. As soon as blood begins to flow, it is drawn into the pharynx by the action of cibarial and pharyngeal pumps. It has been determined that fleas are normally solenophages (Lavoipierre and Hamachi, 1961; Deoras and Prasad, 1967).

Hymenopteran Type. In this type the two general classes of mandibulate and haustellate mouth structures find full development, though the mandibles in many representatives are not involved in the feeding process. The honey bee serves as an example of the fully mandibulate and suctorial condition, details of which may be seen in Snodgrass (1935). A similar full development of mouth parts is found in vespoid wasps, which have been suspected of contaminant transfer of pathogens on meat, as occurs with the house fly. In ants the mandibles may be developed into effective biting organs.

Lepidopteran Type. Represented by ordinary butterflies and moths, this type consists of a coiled sucking tube capable of great

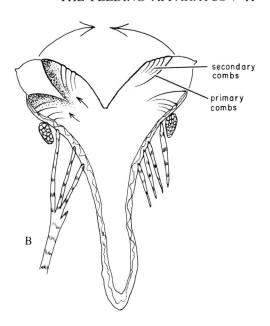

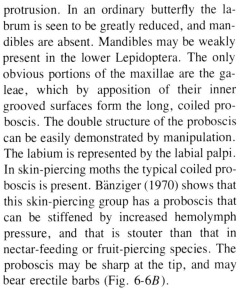

Fig. 3-11 Pharyngeal straining apparatus in mosquito larvae. *A*. Electron micrograph of primary straining combs. *B*. Section across pharynx showing suspended matter strained out on left side.

protrusion. In an ordinary butterfly the labrum is seen to be greatly reduced, and mandibles are absent. Mandibles may be weakly present in the lower Lepidoptera. The only obvious portions of the maxillae are the galeae, which by apposition of their inner grooved surfaces form the long, coiled proboscis. The double structure of the proboscis can be easily demonstrated by manipulation. The labium is represented by the labial palpi. In skin-piercing moths the typical coiled proboscis is present. Bänziger (1970) shows that this skin-piercing group has a proboscis that can be stiffened by increased hemolymph pressure, and that is stouter than that in nectar-feeding or fruit-piercing species. The proboscis may be sharp at the tip, and may bear erectile barbs (Fig. 6-6*B*).

Arachnid Mouth Parts. In his excellent study of the feeding organs of Arachnida, Snodgrass (1948) points out that arachnids come from an ancestral stock that never acquired organs for mastication, and even today have no true jaws, hence subsisting on liquids. A liquid diet requires an ingestion pump, and with all arachnids a highly developed sucking apparatus constitutes an essen-

tial part of the feeding mechanism. The paired CHELICERAE are the first postoral appendages of the arachnid, and although functioning more or less as "jaws," they are not homologous with insect mandibles; they are used for grasping, holding, tearing, crushing, or piercing. In spiders the VENOM GLANDS are associated with the chelicerae (Fig. 17-26). The leglike PEDIPALPS are the second postoral appendages of Arachnida and are the homologs of the mandibles of mandibulate arthropods (Snodgrass). These organs are modified in various ways as organs of prehension, protection, and, in male spiders, as sperm-carrying organs. In scorpions the pedipalps are clawed and serve in catching, holding, and crushing the prey.

FEEDING APPARATUS OF TICKS. According to Snodgrass, the lobes or processes often associated with the distal part of the HYPOSTOME (Fig. 3-13) are the only features that cannot be homologized with structures present in other Arachnida. It is generally believed that the chelicerae serve as cutting structures that gain entrance, and the recurved denticles of the hypostome serve to anchor that structure. Extensive laceration of

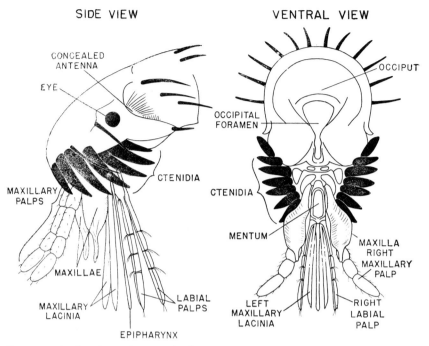

Fig. 3-12 Head and mouth parts of a flea.

blood vessels is characteristic of feeding by ticks, the result being that feeding occurs from a pool of blood (Lavoipierre and Riek, 1955). Feeding ticks can concentrate the blood cells and tissue components ingested (Koch *et al.,* 1974). Soft ticks (Argasidae) generally feed rapidly, hard ticks (Ixodidae) characteristically remain attached for a period of days. Moorhouse (1973) states that more primitive ixodids such as *Ixodes, Amblyomma,* and *Hyalomma* have long mouth parts that are thrust deep into tissue, remaining anchored by the hypostome. They may generally be removed without breaking off the mouth parts in tissue. More advanced ticks such as *Dermacentor, Rhipicephalus, Boophilus,* and *Haemaphysalis* have shorter mouth parts, and they produce a cement substance that aids in sealing them to the host tissue; pulling them from the host frequently causes portions of the mouth parts to break off. Gregson (1960) believed the cement substance to be a special fluid produced by the salivary glands, and Chinery (1973)

showed the cement in *Haemaphysalis spinigera* consists of an inner and outer layer (Fig. 3-13*C*), mainly of protein derived from salivary gland precursor substances.

Gregson (1967), using electrical recording on *D. andersoni* while observing feeding, found that initial feeding was characterized by short periods of sucking followed by bursts of saliva. As days of feeding progressed, sucking, salivation, and intermittent resting intervals lengthened in time. A major blood pool in the host tissue did not form until after about two and a half hours of feeding, and secretory activity reached a peak after about 5 days. These observations were made on restrained hosts, although Sweatman and associates (1976) have since developed a telemetry technique for continuous recording of tick feeding activities, and the temperature of ticks, on free-ranging cattle.

FEEDING APPARATUS OF MITES. The mouth parts of mites resemble those of ticks. A hypostome is usually lacking, but in mites possessing one that structure is not anchoring

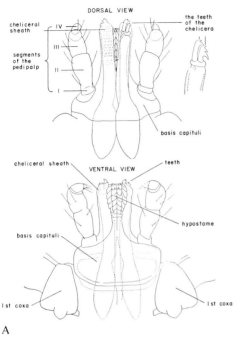

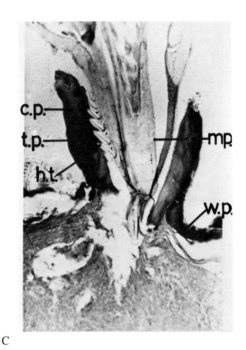

A

C

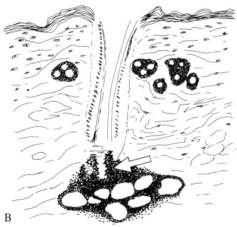

B

Fig. 3-13 Mouth parts of tick. *A.* Overall view. *B.* Section of mouth parts in host's tissue, showing feeding on a pool of blood at arrow (after Lavoipierre and Riek). *C.* Section through the cementing substance and mouth parts of *Haemaphysalis spinigera* attached to the skin of a rabbit. *c.p.,* conical projection of cement; *h.t.,* hypostomal teeth; *m.p.,* mouth parts; *t.p.,* toothlike processes of internal cement; *w.p.,* winglike lateral extension of external cement. (*C* courtesy of Chinery, 1973, *J. Med. Entomol.,* **10:**355–62.)

in nature because it is not armed with barb-like teeth. Among trombidiform mites the chelicerae become progressively adapted for piercing by a transformation of the movable digits into hooks or stylets. Mites tend to feed on lymph and tissues other than blood. Larval trombiculid mites remain attached for some time, and around the inserted mouth parts a feeding tube or STYLOSTOME forms.

The stylostome is thought to be developed in part from host tissue reactions, but by feeding through artificial membranes its source is shown to be a fluid (salivary?) introduced by the mite (Schumaker and Hoeppli, 1963; Cross, 1964). The adult female macronyssid mite *Chiroptonyssus,* feeding in bat wing membrane, teases a hole in the wall of a small venule and feeds rapidly (Fig. 3-14);

as the protonymph it imbides tissue fluids for a prolonged time before likewise feeding from a venule (Lavoipierre and Beck, 1967).

DIGESTIVE TRACT

A knowledge of the digestive tract is of particular importance for understanding the development of pathogens in vectors. In species refractory to development of pathogens the gut is frequently the site of resistance, a phenomenon referred to as the "gut barrier" (Fig. 10-22). Thus it is well known that many species of mosquitoes that do not ordinarily serve as a host for the development of given pathogens, particularly viruses, can be readily infected if the viruses in question are injected directly into the hemocoel. In addition, virus replication at high levels can be accomplished in tissue cultures of several arthropod cell lines representing species that do not support development of these pathogens upon ingesting infectious blood.

The general features of the arthropod gut are reviewed in Chapter 2. The purpose here

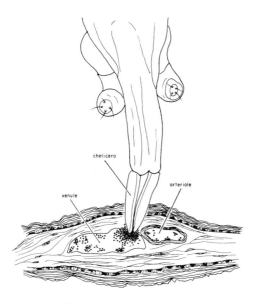

Fig. 3-14 The mite *Chiroptonyssus* feeding in bat wing membrane. (After Lavoipierre.)

is to discuss specific points in relation to pathogen development, and more particularly to present information regarding specific important gut structures and associated glands. Suffice it to say that basically the gut is a tube of epithelial cells divided into the foregut, midgut, and hindgut. The foregut and hindgut are lined with cuticular derivatives and are resistant to the passage of nutrients and pathogens. The midgut is the main site of digestion and absorption, and it is here that pathogens gain entry to the arthropod body (Fig. 4-12). Richards (1975) discusses our knowledge of midgut ultrastructure in blood-feeding insects, reviewing the literature and pointing out deficiencies in knowledge on the functioning of cells and membranes, and major differences between groups of insects.

Peritrophic Membrane. The PERITROPHIC MEMBRANE is a layer between the midgut contents and the epithelial cells of that gut region. This membrane is particularly well developed in chewing insects that feed on coarse foodstuffs, and it may be lacking in liquid feeders. Nonetheless, in blood feeders such as mosquitoes and tsetse, a peritrophic membrane is present and it is believed to comprise an important barrier to the passage of pathogens such as microfilariae in the former and trypanosomes in the latter. A general review of insect peritrophic membrane is provided by Richards and Richards (1977), and Le Berre (1968) reviews information to that time concerning the role of that structure in digestion and its influence on the development of parasitic organisms.

The peritrophic membrane of *Glossina* has been studied by Moloo and coworkers (1970) and Freeman (1973) in terms of its ultrastructure, formation, and site of passage of trypanosomes. In this insect it is a bilaminate structure, the mucopolysaccharide and polysaccharide precursors of which are secreted by two types of cells at the anterior end of the midgut. Crossing of the membrane by trypanosomes appears to take place at the anterior portion where this structure is still soft during the process of formation.

In the mosquito, and other nematocerous adult bloodsucking Diptera, the peritrophic membrane seems to be lacking until a blood meal is taken. The role of this structure and pathogen penetration is reviewed by Orihel (in Maramorosch and Shope, 1975) for a number of arthropods. In adult mosquitoes, representing bloodsucking Nematocera in general, the membrane is formed shortly after a blood meal from the cytoplasmic matrix of epithelial cells and consists of condensed granular material, loosely layered. Though it can be demonstrated as a delicate structure present within five minutes after a blood meal, it may already be penetrated by microfilariae using an anterior hook, and these pathogens can be found within the hemocoel within thirty minutes after blood feeding. In a mosquito with a peritrophic membrane that has aged and hardened for 20 to 30 hours this structure may serve as a significant barrier to the ookinetes of bird malaria. The rapidly forming membrane in *Simulium* may act as a substantial barrier to microfilariae of *Onchocerca*. Species of *Phlebotomus* that are competent vectors of *Leishmania* develop a peritrophic membrane that loses its integrity in 3 to 4 days.

Salivary Glands. These glandular structures are prominently present in hematophagous arthropods. They serve an important role in the concentration and dissemination of some pathogens, yet their normal functions remain obscure. Studies of mouth parts in tissue have demonstrated that salivation is a commonplace occurrence, often in quite copious amounts and therefore apparently of important purpose. But various authors assert that saliva functions as a lubricant to permit ready sliding of the stylets in the fascicular bundle, the introduction of anticoagulins to cause a steady flow of blood to be ingested, anesthetic properties to prevent vertebrate hosts from reacting vigorously and dislodging the offending

arthropod, and digestive functions. In a review of the digestive properties of blood-feeding insects, Gooding (1972) points out that anticoagulins have frequently been demonstrated in the saliva; agglutinins are also found rather often, but digestive enzymes rarely. The copious salivary secretion of ticks enhances the introduction of pathogens and may be a means of excreting excess water (Tatchell, 1967). Anesthetic properties seem to be characteristic, as many competent vector arthropods possess a bite that gives no or very little sensation; related pest species that are not vectors can be very annoying when biting. Hudson and associates (1960) showed that severing the mosquito salivary duct caused the bite of operated females to be significantly more painful than the bite of normal females, suggesting anesthetic qualities in the saliva.

Because salivary glands are the particular site of concentration for some pathogens, these structures and salivary fluid have been thoroughly studied. That the salivary glands are not necessarily simple structures is indicated by an electron microscope study of the glands of the mosquito *Aedes aegypti* by Janzen and Wright (1971). They describe two lateral lobes of the proximal, intermediate, and distal portions of the salivary glands, and a median lobe consisting of intermediate and distal portions. Branches occur in the lateral lobes and to a lesser extent in the median lobe, resulting in varying degrees of duplication of these sections. Each portion has a distinctive cell type, the proximal and distal areas of the lateral lobes and the distal area of the median lobe being glandular in function. Intermediate sections appear to be involved in water transport, possibly to hydrate dense secretory materials in the distal zones.

In order to study salivary secretions, special techniques have been developed for their collection; for mosquitoes the fascicle can be separated from the labium and inserted into a capillary tube filled with mineral oil, thus permitting the measurement of salivary fluid and calculation of infectious virus doses (Hurlbut, 1966); for tsetse, confinement on the stretched wing of a bat causes hungry flies to probe and salivate, and the saliva is readily collected (Youdeowei, 1975); the glass capillary technique has been used to collect tick saliva.

Process of Blood Engorgement. Host seeking involves a behavioral repertoire that is enhanced by hunger and at the time of day for normal feeding activities of a given bloodsucking arthropod. Along with this behavioral sequence of events there is a sensory input, the component parts of which have been studied extensively in a variety of blood feeders. Certain aspects of these findings are mentioned in Chapter 4, and general reviews of bloodsucking behavior of arthropods are provided by Hocking (1971) and Galun (in Shorey and McKelvey, 1977). At this point the processes of probing and engorgement will be considered.

Friend and Smith (1977) have reviewed factors known to affect feeding by bloodsucking insects. In a number of feeding tests on artificial diets through membranes it has been demonstrated that a temperature above ambient and similar to that of a normal host animal will elicit probing, though of course this does not apply in the case of species that feed on cold-blooded hosts. In studying interacting stimuli with the stable fly, *Stomoxys calcitrans,* Gatehouse (1970) found that a substrate with low reflectance increased probing (visual stimulus), rough surfaces lengthened the duration of probing (contact stimulus), and chemical stimuli (olfactory) provided by ammonia and high humidity were also involved.

In several studies specific chemical stimuli, alone or in combination, have proved important in causing engorgement once probing starts. Chemical receptors are present in the buccal and pharyngeal linings, and it is noteworthy that even some extremely fine stylets, like those of mosqui-

toes, have nerves and sensory structures. Nucleotides, particularly ATP (adenosine triphosphate), have frequently been shown to be chemicals present in blood that cause engorgement, as noted for mosquitoes, black flies, tsetse, tabanids, the biting Hemiptera, and ticks. The ATP receptor has been identified in tsetse (Mitchell, 1976). Various amino acids also have some stimulatory effect.

Measuring the size of the blood meal is important in determining the effect of excessive attack on blood loss of hosts, and attempting to estimate the infectious dose of pathogens obtained during feeding. The usual method for measuring blood engorgement is to weigh arthropods before feeding and immediately after blood satiation, but this results in an underestimation because most blood feeders void clear fluid from the anus during feeding as a means of water regulation; some feeders even void whole blood. It has been possible to overcome this problem in studying blood feeding of mosquitoes by feeding them on blood labeled with the radioisotope ^{144}Ce, the isotope remaining with the blood meal and not passing in the clear excreta (Redington and Hockmeyer, 1976). Gooding's review of digestive processes in hematophagous insects lists blood meal size as 0.1 mg in *Phlebotomus papatasi,* 0.23 mg in *Leptoconops kerteszi,* 1.08 to 3.26 μl in Simuliidae, 344.0 mg in *Tabanus sulcifrons,* ca 1.0 μl in *Pediculus humanus,* 6.48 μl in *Cimex lectularius,* 307.0 μl in *Triatoma infestans,* and 1.0 to 25.0 mg in Culicidae.

The size of the blood meal is regulated by abdominal stretch receptors and central nervous system integration, at least in the mosquito, for if the ventral nerve cord is severed anterior to the second abdominal ganglion, massive hyperphagia results (Gwadz, 1969). Likewise there is regulation of rate of blood digestion aside from that affected by ambient temperature. Inseminated female mosquitoes, or unmated females injected with an extract of male accessory glands (matrone), digest blood meals more rapidly than unaltered virgin females. A brain humoral factor is also involved, because cervical ligatured mosquitoes digest blood more rapidly if they received a brain implant from a blood-fed donor (Downe, 1975).

When arthropods fill up with blood, the abdomen is greatly expanded, indicating a cuticular adaptation especially suited for stretching. This is particularly noteworthy in the Hemiptera and ticks (Fig. 16-8). Hackman (1975) found in the bug *Rhodnius* and the tick *Boophilus* that stretching is accomplished by the unfolding of the highly folded epicuticle, and by the stretching of the underlying procuticle so that its thickness is halved. A special type of cuticle is present, differing from flexible but nonstretching cuticles in the nature of the cuticular proteins, and low chitin content.

In reviewing the mouth parts, gut, salivary glands, and control of blood feeding in hematophagous arthropods, it can be seen that a number of unique adaptations are present, remarkable in themselves but also providing clues to how arthropods have developed the capacity to acquire, cultivate, and transmit pathogens of vertebrates.

4

EPIDEMIOLOGY

EPIDEMIOLOGY is here defined as the study of the distribution and determinants of disease prevalence in populations of organisms. It includes the causes of epidemics, the ways that diseases are maintained at low or stable levels, and the factors that are important in the decline of disease incidence. This is a broader scope than the one originally and generally accepted, but it is more adequate for our purposes. The objective for gathering and studying this information is to provide practical solutions for reducing the number of clinical cases of disease.

Epidemiology, as applied to arthropod-associated diseases, covers situations where arthropods transmit pathogens that affect man and animals, as well as circumstances where arthropods are in themselves causal agents of disease. This latter situation will receive little attention here, though infestations where arthropods are pathogens comprise the main substance of Chapter 13, and damage due to arthropod products is the substance of Chapter 17. The procedure here is to discuss the characteristics of vectors, to list pathogens and their associations with vectors, and to review the ways that vertebrate populations influence the course of these disease relationships. Finally, some attention is paid to an amalgamation of the ways that these three distinct groups of organisms interact and the ways that such interactions may be studied coherently.

Some Parasitological Terminology. In the field of parasitology certain terms have been developed to categorize the variety of relationships between two different organisms. The term PARASITISM merely denotes that two species of organisms are living in close association; generally the larger organism is the HOST and the smaller one the PARASITE. Where parasitism is the only means of existence it is called OBLIGATE, for example, the parasitism of lice and parasitic mites; where a free-living form, such as the larva of a scavenger fly, infests a host the relationship is termed FACULTATIVE, or accidental. TEMPORARY PARASITES visit the host for only a short time to feed, for example, most biting flies, most ticks, and bed bugs; CONTINUOUS PARASITES usually infest one host throughout life, for example, scabies mites, and sucking and chewing lice. Continuous parasitism with change of hosts occurs with filarial worms and malaria parasites. The site of parasitism is classified as ECTOPARASITISM for parasites living on the host's body surface, and ENDOPARASITISM for parasites not visible there except, sometimes, through an opening needed for respiration. Regarding arthropod-associated diseases, those caused by a pathogen requiring a vector are OBLIGATE TRANSMISSIBLE DIS-EASES (e.g., malaria, filariasis), those where a vector is only one route of dissemination of a pathogen are FACULTATIVE TRANSMISSIBLE DISEASES (e.g., anthrax, tularemia).

VECTORS

VECTORS, used here in a somewhat restricted context, are those arthropods that are capable of transmitting organisms that cause disease in a vertebrate host. Mere capability, however, is of little significance in understanding how an arthropod-transmitted pathogen is maintained in any specific place at times of high or low incidence.

The common practice is to classify vector arthropods, in a given region and season, as (1) PRIMARY, if they are proved to be transmitting a pathogen to man or other animals, and (2) as SECONDARY, if in these same circumstances they play a supplementary role in transmission but would be unable to maintain a disease in the absence of primary vectors. The status of vectors within these definitions is often difficult to ascertain, and some investigators prefer such terms as *important vectors*, and *vectors of minor importance.*

In addition to simply verifying that many arthropods serve as vectors, further studies have revealed a number of factors useful for evaluating the potential of a given species; that is, those circumstances influencing the actual role played by a vector species. Some of these factors bear discussion.

Incriminating a Vector

In instances where a disease of unknown cause is occurring, certain general characteristics help identify it as arthropod-borne. The time of occurrence, nearly always closely following a time of increase in a suspect arthropod population, is a common clue (Fig. 4-1). In situations where an arthropod

harbors a pathogen throughout the year, there will be no transmission without active feeding, and health problems resulting solely from the bites, stings, or presence of arthropods occur at high rates only when the stages proximal to humans are, or have been, at peak numbers.

The environment in which a disease occurs provides further clues. Many arthropod-associated diseases are connected with wilderness areas, or areas disturbed by agriculture or deforestation, though urban concentrations of humans near such areas may be affected. Malaria, scrub typhus, African trypanosomiasis, and diseases transmitted by ticks fit within this general description. By contrast, however, diseases such as urban plague, urban yellow fever, and rickettsialpox have characteristic urban foci.

In most cases prior information amassed in the field of medical entomology provides an investigator with general clues as to the likely vectors of pathogens causing a disease. The investigator may recognize the pathogen as one known to be arthropod-transmitted. The site of feeding may be diagnostic; marks in the form of characteristic welts or more long-lasting lesions at the site of attack, as occur with the bites of triatomine bugs, chigger mites, ticks, and black flies, can provide important clues. A persistent lesion caused by the pathogen itself at the site of feeding is characteristic of certain diseases, such as cutaneous leishmaniasis, anthrax, and scrub typhus (Fig. 15-24).

In most arthropod-borne diseases the interval between successive vertebrate infections includes two incubation periods: a period in a vector during which the disease-producing organism increases or transforms to the point where it can be transmitted (EXTRINSIC INCUBATION PERIOD), and a period in the vertebrate host before disease is expressed clinically (INTRINSIC INCUBATION PERIOD).

Spatial coincidence relates a characteristic

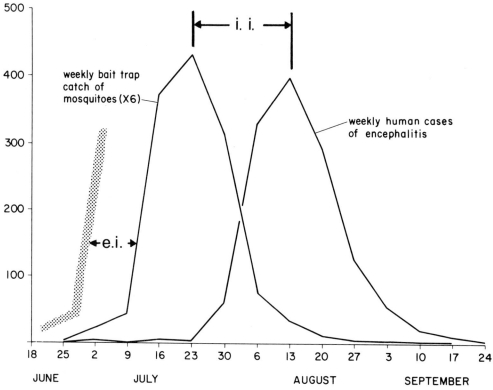

Fig. 4-1 Relationship between mosquito vector populations and human cases of Japanese encephalitis. The extrinsic incubation period (*e.i.*) is a temperature-dependent time for the vector to become infectious by proliferation of the virus into the salivary glands. The intrinsic incubation period (*i.i.*) is the time in the vertebrate host from introduction of the pathogen to expression of clinical symptoms. (Modified from Barnett.)

zone of activity of the vector with its host, such as tsetse in a thicket transmitting trypanosomes between mammalian hosts found there, arboreal mosquitoes transmitting yellow fever virus among monkeys in the tree canopy, or *Anopheles* mosquitoes entering dwellings to transmit malaria.

Criteria similar to what microbiologists term Koch's postulates, regarding pathogens, must be satisfied to accurately connect a specific arthropod with transmission of the causal agent of a disease. Briefly stated these are (1) ASSOCIATION—the demonstration of feeding or other effective contact with the host under natural conditions; (2) SPECIFIC CONNECTION—a convincing biological association in time and/or space of the suspected

arthropod species and occurrence of clinical or subclinical infection in the host; (3) CONSISTENCY—repeated demonstrations that the arthropod, under natural conditions, harbors the infectious agent in the infective stage; (4) TRANSMISSION—ability to transfer the infectious agent under controlled conditions; (5) BIOLOGICAL GRADIENT—low populations and high populations of the suspect vector relate to low incidence and high incidence of cases in susceptible hosts.

Experimental Transmission. Transmission of a pathogen by an arthropod under laboratory conditions to any susceptible animal is another means of incriminating vectors. However, transmission under such conditions really only shows those potentially capable of transmission. Suspect vectors may be exonerated if transmission proves impossible experimentally, though for leishmaniasis it has frequently been difficult to obtain experimental transmission. Difficulty in transmission as suggested by low titer of pathogen, or prolonged developmental time before transmission will occur, can reveal that suspect arthropods probably have an insignificant role in nature. When experimental transmission occurs readily with a vector not found where the disease occurs, there is evident danger in introducing an infected host into its range. Insight into various complications of arthropod transmission of pathogens can be gained through reading the review by Chamberlain and Sudia (1961).

In ordinary transmission tests a naturally infected or inoculation-infected animal is fed on by suspect arthropods. These presumably infected arthropods are then permitted to feed on other susceptible animals, first allowing the pathogen time to develop to the extent that it is in an infective stage or in sufficient quantity to be infective. Since arthropods do not regulate their body temperature, environmental temperature during the incubation period greatly influences the length of extrinsic incubation. Once the presumed infectious arthropod has bitten or otherwise infected the host, the latter is then observed for the appearance of characteristic clinical or other evidence of infection.

Epidemiologists refer to vertical and horizontal modes of transmission. VERTICAL TRANSMISSION means hereditary types of transfer, the direct transfer of infection from a parent organism to its progeny. In arthropod vectors this usually consists of infection of the maternal oocytes by pathogens, with passage to progeny called TRANSOVARIAL TRANSMISSION. There are a number of complications in transovarial transmission, including familial lines more prone to this type of transfer, and the fact that in many cases the level of infected progeny is low and supplemental routes of infection are required

(cf. Fine, 1975). HORIZONTAL TRANSMISSION refers to all types of transfer of infection between host individuals except transfer that occurs directly from parents. An arthropod vector usually acquires pathogens from feeding on an infected vertebrate host, but a more indirect route has been observed in triatomine bugs and other bloodsucking arthropods, which may obtain infested vertebrate blood by feeding on a freshly engorged arthropod of their own or of a different kind.

In those instances where the vector does not transmit by inoculation, as in epidemic typhus or Chagas' disease, after a suitable extrinsic incubation period the infectious vector feces or excreta are tested. This is accomplished by scratching these infectious products into the skin of a susceptible animal, or by using microbial culture techniques to isolate pathogens.

Methods have been developed to overcome situations where the arthropod will not feed readily on an infected host in the laboratory. Thin membranes derived from animal structures, fine cloth fabrics, or latex sheeting have been found suitable for feeding arthropods upon infectious materials. The infectious materials are mixed with attractive substances, such as blood or sugar solutions, and are usually made more acceptable by warming to simulate the temperature of a mammal or bird. A variant of this procedure has been to place a capillary tube of infectious fluid over the stylets of phlebotomine sand flies or mosquitoes (Hertig, 1927), or the hypostome and chelicerae of ixodid ticks (Burgdorfer, 1957). Introduction of pathogens into the posterior end of the gut has been used to infect *Glossina* with trypanosomes, and various arthropods such as ticks with arboviruses (cf. Stelmaszyk, 1975).

Almost certain acquisition of viruses and bacteria is possible by direct injection into the hemocoele. This technique is likely to be used if a test vector is difficult to handle under rearing and feeding conditions, and it may be mechanized (Boorman, 1975).

Direct injection bypasses the gut and there-fore, even if successful, does not really in-criminate a vector.

Relationship of Vertebrate Pathogens to Vectors

The large number and variety of vertebrate pathogens transmitted by arthropods suggest a long-standing relationship. These patho-gens are generally well adapted to the vector, causing no obvious adverse effects, but arthropod pathogenicity is noted in some in-stances. Subtle effects, such as slightly re-duced reproductive rates, or shortened life span, are probably fairly common but have not been systematically looked for in many vectors. Nonetheless, even though effects appear subtle in laboratory evaluations, they probably have a measurable influence on vector populations, and thus on transmission levels in nature.

There is increasing interest in identifying those factors that make blood- or tissue-feed-ing arthropods suitable for development of a specific pathogen. Anatomical features are easily conceived of as being significant, yet in few cases are structural characteristics known to be important. Physiological dif-ferences, whereby the vector host is specifi-cally suited to meet nutritional, respiratory, and other requirements of the pathogen, are readily conceptualized, but also difficult to prove. By and large the available evidence is suggestive rather than irrefutable. The physi-cal and chemical characteristics of the arthropod gut and its contents are involved, and this subject is reviewed in Chapter 3.

A further block in pathogen development may occur, once the gut is bypassed, when inadequate transformation or multiplication occurs in general body tissues such as he-molymph, muscles, or reproductive system. Finally, if a pathogen such as an arbovirus or the causal protozoan of malaria is to concen-trate in the salivary fluid, the salivary glands must be freely permeable to pathogen entry.

Variability of Vectors in Acquiring and Maintaining Vertebrate Pathogens

A species of vector arthropod may be highly variable with respect to its importance as a vector within different geographic areas. Close observation has generally proved such variations to be due to behavioral differ-ences, or basic differences in susceptibility to establishment of the vertebrate pathogen. Ultimately genetic differences have been demonstrated, often to the point of differen-tiating subspecies or cryptic species. But even if genetic isolation is not proved, it is possible to show natural variation in suscep-tibility and to develop changed susceptibility by selection. The whole problem of genetics of vector susceptibility to establishment of vertebrate pathogens is reviewed by Mac-donald (1967) and Barr (in Maramorosch and Shope, 1975), and refractoriness in vector strains is one of the characteristics sought in genetic approaches to controlling pathogen transmission.

Behavioral changes of vectors, some of which are discussed in Chapter 5, may occur in response to selection. The variety of ver-tebrates fed upon by a vector species is highly dependent upon the kinds of hosts available, and behavioral characteristics of a vector strain greatly affect its efficiency as a vector. Insight into such behavioral rela-tionships may be gained from Hocking's (1971) review of bloodsucking behavior of terrestrial arthropods.

Physiology of the Vector

Many physiological characteristics of arthropods have been studied in those of medical importance. In fact the blood-feed-ing habit provides opportunities to study such features as protein digestion, ni-trogenous excretion, and hormonal mecha-nisms. Certain physiological findings dis-cussed here have already contributed to an understanding of the importance of vector species.

Age Estimation Studies. Longevity during the blood-feeding stage of an arthropod is an important consideration in determining its vector potential. In general the longer the life span of a species, the better its chances of acquiring and transmitting pathogens.

Various external signs, though of poor specific accuracy, provide evidence of longevity. These are signs of wear that can be used to advantage in conjunction with more accurate methods. As examples, a recently emerged insect is virtually perfect in its scale and setal covering, and its color pattern is distinct. Within a few days of active life many scales and setae are lost, and colors may become comparatively dull. Further aging is characterized by wear of the tips and back edges of wings in flies and mosquitoes, and the spurlike ctenidial combs of fleas may become worn or broken off. Corbet (1960) has discussed some of the readily utilized external characteristics of mosquitoes, showing that age estimation can be greatly accelerated with little loss of accuracy if some screening by external characteristics precedes techniques requiring dissection.

Internal evidence may also be used. Though burdensome to carry out, a technique of staining and counting the daily growth layers in thoracic phragmata (internal cuticular braces) is accurate and could be applied up to 13 days in anopheline mosquitoes (Schlein and Gratz, 1973). This technique could presumably be applied to other vectors.

Except in cases of transovarial carry-over or venereal transmission, a bloodsucking vector must feed more than once to transmit a pathogen. Certain arthropods, such as many ticks and lice, feed on blood as a sole source of food, in all postembryonic development, so that the achievement of growth after hatching automatically implies blood feeding has occurred. However, blood feeding in the adult stage, generally for the purpose of egg development, is characteristic of vectors such as most bloodsucking Diptera. It is this relationship of blood meals to egg development that has proved of specific value in age estimation. After a full blood meal the ovaries of most adult female insects increase greatly in size, with a return to small size immediately after oviposition. The ovarian tissues thereby undergo changes that are permanent and diagnostic in determining whether any egg development (PAROUS) or no egg development (NULLIPAROUS) has occurred. Recent studies indicate that partial blood meals may be quite frequent in nature, due to vertebrate host defensive behavior. Partial blood meals may be sufficient for pathogens to be acquired, yet insufficient for ovarian development.

Permanent changes in fine endings of tracheoles coincide with advanced ovarian development. In common with insect tissues in general, the ovaries are supplied with tracheoles to meet the oxygen requirements of cellular respiration. Tracheolar proliferation into the cellular sheath of the ovaries is particularly rich to meet increased oxygen requirements during the growth of eggs. The increase in size of oocytes to form fully developed eggs permanently stretches the knotlike endings of fine tracheoles within the ovarian tissue. Thus, unraveled or stretched tracheolar endings indicate the parous condition, whereas knotted endings are typical of nulliparity (Fig. 4-2). Another indication of parity that is infrequent, but quickly recognized when it occurs, is the finding of eggs that are developed but have failed to be laid and may be partially or entirely unresorbed (RELICT EGGS) (cf. Bellamy and Corbet, 1974).

A complication modifying the interpretation of blood feeding is AUTOGENY, a condition whereby a proportion of a vector population can develop a first batch of eggs without a blood meal. This phenomenon is known for many bloodsucking arthropods; it

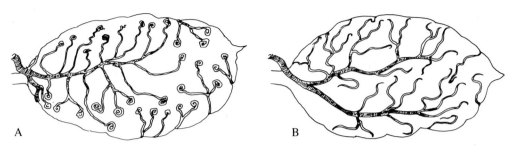

Fig. 4-2 Tracheole changes in mosquito ovaries as a consequence of ovarian development and egg laying (after several authors). *A.* No ovarian development has occurred (nulliparous) and tracheole endings remain in tight coils. *B.* Eggs have been laid (parous) and tracheole endings are permanently stretched.

is a genetically controlled characteristic that requires ideal nutritional and environmental conditions for maximum expression, and it is controlled by hormonal factors (cf. Clements, 1963, and more recent authors).

Numbers of Blood Meals. Information on the effectiveness of a vector population can be gained by learning how many times individuals feed on blood. Again specific evidence, not applicable in cases of autogeny or small blood meals, is provided by the ovaries. Oocytes develop sequentially within ovarian tubes (OVARIOLES). In most blood-feeding insects only the basal oocyte within each ovariole develops after a blood meal, and generally a single full blood meal will mature one series of basal oocytes; there are cases where more than one blood meal between ovipositions is common. As the oocyte expands to form a mature egg it stretches the epithelial sheath surrounding the ovariole. After oviposition the flaccid epithelium of the sheath in each recently functional ovariole shrinks, leaving a small knotlike dilatation, termed the CORPUS LUTEUM (reminiscent of the same name in the vertebrate ovary, though not known to have a hormonal function). The number of dilatations on the ovariole corresponds to the number of previous ovipositions (Fig. 4-3). By inference the number of ovarian dilatations per ovariole also denotes the number of prior blood meals, and thus the number of opportunities to acquire and transmit pathogens. The number of dilatations per ovariole in females sampled will also provide an estimate of the longevity and stability of the population. Thus in a vector control campaign such measurements over a period of time can indicate the level of success of the program.

Colless and Detinova have been instrumental in the development of these methods and the latter has prepared a monograph (1962) and a review article (1968) that

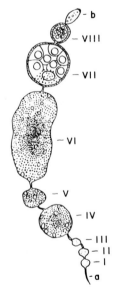

Fig. 4-3 Diagram of ovariole from a multiparous mosquito. Dilatations I through VI are decreasingly old tissue remains from prior ovipositions. Items VII and VIII are undeveloped oocytes. *b* is the germarium. (Courtesy of WHO.)

analyze application of these techniques in studying Culicidae, Simuliidae, Tabanidae, Ceratopogonidae, and other insects.

In addition to the proven utility of these earlier findings, changes in ovarian structure, abdominal wall pigmentation, fat body cells and pigmentation, digestive tract cells, and clarity of the halteres have been shown to indicate parity or nulliparity in many groups of medically important arthropods, including some that do not feed on blood. A representative listing includes ticks, cockroaches, ceratopogonids, cyclorrhaphous Diptera, Hippoboscidae, and fleas.

Survival Under Difficult Conditions. The great variety and abundance of insects and related arthropods attest to their many adaptations for survival under often harsh environmental conditions. Certain of these adaptations, namely cold and drought resistance, and general longevity during starvation, permit the survival of pathogens in environmental extremes. The use of pesticides has in recent years been an important limiting factor of the vector's environment, and adaptations through resistance to pesticides are mentioned in Chapter 5.

Though hematophagous arthropods (and the pathogens they transmit) are often thought to be characteristic of tropical zones, they may also be abundant at latitudes well removed from the equator. In such cases feeding activity is restricted to warmer months of the year, or infrequently occurs in the warm habitations of man in winter. The problem is rarely present, of course, in parasites that continuously infest their hosts. Winter survival of free-living arthropods is generally accomplished by cessation of growth, a condition known as DIAPAUSE. Diapause may be typified by a prolonged period (overwintering; summer aestivation) of an immature stage, or by reproductive cessation in the adult condition. In cases where only a single generation per year is found, as at extreme north and south latitudes (cf. Downes, 1965), diapause is said to

be OBLIGATORY. At midnorthern and midsouthern latitudes (approximately 35 to 65 degrees) multiple generations are possible, and diapause after a few generations in order to survive through winter is referred to as FACULTATIVE. In these latter cases the arthropod responds to environmental cues such as lowered temperatures of autumn or shortened day length. Some characteristics of facultative diapause are discussed under mosquito biology in Chapter 10.

A vector arthropod may have quite different environmental adaptations in different regions. Büttiker (1958) found that in the cold climate of Afghanistan *Anopheles culicifacies* precedes winter inactivity (hibernation) by blood feeding resulting in fat body development and cessation of ovulation; in eastern India, where winters are mild, there is a semihibernation with repeated blood meals converted to fat; in northern Ceylon and central Burma, summer inactivity (aestivation) begins at the start of the dry season. Other similar examples expressed by season of adult activity, as in *Culiseta inornata,* are known (Horsfall, 1955).

Behavioral and physiological changes occur among arthropods in hot and dry environments because their small size and relatively large surface area make desiccation an acute problem. A few examples will suffice as illustrations. The mosquito *Aedes vittatus* breeds in rock holes. In northern Nigeria it was found that its eggs resist desiccation at normal temperature for at least 10 weeks, indicating sheltered rock holes should permit survival in a relatively dry season (Boorman, 1961). Rozeboom and Burgess (1962) studied survival of several mosquito species during extended drought conditions in Liberia. In their investigation of plant cavities and banana leaf axils there was evidence that *Aedes simpsoni* could survive as larvae in the minute film of water on opposing surfaces of the leaf base and trunk. In Tanzania the mosquitoes *Anopheles gambiae* and *An. pharoensis* rest during the day in low vegetation, and in soil cracks, especially during the dry season (A. Smith, 1961). The desert-inhabiting soft tick *Ornithodoros savignyi* of South Africa is very susceptible to desiccation, but digs itself into deep sand under large, shady trees (Theiler, 1964).

Hematophagous arthropods for the most part require frequent blood meals. However, many of those associated with mammalian burrows, dens, or bird roosts have evolved an ability to withstand prolonged starvation. Examples may be noted among fleas and soft ticks in the chapters dealing with these arthropods.

Blood Meal Sources. One means of determining the significance of a potential vector in maintenance and spread of a given disease is the identification of its blood meal sources. Where the disease is present such information can verify that a potential vector feeds on a suspect host animal, and other identified blood meals may suggest where the pathogen is being acquired or maintained (Boreham, 1975). This is not necessarily a static relationship, as seasonal or regional changes in blood meal sources may reflect changed availability of vertebrate hosts to vectors.

Both general and specific techniques are required to determine blood meals. It is first of all desirable to obtain fresh blood, generally recognizable as still being red rather than brownish to blackish. As blood is digested it progressively loses its distinctive morphological and antigenic properties. The morphological characters of blood cells are of only general use; nucleated red blood cells are, with rare exceptions, those of birds, reptiles, or amphibians, and enucleated erythrocytes are from a mammalian source. Specific crystal patterns formed by coagulating hemoglobin of blood with oxalate (Fig. 4-4) provide a promising approach for identifying blood meal sources to species or larger groupings (Washino and Else, 1972). This technique may have to be combined with more sophisticated methods for com-

Fig. 4-4 Coagulated hemoglobin crystals from (a) blood of horse and (b) goat. (Courtesy of R. Washino, University of California, Davis.)

plete accuracy, but it has the advantage of speed and can be carried out under field conditions.

Proteins in the blood serum contain specific antigens. According to the sensitivity of the type of the test employed, as well as the specificity and diversity of the antisera prepared, it is possible to identify blood by immunological reactions to a closely related group of animals, or in some cases even to species of origin. The general reliability of immunological methods is indicated by their increasing use by systematists to determine relatedness of organisms, and by the flourishing field of immunochemistry.

A frequent procedure for blood source determination is to inject blood serum from a donor animal into a host animal whose immunity mechanisms will respond (become

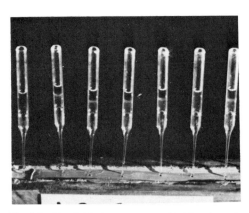

Fig. 4-5 Precipitin test on mosquito blood meals for the presence of cattle blood using general bovid antisera. Reading *left to right*, tubes 1 and 6 are negative, and the remainder are positive as indicated by the ring of whitish precipitate a small distance below the surface. (From Boreham, *PANS*, **18**[2]:205–209, Crown copyright, 1972.)

sensitized) by forming immune serum with antibodies. The immune serum then reacts to specific antigens in a characteristic way, such as precipitation (Fig. 4-5) when mixed with even small amounts of blood serum from the same or a closely related sensitizing species of animal. Experience is required to distinguish nonspecific responses that may occur. Procedures and their interpretation require more background than can be appropriately dealt with here, though the interested student can consult immunology texts or review articles.

Other methods include the "inhibition test" procedure (Weitz, 1956) (Fig. 4-6), and agar gel immunodiffusion techniques for identification of host bloods (cf. Crans, 1969; Foot, 1970).

One particular problem to be avoided is bias in collecting arthropods with a fresh blood meal. A freshly engorged arthropod usually seeks secluded resting places, often near its recent blood meal, for the blood to digest and eggs to mature. It is a natural human failing to collect engorged arthropods where they are most easily obtained, but more diligent generalized searching is required to identify all types of hosts fed on in an area. For example, a preponderance of engorged mosquitoes resting within a cattle shelter will usually contain cattle blood, those within houses will contain the blood of humans or of household pets.

Terms have been developed to identify the preferred blood source of vectors. ANTHROPOPHILOUS (Gr. *philos*, "loving") or ANTHROPOPHAGOUS (Gr. *phagein*, "eat") species of arthropods are those that are attracted to feed on man, ZOOPHILOUS forms (generally meaning mammal-feeders) prefer other animals, and ORNITHOPHILOUS arthropods prefer birds. EXOPHAGOUS species usually feed outdoors, and ENDOPHAGOUS species enter houses or other man-made shelters to feed.

Population Studies

Awareness of the kinds and numbers of medically important arthropods within a given study area is of paramount importance. The objectives are (1) to identify all suspect vectors, (2) to gain some idea of the relative numbers of each species, and (3) to determine whether populations are relatively stable or in a state of increase or decline. Increasing numbers of vectors may alert one to the imminent development of an epidemic. In vector control programs, and particularly during eradication campaigns, a fairly continuous estimate of population size can indicate the degree of success.

Collecting Methods. Many techniques, each with particular utility for specific situations, have been developed for collecting medically important arthropods. In the case of ectoparasites it is generally sufficient to capture and examine vertebrate hosts; ectoparasites may then be obtained by brushing them off their hosts, causing them to let loose by using anesthesia, such as ether, or by vigorously washing them off with detergent solution. For actively flying or crawling arthropods a standardized trapping pro-

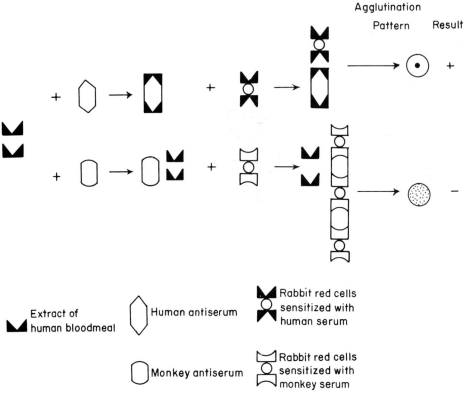

Fig. 4-6 Diagrammatic representation of the hemagglutination inhibition test to distinguish blood from closely related hosts, such as man and monkey in this example. The blood meal extract is mixed with appropriate antisera, and after incubation rabbit red cells sensitized with the test serum are added. If the blood meal prevents the direct agglutination of sensitized cells and antiserum, the meal was derived from that host (man in this example). (From Boreham, *PANS, 18*[2]:205–209, Crown copyright, 1972.)

cedure, such as light trapping, may be used but has the drawback of capturing an excess of irrelevant insect material. Methods that capitalize on specific behavior or attraction to a host are more discriminating but require time and effort during capture, or require maintenance of vertebrate hosts. See Galun (in Shorey and McKelvey, 1977) regarding host finding in blood-feeding arthropods.

Surveys of stages of development other than the bloodsucking stage can be useful. For example, in mosquito control it is common practice to determine the presence of the aquatic larvae, particularly as this is the stage against which control measures are most frequently directed. Knight (1964) has

evaluated mosquito larval survey procedures. Gillies (1974) reviews methods for assessing the density and survival of blood-feeding Diptera, and particularly noteworthy is the comprehensive review of Service (1976) covering surveys on all stages of mosquitoes.

The specific behavior of arthropods is sometimes exploited. Hard ticks (Ixodidae) frequently climb onto grass and low shrubs to transfer to passing hosts (Fig. 4-7). They may be collected by FLAGGING, dragging coarse cloth, such as flannel, to which they cling, over the ground vegetation. Many mosquitoes rest in shaded environments during the day and can be surveyed by placing

Fig. 4-7 A questing tick, *Dermacentor andersoni,* waiting on a blade of grass to attach to a host. Orientation may also be upside down. (Photograph by R. D. Akre, Washington State University.)

artificial shelters and capturing the resting individuals. Several shelter types have been developed, such as wooden nail kegs, artificial huts, earth-covered barrels, and artificial pit shelters. A foot-square wooden box painted red and with one open side has proved simple and effective for many species.

In surveying for the presence of *Aedes aegypti,* particularly where the populations are low, placing ovitrap containers with water where they can oviposit is helpful. By this means fewer man-hours are needed than by trying to capture adults, and the presence of eggs or larvae in the placed containers indicates the need to intensify the search for obscure natural breeding sites. Likewise, artificial containers have been developed to determine numbers of treehole-breeding mosquitoes.

A variety of light traps has been designed to sample flying insects. In mosquito surveys the New Jersey light trap (Fig. 4-8) is widely

used. It has an incandescent bulb as attractant and a fan to draw nearby insects down into a killing chamber. As with insects in general, bright moonlight is competitive and will lower light trap catches. Black light traps employing a fluorescent tube with high ultraviolet output are very successful with many insects, notably some ceratopogonid bloodsucking midges and black flies, but have little advantage for mosquitoes (but see Wilton and Fay, 1972). Light traps serve a useful purpose in demonstrating population trends rather than absolute numbers, and they are often used by mosquito abatement districts to show seasonal trends and to provide population comparisons between years of operation.

A portable battery-operated modification of the New Jersey type of light trap has proved to be extremely useful for arbovirus surveillance (Sudia and Chamberlain, 1962). This particular design has a live trapping chamber and finds increasing use in collecting flying bloodsucking insects to survey for the presence of vertebrate pathogens.

Traps for flying hematophagous insects may be baited with various hosts to yield specific information on host preference, as well as data to develop an index of the popu-

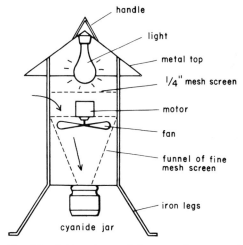

Fig. 4-8 A New Jersey light trap. (Courtesy of WHO.)

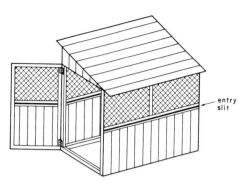

Fig. 4-9 Magoon trap, usually baited with a calf or other large animal. (Courtesy of WHO.)

Fig. 4-10 A trap for tabanids and similar biting flies. The biting flies are visually attracted to the black sphere and then fly up through a trapping funnel into the top collection chamber. (Courtesy of T. R. Adkins, Jr., Clemson University.)

lation attracted. One widely used design is the Magoon or stable-type trap (Fig. 4-9), which is essentially a large screened cage having side entrance baffles. Magoon traps are often baited with a calf or donkey. Smaller baffle entrance traps on the same principle have proved useful in studying host preferences in the wild. Lumsden (1958) developed a trap that permits insects to be attracted to a host and then at predetermined intervals draws them by suction into a trapping chamber. A striped and moving black and white visual pattern with suction is useful for species attracted largely by vision (Haufe and Burgess, 1960).

Carbon dioxide, a ubiquitous respiratory gas, has been used, most commonly in the form of dry ice, to attract many bloodsucking arthropods. The amount of carbon dioxide released can be a selective factor determining which species will predominate in the catch.

Specific methods permit assessment of the presence and numbers of flies that breed in organic matter. For filth-breeding flies an attractant such as fresh or decomposing meat is used, over which a screen cone funnels departing insects into a trap chamber. Fly resting sites called grids are another means of rapidly assessing populations of house flies and filth-infesting flies (Scudder, 1947).

All trapping procedures have some bias with respect to where and how the trapping is done. Such bias may be intentional, as with malaria studies that employ traps to capture mosquitoes entering huts to determine endophily of vectors; platform towers in forest environments to compare the species of vectors biting at ground level and in the canopy; fly rounds of human collectors, perhaps with oxen added as bait, traveling over prescribed routes to determine biting activity of tsetse flies; silhouette traps of dark color and approximate animal size to sample adult tsetse, tabanids, and black flies (Fig. 4-10).

If general information surveys are intended, less discriminating traps or procedures should be used. In comparing relative attractiveness of various hosts or baits, the traps used may be evenly spaced on a slowly rotating turntable, thus eliminating the bias of immediate position. Turntable procedures have been profitable for studying mosquitoes and eye gnats (cf. Mulla et al., 1960). The Malaise trap, a trap that has sidewalls and a funneling top of fabric or screen, has proved extraordinarily successful with a great variety of flying insects, but especially Diptera (Marston, 1965; Breeland and Pickard, 1965) (Fig. 4-11). Several modifications of Malaise traps, operating on the same general principle, have been devised. Malaise trap catches of many flying bloodsucking insects can be greatly enhanced by the addition of

Fig. 4-11 Malaise-type trap for flying insects. Insects strike cross walls under canopy and fly or crawl up into collecting chamber. (Courtesy of R. D. Akre, Washington State University.)

dry ice as an attractant (Roberts, 1970). Moving nets indiscriminately sampling insects flying in a given zone have been used on a rotating axis to capture mosquitoes, or mounted on a driven vehicle to sample ceratopogonid gnats, mosquitoes, and other flying insects (cf. Bidlingmayer, 1961).

As can be noted from the variety of trapping methods mentioned, the procedure employed depends on behavior of the kinds of arthropods sought and the objectives of the capture. Effective attractants and traps are central to most epidemiological investigations involving arthropods, and if an analysis of infected vectors is sought, it will be necessary to use methods that obtain live specimens.

When sampling data have been collected, various statistical procedures may be employed if the goal is to indicate population size and trends. One common objective is to obtain a relative index of population size, a sort of running seasonal and yearly account of trends for which statistical procedures are mainly designed to smooth out wide fluctuations characteristic of daily or weekly trap catches. Sometimes collections are made to map the geographic distribution and relative abundance of vectors, particularly with disease-associated species having very wide distribution (see WHO maps of *Aedes aegypti* distribution presented by Brown, in Howe, 1977).

Absolute population size estimates are sometimes desired, and they may be obtained with considerable accuracy by marking and releasing a known number of insects, recapturing and identifying the proportion of marked to unmarked individuals to calculate the absolute size of the population. This is a simplified description of what is known as the LINCOLN INDEX, widely used in estimating absolute numbers of all kinds of animals. Statistical methods for population estimates may be referred to in Watt (1962), and several WHO publications are applicable for mosquitoes. Marking is generally accomplished with dyes or radioactive isotopes.

Mobility and Rate of Spread. In determining the spatial extent of coverage required in control programs to interrupt transmission, and for quarantine purposes, the mobility of insect vectors must be determined. The problem of controlling transport of vectors by aircraft and other vehicles is discussed in Chapter 5.

Information on natural spread is gained by inference (for example, by knowing certain insects are outside their breeding sites) or by marking and releasing from a specific area and then recapturing at known distances from release. Both methods have yielded greater flight distances than originally estimated. One must distinguish between these distance records, which are rather like Olympic sports records, and average dispersal that is more relevant when attempting to control arthropod-associated diseases.

Inferential methods have yielded some surprising distances of spread. W. D. Haufe (Agriculture Canada, personal communication to R. F. H.) has found species of *Aedes* mosquitoes far from known breeding grounds as a consequence of atmospheric convections; likewise, records for *Culiseta alaskaensis* at Fort Churchill, Manitoba, indicate dispersal at least 480 km in association with weather disturbances along the polar front. Horsfall observed *Aedes vexans* to migrate 144 to 468 km from Wisconsin to Illinois, USA, in a weather front situation. Garrett-Jones attributed a 1959 malaria outbreak in widely separated Israeli settlements to migrations of *Anopheles pharoensis* up to 280 km from the Nile delta. Migrations of salt marsh mosquitoes in the southeastern seaboard of the USA are well known, *Aedes taeniorhynchus* in Florida dispersing at least 96 km into populated areas.

To understand the potential range between breeding sites and zones of pathogen transmission, one needs to know what distances freshly fed or parous vectors can fly. If isolated breeding sites can be identified, it is possible to collect vectors by whatever means available at increasing distances from breeding sites, and to dissect these in order to determine their parous condition. One normally assumes that recently blood-engorged vectors will not fly great distances, but Edman and Bidlingmayer (1969) found several species of recently blood-engorged mosquitoes to fly at least 1.6 km. By marking, it has been shown that spread by flight may be rapid. In a study in Lake County, Oregon, USA, marked muscoid flies were recaptured 8 km from release in less than 2 hours. An artificial but comparative technique for determining flight capacity of vectors under a variety of conditions is to check flight performance on a flight mill (Rowley *et al.*, 1968; Hockmeyer *et al.*, 1975).

Among flying insects one can distinguish between normal purposive or appetential flight—for example, to find a mate, food, or an oviposition site—and migratory or consummatory flight. This whole subject is reviewed by Johnson (1969).

Determining the Proportion of Infected and Transmitting Vectors in a Population. In epidemiological studies of specific diseases it is necessary to know what proportion of naturally occurring vectors is in-

fected. Except for viruses and highly cryptic pathogens, the vectors may be dissected and examined microscopically, provided that pathogens are present in sufficient numbers. Thus, in studying malaria the salivary glands and midguts of *Anopheles* mosquitoes are examined for sporozoites and oocysts, respectively (Figs. 10-13, 10-14), trypanosomes are sought in the salivary glands and digestive tract of tsetses (or digestive tract of triatomine bug vectors); *Leishmania* are looked for in the salivary glands and digestive tract of phlebotomine sand flies; and filarial larvae are searched for throughout the body of their dipteran vectors (Fig. 11-16). Dissection and examination of fresh specimens are preferred, but since this is sometimes impossible, methods have been devised for examining dried, refrigerated, or frozen specimens (Ward, 1962; Minter and Goedbloed, 1970; Ungureanu, 1972).

Pathogens not requiring a specific stage of development to be infectious may be demonstrated by introducing them into susceptible hosts or culture systems. Individual or pooled vectors are ground in diluents, and suspensions are injected into hosts such as suckling mice, guinea pigs, or chick embryos, which are then observed for typical clinical evidence of disease. Similarly, such suspensions may be introduced into susceptible vertebrate tissue culture systems that are observed for the development of cytopathological effects. Further developments include the use of invertebrate tissue cultures to identify and to multiply pathogens (cf. David-West, 1974; Řeháček *et al.*, 1973; Trager, 1975; Vago, 1972; Weiss, 1971; Yunker and Cory, 1975). These techniques have been most helpful in determining if vectors are infected with viruses and rickettsiae (cf. Sudia and Chamberlain, 1974). It is beyond the scope of the present discussion to indicate differences found by site of infection, use of splenectomized hosts, titer of infectivity, and other complications that must be understood to utilize these methods properly. The interested reader should consult articles and textbooks in immunology and virology, such as the review of serodiagnosis by Schmidt and Lennette (1973).

In surveying for numbers of infected vectors, particularly in studying viruses, wild-caught vectors are collected and prepared while fresh, or stored at temperatures well below freezing until processed. The number of vectors in a group, known as a POOL, is recorded. Each pool is preferably composed of a single species, or species as closely related as identification in fresh condition at the time of collection will allow. A pool of vectors is triturated in a diluting fluid of physiological saline and serum or other blood protein product. Antibiotics generally are added to prevent bacterial contamination, and the suspension is injected into susceptible hosts or dispersed in tissue cultures. Statistical treatment, as outlined by Chiang and Reeves (1962) for mosquito vector populations, permits estimation of infection rates.

The actual transmission rate in nature is a more important immediate parameter to be known than the infection rate alone. Reeves and coworkers (1961) describe a method for determining the transmission rate of viral encephalitides, using single-night exposures of susceptible chickens (sentinel animals) in traps. By this method the risk of virus infection of susceptible hosts, such as man, can be estimated.

Where direct visual observation of pathogens within a vector is difficult, as with rickettsiae and viruses, the fluorescent microscope and specific FLUORESCENT ANTIBODY conjugates (immunoflurorescence) have proved useful (Burgdorfer, 1970; Gaidamovich *et al.,* 1974). Although these methods are promising (see Figs. 15-25, 16-20), they require further development before they will find routine use.

Methods used for determining salivary gland sporozoite infection rates of *Anopheles* are treated in books on parasitology or malariology. Because individual mosquitoes must be dissected and examined, and actual infection rates may be very low, as many as two to three thousand dissections may be required for an accurate assessment. Transmission rates usually vary seasonally, making it advisable to examine collections made several times in a year. The World Health Organization has prepared a number of pertinent publications, the two-volume treatise entitled *Practical Entomology in Malaria Eradication* (WHO/PA/62.63) being very useful.

Vector Effectiveness. In all natural situations where arthropods transmit pathogens there are major or important vectors, sometimes limited, to be sure, to a specific geographic area or seasonal circumstance. Through epidemiological and laboratory investigations it has been possible to identify factors that predilect populations of a vector to play a primary role. Factors of known importance are as follows:

1. PATHOGEN RECEPTIVITY. Except in cases of mechanical transmission, a vector must be able to support the pathogen, though such support may not always be efficient. The pathogen must develop (filarial worms) or multiply (most pathogens), and be suitably concentrated to cause infection when introduced into a vertebrate host. The vector's contribution to pathogen development is to provide a suitable physical and chemical milieu. The vertebrate host also provides a developmental medium, but in the vertebrate complete or partial immunity develops, whereas infection is lifelong in most vectors. As examples the efficient malaria vector, *Anopheles gambiae,* is particularly receptive to *Plasmodium* species causing human malaria; the genus *Anopheles* includes the only vectors for the pathogens of human malaria; tick vectors of relapsing fever spirochetes are often suited for the development of only specific strains of the pathogen. In a sense even vectors that transmit pathogens solely by mechanical means are specially suited for their role, with easily contaminated mouth

parts or body surfaces and often with restless feeding behavior.

2. HOST SPECIFICITY. Pathogens causing diseases exclusive (or nearly so) to a specific vertebrate host, such as human malaria and human filariasis, are best transmitted by vectors that feed preferentially on such hosts. There are many cases known where *Anopheles* mosquitoes will readily develop the pathogens of human malaria, but they do not serve as natural vectors because of their preference for feeding on livestock. Also, in several North American mosquito-transmitted arboviruses (WEE, SLE) the normal hosts are birds, and man or other susceptible mammals are only tangentially involved when fed on by infectious vectors. Conversely, pathogens that develop in a number of vertebrate hosts are best served by vectors with little host specificity. The hard ticks *Ixodes persulcatus* and *Ixodes ricinus* have proved to be potent vectors of tick-borne encephalitis-complex viruses, which they transmit to a great variety of rodents, birds, intermediate-sized and large mammals, and man.

3. LONGEVITY. Except in cases of transovarial transmission, as is common with the transmission of several rickettsiae by ticks and chigger mites, a vector must feed more than once to transmit pathogens. Basic to this requirement is the need for a vector to live a sufficient period of time. The importance of vector longevity in malaria transmission has been stressed by MacDonald (1957) and by Garrett-Jones and Grab (1964). Aside from a minimal longevity required for the extrinsic incubation period prior to transmission, maximal longevity permits vectors to serve as essential parts of the reservoir of infection, well illustrated by relapsing fever and soft ticks (see Chapter 16).

4. FREQUENCY OF FEEDING. Vector-host contact of a frequent nature may increase the effectiveness of a vector, though frequent contact can adversely affect vector survival. In mechanical transmission of the etiological agents of anthrax, tularemia, and anaplasmosis by tabanid flies, the frequent interruption of feeding by restless movement of the host, in response to irritation caused by the feeding vectors, aids the spread of the pathogens. Likewise, a flea with digestive tract blocked by plague organisms will bite many times in attempting to feed, increasing its chances of transmitting the pathogen to more than one host. One factor making *Anopheles gambiae* an effective vector is the characteristic of this mosquito to seek a blood meal the same night it oviposits, whereas many mosquitoes wait a day or more before refeeding after oviposition.

5. MOBILITY. The ease with which a vector makes contact with a number of hosts is significant in determining its effectiveness. Superior mobility aids in the rapid dissemination of pathogens over a wide area so that their associated diseases are not limited and focal in nature. Mobility is obvious in insects with good flying ability, but may also be a characteristic of wingless ectoparasites such as fleas, lice, mites, and ticks that are distributed by the relative mobility of their hosts.

6. NUMBERS. Sheer population density, with enormous numbers making contact with susceptible hosts, will permit some vectors that are otherwise poor hosts for a pathogen to be of significance. Such is clearly the case where *Anopheles culicifacies* in south India and *An. albimanus* and *An. aquasalis* in Central America are principal vectors of malaria plasmodia through the density of their populations, even though salivary gland infections are frequently less than one per thousand dissections during active seasons of transmission. Additionally, when large numbers of a vector are present the chance of feeding on, and infecting other than, the preferred hosts is increased; this is of great consequence in mosquito transmission of those North American encephalitis viruses in which the vector mosquitoes prefer to feed on birds, but will bite man and other mammals when the opportunity arises.

7. Physiological and Behavioral Plasticity. Under the pressure of insecticides used extensively in control schemes, successful vectors are those that develop the ability to resist destruction by physiological (biochemical) and behavioral means. Such ability generally is found to be under genetic control. Resistance to insecticides is dealt with more thoroughly in Chapter 5. Adaptation of the mosquitoes *Aedes aegypti* and *Culex pipiens* to urban environments has increased their importance as vectors of pathogens affecting humans.

The applicable factors listed above may occur in combination. This composite of vector capability has been termed VECTORIAL CAPACITY in epidemiological investigations. Vectorial capacity summarizes the effectiveness of a vector population at a specific location and at a particular point in time. This status may be described verbally, but increasingly there is a desire to convert the concept to quantitative terms (Garrett-Jones and Grab, 1964).

In analyzing the effectiveness of vectors we may study individuals, yet from an epidemiological viewpoint it is the characteristics of a vector population that are important. Insight into methods used to characterize vector populations may be gained by referring to Muirhead-Thomson (1968) and WHO (1972c).

ARTHROPOD-TRANSMITTED PATHOGENS

Parasitism by arthropods places them in frequent, intimate, and often dependent association with vertebrates. This has provided pathways for the transfer of a variety of pathogens to other organisms. Each type of pathogen is responded to by the immunological systems of vertebrate hosts in ways that affect its future infectivity in a population of such hosts. Representative agents transmitted by arthropods, without being inclusive, can be grouped as follows:

1. Protozoa. Vertebrates do not generally acquire strong immunity to these pathogens, and repeated infection is possible. However, partial immunity to specific strains may develop, and this is characterized by slow development of the pathogen in the vertebrate host, and lower numbers of circulating pathogens.

a. *Entamoeba histolytica,* the dysentery ameba, which along with other intestinal protozoa, may be transmitted by contamination between fecal sources and foodstuffs by cockroaches and muscoid flies. These and other contaminative pathogens are usually transferred by a variety of routes involving poor sanitation, and arthropods play mainly a minor additive role.

b. Sporozoa blood parasites of man and many other higher vertebrates, such as *Plasmodium* species transmitted by *Anopheles* mosquitoes to man, and by many mosquitoes to other vertebrates. Others include *Babesia* and *Theileria* transmitted to large domestic animals, pets, and rodents by ticks; *Hepatocystis* of monkeys transmitted by *Culicoides; Leucocytozoon* species of fowl transmitted by black flies and ceratopogonid midges; and *Haemoproteus* of birds transmitted by hippoboscid, simuliid, and ceratopogonid flies.

c. *Trypanosoma* blood flagellates of man, domestic animals, and other vertebrates transmitted by tsetse and by triatomine bugs, tabanid flies (mechanically), by fleas to rodents, and by the sheep ked to sheep.

d. *Leishmania* flagellates transmitted by phlebotomine sand flies to man, dogs, rodents, and other wild vertebrates.

2. Helminths. Immunity responses in vertebrates are weak, and repeated infections are possible.

a. There are various insects, mites, and crustaceans that serve as developmental hosts of tapeworms (Cestoda), flukes (Trematoda), roundworms (Nematoda), and spiny-headed worms (Acanthocephala). Filth-associated flies and cockroaches may convey helminth eggs on their body surfaces or via their digestive tract.

b. Filarial worms affecting man and other vertebrates. Mosquitoes and *Wuchereria* and *Brugia;* tabanid flies and *Loa;* simuliid and ceratopogonid flies and *Onchocerca.*

3. Bacteria. (Terminology according to Buchanan and Gibbs, 1974) In most instances vertebrates recovering from arthropod-transmitted bacterial infections develop a strong immunity. In some cases, however, they may be reinfectible with a different strain of the pathogen.

a. The majority of common bacteria are transmitted by mechanical contamination, or the pathogens multiply within a vector and are transmitted when the arthropod takes a blood meal. Examples of mechanical transfer include the contamination of foodstuffs by filth-associated flies with foodpoisoning *Shigella* and *Salmonella;* transfer of the spirochete *Treponema pertenue* of yaws by *Hippelates* gnats and house flies; bites by tabanid flies mechanically transmitting *Bacillus anthracis* of anthrax and *Francisella tularensis* of tularemia. Examples of bacteria multiplying within a vector and transmitted during the act of blood feeding include the plague organism, *Yersinia pestis,* transmitted by fleas; some ten "species" of human-infecting *Borrelia* spirochetes of relapsing fever transmitted by *Ornithodoros* ticks; tick transmission of tularemia organism.

b. Certain groups of bacterial agents having distinct characteristics, and relationships to vectors, are categorized as follows. The order Rickettsiales is a group of very small pathogenic bacteria that are mainly dependent on living cells for growth. Out of past custom these are discussed in the text as the rickettsiae.

1) *Rickettsia,* grow only in host cells and mainly in the cytoplasm.

a) Typhus group. Louse-borne or epidemic typhus, caused by *Rickettsia prowazekii,* transmitted to man by the louse *Pediculus humanus;* murine typhus, caused by *R. typhi,* transmitted by fleas from rats to man.

b) Spotted fever group. Rocky Mountain spotted fever, caused by *R. rickettsii,* transmitted by ticks; North Asian tick typhus, *R. sibirica,* transmitted by ticks to man and domestic and wild animals; boutonneuse fever, *R. conorii,* transmitted by ticks to man; rickettsialpox, *R. akari,* of domestic and wild mice and man, and the mite *Liponyssoides sanguineus;* Queensland tick typhus, *R. australis,* transmitted to man by ticks.

c) Scrub typhus group. The agent of scrub typhus, *Rickettsia tsutsugamushi,* transmitted to man and rodents by trombiculid mites.

2) *Rochalimaea,* usually extracellular in the arthropod host; culturable in certain bacteriological media. *R. quintana,* the etiological agent of trench fever, found in man and the body louse.

3) *Coxiella,* grow preferentially in host cell vacuoles, highly resistant to physical and chemical destruction in extracellular environment. *C. burnetii,* the causal pathogen of Q fever, primarily a zoonosis acquired by various contaminant routes but also transmitted to man by ticks and known to be transmitted by other arthropods.

c. Tribe Ehrlichieae, includes *Ehrlichia* spp. transmitted by ticks to dogs and domestic ruminants, and *Cowdria* sp. of ruminants transmitted by ticks.

d. Family Bartonellaceae, including *Bartonella bacilliformis,* etiological agent of Oroya fever and verruga peruana, transmitted by phlebotomine sand flies.

e. Family Anaplasmataceae, containing the tick-transmitted *Anaplasma* pathogens that infect ruminants; *Aegyptianella pullorum,* affecting several species of birds and transmitted by *Argas* ticks; *Haemobartonella muris,* transmitted to rodents by the rat louse, *Polyplax spinulosa; Eperythrozoon* spp., blood parasites of mice, sheep, pigs, and cattle, transmitted by lice, horse flies, or unknown vectors.

f. Order Chlamydiales, including *Chlamydia trachomatis,* the causal agent of trachoma, which is transferred by ordinary contact routes that may include the house fly and related flies.

g. Mycoplasmas, very small bacterial pathogens of uncertain affinities. Several of these cause "yellows" and similar diseases of plants. A *Spiroplasma* causing cataract development when injected into rats has been isolated from ticks.

4. Viruses. These exceptionally small pathogenic agents are made up of nucleic acids with a protein coat. Vertebrates recovering from infections generally develop a strong and lasting immunity, though in some instances they can be infected again by closely related virus strains. Immunity developed after infection by one virus may provide some protection against another virus in the same family or group. A complete listing of arthropod-transmitted viruses, which will continue to be updated, may be found in the 1975 *International catalogue of arboviruses including certain other viruses of vertebrates* (Berge, 1975). Current attempts to develop a nomenclature system for viruses are based on combinations of antigenic properties, nucleic acid type, virion morphology, and site of development in living cells. The system is discussed by Murphy (1974), and the family and genus listings of ARBOVIRUSES (arthropod-borne viruses) used here are from the second report of the International Committee on Taxonomy of Viruses (Fenner, 1976). Names of viruses are usually derived from area of first isolation, native terms, or a clinical description of the disease they cause.

a. Non-arboviruses. Contaminative transfer of Coxsackie virus and the causal virus of poliomyelitis may be possible by filth-associated flies and cockroaches. In addition certain viruses have developed durabil-

ity that enhances their mechanical transfer on the mouth parts of biting arthropods. In this latter group is the family Poxviridae, which includes *Avipoxvirus* of fowl or avian pox, transmitted by mosquitoes and probably by *Argas* ticks, and Leporipoxvirus, the myxoma virus causing myxomatosis of rabbits, transmitted by mosquitoes, black flies, fleas, and other biting arthropods.

b. Reoviridae. The genus *Orbivirus* containing the etiological agents of blue tongue and African horse sickness, transmitted by ceratopogonid biting gnats; Colorado tick fever, transmitted by hard ticks; Changuinola, isolated from phlebotomines.

c. Bunyaviridae. The genus *Bunyavirus* is separated into several groups with antigenic similarities. Mainly groups with representatives that have been isolated from humans or domestic animals are listed here:

1) Bunyamwera group: The agents of Bunyamwera, Germiston, Main Drain, Tensaw, Wyeomyia—mosquito-transmitted.

2) Bwamba group: Bwamba, Pongola—mosquito-transmitted.

3) C group: Apeu, Caraparu, Itaqui, Madrid, Oriboca, Ossa—mosquito-transmitted.

4) California group: Bocas, California encephalitis, Inkoo, Jamestown Canyon, Jerry Slough, Keystone, LaCrosse, Melao, San Angelo, Tahyna, Trivittatus—mainly mosquito-transmitted.

5) Guama group: Catu, Guama—mosquito-transmitted.

6) Simbu group: Buttonwillow, Ingwavuma, Oropouche, Sathuperi, Shuni, Simbu—mosquito- and *Culicoides*-associated.

7) Possible additional groups involving mosquito, tick, phlebotomine or *Culicoides* vectors include Uukuniemi, Anopheles A, Anopheles B, Bakau, Crimean-Congo hemorrhagic fever, Nairobi sheep disease, Phlebotomus fever, Turlock.

d. Rhabdoviridae. Members of *Vesiculovirus* include Chandipura, Vesicular stomatitis (VS)-Argentina, -Brazil, -Cocal, -Indiana, -New Jersey; possibly also bovine ephemeral fever associated mainly with mosquitoes and phlebotomines.

e. Togaviridae.

1) Alphavirus (Group A), mainly mosquito-borne: Aura, Chikungunya, Eastern equine encephalitis, Everglades, Middelburg, Mucambo, O'Nyong-nyong, Pixuna, Ross River, Semliki forest, Sindbis, Venezuelan equine encephalitis, Western equine encephalitis.

2) Flavivirus (Group B)

a) Mosquito-borne; Bussuquara, dengue 1,2,3,4, ilheus, Japanese B encephalitis, Murray Valley encephalitis, Spondweni, St. Louis encephalitis, Wesselsbron, West Nile, yellow fever, Zika.

b) Tick-borne: Abbsettarov, Kyasanur forest disease, louping ill, Omsk hemorrhagic fever, Powassan, Royal Farm, Russian spring summer encephalitis.

Pathogen Development in Arthropods. Basically, as Chamberlain and Sudia (1961) assert, vectors transmit pathogens mechanically or biologically. In MECHANICAL TRANSMISSION the vector is no more than a carrier that transmits pathogens with contaminated mouth parts or legs, or by passage of agents through the gut without change. Rather obviously the act of feeding should in itself permit some degree of success in mechanical transmission by any blood-feeding arthropod, provided a vertebrate-infective stage of the pathogen is acquired. Arboviruses present excellent examples of mechanical transmission as a consequence of interrupted feeding. Chamberlain and Sudia showed that the mosquito *Aedes triseriatus* infected chicks with eastern equine encephalitis virus, after brief feeding on viremic chicks, about as successfully as a jab by an infective pin. Although this is an artificial situation, it points out the consequences of disrupted feeding. Mechanical transmission is the normal, and apparently only, method

of transfer for *Leporipoxvirus* (myxoma) among lagomorphs (rabbits and hares) by a variety of bloodsucking vectors, and for *Avipoxvirus* (avian pox) by mosquitoes.

The BIOLOGICAL TRANSMISSION of pathogens is by far the commonest method of pathogen development among arthropod-associated diseases of vertebrates. This clearly suggests a long-standing evolutionary partnership between arthropods, which in themselves became parasitic on vertebrates, and parasitic pathogens that exploited this link to transfer into vertebrates. Considering the variety of parasites and microbial pathogens involved in this relationship, and the many kinds of arthropods utilizing vertebrate tissues or secretions as important or essential food sources, it is not surprising that the pathogen strategies evolved have included some very complicated types of development in both arthropod and vertebrate hosts.

Broad classification of types of biological transmission, centering on pathogen development in the vector, has been listed by Huff (1931). This classification helps to simplify our understanding of relationships, but it is not certain where the development of bacterial and viral pathogens in vectors should be placed. Since sexuality or specific developmental stages are not clearly evident for these pathogens, they are simply grouped under propagative transmission:

1. CYCLOPROPAGATIVE TRANSMISSION. The causal organisms "undergo cyclical changes and multiply" in the body of the arthropod, as in the transmission of malaria plasmodia by anopheline mosquitoes and many other instances of development in sporozoan protozoa utilizing arthropod vectors. There is great multiplication in numbers within the vector, and there are distinct intermediate growth stages of the pathogen during which it is incapable of infecting the vertebrate host; only the final growth stage is infective.

2. CYCLODEVELOPMENTAL TRANSMISSION. When the causal organisms "un-

dergo cyclical change but do not multiply'' in the body of the arthropod, transmission may be classified as cyclo-developmental, as in mosquito transmission of filarial worms such as *Wuchereria bancrofti,* the causal organism of bancroftian filariasis. Note that there is development of the pathogen to a vertebrate-infective stage, but no multiplication. This type of development is characteristic of helminths using arthropods as an agent of transfer, as for example when the larval stage of poultry-infesting tapeworms develops in beetles. The basic difference between the mosquito and beetle examples cited, however, is that the first of these is a vector that actively seeks a vertebrate blood source, whereas in the latter case the vertebrate consumes the vector or transfer host that bears infective larvae.

3. PROPAGATIVE TRANSMISSION. When ''the organisms undergo no cyclical change, but multiply'' in the body of the vector, transmission is said to be propagative only. This term is used to describe bacterial and viral multiplication in vectors, for example, the plague organism in fleas, or the relapsing fever organism in ticks. Any stage of these pathogens can infect a vertebrate host.

The diagrammatic representation in Figure 4-12 illustrates the variety of pathways involved in arthropod acquisition, development and dissemination of vertebrate pathogens.

For the most part vertebrate pathogens show a benign adaptation to their arthropod hosts, though *Rickettsia prowazekii* of epidemic typhus is pathogenic to its body louse vector, and *Yersinia pestis* of plague needs to block the digestive tract of flea vectors in order to be transmitted by the usual route. A number of instances of defense reactions of arthropods to parasites are known (Jackson *et al.,* 1969), indicating that vertebrate pathogens that utilize arthropods have generally compromised with their carriers so as not to trigger cellular defensive responses or to excessively damage their invertebrate hosts.

The most neutral relationships occur when transmission is simply by mechanical transmission, yet a pathogen such as myxoma virus is specifically adapted by high resistance to degradation, and can thus survive prolonged periods on a variety of biting insect mouth parts (Fig. 10-24). The general evidence suggests that most of the vertebrate pathogens we are concerned with were first arthropod parasites, based on the fact that complex developmental cycles predominate in arthropod hosts. There are numerous examples where these pathogens are at least mildly detrimental to their invertebrate hosts (review by Maier, 1976; selected studies by Hacker and Kilama, 1974; Hockmeyer *et al.,* 1975; Srivastava, 1975).

VERTEBRATE HOSTS

Vertebrates are hosts to arthropods and to the pathogens they disseminate, either in a normal parasitic relationship or by chance. In medical entomology the infections of prime concern are those in man, but other vertebrates such as domestic animals and wildlife may show the same disease or discomfort responses. By far the majority of pathogens transmitted to man by arthropods are ZOONOSES; that is, these pathogens of man are maintained in other vertebrate hosts. A satisfactory discussion and analysis of the features of zoonoses can be found in Garnham (1971). ANTHROPONOSES are those pathogen-caused diseases in which man is the only known vertebrate host (i.e., *vivax* malaria, bancroftian filariasis, epidemic typhus). Even in cases where other vertebrate hosts of the pathogen are unknown, there are generally other arthropod-pathogen-vertebrate host cycles recognizable as being very closely related by virtue of the same variety of vector(s), and pathogens with close phylogeny. Likewise, the transmitted pathogens causing diseases in domestic animals show very close affinities to diseases of wild

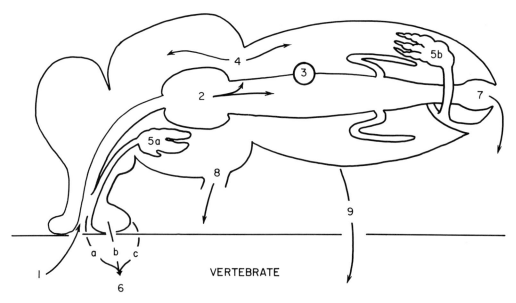

Fig. 4-12 Diagram of acquisition, development, and transmission of vertebrate pathogens by arthropods.

Arthropod Acquisition and Development of Pathogens:

1. Pathogen and blood or tissues of vertebrate host ingested. Ingestion of helminth eggs in vertebrate feces also possible.

2. Pathogens in lumen of gut may be "inactivated" (digested or adsorbed), or multiply here as with fleas and plague bacillus or *Leishmania* in phlebotomine flies.

3. Passage directly though gut wall as with filarial worms in biting flies and *Borrelia* in lice and ticks, or development in the epithelial layer as probably occurs with most rickettsiae and arboviruses, or development in cysts on the epithelial layer as in malaria.

4. Transport by hemolymph to tissues. There is evidence that some arboviruses can multiply in hemocytes.

5. Tissue concentration such as (*a*) salivary gland concentration of many arboviruses and the sporozoites of malaria and (*b*) invasion of reproductive system for transovarial transmission as with tick vectors of spirochetes and viruses, tick and mite vectors of rickettsiae, mosquito vectors of some viruses in California encephalitis complex; or sexual transmission of African swine fever virus by ticks, La Crosse virus by mosquitoes.

Acquisition of Pathogens by Vertebrate:

6. Introduction into the vertebrate host by (*a*) back pressure of digestive tract as with plague bacillus by fleas and *Leishmania* by phlebotomids, or introduction of saliva in act of feeding (see *5a*), or (*b*) escape of pathogens through the body wall as in transmission of filariae, or (*c*) by contaminated mouth parts or other body surfaces; tularemia and *Chrysops,* avian pox virus and mosquitoes, myxoma virus and various bloodsucking insects, dysentery organisms and filth flies.

7. Infected feces egested onto skin of host where pathogens enter cuts, are scratched in, or rubbed into conjunctiva (*Trypanosoma cruzi* and triatomine bugs; *Rickettsia prowazekii* and the human body louse).

8. Infective fluids may be excreted from glands, for example, viruses and spirochetes from the coxal glands of argasid ticks.

9. Vertebrate may acquire pathogen by ingesting or crushing infected arthropod. Examples include ingestion of arthropod hosts of helminths; *Yersinia pestis* acquired by rats through the ingestion or crushing of fleas, or crushing of *Borrelia*-infected lice by vertebrates.

hosts. From these epidemiological links we realize that common systems have evolved, and comprehension of the vertebrate portion of the cycle—whether humans, their domestic animals, or wild vertebrates—is an essential factor in understanding disease cycles.

Vertebrate hosts respond to pathogens in a number of ways that affect further spread of these agents by arthropods. Certain descriptive terms have been developed to categorize these relationships from an epidemiological standpoint. A DEAD END HOST is a vertebrate (usually) that may harbor the pathogen being considered, may even be severely affected by it, yet the level of pathogen in the blood and other peripheral tissues may be too low for a bloodsucking arthropod to become infective after feeding on this host. An AMPLIFYING HOST is one in which the level of pathogen is high enough that a feeding vector will likely become INFECTIOUS (capable of transmitting the pathogen after a suitable extrinsic incubation period). A SILENT HOST is one that harbors the pathogen but shows no obvious signs of disease. To illustrate the above designations, in western equine encephalitis man and horses are dead end hosts to a vector mosquito, even though they may be severely affected by the virus, while several species of nestling birds are silent and amplifying hosts. A RESISTANT HOST generally refers to a vertebrate that is not naturally affected by a pathogen (even in the absence of prior exposure, recovery, and development of an immune state), a PARTIALLY RESISTANT HOST is one that harbors the pathogen for a long period before being overcome by it or eventually recovering, and a SUSCEPTIBLE HOST is a victim of the pathogen. In plague the black or roof rat, *Rattus rattus*, is a susceptible host, certain gerbils are partially resistant hosts, and other wild rodents are resistant hosts that are refractory to infection by the pathogen.

Human Habits and Disease Prevalence. Various characteristics of man have created public health problems, no less so in vector-borne than in other disease situations. Man is unquestionably the greatest environmental disrupter of all living things and consequently may unintentionally create problems of great complexity and gravity. As with other forms of life, any tendency to overpopulate finally meets with population-limiting controls; the medically important arthropods and the pathogens they transmit are simply one of many biotic population-limiting forces. Humans view population-controlling diseases with alarm and as unnatural, but such diseases have no doubt influenced the evolutionary course of man into his present state. One well-known evolutionary force influenced by an arthropod-transmitted disease is sickle cell trait, a genetically controlled hemoglobin variant of man that is harmful when homozygous but has some protective value against malaria when heterozygous.

Historical aspects of arthropod-transmitted diseases are referred to in Chapter 1. Especially noteworthy is the role of arthropod-associated diseases such as typhus, plague, relapsing fever, and malaria in early military campaigns, and plague as a factor that reshaped society.

An understanding of human characteristics that promote or prevent arthropod-borne diseases is vital, for often a relatively minor change in existing practices can significantly reduce or increase a problem. It has been stated that every properly aware traveler, and many similarly knowledgeable residents, can avoid infection by all the parasitic diseases of warm climates (Garnham, 1971). Each disease situation has such human elements, but only a few examples will be cited here as illustrations.

Disposal of waste and unwanted articles creates acute problems. Mosquitoes that normally breed in the water of rock holes and tree cavities are found in vessels for storing water, in car and airplane tires, in tin cans, and in other artificial containers. This is clearly the case with *Aedes aegypti,* the urban vector of dengue, chikungunya, and yellow fever viruses, in which man-associated strains have developed and spread throughout the world. Improper disposal of garbage provides breeding sites for filth-breeding flies as well as rats and their ectoparasites. Urbanization as a problem promoting mosquito-borne diseases in tropical countries is reviewed by Gratz (1973). Further examples of human failures include villagers in savannah regions of Africa who permit bush vegetation that shelters tsetses to grow back in clearings next to their villages, and careless omission of chemoprophylaxis among travelers in malarious regions.

Water management problems are also a failing of man. Irrigated agriculture has increased, often with little thought for disposal of excess water. Some crops require extensive periods of standing water, and widespread world production of rice is a factor closely linked to the prevalence of mosquito-associated diseases (National Academy of Sciences, 1976). To an increasing degree, land use in countries with advanced economies includes the development of artificial lakes in housing areas for aesthetic and recreational purposes, often with serious increases in mosquito and other biting fly populations, and great hatches of chironomid midges and other aquatic insects, resulting in allergic sensitization of nearby humans.

Another type of human water use problem concerns the development of fish ponds for providing a source of food. In the Hollandia area of New Guinea it was found that in poorly managed ponds, *Anopheles farauti* and three other known vectors of malaria plasmodia were prevalent. Many examples of arthropod-associated health problems connected with man-made lakes may be found in Stanley and Alpers (1975).

Nomadic habits of less-developed human cultures have complicated the control and eradication of malaria. Infected persons in nomadic tribes of Africa and Afghanistan have reinfected stable human populations through vectors in regions where the pathogen had been eliminated. Likewise, people of remote mountainous areas in Asia and New Guinea have served as residual reservoirs of the parasites. Jungle agricultural communities are characterized by frequent moving, so that any vector control program dependent on treatment of dwellings with residual insecticides is hampered, because new huts are built but unrecorded, and so are missed in treatment schedules.

Man's activities are directly responsible for the introduction of vector arthropods, along with the pathogens they transmit. This problem is further dealt with in Chapter 5. It is worth noting that of the six internationally quarantinable diseases, flea-borne plague, louse-borne typhus, louse-borne relapsing fever, and yellow fever are arthropod-borne. It is also well to know that in temperate zones where malaria has been eradicated, vector *Anopheles* mosquitoes are often still abundant.

The very fact that man characteristically lives in close groups favors arthropod-associated diseases, and the same statement may be made regarding herd behavior among domestic and wild animals. Examples include tsetse-transmitted trypanosomiases of man and cattle, mosquito-transmitted bancroftian filariasis of man, tabanid-transmitted anaplasmosis of cattle, and all of the ectoparasitic louse and mite infestations of man and animals. Social grooming and self-grooming, however, help to reduce numbers of ectoparasites (Fraser and Waddell, 1974; Weisbroth *et al.*, 1974). In addition, the defensive behavior of hosts against feeding by arthropods can greatly influence how frequently they will be fed on, and no doubt similarly affects their suitability to serve as reservoir hosts of pathogens (Edman and Kale, 1971).

Response of Vertebrate Hosts to the Pathogen. Once infected by pathogens, vertebrates commonly develop complete or partial immunity. For this reason the number of unexposed susceptibles is a major factor determining whether an epidemic can occur. Reeves and associates (1962, 1964), analyzed factors involved in epidemics of western equine and St. Louis encephalitis. In these diseases there are many instances where a population is found to have high specific antibody rates, although the proportion of immunes must be extremely high to prevent epidemics. Even a disease eradication program has inherent dangers because a large unexposed and therefore nonimmune vertebrate host population develops. As an example, a hurricane on the southern peninsula of Haiti occurred in October, 1963, in the fourth 6-month cycle of spraying to control *falciparum* malaria transmitted by *Anopheles albimanus*. The parasite rate before control had been 10 percent, was down to 0.8 percent in the course of the control program, and rebounded to 17 percent after the hurricane, with the occurrence of 75,000 cases in a 3- to 4-month period (Mason and Cavalie, 1965).

Diseases of Wild Animals. Similar groups of arthropod-transmitted pathogens causing diseases of humans are serious problems for domestic animals and wild vertebrates. Diseases of domestic animals due to arthropod transmission of pathogens will be dealt with more fully in discussions of specific groups of vectors and the diseases with which they are associated.

From an epidemiological standpoint wildlife without evident disease may be very important sources of vector infection with human pathogens. Thus, certain species of mice and gerbils can serve as partially resistant hosts to maintain for a considerable period the plague bacilli that rapidly affect rats and man, and snowshoe hares are rarely affected by the tularemia pathogen in marked contrast to its effect on cottontail rabbits and man. Birds are well adapted to western equine and St. Louis encephalitis, but man may be seriously affected. Myxoma virus kills the European rabbit but does not kill most American rabbit hosts. These illustrations reveal that man or other affected animals are often merely tangential hosts in a cycle normally involving vertebrates that are little affected by arthropod-borne pathogens.

We are not very aware of the effect of arthropod-transmitted diseases on wildlife. Myxomatosis, a virus disease of rabbits and hares, is an example of the known serious effects of disease on wild animal populations (Fenner and Ratcliffe, 1965). There is little doubt that diseases of animals transmissible to man, termed ZOONOSES, have had a significant effect on the survival and evolutionary development of wild vertebrates, though only the most extensive epidemics in wild animal populations (EPIZOOTICS) tend to be recorded. A few illustrations will suffice.

Plague, caused by a flea-transmitted bacillus, has frequently decimated wild rodent populations in addition to its well-known effect on domestic and commensal rats. Hoogstraal (1966) cites a paper describing an epizootic in muskrats (*Ondatra*) in the Omsk area of the Soviet Union. These animals had been introduced and became numerous along streams, but large numbers died off between 1960 and 1962 due to Omsk hemorrhagic fever virus and the bacterium of tularemia, the former apparently being transmitted between muskrats and other rodents by gamasid mites.

Introduction of nonindigenous fauna, for example, exotic game birds such as the Chinese ring-neck pheasant and the Asian red-legged partridge (chukar) to the USA, has been attended by epizootics caused by eastern equine encephalitis virus while native bird populations appear to be unaffected. Such epizootics are especially noted on game farms and are not readily observed in nature. In fact, it is so much more difficult to observe the natural decimation of wild populations that we usually cannot be certain epizootics are absent.

Bird malarias likely have a very significant effect on avian populations. In the

United States it has been noted that annual Audubon Society bird census figures have increased in recent years, especially swamp- and salt-marsh-inhabiting species such as blackbirds. There has been some conjecture that improved control of marsh-breeding mosquitoes has resulted in reduced avian malaria and filariasis, permitting better survival of this group of birds. Warner (1968) has developed a hypothesis, well supported with experiments and observations, that the extinction of nearly half the native Hawaiian bird species (drepaniids), and present restriction of the remainder remnant populations to a few high mountain forest sanctuaries, is a consequence of introduction of the mosquito *Culex pipiens fatigans,* which transmits avian pox virus and bird malarias from considerably more resistant introduced avian species.

Contributions of Laboratory Vertebrates as Model Arthropod-Transmitted Disease Systems. In addition to their rather obvious use in transmission studies, laboratory and wild vertebrates have been utilized to gain an understanding of approaches in controlling arthropod-transmitted diseases. One noteworthy practice is their use in chemotherapy investigations. Chicks and canaries infected with plasmodia have aided in the screening of thousands of compounds for antimalarial activity, and promising drugs thus discovered have been tested on malarious monkeys and apes before final clinical tests with humans. Candidate trypanocides may be screened initially against *Trypanosoma lewisi* in the common laboratory rat, and antifilarial agents may be tested against microfilariae of *Litomosoides carinii* in the cotton rat *Sigmodon hispidus.*

Vertebrates previously unexposed to a pathogen under study may provide information on existing transmission rates by vectors under more or less normal conditions. Animals used in this manner are referred to as SENTINEL HOSTS; for instance, exposed chickens are termed sentinel flocks. Because the vertebrate used may be more attractive (or under greater exposure) to a vector than man is, routine use of sentinel animals can warn of a potential epidemic before the first human cases occur. Such methods have found greatest use in studying arboviruses; routine sampling of the serum of exposed sentinel animals may indicate acquisition of a virus by VIREMIA (virus actively circulating in the peripheral circulation) or by the appearance of pathogen-specific antibodies. Pathogens circulating among wild vertebrates can be detected near human habitations even when clinical cases in man are very rare, as with Venezuelan equine encephalitis in Florida, USA (Ventura and Ehrenkranz, 1975).

RESERVOIRS

The concept of reservoirs of infection for diseases caused by arthropod-transmitted pathogens has been introduced to explain the maintenance of pathogens during times when active transmission is not occurring. As Philip and Burgdorfer (1961) note, reservoirs are more a feature of temperate climates where vectors become dormant or hibernate; in the tropics there may be continuous passage between susceptible hosts and vectors. Basically, the term RESERVOIR is applied to maintenance infections, that is, to any animal system, whether vector or vertebrate, capable of maintaining a pathogen for considerable periods of time. Such a system may involve vertebrates that show no evidence of serious disease (i.e., prolonged viremia of birds, reptiles, or amphibians infected with western equine encephalitis virus), or a reservoir of infection may be maintained by continuous transmission among a group of severely affected animals (i.e., forest monkeys and sylvan yellow fever). Most arthropod vectors are short-lived (but see soft ticks, Chapter 16) and would not serve as important components of reservoirs, except possibly during periods of survival during harsh weather conditions under aestivation or hibernation.

An interacting and self-regulating unit in an ecological community is referred to as a BIOCENOSIS (pl. -noses). This term has been widely adopted in disease relationships to refer to the interacting organisms involved; in a disease caused by a pathogen transmitted by arthropods the biocenosis might include pathogen, vector(s), susceptible host(s) and nonclinical maintenance host(s). To illustrate with a simple specific case, in epidemic typhus the biocenosis includes only the rickettsial pathogen, human lice (particularly the body louse), and man.

COMBINING INFORMATION IN EPIDEMIOLOGICAL INVESTIGATIONS

Emphasis on epidemiological discussions, thus far, has been on how the characteristics of vectors may be studied. In addition, a more cursory glimpse of the pathogen and vertebrate portions of biocenoses has been provided. But in any arthropod-associated disease condition of man or animals all of these factors interact, often in quite intricate ways. Epidemiology, properly applied, should attempt to integrate these interactions, so that through an understanding of the system it may be possible to achieve such objectives as predicting number of clinical cases to be expected in man and animals, determining most effective means to prevent the occurrence of cases, and predicting the degree of disease control one might expect if certain remedial practices are carried out.

The interactions may be studied by a team. For example, in malaria control studies the individuals in a given control team frequently include persons with specialty training in entomology and in identification of blood parasites, and an epidemiologist who specializes in the quantitative aspects of disease incidence. An alternate approach is to develop mathematical descriptions of the important interactions that will mimic the circumstances found in nature. These descriptions are termed models, of which there are many kinds, each of which may be as limited or inclusive as desired, dependent mainly on how complex the interactions are and how detailed the information sought is.

Epidemiological modeling has in recent years been primarily carried out by Macdonald and associates studying malaria and schistosomiasis, and fortunately the papers published by these investigators have been assembled in one collection by Bruce-Chwatt (1976). Macdonald acknowledges that the baselines for present studies were provided by Ronald Ross and Christophers in 1908–11, in their attempting to explain why insect-associated diseases, and particularly malaria, occur in endemic and sometimes in epidemic form. The first assumption was that for any given set of malariological circumstances some minimum number of mosquitoes, above zero, was required to keep transmission going. They, and subsequent investigators, developed a generally accepted theory that interrupted transmission and low endemicity lead to the production of unstable (vertebrate) immunity, which in turn occasionally precipitates epidemics. Steady transmission yearly leads to the production of a firm immunity, which protects the community against major disease outbreaks. The essential features of the epidemiology of malaria and of any disease carried from man to man by arthropods, and in which superinfection occurs, are believed by Macdonald to be (1) the duration of the extrinsic cycle of the parasite, (2) the frequency with which the vector feeds on man, and (3) the vector's normal longevity.

It is beyond the scope of this book to go into the details of modeling. The interested reader may refer to general modeling references, such as Jeffers (1972), which includes an article by Conway and Murdie on the man-mosquito-man cycle in urban yellow fever transmission. Representative modeling exercises concern the testing of a malaria model (Dietz *et al.,* 1974), and studies on anopheline populations (Yanes de Ramirez, 1974) and *Culex* populations (Juricic *et al.,* 1974), on an arthropod-virus transmission cycle (Moor and Steffens, 1970) and on a plague epizootic (Soldatkin *et al.,* 1975). In addition to describing major systems, more limited models may be developed in order to understand such phenomena as transmission and the theory of control (Cuellar, 1973), and vector population dynamics related to control technology (Weidhaas, 1974) and pest assessment (Wright, 1975). The validity of models can be tested in the field (Najera, 1974), providing opportunities to modify parameters so that their simulation better describes real systems, or mathematical techniques can be used for model improvement when real life tests are too difficult (Miller, 1974).

Modeling is resorted to in order to integrate systems too complex to be assessed mentally, and to change systems more accurately and economically than is possible in actual circumstances. Although a main purpose is to accurately describe the effects of complex interactions, one substantial benefit is to identify those areas where information critically necessary for understanding the interplay of factors controlling numbers of cases is missing or inadequate.

5

CONTROL OF ARTHROPODS AFFECTING MAN AND ANIMALS

This discussion centers on control principles, occasionally illustrated by specific examples. For information regarding control of a specific disease, or group of arthropods, consult the chapter dealing with the relevant arthropods. Comprehensive general discussions on the philosophy of vector control, as well as information on specific pests, may be found in publications by Cherrett and colleagues (1971), WHO (1972), Vinogradskaya (1972), Beesley (1973), and Pal and Wharton (1974). The use of pesticides, and ectoparasite control, on small animals and livestock are covered by Beesley (1974b) and Kirk (1974). Mallis (1969) and Ebeling (1975) provide comprehensive coverage of household-associated pests, including those directly attacking man.

The control of medically important arthropods employs many principles used against pests of agricultural importance, but there may be quite different reasons for control. Bear in mind that this discussion concentrates on arthropod control, not disease control. Such techniques as immunization are often the simplest and most reliable means of controlling arthropod-borne diseases, but in many circumstances the control of arthropod vectors may be the chief means of counterattack. Quite obviously, controlling pest arthropods is the main means of dealing with simply annoying species, or species that cause health problems by themselves.

GENERAL CONSIDERATIONS

Unique Aspects of Control in Medical Entomology. The basic purpose in controlling medically important arthropods is to preserve the health and well-being of man, whereas control of arthropods affecting crops and livestock is fundamentally guided by economic principles. The protection of human lives and the promotion of human comfort cannot be measured by monetary considerations alone because man views his own welfare as priceless. Economics enter the picture, however, since limited resources tend to place most emphasis on diseases of major importance. Widespread debilitating diseases, such as malaria and Chagas' disease, which cause considerable losses in

77

human productivity, can be important factors retarding economic improvement in underdeveloped countries.

The economic benefits gained from controlling vector-borne diseases of man are essentially the value of increased production of manufactured and agricultural goods realized through control, minus the costs of control measures applied. Because of adequate control practices in temperate regions of the world, arthropod-associated problems of human health are presently too minimal to make estimates profitable. In these regions control measures tend to be applied prophylactically to prevent health problems, or simply to free man from annoyance and anxiety. Such is not the case in developing countries of the tropics where malarias, filariases, and trypanosomiases cause widespread loss of working days and significant mortality. In some regions these diseases completely prevent the use of lands for human habitation and the production of food.

If satisfactory control measures are available, their costs are readily calculated, but a true economic measure of the benefits derived is far more complex. Galley (1973) argues that in developing countries the case for control measures against arthropod-associated diseases must be based mainly on ethical and humanitarian grounds, though the kinds of benefits realized from malaria control are enumerated.

Complex ethical and emotional considerations arise when control practices affect a whole region, and in order to be effective the application of pesticides may involve public and private property and agricultural, urban, and wild lands. Greater complexities arise when pests are produced through faulty practices in one area, but the resultant pests are not bothersome at the source; instead they fly great distances and affect highly populated regions. An example is the production of vast numbers of stable flies, *Stomoxys calcitrans,* in conjunction with poultry production in parts of north-central Florida, USA, and flight of these pests to recreational beaches many miles away. With this type of problem the pests produced do not seriously affect the industry producing them, the human population affected is far removed from the source of pest production, and even though control of the pest is probably most readily carried out at the source there are no economic incentives there to do so; who then should bear the cost?

One sees from these examples that, quite aside from economic returns, the available resources, political influences, and public relations may strongly influence what will be controlled and at what costs.

Even the reduced productivity of animals annoyed or debilitated by the attack of arthropods may be difficult to determine. The effectiveness of livestock management has some bearing on whether losses are experienced. As an example, on dairy cattle Cheng and Kesler (1961) found that controlling biting and merely annoying flies with sprays and aerosols did not significantly increase the milk production of well-managed herds provided with supplementary feed, though a poorly managed herd provided with only minimal pasture responded favorably. Where biting fly populations are very high, increased income above the cost of control by sprays may be realized for dairy herds (Granett and Hansens, 1957). The subject of losses in livestock productivity due to arthropod attack has been reviewed by Steelman (1976), in which the importance of knowing precise levels of infestation, and the general nutrition and health conditions of the animals affected, is pointed out. Losses are far more serious when quarantines preventing movement of animals are enacted, as with psoroptic mange outbreaks or Texas cattle fever.

Large-Scale Campaigns. Thorough area control of arthropods is required in many disease-control programs, but this imposes certain problems of management. Comprehensive areawide measures are seldom achieved through the voluntary cooperation of individuals. Supervision, coordination, and implementation have evolved from local community governments to somewhat larger county or state units, to national governments, to hemispheric multinational units like the Pan American Health Organization (PAHO), and to ultimate coordination by an international agency, such as the World Health Organization (WHO). Objectives have also expanded to envision regional or world eradication of important diseases or pests.

The mere alleviation of annoyance has been a prime motivating factor responsible for many control programs. This has been an important factor promoting the development of many of the mosquito-abatement districts that include coastal salt marshes of the USA. Increased use of wetlands for recreational, residential, and agricultural purposes, and attendant increases in property values, have been possible only with the development of effective mosquito-control schemes. Examples for the USA include the states of New Jersey, Delaware, Florida, and California. With such developments malaria has disappeared, though *Anopheles* control was seldom the main impetus for starting control districts. The organizational resources of such agencies can be rapidly mobilized to reduce vector populations in the event of arthropod-associated disease outbreaks. Annoyance occasioned by household contaminators such as house flies, cluster flies, and cockroaches has provided justification for employing entomological advisory personnel in local public health agencies, such as city or county health departments.

Some mention of mosquito abatement districts (MADs) seems warranted. These units are prevalent in the USA; in 1971 that country had some 251 abatement districts (National Academy of Sciences, 1976). MADs are initiated through public referenda that are a response to the annoyance and disease potential associated with mosquitoes. In order to be effective, benefits of con-

trol must be discernible to the majority of rural and urban property owners over a rather large area, often a county or several counties. Their main activities are to control mosquito production at the source, often in rather sparsely populated sections of a district. This concept requires effective education of the people served. A carefully planned public information program is needed to initiate a MAD, for voters are not only authorizing the formation of a district, they are also authorizing annually recurring tax assessments for district operations. A district requires personnel of various skills, and it should be managed by someone who not only has the scientific and technical background required, but who can also educate the public, as well as be sensitive to widely disparate viewpoints of the people served. Organization of MADs has been described by the American Mosquito Control Association (AMCA, 1961), and other information may be found in Ebeling (1975), Hatch and associates (1973), Mulhern (1973), and Axtell (1974).

METHODS OF ARTHROPOD CONTROL

Controlling arthropods generally implies reduction in their numbers, but as used here it also includes methods that deny these pests physical access to humans or domestic animals.

Personal Protection

Three common categories of personal protection may be distinguished: (1) physical barriers between a vertebrate and arthropods; (2) chemical barriers that repel arthropods from actually biting; and (3) arthropod toxicants that are applied directly to or within a vertebrate. The subject of arthropod toxicants is dealt with in the discussion of insecticides (pages 88–93).

Physical Barriers.   Screening fabricated of metal or durable synthetic fibers, and covering potential entrances to man's dwellings and those of domestic animals, effectively reduces arthropod attack. The dimensions of the mesh in common screening will exclude house flies, mosquitoes, and similar-sized insects, but in some regions fine screening is needed to prevent passage of very small biting flies, such as phlebotomine and ceratopogonid biting gnats. Screens on windows and doors to exclude flying insects contributed decidedly to the decline of malaria in the USA. This would be the case in much of the world because the most important mosquito vectors of human malaria transmit the causal pathogens when they enter dwellings and bite. Bacillary dysentery outbreaks and intestinal myiasis are also undoubtedly reduced when screening denies access of house flies and other filth flies to food in homes and at the point of processing and vending. Considering the long-term benefits, the basic costs of screening are reasonable, but in many underdeveloped economies these costs are still too high. Cultural acceptance is also required, and, particularly in the tropics, the construction of dwellings would often have to be altered to screen openings at all sites of air exchange.

Where circumstances make house screening difficult, fabric netting may be practical. In the tropics, sleeping under bed nets reduces malaria, filariasis, and dengue where these diseases are transmitted primarily by nocturnally active mosquitoes. Likewise, protection against hordes of Arctic tundra mosquitoes and other biting flies is possible with appropriate clothing, and a head net that is kept away from the skin.

Clothing may be useful for diminishing bites by temporary ectoparasites that attack when a person is actively moving outdoors. Boots and other items of apparel that overlap the upper garments can markedly reduce bites by ticks and chiggers.

Other physical barriers have been used in specific situations, and doubtless still more will be developed. The prevalence of air conditioning in some regions, primarily to improve human comfort at high temperatures, has reduced insect attack within buildings because windows and doors must remain closed for these systems to operate efficiently. The Congo floor maggot, *Auchmeromyia luteola,* feeds on warm-blooded animals in contact with the floor or ground; thus protection of man from this insect can be achieved merely by using a bed, a sleeping platform off the floor, or a hammock. Experimental "curtains" of air jets at doorways have proved to be rather effective in excluding a variety of flying insects.

Small electronic sound generators that can be carried on the body and allegedly produce a sound that repels attack by mosquitoes have been produced commercially. However, personal testing of this type of oscillator (by R.F.H.), and careful analysis by investigators acquainted with repellancy tests, have thus far indicated that such devices are essentially worthless (Kutz, 1974; Rasnitsyn *et al.,* 1974).

Chemical Barriers.   Chemical barriers are preparations that repel attacking arthropods when applied to a vertebrate host or to clothing. These substances are commonly called REPELLENTS.

Repellency is generally not effective at a great distance, and there may still be considerable annoyance through persistent flying or crawling. The main feature of repellency is that arthropods leave without actually biting, and these compounds irritate and disorient biting arthropods, but are of very low toxicity to them. By comparison, conventional insecticides have a much higher acute toxicity to arthropods, and though they may have some repellent characteristics, this is not one of their necessary features. However, some insecticides, particularly pyrethrins and various synthetic analogs, cause marked irritation to arthropods upon contact.

Certain features are essential in repellents.

They must be (1) highly repellent to blood-sucking arthropods but not unpleasant to humans; (2) long-lasting; (3) nontoxic and non-irritating to skin (though many repellents irritate mucuous membranes); (4) stainless. Additional desirable features include innocuous tactile qualities (for example, not feeling sticky or greasy), low cost, and no solvent action on plastics or synthetic fibers.

Although initial tests for developing repellents may be quite simple, several thousand compounds have now been tested, making the task of discriminating more effective compounds rather complicated. Most tests have been against bloodsucking insects and acarines, and proceed from evaluation and primary selection in the laboratory to exposures under field conditions. Consult Schreck (1977) for a more inclusive review of repellency evaluation, and Khan (in Shorey and McKelvey, 1977) on the nature of repellents and extender formulations.

The rate of loss of repellents varies; loss is mainly due to absorption and abrasion (Smith *et al.,* 1963), and high temperatures (Khan *et al.,* 1973). Repellents currently used on human skin may remain effective up to about 6 hours in temperate climates after thorough application, yet these same formulations are effective for only an hour or 2 under humid tropical conditions owing to increased absorption, volatilization, and loss from sweating.

Repellents found effective against bloodsucking arthropods are mainly ineffective against merely annoying species, such as filth flies or *Hippelates* eye gnats, but some success with repellent tests has been reported for house and face flies, the bush fly, yellow jacket wasps, honey bees, and cockroaches.

Some of the commercial repellent compounds listed in Kenaga and End (1974) are used to protect livestock, particularly dairy cattle. Two compounds commonly in use for humans throughout much of the world are mentioned here because of their effectiveness in preventing attack by mosquitoes and other biting flies. The compound called *deet* (N,N-diethyl-m-toluamide) is the active ingredient in many proprietary products. It is noted for its essential odorlessness, waterlike texture, and long-lasting effectiveness. It causes a burning sensation in eyes, cuts, and membranous areas, and will damage some plastic and synthetic fibers. Another well-known product is ethyl hexanediol (2-ethyl-1,3-hexanediol; also known as 6-12). Its qualities are similar to *deet,* though it is generally not as long lasting. In addition, investigations in the Soviet Union indicate that carboxide [1,1-carbonylbis (hexahydro-1H-azepine)] is promising against mosquitoes and black flies (Dremova *et al.,* 1973).

Because repellent effectiveness needs to be improved, there continues to be an active research program, particularly in North America and the Soviet Union, to find new compounds, mixtures of compounds effective against a wide complex of biting arthropods, and improved formulations.

An especially durable method of personal protection employs repellent compounds in an open mesh fabric outer garment. It has been known for some years that ordinary cotton fabrics can be successfully treated with repellents, and that shirts, pants, and leggings so treated remain effective for some weeks; however, under tropical conditions ordinary garments feel too warm. It has now been shown that repellent-impregnated looser-fitting outer garments, window nets, or bed nets of mesh with up to 1.25 cm spacings are also highly effective (Fig. 5-1). These mesh fabrics generally have a synthetic fiber matrix for strength and durability, with cotton fiber incorporated for absorption characteristics. Field tests with repellent-treated wide mesh jackets have been very successful against mosquitoes, tabanids, black flies, ceratopogonid midges and tsetse; bed nets of similar fabric and repellent characteristics have worked well to protect against mosquitoes and smaller biting

Fig. 5-1 A wide-mesh fabric jacket impregnated with repellent. (USDA photograph; courtesy of C. E. Schreck.)

flies. Repellency has been retained for from 6 weeks to as long as 8 months under demanding environmental conditions, and the fabrics are readily reimpregnated with repellent(s) (Gouck et al., 1971; general review by Grothaus, Haskins, Schreck, and Gouck, 1976).

One approach in which there may be future promising developments involves the use of space repellents to protect man and animals from arthropod bites over a considerable area, either indoors or outdoors. As examples, protection might be sought in dairy parlors or stables, or on patios, porches, and picnic areas. Protocols for testing space repellency are available (Bar-Zeev and Sternburg, 1970; Schreck et al., 1970), and field tests of a granular formulated repellent mixture have demonstrated marked reduction of attack by mosquitoes and black flies for 24 hours in temperate climates (Means, 1973).

Environmental Manipulation

Modification of the often specific breeding habitat of an arthropod can provide effective control. Environmental modification may consist of complete and permanent change—for example draining or filling mosquito-producing waters—or less-permanent change, such as the change that occurs when a salt marsh is diked and kept flooded for several months to control mosquitoes and *Culicoides* (Rogers, 1962). Thorough knowledge of the life history and biology of the vector in question is required for a successful program, and one must be certain that the changes will not lead to increases of other pests.

Where otherwise compatible with man's uses, environmental modification is desirable because of its permanent nature. It is also termed SOURCE REDUCTION when it denies arthropod pests in the growth stage a place to develop. Initial costs of source reduction programs, such as costs for drainage ditching or landfill, are generally high, but the permanent effectiveness of such programs (perhaps with some annual maintenance) reduces the need for seasonal temporary measures. It should be pointed out that unwanted environmental disturbances, particularly in aquatic environments, may result from controlling the vectors of major arthropod-borne diseases, but careful evaluation often suggests methods more compatible with the existing biota (Provost, 1972).

Permanent source reduction is often consistent with good agricultural practice (Fig. 5-2). In fact, poor agronomic practices that leave standing water that drowns out crops also provide sites of mosquito production (Davis, 1961). Drainage of marshy areas in fields will remove mosquito breeding and will frequently increase crop yields; it also allows equipment more ready access for tillage.

One type of effective environmental manipulation consists of pasture rotation to reduce populations of certain ticks attacking livestock. The object is to remove stock so that ticks leaving the host to molt will not obtain a blood meal at the next stage, and will experience heavy mortality from starvation or dehydration before stock is returned. This

Fig. 5-2 Storage sump collects extra irrigation water to be used again. This prevents water from standing in fields in shallow puddles. (Photograph by R. McCarrell; courtesy of the Delta Mosquito Abatement District, Visalia, California.)

method can be particularly effective in arid areas, and periodic "spelling" of grazing lands is consistent with sustained production of livestock because it permits good regrowth of forage. PASTURE SPELLING has been widely practiced in Australia, and general features of pasture rotation for tick control may be consulted in Wilkinson (1957) and Waters (1972).

Certain other types of environmental modification can be listed as examples. The removal of nests and nesting places for pigeons and other birds is an effective means of preventing annoying numbers of bird mites from entering homes and biting humans. Water-level management in reservoir systems as practiced in the Tennessee Valley Authority (USA) has been long recognized as an effective means of reducing *Anopheles* breeding. Various mosquito breeding problems in the tropics, and their correction, using specific examples, are discussed by Carmichael (1972), and Stanley and Alpers (1975) provide a review of the management of man-made lakes to prevent or reduce diseases.

There is increasing interest in the manage-ment of pest insects and parasitic helminths associated with the soil or in substrates at ground level. Cultivation and addition of urea fertilizer to soil in California had variable effects on the eye gnat, *Hippelates collusor,* but caused around 50 percent reduction in emergence of the biting ceratopogonid midge, *Leptoconops kerteszi* (Legner, Sjogren, Olton, and Moore, 1976). Dung and soil management through biological means, using imported species of scarabaeid beetles to bury dung before it can be effectively utilized by the bush fly, *Musca vetustissima,* has met with much success in Australia and is being tested elsewhere.

Barrier Zones and Quarantines

An area free from certain vectors, either naturally or as a consequence of control programs, may need protection from invasion. This protection is recognized to be of increasing importance with the expanding amount and speed of air traffic. Test protocols for aerosols and micronized dusts have been developed to control insects in aircraft (Sullivan, Pal, Wright, Azurin, Okamoto, Mcuire, and Waters, 1972; Sullivan, Schechter, Amyx, and Crooks, 1972). The routine control of vectors in international health efforts at ports and airports and in ships and aircraft is recommended by WHO, and this agency has sponsored a manual on the subject (WHO, 1972b). To prevent transport of tsetse, it has been necessary to place vehicle check points on roads entering control zones. At these points vehicles are sprayed with insecticide to prevent the flies from traveling on them.

A few examples will indicate how serious recent problems of vector introduction are. The best known was the introduction of *Anopheles gambiae sens. lat.* to Brazil from Africa, presumably by aircraft, and its discovery in 1930 with an intensive and successful follow-up program of eradication (Soper and Wilson, 1943). Eradication of *Aedes aegypti* was achieved in French Guiana in 1952, but a reinfestation was discovered in 1959. This reinfestation extended 125 miles in 2 months, nearly 3,000 square miles were affected in 7 months, and the evidence suggested spread by land vehicles. With the cessation or relaxation of *Aedes aegypti* eradication programs in many countries, this mosquito has reinvaded a large amount of freed territory (see Brown, in Howe, 1977). Interceptions of traffic arriving at international airports in the United States and Puerto Rico for the years 1947–60 revealed 20,000 mosquitoes of 87 species; 48 were indigenous and four species were alive when found (Hughes, 1961). Similarly, *Aedes nocturnus,* a potential vector of the virus of Japanese encephalitis, became established in Hawaii (Joyce and Nakagawa, 1964). The tsetse, *Glossina palpalis,* was eradicated from the island of Principé off the West African coast, but was later reintroduced, presumably from the island of Fernando Po some 200 km distant (Azevedo *et al.,* 1956).

Biological Control (Biocontrol)

All animal populations, including arthropods affecting man and animals, are reduced in numbers by certain other forms of life. For arthropods these control agents are categorized as predators (both vertebrate and invertebrate), parasites (generally meaning metazoan arthropods or nematodes), or pathogens (viruses, rickettsiae, bacteria, fungi, protozoa). Even when biocontrol agents drastically reduce vector populations, they do not ordinarily increase their own numbers to an effective level until after the host vectors have reached an undesirably high level. For this reason biological control agents are manipulated to more effectively reduce arthropod populations.

Principles and procedures involved in using biological agents to control medically important arthropods are the same as those used in controlling agricultural pests; such methods are discussed at length in Steinhaus (1963), Huffaker (1971), and DeBach (1974). Further reviews on biocontrol of medically important arthropods are in Chapman (1974) and Legner and coauthors (1974).

Because of the lag in development of effective populations of biological control agents, major effort may be directed toward their rearing and release. Other means of improving biocontrol include better distribution of naturally occurring agents, or the provision of environments that favor their survival. An awareness of naturally present biocontrol agents is important in order to avoid, as far as possible, their destruction when any control procedures are used.

Biocontrol is frequently criticized, when compared with the use of pesticides, in that it seldom achieves rapid control. Then, too, some notoriously important vectors or vector groups have a relatively large number of biocontrol agents. For larval mosquitoes, as an example, at least 300 species of biocontrol agents have been listed, many of these often inflicting high mortality. Yet despite the exuberance of pathogenic viruses, microsporidian protozoa, and fungal pathogens, as well as a variety of invertebrate and fish predators, mosquitoes still occur in many areas in numbers sufficient to maintain diseases and cause epidemics. But application of pesticides, particularly when these affect insect and fish predators, can throw biotic restraints out of balance and thus require more frequent application of pesticides. Huffaker (1971) points out that permanent establishment of predators or parasites in a pest population not only generally results in lower average absolute levels, but also often dampens the population oscillations that can carry vector numbers above the epidemic threshold.

Biocontrol has some of the same advantages as environmental manipulation, namely, once controlling agents have been successfully established in an area, they may remain as permanent suppressants of pest species. Furthermore, biocontrol agents such as predators or parasites actively seek prey in the host's habitat, and their life history may be so intimately coordinated with that of their host that they respond to the same environmental factors that cause renewed growth and development. Examples of developmental synchrony occur with mermithid nematodes that parasitize *Culicoides,* and certain South American egg-laying fish predators of mosquito larvae that lay resistant eggs that hatch with the flooding that initiates development of the resistant eggs of their floodwater mosquito prey. Mutual adaptation, unfortunately, often causes newly introduced biocontrol agents to become less destructive; one common feature of parasitism is that long-term association of host and parasite tends to result in reduced pathogenicity of the latter.

Environmental modification and improved biological control can be linked effectively; as an example, keeping waterways free of emergent vegetation provides fish and predacious aquatic insects ready access to mosquito larvae and pupae. Then too, certain types of biocontrol are functionally a sort of environmental modification, as when dung beetles bury and utilize cattle dung, thus denying it as a breeding site for various pestiferous higher Diptera.

Types of Biological Control Agents. Some review of types of biological control agents seems warranted, with further examples added when control of specific kinds of medically important arthropods are discussed in later chapters. At the outset one should not assume that all biocontrol agents are safe to use around man, his pets, and domestic animals. Various pathogens, particularly entomogenous fungi and viruses, have been of specific concern. To study some of the issues involved, the World Health Organization (WHO) convened a conference to review the safety of biocontrol agents (Brown, 1974). Continuing review and discussion of these issues should accelerate the development and utilization of more effective microbial control agents against medically important arthropods.

MICROBIAL PATHOGENS OF ARTHROPODS. Viruses, bacteria, fungi, and protozoa are widespread in nature and can on occasion be seen to decimate populations of their arthropod hosts. When it is possible to culture large numbers of pathogens they may be dispersed in application systems similar to those used to apply insecticides, and indeed they are referred to as microbial insecticides, requiring proof of vertebrate safety and effectiveness against the intended arthropods by governmental agencies that regulate the use of pesticides. Some idea of the widespread occurrence and variety of pathogens affecting insects and acarines may be gained by consulting the atlas of insect diseases developed by Weiser (1969), and principles on the use of microbial pathogens are in the book edited by Burges and Hussey (1971). Entomoviruses are especially damaging to

Lepidoptera, and are therefore of great interest in crop and forest protection. Irridescent viruses affect Diptera, though not necessarily with a high degree of pathogenicity, and viruses of this type are known from mosquito larvae. Even though viruses affecting insects and acarines seem to be distinctly different from vertebrate pathogens, and they are very widespread during insect epizootics and must therefore be frequently ingested by man, there is great apprehension over permitting their use in control. Such reluctance is understandable when one realizes that over 300 viruses are known to be transmitted by arthropods to man and other higher vertebrates, and most of these replicate in both their vertebrate and invertebrate hosts. Even though elaborate test protocols can indicate safety for use around humans, it is well known that viral mutations occur and the consequences are unpredictable. But since entomoviruses have been shown to be highly effective in several pilot experiments with agricultural pests, efforts continue to solve the safety and effectiveness problems associated with their use. The subject of potential uses is reviewed in an FAO/WHO report (WHO, 1973d), and Falcon (1976) discusses problems associated with the use of arthropod viruses.

Bacteria, fungi, and protozoa are utilized, or are under intensive study, with special concern that some fungi pathogenic to invertebrates are not very host specific. Suffice it to say that spores of the bacterium, *Bacillus thuringiensis,* are commercially available as a microbial pathogen, and are particularly effective against Lepidoptera larvae and therefore mainly used in agriculture and forestry. A number of fungal pathogens, such as species of *Coelomomyces, Lagenidium, Metarhizium,* and *Entomophthora* appear to have great promise against larval mosquitoes. Of particular note, in culturing the first of these and in understanding how they may flourish in the field, is the discovery that they depend on crustacean hosts as well as their insect host for development (Whisler *et al.,* 1975). The entomopathogenic protozoa are mainly sporozoan microsporidia such as species in the genera *Lankesteria, Nosema, Stempellia,* and *Thelohania.* Many of these protozoans do not cause very high host mortality, but if they infect a significant proportion of a vector population, their importance should not be discounted, since they affect the reproductive capacity of their hosts. Control specialists have been prone to look only at direct mortality in assessing the effectiveness of control agents.

ALGAE AND AQUATIC HIGHER PLANTS. Certain of the aquatic plants reduce breeding by mosquitoes. Filamentous algae, *Cladophora* spp., and branched algae, *Chara,* have been encouraged in rice field and pond situations because they apparently release toxins inhibiting larval mosquito growth. Yet these algae may become too abundant and cause problems requiring correction (Yeo, 1972). Common duckweed, *Lemna minor,* may develop dense surface mats that control mosquito larval development.

NONARTHROPOD INVERTEBRATES. Nematodes appear to have the particular promise for further exploitation to control medically important arthropods. General information on the use of nematodes as biocontrol agents may be found in Poinar (1972, 1975), Gordon and Webster (1974), and Shephard (1974). *Heterotylenchus autumnalis* (Fig. 5-3) has been highly effective in reducing annoying populations of the face fly, *Musca autumnalis.* Much progress is being realized in controlling several species of mosquito larvae by mass rearing and release of the facultatively parasitic mermithid nematode, *Romanomermis culicivorax* (=*Reesimermis nielseni*) (Fig. 5-4) (Chapman, 1974, Petersen, 1973). Unfortunately, this nematode has been shown to harbor an arbovirus (Poinar and Hess, 1977), illustrating the potential for serious problems without careful evaluation of biocontrol agents. Among nonarthropod invertebrates, flat-

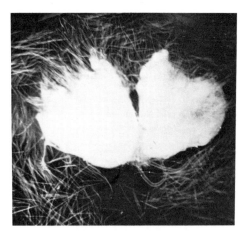

Fig. 5-3 A parasitic nematode, *Heterotylenchus autumnalis,* emerging in large numbers from the ovaries of the face fly, *Musca autumnalis.* (Courtesy of J. Stoffolano, University of Massachusetts.)

worm planarians appear to be quite effective against mosquito and chironomid larvae (cf. Yu and Legner, 1976), and freshwater hydra have also shown promise.

Use of insect parasites and predators in biological control is a highly specialized and documented subject in itself that will not be dealt with here, but one may consult DeBach (1974) and Huffaker (1971) for comprehensive accounts on this subject.

PREDACIOUS FISHES. These may be extraordinarily effective predators of medically important insects. A very comprehensive annotated bibliography on the use of fish in mosquito control has been provided by Gerberich and Laird, 1968. Best known is a livebearing South American top minnow, the mosquitofish *Gambusia affinis* (Fig. 5-5*a*),

Fig. 5-4 Larvae of a parasitic nematode, *Romanomermis culicivorax,* in the thorax region of the fourth stage larvae of *Culex pipiens quinquefasciatus.* In the center a nematode is emerging. (USDA photograph; courtesy of J. J. Petersen.)

Fig. 5-5 Live-bearing fish used as predators on the aquatic stages of mosquitoes. *a.* The mosquitofish, *Gambusia affinis. b.* The common guppy, *Poecilia reticulata.* (Courtesy of E. C. Bay, Washington State University.)

which has been introduced to all continents. In California reasonable control of *Culex tarsalis* and *Anopheles freeborni* in rice fields was achieved by stocking at the rate of about 740 mature females per hectare (300 per acre) early in the season (Hoy *et al.,* 1971). Cold-resistant strains of this fish have been developed in North America, thus increasing their usefulness in temperate zones. In tropical areas the common guppy, *Poecilia reticulata* (Fig. 5-5*b*), seems particularly effective for controlling mosquitoes, such as *Culex pipiens quinquefasciatus,* in waters with high organic pollution (Bay and Self, 1972; Kurihara *et al.,* 1973). Intensive culture programs have been developed for exotic fish used in mosquito control programs (Challet *et al.,* 1974), and there is considerable interest in studying the use of indigenous predacious fishes for mosquito control (examples in Dixon and Brust, 1971; Legner and Medved, 1974; Lomax, 1970). Care is required in choosing pesticides for control of rice pests, to ensure that toxicants do not destroy fishes useful in controlling mosquitoes (cf. Shim and Self, 1973).

It is well to terminate this discussion on biocontrol by citing an example where arthropods may find use against hosts of other parasitic diseases of man and animals. Flies in the family Sciomyzidae parasitize molluscs, and may effectively reduce snail hosts of blood flukes and liver flukes. Berg (1973) provides a review on the subject. In Thailand the biology of *Sepedon plumbellus* has been studied, and it is stated that infestation of man and animals with trematode liver flukes occurs only in areas where this insect has not been found (Bhuangprakone and Areekul, 1973). The rapid utilization of cattle dung by scarabaeid beetles reduces the number of parasitic nematodes in this substrate and in pasture herbage (examples from Australia and USA in Bryan, 1973; Fincher, 1973). Doubtless other instances of arthropod predation or parasitization of invertebrate hosts to diminish the incidence of parasitic diseases will be discovered and studied.

Insecticides and Acaricides

Effective, long-lasting contact insecticides were developed during World War II. During that conflict DDT was found to be spectacularly effective against human body lice, mosquitoes, and various muscoid flies, and was credited with controlling outbreaks of malaria, typhus, and dysentery. The historical aspects of the development of DDT and its use in human and veterinary medicine are covered in Simmons (1959).

Since World War II there has been a vast proliferation of many kinds of organic chemical control agents, until, by 1974, some 250 compounds actually used or extensively field tested were listed for the USA (Kenaga and End, 1974). This number includes plant and synthetic products, insecticides, acaricides (effective against ticks and/or mites), growth modifiers, repellents, attractants, fumigants, synergists, and chemosterilants. The listing by Kenaga and End is used here because it provides common names of compounds approved by the *Entomological Society of America* (ESA) and by *Chemical Abstracts,* and includes other names and trademarks, structural formulas, manufacturers, and levels of mammalian toxicity. This listing, sponsored by ESA, undergoes frequent revision. Other useful governmental or agency listings in the English language include frequently updated WHO publications and occasional editorials on pesticide nomenclature in the *Review of Applied Entomology, B*.

The most frequently used insecticides and acaricides with acute toxicity characteristics against arthropods have potentially dangerous toxicity to man and other vertebrates. Acute toxicity hazard in vertebrates has been highly dependent on whether uptake is mainly by way of the dermal, respiratory, or digestive systems. Many earlier toxicants achieved a differential toxicity between arthropods and larger vertebrates mainly because of the difference in size between these two groups of organisms; that is, on a comparable weight basis both mammalian and insect toxicity were high, but arthropods are small and they require a small amount of toxicant to be killed.

Because pesticides may endanger a variety of animal life, manufacturers are required to carry out quite elaborate tests on toxicity, mutagenicity, or oncogenicity of compounds. Initially tests concerned human safety, but in recent years procedures have been included to assure relative safety for many forms of life. It is beyond the scope of this discussion to deal with the complexities of this approach, but the interested reader is referred to Hayes (1975) and Matsumura (1975). Since public health applications tend to involve a great variety of human ecosystems and wild environments, the program of pesticide development sponsored by WHO is of interest.

The WHO program for evaluating and testing new insecticides consists of a progression in seven stages, from initial laboratory screening of large numbers of candidate compounds to reduction in number of compounds and increasingly larger field trials (Wright, 1971). Pesticide manufacturers develop new compounds, which are submitted for WHO tests, perhaps after they have shown initial promise against agricultural pests. Stage I, conducted at a university laboratory in the USA (1,265 compounds through 1970), involves acute toxicity tests on adults and larvae of a susceptible strain of the southern house mosquito, *Culex pipiens quinquefasciatus*(=*C. fatigans*) and a dieldrin-resistant strain of the Central American malaria vector, *Anopheles albimanus,* with cross-resistance tests on the house fly, *Musca domestica.* Stage II consists of further testing for contact and residual toxicity in two US and two British laboratories, using

larvae and adults of additional mosquito species. Promising materials are evaluated in stage III (305 compounds) in these same laboratories under simulated field and normal weathering conditions, using reference strains of mosquitoes resistant to the major groups of insecticides currently used. Many promising insecticides are eliminated at this stage because of arthropod resistance. In stage IV promising toxicants (82 compounds) are subjected to limited field tests by two facilities in the USA and two each in Africa and Asia, and against various vector arthropods. This might include experimental huts for mosquitoes, part of a stream for *Simulium* larvae, and plots naturally infested with ticks. Stages V and VI (20 compounds and 6 compounds) consist of expanded natural test areas. In stage VII (3 compounds) full epidemiological evaluations are conducted, such as measurements on the disruption of malaria transmission in problem areas. As humans or domestic animals are included in later tests, they are monitored for signs of intoxication. Tests by manufacturers must also meet registration protocols required by registering governmental agencies. Somewhat similar testing procedures are conducted against arthropods affecting pets and livestock, but in the latter group food products derived from these animals must be monitored to ensure that with suitable intervals between treatment and marketing there will be no significant levels of toxicants remaining. Ultimately, specifications are developed for pesticides used in public health (WHO, 1973b).

Unfortunately, even though many pesticides registered for human and animal uses can be safely used with adequate precautions, methods of application and general carelessness or ignorance regarding toxicity have encountered unexpected problems, particularly in less well-developed countries. Furthermore, all toxicity hazards during use under actual field conditions cannot be anticipated. Examples of unanticipated problems include dieldrin intoxication in several tropical countries among applicators using small portable sprayers, poisoning of applicators by malathion in Pakistan because of excessive amounts of highly toxic contaminants in the formulation used, and death or impaired coordination of cats in dry environments wearing collars that slowly release DDVP to control fleas (Bell *et al.,* 1975). Examples of animal deaths due to improper applications to control ectoparasites are represented by McCurnin (1969) and Ray and colleagues (1975).

One technique for early screening of environmental interactions is the development of model ecosystems in which distribution, accumulation, metabolism, and breakdown may be studied on a limited scale (Metcalf, 1977). Concern over many earlier toxicants has heightened interest in producing more selective chemicals that control vectors by disrupting behavior, preventing reproduction, or inhibiting normal growth.

Pesticides have become a highly effective means of control, but they have introduced complications concerning public domain and the rights and well-being of individuals. Because of potential environmental contamination as a consequence of improper use of pesticides, much research and public education are needed before embarking on schemes requiring the widespread use of these toxicants. Agencies conducting programs to control arthropods in recreational areas, forests, marshlands, and other public lands must have the responsible use of pesticides as one of their main objectives. In addition, arthropods affecting man and animals are often associated with agricultural areas where crops are produced for human and animal consumption. Insecticides applied to control these pests may not be registered on such crops, and contamination could affect the safety and marketability of foods.

Problems of toxicity and environmental dissemination of toxicants are alleviated in several ways: (1) formulating toxicants at low concentrations and only in formulations that will reduce the likelihood that applicators will absorb toxic quantities; (2) protective clothing and equipment; (3) restricting application to appropriately trained personnel under close supervision; (4) frequent monitoring of applicator personnel for physiological evidence of excessive exposure and not permitting them to work with pesticides when safety thresholds are exceeded; and (5) limiting application sites to those with minimal exposure hazard to man and domestic animals, and with reduced likelihood of dispersal throughout the general environment. Examples in this last category include treatment of room walls with a contact residual toxicant to control house-entering vector *Anopheles* of human malaria, or applying similar toxicants to the restricted specific resting sites of tsetse.

Obviously, the objectives of a control program govern the kind of chemicals used, the type of application equipment, the formulation of insecticide, and the extent of the area to be treated. Applying pesticides to sites where immature forms of pest arthropods are developing is generally thought to be more efficient as a control measure than applications aimed at adults, especially if the adults are flying insects. The nature and extent of the source, however, are of prime importance. Where breeding sites are aquatic environments harboring fish and wildlife, or are difficult to identify, as in jungle environments, or are at some distance, as may be the case with flying insects, it is more appropriate to control the adult stage. Insecticidal control of flying vectors is generally practiced by (1) treating a limited area on which the vector habitually rests, such as walls of houses, or (2) laying down a barrier zone around a community in the knowledge that flying insects frequently rest on vegetation during their flight dispersal, or (3) using atomized sprays to treat large portions of the aerial environment.

CONVENTIONAL ACUTE TOXICANTS

The attempt is made here to provide an overview of the general types of conventional insecticides that have been used to control arthropods affecting man and animals. Detailed recommendations are not advisable because these change frequently owing to the development of resistance by arthropods, or the advent of more effective compounds or application methods, or cancellation of uses due to mammalian or environmental hazard. Broad groupings are discussed, but further details regarding specific compounds may be obtained by consulting Kenaga and End (1974).

Botanicals and Derivatives. Insecticides extracted from plants have very low mammalian toxicity by contact. These toxicants include rotenone and pyrethrum. Rotenone is an effective ectoparasiticide but finds little present use. Pyrethrum is in reality a mixture of four compounds termed pyrethrins. These have extremely low dermal mammalian toxicity and rather low oral toxicity. They possess some arthropod repellency and have rapid knockdown characteristics. Pyrethrins find use against ectoparasites and household pests.

Many analogs of pyrethrum have been synthesized: allethrin and resmethrin are examples. Unfortunately some analogs may have high mammalian chronic toxicity. Pyrethrum or its derivatives applied over a wide area are not very selective, tending to disrupt existing levels of biocontrol owing to arthropod predators and parasites.

The burning of punk coils containing pyrethrum is an established and convenient practice to reduce mosquito annoyance indoors or around sedentary humans outdoors (Winney, 1975).

Synergists. Certain compounds, essentially nontoxic in themselves, greatly increase the toxicity of pyrethrum and derivatives to arthropods. These synergists are commonly used with pyrethrum to reduce costs. The three commonest synergists are piperonyl butoxide, sesamex, and sulfoxide.

Chlorinated Hydrocarbon Insecticides. The earliest used group of synthetic insecticides, chlorinated hydrocarbons, initially enjoyed widespread use because of their low to moderate acute human toxicity coupled with extraordinary toxicity to many arthropods. One of their characteristics is prolonged contact toxicity, a feature that proved of great value in controlling adults of many important mosquitoes, filth-associated flies, and lice. Durability proved to be a liability, as it was soon apparent that treatment of plants and animals would result in toxic residues within foodstuffs, and the extraordinary success of these compounds destroyed susceptible arthropod populations so effectively that extreme selective pressure caused the appearance of resistance. Ultimately this class of insecticides, with few exceptions, was shown to distribute widely in the environment and to accumulate at increasing rates within higher trophic levels of food chains. This accumulation was believed to be causal in the reproductive failure of some animals, particularly certain fishes and birds. The net result has been banning, severe restriction, or careful review of this class of toxicants in the USA and much of the world. Nonetheless they are mentioned here because of their key role at one time in the control of arthropod vectors and pests of man and animals, and because current reassessment of this group is underway to find compounds with limited drawbacks, namely, with a tendency to degrade after a reasonable time period rather than to persist in foodstuffs and in organisms.

DDT is probably the best known of all synthetic insecticides. It was the first chlorinated hydrocarbon seriously questioned in the USA, and it has been banned or restricted in many countries. DDT was especially useful in treating walls of houses for *Anopheles* mosquito control, but in due time resistance developed in most of the important species of these mosquitoes. A close relative of DDT is methoxychlor, which has very low mammalian toxicity and is rapidly excreted by mammals. The effects of this compound on the environment have been reviewed by Gardner and Bailey (1975).

Polychlorinated hydrocarbon compounds have six or more chlorine atoms. They have relatively long residual toxicity, and are for the most part contact in action. The compounds lindane, heptachlor, toxaphene, dieldrin, and endrin have quite high acute mammalian toxicity; chlordane has rather low acute toxicity. Best known of these compounds, for controlling medically important insects, is dieldrin. This compound once found widespread use as a residual application on resting sites for mosquitoes, but resistance developed rapidly in some cases.

Organophosphorus (OP) Compounds. More organophosphorus products have been synthesized as insecticides and acaricides than any other group of compounds. They are characterized as nerve poisons; some members have extremely high mammalian toxicity whereas others are quite nontoxic; residual activity is shorter as a group than is characteristic of chlorinated hydrocarbons, and buildup in food chains is less likely. Some members are animal systemics.

The more utilized OP compounds include coumaphos, diazinon, dichlorvos, dimethoate, dursban, fenthion, malathion, methyl parathion, naled, ronnel, and trichlorfon. Dichlorvos (Vapona®) has high vapor toxicity and can be used to reach into inaccessible areas. This compound can be formulated in plastic resins that delay vaporization and thereby provide long-lasting control. Ronnel and coumaphos may be used as systemic insecticides for livestock pest control.

Three aziridinyl compounds, *tepa, metepa,* and *apholate* have found extensive experimental use as chemosterilants.

Carbamates. Carbamates have been developed for arthropod control more recently than organophosphorus compounds. They are also nerve toxicants. Some have extremely high mammalian toxicity; however *carbaryl* is noted for very low acute mammalian toxicity.

Inorganic Compounds. Some inorganic compounds have found limited use in controlling medically important arthropods. Powdered *sulfur,* added to an inert carrier such as talcum, helps to control ectoparasites of poultry when placed in dusting boxes. Boric acid is an effective toxicant against cockroaches (Ebeling *et al.,* 1975). Various arsenicals find use; notably *Paris green* for controlling mosquito larvae, and *lead arsenate* as a livestock dip against ectoparasites.

One group of inorganics causes the death of arthropods by water loss. Various formulations of *silica gel* have this characteristic, and have proved useful under some conditions for ectoparasite and household pest control because they are virtually nontoxic to warm-blooded animals (Ebeling, 1971).

Larviciding Oils. Petroleum-based oils are effective mosquito larvicides. When applied to waters these oils form a thin surface film that suffocates mosquito larvae by blocking the respiratory tubes, and by direct entry through the cuticle to affect other internal systems. Diesel type oils may be used, and special distillates such as Flit® MLO have been developed specifically for this purpose.

SPECIAL APPLICATIONS

Systemic Insecticides. Specific mention is due systemic insecticides that provide outstanding control of livestock pests, most notably of cattle grubs and other warble flies. Systemic insecticides, as the term suggests, are toxicants that can be absorbed by an animal to circulate through its body and control pests internally or at some distance from the area of absorption. Systemics work well on internal parasites and on ectoparasites that require frequent feeding upon the blood and tissues of the host. Some requirements for good systemics are (1) high toxicity for arthropods and low toxicity for the vertebrate host; (2) elimination of the toxicant from the vertebrate so that its meat and products do not contain harmful residues when consumed by humans; (3) ease of application, particularly for animals on range. Systemic insectices are administered by (1) forced feeding of a bolus (large pill), not used much because of the amount of handling required; (2) thorough spraying or dipping to cause absorption through the skin; (3) pouring a concentrated formulation along the back of the animal (pour-on-treatment); (4) addition to feed or mineral salt. A largely historical review of methods used in the development of animal systemics is provided by Bushland and associates (1963), and reviews on systemics are presented by Khan (1969) and Khan and Haufe (1972).

Fumigation. Toxic gaseous compounds are used mainly for disinfestation of stored agricultural commodities. Nonetheless this method is enlisted to destroy arthropods in bedding or clothing, and to rid households of pests such as cockroaches and bed bugs. The techniques, and particularly the precautions required, are the same for agricultural or household applications, and these may be reviewed in Monro (1969).

Disinsection of Aircraft. The need to prevent dispersal of medically important arthropods in vehicles, and particularly in jet passenger aircraft, has been mentioned. For this purpose specific application methods of toxicants by aerosols or micronized dusts have been developed and evaluated (Burden, 1972), and the search for more effective and safe techniques continues.

Pesticides Disrupting Arthropod Growth and Reproduction. There is great interest in compounds that mimic the hormones that affect arthropod growth and reproduction, or chemicals that disrupt the arthropod endocrines that control these natural processes. These chemicals do not usually kill arthropods rapidly, though affected individuals may cease to act normally

and therefore may not bite. Usually only a specific growth stage of an arthropod is affected by hormonal modifiers. Natural compounds may be identified in arthropods, and analogs synthesized, or natural products may be sought in plants or other organisms. Since desirable arthropods, such as spiders, beneficial insects, and a variety of Crustacea have similar endocrine control systems, it has been necessary to apply such compounds in highly selective ways, or to study analogs for specificity.

The main compounds used have been isoprenoid juvenile hormone (JH) mimics, with methoprene (Altosid) being the first such chemical to be commercially available. Unfortunately, resistance occurs to JH-like compounds as well as to conventional insecticides (Cerf and Georghiou, 1974). The subject of hormonal analogs and regulators may be reviewed through selected examples (Chamberlain, 1975; Steelman *et al.*, 1975).

NATURAL POPULATION REGULATORS

It is readily observable that animals living in crowded conditions may limit their numbers through the production of toxic substances. This has been noted among the larvae of mosquitoes and chironomids, and the nature of these toxicants has been investigated for possible application. Fletcher (in Shorey and McKelvey, 1977) has reviewed dipteran responses to natural attractants, including the behavior of mosquitoes to oviposition attractants, and inhibition of their growth by retardants produced by larvae.

RESISTANT ANIMAL VARIETIES

Varieties of plants resistant to arthropod attack, and to the effects of plant pathogens, are well known. This specialty area of plant breeding provides potent and lasting answers in pest control. There has been little exploitation of similar techniques in animal breeding, though two general types of animal resistance to arthropod attack can be noted: (1) behavioral resistance wherein birds protect themselves from mosquitoes while resting, or grooming behavior of animals that reduces numbers of ectoparasites such as lice; (2) physiological resistance in which cellular or chemical factors within the blood or other tissues deter the development of parasites on a host. It is the latter process that has been exploited in selecting resistant cattle in Australia and elsewhere. Interest in such a program has been heightened by the development of tick resistance to pesticides.

The use of tick-resistant cattle is further documented in Chapter 16, but suffice it to mention here that resistance to ticks seems to be an acquired phenomenon. Hewetson and Nolan (1968) provide a good quantitative account of this type of resistance to the tick *Boophilus microplus,* and inheritance of such resistance characteristics. Bulls of half-bred zebu stock acquired marked resistance to this tick once they had been challenged by the feeding of large numbers of larvae, even to the point where on one resistant bull female ticks engorged to only half their expected size and died. Quarter-bred zebu progeny of these bulls showed inherited resistance characters following challenge by larval ticks, expressed by reduced weight in engorged female ticks, and reduced number and viability of their eggs.

It seems likely a rather widespread general phenomenon is involved in the physiological resistance of animals to blood-feeding arthropods, and this phenomenon should be further sought and exploited. Additional examples of observations include an acquired level of resistance in sheep after extensive feeding by the sheep ked, *Melophagus ovinus* (Nelson, 1962), and increased death rate of *Anopheles* mosquitoes fed on rabbits immunized with mosquito antigen (Alger and Cabrera, 1972).

INSECTICIDE RESISTANCE

The high degree of selective pressure placed by modern organic insecticides upon arthropod populations has resulted in the de-

velopment of many cases of marked resistance. By 1971 it was reported that about 105 species of insects and acarines of public health and veterinary importance had developed resistance to one or several groups of insecticides, and the number of resistant orders, families, and species continues to grow. Resistance is best known for conventional acute toxicants, but is also known for chemosterilants and hormonal analogs (cf. Vinson and Plapp, 1974).

A comprehensive review on all aspects of resistance in vectors and merely annoying arthropods has been prepared by the World Health Organization (WHO) (Brown and Pal, 1971), and FAO has published methods for detecting and measuring resistance in agricultural pests, including those affecting animal production, such as sheep blow flies, *Phaenicia,* and others (FAO, 1974). WHO plays a leading coordinating role among several laboratories throughout the world in studying and countering resistance (Pal, 1972), with updating summaries developed in reports by the WHO Expert Committee on Insecticides (as example see WHO, 1976).

The term RESISTANCE requires clarification. In the context of this discussion this term is applied when arthropods are no longer controlled by a formerly effective pesticide, for our purposes meaning that at least four to ten times (and often as much as 100 times) as much toxicant as was initially used is required to achieve the same degree of control. Prior to the actual development of resistance there may be some slight increase in required dosage, but this condition is due to a more vigorous population rather than a specific mechanism and is termed VIGOR TOLERANCE (Hoskins and Gordon, 1956).

The underlying process controlling the appearance of resistance is genetic selection. Factors favoring the development of resistance in a natural population of arthropods include (1) The selection pressure; resistance to a given pesticide develops with maximum rapidity the more effective the control has been. (2) The generation time of an arthropod; more rapid selection for resistance is likely in an arthropod exposed to pesticides over several generations per year than in a species having only a single annual generation. (3) Complexity of the gene pool governing resistance; rapid development of resistance is most likely where a single gene governs the resistance mechanism, and it is slow where several genes must be selected out.

Two general types of resistance have been distinguished. PHYSIOLOGICAL RESISTANCE refers to any of several methods of detoxifying pesticides into less toxic metabolites, or other physiological processes, such as increased impermeability of the integument or around critical target systems. BEHAVIORAL RESISTANCE refers to increased avoidance of treated surfaces by arthropods. It is quite well documented now that many instances of "behavioral resistance" are not due to the selection of strains but due to a cryptic susceptible species' being locally eliminated and replaced by a coexisting morphologically indistinguishable one having different behavior. Such were the circumstances involved in revealing that *Anopheles gambiae* was not a single entity, but rather a species complex (see Chapter 4). Nonetheless, clearcut cases of strain selection for increased irritability and avoidance do seem to occur.

Resistance has occurred not only as a direct consequence of control programs against arthropods affecting man and animals, but also through interactions between public health and agriculture. Thus in El Salvador the use of agricultural pesticides, particularly on cotton, hastened resistance to insecticides by *Anopheles albimanus* (WHO Tech. Rept. Ser. 585). Rice culture is an important worldwide source of mosquito production (National Academy of Sciences, 1976), and the use of insecticides to control rice pests will also affect mosquitoes (cf. Self, *et al.*, 1973), thus hastening the development of resistance in these important vec-

tors. Quite likely the control of vector arthropods by insecticides has also accelerated the appearance of resistance in agricultural development programs.

COMMON INSECTICIDE FORMULATIONS

Most organic insecticides have high solubility in organic solvents and low solubility in water. They are formulated in a number of ways for purposes of safety, effectiveness, and ease of application. The main formulations are described here:

1. Technical Grade Insecticides (tech.). This type of insecticide is the greatest purification of toxicant that can be practically manufactured on a bulk scale; it generally consists of 90% or more pure toxicant plus other byproducts of synthesis. Technical grade may be purchased for addition to oil carrier in larviciding mosquitoes, for dilution with oil for use in fogging against adult flying insects, or in some cases for direct application at ULTRALOW VOLUME (ULV). ULV applications usually consist of very uniform aerosol droplets dispersed in such small quantities that a pilot has difficulty observing coverage during aerial application (Fig. 5-6). The spray swath readily drifts downwind, which in very weak air movement advantageously increases coverage when applying from fixed roadway pe-

Fig. 5-6 Aerial spraying by ULV technique to provide rapid and effective coverage in mosquito control. This application occurred during the 1971 outbreak of Venezuelan equine encephalitis in Texas, USA. (Courtesy of USDA, APHIS, Washington, D.C.)

rimeters with ground-based vehicles (Fig. 5-11), but can be especially detrimental if drift lands on crops for which the pesticide has no registration. Strong breezes cause severe problems of poor coverage and excessive drift. ULV applications find greatest use in covering very large areas, where they can rapidly destroy flying insects. They leave very low levels of residue on surfaces and therefore tend to be short-lived in effect. Lofgren (1970) has summarized ULV applications to control pests of man and animals.

2. Solutions (s). These are composed of technical toxicant diluted with suitable solvents. Oil solutions applied to water stay on the surface, where most mosquito larvae breathe, but the organic solvent may damage plants.

3. Emulsifiable Concentrates (ec). These are a technical grade toxicant in a suitable organic solvent plus emulsifier. The latter makes it possible to mix the concentrate with water to provide a cheap carrier during spraying.

4. Wettable Powders (wp). These are inert carriers impregnated with insecticide. A wetting agent is added, and this formulation can be kept in suspension in water by agitation. When sprayed on objects, wettable powders stay on the surface, are less harmful to plants than are emulsions or oil solutions, and frequently have longer residual effectiveness. They may leave visible powdery deposits.

5. Dusts (d). Dusts are finely ground inert carriers impregnated or mixed with insecticide and applied in dry form. The dust form of Paris green floats on the water surface where *Anopheles* larvae feed. Dusts are less often used than formerly because of their tendency to drift beyond the intended application target. Special dust formulations have been developed for direct application on pets, or in poultry dusting boxes or cattle dust bags (Fig. 5-7) to control ectoparasites.

6. Granules (g). Granules are much like impregnated dusts, but the particles are larger and therefore settle rapidly when

Fig. 5-7 Use of dust bags with rangeland cattle, to apply pesticides controlling biting flies and ectoparasites. (Courtesy of R. W. Portman, Washington State University.)

broadcast, resting on the ground or sinking in water.

7. Baits (b). Baits are comprised of toxicant mixed with a material upon which insects like to feed; for example sugar syrup or granules for flies, or peanut meal granules for ants and cockroaches.

8. Special Formulations. Some formulations have been developed for limited specific applications.

BRIQUETTES AND CAPSULES. For continuous slow release, or convenience in difficult placement, it has been possible to formulate pesticides in a solid matrix such as plaster of paris, or organic solvent and pesticide in a water-soluble gelatinous packet. The first of these methods has been tested extensively against black fly larvae in streams. Both formulations have been hand tossed to control mosquito larvae in such habitats as sewage oxidation lagoons and thick beds of marsh vegetation.

ENCAPSULATED FORMULATIONS. Encapsulation of technical insecticide in microscopic capsules of plastic with prolonged or brief duration characteristics, or with proteinaceous polymers having adjustable degrading time, has completely changed the effective period of some insecticides. This technique increases the effective period of compounds with low mammalian toxicity, and shows promise for durable residual insecticide applications to surfaces likely to be contacted by vectors.

PLASTIC SLOW-RELEASE FORMULATIONS. Binding pesticides in a plastic resin base may result in slow release over an extended period. Resin incorporation has been tested as granules fed to livestock to control intestinal parasites, or as boluses placed in the digestive tract of ruminants to render their feces toxic to the larvae of Diptera. There are also plastic collars or ear tags that slowly release the vapor phase of toxicants to control ectoparasites of pets and livestock.

SOAPS, SHAMPOOS, LOTIONS, AND CREAMS. These have been formulated, with pesticide incorporated, to control ectoparasites on man and pets.

MINERAL BLOCKS. Pesticides have been added to mineral blocks licked by cattle. The toxicants may be systemics that control ectoparasites, or they may render the feces unsuitable for the development of dung-breeding flies.

APPLICATION EQUIPMENT

Equipment to apply insecticides has been developed for many of the formulations listed above. Application equipment may consist of devices that can readily be carried and operated by one person, often a necessity in remote and inaccessible areas. On a worldwide basis the type of hand-carried

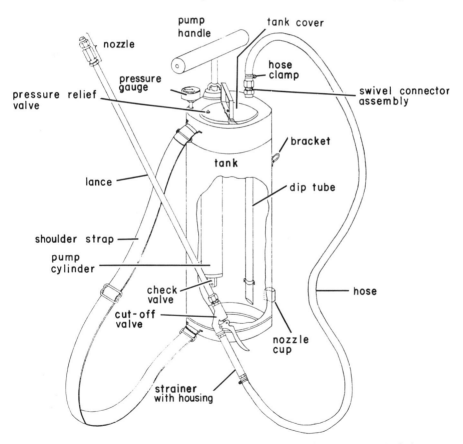

Fig. 5-8 Hand pump compression sprayer. This item of equipment is particularly useful because it is easily carried and can be used to apply a residue to walls or vegetation to control adult vectors, or it readily treats waters that breed mosquito larvae. (Courtesy of WHO.)

sprayer shown in Figure 5-8 is probably used more than any other one piece of equipment for insecticide application to control vectors, and is often the only type of equipment used in underdeveloped countries. Motorized application devices are more characteristic where large areas are being covered. The World Health Organization has prepared a publication on the subject of application equipment for vector control that is occasionally revised and updated (WHO, 1974).

Vehicles used during application depend on the type of terrain and the amount of area to be covered. For general distribution upon a terrestrial environment conventional vehicles such as bicycles, motorcycles, or light trucks are used. In marshy environments four-wheel-drive or tracked vehicles prove necessary, and amphibious all-terrain vehicles can operate in deep water also. On water, boats with conventional propeller propulsion find use, though in very shallow water choked with vegetation, a shallow-draft, flat-bottomed craft with aerial propeller is ideal. Aircraft, either rotor or conventional fixed wing, can provide rapid coverage of terrestrial or aquatic environments (Fig. 5-9). Since the principles of aerial application in

agriculture and for control of medically important arthropods are similar, the publication of Akesson and Yates (1974) may be consulted.

One drawback of aerial application has been the relatively small amount of territory that could be covered when insecticides were diluted, necessitating frequent returns for more pesticide. Ultra low volume (ULV) spraying (often only a fraction of a kilogram of technical insecticide per hectare) has changed this situation whenever it could be used, making returns to base a necessity more for refueling than for reloading.

In addition to the equipment traditionally used to control agricultural pests, devices for applying aerosols of very fine droplets have been developed to control adult flying insects of medical importance. For use indoors aerosols are available in the form of pressurized cans or cylinders, the released compressed propellant gas forcing insecticide and solvent through a narrow orifice that breaks it into very fine droplets. For outdoor application a common aerosol disperser is a THERMAL FOGGER (Fig. 5-10); there are a number of devices that heat diesel oil to the vapor point and release it along with insecticide so that a dense smokelike or foggy cloud of very fine

Fig. 5-9 Use of airplane in spraying operations. (Courtesy of Mosquito Abatement District, Kern County, California.)

Fig. 5-10   A thermal fogger application of insecticide in aerosol form to control mosquitoes and other flying insects. (Courtesy of Tifa, Ltd., Stirling, N.J.)

particles of oil and insecticide is released. COLD FOGGERS that mechanically atomize very small amounts of pesticide are increasingly used, as they do not produce as visible emissions and are generally quite effective (Fig. 5-11).

Attractants and Traps

Various attractant principles are used mainly as monitoring tools to assess the level of arthropod populations. Assessing population levels is highly desirable during control programs or epidemiological investigations, and this general subject is covered in Chapter 4. Although sex pheromones and other

Fig. 5-11   Cold fogging application to control mosquitoes and other flying insects.

chemical attractants are undergoing widespread investigation as control tactics for agricultural applications, as a group these methods are not sufficiently attractive or specific to control arthropod pests of man and animals. One exception is a group of organic acid esters highly attractive to yellowjacket wasps (see Chapter 17; control of bees and wasps). Certain trapping techniques may be useful indoors to lower numbers of pest arthropods. Various traps have been devised to catch cockroaches, and electrocutor traps combined with attractive lights are used to destroy flies and other flying insects in areas where meat products and other foodstuffs are being processed or sold (Hienton, 1974).

Genetic Methods

Genetic methods have been field tested successfully, but in most instances they are still very much under development. The first notable application of genetics concerned eradication of the primary screwworm fly, *Cochliomyia hominivorax*. In this program sterilization was achieved by exposing male flies to gamma irradiation and overwhelming native populations locally over several generations by releasing vast numbers of sterilized flies. It was first successfully utilized on the island of Curaçao, thereafter progressing from the southern tip of Florida north and westward along the southeastern tier of states. An excellent appraisal of this program has been prepared by Baumhover (1966).

The impact of the remarkably successful screwworm eradication scheme has inspired numerous studies on sterilization procedures for arthropod control. These studies have included population models to determine the amount of overflooding needed in releases; genetic manipulation to develop predominantly male-producing strains, or to take advantage of cytoplasmic sterility factors. Examples of genetic methods, including more sophisticated applications of heritable partial sterility, are available in several reviews (Knipling *et al.*, 1968; LaChance, 1974; Proverbs, 1969; Pal and LaChance, 1974; Whitten and Foster, 1975).

Chemosterilants have been investigated extensively (Bořkovec, 1966; LaBrecque and Smith, 1967) but these chemicals are presently prone to affect several forms of life and they cannot be put into general use. Reviews and bibliographies are available on the subject (Campion, 1972; Fye and La-Brecque, 1976). Arthropod resistance to chemosterilants has been demonstrated.

Pest Management

Because single approaches to the control of arthropods are seldom satisfactory there has been a growing interest in the integration of several tactics, a program called pest management. The pest-management strategy is very compatible with control requirements in medical and veterinary entomology, though in the latter case the options available are often limited, and a single approach predominates. That is, when a farm animal or pet is infested with ectoparasites, the most apparent option for control is the application of a pesticide. Even here, however, prevention of initial infestation by reducing contact between animals will lower the possibility of infestation in the first place. General princi-

ples of pest management are described in Metcalf and Luckmann (1975), which includes a chapter on applying these strategies against insects affecting man and animals.

When approaching the control of diseases in which the pathogens are transmitted by arthropods, the main objective consists of permanently reducing the number of clinical cases. The key lies in maintaining contacts between vertebrate hosts and infectious vectors at a low level. For irregularly occurring diseases where man is an accidental host, such as North American mosquito-borne encephalitides, major control efforts may be limited to those seasons when human-vector contacts are predicted to be high, in which case it may be best to carry out source-reduction practices in seasons of low to moderate mosquito populations, and to add aerosol sprays against adults and the use of repellents on people exposed frequently outdoors when vector-human contacts will be high and an epidemic threatens.

In diseases where man is the normal pathogen host, such as malaria or filariasis, diverse strategies need to be applied constantly. Here the primary goal is to reduce numbers of human cases, and preventive and remedial measures can also be applied to the vertebrate host. Pest-management principles to lower the proportion of infectious vectors may include reduction of the sources where vectors can acquire the pathogen, by administering drug prophylaxis to prevent new cases, and intensive treatment to cure those having the disease; reduction of vector populations by source reduction, biocontrol with larvivorous fishes or the encouragement of natural predators and parasites, and possibly by larviciding sites where vectors develop; prevention of vector-host contacts by screening of homes, application of pesticides to surfaces of homes, and the use of repellents on humans exposed excessively.

Pest management is a dynamic process that adjusts options to particular circumstances. The methods are thus a combination of techniques developed in response to the gravity of the disease or potential threat of disease in the human or animal population, and practical limitations of combined tactics in the environment under study.

As specific vector or disease control options are discussed in succeeding chapters, there is a tendency to list tactics as though they might be successful when used singly. Pest management is designed to provide long-term amelioration, and to do so a single tactic seldom provides sustained stability at a low level. For this reason it is important to understand and apply the principles of pest management.

6

COCKROACHES, BEETLES, AND ORDERS OF MINOR IMPORTANCE

ORDER DICTYOPTERA

COCKROACHES

The cockroaches are an ancient group, extending back to the Silurian and showing little change in general structure since the Devonian, some 320 million years ago. They occurred abundantly in the Carboniferous swamps, as indicated by their fossil remains in the coal deposits of that period. Their taxonomic position has been subject to considerable disagreement, but currently they are either placed in an order by themselves or, along with the mantids, in the Dictyoptera as a separate order or as a suborder of the Orthoptera. Evidence strongly indicates a close relationship, probably ancestral, to the termites.

Accounts of cockroaches such as one usually finds in text and reference works on sanitary, medical, veterinary, and pest control entomology are definitely skewed toward a consideration of the man-associated (synanthropic, domestic, domiciliary) forms, and that will be the treatment here. However, fewer than 1 percent of the 3500 to

4000 known species fall in this category. Though pest species are for the most part cursorial (running) insects and nocturnal, many others are active diurnal fliers, inhabiting tropical forests. Others live in the ground or under stones, boards, or various types of rubbish; some are commensal or suspectedly so in nests of ants, termites, or wasps; some inhabit rodent burrows or live in caves in association with bats; some are even aquatic or bore into decayed wood (Cornwell, 1968).

Cockroaches are usually flattened dorsoventrally with a smooth (sometimes pilose) integument, varying in color from chestnut brown to black in the more pestiferous house-invading species, but are frequently green, orange, or other colors, especially in the tropical species. The prominent antennae are filiform and many-segmented. The mouth parts are of the generalized biting-chewing type (orthopteran). There are two pairs of wings in most species; in some the wings are vestigial; in others, for example, *Blatta orientalis,* they are well developed in the male and short in the female. The outer pair of wings (TEGMINA) is narrow, thick and leathery; the inner pair is membranous and folds fanlike.

The name cockroach, supposedly derived from the Spanish name for the insect (*cucaracha*) is preferable to the commonly used "roach," which properly should be applied to certain species of cyprinid fishes.

Important references on medically important cockroaches and on these insects in general are Roth and Willis (1957), Cornwell (1968), Guthrie and Tindall (1968), and Hickin (1977). A discussion of particular value to the sanitary entomologist and pest-control operator, with much basic as well as applied information, is included in Ebeling (1975, pp. 217–244).

Life History. The eggs of cockroaches are lined up vertically, two by two, in the vestibule or oothecal chamber, covered with secretions from the accessory glands, and the leathery, bean-shaped ootheca is then extruded to the outside. Some species, such as *Blattella germanica,* may carry the ootheca for several weeks, but most cockroaches will drop or deposit it within a day or two. In the so-called ovoviviparous species the ootheca is retracted into the uterus or brood sac for incubation until the young are born. In some species parthenogenesis may occur.

The number of eggs in each capsule varies with the species. It is 16 ± 2 for *Periplaneta americana,* about 24 in *P. brunnea,* about 16 in *Blatta orientalis,* and 16 to 18 in *Supella longipalpa;* in *Blatella germanica* it averages about 40 but may range from fewer than 30 to more than 50. Oothecae are quite distinctive and may be used as an aid in determining what species are infesting properties. They also serve as a common means of establishing a pest species in new premises. Many egg capsules are produced during the lifetime of the female—for example, as many as 90 by the American cockroach, 18 by the Oriental, and about 4 to 6 by the German.

The length of incubation period varies with temperature and humidity. For the American cockroach it is about 32 to 53 days, for the Oriental about 42 to 81 days, and for the German about 28 days (Gould and Deay, 1940). The brown-banded cockroach requires 90 days at 23° C and 49 at 28° C.

On hatching, the young cockroaches are quite wingless; they are then, and immediately following each molt, almost all white. Metamorphosis is simple. The number of instars may be variable within a species and is sometimes difficult to determine, partly because of the habit of the insects of devouring the cast skins. The skin is cast first on emergence, with a second molt in 3 or 4 weeks, followed by other molts until maturity is reached. The American cockroach may have as many as 13 molts, maturity being attained in 285 to 642 days, though development requiring 971 days has been reported. The period for the Oriental cockroach is about 1 year. For the German cockroach it is shorter, 2 to 3 months, thus permitting the development of two or more generations a year for that species; normally six molts occur, but there may be seven under adverse conditions.

The longevity of the adult American cockroach is reported by Gould and Deay to range from 102 to 588 days under room conditions, and the complete life span of three females is reported to have been 783, 793, and 913, respectively. The mean length of life of female German cockroaches is reported by these authors to be 200 days, with a maximum of 303 days.

Habits. All immature instars may be found aggregated in association with the adults; in some species, such as the German and the American cockroach, an aggregation pheromone, present in the feces and on the body of the insect, is responsible for this aggregation. Cockroaches are omnivorous, feeding on a great variety of foods, with preference for starchy and sugary materials. They will sip milk, nibble at cheese, meats, pastry, grain products, sugar, sweet chocolate—in fact, no edible material available for human consumption is exempt from contamination by these insects, which also feed freely on book bindings, ceiling boards containing starch (Burden, 1976), the sized inner lining of soles, their own cast-off skins, their dead and crippled kin, fresh and dried blood, excrement, sputum, and the fingernails and toenails of babies, sleeping, sick, or comatose persons, and corpses. They habitually disgorge portions of their partially digested food at intervals and drop feces wherever they go. They also discharge a nauseous secretion both from their mouths and from glands opening on the body, thus imparting a persistent and typical rather musty "cockroach" odor, more offensive in some species than in others, to food and dishes with which they come into contact.

Most pest cockroaches are nocturnal, prowling around and feeding at night and hiding in dark places in the daytime. Consequently, many persons are unfamiliar with their disgusting and dangerous feeding habits. In heavy infestations, these insects may be much in evidence by daylight, or at least the telltale long slender antennae may be seen waving outside places of concealment. Crawl spaces, narrow crevices in damp, dark areas, are especially suited to the flattened body of the insect. Recesses out-of-doors may furnish temporary hiding places from which entries may be made into homes, stores, restaurants, and other buildings. One or more of the pest species may establish themselves in sewer systems. Mass migrations, resulting from overcrowding and often involving immense numbers of these insects, may occur in such species as the German, Oriental and American cockroaches, the last mentioned often resorting to flight as well as crawling in their move into a new area (consult Ebeling, 1975). Various species may also be transported by man from one human edifice to another on furniture or appliances, cartons and empty food containers especially when uncleaned or soiled with spilled

syrups, yeasts, and the like, in packages of groceries, and by other means.

Species of Sanitary Importance. These are the so-called domiciliary, domestic, or synanthropic species that have become adapted to living in close association with man in homes, restaurants, hotel kitchens, grocery stores, rest homes, damp basements where food is available, sewer systems connected with any of the above, or other manmade structures that provide sufficient moisture, food, and hiding places; they carry contaminants to human food, pollute the air with their allergens, produce their characteristic disagreeable odors, and degrade the environment aesthetically. Like man, cockroaches are tropical in origin; when man moved into temperate zones and created his own localized "tropics," certain cockroaches moved along with him. Pest species fall chiefly into three families: The Blattidae (*Blatta, Periplaneta*), the Blattellidae (*Blattella, Supella*), and the Blaberidae (*Blaberus, Pycnoscelus, Leucophaea*). These have been widely distributed by maritime trading from the warm and moist tropics into temperate zones; holds of vessels, the galleys, and the crew's sleeping quarters are often overrun with them. Some of the most important species are the following.

Blattella germanica, the German cockroach, (synonyms: Croton bug, steamfly), is the best known and probably the most widely distributed species (Fig. 6-1). Despite its popular name, it is probably African in origin. It is a small species, adults 12 to 16 mm in length, pale yellowish brown with two dark brown longitudinal stripes on the pronotum. Both sexes are fully winged. The female carries the egg capsule partly protruding from the abdomen until hatching time. It is usually the most abundant pest cockroach in kitchens of homes, especially under sinks and in the dead spaces between sinks and walls, but it may infest other parts of the home where warmth, moisture, and food are adequate. As in the case of the brown-

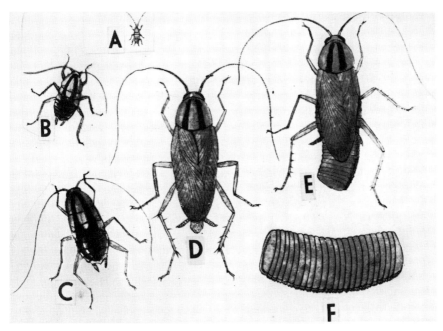

Fig. 6-1 The German cockroach, *Blattella germanica. A.* First nymphal instar. *B.* Third instar. *C.* Fourth instar. *D.* Adult female. *E.* Female with egg case. *F.* Egg case. Egg case, × 5; others, × 3. (USDA drawing.)

banded cockroach, television sets may harbor it. In warmer climates suitable out-of-doors areas may become infested. On a worldwide basis this is one of the most important species; it is probably second to none in the United States.

Blatta orientalis, the Oriental cockroach (synonyms: black beetle, waterbug), is very much darker and larger than the German cockroach. It is dark brown to black, about 22 to 27 mm in length; the wings are rudimentary in the female, and they do not quite reach the tip of the abdomen in the male. This species is more tolerant of cooler temperatures than the German cockroach; consequently it is more often found in cooler climates and out of doors. Its origin is uncertain, possibly African, though Princis (1954) believes it to be South Russian. Its slower reproductive rate, with one generation a year or one in two years, tends to keep its populations to a low level, though in some parts of the world, in Great Britain, for example, it is the most important pest species.

Periplaneta americana, the misnamed American cockroach (actually native of Africa) (Fig. 6-2) is large, 30 to 40 mm in length, and chestnut brown. Both sexes have long wings and may fly short distances. It is a widespread tropicopolitan species, and in

Fig. 6-2 The American cockroach, *Periplaneta americana.* × 1.3.

some parts of the world it becomes the predominant pest species, even in homes; in the warmer parts of the United States and as far north as Seattle, Washington, on the Pacific coast, it may be common in sewer systems and in larger buildings. A closely related household pest, *Periplaneta fuliginosa,* the smoky-brown cockroach, is smaller than *americana* and is dark brown to mahogany black in color; in some parts of the southeastern United States it is second only to the German cockroach as a domestic pest. *Periplaneta brunnea,* the large brown cockroach, is a tropical and subtropical species that is common in some parts of the southern United States westward to Texas.

Periplaneta australasiae, the misnamed Australian cockroach (another African native) is, like *americana,* a tropicopolitan species. It is reddish brown, like *americana,* and about the same size, but it has an evident straw-tan streak extending about one-third of the way down the outer margin of the wing covers as well as a yellowish area around the margin of the pronotum, leaving a double dark area on the dorsum. The wings are well developed in both sexes. Among the species of *Periplaneta,* it is second only to *americana* as a cosmopolitan pest in buildings.

Supella longipalpa, the brown-banded cockroach (synonyms: TV cockroach, furniture cockroach) (Fig. 6-3), is almost certainly of African origin. It resembles the German cockroach in appearance and size but has two cross bands, one at the base of the wings and the second about one-third of the way back. The tegmina do not quite reach the tip of the abdomen in the female; the male has longer tegmina and is more slender. The difference in appearance in the two sexes may confuse one into thinking that two species are involved in an infestation. Adults fly readily when disturbed. The species is decidedly gregarious. This cockroach hides in cupboards and pantries, invades all rooms of the house, and frequently occurs in high locations, such as on shelves in closets or behind picture moldings. The egg capsule is fairly regularly stuck with an adhesive to surfaces, often in furniture such as radio or TV cabinets. The insect is consequently easily transported with furniture.

Pycnoscelus surinamensis, the Surinam cockroach, is a dark brown to black tropicopolitan species, 18 to 24 mm in length. It is probably native to the East Indies or the eastern Oriental Region mainland. It is a burrowing species in piles of debris, leaves, and

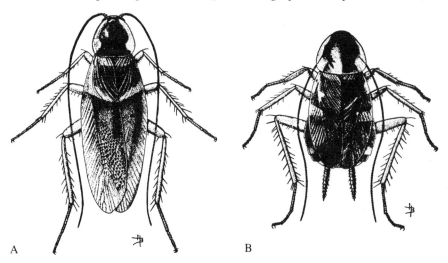

Fig. 6-3 The brown-banded cockroach, *Supella longipalpa.* A. Male. B. Female. (After Back.)

other such materials, but it will invade houses and has become a household pest in some areas, including parts of the southeastern United States. It is parthenogenetic, though it has been confused with the closely related, bisexual *P. indicus* (Roth, 1967).

Leucophaea maderae, the Madeira cockroach, is a widespread tropical and subtropical species. It is large, 4 to 5 cm in length, and long-lived, reported to live up to 2½ years. There are indications that its sanitary importance in tropical America may be greater than has been previously realized (cf. Chinchilla and Ruiz, 1976).

Species of other genera, including *Blaberus, Nauphoeta, Polyphaga, Neostylopyga,* and *Ectobius,* are important pests in certain parts of the world.

Cockroaches as Vectors. Cockroaches have never been irrefutably incriminated in the natural transmission of pathogenic organisms to man. However, that they potentially and, under proper circumstances, actually act in this capacity is virtually undeniable, in light of the studies of Roth and Willis (1957) and other workers. The evidence, though circumstantial, is as strong as much of that generally accepted in other instances of mechanical transmission. To paraphrase Cornwell (1968), who has well summarized the situation, (1) cockroaches favor environments where both human pathogens and human food are found and they pass readily from one to the other; (2) they may carry pathogens in and on their bodies, and these may remain viable on the cuticle and in the digestive tract and feces to the extent that the insects may even be chronic carriers; and (3) the evidence, though circumstantial, is strong enough to justify any cockroach control programs where human health is endangered.

A particularly convincing bit of evidence is cited by Roth and Willis. In a pediatric hospital in Brussels, Belgium, an epidemic of *Salmonella typhimurium* in human infants persisted in spite of quick isolation of the patients, the absence of healthy carriers, and the suppression of direct or indirect contact, other than through cockroaches. It was discovered, however, that the cockroaches were running over clothing, covers, and bodies of the babies by night, and the bacterium was isolated in considerable numbers from the bodies of the insects. The epidemic ceased immediately after cockroach control was effected.

Herms and Nelson, as early as 1913, showed that *Blattella germanica* can acquire specific bacteria by crawling over cultures and then depositing the bacteria on food, for example, on sugar. Rueger and Olson (1969) have shown that the feces of the American cockroach infected with *Salmonella oranienberg,* when spread on human foods and on glass, still contained live bacteria after 3.25 to 4.25 years, and that mice placed in jars containing minute quantities of infected feces themselves became infected within as short a period as one day.

Roth and Willis list eighteen species of domiciliary cockroaches, including most of the common pest species, which have been incriminated by the recovery from their bodies of organisms pathogenic to man or by their experimental exposure to such organisms, or which have been known to bite man. A major consideration relating to the actual role of cockroaches in the transmission of disease agents concerns the likelihood of these insects' passing from contaminated areas and media to homes. Though largely confined to buildings in cooler climates, domestic cockroaches may freely leave such structures under tropical and warm temperate conditions or during the warmer parts of the year; they may frequently migrate into buildings from sewers, cesspools, septic tanks, privies, and dumps. Most domestic species feed on both human feces and human food. The abundance of cockroaches in some areas where poor hygienic conditions prevail is beyond the imagination of persons who live in civilized, highly sanitary areas. The true nature of the problem of cockroaches relative to vectorship of pathogens can only be judged in reference to those areas where these insects afford the greatest threat to human health, complicated by the fact that the threat tends to be hidden from view because of the nocturnal habits of the common pest species.

Natural isolations from wild-caught cockroaches include four strains of poliomyelitis virus, about 40 species of pathogenic bacteria, largely Enterobacteriaceae but including what is probably the leprosy bacterium, two pathogenic fungi (*Aspergillus*), and the protozoon *Entamoeba histolytica*. Other pathogenic organisms cockroaches have been shown to harbor under experimental conditions include the Coxsackie, mouse encephalitis, and yellow fever viruses; the bacterial agents of cholera, cerebrospinal fever, pneumonia, diphtheria, undulant fever, anthrax, tetanus, tuberculosis, and others; and the Protozoa *Trichomonas hominis, Giardia intestinalis,* and *Balantidium coli,* all suspected or proven agents of diarrhea or dysentery.

A protozoon that deserves special mention is the Sporozoan *Toxoplasma gondii,* which causes toxoplasmosis in man and in a wide range of mammals and birds. The disease is common in man, though usually asymptomatic; however, it can cause congenital defects in an unborn child even though it may be asymptomatic in the mother. An amazingly wide range of domestic and wild mammals and birds, even some cold-blooded vertebrates, are susceptible to it, and sometimes the acute form is fatal. It has recently been shown that the definitive cycle is limited to domestic cats and other felines (Kean, 1974). Cats may acquire the parasite by feeding on infected birds and rodents and then pass the organisms in their feces; cockroaches that feed on these feces may pass them on to man. Chinchilla and Ruiz (1976) have demonstrated that in Costa Rica

Periplaneta australasiae, P. americana, and particularly *Leucophaea maderae* readily accepted cat feces when offered with other foods and concluded that these and other cockroaches were potential transport hosts for *Toxoplasma.*

Cockroaches as Developmental Hosts of Parasites. Cockroaches may serve in the above capacity for several groups of parasitic worms. The eye worm of poultry, *Oxyspirura mansoni,* has the Surinam cockroach as one of its intermediate hosts. Cockroaches may become infected with the nematode *Spirura gastrophila* of the rat by feeding on rat feces, and other rats may become infected in turn by feeding on infected cockroaches. At least four species of cockroaches serve as developmental hosts of the nematode *Gongylonema neoplasticum,* another rat parasite that was at one time thought erroneously to produce malignant tumors.

Roth and Willis (1957) summarize the association of cockroaches with helminths as follows:

The eggs of 7 species of pathogenic helminths have been found naturally in cockroaches 11 times. The eggs of 4 of these species and of 5 additional species have been fed experimentally to cockroaches 19 times. Cockroaches have been found to serve naturally as the intermediate hosts of 12 species of helminths in about 43 observations. Cockroaches were used successfully as intermediate hosts for 11 of these species and also for 11 other species in about 44 experiments.

A pentastomid, tentatively identified as *Raillietiella hemidactyli,* has been found to use *Periplaneta americana* as an intermediate host (Lavoipierre and Lavoipierre, 1966). The definitive hosts were geckoes. The acanthocephalans *Hormorhynchus clarki,* a rodent parasite, and *Prosthenorchis spirula,* a parasite of mammals, including primates, have *Blattella germanica* among their intermediate hosts, and *Moniliformis moniliformis,* a parasite of the dog and several wild mammals, has the American cockroach as one of its hosts.

Cockroaches and Allergy. This subject is discussed in Chapter 17.

Control of Cockroaches. Cockroach control is basically easier in temperate climates where populations cannot persist outdoors in winter, than in humid and warm zones. In warmer areas where populations of cockroaches outdoors may readily enter buildings, control within living quarters and commercial establishments requires some tolerance toward a few cockroaches. Isolated homes are easier to achieve control in than apartment dwellings, in the latter circumstances particularly because turnover occupation of adjacent quarters permits easy reinfestation and invasion of surrounding cockroach-free quarters. Cockroaches may be a persistent and annoying problem in ships.

The key to cockroach control is cleanliness so that no substances that could conceivably serve as food for these omnivorous insects is left accessible to them. This task is more difficult where pets and children also occupy living quarters. Reinfestation occurs from outdoors in warmer areas, or along heating ducts and water pipes in apartments, or from bringing in groceries or luggage that has been where cockroaches are abundant. Homeowners are often sensitive about the presence of any cockroaches, because of the connotation that their presence is due to poor housekeeping habits, yet they should understand that the presence of a few cockroaches in certain circumstances is nearly inevitable. Excellent reviews on cockroach biology and control may be found in Ebeling (1975) and Cochran and colleagues (1975).

Surveying for cockroaches is relatively easy. Where numbers are high they are readily seen by use of light during the dark period, when these essentially nocturnal insects are moving about freely. A search behind baseboards and in back of boxes, furniture, or other objects may reveal these

insects or their characteristic oothecae. Spraying a pyrethrum or synthetic pyrethroid aerosol into crevices will flush out these pests. Baited traps are also useful to assess their presence.

Cockroaches are difficult to control permanently with insecticides. This is the case because these insects readily become resistant to common toxicants (cf. WHO, 1975a), and most toxicants are repellent to them and are therefore avoided. Insecticides are usually sprayed or dusted into places of concealment. Slow-release plastic tapes of contact toxicants sandwiched between an uppermost film to retard vapor loss, and a base layer with adhesive, may be placed in concealed areas and remain effective for long periods. Promising results have been obtained by the use of silica aerogel desiccant that kills insects by causing water loss, and by boric acid that is either ingested or enters into cockroaches through the general body surface. Both substances are low in repellency, but they may be slow in effect; they may be combined with other toxicants. Analogs of insect growth hormones also show promise but are slow in action (Das and Gupta, 1974).

Baits consisting of peanut meal, dog food, maltose, traces of methyl myristate, or other attractants combined with a toxicant may be effective (cf. Tsuji and Ono, 1970). Such baits embedded in paraffin and combined with a mold inhibitor, so that they can be used in areas of high moisture, have provided long-term reductions of cockroaches in sewer lines (Wright *et al.,* 1973).

Where control rather than eradication is the goal, good results are possible. There is increasing interest in preventive construction of new apartment complexes, providing as much isolation as practical between units to retard the spread of cockroaches. Silica aerogel, dry boric acid, or combinations of these may be placed liberally in wall compartments during or after construction to provide long-term control. Where contact insecticides are not used, a parasitic eulophid wasp, *Tetrastichus hagenowii,* that oviposits in oothecae, may be introduced and will help to lower cockroach populations. Use of this wasp in conjunction with trapping procedures yields quite satisfactory control.

Restaurants, food-processing plants, and a variety of commercial establishments generally rely on contract control of cockroaches by commercial applicators. These establishments can do much to minimize their own problems by keeping food inaccessible and through care in handling incoming supplies to prevent reinfestation. Cockroach repellents are under development (cf. McGovern *et al.,* 1975), and these might be employed to treat packagings and containers, thus preventing reinfestation when food and supplies are moved into uninfested premises.

ORDER COLEOPTERA

BEETLES

The Coleoptera constitute the largest insect order, comprising more than 270,000 described species. Relatively few beetles concern the medical or veterinary entomologist, but because of their abundance and successful invasion of many types of environments, certain relationships have developed that require discussion here.

Beetles are readily distinguishable from all other insects. Their integument is horny or leathery; their mouth parts are strongly mandibulate, that is biting-chewing. Although wings are absent in some species, usually at least the fore pair is present; these, which are called ELYTRA, are not used in flight, are horny, and when at rest usually meet in a straight line down the dorsum; the hind wings, when present, are membranous and functional, often folded both horizontally and vertically. Metamorphosis is complex. The larvae are of various forms. Most of them have three pairs of well-developed

legs, although those of weevils and of some other groups are legless.

Beetles as Mechanical Transmitters of Pathogens. All scavenger beetles, of which there are several families, are potentially of some public health importance because of their habits of feeding as larvae or adults, or both, on dead animals, hides, or other animal matter that may accidentally bring them into contact with pathogenic organisms. They may carry such pathogens either mechanically on their legs, mouth parts, or body, or in their excreta after feeding on contaminated materials. Of the families so involved, the Dermestidae deserve particular mention. This family includes the hide beetles, larder beetles, museum pests (Fig. 6-4), and related forms. The hairy larvae as well as the adults feed on dead animals, museum specimens, hides, wool, cured meats, cheese and many other animal products, as well as some vegetable materials. *Dermestes lardarius* is known as the larder beetle; *Dermestes maculatus,* the hide beetle, is commonly used to clean flesh from bones for museum use; *Anthrenus scrophulariae,* the carpet beetle, is also a museum pest.

The relationship of dermestids to transmission of the anthrax bacillus was pointed out by Proust as early as 1894. According to Nuttall, writing in 1899,

Proust in examining goatskins taken from anthracic animals, found quantities of living

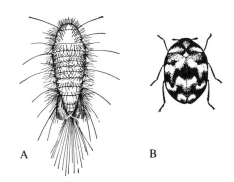

Fig. 6-4 The varied carpet beetle, *Anthrenus verbasci. A.* Larva. *B.* Adult.

[*Dermestes maculatus*] upon them. He found virulent anthrax bacilli in their excrements, as also in the eggs and in the larvae. It is evident from this that these insects which feed on the skins permit the anthrax spores to pass uninjured through the alimentary tract. Heim (1894) also had occasion to examine some skins which were suspected of having caused anthrax in three persons engaged in handling the leather. He found larvae of [*Attagenus pellio, Anthrenus museorum,* both Dermestidae, and *Ptinus,* family Ptinidae]; also fully developed insects of the latter . . . on the skins. All these insects had virulent anthrax bacilli (spores) on the surface and their excreta, from which Heim concludes that they might spread the disease.

Another family of scavenger beetles that might be so involved is the Staphylinidae, the rove beetles (Fig. 6-5). These commonly feed on dung, carrion, and other decaying animal matter, and one often sees them under carrion, hides, heaps of bones, and such. Still another such family is the Silphidae (Fig. 6-5), the carrion beetles, sexton beetles, and burying beetles. They as a rule are attracted to dead animals, which they undermine and, in the case of small carcasses,

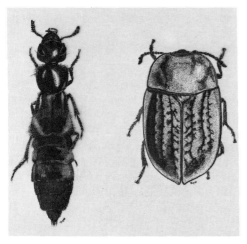

Fig. 6-5 An example of a rove beetle (Staphylinidae, *left*) and a sexton beetle (Silphidae, *right*). Illustrated at same scale; sexton beetle is 13 mm in actual length.

may even bury. They deposit their eggs on the dead bodies, and the larvae feed on the decomposing flesh. According to Théodoridès (1950), three silphids, *Necrophorus vespillo, Silpha obscura,* and *Silpha atrata,* have been recorded in the transmission of the anthrax bacillus.

Some beetles of other families are potential or actual transmitters of pathogens. Čuturić and Topolnik (1975) isolated *Salmonella eimsbuettel* from chicken and pig food heavily infested by the drugstore beetle, *Stegobium paniceum,* an anobiid, and transferred it to sterile samples, where the bacteria survived for several years. Théodoridès (1950) cites experiments in which the tubercle bacillus, injected into the yellow mealworm, the larva of the tenebrionid *Tenebrio molitor,* persisted and remained infective in the developing beetle from stage to stage. Another tenebrionid, the lesser mealworm *Alphitobius diaperinus,* a stored products pest that is common in poultry litter, is often implicated in the transmission through the litter of the virus of Marek's disease, a leucosis of poultry.

Beetles as Developmental Hosts of Helminths. Many species of beetles serve as developmental hosts of helminth parasites of man and of wild and domesticated animals. The common relationship is no doubt due to the variety of feeding habits of beetles that enables them to ingest fecal matter in which eggs of intestinal parasites of animals commonly occur; thus many soil- and dung-inhabiting beetles may readily lend themselves to this role. The helminth groups commonly involved are the nematodes, acanthocephalans, and cestodes.

Théodoridès (1950) cites Hall as recording more than 40 species of vertebrate-infesting nematodes utilizing various Scarabaeidae and Tenebrionidae as hosts, and others have been added since. The infective stage of the nematode *Gongylonema* commonly occurs in scarabaeids such as *Aphodius* and in darkling beetles (Tenebrionidae), including *Tenebrio molitor*. *Gongylonema pulchrum* parasitizes ruminants, sometimes swine, and occasionally man; it invades the tissues of the oral cavity and esophagus, causing *gongylonemiasis*. The swine parasite *Physocephalus sexalatus* was reported by Prokopič and Bilý (1975) to have been recovered from two species of Silphidae, *Oeceoptoma thoracica* and *Phosphuga atrata*, in Czechoslovakia.

De Giusti (1971) has reviewed parasitism of wild and domestic mammals by Acanthocephala. Two essential hosts are required for the development of these worms, a vertebrate for the definitive host and, as a developmental host, an insect for terrestrial and a crustacean for aquatic cycles. The adult worm is highly nonspecific in its host; for the swine parasite *Macracanthorhynchus hirundinaceus* twelve other hosts, including man, are known; *Echinorynchus salmonis* parasitizes 57 species and subspecies of fish, and *Polymorphus minutus* 84 species of birds.

Cockroaches have already been mentioned as hosts. However, in the terrestrial cycles scarabaeid beetles appear to be the most important. *Macracanthorhynchus hirundinaceus* has as developmental hosts scarabaeids of the genera *Cetonia, Lachnosterna, Melolontha, Phyllophaga, Scarabaeus, Xyloryctes,* and others, and occasionally other Coleoptera. These soil-inhabiting larvae may be ingested in large numbers by swine. In the mammalian host the parasite may cause a simple enteritis, with or without bacterial infection, but in more severe cases there may be deep penetration and peritonitis, which may terminate in death. *Moniliformis moniliformis* has, in addition to the American cockroach, the scarabaeid *Scarabaeus sacer* and the tenebrionids *Blaps halofila, B. mucronata,* and *B. gigas* as recorded hosts. It is a parasite of the domestic dog, rodents, hares, and the European hedgehog, and man has been experimentally infested with it.

Some Tenebrionidae, Scarabaeidae, and Carabidae serve as developmental hosts of tapeworms. One tenebrionid, two scarabaeids, and at least 38 carabids of the subfamily Harpalinae, have been reported as developmental hosts of the fowl tapeworm, *Raillietina cesticillus*. The genus *Amara* proved to be particularly favorable as hosts, though the highest degree of parasitemia, 626 cysticercoids, was reported from an individual *Pterostichus*. Several species of *Hymenolepis,* mostly parasites of ducks and chickens, were reported by Théodoridès as having tenebrionids and scarabaeids of several genera as developmental hosts; Théodoridès gives man as an abnormal host for one unidentified species of that genus. *Hymenolepis diminuta,* a common parasite of rats, is occasionally found in man (K. G. V. Smith, 1973). Its developmental hosts include the yellow mealworm, *Tenebrio molitor*.

Annoyance by Beetles. Aside from any economic or other injury for which beetles may be responsible, the sheer abundance of certain species at times causes annoyance. Herms reported considerable annoyance caused by the copra "bug," or redlegged ham beetle, *Necrobia rufipes,* in the Philippine Islands, as these insects swarmed over him while he was trying to do desk work. These small greenish-blue beetles originated in vast numbers in copra stored in neighboring sheds.

The tiny (3 mm long) sawtoothed grain beetle, *Oryzaephilus surinamensis,* family Cucujidae, may invade bedchambers in great numbers, crawl over the bodies of occupants, and nibble the skin. One infestation of this kind was traced to the bathroom and thence out of the house through the yard and into an old barn where, under the stalls, grain from the manger had collected, affording a breeding ground. Extreme dryness had apparently driven the insects to the bathroom for moisture, and the annoyance of the adjoining bedchamber was merely accidental.

Minute Staphylinidae, such as *Atheta occidentalis,* a blackish species 3 mm long, are often encountered on the wing in the late autumn and may accidentally enter the eyes, causing a severe burning sensation and temporarily blinding the victim. Such a mishap to a person driving an automobile might lead to a serious accident. These minute insects breed in cow dung and decomposing plant refuse.

Many species of Carabidae possess vile odors. One of these, the tule beetle, *Agonum maculicolle* of California, is normally a beneficial predator, but when its natural habitat in the marshes becomes dry in summer it commonly leaves in search of moisture and may invade homes in the neighborhood. Heavy winter rains and cold weather may likewise induce invasions. The nauseous odor of this beetle is almost intolerable.

Another beneficial beetle, the convergent lady beetle, *Hippodamia convergens,* a coccinellid, is well known for its habit of forming enormous hibernating aggregations in the adult stage. It can become a definite nuisance when these aggregations are forming in late summer or are breaking up after the hibernating season is past. One of us (James) observed a situation in a mountain valley in northeastern Washington when these beetles filled the air densely for several days and seriously annoyed bathers and vacationers in the area, nipping exposed areas of skin in a very annoying fashion.

Both adult and larval beetles have been known to crawl into the auditory meatus of a sleeping person, sometimes causing considerable pain and discomfort and even further complications. The larvae of the carpet beetle, *Anthrenus scrophulariae,* have been recorded as doing this. A notable example was described by Maddock and Fenn (1958) when, at the annual Boy Scout jamboree at Valley Forge, Pennyslvania, the ears of 186 boys who were sleeping on the ground were invaded by adult scarabaeids, *Cyclocephala borealis,* the northern masked chafer, and

Autoserica castanea, as these insects emerged from the ground. Extraction in some cases was difficult, and some of the boys experienced considerable pain because of the tibial spines of the beetles.

Canthariasis and Scarabiasis. These terms designate rare beetle infestation or invasion of organs of the body (comparable to the much more common myiasis caused by the Diptera). The first term has long been used to designate larval infestation; the latter was coined by Théodoridès for infestation by adult beetles.

Though in comparison with myiasis (pseudomyiasis) canthariasis rarely occurs, cases of it should not be ignored. Théodoridès (1950) reviews a number of these, and Perez-Inigo (1971), who considers these to be "pseudoparasites" that do little or no harm, lists, among others, the yellow mealworm, *Tenebrio molitor,* and the dark mealworm, *T. obscurus.* In Europe infestation of the alimentary canal has reportedly been due to ingestion of larvae of the churchyard beetle *Blaps mortisaga,* a tenebrionid.

Ebeling (1975) records the case of a baby suffering considerable pain after eating cereals containing fragments of *Trogoderma ornatum,* the discomfort resulting from irritation by the larval setae. Actually, any living larva or adult ingested with food or in some other manner might conceivably live for a time and cause discomfort, although as a rule it cannot survive mastication and the action of the digestive fluids.

Very rarely, canthariasis and scarabiasis may be other than enteric. Adult and larval beetles have been found alive in the nasal sinuses, probably crawling there from the pharynx after being inhaled directly or ingested with food (T. J. Spilman, personal communication). Leclercq (1969) lists (1) attacks on the conjunctiva of the eye, sometimes involving encystment, and (2) two curious cases of urinary canthariasis, one in the Sudan involving a clerid beetle and the other in Finland involving the larva of *Nip-*

tus, a ptinid. Spilman (personal communication; see also Senior-White and Fletcher, cited by K. G. V. Smith, 1973) says that adult scarabs, such as *Onthophagus bifasciatus, O. unifasciatus,* and *Caccobius mutans,* occasionally live in the rectum of children in India, Ceylon, and South Africa, causing diarrhea and debilitation. The beetles undoubtedly crawl into the anus when the children sleep out-of-doors on the ground or defecate out-of-doors where the beetles are common.

Théodoridès (1950) records several types of injury to domestic animals, namely, a scarabaeid, *Onthophagus granulatus* perforating the stomachs of horses and calves in Queensland; larvae of a cerambycid, *Ergates faber,* in nostrils of camels in Yemen; larvae of *Tenebrio molitor* recorded as biting chickens to death; and *Dermestes* and *Necrophorus* attacking young pigeons in their nests, burrowing galleries in their skin, and causing death. The last mentioned is somewhat analogous to dermal and bloodsucking myiasis in nestling birds.

Poisonous Beetles. The rose chafer, *Macrodactylus subspinosus,* is poisonous to chickens, ducklings, goslings, and young turkeys. Death may occur within 5 to 24 hours after feeding; if the bird survives that long, it usually recovers, though it does not return to normal for several days following the poisoning.

Vesicating beetles and beetles poisonous to man are dealt with in Chapter 17.

Beetles as Ectoparasites. The beaver parasite, *Platypsyllus castoris,* of Europe and North America, is an obligate parasite, in all its stages, of the beaver. Three known species of family Leptinidae occur. These are *Leptinus testaceus,* parasitic on mice and shrews in Europe and North America; *Leptinus validus,* found on North American beavers; and *Leptinillus aplodontiae,* on *Aplodontia,* the mountain beaver of the Pacific Coast States.

Control of Beetles. From the health standpoint it is seldom necessary to control beetles. Ebeling (1975) reviews methods effective against stored products species found in homes. A variety of toxicants can successfully knock down beetle populations serving as developmental hosts for helminths in the vicinity of poultry yards. The lesser mealworm, *Alphitobius diaperinus,* a dispersal and transmission agent for avian leucosis virus, may develop in enormous numbers in poultry litter. Frequent and thorough changing of the litter will help to reduce its numbers, and pesticides applied to fresh litter materials may provide control for as long as 2 months (cf. Lancaster *et al.,* 1969).

ORDER LEPIDOPTERA

MOTHS AND BUTTERFLIES

The extremely rare accidental ingestion of moth and butterfly larvae is called SCOLECIASIS. The relationship is purely an accidental one, and if there is any medical importance attached to it, this has been poorly documented.

Of much greater significance is the habit of certain adult moths of feeding on lacrimal secretions of wild or domestic mammals, or even of man. Bänziger (1971) traces the hypothetical development of this habit through three steps, each documented by field observation: (1) the consumption of animal urine, dung, and skin secretions of large mammals left on plants; (2) the sucking of sweat, sebum, and blood directly from the skin of these mammals; and (3) attraction to the eyes, where there are proteinaceous deposits. Such eye feeding has been observed in Thailand and Cambodia, involving six noctuids, including *Arcyophora sylvaticus* and *Lobocraspis griseifusa,* and a lycaenid (Büttiker, 1959, 1962, 1964); in Switzerland, involving the hesperiid *Pyrgus malvae malvoides;* and in Germany and Switzerland, involving unidentified lepidotera. Some of the Asiatic species will also feed in the eyes of man (Fig. 6-6).

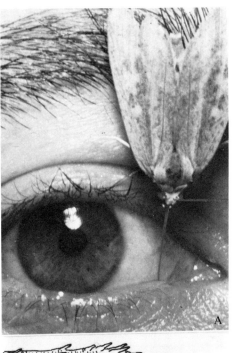

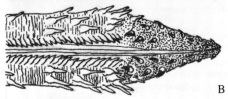

Fig. 6-6 *A*. An eye-frequenting lepidopteran from Thailand. *B*. Tip of a proboscis of the skin-piercing, bloodsucking moth, *Calyptra eustrigata*. (*A* is a photograph by H. Bänziger; courtesy of WHO. *B* is redrawn from Bänziger, 1970.)

If the proboscis of the moth is contaminated with pathogenic bacteria, mechanical transmission may occur. In Uganda a noctuid moth, *Arcyophora longivalvis*, was seen to feed in large numbers on lacrimal secretions of cattle and was thought to be the primary agent disseminating *Moraxella bovis*, the causal organism of infectious keratitis (keratoconjunctivitis) that blinds many cattle and is widespread in tropical Africa and parts of Asia (Guilbride *et al.*, 1959). In the Ivory Coast Nicolet and Büttiker (1975) have implicated that species

and *A. patricula* as probable mechanical vectors of the same pathogen.

Some Lepidoptera can prick the skin, causing mild annoyance, while probing for liquids (e.g., an unidentified lycaenid in Colorado, personal observation, James), but the development of a proven bloodsucking habit, involving modification of mouth parts for such a purpose, as described by Bänziger (1971, 1975) is surprising. The noctuid, *Calyptra eustrigata*, attacks the Malayan tapir, black rhinoceros, Indian elephant, sambar deer, nilgai antelope, and water buffalo, in southeast Asia. Though hairless scars, excoriations, fissures in the skin, and fresh and old sores are preferred to healthy skin, the moth always pierces tissues to obtain blood, even when blood is otherwise available. There is no evidence that the moth will feed on man naturally, although it has been induced experimentally to do so.

Many moths and butterflies have urticating larva and some produce allergens. These will be discussed in Chapter 17.

ORDER HYMENOPTERA

ANTS

The primary medical importance of ants is concerned with their stings (see Chapter 17) and bites, but their role as mechanical vectors of pathogens and as developmental hosts of helminths of veterinary importance needs also to be considered.

Ants as Mechanical Vectors. Since certain species of ants readily enter houses and are attracted to human food, they are capable of contaminating such foods with viable pathogens on their bodies or in their digestive tracts or mouth parts. Smith has pointed out that *Solenopsis* workers may carry viable dysentery germs on their bodies for at least 24 hours, and Donisthorpe recorded mechanical transmission of smallpox in a hospital in Egypt. In Germany, the Pharaoh ant, *Monomorium pharaonis*, imported from the subtropics, shows a marked predilection for hospitals, where it is suspected of contaminating surgical objects and other aseptic items.

Beatson (1972) made a study of long-standing infestations of the Pharaoh ant in nine hospitals in the United Kingdom in which she isolated *Pseudomonas aeruginosa* and species of *Streptococcus, Staphylococcus, Salmonella, Clostridium,* and other bacteria. The ants would readily visit feverish and sweating patients and those with suppurating lesions, as well as bedpans, toilets, sluices, drains, food preparation areas, and sterile packs in stores, which they would invade. That they could easily transport pathogens was indicated by a study involving piglets in an isolated unit of a veterinary hospital. This ant species must definitely be considered a health hazard, and other regular house-invading species should not be dismissed lightly from consideration.

Ants as Developmental Hosts of Parasites. Passera (1975) has reviewed the literature on ants as developmental hosts of helminths. Both trematodes and cestodes are involved. The trematode of concern to us is *Dicrocoelium dendriticum,* the little liver fluke of sheep, for which ants of the genus *Formica* serve as a second developmental host, the first being a snail. Passera lists 11 species of *Formica* known to be thus involved. The cercaria enters the ant and transforms into the metacercaria in the ant's abdomen. Parasitism renders the ant torpid, thereby adding to the likelihood of its being ingested by sheep as it clings to plants on which the latter are grazing (Salimov, 1971).

Of the cestodes, nine species of *Raillietina,* all parasites of the domestic fowl and of the domestic pigeon, are recorded by Passera. Ant genera involved are mainly *Pheidole* and *Tetramorium,* but other genera of carnivorous ants are well represented. Host specificity is very low; as extremes, one tapeworm species (*Raillietina tetragona*) may have as many as 23 ant hosts, whereas one ant species (*Tetramorium caespitum*) may serve as host for as many as six cestodes.

Control of Ants. Pest ants that are attracted to foodstuffs in buildings are best kept at a minimum by making food inaccessible. Pharaoh ant, *Monomorium pharaonis,* is particularly invasive and a problem in sterile areas and within sterile packaging in hospitals. Cabinets, counters, and tables for holding sterile materials may have to be isolated from ant access by setting their legs in cups filled with a low volatility mineral-base oil. Highly attractive toxic baits have been developed for control of this ant in Germany (Hudemann *et al.,* 1972; Schröder, 1975). Baits containing a juvenile hormone analog have been experimentally effective against Pharaoh ant (Edwards, 1975) and would likely work well against other ants.

Ants outdoors that serve as developmental hosts for helminths, or that are harmful because of their stinging propensities, pose different kinds of problems. Approved pesticides may be sprayed around poultry houses and runs, and if not too rapidly toxic, these tend to be distributed back to the colonies to cause their annihilation. Pasture or rangeland species, such as imported fire ants, *Solenopsis,* or harvester ants, *Pogonomyrmex,* may be controlled with pesticides applied directly into the mounds. This method is burdensome for extensive areas, so granular baits with toxicant can be aerially broadcast. Such baits are taken by the workers into colonies and can provide excellent control. However, this type of widespread application has been criticized because of the dangers of side effects, so great care must be taken in choosing the insecticide to be incorporated into baits.

7

THE BUGS
Bed Bugs, Assassin Bugs, Conenoses

ORDER HEMIPTERA

The order Hemiptera is used here to include and consist of the true bugs, the Heteroptera of those workers who use the ordinal name in a broader sense so as to include within it the Homoptera. The latter are phytophagous and of slight medical importance, though a number of them, as well as some phytophagous bugs, are known to bite man occasionally. These will be discussed briefly in Chapter 17. The present chapter concerns mainly two families that, at least in a significantly large part, feed obligatorily on the blood of vertebrates, including man. The insectivorous Notonectidae and Belostomatidae, which may on occasion attack and inflict a painful wound on man, are likewise discussed in Chapter 17.

The true bugs are, for the most part, easily recognizable as such in the adult stage. The fore wing, the HEMELYTRON, is usually divided into a coriaceous or leathery basal part and a membranous apical portion, the latter overlapping the membranous part of the opposite fore wing. The proboscis is suctorial and segmented, attached anteriorly and, when not in use, flexed under the head.

Metamorphosis is simple. All instars take liquid food, animal or plant. Usinger (1934) remarks that the change

. . . from plant feeding to bloodsucking, is not such a profound one as would at first be supposed. This is evidenced by a comparison of the composition of plant juices and blood and by the various plant-feeding groups, some members of which have adapted themselves to a predaceous habit or have shown their ability occasionally to suck the blood of mammals.

Family Cimicidae

BED BUGS

The family Cimicidae includes the bed bugs, swallow bugs, poultry bugs, bat bugs, and others. The wings are reduced to inconspicuous pads. The bodies are broad and flat, enabling the bugs to creep into narrow crevices. A disagreeable pungent odor is present in the group as a whole, with few exceptions. These insects are night-prowling and bloodsucking, feeding on birds and bats, but some either regularly or occasionally attack humans. Peculiar to these bugs is the organ of Ribaga located in the fourth and fifth abdominal segments. The presence or

117

absence of this organ and its particular location when present provide characters useful in the identification of species.

In Usinger's (1966) monograph 74 species were listed, representing 22 genera and 6 subfamilies. Others have been described since then. Many are local in distribution and are of little medical importance, but two have followed man, a host unusual for the family, over a large part of the world. These are *Cimex lectularius* the common bed bug, a cosmopolitan species of both hemispheres but particularly occurring in the temperate regions, and *C. hemipterus,* also widespread in both hemispheres but essentially a species of the tropics. A third human parasite is *Leptocimex boueti* restricted to tropical Africa, where it inhabits native huts. The genus *Oeciacus,* common parasites of swallows, is represented by one species, *O. hirundinis* in the Old World and by another, *O. vicarius,* in the New. Both of these infest man occasionally.

The monograph of Usinger (1966) deals not only with the taxonomy of the Cimicidae of the world but also, extensively, with the ecology, external and internal morphology, reproduction, cytology and cytogenetics, medical importance, and control of these bugs.

Common Bed Bug. The adult *Cimex lectularius* (Fig. 7-1) is reddish brown in color, whereas the young are yellowish white. Many local names are applied to it;

Usinger has given an interesting list of more than 50 in various languages that have been used for this insect.

Bed bugs, like lice, have been the constant companions of man for centuries. The earliest writings on natural history (Aristotle and Pliny) mention them. Almost certainly prehistoric man had to deal with them; it is interesting to speculate that the cave dwellings of early man may have been the site of adaptation to the human host of an insect, previously foreign to man but suited to bats in the cave habitat. Just as bed bugs of swallows and bats may occasionally infest man, the human species may in like manner attack such animals as guinea pigs and white rats, upon which they feed readily.

The bed bug is nocturnal in its feeding habits, hiding in crevices during the day. In its nighttime activity it leaves its hiding place, often traveling considerable distances to attack its victims. Ordinarily, the bugs stay closer to wooden bedsteads. Seams in mattresses and box springs commonly afford harborage. Bed bugs are gregarious; great assemblages may be found in some convenient crevice or beneath some nearby loose wallpaper, where the eggs are deposited and the tarry black excrement may collect.

The female deposits eggs in batches of 10 to 50, totaling 200 to 500, spread out in a yellowish patch. The eggs are relatively large and yellowish white in color. The young hatch in 4 to 21 (usually about 10)

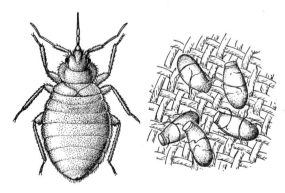

Fig. 7-1 The bed bug, *Cimex lectularius.* Eggs shown at right.

days. The time required for development from egg to maturity is 37 to 128 days, under favorable laboratory conditions; these, however, are seldom, if ever, realized in nature. All instars, particularly the later ones, can endure prolonged starvation (80 to 140 days, depending on the instar and sex, the higher figure being for the adult male); this, of course, lengthens the life cycle.

Bed bugs are sensitive to high temperatures. The thermal death point for *C. lectularius* is 44 to 45° C, but even a temperature of 36 to 37° C with fairly high humidity may kill large numbers of them. At the lower level, activity ceases below about 13 to 15° C. The life span may be influenced considerably by temperature. Johnson, cited by Usinger, calculates that in an unheated bedroom in London, where conditions for development are limited to a period extending from the second week of May to the second or third week of October, only one complete generation and partial second could be produced in 1 year's time.

Bed bugs molt five times (only four in *Haematosiphon inodorus;* Lee, 1955), and the minute wing pads characteristic of the adult insect make their appearance with the last molt. Ordinarily one meal is taken between each molt and one before egg deposition; an average period of 8 days is required between each molting.

METHODS OF DISTRIBUTION. Infestations by bed bugs are traceable to introductions of eggs, young, or adults. One impregnated female might furnish the nucleus for a well-developed colony within a few months. The best regulated household is not exempt from invasion, though cleanliness is the best preventive against the multiplication of any household pest.

Public conveyances and gathering places are common avenues for the dissemination of bed bugs. Furthermore, migration from one living unit to another by way of water pipes, walls, and the like is not at all unlikely when infested houses are vacated and the food

supply cut off. Bugs in all stages are also easily carried in clothing, traveling bags, suitcases, or laundry, and they may be introduced with secondhand beds, bedding, and furniture. They may pass from the clothing of one person to another on crowded public vehicles.

BED BUG BITES. Persons bitten by bed bugs are affected in various ways. In some the bite produces marked swellings and considerable irritation; in others, not the slightest inconvenience is caused. Some persons suffer from broken sleep as a result of the annoyance caused by the bug. The hemoglobin count may be reduced significantly as a result of persistent feeding.

The bite of the bed bug is produced by piercing organs of the hemipteran type. It is probable that punctures by these stylets, unattended by contamination or specific poisons, would produce little pain. The welts and local inflammation are caused by allergic reactions to the saliva that is introduced early in the act of feeding. The bed bug is able to engorge itself completely with blood in from 3 to 10 minutes. Although persons are usually bitten at night while in bed, the bite may be freely administered in subdued daylight.

PATHOGEN TRANSMISSION. The fact that bed bugs must feed at least four and usually five times, upon either the same or a different host, to reach maturity, has placed these insects under grave suspicion as potential vectors of disease-causing organisms. Ryckman (personal communication) points out that bed bugs fulfill all the usual requirements of being good vectors, namely, (1) they are obligate haematophagous species; (2) they feed repeatedly throughout their immature and adult lives; (3) in hotels and other such harborage they may feed on different human hosts with great frequency; and (4) in the laboratory they have been readily infected with various pathogens. In addition, the fact that defecation takes place during the feeding process makes contamination from this source possible. Though many workers have tried to incriminate the bed bug in the role of vector (see the review of Burton, 1963), virtually all evidence has been negative or inconclusive. The bed bug, therefore, appears to be relatively unimportant in this respect, in spite of the fact that it has been shown experimentally capable of harboring the pathogens of hepatitis, Oriental sore, Chagas' disease, and other diseases. The known medical importance of this insect therefore lies in its pest value, the social stigma of its presence in the home, and the effects of its bite.

Other Species That Attack Man. The bed bug of the tropics, *Cimex hemipterus,* is similar in its biology to *C. lectularius,* but it is adapted to a warmer climate. It is widespread in tropical America, Africa, and Asia, in the East Indies, and in some islands of the Pacific. It is essentially a human parasite but, like *C. lectularius,* also attacks the domestic fowl and certain bats. It is difficult to distinguish from *C. lectularius,* but its pronotum is only a little more than twice as wide as long, whereas that of *C. lectularius* is about two and a half times as broad as long. The short hairs on the sides of the pronotum will distinguish both *C. lectularius* and *C. hemipterus* from such bat-infesting species as *C. pilosellus* and from the swallow-infesting *Oeciacus* species.

Leptocimex boueti has a narrow thorax, not much wider than the head, short, flaplike hemelytral pads, and a greatly elongated third antennal segment, which is twice as long as the fourth and almost four times as long as the second. It will also feed on bats, the normal hosts for other members of the genus.

The genus *Cimex* includes two species of bat parasites, namely, *C. pilosellus* of America and *C. pipistrelli,* of Europe. These bugs, as well as the *Oeciacus* species, will bite man if he disturbs them but will not utilize him as a regular host. The swallow bug, *Oeciacus vicarius,* has been known to invade houses where swallows have been nesting close by or under the eaves; it has been reported to occur in such numbers as to drive out the human occupants. Infestation of human premises, though at times severe, is only temporary.

Poultry bugs may also bite man, but only incidentally. These belong to two genera of the subfamily Haematosiphoninae. The Mexican chicken bug, *Haematosiphon inodorus,* has an interesting host range, as it feeds on the California condor, owls, eagles, and chickens. Lee (1955), who studied the biology of this species, reports infestation of barn owls as high as 1,778 bugs on one bird. Reports of this species infesting houses, however, are rare, though the infestation may be severe while it lasts. *Ornithocoris toledo,* the Brazilian chicken bug, once a pest of considerable importance, has been brought under control by the use of residual insecticides. Another species of this genus, *O. pallidus,* feeds on swallows and poultry in Brazil and also in the southeastern US, where it was probably introduced.

Family Reduviidae

ASSASSIN BUGS AND KISSING BUGS (CONENOSES)

Reduviidae are typical Hemiptera. About 2,500 species, in approximately 20 subfamilies, are known. Most of these are true assassin bugs, predacious on insects, which they "assassinate" or kill. These can be beneficial if their prey consists largely of harmful insects; however, some entomophagous species occasionally bite man. A relatively small but highly important group, constituting the subfamily Triatominae, feeds exclusively on the blood of vertebrates. They are commonly known as conenoses, but various vernacular names are applied to them: kissing bugs in the United States; *vinchuca, pito, chupon, chirmacha,* and *chinche* in Central and South America.

The last-mentioned name is often used with a modifier, for example, chinche de monte or chinche tigre. The term "assassin bug," when applied to the Triatominae, is of course a misnomer.

Life History. The rather large, more or less barrel-shaped eggs of reduviids (often with stellate or fringed caps) are generally deposited in areas where adults occur; that is, the ground-inhabiting forms deposit their eggs on the ground, arboreal forms lay their eggs on leaves and stems, and house-inhabiting forms oviposit in dusty corners, crevices, among debris, or in the leaves or stems of palm-thatched roofs. Eggs are commonly deposited singly, either free or glued to the substrate; sometimes they are laid in small clusters. The total number of eggs per female varies considerably from one species to another, from a few dozen to about 1,000 in *Triatoma dimidiata.* The incubation period varies considerably, depending upon the species and temperature. The usual number of nymphal instars is five, though Readio states that *Melanolestes picipes* passes through only four. Mature nymphs often camouflage themselves by covering their bodies with dust particles or other bits of debris (Fig. 7-2). In temperate regions some species overwinter in the egg stage, others as adults, and still others as nymphs. One generation a year seems to be the rule, but this varies from species to species and even within a species. According to Zeledón (1974b), the range, depending on the species, may be from two generations a year to one in two years. Within a species, temperature, host, and the frequency with which the insect takes its blood meal may affect the duration of a generation; in nature longer cycles may occur if the bug lacks aggressiveness or has to endure prolonged starvation.

Assassin Bug Bites. Many species of assassin bugs inflict a painful bite when carelessly handled. The masked hunter, *Reduvius personatus,* a European species that has been introduced into the US, where it is now widespread, is one of these. Its popular name is derived from the habit of "masking" itself under debris that accumulates upon it. Another reduviid that has a bad reputation as a biter is the wheel bug, *Arilus cristatus,* a common species in the southern and eastern US.

Bites of the two-spotted corsairs, *Rasahus biguttatus* and *R. thoracicus* (Fig. 7-5) subfamily Piratinae, are often mistaken for spider bites. The former is common in the

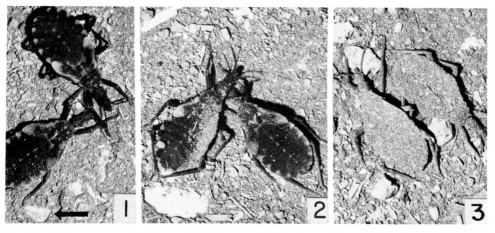

Fig. 7-2 Mature nymphs of *Triatoma dimidiata* camouflaging themselves with debris. (Courtesy of R. Zeledón, Louisiana State University, University of Costa Rica, International Center for Medical Research and Training.)

southern United States, Cuba, and South America; it gives way to the latter in the western United States. The black corsair, *Melanolestes picipes,* inflicts a very painful bite.

The Triatominae. In this subfamily the head is more or less elongated or cone-shaped and has a remarkably free movement. The sturdy, three-segmented proboscis can be thrust forward, but when in repose it lies beneath the head. Most species are good fliers.

The feeding bites of species of *Triatoma* and *Panstrongylus* are in themselves benign; they are hardly, if at all, felt by their hosts. The painless bite is an adaptation to the habitual bloodsucking behavior, in contrast to the painful bite of the assassin bug. On the other hand, the thrust produced in self-defense (but not the feeding bite) of *Triatoma rubrofasciata* in Hawaii is sharp and painful and is accompanied by the emission of an offensive musk, according to Zimmerman (1948). Allergic responses to triatomine bites are discussed in Chapter 17.

Triatominae and Disease. Chagas' disease, or American trypanosomiasis, one of the most important arthropod-associated diseases in tropical America, is caused by *Trypanosoma (Schizotrypanum) cruzi*. The vectors are various species of Triatominae.

CHAGAS' DISEASE

Disease: Chagas' disease (American trypanosomiasis).

Pathogen: Flagellate protozoon, *Trypanosoma (Schizotrypanum) cruzi.*

Victims: Man; dogs and, to a varying extent, other domestic and wild animals; about 100 recorded mammalian hosts.

Vectors: Mostly conenose bugs (Reduviidae, Triatominae), especially *Panstrongylus megistus, Triatoma infestans, T. dimidiata,* and *Rhodnius prolixus.*

Reservoirs: Man, often asymptomatic; numerous mammals, particularly opossums, armadillos, rodents (various, including rats and guinea pigs), carnivores, monkeys.

Distribution: American tropics and subtropics, northern Mexico (about 25° N) to Rio Negro, Argentina (about 38° S); infected mammals and some human cases beyond those limits (see map, Fig. 7-3).

Importance: One of the most widespread and important insect-associated diseases in America; currently an estimated 10 million or more human cases in South America.

The causative organism of Chagas' disease was discovered by Carlos Chagas in

Fig. 7-3 Distribution of Chagas' disease in humans and of its four principal vectors. (Courtesy of US Armed Forces Institute of Pathology. Neg. No. 65-5015.)

Brazil and was named *Trypanosoma cruzi* by him in 1909. The generic name *Schizotrypanum* was later used for this species by Chagas and other workers, but that name is now used only in a subgeneric sense (Hoare, 1972). The life cycle of *T. cruzi* involves both a mammalian and a hemipteran host. Multiplication, by binary fission, takes place in mammalian tissues in a nonflagellated form, called an AMASTIGOTE; only the flagellated (TRYPOMASTIGOTE) form occurs in the blood stream. In the bug, development is in the hind gut; it is rather complicated and involves both metamorphic changes and multiplication. The parasite is finally passed through the feces of the bug (see discussion of transmission).

Except in the acute form of the disease, which is not always clinically discovered, the trypanosomes occur sparsely in human blood; therefore diagnosis by recovery of the organism from the blood is difficult. XENODIAGNOSIS, the method proposed by Brumpt in 1914, is now widely used. Essentially, this involves the use of noninfected (clean) bugs, which are then fed on the suspected individual; after incubation in the body of the insect, the trypanosomes, if present, may be recovered very easily from the digestive tract or by microscopic sampling of feces taken from the rectum by means of a slender pipette. In all but obvious cases complement fixation alone or supplemented by other serological methods is usually used along with xenodiagnosis as a diagnostic tool. Evidence of the disease present in a community may be obtained through the discovery of infected bugs in or near human habitations.

The most apparent external sign of Chagas' disease in some cases is the unilateral swelling, or "chagoma," on the face or eyelid, known as the SIGN OF ROMAÑA (Fig. 7-4). This swelling marks the site of the initial infection (but see page 434). When entry is made other than near the eye, a similar chagoma sometimes marks its site.

After a period of 1 or 2 weeks the trypano-

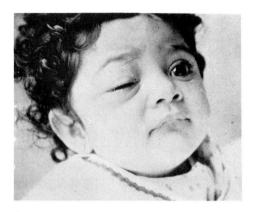

Fig. 7-4 Child showing sign of Romaña of right eye (left side of picture). (Courtesy of US Armed Forces Institute of Pathology. Neg. No. 62-3934-6.)

somes enter the blood stream and initiate the acute stage, during which there is a marked parasitemia, and the parasites penetrate the body cells. The acute form is characterized by a high or moderate fever, edema of the face or other parts of the body, adenopathy, and sometimes nervous disorders. This phase is more common in small children and, in general, the younger the child, the more severe the symptoms and the higher the mortality. The disease runs its course rapidly. Within about 4 weeks of the initial infection, if the patient recovers, the disease becomes more stabilized and enters the chronic phase, in which an equilibrium is set up between the parasite and its host. In some patients the acute phase may be asymptomatic and the chronic phase may be the only one observed, particularly in older children and adults. In this phase the infection may last for many years. At least one case of 50 years' duration, that of a convalescent carrier who, as a small girl, was one of the patients originally examined by Chagas, has been documented (Salgado *et al.,* cited by Hoare, 1972). Cardiac lesions may appear 10 or more years after the first contact with the parasite. Undoubtedly, many asymptomatic cases exist, without any evidence of disease other than that obtained through special diagnostic methods. Also, virulence may vary from one strain to another, and, as a consequence, the disease may be more serious in one area than in another.

In chronic cases, nevertheless, death may occur at any time, often suddenly. The trypanosome invades and destroys cardiac and other body cells; it may produce loss of nervous control, resulting from damage and destruction of nervous tissue. Chronic cases may be marked by cardiopathy and disturbances of cardiac functions and by injury to the alimentary tract; death may result from cardiac failure or from intestinal obstruction resulting from loss of nervous control of peristalsis.

The disease is usually considered rural, though it may be of importance in suburban and urban areas where poorly constructed domiciles are common. It occurs throughout most of South and Central America from Argentina to Mexico. According to World Health Organization estimates made in 1960, about 7 million cases occurred in Latin America at that time; the number of current cases is probably in excess of 10 million (Symposium, First Argentine Congr. Parasitol., 1972 [1974]). It has been recorded from humans in Texas and it has probably occurred, though unreported or not diagnosed, in areas of the southwestern United States near the Mexican border. In reservoir animals other than man the infection occurs rather widely in the southern United States at least as far eastward as Alabama.

So far as is known, Chagas' disease is limited to the Americas. It has been reported in macaque monkeys, in zoos and laboratories, originating from southeastern Asia; though triatomines, notably *T. rubrofasciata,* occur in that part of the world, evidence indicates that the parasite affecting these monkeys had other origins. The matter is discussed thoroughly by Hoare (1972).

Transmission and Epidemiology.
Chagas reported successful transmission through the agency of *Panstrongylus meg-istus* but believed that it was effected through the bite of the insect. In 1912–13 Brumpt disproved the theory of salivary transmission by demonstrating that the infectious stage of the trypanosome resides in the hind gut of the insect and that the infection reaches the victim through the feces of the bug, which almost invariably defecates on the skin of its victim while in the act of sucking blood. From the soiled skin the trypanosomes are readily transferred by the fingers or otherwise to the highly receptive conjunctiva of the eye or the mucosa of the nose or mouth, where entry of the infectious agent takes place. Inoculation may also be effected by rubbing in the organism through the excoriated skin, for example, by scratching. The incubation period in man before fever and edema are noted is 1 to 2 weeks; it varies as to the virulence of the strain of parasite, the location of entry, and the size of the inoculum.

Transmission to man is usually associated with the bite of the bug, which, in turn, usually obtains the parasite from man or from a reservoir host. Sometimes "cannibalism" occurs, in which a nymph may feed upon an engorged nymph, apparently not to the detriment of the latter. Sometimes transmission is not associated with the bite of the bug (cf. Zeledón, 1974b). It has been well documented that congenital transmission may take place through a damaged placenta. Also, infection may be achieved through blood transfusion; the parasites may be viable in refrigerated blood at least for several weeks. Dogs, cats, raccoons, and other carnivores may become infected by eating infected prey or even the bugs themselves or their feces. Other avenues of human infection include contact with feces dropping upon a sleeping person, and infection through mother's milk. Whatever the mechanism, however, one ultimately has to come back to the bug as the vector.

Panstrongylus megistus, a bug with distinctly domestic habits, is the chief vector of

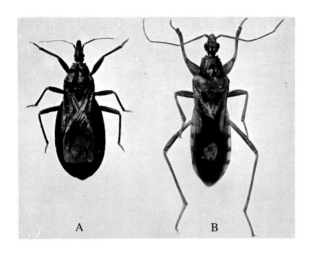

Fig. 7-5 Members of the family Reduviidae. *A. Triatoma protracta. B. Rasahus thoracicus.*

T. cruzi in part of Brazil. This is a rather large bug, 21 to 34 mm in length, blackish in color, with four reddish spots on the pronotum and a series of six reddish spots on each side of the abdomen. In southern Brazil, Uruguay, Paraguay, Bolivia, Argentina, and Chile this species is replaced in importance by *Triatoma infestans,* also a highly domestic species. *T. infestans* is a little smaller, its head is longer, and the red spots on the pronotum are lacking. In Venezuela and Colombia *Rhodnius prolixus* (Fig. 2-4) becomes the most important vector, with *T. maculata* also of considerable importance. Through Central America into Mexico *Triatoma dimidiata* along with *Rhodnius prolixus,* assumes primary importance, and *T. barberi* appears to be important in Mexico. In Panama the main vector is *Rhodnius pallescens* (Fig. 7-6).

Apparently, *Trypanosoma cruzi* is capable of developing in any triatomine bug (cf. Figs. 7-5, 7-6). Of the 90 species of Triatominae known from the Americas, 53 have been found naturally infected, although only 10 or 12 of these are important enough in domestic situations (i.e., as synanthropes) to serve as vectors of *Trypanosoma cruzi* in man or domestic animals (Zeledón, 1974a). In California, S. F. Wood found a high natural infection in *Triatoma protracta* (Fig. 7-5), namely, 25 percent of 816 bugs examined.

Domesticity (synanthropism) on the part of triatomines is, of course, necessary if they are to become significant vectors of *T. cruzi* to the human host. These insects, being nocturnal, seek refuge by day in the cracks and crevices in poorly constructed houses or huts (Fig. 7-7) and in loose thatching of the roofs; by night adults of both sexes and nymphs of

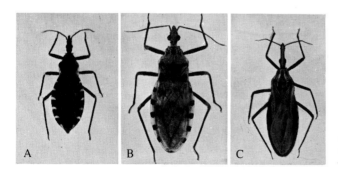

Fig. 7-6 Examples of Reduviidae. *A. Triatoma sanguisuga. B. Panstrongylus geniculatus. C. Rhodnius pallescens.*

Fig. 7-7 Houses like the above (in Central America) are good haunts for synanthropic triatomines. (Courtesy of US Armed Forces Institute of Pathology. Neg. No. 75-4181.)

all stages leave these places to seek food. Obviously, proper construction of human habitations would be an important preventive measure against trypanosome infection. Species such as those of the *Triatoma protracta* complex, which do not frequent human habitations, maintain the parasitemia in nonhuman reservoirs, and it is conceivable that in occasional attacks they may also transmit the trypanosome to man.

Another important aspect of the epizootiology and epidemiology of Chagas' disease is the differential susceptibility of triatomine bugs to different strains of *Trypanosoma cruzi*. This was demonstrated by Ryckman and colleagues (1965), and by Little and associates (1966), who found that the North American *Triatoma barberi* and *T. protracta*, respectively, are much more susceptible to Mexican strains of *Trypanosoma cruzi* than is the Argentine *Triatoma infestans*, although the latter is the most important vector of the parasite in its own area.

Obviously, the triatomine bugs were sylvestral before the advent of man to the Americas. Zeledón (1974a, 1974b) gives an interesting discussion of their adaptation to the human ecotope. Entomological factors involved in this process are the aggressiveness of the bug, its physiologic adaptability, its capacity for feeding on man or

domestic animals, the length of its life, and its reproductive potential.

Environmental factors include sanitary conditions and materials used in construction of human dwellings that triatomines might occupy, climatic conditions (temperature and humidity), and natural enemies and competitors. On the basis of the degree of success in this adaptation Zeledón recognizes five categories: (1) triatomines well adapted to the human habitation, a close synanthropic relation and probably an ancient one, in which the bugs now find relatively few natural habitats and in which distribution of the insect is through the medium of man himself; (2) triatomines adapted to human habitations or in process of becoming so, but still with numerous natural habitats; (3) essentially wild triatomines that are attempting to invade human habitations, in which some breeding occurs; (4) essentially wild triatomines, adults of which occasionally are found in houses, especially when attracted by lights, but that breed elsewhere; and (5) wholly wild species. Categories (1) and (2) are, of course, of most importance, but transfer of *Trypanosoma cruzi* to reservoir hosts, and, occasionally, to man, may occur in bugs that fall into other categories.

Reservoirs. Reservoirs may be domiciliary or sylvatic. Dogs and cats, especially in areas where *Triatoma infestans* is the chief vector, are important domiciliary reservoirs. In Panama and Costa Rica the black rat, *Rattus rattus*, is the main one (Zeledón, 1974a). Sylvatic reservoirs include mammals belonging to seven orders: Marsupialia, Chiroptera, Rodentia, Lagomorpha, Edentata, Carnivora, and Primates. Probably the most important sylvatic reservoir is the opossum, *Didelphis marsupialis*, a prolific animal with a great adaptive capacity and widely distributed from the southern United States to Argentina. In this animal indices of *T. cruzi* infection are usually high, the parasitemia persisting possibly for life, and there is a close association with several *Triatoma* species on which it feeds at times (Zeledón, 1974b). Armadillos are also an important reservoir. Others include guinea pigs, wild rats and mice, squirrels, bats, sloths, and others. Ryckman and his coworkers in the United States have shown that the southern alligator lizard, *Gerrhonotus multicarinatus*, and the whiptail lizard, *Gnemidophorus tesselatus multiscutatus,* can become infected and can pass the parasite on to triatomine bugs; moreover, there is evidence that a given strain of *T. cruzi* may increase in virulence by passing through these hosts.

Several species of bed bugs and ticks are capable of acquiring the parasite experimentally. *Ornithodoros turicata* has been proven to be an experimental vector of the Brazilian strain. Other arthropods that have been experimentally infected include the sheep ked, *Melophagus ovinus,* and a caterpillar, the wax moth, *Galleria mellonella,* listed by Usinger (1944).

Other Triatomine-Transmitted Trypanosomes. Other trypanosomes of mammals are transmitted by Triatominae; so far as known, they have no medical importance. A discussion of them may be found in Hoare (1972).

Trypanosoma rangeli is often associated with *T. cruzi* and discussed in connection with that species. It is not of the *cruzi* complex but belongs to a different subgenus (*Herpetosoma*); it is readily distinguishable from *cruzi* morphologically and in other ways. It is transmitted by *Rhodnius prolixus* and other triatomine bugs, but transmission is through the bite of the bug, not through its feces. Development in the bug, unlike that of *T. cruzi,* involves not only the hind gut but also the haemolymph and salivary glands. *T. rangeli* is generally considered nonpathogenic to man; it obviously is well adapted to its vertebrate host. On the other hand, it is often pathogenic to its insect host, resulting in high mortality and interference with the molting process; it is unique among the mammalian trypanosomes in this respect (Hoare, 1972). There is speculation that the presence of *T. rangeli* in a mammalian host might modify the pathogenic response to *T. cruzi.*

Bugs as Nuisances

Occasionally bugs that do not attack man or animals may enter houses or other manmade structures to such an extent as to cause considerable annoyance. A common example in parts of the USA is the boxelder bug, *Leptocoris trivittatus.* Prior to hibernation, all immature instars and adults may, in a mass, invade houses, much to the discomfiture of their occupants.

The family Pentatomidae is especially objectionable in this respect, because of the offensive odors produced by these bugs. In Tucumán Province, Argentina, *Antiteuchus variolosus* is well known as a nuisance insect (Barrera, 1973); contact with the foulsmelling fluid that it ejects is known to cause dermatitis and pruritus. In Nigeria mass invasions of houses by pentatomids, violent enough to require temporary vacation of the premises, have been reported (Azeez, 1972).

Control of the True Bugs (Hemiptera)

The true bugs, or Hemiptera, are not ordinarily controlled for public health purposes, with the exception of bloodsucking forms, such as the bed bugs, *Cimex* spp., and the Triatominae. The presence of bloodfeeders is generally suspected when welts caused by their bites are in evidence, though some triatomines are more overtly aggressive and their bites may be noticeably painful. Reduviids indoors, such as *Reduvius personatus* and its relatives, may be observed quite readily, though the nymphs are cryptic and seldom seen. Predatory and phytophagous Hemiptera that may bite or be disagreeably odoriferous can be controlled by several pesticides, and natural or synthetic pyrethroids are quite satisfactory and safe to use indoors.

Domiciliary species of triatomines can be reduced in number by several means. House construction can have a decided influence on populations; thatch roofs harbor many more triatomines than sheet metal roofing, and whitewashing of mud walls also tends to reduce their numbers (Gamboa, 1973). Reduction in numbers of domestic and peridomestic animals such as opossums, birds, rats and dogs reduced the level of infection of these vectors with the Chagas pathogen (Rocha e Silva *et al.*, 1975), and should also help to lower the population level of house-associated triatomines.

Biological control agents may be quite effective against Chagas' vectors. In Venezuela the absence of *Rhodnius prolixus* in some areas is believed to be due to heavy predation of these bugs by an ant *Tapinoma melanocephalum,* and ants are known to be especially predacious on blood-filled triatomines in Uruguay and Venezuela (Gómez-Núñez, 1971; Jenkins, 1964). A scelionid wasp egg parasite, *Telenomus fariai,* may cause quite high rates of parasitism under natural conditions where pesticides are not used; a similar wasp in the genus *Gryon* parasitizes triatomines in India, and it has been suggested for introduction into other areas (Sankaram and Nagaraja, 1975). *Telenomus* has been reared and released in urban areas with some success, and a simulation model has been developed to study strategies employing this parasite alone or in conjunction with other methods (Rabinovich, 1971).

Residual insecticides, initially used to control *Anopheles* mosquitoes entering houses, were found to be very effective against triatomines. These were then applied to crevices and other areas of concealment, practically eliminating house-infesting bugs in some situations. In time, however, resistance developed, and tests have been devised to determine the level of reduviid susceptibility to insecticides (WHO, 1975b). In addition to conventional toxicants, juvenile hormone mimics have been tested and show promise (Patterson, 1973).

Bed bugs, *Cimex* spp., have been very successfully controlled with residual insecticides applied to crevices and other places of concealment, but these parasites have also developed resistance to some toxicants. The martin bug, *Oeciacus hirundinis,* may enter homes in England and bite humans. Control of this pest has been accomplished by removing nests in the fall after the young birds have left in order to prevent overwintering of the bugs, and then applying insecticides in the general surroundings but not into nest entrances, in order to safeguard the birds (Beatson, 1971). For control against biting aquatic Hemiptera such as notonectids or corixids in outdoor swimming pools, pyrethrin-type sprays have been suggested.

8

THE LICE

General Characteristics and Classification of Lice. The common practice is to consider the lice as comprising two orders, the Anoplura, or sucking lice, and the Mallophaga, or chewing lice. The term "biting lice," often used for the latter, is ambiguous and consequently objectionable. Since our concern is not primarily with the taxonomy of these insects, the purposes of the present work are best served by maintaining the traditional two orders. Königsmann (1960), Clay (1970), and others believe that the "Mallophaga" are not a monophyletic group; Clay proposes discarding that name and considering all lice as forming a single order, the Phthiraptera, with four suborders, Anoplura, Amblycera, Rhynchophthirina, and Ischnocera. Clay's study is an important reference for the serious student of louse taxonomy.

The Anoplura are bloodsucking parasites of eutherian mammals. For a discussion of their mouth parts, see Chapter 3. Mallophaga feed on sloughed epidermal tissues, parts of feathers, and sebaceous secretions of the host, although in the Amblycera blood constitutes at least part of the diet of many species (cf. Seegar *et al.,* 1976). In all lice wings are absent, the body is flattened in a dorsoventral axis, and the legs are in part adapted for clinging to hairs and feathers. Metamorphosis is simple.

The literature on lice is very extensive. For a bibliography on human lice prior to 1943 (961 entries) see Grinnell and Haws (1943). A valuable foundation work on Anoplura and Mallophaga, with a summary of host records and extensive information on the phylogeny and biology of lice, is that of Hopkins (1949).

ORDER ANOPLURA

SUCKING LICE

Classification. An important foundation work on the classification of the sucking lice is the monograph of the species of the world by Ferris (1951). It is, of course, outdated, and it was incomplete even at the time when it was written, as Ferris could not include the work that was then being done in eastern Europe and China. Subsequent studies have advanced our knowledge of louse taxonomy considerably. Ferris listed 225 species; this number has been increased to 486 at the time of this writing (Kim and Ludwig, 1978).

Ferris recognized six families of Anoplura, four of which included species of medical and veterinary importance. Kim and Ludwig now recognize fifteen, the increased number resulting largely from the breakup of the admittedly heterogeneous Hoplopleuridae of Ferris into its more logical elements. The

129

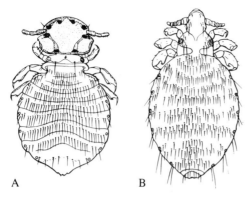

A   B

Fig. 8-1 Comparison of a biting louse, *Trichodectes canis* (A) and a sucking louse, *Linognathus setosus* (B). Both are parasites of the domestic dog. (Courtesy of US Public Health Service.)

six families of greater medical and veterinary importance, according to Kim and Ludwig, are as follows:

1. Haematopinidae, consisting of one genus, the common *Haematopinus*.

2. Linognathidae, including two genera of veterinary importance, *Linognathus* (Fig. 8-1) and *Solenopotes*.

3–4. Pediculidae and Pthiridae, consisting of *Pediculus* and *Pthirus*, respectively, and containing the lice that normally infest man.

5–6. Polyplacidae and Hoplopleuridae, including, among others, the genera *Polyplax* and *Hoplopleura*, respectively, rodent parasites that may be involved in the transfer from rodent to rodent of such human and animal pathogens as those of tularemia and murine typhus.

The genus *Pediculus*, according to Ferris (1951), includes only three or four species: (1) *Pediculus humanus*, the head louse and body louse of man; (2) *Pediculus mjoebergi*, on New World monkeys; (3) *Pediculus shaeffi*, on the chimpanzee; and (4) *Pediculus pseudohumanus*, which Ferris thinks may or may not be valid, described from a monkey, *Pithecia monachus*, and subsequently recorded by Ewing from aboriginal man in tropical Africa and Polyne-

sia. The nomenclature affecting the species considered here has been confused in the literature. Ferris has discussed this subject in detail. The names adopted here, in accordance with current usage, are *Pediculus humanus humanus* for the body louse and *Pediculus humanus capitis* for the head louse.

The genus *Pthirus* (often incorrectly spelled *Phthirus* and *Phthirius*) includes the crab louse, *Pthirus pubis*, of man, and *Pthirus gorillae*, of the gorilla.

Crab Louse. The crab louse, *Pthirus pubis* (Fig. 8-2), also called the pubic louse,

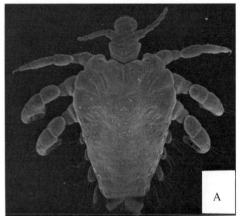

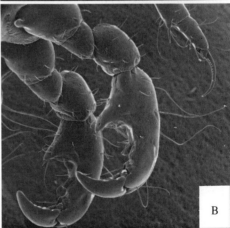

Fig. 8-2 The crab louse, *Pthirus pubis*, scanning electron micrograph. *A.* Dorsal view. *B.* Adaptation of meso- and metathoracic legs for grasping mammalian hairs by hooklike tarsus approximating "thumb" of tibia. (Courtesy of D. Corwin and C. Clifford, Rocky Mountain Laboratory, NIH.)

is easily recognized by its crablike appearance. It is 1.5 to 2.0 mm long, nearly as broad as long, and grayish white. Its middle and hind legs are much stouter than those of the head louse and body louse. It infests the pubic regions particularly but also the armpits and more rarely other hairy parts of the body, such as the mustache, beard, and eyelashes. These lice are remarkably stationary in their habits, often remaining attached for days at one point with mouth parts inserted into the skin. Continued defecation during this time results in accumulation of excrementous materials around the insect. The pruritus caused by their bites is very intense, and a discoloration of the skin usually results if the infestation continues for some time.

The female louse deposits her eggs on the coarser hair of the body where the parasites occur. The number of eggs per female is apparently quite small, usually not more than 30. The life cycle requires not more than a month under usual conditions.

Crab louse infestation, known as pthiriasis, is characteristic of human adults, children under the age of puberty not usually being affected. It is, in a sense, a venereal disease, often showing a marked increase in incidence in subcultures where sexual laxity prevails. The louse is spread by contact with infested humans or objects used by them, such as blankets. Because of the sedentary nature of the parasite, spread is usually relatively slow. It is a human louse, though very rarely dogs are reported to harbor it.

Head Louse.   The head louse, *Pediculus humanus capitis,* is gray in color but tends to resemble the color of the hair of the host. The male averages 2 mm in length and the female 3 mm. This form occurs on the head, about the ears and occiput, but in heavy infestations it may establish itself on other hairy parts of the body. In severe infestations the hair may become matted with eggs (NITS), parasites, and exudate from the pustules that originate from the louse bite; a

fungus may grow in the whole fetid mass, forming a carapacelike covering under which large numbers of lice may be found.

The number of eggs deposited by the female ranges from 50 to 150. These are glued to the hair and hatch in five to ten days. Development is very rapid. There are three molts and 3 weeks usually covers the entire life cycle from egg to egg. Lice are easily disseminated by physical contact, stray hairs, and similar means; therefore, slight infestations may occur under the best of sanitary conditions, particularly among school children. As in the case of the body louse, crowding under unsanitary conditions aids in the development of massive infestations.

Both the crab louse and the head louse are capable of serving as hosts in which the rickettsiae of typhus and the spirochetes of relapsing fever can proliferate. The role of these two insects in transmitting these pathogens is insignificant if, indeed, it exists at all, although Murray and Torrey (1975) have shown that the head louse has a high potential for transmitting the typhus organism.

Body Louse.   The body louse, *Pediculus humanus humanus* (Fig. 8-3), is most common where the clothing comes in contact with the body rather continuously, for example on underwear, the fork of the trousers, the armpits, the waistline, neck, and shoulders. Usually it stays on the clothing and makes contact with the body while feeding; thus, it has a very suitable private environment, which its host has created for it. In heavy infestations some lice may remain on the body after all clothing has been removed. Eggs are deposited by preference in the seams of clothing. Only rarely may the louse attach its eggs to the coarser hairs of the body.

Nuttall states that a female body louse may lay 275 to 300 eggs during her lifetime, but 50 to 150 seems to be a more realistic estimate. The incubation period varies from 5 to 7 days when eggs are laid near the body; it is longer at lower temperatures, and eggs do

Fig. 8-3 Human body louse, *Pediculus humanus humanus,* adult louse and eggs. (Courtesy of Oregon State University and Ken Gray.)

not hatch above 38° or below 23° C. After hatching, the young lice begin to suck blood at once, and throughout their development they feed frequently both day and night, particularly when the host is quiet. Maturity is reached in 16 to 18 days after oviposition. There are three molts. Females begin to lay eggs a day or 2 after reaching maturity. The egg-to-egg cycle averages about 3 weeks. Unfed lice soon die; probably ten days would cover the longest period of survival without food. However, if fed, lice may live 30 to 40 days. Moist fecal matter in masses of spiral threads is extruded as the louse feeds; the feces dry quickly in the air.

Suitable temperatures are essential for the continued existence of louse populations. The optimum for the adult body louse is approximately the temperature of the normal human body. A rise of 4 to 5 degrees is fatal to them within a few hours. Temperatures below the optimum are much less critical, although prolonged exposure to 20° C or lower

may result in death. The egg is the most resistant stage, but even it will not withstand 5 minutes at 51.5° or 30 minutes at 49.5° C, according to Buxton. A temperature of – 20° C for 5 hours or – 15° for 10 hours is lethal.

DISSEMINATION OF BODY LICE. Lice normally live on the surface of the body or in clothing being worn. They thrive best in temperate regions where at least moderately heavy clothing is worn, thus producing an area next to the body where the conditions of temperature and relative humidity are close to optimum. They do not voluntarily leave unless the body grows cold in death or becomes hot with high fever. Even then they cannot travel far, but are easily dislodged. They will quickly invade a nearby new host. Louse infestation is mainly the result of contact with lousy persons or their infested clothing; a person free of them may acquire them merely by walking among heavily infested individuals or by contact with them in crowds, public transportation vehicles, or otherwise.

It has been estimated that under favorable conditions the progeny of a single female could increase to 4,000 to 5,000 in 3 months time. Very heavy infestations, numbering into the thousands, have been reported, but even under unsanitary conditions 400 from one person is an unusually large number. Busvine (in PAHO, 1973) states that in England, even under squalid conditions of general lousiness, most persons carry no more than about a dozen lice. Therefore, though having an abundance of food, equable environmental conditions, and virtually no arthropod natural enemies, lice on the body fall far short of realizing their reproductive potential.

Pediculosis. The presence of lice on any part of the body is called PEDICULOSIS. Louse bites may produce certain systemic disturbances such as general tiredness, irritability, depression and pessimism, and body rash. The intense discomfort some persons feel when lice are biting may persist for several days. Typically a red papule will develop at the site of each feeding puncture. The skin may "weep" and swellings may occur. In time sensitization may develop. The skin of persons who continually harbor lice becomes hardened and deeply pigmented, a condition designated as VAGABOND'S DISEASE or MORBUS ERRORUM.

LOUSE-BORNE (EPIDEMIC) TYPHUS

Disease: Louse-borne typhus (epidemic typhus); Brill-Zinnser disease, classical typhus, European typhus, tabardillo, war fever, jail fever.
Pathogen: Rickettsia, *Rickettsia prowazekii.*
Victims: Man; often pathogenic and fatal to the louse.
Vector: Louse, *Pediculus humanus humanus.*
Reservoir: Man, nonhuman reservoirs doubtful.
Distribution: African highlands, especially Burundi, Rwanda, and Ethiopia, with scattered foci elsewhere, especially in Bolivia and mountainous areas of Ecuador; scattered elsewhere in Africa, Europe, Asia, and Central and South America (cf. Fig 8-4).
Importance: In 1971, 10,272 cases and 106 deaths reported for the world, many fewer in 1972. Devastating, widespread epidemics occurred during the two world wars and lesser, though great, epidemics between and since these wars, under conditions of stress, poverty, crowding, and mass migration. Fatality high

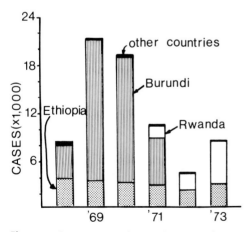

Fig. 8-4 Occurrence of louse-borne typhus in the African highlands, 1968–1973. (From WHO, 1974, *Wkly. Epid. Record,* **49**[18]:109.)

under epidemic conditions, sometimes nearly 100 percent.

This disease is of ancient origin and wide distribution, chiefly in Europe, Africa, Asia, and higher altitudes of Mexico and Central and South America. Brill-Zinnser disease is a recrudescent form of it. The causative organism is *Rickettsia prowazeki.* Whenever humans are concentrated in close quarters, and whenever excessive poverty or mass migrations prevail, especially in times of war and famine, the disease may become rampant. It is chiefly a disease of cool weather or, in the tropics, of the cooler high mountain areas. Mortality varies; under epidemic conditions, especially when malnutrition prevails, it may reach 100 percent.

The disease is characterized by a high fever continuing about 2 weeks, backache, intense headache, bronchial disturbances, mental confusion, stupor, a congested face (designated as a "besotted expression"), and, on the fifth or sixth day, by a red macular eruption on chest and abdomen, later spreading to other parts of the body, even to the hands, feet, and face. In early investigations in Mexico, "tabardillo" was mistaken for Rocky Mountain spotted fever.

The first record of a typhus epidemic was probably the great "plague" in Athens, during the Peloponnesian War, in 420 B.C., which Thucidides described as being without record elsewhere or with such a great destruction of human life (Marks and Beatty, 1976). This probably was typhus, though we cannot be sure of it. Prior to 1970 typhus was widespread and endemic throughout much of Europe, but since then, except during times of major emergencies, it has declined to the point that few deaths from it occurred during the period between the two world wars. During World War I, severe epidemics occurred in Russia, Poland, and the Balkan States. Russia alone lost 2 to 3 million of her people. During World War II, in 1942, there were about 80,000 cases in North Africa.

When Allied forces landed in Italy, in September 1943, a typhus epidemic in Naples was threatening virtually to wipe out that city of a million population. Under the crowded, highly unsanitary conditions, the death rate reached as high as 81 percent. An effective delousing campaign, using chiefly DDT, saved the situation, as it did also in Cologne, Germany. For an excellent account, see Cushing (1957).

Transmission by Lice. *Pediculus humanus humanus* is probably the sole agent in the transmission of the typhus organism from man to man, although both the head louse and the crab louse can serve as hosts in which it can multiply. The typhus patient is infectious for the louse during the febrile period. The louse acquires the parasite by way of the blood meal. The rickettsiae multiply enormously in the epithelial cells of the midgut of the louse; these cells become so distended after a few days that they rupture and release enormous numbers of rickettsiae into the lumen of the digestive tract; these then appear in the louse's feces. The rickettsiae are pathogenic to the louse; most infected lice die in 8 to 12 days because of damage to the gut epithelial cells. If a louse survives the infection it remains infective for life.

The usual route by which man becomes infected is through fecal contamination, though it may be brought about through the crushed body contents of the louse. The bite of the louse is not directly involved; the rickettsiae do not occur in the salivary glands. The parasites may remain alive and virulent in the louse's feces at room temperatures for more than 60 days; thus, infection may be acquired through the respiratory passages by inhalation of minute particles of louse excrement. The usual means of transmission is through scratches and through contact with the conjunctivae or mucous membranes by fingers contaminated with louse feces.

Man is the reservoir of the pathogen. Asymptomatic human carriers may be capable of infecting lice for many years; either changes in local conditions that might favor excessive multiplication of lice, or the passing of a carrier through an area where a susceptible human population resides, might lead to an infection of the louse population and the beginning of an epidemic. A characteristic of louse-borne typhus is its tendency to recrudesce, producing a mild form of the disease, known as the Brill-Zinnser syndrome, which may maintain the pathogen and introduce it into a susceptible population for as many as 50 years after the original onset of the disease. Thus, the reservoir may be maintained long after any overt cases have ceased to occur. Epidemics or local outbreaks of unexplained origin could result from such sources.

Animal reservoirs have been investigated by many workers, notably Reiss-Gutfreund, in Ethiopia, who demonstrated a higher titer by agglutination tests in domestic sheep, goats, and zebu than in the human population, and who infected lice from strains of the parasite isolated from ticks that were feeding on zebu. These studies, which have been summarized by Ormsbee, Burgdorfer, and McDade and Wisseman (PAHO, 1973) lend little support to the hypothesis of an extrahuman reservoir.

Louse-borne typhus is a poverty-associated disease. Unsanitary conditions, in cooler climates, with infrequent bathing and ineffective laundering, lead to multiplication of lice, and consequent opportunities for transmission. Likewise, susceptibility and pathogenicity increase with malnutrition, particularly protein deficiency. Current standards of living in well-developed countries have largely eliminated the disease there, but its cause lies smoldering, ready to erupt quickly and violently under conditions favorable to it. The great epidemics of World Wars I and II originated from latent foci; the possibility of similar future ones still exists.

Murine (Flea-Borne) Typhus. This is a much milder disease, maintained in nature in rats and transmitted to man by rat fleas. The use of the term "endemic" to distinguish the flea-borne from the louse-borne strains of typhus is open to criticism, since both must have an endemicity to survive and both may be epidemic at times. The relation of fleas to murine typhus is discussed in Chapter 14.

TRENCH FEVER

Disease: Trench fever; five-day fever, Wolhynia fever, shank fever, His-Wernerische Krankheit.
Pathogen: Rickettsia, *Rochalimaea quintana.*
Victim: Man.
Vector: Louse, *Pediculus humanus humanus.*
Reservoir: Man.
Distribution: Widespread, but latent, local foci in Europe, Africa, Asia, the altiplano of Mexico and Central and South America. Explosive under proper conditions, but no cases have been diagnosed anywhere in recent years.
Importance: A nonfatal but highly debilitating disease.

Trench fever is a nonfatal disease characterized in overt cases by sudden onset of fever, headache, dizziness, pain in the muscles and bones, particularly in the legs, with especial tenderness of the shins, and lasting 24 to 48 hours or more, followed at intervals of about 5 days by other attacks of fever of diminishing severity. Diagnosis is difficult; the disease is very often asymptomatic or very mild and may be confused with influenza, dengue, and several other diseases. Man is the only known reservoir.

Epidemics, as in the case of typhus, depend upon heavy louse infestations in a susceptible human population under low socioeconomic conditions. The disease was first noticed during World War I when, under conditions of trench warfare, it became of considerable importance, involving at least a million men. It reappeared in Yugoslavia and the Ukraine during World War II. Between the two great wars and since then, mostly sporadic cases have occurred. However, recovery of the rickettsiae from lice and serologic evidence indicates that the disease is widespread, even though in an asymptomatic or suppressed form, in louse-infested areas of Europe, Asia, Africa, and the highlands of Mexico and Central and South America.

The causative organism is *Rochalimaea quintana.* Unlike *Rickettsia prowazekii,* this organism multiplies freely in the lumen of the digestive tract, not in the epithelial cells, and it is not pathogenic to the louse. An infected louse may live a normal span and is infective for life. As in the case of louse-borne typhus, the louse acquires the rickettsia in its blood meal and passes it on to man through the feces or through the body contents of a crushed louse.

EPIDEMIC RELAPSING FEVER

Disease: Epidemic relapsing fever.
Pathogen: Spirochete, *Borrelia recurrentis.*
Victim: Man.
Vector: Louse, *Pediculus humanus humanus.*
Reservoir: Man; questionably rodents and ticks.
Distribution: Africa, especially Ethiopia; only scattered foci and sporadic occurrence elsewhere. Great epidemics occurred in Europe during World Wars I and II.
Importance: In 1971, 4,972 cases and 29 deaths in the world (4,700 and 29 respectively in Ethiopia).

This is another of the three important diseases of man that are associated with lice. It has occurred in many parts of the world and was probably once cosmopolitan. There were frequent epidemics of the disease in Europe during the eighteenth and nineteenth centuries. Major ones have occurred as recently as the period of World War I and its aftermath in Russia, Central Europe, and North Africa, and during and after World War II, in 1943–1946. According to Felsenfeld (PAHO, 1973) an epidemic occurred in the Democratic People's Republic of Viet Nam during the conflict there. Except during wartime, louse-borne relapsing fever has in recent years been restricted largely to Africa, though imported cases and local foci have been reported elsewhere. Between 2,000 and 5,000 cases were reported annually from Ethiopia, and a smaller number from Sudan, from 1967 through 1971.

The name currently accepted for the pathogen by microbiologists is *Borrelia recurrentis*. The role of lice in its transmission, though long suspected, is now clearly established. Not only *Borrelia recurrentis,* but also most other members of the genus, will survive and multiply in the human body louse. The insect can acquire the pathogen by a single feeding on an infected person but cannot pass it on to a second human in this way. Man acquires the parasite by crushing the louse, usually in the act of scratching to alleviate the irritation caused by the bite, and in this way releases the spirochete, which then enters the excoriated skin or mucous membrane.

After being ingested by the louse, the spirochetes pass through the stomach wall into the hemolymph. The digestive tract of the insect seems to be a hostile environment, but the ability of a sufficient number of spirochetes to survive there and to pass into the hemolymph, where they will multiply without being affected themselves and without damaging the host, indicates a highly successful adaptation of the parasite to its host. Once infected, the louse remains so for life. Multiplication of the spirochetes in the hemolymph is rapid, becoming even more so after 5 to 7 days and reaching a maximum after 10 to 12 days. The spirochetes do not invade the gonads, salivary glands, or malpighian tubes and are not found in the feces. Consequently, transovarian transmission and transmission by fecal contamination are not possible.

Whereas in tick-borne or endemic relapsing fever endemicity is maintained in nonhuman reservoirs, such as rodents, man is the usual reservoir of the louse-associated disease. The theory that ticks preserve and lice propagate the *Borrelia* has been proposed, but this is subject to serious question. Transmission from man to man is through transfer of lice.

The incubation period in the human is 3 to 10 days. The onset of the disease is sudden, with headaches, chills, and fever, and generalized pains. The fever remains high for several days (an average of 4) and subsides abruptly, with an afebrile period of 3 to 10 days, followed by one or more relapses. Mortality is usually low to negligible, but it may vary up to 50 percent or even higher in crowded poverty-stricken and louse-infested populations.

OTHER LOUSE-RELATED INFECTIONS

The only known microbial agents pathogenic to man that are adapted to lice and, consequently transmitted by them, are rickettsiae and spirochetes. Lice have been suspected, however, in the transmission of other human and animal pathogens. In general, louse feces have proven to be a good medium for the preservation of several pathogenic bacteria.

Weyer (1960) has concluded that natural transmission of other pathogens by lice is probably insignificant, although an extensive louse infestation may play some role in spreading *Salmonella* infections. Such occurred, for example, in Russia in 1920–1922 in association with epidemics of typhus and relapsing fevers. Milne and associates (1957) produced infections of *Salmonella enteritidis* by feeding lice on infected rabbits. The salmonella remained infective in lice for more than a year, in one extreme case for more than 4 years. Similar results have been obtained with typhoid, paratyphoid, and colon bacilli. Price (1956) obtained experimental infection with tubercle bacillus in the body louse, with multiplication in the louse. Experiments with other microorganisms, including *Bartonella bacilli-*

formis and the protozoon *Toxoplasma gondii,* have proven less conclusive.

ANOPLURA AFFECTING ZOO AND DOMESTIC ANIMALS

Among the Anoplura that attack wild mammals, *Pediculus mjobergi* parasitizes New World monkeys. Ronald and Wagner (1973) report a fatal infestation of two spider monkeys *Ateles geoffroyi,* in a St. Louis, Missouri, zoo, death apparently resulting from anemia. Men who cared for these monkeys also became infested. *P. mjobergi* may be merely a subspecies of *P. humanus.*

The important sucking lice of domestic mammals belong to one of three genera, *Haematopinus, Linognathus,* and *Solenopotes.* Significant weight losses and anemia may result from heavy infestations by these lice. A monetary value is hard to estimate, but the USDA estimated in 1965 that losses in the USA that year amounted to $47 million each for cattle and sheep (Steelman, 1976); these estimates included both Anoplura and Mallophaga.

Swine have one species, the hog louse, *Haematopinus suis* (Fig. 8-5). This is the largest species of the entire group, measuring as much as 5 or 6 mm in length. It is cosmopolitan in distribution. According to Steelman (1976), it is considered, next to hog cholera, the worst enemy of swine. Hog lice feed readily on man and, under experimental conditions, on some rodents, but not on guinea pigs.

Five species of Anoplura infest cattle. (1) *Linognathus vituli,* the long-nosed cattle louse, is characterized by its small size (about 2 mm in length), its slender body, and its long snout. (2) *Solenopotes capillatus,* known as the little blue cattle louse in the USA and the tubercle-bearing louse in Australia, is smaller, 1.2 to 1.5 mm in length; it generally resembles the previous-mentioned species.

The confused situation concerning the three species of *Haematopinus* on cattle has

A

B

Fig. 8-5 *A.* Hog louse, *Haematopinus suis.* × 7. *B.* Nits (eggs) of the hog louse attached to the hairs of the host. One of the eggs has hatched. × 10.

been clarified by Meleney and Kim (1974), who discussed them thoroughly and presented a key and numerous illustrations as aids in their separation. (3) *Haematopinus eurysternus,* the shortnosed cattle louse, is generally considered economically the most important louse infesting domestic cattle, *Bos taurus,* its normal host, through which it has been introduced in all parts of the world, particularly in cold and temperate areas. It is found widely distributed over the body of the host. (4) *H. quadripertusus,* the cattle tail louse, is found in the long hair around the tail, and sometimes around the eyes and in the ears. Its normal host is the zebu or humped cattle, *Bos indicus,* but it also infests hybrids between this species and *Bos taurus.* It is predominantly tropical or subtropical in distribution. (5) *H. tuberculatus,* the buffalo louse, is primarily a parasite of the water buffalo, *Bubalus bubalis.* It has been recorded from many parts of the Old

World where the buffalo has been introduced; it will infest domestic cattle, usually when in association with the water buffalo. *Haematopinus* species are larger than those of *Linognathus* or *Solenopotes,* usually 3 to 5.5 mm in length, the buffalo louse being the largest of the three.

Horses, mules, and asses are frequently infested with the horse sucking louse, *Haematopinus asini.* It resembles the hog louse except that the head is relatively longer and more robust. Sheep may be affected by the foot louse, *Linognathus pedalis.* This species occurs not only in the United States but also in parts of South America, New Zealand, Australia, and South Africa. Two other species of *Linognathus* occur on sheep, *L. ovillus* and *L. africanus,* the three sometimes occurring on the same animal though on different parts of its body (Meleney and Kim, 1974).

Several studies have shown that nutrition is of importance in keeping populations of both sucking and chewing lice under control. Decreased ability of poorly nourished animals to groom themselves and interference with normal seasonal shedding of hair appears to favor survival of their lice.

ORDER MALLOPHAGA

Chewing Lice

The Mallophaga, in the sense accepted here, are divided into three suborders. (1) The Rhynchophthirina consists of one family, the Haematomyzidae, which in turn includes but two species, *Haematomyzus elephantis,* a parasite of elephants, and *H. hopkinsi,* a parasite of wart hogs. (2) The Amblycera includes several families, of which one, the Menoponidae, is of importance because of the genera *Menacanthus* and *Menopon,* which attack domestic birds. Some Amblycera, notably the Gyropidae, attack mammals; at least two of these infest guinea pigs. *Heterodoxus spiniger,* family Boopidae, a parasite of coyotes and wolves in the New World, may infest dogs. Amblycera are body lice. (3) The Ischnocera also contain both avian and mammalian parasites, but unlike the Amblycera they are found fixed to fur or feathers. Such genera as *Columbicola,* which feeds on the domestic pigeon, and *Chelopistes, Cuclotogaster, Goniocotes, Goniodes, Lipeurus,* and *Oxylipeurus,* which parasitize domestic fowls, belong to the family Philopteridae, and three genera of mammalian lice of the family Trichodectidae, namely *Bovicola, Felicola,* and *Trichodectes* infest domestic mammals.

The chewing lice, of which there are about 3,000 described species in the world, are much more numerous than the Anoplura but are of relatively little medical or veterinary importance. The injury done by them is largely to poultry, although some trouble may result when mammals are badly infested. The possibility of mechanical transmission of several pathogenic bacteria exists (cf. Seegar *et al.,* 1976). Man is attacked only by accident, if at all. Poultry become irritated by the creeping insect and its incessant gnawing at the skin. Some species, such as the chicken body louse, *Menacanthus stramineus,* frequently obtain blood by gnawing through the skin and rupturing the quills of pinfeathers. Parts of the feathers, particularly the barbs and barbules, constitute a major part of the food of this and certain other species. The irritation from the feeding of the louse causes the host to become exceedingly restless, thereby affecting its feeding habits and digestion; young birds are particularly vulnerable. Egg production in fowls is greatly reduced and development retarded. Lice tend to be abundant where uncleanliness and overcrowded conditions exist.

Important literature on chewing lice includes the work of Hopkins (1949), the useful checklists of Emerson (1962b, 1962c, 1964), and the same author's reviews of Mallophaga occurring on the domestic chicken

(Emerson, 1956) and turkey (Emerson, 1962a).

Chewing Lice of Domestic Fowls. More than 40 species of lice occur on domestic fowls. The most common lice of chickens (Fig. 8-6) are the following. (1) The chicken body louse, *Menacanthus stramineus,* is a very active species occurring on all parts of the fowl; it is probably the most damaging species to its host because it frequently occurs in large numbers (up to 35,000 or perhaps more per bird). (2) The shaft louse, *Menopon gallinae* (Fig. 8-6), a similar but smaller species, occurs mainly on the shafts of the feathers. It may infest turkeys, ducks, and guinea fowl, especially when they are housed with chickens, and it sometimes infests horses that are stabled nearby. Other species of major importance on chickens are the fluff louse, *Goniocotes gallinae;* the brown chicken louse, *Goniodes*

Fig. 8-7 The large turkey louse, *Chelopistes meleagridis.* A. Adult. B. Eggs at base of feather. (Courtesy of Pacific Supply Coop and Ken Gray.)

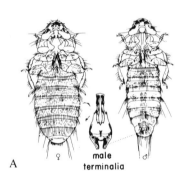

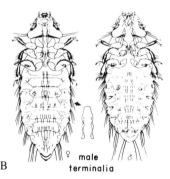

Fig. 8-6 Common biting lice of poultry. A. *Menacanthus stramineus.* B. *Menopon gallinae.* (Courtesy of US Public Health Service.)

dissimilis, the wing louse, *Lipeurus caponis;* and the chicken head louse, *Cuclotogaster heterographus.* All these species are worldwide, wherever their hosts occur.

Turkeys may likewise be attacked by the chicken body louse, but they have their host-specific parasites as well, as would be expected in consideration of the New World origin of the turkey. Chief among these are the large turkey louse, *Chelopistes meleagridis* (Fig. 8-7) and the slender turkey louse, *Oxylipeurus polytrapezius.*

Pigeons are often abundantly infested with the slender pigeon louse, *Columbicola columbae,* a very slender species measuring 2 mm in length, and with the small pigeon louse, *Campanulotes bidentatus compar,* about 1 mm in length, whitish in color, with the head rounded in front.

A study by Ledger (1970) indicated the importance of preening to the ability of the bird to rid itself of lice. A laughing dove, *Streptopelia senegalensis,* with a badly deformed mandible, though able to feed normally, removed all lice from its body except one species that attacks only the head and neck region of the host where the abnormal bird could not grasp hold of them; other birds in the study were able to remove these lice of the head and neck region.

Chewing Lice of Domestic Mammals. The chewing lice of domestic mammals are for the most part rather easily identified by their presence on a given host, since commonly not more than one species of Mallophaga is found on each host species. Thriftiness in herds, promoting better grooming and normal shedding of winter coats, reduces the louse infestations. Spread from one animal to another is commonly by large-scale direct contact or by introduction of uninfested animals into quarters that have recently been vacated by infested ones, though phoresy, for example, on mosquitoes and hippoboscids, sometimes occurs.

Cattle are often heavily infested on the withers, root of tail, neck, and shoulders by the cattle-biting louse, *Bovicola bovis* (Fig. 8-8), a little red louse about 1.5 mm in length, with transverse bars that produce a ladderlike marking on the abdomen. Horses, mules, and asses, but horses more particularly, when poorly or irregularly groomed, may suffer from the horse-biting louse, *Bovicola equi*. Sheep at times may show severe infestation of the sheep-biting louse, *Bovicola ovis*. Goats are commonly very heavily infested with chewing lice. Several species from goats have been described, about which there is some confusion, but the common species is the goat-biting louse, *Bovicola caprae*. Dogs, particularly puppies, may suffer much irritation from the dog-biting louse, *Trichodectes canis* (Figs. 8-1, 8-8), a broad, short species measuring about 1 mm in length, and by *Heterodoxus spiniger*. Cats may become heavily infested with the cat louse, *Felicola subrostratus*. Elephants may suffer from dermatitis caused by the elephant louse *Haematomyzus elephantis;* in one zoo animal reported by Raghavan and associates (1968) the dermatitis was severe and involved deep penetration of the skin. *Bovicola lipeuroides* and *Bovicola parallela* infest American deer (*Odocoileus*) and other genera are found on other wild mammals.

Chewing Lice as Developmental Hosts of Parasites. *Dipylidium caninum,* the double-pored dog tapeworm, is a common parasite of the dog and is occasionally found in humans, especially children. Although fleas are of greater importance as developmental hosts of this cestode, lice may also serve as such. Since many chewing lice subsist on epidermal scales, skin exudations, and other matter on the skin of the animals, it is comparatively easy for the lice to become infected by swallowing egg capsules. The dog, on the other hand, readily infects itself by devouring the lice (or fleas) that irritate its skin. It may pass the infected insects onto persons, particularly children, as they fondle their pets.

At least two filarial nematodes of the family Dipetalonematidae are known to have chewing lice as developmental hosts. Pennington and Phelps (1969), investigating causes of canine filariasis that was rendering dogs unfit for guard services in Okinawa, found the developing forms of *Dipetalonema reconditum* in the chewing louse *Heterodoxus spiniger*, the anopluran *Linognathus setulosus*, and three species of fleas. Seegar and coworkers (1976) have shown that *Trinoton*

Fig. 8-8 *A.* The biting ox louse, *Bovicola bovis.* × 26. *B.* The biting dog louse, *Trichodectes canis.* × 35.

A   B

anserinum, suborder Amblycera, is a natural cyclodevelopmental host of *Sarconema eurycera,* a heartworm of the whistling swan *Cygnus colombianus columbianus* and at least six other species of swans and geese. This host-vector relationship is possible because of the habitual blood-ingesting habit of the louse.

CONTROL OF SUCKING AND CHEWING LICE

The head louse, and crab louse of humans are all controlled by similar treatments. Prevention of lousiness is accomplished mainly by reducing intimate contacts. For the body louse, control may be achieved by frequent changes of clothing, treating infested garments with hot water or steam, or fumigation. Lotions such as Kwell® or shampoos containing a persistent insecticide, or an insecticide with short residual activity plus an ovicide, are highly successful. For mass treatments, such as may be necessary in conditions where typhus threatens, insecticide dusts may be applied in clothing and onto the body. Useful discussions are found in PAHO (1973) and Keh and Poorbaugh (1971), and instructions are available for determining the susceptibility status of body lice to various pesticides (WHO, 1975c).

Control of chewing lice on birds, and of sucking and chewing lice on laboratory animals or pets, is usually accomplished with insecticidal dusts. In addition, insecticidal soaps are available for use on pets.

Lice on cattle may be partially controlled if animals have free opportunity to groom themselves; it has been observed that infestations become particularly high on animals in stanchions. Sprays of contact or systemic insecticides, or dips, are used on cattle; for sheep, generally dusts or dips are used. Animals on poor nutrition are more likely to develop serious infestations.

9

GNATS, BLACK FLIES, AND RELATED FORMS

ORDER DIPTERA

The several families of insects discussed in this and the following four chapters are members of the order Diptera, the two-winged or true flies. Eighty to a hundred thousand species, in approximately 140 families, have been described, and additional ones are being added at a rapid rate. Members of this order are involved more abundantly in the transmission of human and animal pathogens than any other group of arthropods; moreover, many species in a wide range of families are important because of their bites, production of allergens, attacks as larvae, and in other ways. The medical or veterinary entomologist, therefore, must be extensively familiar with this order.

As the name implies, all winged members of this order have only one pair of wings; instead of the posterior pair, in nearly all species, even the few wingless ones, there is a pair of short knobbed structures known as HALTERES (singular, HALTER) (cf. Figs. 10-3, 12-2). Conspicuous compound eyes are usually present, and most species possess three simple eyes, the ocelli, set in a triangle at the top of the head. The mouth parts, as previously described, are subject to great variation, though, except when vestigial, they are all suctorial; many species are provided with piercing or lacerating stylets that enable them to penetrate flesh and either to suck blood through a tube formed by these stylets or to lap up the blood caused to flow from the wound.

Metamorphosis is complex. Most flies are oviparous. A number of species are larviparous, that is, they deposit active larvae; these are sometimes retained within the body of the female through the second instar, as in some muscids and calliphorids, or through complete development, as in tsetse and the so-called Pupipara (Hippoboscidae and related families), pupation then taking place very soon after deposition.

Much attention should be given to the larval stages because the larvae, especially those of the higher Diptera, frequently invade the tissues and organs of man and animals, causing myiasis (see Chapter 13); also, a wider knowledge of aquatic larvae, of which there are many species, is important in pursuing work with mosquitoes, horse flies and deer flies, black flies and others, as well as in the study of the biology of water supplies. Unfortunately, the study of these immature forms is often fraught with severe difficulties because of the paucity of good characters for their identification, and rearing or the use of nonmorphologic information is often necessary or at least a helpful adjunct.

Diptera as a whole have a wide range of breeding habits. Larvae of most species of medical importance require a high humidity, being either aquatic, endoparasitic, or living in a wet medium such as carrion, live or decaying vegetable tissues, or moist or wet manure or soil. Some diptera are highly resistant to adverse environments. The integument may be such as to permit desiccation to a considerable extent. Some may live under almost unbelievably adverse conditions; the petroleum fly, an ephydrid, *Psilopa petrolei,* breeds in oil or tar pits; certain Ephydridae live in highly, even saturated,

saline waters; some blow flies have been known to survive for extended periods in alcohol- or formalin-preserved tissues.

Classification and Identification of the Diptera. In the classification of the Diptera, knowledge of the wing venation (Fig. 9-1) is important. The great diversity of antennal structure, three types of which are illustrated in Figure 9-2, provides useful basic characters, as does the arrangement of bristles on the body of such species as the blow flies (see Chapter 12). The genitalia furnish important taxonomic characters in many Diptera. Structures of special importance in the various groups will be discussed in connection with the individual groups involved.

Division into suborders has varied according to the authority, but the most commonly accepted now recognizes three suborders. (1)

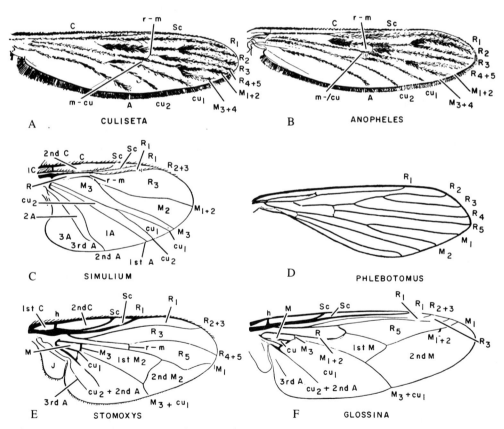

Fig. 9-1 Wings of Diptera. For explanation of venation, see Figure 2-2.

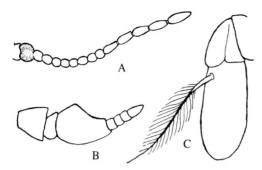

Fig. 9-2 Three antennal types in the Diptera; in outline, setation and hairs omitted. *A.* The many-segmented type, *Culicoides* (after I. Fox). *B.* The stylate type, *Tabanus. C.* The aristate type, Sarcophaga.

In the NEMATOCERA, the larva has a well-developed head with horizontally biting mandibles, the pupa is free, and the adult has a many-segmented antenna, usually longer than the head, and never aristate or stylate. Families of Nematocera treated here are the Simuliidae, Psychodidae, Ceratopogonidae, Chironomidae, Dixidae (not affecting man or animals), Chaoboridae, and Culicidae. (2) In the BRACHYCERA, the larva has an incomplete, usually retractile head, the pupa is free or, uncommonly, enclosed in the last larval skin, and the adult has a shortened, three-segmented antenna, although the third segment may be subdivided and may appear to be many-segmented, some of the subsegments often forming a STYLE. Brachycerous families treated here are the Tabanidae, Athericeridae, and Rhagionidae. (3) the so-called higher flies constitute the CYCLORRHAPHA. In these the larva is a headless MAGGOT or GRUB with a pair of mouth hooks that operate side by side, or with only one such hook; the pupa is enclosed in a PUPARIUM formed from the last larval skin; and the adult has a short, three-segmented antenna, the last segment usually bearing an ARISTA or STYLE.

Although the above classification is the one most generally in use, many dipterists are dissatisfied with it. An alternate one, proposed by Oldroyd (1977) and based ex-

tensively on the paleontological studies and phylogenetic conclusions of Rohdendorf, along with a morphological consideration of mouth parts structure, deserves serious consideration.

Identification of flies and their immature stages must often be the task of specialists in the various families. However, in many cases the nonspecialist can make an identification that is satisfactory for certain needs. A very useful reference is K. G. V. Smith (1973), more than three-fifths of which is devoted to the Diptera, both adult and immature stages being included. Lindner's *Die Fliegen der Palaearktischen Region* (Handbuch, Lindner, 1925–1949, and individual parts by various authors at various dates) is invaluable for temperate Eurasia, and Cole (1969) will greatly aid in the identification of flies of western North America. A new manual of the Nearctic families and genera of Diptera, to replace the outdated 1934 manual of Curran, is in preparation (Peterson *et al.,* ?1979). Important works of a broad nature on dipterous immature stages are Hennig, 1948–1952; Peterson, 1951; and Johannsen, 1933–1937. Important identification guides for more restricted areas or for specialized groups will be cited in the proper places. Catalogs of the entire order for entire zoogeographic regions are the following: Stone *et al.,* 1965, Nearctic Region; Papavero *et al.,* 1966– , Neotropical Region; Delfinado and Hardy, 1973–1977, Oriental Region; and Crosskey *et al.,* to be published, Ethiopian Region. The Neotropical catalog is currently incomplete; the Ethiopian one is scheduled for early in 1980.

Family Simuliidae

BLACK FLIES

Characteristics. These flies, of which there are more than 1,000 known species, are small (1 to 5 mm in length), stout bodied, and variable in color; the term "black fly" is

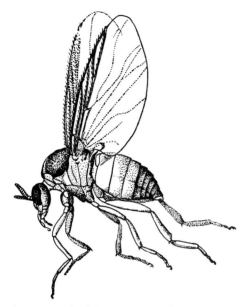

Fig. 9-3 A buffalo gnat, *Cnephia pecuarum.* (After Garman.)

somewhat of a misnomer as certain species may be gray or even predominantly tannish yellow. The mouth parts include bladelike piercing stylets in the female (Fig. 3-6) but are reduced in the male. The thorax presents a strong development of the scutum and a reduction of the prescutum resulting in a prominent hump (Fig. 9-3). The antennae are 9 to 12, usually 11-segmented; all but the basal 2 segments form the flagellum. Eyes of the female are distinctly separated, those of the male are, with rare exceptions, contiguous above the antennae. Ocelli are absent. The palpi are 5-segmented. The wings are broad and irridescent, with distinct alulae; the venation is characterized by a strong development of the anterior veins (Fig. 9-3).

Biology. Black flies often occur in enormous swarms where strong or swiftly flowing streams provide well-aerated water for larval development. They are particularly abundant in the north temperate and subarctic zones, but many species occur in the subtropics and tropics where, of course, factors other than seasonal temperatures affect their developmental and abundance patterns.

Larvae are nearly always found in running water, shallow mountain torrents being favored breeding places. Some species, including some notable pests, breed in the larger rivers; others live in temporary or semipermanent streams. Larvae attach themselves to rocks or other solid objects in the stream; sometimes they cling to aquatic or emergent vegetation. The sides and concrete drop structures in irrigation canals, concrete dams, and other such situations may serve as places of attachment for large numbers of larvae. Some attach themselves to aquatic animals, such as bivalve molluscs, crustaceans, and the naiads of Odonata and Ephemeroptera. The African *Simulium neavei,* an important vector of *Onchocerca volvulus,* and several other species, have an obligate phoretic relationship with river crabs of the genus *Potamonautes.*

Adults may fly from 12 to 18 km from their breeding places, but migrating windborne swarms may go much farther. *Simulium arcticum* may travel 150 km or more in this way in the Saskatchewan Valley of Canada, and other species are reported to travel 250 km or more. *Simulium colombaschense* may travel 200 to 450 km with wind currents in the Danube Valley.

Nectar from flowers provides both males and females with carbohydrates for flight energy, and females, in addition, usually require blood for ovarian development. As in the other Nematocera, males never suck blood. Most females are bloodsuckers, often vicious ones, but in some species the mouth parts are so feeble that they cannot penetrate the skin of vertebrates. Nonbloodsucking species are autogenous, but autogeny may occur in some bloodsucking species as well. Of the hematophagous species, only certain ones attack man. A number of birds and mammals, however, have been reported as hosts.

Males form hovering swarms, and mating may occur when females fly into or near such swarms. Parthenogenetic reproduction is known to occur, but apparently as an extreme rarity.

Life History. Eggs, usually 200 to 500 per female, but sometimes as many as 800, are deposited on the water surface, on aquatic plants, or on logs, water-splashed rocks, or other solid surfaces in or at the edge of the water. Sometimes they sink and become scattered loosely on stream bottoms. Commonly the female drops eggs while flying over the water surface; some species will hover and oviposit through a thin film of water that covers sand, rock, or vegetation; others will settle and oviposit on water-lapped surfaces at the water's edge.

In the multivoltine species the time required for hatching is usually 3 to 7 days, except that eggs laid in late summer do not hatch until the following spring. When diapause takes place in the egg stage, as is usually the case in univoltine species in temperate regions, the time from egg deposition to hatching is, of course, much greater. *Simulium venustum* requires 4 days at 24° C, 5 days at 18° C, and 27 days at 7° C.

The newly emerged larva attaches itself by means of a caudal sucker and silken threads to the submerged object that is to form its support and resting place. Movement from place to place is achieved by shifting anchorage. In some favorable locations, such as rills on the downstream side of an old log partially damming a stream, there may be thousands of these tiny, black, spindle-shaped larvae. The larvae, as well as the pupae, are provided with gill filaments and usually remain submerged or partially so. The larvae are largely indiscriminate filter feeders. Their food consists of small suspended crustacea, protozoa, algae, bacteria, parts of animals and plants, and decaying organic matter. Some may graze in the silt or on submerged surfaces; some even resort to cannibalism. The larval period may require 7 to 12 days under favorable conditions; however, it may be extended to weeks or months, depending on the water tempera-

ture, food available, and the species involved. Some species overwinter as larvae. There are six or seven larval instars. Pupation takes place in a basketlike cocoon that is firmly attached to shallowly covered objects, such as rocks.

The pupal period may require only 2 to 6 days or it may last 3 to 4 weeks. Cooler weather retards the emergence of adults. The fully formed adult, in about a minute's time, breaks through the T-shaped emergence slit and rises quickly to the surface in a bubble of gas. In the multivoltine species there is continual breeding from early spring to late autumn, with overlapping generations; in the univoltine species there is one sudden brood coming early in the spring, with stragglers following. The life cycle, egg to adult, ranges from 60 days to 15 weeks or over; there are one to five or six generations a year, depending on the species and climatic conditions. Adult female longevity is apparently short, usually 2 or 3 weeks, although *Simulium metallicum,* according to Dalmat (1955), may live up to 85 days; in some species there is evidence of aestivation during periods of extended drought, which may extend the longevity further (Crosskey, in K. G. V. Smith, 1973).

Larvae. The light brown to black larva is cylindrical, slightly thinner in the midregion, and when fully grown 10 to 15 mm in length (Fig. 9-4). Fan-shaped filamentous structures located on the head are used for filtration of food from the water. The posterior end of the body is provided with a toothed disklike sucker. The anterior pseudopod, or proleg, is also modified as a toothed disk. The larva is one of the most characteristic creatures that occur in running water.

Pupae. The form of the pupal cocoon, as well as the pupal site, varies with the species. The pupa is provided with respiratory filaments attached anteriorly to the dorsal part of the thorax (Fig. 9-4). The filaments are often quite numerous. Because of

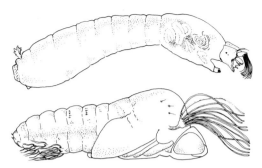

Fig. 9-4 Black fly larva (*above*) and pupa (*below*). (After Dinulescu.)

their constancy in form and number, they are of great taxonomic importance.

Natural Enemies. Black flies have numerous parasites and predators, and these may act at times as important natural controls of black fly populations. Useful reviews of this subject are given by Dinulescu (1966) and in the Bull. WHO, suppl. 1 to no. 55, 1977. The systematic utilization of this information for control purposes, however, is quite a different matter.

Dinulescu discusses the following natural enemies of the Simuliidae. Invertebrate predators on immature stages include hydras, planarians, crustaceans (*Rivulogammarus*), and many aquatic insects. Vertebrate predators include numerous fish as well as three species of birds of the genus *Cinclus*. There are numerous predators on adult black flies, as well as the insectivorous plant *Pinguicula vulgaris*. Parasites on either the larval or adult simuliids include fungi, trypanosomes, infusoria, sporozoa, spirochetes, and mermithid nematodes. Natural enemies mentioned by other authors include spiders, Coleoptera, and possibly leeches. Certain muscoid flies (Coenosiinae) may exercise considerable control when black fly populations become depleted by other methods.

Mermithid parasitism may be especially important. Phelps and De Foliart (1964) have pointed out that, in Wisconsin, parasitism resulting in castration or death of adults amounted to 37 to 63 percent, with an additional conservative estimate of 50 percent larval mortality. The effect of such high parasitism on black fly populations is evident. Mermithid parasitism cannot be expected to eliminate a species from any given area, but it certainly can hold down the numbers.

Classification. The classification of the Simuliidae is at present in need of considerable study. Authors differ as to what genera should be recognized. According to the conservative classification adopted by Crosskey (*in* K. G. V. Smith, 1973), most species, about 810, belong to the genus *Simulium*. *Cnephia, Prosimulium,* and *Austrosimulium* also contain species of medical importance (the first two of these are combined under *Prosimulium* by Crosskey). The occurrence of sibling species, suspected in some cases and clearly demonstrated in others, complicates species-level taxonomy. For example, the so-called American *Simulium vittatum* bites man in parts of its range but not in others. What was formerly considered to be one species, *Simulium damnosum*, in Africa, was reported by Crosskey (*in* K. G. V. Smith, 1973) to consist of 17 different cytotypes, and additional ones are now known to occur. Several of these have been given specific names and are recognized as having specific status. Likewise, so-called *Simulium neavei* is clearly a complex.

Two important taxonomic works on a broader scale are those of Rubtsov (1964) for the Palaearctic Region and Crosskey (1969) for the Ethiopian. Some good studies on a more restricted basis are available, among them Dinulescu's very informative work on the Romanian species (Dinulescu, 1966, in Romanian), the publications by Davies and coworkers (1962) and Wood and associates (1963) on the Ontario species, Stone (1964) in the Diptera of Connecticut series, and Dalmat's (1955) monograph of the Guatemala species. Additional taxonomic references, along with many others dealing with the medical importance, biology, control, morphology, and collecting and rearing of black flies, are included in the extensive list of bibliographic references in Crosskey (in K. G. V. Smith, 1973).

Important Species. Only a few of the many important species of black flies can be mentioned here. *Prosimulium mixtum* can be seriously annoying to man and animals in the USA. *Cnephia pecuarum,* the southern buffalo gnat, has been known as a great scourge to livestock as well as to man in the Mississippi Valley, although ecological changes such as increased silt and organic pollution in recent years have reduced the importance of this fly. The genus *Austrosimulium* is represented by a number of species in Australia, New Zealand, and nearby areas; *A. pestilens, A. bancrofti,* and *A. ungulatum* bite man regularly.

Most of the important species of black flies belong to the genus *Simulium* (used in a moderately broad sense and including *Eusimulium*). *S. vittatum* is widespread in North America. It may be annoying to livestock because of its constant crawling over the skin and probing. *S. meridionale,* the turkey gnat, is also common in spring in North America, particularly in the Mississippi Valley and southern USA. It attacks poultry, biting the wattles and combs. *Simulium venustum* is one of the most annoying and widespread species. It torments fishermen and campers in the northern United States, Canada, and Alaska. The gnats occur in the greatest numbers during June and July but may persist through the summer. *Simulium arcticum* is a plague to livestock in western Canada (Fig. 9-5) where vast numbers of this gnat frequently attack and kill livestock and seriously annoy man (Fredeen, 1969).

Simulium colombaschense, the infamous golubatz fly of middle and southern Europe, has caused heavy mortality to domestic and wild mammals. Other European livestock

Fig. 9-5 Cattle attacked by swarms of black flies, *Simulium arcticum,* in central Alberta Province, Canada. The bunching of the cattle is a protective behavior against black fly attack. (Courtesy of J. A. Shemanchuk, Agriculture Canada.)

pests include *S. kurenze* in the Soviet Union and *S. erythrocephala* in western Europe.

In Africa, members of the complexes grouped around *S. damnosum* and *S. neavei* are important as vectors of *Onchocerca*. In Central America *S. ochraceum, S. metallicum, S. callidum,* and *S. exiguum* are important *Onchocerca* vectors. *S. ochraceum* and *S. metallicum* can likewise be vicious biters.

The Bite. Simuliids are daytime biters and are rarely found indoors. The mouth parts are similar to those of the horse fly but differ in detail; the mandibles and maxillae are flattened and usually serrate.

Humans may be viciously attacked. In addition to local reactions, involving reddened, itching wheals, there may be general conditions that vary in intensity with the sensitivity of the individual and the number of bites. A syndrome known as "black fly fever" and involving headache, fever, nausea, and adenitis, may occur, as well as general dermatitis and allergic asthma. In some individuals the face, arms, and other exposed parts may be greatly swollen as a result of the bites; in others, effects other than blood loss may scarcely be noticeable.

Immense numbers of the flies at times attack livestock with resulting heavy damage and high mortality (Fig. 9-5). In Canada between 1944 and 1948 more than 1,100 animals perished annually, and the toll in the Balkan States in 1923 and 1934, resulting from plagues of *Simulium colombaschense,* were 16,000 and 13,900 respectively. Death apparently results from a toxin in the saliva, the effect of which is to increase the permeability of the capillaries, thus permitting the fluid from the circulatory system to ooze out into the body cavity and the tissue spaces. The animal rapidly succumbs to a mass attack, but it can recover quickly if protected from further onslaughts by the fly. Reductions in meat, milk, and egg production may result from less extensive attacks.

There is some correlation between the habit of larval breeding in large rivers and adult viciousness, particularly where attacks on livestock are concerned. *Cnephia pecuarum* has, in past times, killed horses, mules, and cattle in numbers within a few hours in the Mississippi Valley. *S. arcticum* is an outstanding pest where it is associated with large river breeding conditions, as in the Saskatchewan, and outbreaks of *S. colombaschense* have been associated with breeding in the Danube. On the other hand, mountain stream species are often seriously annoying to man. *S. ochraceum* and other smaller stream species become so annoying in the coffee plantations of Guatemala that workmen at times refuse to continue their work (Dalmat, 1955).

In recreational areas where conditions are suitable for their breeding, black flies are second only to mosquitoes in the role of annoying arthropod pests. A low population of an anthropophagous species can be seriously annoying; the immense numbers that occur on occasions may be intolerable. A species that is usually nonanthropophagous may become a pest to man when its usual host is not present. Often the annoyance is seasonal, in temperate regions usually spring or early summer. When the black fly threat prevails, tourist trade, vacationing, or seasonal occupancy of summer homes may be strongly discouraged in otherwise desirable areas near natural simuliid breeding places or those that have become so through man-made structures, water developments, or modifications of the environment.

Pinheiro and associates (1974) have recorded a disease, designated as the hemorrhagic syndrome of Altamira, that has affected immigrants into the area of Brazil along the new Transamazon Highway. Children especially are affected, and some deaths have occurred. No microbial agent has been discovered; the illness, from evidence at hand, seemingly is caused by a toxin associated with intense biting by simuliids. *Simulium amazonicum* is a strongly suspected causative agent.

ONCHOCERCIASIS

Disease: Onchocerciasis; river blindness, blinding filarial disease, Robles' disease (Guatemala), sowda (Yemen).

Pathogen: Filarial nematode, *Onchocerca volvulus.*

Victim: Man.

Vectors: Black flies, genus *Simulium,* especially the *S. damnosum* and *S. neavei* complexes in Africa; *S. ochraceum, S. metallicum,* and others in America.

Reservoir: Man.

Distribution: Tropical Africa roughly between 15° N and 12° S; tropical America from southern Mexico to Columbia, Venezuela, and northern Brazil (Figs. 9-6, 9-7).

Importance: An estimated 20 million cases occur in the world. Blinding is the most common serious effect; visceral involvement, sometimes fatal, and other disorders may occur.

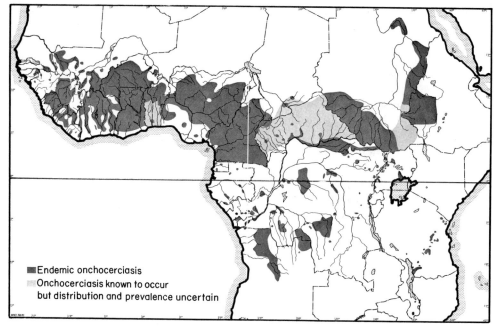

Fig. 9-6   Distribution of onchocerciasis in the Old World. (Courtesy of WHO.)

The most important human parasite transmitted by simuliid flies is a filarial worm, *Onchocerca volvulus,* the causal agent of human onchocerciasis. In the Old World it occurs extensively in tropical central Africa south of the Sahara and in Yemen. The New World species is sometimes considered a separate one, *O. caecutiens,* but evidence to

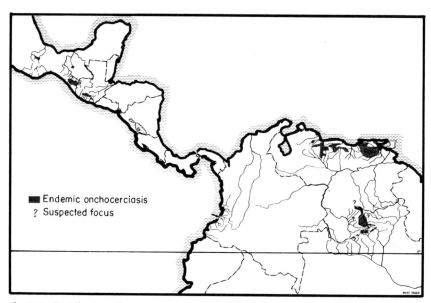

Fig. 9-7   Distribution of onchocerciasis in the New World. (Courtesy of WHO.)

support this separation is inconclusive. Its distribution extends from southern Mexico at least to Colombia, Venezuela, and northwestern Brazil, but probably to Guiana and possibly farther southward; the most important foci are in southern Mexico and Guatemala. It affects mostly rural natives or persons who lack availability of good medical care. In Africa it has severely hindered the human development of the river valleys, where it is most characteristic. The upper Volta Valley is considered the most severely infested area in the world. There are some differences in the manifestation of onchocerciasis in the different areas where the disease occurs.

An important symposium on the disease in America is that of Buck and associates (1974); for important reviews see Nelson (1970) and Duke (1971).

Adult *Onchocerca volvulus* of both sexes occur in subcutaneous nodules just under the skin. Their position on the body varies from one geographical region to another and apparently according to the biting habits of the vector involved; in Africa, where *Simulium damnosum* is the chief vector, they occur mostly on the lower parts of the body, whereas in Central America, where *S. ochraceum* is involved, they tend to be on the upper parts. Furthermore, deep bundles of worms may occur against capsules of joints, between the muscles, and against the periosteum of bones; these may give rise to deep-seated pains or may constitute nidi for the development of deep-seated abscesses (Duke, in Buck *et al.*, 1974). Microfilariae, produced viviparously by the female worm, migrate to the skin, where they are ingested by *Simulium*. Only a small percentage of those ingested go beyond the vector's stomach; the formation of a peritrophic membrane provides a barrier to some, and others are damaged in the feeding process. The successful parasites migrate to the thoracic muscles, where further development takes place, then to the vector's head and proboscis. Repro-duction and multiplication of the worms occur only in the vertebrate host. In man, prolonged and repetitive exposure to black fly biting will progressively build up the parasitemia; frequency of infective attacks and their duration seem to be of greater importance in transmission than the amount of blood taken.

Only the microfilariae, and not the adult worms, produce a pathology of importance to man. The skin may be affected in various ways, the lymph glands may be swollen, and even a form of elephantiasis may occur. Serious systemic involvement, including such organs as the lungs and liver, may develop and may be fatal (Connor, in Buck *et al.*, 1974). The most important effect, however, is blindness, which may occur in as many as 10 to 15% of an affected population. This results from microfilariae migrating from nodules, chiefly those located in the head, through the skin of the face and the conjunctivae, to the eye. Though massive involvement is needed to produce blindness, impaired vision may occur in up to 30% of a population.

The most important vector of human onchocerciasis in Africa is *Simulium damnosum* as more broadly interpreted; *S. neavei* is also of considerable importance in this respect and in some areas is apparently entirely responsible for transmission. The most important vectors in the New World are *Simulium ochraceum, S. metallicum, S. callidum,* and *S. exiguum,* though a number of additional species may be involved. In Guatemala and Mexico *S. ochraceum,* a highly anthropophilic species, assumes the greatest importance. The other three species are strongly zoophilic, though all bite men readily. In Colombia *S. exiguum* tends to become more strongly anthropophilic and assumes greater importance in the disease foci there. Not merely host preference, but also biting habits, can influence the relative importance of a fly as a vector.

BOVINE AND EQUINE ONCHOCERCIASIS

Other species of Onchocerca are transmitted by black flies, although surprisingly few studies of such transmission have been made. Certain *Culicoides* as well as *Simulium* species are sometimes involved. A concise review of onchocerciasis in domestic animals is included in Nelson (1970). *Onchocerca gutturosa* is transmitted by *Simulium ornatum* in England; there it is common and widespread in cattle but is relatively innocuous (cf. Eichler, 1971, 1973; and Eichler and Nelson, 1971). In Australia *O. gibsoni*, a parasite that causes considerable damage to flesh and hides, apparently has both *Simulium* and *Culicoides* species as vectors. Equine ophthalmia, as well as certain other disorders, may be due to the *Culicoides*-transmitted *Onchocerca cervicalis*.

AVIAN LEUCOCYTOZOON INFECTIONS

Infections caused by several species of *Leucocytozoon*, a sporozoan (Protozoa), are known to occur in birds belonging to at least 15 orders (Fallis *et al.*, 1974), but most of them seem to be only mildly pathogenic. However, severe pathogenicity has been attributed to *L. simondi* in ducks, *L. smithi* in turkeys, and *L. caulleryi* in chickens. The disease, which is similar in all three, has been described as malarialike, resulting in loss of appetite, emaciation, drowsiness, enlarged spleen, damaged liver, and congested lungs and heart.

L. caulleryi is transmitted by *Culicoides*; the other species of *Leucocytozoon*, so far as known, develop in and are transmitted by *Simulium*. *S. rugglesi*, *S. anatinum*, and other species transmit *L. simondi*; *S. congareenarum*, *S. slossonae*, *S. nigroparvum*, and *S. occidentale* are hosts and vectors of *L. smithi*. Simuliids are known to transmit at least seven other species of avian *Leucocytozoon*.

Family Psychodidae

MOTH FLIES AND SAND FLIES

The family Psychodidae includes the tiny gnats known as owl midges, moth flies, and phlebotomine sand flies. Those of medical importance belong to two subfamilies: (1) the Psychodinae, the moth flies and owl midges (females not bloodsuckers, wings held rooflike over the body, and larvae commonly aquatic); and (2) the Phlebotominae (females bloodsuckers, wings not held rooflike over the body, and larvae never truly aquatic). Some authorities consider the above as families rather than subfamilies, but the more conservative system, which is used by the majority of workers, is adopted here (Lewis *et al.*, 1978).

Subfamily Psychodinae. The ovate, usually pointed wings and the body are densely covered with hairs; this feature, along with the rooflike position of the wings, present the appearance of a tiny, rather robust moth and suggests the name "moth fly." Only the longitudinal veins are prominent, the cross veins, when present, being restricted to the base of the wing. The antennae are long, and usually each segment bears a whorl of hairs.

Several species of *Psychoda* and *Telmatoscopus* are commonly found in great numbers in sewage disposal plants, cesspools, and washbasins in bathrooms, where larvae may develop in sink drains in spite of hot water and soap. Breeding in such places may occur in such numbers as to constitute a real annoyance to neighboring households. *Psychoda alternata*, a cosmopolitan species that has probably been widely distributed by man, may become annoying in the house, either when it enters from the outside or when it breeds in the surface of the gelatinous material in sink and bathroom drain taps. This species may also be involved in mechanical transmission of parasitic nematodes of livestock (Tod *et al.*, 1971), and its larvae have been reported in pseudomyiasis. The life cycle is short, ranging from 21 to 27 days at room temperatures according to Quate.

Subfamily Phlebotominae. These differ from *Psychoda, Telmatoscopus,* and their allies in that the wings are held upward and outward so that the costal margins form angles of about 60° with each other and with the body. The body is less conspicuously hairy than in the Psychodinae, and the moth-like appearance is not evident. The females have piercing mouth parts and are blood-suckers. Many species feed on cold-blooded animals such as lizards, snakes, and amphibians; others feed on a variety of warm-blooded animals, including man. There is evidence that some species take plant juices; as Adler and Theodor (1957) have shown, this is of importance, since the capacity of at least one species to transmit *Leishmania donovani* is enhanced by feeding on raisins. The males take moisture from any available source and are even said to suck sweat from humans. Females usually require blood meals for ovarian development, but some are autogenous.

These gnats are active only when there is little or no wind. Usually they bite only by night, though at least one species, *Lutzomyia wellcomei,* a probable vector of *Leishmania braziliensis* in Amazonian Brazil, is a daytime biter (Fraiha *et al.,* 1971). By day they seek protection in shelters either in dark corners of buildings, or out-of-doors, where they may hide in crevices and caves, among vegetation, and in buttresses of trees, rodent and armadillo burrows, termite mounds, and such. Their weak, noiseless flight is usually in short "hops" when they are disturbed; in longer flights their progress is slow and steady and can be followed with the eye. Movement as far as 1 km in still weather has been recorded.

The phlebotomine sand flies of medical importance were formerly considered as constituting a single genus, *Phlebotomus.* In the current literature, this genus is restricted to the Old World, along with a second genus of some medical importance, *Sergentomyia,* and the American forms are referred to *Lutzomyia* (Lewis, 1974). Some authors will divide these genera still further. Although *Lutzomyia* is known to occur as far north as southern Canada, species of medical importance in the New World are confined to the tropical, subtropical, and some montane regions of Central and South America, and northward into Texas. On the other hand, considerable pathogen transmission by Old World sand flies occurs also in the warmer temperate regions. Many species never attack man and consequently are of no medical importance. Many of these feed on reptiles.

Valuable taxonomic references to the Phlebotominae are the studies of the Palaearctic species by Theodor (1958), of the Ethiopian species by Abonnenc (1972), and of the New World species by Theodor (1965) and Forattini (1971, 1973).

Life History of Sand Flies. Typical breeding places for sand flies, according to Hertig, are "under stones in masonry cracks, in stables, poultry houses, etc., in situations combining darkness, humidity, and a supply of organic matter which serves as food for the larvae. In no case is the breeding place aquatic." To the situations mentioned might be added surfaces under dead leaves on the forest floor, hollow trees, tree buttresses, and animal burrows. Rodent burrows, which fulfill very well the three requirements stipulated by Hertig, have been shown to be an extensive habitat for these larvae in various parts of the world. In addition Thatcher (1968) has shown that dead leaves and detritus in trees, as much as 12 meters above the ground, may provide breeding grounds for certain species.

The eggs are deposited in small batches (Fig. 9-8). The incubation period is from 6 to 17 days. The minute, whitish larvae have long anal setae; the mouth parts are strongly mandibulate. The larvae feed on organic debris, such as moist excrement of lizards and mammals, insect debris, decaying plant ma-

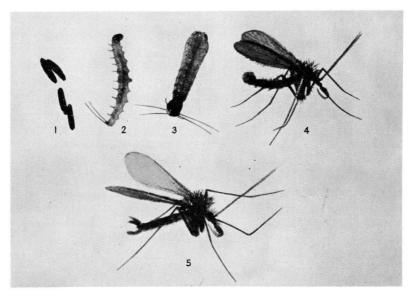

Fig. 9-8 Life cycle of *Phlebotomus*. *1.* Eggs. *2.* Larva. *3.* Pupa. *4.* Adult female. *5.* Adult male. (After Hertig.)

terials, and fungi. There are four instars; the duration of the larval stage is usually 4 to 6 weeks, with extremes ranging from about 2 to 10 weeks. The pupa requires about 10 days for development. The female usually lays eggs in 5 to 7 days under microclimatic conditions of virtually 100% relative humidity. Refeeding habits of females in relation to oviposition vary greatly with the different species, and this fact has considerable significance in respect to transmission of pathogenic organisms. The egg-to-egg cycle requires 7 to 10 weeks; however, where winters are cold, sand flies are subject to diapause in the fourth instar, which may last from several weeks to nearly a year. Tropical species may undergo diapause in dry seasons in certain areas.

Medical Importance of Sand Flies. Phlebotomine sand flies are of known medical importance in transmitting the pathogens of several leishmaniases, Carrión's disease, and sand fly fever and other viral infections. The ability of these insects to harbor and transmit other pathogens of man should be taken into consideration. For example, two

California species of *Lutzomyia* have been reported as transmitting seven protozoan parasites of reptiles and amphibians (Ayala, 1973), including that of lizard malaria. These small, fragile flies have a surprising capacity for serving as vectors and intermediate hosts for a variety of pathogens.

THE LEISHMANIASES

Disease: Leishmaniasis, visceral (kala-azar), and dermal.

Pathogens: Flagellate protozoa in the genus *Leishmania; Leishmania donovani* and *L. infantum* (visceral); *L. tropica, L. braziliensis,* and *L. mexicana* (cutaneous).

Victims: Man, dogs.

Vectors: Sand flies, *Phlebotomus* species (Old World) and *Lutzomyia* species (New World).

Reservoirs: Man, dogs and other canids, sloths, rodents.

Distribution: Forested and arid tropics of New World; mostly savanna and steppe, subtropical and warm temperate, of the Old World (Fig. 9-9).

Importance: Of global significance, wherever adequate vectors occur; with a potential, in its various forms, for high fatality unless treated, and from mild to serious disfiguration.

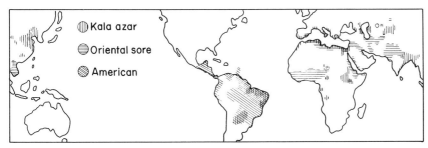

Fig. 9-9 Distribution of leishmaniases. (Adapted from WHO Map VBC 71.255 and US Armed Forces Institute of Pathology Negative No. 76-1380.)

The leishmaniases are caused by parasitic protozoa belonging to the genus *Leishmania,* round or oval intracellular bodies that develop the flagellated promastigotes in the digestive tracts of insects. Several types of human leishmaniasis result from phlebotomine-transmitted pathogens. These diseases are commonly classified as either visceral (kala-azar) or cutaneous infections, although the division is not always distinct; kala-azar has an initial cutaneous lesion and may later develop an extensive cutaneous involvement, whereas the cutaneous form may spread to the viscera in dogs and in experimental animals.

For valuable references on the Phlebotominae in respect to the leishmaniases, see Lainson and Shaw (1974), Lysenko (1971), and Lewis (1974).

Kala-Azar. Kala-azar, dumdum fever, or tropical splenomegaly is a visceral leishmaniasis caused by *L. donovani.* It is widespread, occurring in all countries on the shores of the Mediterranean, south Russia, India, China, Manchuria, equatorial Africa, Brazil, and other parts of tropical America, from Argentina to Mexico. In man there is a progressive enlargement of the spleen and later of the liver. As the disease progresses, the skin becomes grayish in color. In untreated cases it is usually fatal, death resulting in acute infections within a few weeks and in chronic cases in from 2 to 3 years. Medication, however, has reduced fatality to a very low level.

Visceral leishmaniasis is a complicated disease subject to considerable variation and with a variety of reservoir hosts. The brief account that follows deals with generalizations, to which exceptions may occur. There are, however, four recognized epidemiological types.

1. In India, kala-azar occurs in both endemic and epidemic forms; all age groups, but mostly young adults, are attacked. No animal reservoir is known, but the epidemiologic picture can be explained by assuming that man himself constitutes the chief reservoir. This assumption is supported by the facts that the parasite can easily be demonstrated in the peripheral blood, thereby being readily available to the vector, and that the chief vector, *Phlebotomus argentipes,* is a strongly domestic species.

2. In the Sudan and other parts of tropical Africa, cases are sporadic, but epidemics occur in which cases may be unevenly distributed. Again, all age groups, but principally young men, are victims. The status of the disease here seems to be that of a zoonosis, the reservoirs being chiefly rodents that have acquired a well-established tolerance to the parasite, an indication that rodents probably were the original vertebrate hosts.

3. In the Mediterranean and some other areas within the range of the disease, it attacks to a high degree, though not exclusively, children under the age of 5 years. Here dogs are highly susceptible; in fact, the incidence among dogs usually far surpasses

that among humans. This form of the disease is often attributed to a separate pathogen, *Leishmania infantum,* though many consider this a form of *donovani.* The Mediterranean form also occurs in parts of Asia.

4. In the New World kala-azar is known to occur in various areas from Mexico to Argentina, but most importantly in the dry areas of northeast Brazil. As in the Mediterranean form dogs and particularly the wild fox, *Lycalopex vetulus,* are reservoirs, and human cases occur sporadically. Lainson and Shaw (1974) believe that the parasite has been long established in the New World. Its chief vector is a domestic species, *Lutzomyia longipalpis.*

A fifth type, in Northeast China, is sometimes recognized. Like the Mediterranean type, it is a zoonosis and is probably a derivative of that type, according to Lysenko.

The task of determining which species of sand flies are true vectors of visceral leishmaniasis and their relative importance has been a difficult one. This has been complicated by the variation of the epidemiology with place and time. Moreover, progress was made difficult by the low susceptibility of laboratory animals. The discovery that hamsters were highly susceptible and the successful transmission of the disease to human volunteers by Swaminath and associates in 1942 established *P. argentipes* as an important vector. Lewis (1974) lists 15 species and subspecies of Old World *Phlebotomus* and 2 of New World *Lutzomyia* as known or probable vectors. In addition to *argentipes,* the more important established vectors seem to be *Phlebotomus martini* and *P. langeroni* in tropical Africa, *P. longicuspis* in Algeria, *P. major* and *P. perniciosus* in the Mediterranean, *P. simici* in the eastern Mediterranean, and *P. chinensis* in China.

Dermal Leishmaniases. These include a complex of diseases known by various names, particularly Oriental sore in the Old World and uta, espundia, and chiclero ulcer in the New. Oriental sore is a cutaneous leishmaniasis with wide distribution in Med-

iterranean areas, Asia Minor, Arabia, India, the southern part of Asiatic USSR, and parts of Africa. Unlike kala-azar, the leishmanias inhabit the skin and do not ordinarily invade the viscera.

Two types of this leishmaniasis are recognized in the Near East. The wet rural form is a zoonosis with gerbils, particularly *Rhombomys opimus,* and ground squirrels as reservoirs. *R. opimus* is of particular importance in the Turanian deserts because its deep burrows, with high humidity and ample food, provide ideal breeding conditions for the sand fly (Dubrovsky, 1975). *Leishmania tropica major* is the etiological agent; the chief phlebotomines are *Phlebotomus caucasicus,* an enzootic vector that maintains the infection among gerbils, and *P. papatasi,* which transmits it to man. Other species of *Phlebotomus,* however, are known to be involved. The dry urban form, a disease of man that also affects dogs, is caused by *Leishmania tropica tropica* and is transmitted chiefly by *P. papatasi* and *P. sergenti.*

An African form, occurring in high elevations in Ethiopia and Kenya, is a zoonosis, the animal hosts being hyraxes and possibly other rodents. The vectors are *P. longipes* and *P. pedifer.* The etiological agent is generally considered to be *L. t. tropica,* although Bray, Ashford, and Bray (1973) believe it to be a distinct species, which they call *L. aethiopica.*

American dermal leishmaniasis is widely distributed in tropical and subtropical America, from Texas, USA, to Argentina. Prior to discovery of overt human cases and of the presence of antibodies in dogs and humans in Gonzales and Kenedy Counties, Texas (Shaw *et al.,* 1976), the northern limit was thought to be southern Mexico. The taxonomy of the leishmanias involved is unsettled (see Bray, 1974), but this disease apparently is attributable, in its three best-marked forms, to three causal agents, *Leishmania braziliensis, L. peruana,* and *L. mexicana.* The entire epidemiological picture, however, is complex, much more so than that of the Old World, and the following account is a necessary simplification.

The most serious American form, known as espundia, involves the nasopharyngeal region and may result in horribly disfiguring ulcers that may destroy the nasal cartilage and bone and may terminate fatally. The pathogen is *Leishmania braziliensis,* and the disease is best known from southern Brazil. A much milder form, caused by *L. peruana,* occurs in nonforested areas at high elevations in the Peruvian Andes, where it is known as uta. Unlike the other American cutaneous leishmaniases, it is not known to occur in wild animals, although dogs are commonly infected. The disease in man is usually mild, producing a single ulcer. *L. mexicana* likewise produces a relatively mild form known as chiclero ulcer, in southern Mexico, Belize, and Guatemala; like uta, it generally results in a single sore with no nasopharyngeal involvement, but, unlike uta, it is a disease of the low-elevation rain forest.

Except for uta, the American dermal leishmaniases are essentially zoonoses. A number of animal hosts serve as reservoirs, including the spiny rat, opossums, sloths, kinkajous, marmosets, and ocelots. The pathogen may be acquired at ground level, or it may be maintained among forest mammals and brought to man by sand flies that have a broad vertical distribution.

Lewis (1974) tabulates 14 species of *Lutzomyia* associated with the transmission of leishmanias of dermal leishmaniases in the Americas, and Ward (1977) lists other suspected vectors. Important vectors include *Lu. verrucarum* and *Lu. peruensis* in connection with uta in Peru, *Lu. flaviscutellata* and *Lu. olmeca* as vectors of *Leishmania mexicana,* and *Lu. trapidoi* and *Lu. ylephiletrix* and others as vectors of *Leishmania braziliensis.* Other species of *Lutzomyia* maintain the parasites in enzootic areas in various parts of tropical America and transmit them to man when proper contacts are made. Not all sand flies have equal capacity for transmitting leishmanias. An efficient vector must, of course, contact and bite both man and the reservoir host, if one other than man is involved. Moreover, there must be high infection rates in the foregut and mouth parts of the insect; some would-be vectors develop only midgut and hindgut infections.

CARRIÓN'S DISEASE (BARTONELLOSIS)

Disease: Carrión's disease (bartonellosis); two clinical forms, oroya fever and verruga peruana.

Pathogen: Bacterium, *Bartonella bacilliformis.*

Victim: Man.

Vectors: Sand flies, *Lutzomyia verrucarum* and probably *L. colombiana.*

Reservoir: Unknown, probably man.

Distribution: Andean region of Peru, Ecuador, and southern Colombia (Fig. 9-10).

Importance: Visceral and highly fatal (oroya fever) or cutaneous and mild (verruga peruana). Importance restricted to endemic areas.

Carrión's disease is known also by its two clinical forms, oroya fever and verruga peruana. These were proven to be one and the same by Daniel Carrión, a medical student of Lima, Peru, who inoculated himself with the pathogens obtained from a verruga patient and developed as a result the classical symptoms of the much more virulent oroya fever. Carrión, who died from the disease, had the perspicacity to recognize the identity of the two forms, though he did not, as is usually stated, set out to prove this (Schultz, 1968).

The etiological agent, the bacterium *Bartonella bacilliformis* (Fig. 9-11), is a minute rodlike or coccoid organism that "occurs in or on the red cells and intracellularly in a number of organs, notably in the endothelial cells of the lymph glands" (Hertig, 1948). Verruga is a mild, cutaneous form; oroya fever is visceral, a highly fatal form accompanied by bone, joint, and muscle pains, with rapidly developing anemia and jaundice.

The disease occurs in the mountain areas of Peru, Colombia, and Ecuador at elevations between 800 and 3000 meters. The

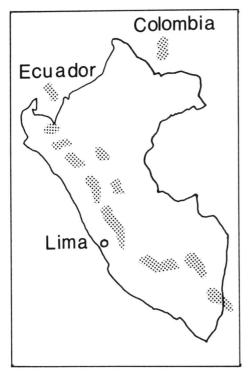

Fig. 9-10 Distribution of Carrión's disease, occurring mainly in Peru. (Redrawn from US Armed Forces Institute of Pathology Negative No. 74-13046-6.)

range is determined by the breeding habits of its vector, the night temperatures above this zone being too cold and the rainfall below it insufficient for successful development. The

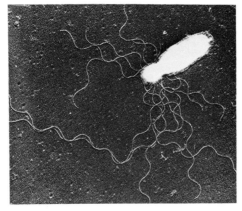

Fig. 9-11 *Bartonella bacilliformis.* (Electron micrograph, courtesy of US Armed Forces Institute of Pathology. Negative No. 75-8592.)

disease is not known to occur in lower animals. However, the pathogen can be recovered from patients who no longer show any symptoms of the disease and even from individuals who have no clinical history. Consequently, man himself appears to be the source of infection.

Lutzomyia verrucarum is at least the chief, and perhaps the sole, vector in Peru. The relationship of this fly to the transmission of the infective agent was indicated by Townsend in 1913 and confirmed by the work of Noguchi and his coworkers in 1929. *Lu. noguchii,* at one time thought to be a vector, must be eliminated from consideration as such; its feeding is apparently limited to field mice. In epidemic areas outside the range of *Lu. verrucarum,* the closely related *Lu. colombiana* is probably the vector.

Many aspects of the transmission mechanism need to be investigated. However, in 1939 Hertig discovered an important clue when the verruga organism was recovered in pure culture from the extreme tip of the proboscis of the fly, the piercing stylets being thoroughly contaminated.

SAND FLY FEVER

Disease: Sand fly fever; also papatasi fever, three-day fever, *Phlebotomus* fever.
Pathogen: Virus, tentatively family Bunyaviridae.
Victims: Man.
Vectors: Sand flies, *Phlebotomus papatasi, P. sergenti,* and other species.
Reservoirs: The sand fly vectors, by virtue of transovarial transmission.
Distribution: Southern Europe and northern Africa, eastward to central Asia, southern China, and India.
Importance: Nonfatal but highly debilitating, occurring spasmodically in great epidemics when introduced into susceptible populations.

Sand fly fever is a seasonal (May to October) febrile viral disease of short duration, occurring in the *Phlebotomus*-infected regions of the Mediterranean, south China, parts of India, Ceylon, the Near and Middle East, and central Asia. It is a nonfatal infection that because of its clinical similarities, especially in sporadic cases, may be confused with dengue and other febrile infections. At least two types of the virus, designated as the Naples and Sicilian, can produce the disease. The epidemic form may occur either when a large number of susceptible individuals enter an endemic area or when the disease becomes extended into a previously uninfected area where there are large numbers of susceptible individuals. As an example of the latter, when the disease was introduced into Jugoslavia (Serbia) in 1948, three-fourths of the population of the new area, or 1.2 million persons, acquired it.

Epidemiological evidence and the fact that the virus has been isolated from *Phlebotomus* males, which do not suck blood, indicate that the infective agent is transmitted transovarially. Consequently, the insect itself is the reservoir. *Phlebotomus papatasi* in the Mediterranean region becomes infective about 1 week after an infecting blood meal. The virus is present in human blood for 24 hours prior to onset and for the first 24 hours of the disease; hence the infective period for the sand fly is limited to that length of time. Though the course of the clinical disease is short, there may be a long period of convalescence. *Phlebotomus papatasi* and *P. sergenti,* both urban and domestic in habit, are the known vectors in the areas where the disease is endemic. Other species serve as vectors in such areas as China where *P. papatasi* is not known to occur.

OTHER VIRAL INFECTIONS

Sand fly transmission of viruses of the Changuinola (*Orbivirus*) and *Phlebotomus* fever complexes has recently been demonstrated in the Americas. Tesh and associates (1974) reported obtaining 269 virus strains, representing nine different virus types, from phlebotomines in Panama. One of these types was the VSV-Indiana strain of vesicular stomatitis virus, causing an important disease of cattle, horses, and pigs in the West-

ern hemisphere. Neutralizing antibodies were found to a significant extent in the human population of the area investigated, and some veterinarians and other persons caring for sick animals contracted an influenzalike disease, sometimes severe enough to require hospitalization (Shelokov and Peralta, 1967). Three viruses of the Phlebotomus fever group, the Punto Toro and Chagres types from Panama, and the Candiru from Brazil, have been recovered from persons sick with an illness indistinguishable from classical sand fly fever, and there is serological evidence for four others (Tesh and associates, 1974). An additional virus of this group, so far known only from the wood rat *Neotoma,* has been isolated near Brownsville, Texas, the first phlebotomus fever group virus known from the United States (Calisher *et al.*, 1977). Apparently, phlebotomine-related infections in America are more prevalent than present information would indicate.

Family Ceratopogonidae

BITING MIDGES

Characteristics. The Ceratopogonidae are very small (0.6–5.0 mm in length), slender gnats resembling some of the smaller nonbiting midges of the family Chironomidae. Many species feed on insects and some attack cold-blooded vertebrates. Among the 50 or more genera comprising the family, 4, namely, *Culicoides* (Fig. 9-12), *Forcipomyia* (subgenus *Lasiohelia*), *Austroconops,* and *Leptoconops,* attack man and other warm-blooded animals. They are popularly known by a variety of names, including "punkies," "no-see-ums," and "sand flies" (not to be confused with the phlebotomine sand flies). In their biting habits, the anthropophagous species resemble the simuliids and are frequently mistaken for them. Some nonanthropophagous species may at times be severe nuisances, for example, an

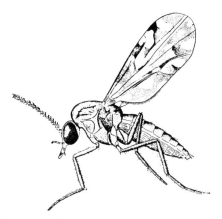

Fig. 9-12 *Culicoides* sp., female. (After Dampf.)

undetermined species of *Dasyhelia* reported by Shemanchuk (1972) as crawling persistently into the eyes, nose, and ears.

The wings, which are narrow, with few veins and no scales, may be clear or hairy, sometimes distinctly spotted, and are folded over the abdomen when at rest; the alulae are slender. The larvae (Fig. 9-13) are aquatic, semiaquatic, or dwellers in moist soil; habitats include fresh, salt, or brackish water, notably salt marshes and mangrove swamps, tree holes, decaying materials such as cactus, banana stems, or plantain, moist or muddy adobe, sandy and alkaline soils, and cattle dung pats.

Only females suck blood, but autogeny occurs in a number of species. One Malayan species, *Culicoides anophelis,* obtains its meal of vertebrate blood from an engorged mosquito. Of the nonbloodsucking species, some feed carnivorously on small insects, whereas others feed on the blood of larger insects. Most species remain within 500 meters of their breeding grounds, though some may go much farther. The flight range

Fig. 9-13 *Culicoides variipennis,* larva. (Drawing by James in W. T. Edmondson, 1959, *Ward and Whipple's Fresh Water Biology.* John Wiley & Sons, Inc., New York.)

of *Leptoconops kerteszi* may exceed 15 km under certain conditions.

The literature on ceratopogonids is extensive. Freeman (in K. G. V. Smith, 1973) gives a number of selected important citations. An important reference source, consisting of a bibliography and keyword-in-context index, of 1,950 papers published between 1758 and the end of 1973, is Atchley, *et al.* (1975). Another useful reference for bibliographical and general information is Kwan and Morrison (1974).

Bites and Biting Species. Many species of *Culicoides* attack man and domestic animals, and may be extremely annoying in doing so. Certain ones constitute a serious economic problem, for example, in coastal summer resort areas, particularly about fresh-water inlets and tidewater pools, and in mountain areas, where they may be so annoying as to drive out campers and fishermen. The bites of these minute flies, which are often blamed on the larger and more conspicuous black flies or mosquitoes, cause itching and, in sensitive individuals, welts and lesions that may persist for several days. These are sometimes complicated by secondary infections resulting from scratching. An instance of a severe attack on livestock was reported by Nielsen and Christensen (1975) in Denmark, where *Culicoides nubeculosus,* breeding in a marsh heavily polluted by raw sewage and liquid manure, attacked nearby dairy cattle to the extent that an estimated 10,000 flies were present on a single cow; they even invaded cow houses and fed on permanently stalled cattle inside.

Culicoides furens may be taken as an example of a vicious biter that occurs in a broad geographical area. This species breeds in salt marshes and similar habitats along the Atlantic and Gulf coasts from Massachusetts to Brazil, throughout the West Indies, and along the Pacific coast from Mexico to Ecuador. It has been recorded repeatedly as biting man furiously, especially in the Caribbean areas. In Europe, *C. pulicaris* seriously interferes with the operations of farmworkers near the areas where the fly breeds. Of the approximately 800 known species of *Culicoides,* many others seriously attack man and animals; many, however, are not man-biters.

Lasiohelea, now considered a subgenus of *Forcipomyia,* feeds on vertebrates. The European *Forcipomyia* (*Lasiohelea*) *velox* and the eastern North American *F.* (*L.*) *fairfaxensis* feed on frogs, and the former transmits a filarial parasite of the frog. Some species of this subgenus are vicious biters of man and other warm-blooded animals in the tropics; *F.* (*L.*) *taiwana* bites man in Taiwan and Japan.

The genus *Leptoconops* occurs throughout a large part of the warmer temperate and tropical areas of the whole world. Many of its species are severely annoying to man and animals. Autogeny sometimes occurs, however. The larvae are found in sandy or clayey soil, often alkaline, that is kept moist because of proximity to aquatic areas or because of a shallow water table that contains enough organic matter for larval subsistence. Feeding is diurnal, and the flies may be especially annoying to bathers on beaches, thereby affecting tourism, or to workmen in fields in infested areas. Biting habits may vary from species to species; for example, in California *L. torrens* tends to bite around the head and arms, rarely attacking the legs, whereas the closely related *L. carteri* attacks low on the body, readily on the ankles and legs, and will move upward inside the clothing.

Wirth and Atchley (1973) have treated the taxonomy of the North American species. This publication is especially valuable, since it presents a subgeneric classification and list of the species of the world, summarizes the known material on biology, economic importance, and control, and provides an extensive list of bibliographic references.

Leptoconops kerteszi is a widely distributed species, occurring in the Near and Mid-

dle East, central Asia, Africa, southern Europe, and western North America. Its bite usually produces transient swellings that may become vesicular; this may result in an open lesion that may exude moisture for weeks. In sheep, dermatitis thus produced around the eyes and ears is known to veterinarians as Leptoconops mange (Howell, 1970). *L. torrens* and *L. carteri* are serious pests in the western United States. These two have been confused with each other, and much of the literature on "*Leptoconops torrens*" and on the "valley black gnat" really refers to *L. carteri.* Other seriously annoying species are *L. irritans* in Mediterranean Europe; *L. spinosifrons,* a beach breeder that interferes seriously with tourism in East Africa, Madagascar, the Seychelles, India, Ceylon, and the Malay Archipelago; *L. bequaerti,* not "*becquaerti*" as sometimes written, widespread in the Caribbean region, from Texas to Panama and in the West Indies; and *L. camelorum* and *L. caucasicus,* both Asiatic in distribution.

Austroconops is known from only one species, *A. macmillani,* which bites man in western Australia. Some species of *Dasyhelia* have been accused of biting man, but this behavior needs confirmation.

Relationship to Disease. The importance of bloodsucking ceratopogonids in relation to human and animal pathogens is perhaps more important than has been previously realized. These flies are known to be involved in the transmission of pathogenic nematodes, protozoa, and viruses. The night-biting *Culicoides austeni* is an intermediate host of the filarial worm *Dipetalonema perstans* (*Acanthocheilonema perstans* in much of the literature). The microfilariae of this worm are found in the peripheral circulation both by day and by night, the diurnal being the more common. The microfilariae undergo metamorphosis in the body of the fly, appearing then on its proboscis. In the vast majority of cases it is nonpathogenic to the human host. *Culicoides grahami* is a natural, though less efficient, host, and both *Culicoides* species transmit *Dipetalonema streptocerca,* another largely nonpathogenic filarial nematode. Both of the above species of *Dipetalonema* are African. In tropical America, *Culicoides furens* transmits still a third largely nonpathogenic nematode of man, *Mansonella ozzardi.*

Onchocerca cervicalis, a common parasite of horses, is transmitted by *Culicoides nubeculosus.* Its life history in both the insect and mammalian hosts is similar to that of *O. volvulus* of man (Mellor, 1975), and some eye involvement may occur. However, its pathogenicity is relatively unimportant, and its supposed relationship to fistulary withers and the so-called sweet itch of horses has recently been discounted. *Onchocerca gibsoni,* which infests cattle in Australia, South Africa, and southeast Asia, has *C. pungens* as an important vector; other known vectors include other species of *Culicoides* and some of *Simulium.* The disease is reported to be of considerable importance in Australia. According to Atchley and Wirth (1975), *Culicoides multidentatus* and *C. lophortygis* harbor filariae of the California Valley quail.

Several protozoan and viral diseases of domestic and wild animals, poultry, and waterfowl are related to ceratopogonid transmission. The sporozoon *Hepatocystis* parasitizes Old World monkeys and other mammals, usually arboreal. Its vectors are either known or presumed to be *Culicoides.* *H. kochi,* a monkey parasite, is transmitted by *C. adersi* and probably by *C. fulvithorax.* There is little pathogenicity resulting from this parasite and, so far as known, from other species of the genus. The related genus, *Haemoproteus,* also *Culicoides*-transmitted, includes several avian parasites, one of which, *H. nettionis,* parasitizes domestic and wild ducks and geese and other wild waterfowl. Wild birds probably serve as the reservoir in epizootics among domestic birds (Garnham, 1966).

Although most *Leucocytozoon* parasites of poultry are *Simulium*-transmitted, *L. caulleryi* (sometimes referred to the genus *Akiba*) is vectored by *Culicoides.* The parasite produces a very important poultry disease in Japan and southeast Asia. Known vectors are *Culicoides arakawae, C. circumscriptus,* and *C. odibilis.* Garnham (1966) points out that *Culicoides* may replace *Simulium* as a vector of the Leucocytozoidae in areas where the latter genus is absent; for example, *C. adersi,* which feeds avidly on chickens, may transmit a *Leucocytozoon* of poultry in Africa.

The virus of bluetongue (*Orbivirus*) (Fig. 9-14), an important disease of sheep, is transmitted by *Culicoides* species, including, as important vectors, *C. pallidipennis* in Africa and Asia Minor and *C. variipennis* in the USA. In warmer areas, where *Culicoides* adults occur throughout the year, the fly itself may harbor the virus during the winter months. However, cattle, in which the disease is usually, though not always, inapparent, probably serve as the main reser-

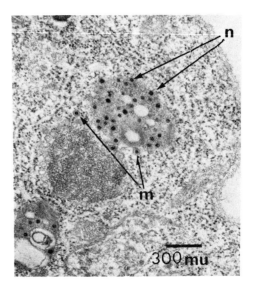

Fig. 9-14 Salivary gland of infected *Culicoides variipennis* 10 days after infection, showing mature (*m*) and nonencapsulated (*n*) virus particles. (Electron micrograph from Bowne and Jones, 1966, *Virology,* **30**:127–33.)

voir. Clinical bluetongue is known from bighorn sheep, mule deer, and white tailed deer, and, by serological evidence, from other ruminants in the wild (Hourrigan and Klingsporn, 1975). Mortality in sheep may occasionally be high, though it usually does not exceed 15% of the infected animals. However, the economic loss due to subsequent poor condition of the animals, interruption of their breeding schedules, long convalescence, and wool loss may be even more important than the mortality (Steelman, 1976). An etiologically similar hemorrhagic disease of deer is also apparently transmitted by *C. variipennis* and several closely related *Culicoides*.

Bovine ephemeral fever, an important viral disease of cattle in Africa and Australia, apparently is transmitted by a complex of *Culicoides* species. In Australia it results in a high milk reduction but has a very low mortality. Species of *Culicoides* are involved in transmitting the virus of African horse sickness, a disease that has extended into Pakistan and other parts of the Middle East, where it has taken a heavy toll of horses and mules (Steelman, 1976). Three encephalitis viruses, JBE, EEE, and VEE, have been isolated from *Culicoides,* but their epidemiologic significance is uncertain. There is, however, strong evidence that *Culicoides arubae* and, possibly, also *Leptoconops kerteszi,* may have been involved in the transmission of Venezuelan equine encephalitis to horses in the 1971 enzootic in Texas and possibly in the introduction of this virus into the United States (Jones *et al.,* 1972). The first incrimination of *Forcipomyia* as a possible vector of a human viral pathogen came from the recovery of Japanese B encephalitis from *F. (L.) taiwana.*

Control of Biting Gnats

Simuliids, ceratopogonids, and phlebotomine sand flies have quite different developmental sites, but as adults their behavior is more alike. Surveys for adults are accomplished by biting collections, netting from moving vehicles, or using traps that employ light or carbon dioxide as attractants. Larvae are sought in aquatic or near-aquatic habitats, and for simuliids plastic cones or fabric strips may be anchored in flowing waters and examined periodically for attached larvae.

The immature stages of these three families of midges often develop in widespread and diverse habitats. These developmental sites may be hard to identify, and even when they are known, environmental manipulation or pesticide applications may be difficult or undesirable because of accompanying changes in habitat; or developmental areas may be so vast or at such great distances that modification or pesticide applications are entirely impractical. Of these three families, black flies are the strongest fliers, in migratory flights traveling 30–300 km and attacking man and animals; no-see-um adults tend to be more local, usually within 2 km; phlebotomines are weak fliers and tend to be quite focal near their developmental sites. Biting activities are mainly outdoors in black flies and no-see-ums, but phlebotomine sand flies tend to be active in protected areas such as forested or scrub zones, or indoors.

Biting gnats generally have distinct periods of activity each day. Simuliids are mainly biters during daylight; ceratopogonids are variable, with *Culicoides* often crepuscular and night feeders and *Leptoconops* biting throughout the day; phlebotomines, except in protected areas with deep shade, are distinctly crepuscular and night biters. With this knowledge it is possible to achieve satisfactory protection for grazing animals by restricting their outdoor exposure to times when locally important biting gnats are not active, and by protecting them in screened or pesticide-treated shelters at other times. For example, *Simulium arcticum* biting activity occurs during daylight hours, and animals might be sheltered during normal attack times and let out to graze at night (also see Kovban *et al.*, 1966). In the case of *Culicoides pulicaris,* which causes a typical dermatitis condition of horses in Britain, the animals can be protected by bringing them into shelter in late afternoon through the evening when attack occurs (Mellor and McCraig, 1974). Since many biting gnats tend to be quite seasonal in their main periods of adult activity, such protective behavioral approaches may be required for only rather short periods.

Repellents have been developed for human protection (examples in Sjogren, 1971; Spencer *et al.*, 1975), and during outdoor exposure protective clothing, including jackets of open fabric impregnated with repellents, is successful (Frommer *et al.*, 1975; Mulrennan *et al.*, 1975). Repellents have been difficult to develop for livestock, particularly because of problems connected with treating the animals frequently enough. For dairy cattle coming twice daily to milking parlors, or range cattle with a restricted and fenced supply of water, entry chutes with automatic spraying devices should be suitable. The applicator devices could spray repellents, or natural or synthetic pyrethroids with short-term insect repellency and rapid knockdown characteristics. Dust bag applicators can also help protect range cattle.

Space spraying or residual spraying can provide temporary relief from adult biting gnats. The application of pesticides in aerosol form, along the perimeter of salt marshes, such as with ULV or thermal fogger techniques, has provided temporary relief from *Culicoides* near recreational and housing areas in Florida, USA. Applications around cattle, or turkey pens, can protect these hosts from simuliid attacks. Residual treatments do not work well with simuliids or ceratopogonids because they tend to spend little time in contact with walls or other surfaces, but residual treatments are very effective against phlebotomines, which are weak

fliers and frequently contact surfaces. Residual applications of DDT to the inside walls of houses, to control *Anopheles* mosquitoes during antimalaria campaigns, virtually eliminated leishmaniasis through control of adult phlebotomus (cf. Seyedi-Rashti and Nadim, 1975; Sharma *et al.*, 1973). The tendency of adult phlebotomines to be in frequent contact with the ground, particularly when replete with blood, makes them susceptible to predation (cf. Christensen and Herrer, 1975).

Predators, parasites, and pathogens all affect the immature stages of biting gnats, though the complex for phlebotomines is not well known. Nematodes have been noted particularly for simuliids and ceratopogonids. Black flies often become very numerous in new irrigation systems, but mermithid nematodes soon thereafter become prevalent and greatly reduce their numbers (cf. Fredeen and Shemanchuk, 1960; Welch and Rubtsov, 1965). *Neomesomermis flumenalis* is a widespread mermithid nematode parasite of black flies, and attempts are underway to utilize this facultative parasite for control (Molloy and Jamnback, 1975). Some 19 species of microsporidian protozoa infect black flies (Maraund, 1975), and even those that are not rapidly pathogenic may reduce adult fecundity. These pathogens need further study before they can be practically utilized. Effective predators of black fly larvae and pupae include several insects and various fishes, suggesting caution is required in treating water with insecticides. Near Mt. Elgon, Uganda, crabs that harbor larvae of *Simulium neavei* were rare in two rivers due to predation by introduced rainbow trout (Hynes *et al.*, 1961). Bacon (1970) has reviewed the natural enemies of ceratopogonids.

Environmental manipulation has been practiced mainly against simuliids, but also has possibilities against ceratopogonids and phlebotomines. Black fly numbers have been reduced unintentionally in large river systems such as the Mississippi River by way of siltation and industrial pollution (also see Rivosecchi and Colombo, 1973). Dams that remove sections of flowing water reduce black fly development, though care is necessary in spillway design to minimize larval development there (cf. Quélennec *et al.,* 1968). Similar attention must be paid to the construction of flow channels in irrigation systems. In contemplating modifications of developmental sites for ceratopogonids and phlebotomines, one must be aware that such aqueous or moist areas as tree holes can be involved. Diking, filling, and flooding have provided a lasting means of controlling salt marsh *Culicoides furens* in Florida, USA, and Panama (Rogers, 1962; MacLaren *et al.,* 1967); annual disking or irrigating, and proper irrigation schedules for crops, can markedly reduce pest species of *Leptoconops* developing in soil crevices (Whitsel and Schoeppner, 1966). Piled mud from dredging operations can cause outbreaks of *Culicoides* (cf. Altman *et al., 1970*). Levels of biting by Panamanian phlebotomines can be substantially reduced by clearing the forest around buildings (Chaniotis and Corrêa, 1974).

Pesticides have been applied to control biting gnat larvae, but this tactic presents many application problems and possibilities of undesirable side effects. For ceratopogonid control vast areas of salt marsh or agricultural lands may be involved. The application of granular formulations of pesticides for these situations has been successful (cf. Wall and Marganian, 1973), but this approach is not usually practical. Phlebotomine developmental zones are too diverse and secretive for much successful general application of pesticides as a means of control. The greatest development of pesticides against biting gnat larvae has concerned control of black flies in flowing waters, and here side effects on other aquatic insects and on fishes cause problems. The largest current

effort relying heavily on insecticides is against *Simulium* vectors of human onchocerciasis filariae in the Volta River and adjacent river systems of West Africa. This program can be reviewed in Waddy (1969), Lee (1973) and Bauer (1974); related aerial applications to control vectors of turkey *Leucocytozoon* are described by Kissam and associates (1975). Dispenser systems for insecticides at upstream locations have also been developed (cf. Fredeen, 1970, 1975). Developmental inhibitors have been tested, in addition to conventional toxicants (cf. Dove and McKague, 1975). Consult Jamnback (1973) for a review of developments in black fly control procedures. Quélennec (1976) has provided instructions for determining the susceptibility of black fly larvae to insecticides.

Family Chironomidae

MIDGES

The family Chironomidae is a large one, comprising more than 2,000 species. Although often mistaken for mosquitoes, these

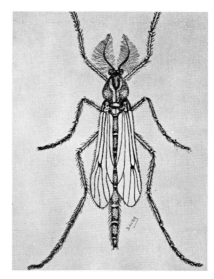

Fig. 9-15 A male midge (Chironomidae), commonly mistaken for a mosquito. × 12. (After Osborn.)

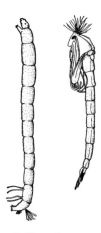

Fig. 9-16 Larva (*left*) and pupa (*right*) of a chironomid gnat (midge). (Larva redrawn from Needham, and pupa redrawn from Grünberg.)

flies bear little resemblance to them under closer examination. Notably, the proboscis is short and not adapted for piercing. Midges (Fig. 9-15) are widely distributed and may be extremely abundant in the vicinity of standing water, since the larvae (Fig. 9-16) of most species are aquatic. Massive adult emergence frequently occurs, sometimes accompanied by a similar large emergence of mayflies. The general adult behavior pattern is to rest by day and fly in the evening, night, or early morning hours. Occasionally great swarms of these insects hover in the air toward evening, and in these swarms they may produce a distinct humming sound. They are often attracted to light in great numbers.

Medical Importance. Chironomids cannot bite, but they frequently constitute a major nuisance. Grodhaus (1963, 1975) has given a bibliography and an excellent summary of this subject, and Beck and Beck (1969) have summarized the situation in Florida, with aids for the identification of the nuisance species. Undoubtedly, chironomids as nuisances are of widespread occurrence; for example, *Chironomus yoshimatsui* is serious in parts of Japan. Comfort and normal activities of persons may be interfered with near areas of heavy midge emergence, and even livestock may be bothered. Attraction of the insects to lights in homes may

result in the accumulation of their dead bodies and the production of a very offensive odor. It may be difficult to keep midges in swarms out of the eyes or to avoid inhaling them. The latter may result in an allergic reaction. Swarms crossing highways may produce a traffic hazard, and their dead bodies may make the highway slippery. The aquatic forms may be troublesome when, as occasionally happens, they enter water distribution systems and occur in the tap water.

So far as is known, mechanical transmission of pathogens by midges is not a probability, though the possibility exists in species whose larval stages are spent in sewage-contaminated water. But, aside from the health and accident hazard, the discomfort and offense to personal tastes cannot be tolerated in a sanitation-conscious community.

Family Dixidae

DIXA MIDGES

Dixa midges are mentioned only because the larvae are frequently mistaken for those of *Anopheles,* being commonly found in similar situations, and because the adults somewhat resemble true mosquitoes. They lack biting mouth parts and are of no known importance to health. Dixa larvae are usually seen at the surface of water among vegetation and debris, moving in a horizontal U-shaped pattern.

Family Chaoboridae

CHAOBORID GNATS

The Chaoboridae, at one time considered a subfamily (Chaoborinae) of mosquitoes but now usually given family status, constitute a group of mostly nonbloodsucking flies that have some importance as predators on mosquito larvae. Some species may at times become nuisances, even to the extent of being dangerous. One such is the Clear Lake gnat, *Chaoborus astictopus* (Fig. 9-17), best known in northern California but becoming more widespread as the result of the develop-

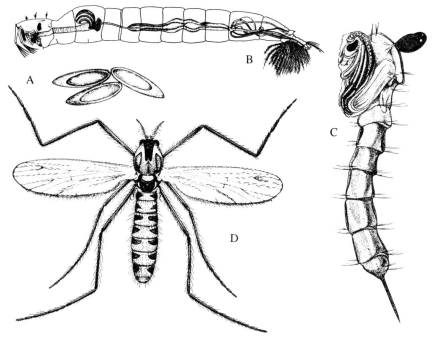

Fig. 9-17 A chaoborid, the Clear Lake gnat, *Chaoborus astictopus*, showing life cycle. *A.* Eggs. *B.* Mature larva. *C.* Pupa. *D.* Adult female. (After Herms.)

ment of impounded waters and other habitats that provide conditions suitable for its breeding. These flies may swarm by the thousands around lights. The larvae, which live in deep water, are almost transparent and are seen with some difficulty, even in clear water, except when in motion; hence, they are called "phantom larvae." The tiny, lead-colored, cigar-shaped eggs are deposited in great numbers on the surface of still water. They soon sink to the bottom. The larvae develop slowly during the summer and overwinter, pupating in the early spring. After about 2 weeks, the pupae quickly come to the surface, where the gnats literally "pop out" of the pupal skins, balance on the water momentarily, then fly shoreward.

Chaoborus edulis is a comparable pest over Lake Victoria in East Africa (Oldroyd, in K. G. V. Smith, 1973), where swarms of these move with the wind. They may be dangerous to persons trapped in their path; a fisherman in a small boat might even be suffocated before escape is possible.

Chaoborids have until recently been considered wholly nonbloodsucking, so far as feeding on vertebrates is concerned. However, Williams and Edman (1968) have shown that *Corethrella blakeleyi* and *C. wirthi* feed on both avian and mammalian blood in Florida, and McKeever (1977) has reported feeding by the same two species on the blood of tree frogs of the genus *Hyla*. Miyaga (1975) has confirmed the presence of hemoglobin in the midguts of female *C. japonica* in Japan. All three species have biting mouth parts with toothed mandibles. It is doubtful whether this bloodsucking habit may assume much importance.

Control of Nonbiting Midges and Gnats

These insects as a group must be considered essentially as beneficial, serving in the biodegradation of wastes and as food for invertebrate and vertebrate predators. Nonetheless, on occasion the massive emergence of adults can be very bothersome to humans, and may cause asthmatic reactions in sensitized persons. Adults are attracted to lights, and during mass emergences it is helpful to turn off lights at night that might be visible outdoors.

Chironomidae (Midges). Considerable emphasis has been directed toward preventing huge swarms of chironomid midges developing in sewage settling lagoons and artificial ponds and recreational lakes. Studies of Anderson and coworkers (1964) revealed that pond rotation, letting settling basins dry 5 to 10 days, is very effective. Carp (fish) at 200 to 400 pounds per acre were particularly useful in test basins, but were not as valuable in large lakes and temporarily disrupted habitats (Bay and Anderson, 1965). The fish, *Tilapia mossambica,* was effective against pond populations of chironomids, but this predator and related tropical forms require supplementation during the early cool season because they do not reproduce at lower temperatures (Legner and Medved, 1973). Several pesticides, including insect growth regulators, give good control, but their effect is short-lived (cf. Polls *et al.,* 1975; Mulla *et al.,* 1974).

Chaoboridae (Chaoborid Gnats). The Clear Lake gnat, *Chaoborus astictopus,* continues to be a difficult problem. A protozoan in the genus *Thelohania* is very damaging at times, and is transovarially maintained and thereby continues to take its toll in succeeding generations (Sikorowski and Madison, 1967). Various fishes exert some control through direct predation and by competition (Cook, 1967). Pesticides have been tested, but there have been significant undesirable side effects from such uses.

Psychodidae (Moth Flies). Moth flies in the genera *Psychoda* and *Telmatoscopus* are seldom a serious problem in homes, but enormous numbers may develop in sewage-trickling filter beds, and the emerging adults can be a significant nuisance to workers. Basically their presence in the beds is considered beneficial. Pesticides have been tested against larvae and adults to provide temporary control, aerosols of natural and synthetic pyrethroids being quite effective against the latter (Hayashi and Hatsukade, 1970). Simple flooding of filter beds for 6 hours was found to reduce *Psychoda alternata* larval populations by about half, without excessively affecting the subsequent functions of the filter (Callahan and Bailey, 1974).

10

MOSQUITOES

ORDER DIPTERA; FAMILY CULICIDAE

Importance

Mosquitoes are the most prominent of the numerous kinds of bloodsucking arthropods that annoy man, other mammals, and birds. Their attacks are not confined to warm-blooded animals, as there are records of feeding on fishes, reptiles, and amphibians, and they are known to transmit pathogens to the latter two groups of cold-blooded vertebrates. Their number is legion, and there are species that are exceedingly annoying in daytime, though most mosquitoes are nocturnal feeders. Great swarms may be produced from practically all sorts of still water, fresh and brackish, foul or clear; water in tin cans, car and airplane tires, hoof prints, tree holes, deposits in leaf cups; the margins of streams, rivers, lakes, and water impoundments. Vast areas of seacoast are at times made unbearable by salt marsh mosquitoes, and agriculture and real estate development may be affected. They can make potential recreational areas unsuitable and interfere with normal living, particularly outside the home. They successfully enter dwellings and transmit serious pathogens of man in many areas of the world, particularly in the humid tropics.

Losses resulting from lowered productivity of industries concentrating on outdoor activities are frequently considerable because of mosquito annoyance. Such losses would alone amply justify the great sums that are spent on mosquito abatement, yet these losses must be added to widespread suffering and death due to mosquitoes as vectors of pathogens causing disease.

Mosquito annoyance is known to affect wild and domestic vertebrate populations. Unusually heavy seasonal development of *Aedes* mosquito populations on the Galapagos islands has been observed to disrupt the successful nesting of birds. Arctic mosquitoes, and other biting flies, influence the migratory activities of reindeer (Terent'ev, 1972) and caribou. Numerous accounts report that mosquito attacks affect weight gains in livestock significantly, and milk production in cattle; some $25 million in the USA in 1965, $10 million of this due to reduction in milk production (Steelman, 1976); an estimated 166 ml of blood loss per animal each night in parts of Queensland, Australia (Standfast and Dyce, 1968). Annoyance from mosquito bites has resulted in behavioral responses in some vertebrates, and particularly noteworthy is the behavioral repertoire of defense developed by some birds (Webber and Edman, 1972).

Mosquitoes are the sole vectors of the pathogens causing human malaria, yellow fever, and dengue, and they are of prime importance in filariases and viral encephalitides of man. It is reported that in an ordinary year in India, before the nationwide antimalaria campaign went into effect, at least 100 million persons suffered from, and a million succumbed to, the direct ravages of malaria. The indirect effect in lowered vitality and susceptibility to other diseases accounted for another million deaths as well as greatly decreased production resulting from work losses.

Through their transmission of the plasmodia of malaria, mosquitoes have had a decided influence on the evolutionary history of mankind. This is evident through the presence of sickle cell anemia, thalassemia, and other suspected related blood disorders among present and derived African populations, and other human groups historically subjected to the devastating selective effects of endemic malaria. Gillett (1972) summarizes the effects of mosquito-borne diseases on human affairs, and Mattingly (1969) reviews several aspects of these diseases.

Classification

Proper identification of mosquitoes is vital in studying and combating mosquito-borne diseases (see Eldridge, 1974). The literature dealing with the family Culicidae is extensive. There are adequate keys to identify adults and larvae for most countries and regions of the world. One source of information dealing exclusively with the classification of culicids is the journal *Mosquito Systematics*. Knight and Stone (1977) have updated the catalog on mosquitoes of the world. This publication is updated by supplements (Knight, 1978). Carpenter and La-Casse (1955) monographed the Nearctic spe-

cies, and this publication has been updated by summarizations of more recent literature (Carpenter, 1968; Darsie, 1973); state and regional studies are available for most of the USA, for example, Gjullin and Eddy (1972) for the Pacific Northwest. Larval and adult characters are described, figured, and keyed by Carpenter and LaCasse and in most of the regional works. Vargas (1972, 1974) provides keys for separating females and fourth-stage larvae of the 20 New World genera, and Diaz Nájera and Vargas (1973) list the species and distribution of Mexican mosquitoes. Characters of other immature stages have not proven of as much value in identification, but some use has been made of pupal structures. Keys to larvae, pupae, and adults of world genera of mosquitoes are provided by Mattingly (1971). *Anopheles* in species complexes have in some instances been differentiated on egg structure, and the eggs of *Psorophora* and *Aedes* can similarly be identified (cf. Horsfall and Craig, 1956; Horsfall *et al.*, 1952).

Classification by Nonstructural Characteristics. Most mosquito species can be separated by fairly straightforward external structural characters, but there is increasing realization that in some important disease situations difficulties in control are due to species complexes where external morphological characters do not suffice to make distinctions. Recent concepts for discriminating species and subspecies are reviewed by Barr (1974b), and identification may require such techniques as studying the sensillae on antennae, cross-mating experiments, characterizing polytene chromosomes, and comparing electrophoretic patterns of enzyme systems (examples in Ismail and Hammoud, 1968; Davidson and Hunt, 1973; Hunt, 1973; Bullini and Coluzzi, 1974).

External Structure. Although some authors include the dixa midges and chaoborid gnats as well as the true mosquitoes in the family Culicidae, only the latter are so classified in this book. But note in Chapter 9 that some chaoborids are blood feeders with mouth parts similar to mosquitoes. Excepting the large, sturdily built American gallinippers, *Psorophora ciliata* and *Ps. howardii* (body length 9 mm and wingspread 13 mm), and a few other species, mosquitoes are small and fragile, ranging in body length from 3–6 mm. The most obvious characteristics separating adult mosquitoes from all other Diptera are a combination of wings with scales on the wing veins and posterior margin, and an elongate proboscis. The antennae are long and filamentous with 14 or 15 segments, hairs in whorls, plumose in the males of most species.

Larvae of mosquitoes are without exception aquatic. They are separable from unrelated aquatic insects by an absence of legs, and by the bulbous thorax, which is wider than the head or abdomen (Fig. 10-1). They differ from other dipterous larvae by a complete head capsule plus only one pair of functional spiracles situated dorsally on the eighth abdominal segment. Larvae of mos-

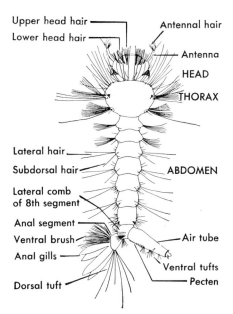

Upper head hair
Lower head hair
Antennal hair
Antenna
HEAD
THORAX
Lateral hair
Subdorsal hair
ABDOMEN
Lateral comb of 8th segment
Anal segment
Ventral brush
Anal gills
Air tube
Ventral tufts
Pecten
Dorsal tuft

Fig. 10-1 Larval characters used in identifying mosquitoes. (After Stage, Gjullin, and Yates; USDA drawing.)

quitoes are collected to (1) provide a positive record in the absence of adult specimens, (2) in some cases give a more positive identification than adults, (3) check on the identification of adults, (4) yield perfect specimens of adults through rearing, (5) provide infection-free stock for experimental transmission studies (but see California encephalitis).

About 3,000 species of mosquitoes have been described worldwide with several additional subspecies, and approximately 150 occur in temperate North America. These ubiquitous insects occur at elevations of 4,300 meters in Kashmir and 1,160 meters below sea level in gold mines of south India. They are abundant in species in the tropics, and almost unbelievably large swarms of them occur in the Arctic. Mosquitoes are divided into three subfamilies: (1) Toxorhynchitinae (Megarhininae), adults large with metallic scales, proboscis strongly curved downward, flower-feeding, and larvae predacious, for example, *Toxorhynchites rutilus;* (2) Culicinae with palpi of the female less than half as long as proboscis, scutellum trilobed, and females hematophagous, for example, *Culex pipiens* and *Aedes aegypti;* (3) Anophelinae with palpi of both sexes as long as or nearly as long as proboscis, scutellum with rare exceptions straplike, and females likewise hematophagous, for example, *Anopheles maculipennis.* Reinert (1975) provides generic and subgeneric abbreviations that have been adopted by the American Mosquito Control Association to avoid confusion when abbreviating, and these are used in this text.

MALE TERMINALIA. Most critical differences used in the identification of adult mosquitoes are found in the male terminalia (Fig. 10-2), although the field collector is more likely to encounter females in the act of biting. The terminalia are prepared for microscopic comparison with illustrations of the same structures in taxonomic publications.

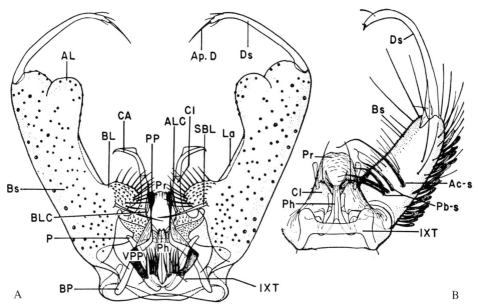

Fig. 10-2 Male terminalia of (*A*) *Aedes squamiger*, (*B*) *Anopheles gambiae*. *AL*, apical lobe; *ALC*, apical lobe of claspette; *Ap.D*, appendage of dististyle; *Bs*, basistyle; *BL*, basal lobe; *BLC*, basal lobe of claspette; *BP*, basal plate; *CL*, claspette; *CA*, claspette appendage; *Ds*, dististyle; *La*, lacuna; IXT, lobe of 9th tergite (= process of 9th tergite); *P*, paramere; *Ph*, phallosome; *Pr*, proctiger; *SBL*, spine of basal lobe; *VPP*, ventral arm of paraproct; *IXT*, 9th tergite. (*B* after Ross and Roberts.)

Internal Structure. The internal anatomy of mosquitoes is that of a relatively uncomplicated insect (Fig. 2-6*C*, and Snodgrass, 1959). In epidemiological studies emphasis is placed on examination of the stomach (midgut) and salivary glands when concerned with malaria, or the thoracic flight muscles, malpighian tubes, and head region when investigating human and canine filariases. The development of ovaries (Fig. 4-3) and fat body is of importance when determining such phenomena as AUTOGENY (ovarian development without a prior blood meal) and PARITY (whether there has been previous ovarian development, most often as a consequence of one or more blood meals). The spermathecae of females are examined, particularly in attempting to rear mosquitoes, to determine whether successful mating has taken place. Electron microscope investigations to reveal the development of pathogens in vectors require a general knowledge of normal and uninfected mosquito tissues at very high magnification. For this purpose the ultrastructure micrographs of Ohmori and Banfield (1974) on *Aedes aegypti* provide a useful background. That there may be changes in the ultrastructure of adult tissues is indicated in an investigation of the salivary glands of *Culex pipiens* during the process of adult maturation, and following blood feeding (Barrow *et al.*, 1975).

Biology

Important books dealing with the biology and physiology of mosquitoes are those of Bates (1949), Horsfall (1955), Christophers (1960), and Clements (1963); more recent general discussions are in Gillett (1971, 1972). Gerberg (1970) reviews techniques that are useful in rearing large numbers of mosquitoes. The serious student will need to be familiar with these works as well as texts

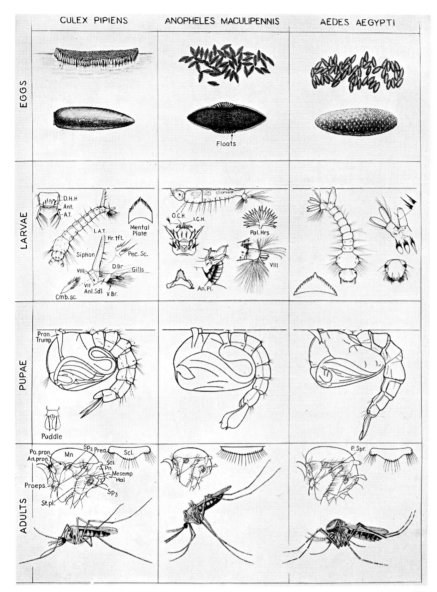

Fig. 10-3 Anatomic details and life history of three genera of mosquitoes, *Culex,* *Anopheles,* and *Aedes.* Explanation of abbreviations: *An.Pl.,* anal plate; *An.pron.,* anterior pronotal setae; *Anl.Sdl.,* anal saddle (dorsal plate); *Ant.,* antenna; *A.T.,* antennal tuft; *Cmb.Sc.,* comb scale; *D.Br.,* dorsal brush; *D.H.H.,* dorsal head hair; *Hal.,* halter; *Hr.tft.,* siphon hair tuft; *I.C.H.,* inner clypeal hair; *L.A.T.,* lateral abdominal tuft; *Mesemp,* mesepimeral setae; Mn., mesonotum (tergum₂); *O.C.H.,* outer clypeal hairs; *Pal.Hrs.,* palmate or float hairs (tuft); *Pec. Sc.,* pecten scale; *Pn.,* postnotum; *Po.pron.,* posterior pronotal setae; *Prea.,* prealar setae; *Pron. Trump.,* pronotal trumpet; *Proeps.,* proepisternal setae; *P.Spr.,* postspiracular setae; *Scl.,* scutellum; *Sp.,* spiracle; *St.pl.,* sternopleural setae; *V.Br.,* ventral brush. (Adapted from various authors.)

on insect physiology that include investigations on mosquitoes, and review articles on aspects of mosquito biology, behavior, and physiology.

Mosquitoes undergo complex metamorphosis—egg, larva, pupa, and adult (Fig. 10-3). The larvae are commonly known as WIGGLERS, and the pupae as TUMBLERS. Water is essential for the larval and pupal stages.

Egg Stage. In some species of mosquitoes eggs may survive long periods out of water, though preferably under humid conditions. Gjullin and coworkers (1950) found eggs of *Aedes vexans* and *Ae. sticticus* to survive in numbers for 3 years when kept moist.

Eggs of all the major groups of mosquitoes are deposited singly, for example, *Anopheles* and *Aedes,* or in rafts, for example, *Culex* and *Culiseta* (Fig. 10-4). Oviposition among most groups is on the surface or along the margins of quiet pools of water. Floodwater and salt marsh species as well as many tree-hole breeders, for example, *Aedes* and *Psorophora,* oviposit in sites subject to inundation by tidal water, seepage, overflow, or rainwater. One feature that all such sites have in common is protection of the ovipositing female from action of wind and wave. It has been shown by a number of authors that substances attractive to ovipositing females may aid in the selection of specific sites for oviposition. Such substances may be produced by the immature forms of mosquitoes in breeding water and be composed of diglycerides or fatty acids associated with egg rafts or bacteria (Starratt and Osgood, 1972; Ikeshoji *et al.*, 1975), or oviposition attractants may be phenolic compounds of plant origin (Ikeshoji, 1975). The egg completes embryonation within a few days, and hatching may occur as soon as this period is completed.

In the case of snow pool *Aedes* mosquitoes, eggs laid during the summer and autumn remain under snow through the winter and hatch with the spring melt; to hatch they require a prolonged chilling period before embryonic diapause is broken, thus one brood is produced annually.

Eggs of floodwater mosquitoes do not all hatch at one time; most of the eggs of one laying will hatch with the first flooding, but some remain for the second or subsequent

Fig. 10-4 A *Culex* mosquito laying an egg raft on the surface of water. The raft conformation forms between the back pair of legs. Mosquito eggs are white when first laid, darkening to a tan-to-black color shortly after. (Courtesy of Oregon State University and Ken Gray.)

floodings. Hatching of eggs of common *Aedes* and *Psorophora* may not be completed until 4 years of intermittent flooding (Breeland and Pickard, 1967; Woodard and Chapman, 1970). In all floodwater mosquitoes studied thus far, essentially *Aedes* and *Psorophora* species, there must be a lowering of the dissolved oxygen level in the water that covers eggs for significant hatching to occur. This oxygen depletion occurs naturally through the growth of microorganisms on freshly flooded nutrients (see discussion in Clements, 1963). In summary, for eggs of *Aedes* and *Psorophora* to hatch they must (1) complete embryonation, (2) not be in diapause, (3) be conditioned, perhaps by alternate drying and moistening of the CHORION (the shell structures), and (4) the water flooding them must have a hatching stimulus of lowered oxygen tension.

Larvae. Larvae of most Culicinae hang suspended diagonally from the water surface by means of a prominent breathing siphon (Figs. 5-4, 10-3). Exceptions occur in species that remain most of the time under water respiring by cutaneous diffusion, and the tribe Mansoniini (*Mansonia, Coquillettidia*) in which the air tube is a short and pointed structure that pierces the roots of aquatic plants. The larvae of Anophelinae lie suspended horizontally just beneath the surface of the water, by means of palmate hairs (Fig. 10-3).

Most mosquito larvae filter out microorganisms and other particulates in water or browse microorganisms present on solid surfaces. Larvae produce a water-soluble feeding stimulant that promotes feeding activity (Dadd and Kleinjan, 1974). The food is carried to the mouth by currents produced through action of long curved hairs borne by the maxillae; it then flows into the pharynx through suction, and filtering of suspended solids occurs there (see Chapter 3). Among at least some species with predacious larvae, the first instar is a filter feeder and the predacious feeding structures are not developed until after the first molt.

The larvae molt four times, the last molt resulting in the pupa. About 7 days, depending on temperature, are required for completion of larval development in commoner Culicinae under optimum food conditions in summer; larvae of the Anophelinae generally require somewhat longer.

Pupae. With the fourth molt the nonfeeding pupa or "tumbler" appears. This stage is quite short, usually 2 to 3 days. There is a pair of breathing "trumpets" situated dorsally on the cephalothorax, and paddlelike flaps on the end of the abdomen (Fig. 10-3). The pupa is remarkably active and sensitive to disturbances, suddenly darting with a tumbling motion to deeper water and after a few moments rising with little motion to the surface. The pupa is held at the surface by buoyancy, and a hydrophilous area where the respiratory trumpets break the water surface aids stability. In diving, in addition to tumbling motion of the whole body, the trumpets are moved backward by muscular activity to permit escape from the surface (Houlihan, 1971).

Food Habits of Adults. Consult Chapter 3 on adult mouth parts. Nourishment of both sexes for flight and ordinary metabolic maintenance is normally derived from nectar and plant juices. With the exception of a few species, such as the plantfeeding Toxorhynchitinae, female mosquitoes pierce the skin of many kinds of animals and feed on blood, the protein from blood being required for eggs to develop.

While most mosquitoes prefer to feed on mammals or birds, there are species that readily feed on fish that live exposed to air, amphibians, reptiles, and even on insect larvae (Harris *et al.*, 1969). The great majority are ZOOPHILOUS, that is feed in nature on animals other than man; species that feed on man by preference are said to be ANTHROPOPHILOUS.

There is a combination of innate behavioral patterns and habitat- or host-associated stimuli involved in finding a host. This subject, as well as a discussion of flight distance, is reviewed in Chapter 4.

Reproduction. The general subject of mosquito reproduction has been reviewed by Barr (1974). Most mosquitoes form crepuscular swarms of males in which females can be observed to enter and mate. Apparently such swarms are not exclusively for mating (Nielsen and Haeger, 1960). Visual and auditory cues appear to be mainly involved during attraction of males to females. The males of *Aedes aegypti* are readily attracted to a vibrating tuning fork that mimics the whining wing beat frequency of females in flight (Roth, 1948). There is evidence that sex pheromones may be produced by either sex in various species to attract the opposite sex (example in Gjullin *et al.*, 1967), but this kind of attraction is weak compared to that found in several other groups of insects. Only the first mating is effective, and frequently females are unreceptive to second matings. Second matings are unsuccessful owing to the polypeptide MATRONE that is introduced into the female reproductive tract from the accessory gland fluid of the first mating partner (Craig, 1967). Newly emerged females of many species are refractory to mating, but they become receptive after a day or 2 owing to the production of juvenile hormone and its stimulation of the last abdominal ganglion in the central nervous system (Gwadz, 1972).

Normally, females of bloodsucking species require a blood meal before oviposition. However, the term AUTOGENY is applied to cases where ovarian maturation occurs without a blood meal; normal ovarian development in most species requires a blood meal and is termed ANAUTOGENOUS. Autogeny was once regarded as something of a rarity in mosquitoes, but further study has revealed that more than 70 species in at least 13 genera are known to have autogenous strains or

to be totally autogenous (Rioux *et al.*, 1975). Autogenous reproduction reduces the opportunities for vector species to acquire an infectious blood meal, though normal egg development follows after the first autogenous egg batch in most autogenous strains. Autogeny is genetically controlled, and its expression may be mediated by larval nutrition, seasonal factors including photoperiod, and mating. The humoral system stimulating egg development during autogeny involves products of the brain median neurosecretory cells, and juvenile hormone produced by the corpora allata (Lea, 1970). During normal or anautogenous development proteins from the blood meal stimulate oocytes in the ovaries to develop yolk and thus grow (Bellamy and Bracken, 1971; Spielman and Wong, 1974), the maturation of eggs being controlled by release of a brain hormone, and vitellogenin stimulating hormone from the ovary (Hagedorn, 1974). The ovarian follicular epithelium secretes the membrane and shell (chorion) surrounding the developing egg (cf. Mathew and Rai, 1975).

In temperate climates, in the fall of the year, females in which adult hibernation occurs (most *Culex, Culiseta, Anopheles*) undergo physiological changes in response to lower temperatures and short daily photoperiods. These females become reluctant to take blood meals, and their ovaries will not mature at that time even if blood feeding occurs. These changes include the development of large amounts of body fat, and thus successful hibernation (cf. Eldridge, 1968; Spielman and Wong, 1973).

Longevity. Male mosquitoes usually remain alive for no longer than a week, although careful laboratory maintenance with adequate carbohydrate and high humidity may yield survival beyond a month. Females with ample food may live 4 or 5 months, particularly under conditions of hibernation. During their period of greatest activity, when summers are hot, female survival may average closer to 2 weeks. The importance of longevity for vector effectiveness has been mentioned in Chapter 4.

Subfamily Culicinae

Characteristics. All adult Culicinae have a trilobed scutellum with each lobe bearing bristles, but with bare areas between lobes. The abdomen is completely clothed with broad scales, which nearly always lie flat. Larvae have a prominent siphon, usually with a well-developed pecten (Figs. 5-4, 10-3), and one to several hair tufts. In most cases the eggs are deposited either in tight raftlike masses on the surface of water, or singly above the water line. Eggs lack the floats characteristic of those in anophelines.

The subfamily includes about 1,700 species distributed among more than 20 genera; roughly two-thirds of the described species belong to *Culex* and *Aedes*.

GENUS CULEX

Culex includes a number of proven and potential vectors of arboviruses and avian malaria. They generally prefer to feed on birds, though narrow host specificity is uncommon. They overwinter as inseminated diapausing females, preparing for hibernation by reduced blood feeding and fat hypertrophy in response to cool temperatures and short day lengths. Abdomens of females are blunt ended, and larvae usually possess a long and slender air tube bearing numerous hair tufts.

Culex pipiens. The northern house mosquito, or rain-barrel mosquito, *'Culex pipiens,* and the southern house mosquito, *Culex pipiens quinquefasciatus* (= *Cx. p. fatigans* of European authors), are considered subspecies of the same polytypic species (Mattingly, 1967). In North America *Cx. pipiens* occurs only north of the thirty-sixth parallel and *Cx. p. quinquefasciatus* only south of the thirty-ninth (except for coastal California); in the area of overlap identification is difficult because of the frequent oc-

currence of hybrids and intergrades (Barr, 1967). Both mosquitoes occur widely in the Old World. Autogeny is of widespread occurrence in this species complex. Autogenous forms within the range of *pipiens* have been designated as *Cx. p. molestus,* but Barr provides evidence that *"Culex molestus"* represents local autogenous populations of *Cx. pipiens.* Sterility barriers called cytoplasmic incompatibility have been noted between strains of *Culex pipiens* from different geographical areas, and this has been considered evidence of an active speciation process. It is now clear that this phenomenon is caused by the presence or absence in various strains of a transovarially transmitted rickettsialike organism, *Wolbachia pipientis* (Yen and Barr, 1973; Irving-Bell, 1974).

This plain brown domestic mosquito lays egg rafts on water in rain barrels, tanks, cisterns, catch basins, and other small collections, tolerating water with high organic pollution. Where breeding conditions are favorable it occurs in enormous numbers. It invades houses freely. Because of its persistence and high-pitched hum continued late into the night, it may be a considerable pest. Although greatly modified by temperature, typically its life cycle requires but 10 to 14 days under warm summer conditions; egg stage 24 to 36 hours, larva 7 to 10 days, pupa about 2 days.

Culex p. quinquefasciatus is increasing in Africa and Asia in response to favorable habitats accompanying urbanization, and open sewage drains and pit latrines in disregard of sanitary measures as a consequence of increased use of persistent insecticides (Service, 1966; Singh, 1967). Sewage oxidation lagoons are particularly attractive for oviposition when coliform bacterial counts increase sufficiently (example in Steelman and Colmer, 1970). This insect is being used to study the population dynamics of a mosquito species in mathematical exercises on sterile insect control (Weidhaas *et al.,* 1971, 1973).

Culex pipiens and *Cx. p. quinquefasciatus* are major developmental hosts and vectors of the human filarial worm *Wuchereria bancrofti*. They transmit the pathogens of bird malaria, heartworm of dogs, and avian pox virus. In the eastern USA, members of the *Cx. pipiens* complex are important vectors of St. Louis encephalitis virus. Western equine encephalitis virus is also frequently isolated from this complex.

Culex tarsalis. This mosquito (Fig. 10-5) is an abundant and widespread species of the semiarid regions of North America, further occurring throughout the southern U.S. and as far northeast as Indiana, southwestern Canada and Mexico. It has been taken at elevations up to 2,750 m. This mosquito is the main vector of western equine encephalitis virus, and it transmits St. Louis encephalitis virus. It is a fairly large and robust *Culex,* generally dark brown to near black, with evident white banding on legs and abdomen and wide ring of white on the proboscis.

Cx. tarsalis breeds in waters both clear and with high organic pollution, but especially sunny ground collections in pools, in ditches, in and around corrals; also in artificial containers. It is considered a summer-active mosquito, but winter adult activity is known for part of its southern range, as reported for the lower Rio Grande Valley of Texas. In cold regions inseminated females hibernate in natural sites, for example, rock piles in the northwestern states. Autogenous strains are known. Domestic and wild birds are the preferred host of this nocturnally active mosquito, although it readily bites man, horses, and cattle, and will feed on amphibians and reptiles.

Culex tritaeniorhynchus. This is the most important vector of Japanese encephalitis virus in the Oriental region. Its distribution includes the Far East, Near East, and several locations in Africa. It is a small mosquito, with brown scaling on the scutum and accessory pale scaling on the lower surface of the proboscis, resembling other species in the *Culex vishnui* subgroup. In Japan its biology includes overwintering as adult females, mainly in brush and woodpiles; about 14 days from adult to adult in summer, larvae in temporary and semipermanent ground water such as hyacinth ponds, streams, swamps; preferred hosts are cattle and pigs, but birds and man are also bitten.

GENUS CULISETA

Culiseta includes eight North American species and subspecies. These are mainly rather large mosquitoes; the postspiracular bristles are absent, and in the females of most species the anterior and posterior cross veins tend to lie in one line; egg rafts resembling those of *Culex* are produced. Many of the common species are univoltine, with larvae found only in early spring. *Culiseta inornata* occurs throughout the contiguous USA and southern Canada. Its wings are broad and clear, the cross veins are scaled, and the very short black palpi have white scales at the tip. Precipitin tests indicate this species feeds chiefly on large mammals such as horses and cattle. *Culiseta melanura* of the eastern and central USA is predominantly a

Fig. 10-5 *Culex tarsalis,* an important vector of the virus of western equine encephalitis. (Photograph by R. Craig.)

swamp-inhabiting bird feeder that rarely bites man or large mammals. In this species there is a winter larval diapause (Morris *et al.*, 1976). *Culiseta inornata* and *Cs. melanura* are involved in the transmission of viral encephalitides, particularly WEE and EEE.

GENUS AEDES

Nearly half of all North American mosquito species belong to the genus *Aedes*. In most the claws are toothed in the female, postspiracular bristles are present, the pulvilli are absent or hairlike, and the female abdomen tends to be pointed and the cerci longer than in other groups; larvae have siphons bearing one pair of posteroventral hair tufts and nearly always a distinct pecten (Fig. 10-1). The eggs are deposited singly on the surface of water, on mud, just above water line in container-breeding species, or even in situations where there may be little moisture but where submergence is likely. The females are often aggressive biters. Many species are diurnal in activity, most of them biting toward evening. The genetics of *Aedes*, including formal genetics, cytogenetics, ability to develop vertebrate pathogens, insecticide resistance, chemosterilization, and several miscellaneous topics, has been reviewed by Rai and Hartberg (in King, 1975). *Aedes* can be conveniently separated by larval habitat into characteristic groups for purposes of discussion.

Salt Marsh Aedes. *Aedes taeniorhynchus* is a typical species of the American coast and inland saline waters from Massachusetts to Brazil and California to Peru, the Antilles, and Galapagos Islands, with a northern US limit imposed by an average annual minimal temperature of 0° F (Knight, 1967). It is the black salt marsh mosquito with distinctly white-banded proboscis. It is a fierce biter by day, and well-documented large migratory flights are preceded by highly synchronous development (Provost, 1972b). Flight and feeding behavior differences have been noted between normal and autogenous

strains (Nayar and Sauerman, 1975). There are monthly broods throughout the summer, and under ideal conditions the larval stage may require but 4 days and adults can emerge in 8 to 10 days. This mosquito has been artificially hybridized with *Ae. mitchellae, sollicitans,* and *nigromaculis* (Fukuda and Woodard, 1974; O'Meara *et al.*, 1974).

Aedes sollicitans of the Nearctic region and Greater Antilles is a pestiferous salt marsh mosquito of the Atlantic coast from Maine to Florida and thence west along the Gulf of Mexico to Texas, with a northern limit imposed by an annual average minimal isotherm of −3° F (Knight, 1967). There are many broods, and in its southern range breeding may be continuous. As with *Ae. taeniorhynchus,* enormous numbers of larvae may develop in pools.

Aedes dorsalis, a fierce biter by day, has a Holarctic distribution that includes Mexico, Formosa, and North Africa. In general the body is tan and white, and this species is readily confused with *Ae. campestris.* Although *Ae. dorsalis* breeds freely and abundantly in fresh water, such as floodwater, rice fields, and drainage from irrigation, it is nevertheless the commonest salt marsh mosquito of the Pacific coast north of Monterey, California. It is here a distinctly brackish water breeder, generally in pools reached only by monthly highest tides. Eggs, deposited singly and mainly in mud along the edge of receding pools, may remain unhatched for many months if not inundated. Development after hatching is rapid, and emergence of the adults occurs within 8 days.

Floodwater Aedes. This group includes species inhabiting riverine flood plains, and species that develop in water accumulations due to irrigation and seepage.

Aedes vexans is a typical floodwater mosquito of Holarctic and Oriental regions, Pacific Islands, and South Africa (Horsfall *et al.*, 1973). The distribution of this species and its subspecies and variants has been reviewed by Zharov (1973). This brown to grayish mosquito with narrowly banded tarsi is a fierce daytime biter and is exceedingly abundant and vexatious. It breeds in greatest numbers along the flood plains of rivers, and like many other *Aedes* it lays eggs along the muddy edges of receding pools where there is shrubby vegetation; here they may hatch the same season when water due to intermittent flooding or freshets reaches them, or they may carry over (Horsfall *et al.,* 1975). There may thus be several broods where flooding occurs from intermittent melting snows or thunderstorms, or only one brood where a single spring flood is characteristic. The species is a rapid breeder and migrates many miles, generally along wooded river valleys. Rainfall and temperature data have been integrated to predict when there will be influxes of adults into urban areas (Clarke and Wray, 1967).

Aedes dorsalis, already referred to as a salt-marsh breeder, is frequently in association with *Ae. vexans* in freshwater habitats in the western USA.

Aedes nigromaculis of the western and central plains of the USA is an important irrigated pasture mosquito, especially in California. Development is extremely rapid, adults appearing as early as 4 days after the eggs have been flooded. Swarms of these fierce daytime biters may bring recreational activities and normal behavior of livestock to a virtual standstill.

Boreal Aedes, or Snow Pool Mosquitoes. An interesting group of *Aedes* consists of the so-called snow pool mosquitoes that appear in early spring in the high mountains and northern ranges of distribution, developing in pools left by melting snow (Fig. 10-6). A low-temperature adaptation is evident; the northern species *Aedes nigripes* and *Ae. impiger* at Fort Churchill, Manitoba, Canada have a low development threshold of 1° C (Haufe and Burgess, 1956). Larvae aggregate in the warmest sector of muskeg pools, and oviposition is concentrated at sites exposed to the sun (Haufe, 1957; Corbet and Danks, 1975) (Fig. 10-7). There is but one generation yearly in snow pool mosquitoes, and adults appear in enormous swarms in higher elevations and northern ranges. The eggs hatch during thawing, and larvae may survive freezing and thawing. Development from larva to adult generally takes more than a month.

Hocking, in a review on northern biting

Fig. 10-6 Typical breeding place for snow pool mosquitoes on floor of Yosemite Valley, California. (Photograph by H. F. Gray.)

Fig. 10-7 Eggs of *Aedes nigripes* massed as a dark line, indicated by arrow, in muskeg. (Photograph by P. S. Corbet.)

Fig. 10-8 Tree hole where *Aedes* larvae are found. (After State, Gjullin, and Yates; USDA photograph.)

flies (1960), points out that arctic snow pool mosquitoes are a direct hazard to life for man and animals, though in these areas the risk of pathogen transmission is small. Snow pool mosquitoes continue to cause major interference with recreational interests, as satisfactory control procedures are very difficult to develop. Facultative autogeny is characteristic of several arctic *Aedes* (Corbet, 1967), and is accompanied in some by autolysis of the main flight muscles. Blood meals are taken from herds of caribou, musk ox, large populations of rodents, nesting waterfowl and other birds (Corbet and Downe, 1966). Grazing animals on the tundra may be driven into restless running by these annoyers. *Aedes nigripes,* strictly a species of the tundra, and *Ae. impiger* (=*nearcticus*) are the most abundant species of the circumpolar far north (Vockeroth, 1954).

Tree Hole Aedes. Although breeding in water-holding tree holes (Fig. 10-8) occurs in various species belonging to other genera, for example, *Anopheles barberi,* there are a number of specialized tree hole breeders in *Aedes,* notably *Ae. sierrensis,* a Pacific Coast USA species; *Ae. triseriatus* of the midwestern and eastern USA; *Ae. luteocephalus,* Ethiopian; *Ae. simpsoni,* Ethiopian; *Ae. seoulensis,* Oriental. These mosquitoes are often very annoying, and their larval habitat may be overlooked unless one is aware of tree holes as a breeding site. The small-container breeding habit seems mainly to be derived from tree hole or leaf axil breeding in response to opportunities provided by man. This appears to be the situation with *Aedes aegypti* and its close relatives.

Aedes triseriatus lacks white rings on the tarsal segments. It is a widespread tree hole breeder east of the Rocky Mountains, USA. Diapause for overwintering may occur in the egg or larval stages in response to short day length (Kappus and Venard, 1967). This species is a major vector of California encephalitis complex viruses in upper midwestern states.

Aedes hendersoni, first thought to be a

subspecies of *Ae. triseriatus,* is the most widespread tree hole *Aedes* in the contiguous USA, possibly being absent from Arizona, California, and Nevada (Zavortink, 1964). Hibernation occurs in the egg stage.

Aedes sierrensis has bright white markings on the legs at bases and apices of the tarsal segments, and many white or silvery scales distributed over the body, giving the vestiture a silver-mottled appearance. It is small but a fierce biter. This species deposits its eggs on the sides of tree holes, notably various oaks. There is an extended larval period of 1 to 7 months, with intermingled broods having pronounced early summer and fall peaks. It overwinters as larvae. Garcia and Ponting (1972) have provided recent details on the life history of this mosquito in a California location.

Aedes aegypti, the yellow fever mosquito, is the most important vector of the viruses of yellow fever, dengue, and chikungunya. It is widely distributed within the limits of 40° N and 40° S latitude, but it is highly susceptible to temperature extremes and does not thrive in dry hot climates. Because of its importance as a vector, mark-release-recapture experiments have been performed with wild populations in Tanzania to determine epidemiological parameters (Conway *et al.,* 1974). Where man-made containers are the principal breeding sites, numbers of emerging adults seem to be mainly determined by larval mortality (Southwood et al., 1972). In locations where this mosquito was eliminated in eradication campaigns now discontinued it has reappeared extensively, including many locations in the mainland USA. Changes in domestic water supplies from open containers to piped systems have been a highly important cause of population reductions (Curtin, 1967).

The adult (Fig. 10-9) is beautifully marked with silvery-white or yellowish-white bands and stripes on a nearly black background. It has a "lyrelike" pattern dor-

Fig. 10-9 The yellow fever mosquito, *Aedes aegypti,* engorged with blood. (Courtesy of Pacific Supply Coop and Ken Gray.)

sally on its thorax; legs are conspicuously banded, and the last segment of the hind leg is white.

A comprehensive account of the yellow fever mosquito has been provided by Christophers (1960). Three varieties are generally recognized (Mattingly, 1957, 1958; Craig and Hickey, in Wright and Pal, 1967): the variety *aegypti* is brownish or blackish, widely distributed but absent from inland Africa south of the Sahara; variety *queenslandensis* is pale and from northern Australia with distribution similar to variety *aegypti;* variety *formosus* is black and confined to Africa south of the Sahara, where it is the only variety except in coastal areas. Outside of Africa the distribution of *Ae. aegypti* is often mainly coastal. In comparing general coloring and abdominal scale pattern of samples from many areas of the world, McClelland (1974) concludes that variation is continuous, with a tendency in any region for populations associated with man to be relatively paler than those less synanthropic. This he believes represents incipient speciation, with a paler synanthropic "species" independent of rainfall found where humans store water in open containers. If this variety and the primitive dark or feral variety occur together, each competitively excludes the other from its specialized habitat. McClelland also reviews other differences, though there is no evidence of significant isolation in crossing experiments. The man-associated form has greater vector potential in human disease than does the feral form.

Eggs of the yellow fever mosquito are deposited singly at or near the waterline, mainly in water storage containers and in discarded or stored objects associated with man in the synanthropic forms, and in rain-filled cavities such as tree holes, leaf axils, or rock depressions in the feral form. Comparatively few eggs are deposited at one laying, and although there may be two or more layings, the total number of eggs produced averages about 140 when *Ae. aegypti* has fed on man.

The eggs can withstand desiccation up to a year. Ordinarily they complete embryonic development and can hatch when flooded after 4 days. The larval stage is passed in about 9 days, dependent on temperatures; the pupal stage in 1 to 5 days.

OTHER GENERA OF CULICINAE

There are many additional genera of culicines. Five genera are of sufficient importance to merit brief inclusion. Even though several tropical forms are vectors of arboviruses, they are mainly forest-associated and are not mentioned except in discussions on specific diseases.

Genus Psorophora. The genus *Psorophora* is a totally American genus of nearly 50 species. It is distinguished by the presence of both prespiracular and postspiracular bristles and by the second marginal cell of the wing, which is more than half as long as its petiole. The larvae of very large species (*Ps. howardii, ciliata*) are predacious on other mosquito larvae and other aquatic animals of similar size in temporary ground pools and their adults are blood feeders called "gallinippers." General habits and life history of *Psorophora* mosquitoes resemble those of floodwater *Aedes*.

Psorophora columbiae is widely distributed in the mid to lower eastern USA and ranges from parts of South America through Cuba and Mexico. It is strikingly speckled in appearance and is a fierce biter. It commonly breeds in rice fields. In 1932 this species is reported to have caused a great loss of livestock in the Everglades section of Florida, and the milk supply was also greatly reduced during the 4 days of the infestation. It has been implicated as a vector of Venezuelan equine encephalitis virus (PAHO, 1972).

Genera Mansonia and Coquillettidia. The genera *Mansonia* and *Coquillettidia* are characterized in large measure by the scales of the wings, which are very broad by comparison with those of other mosquitoes. These were formerly combined in the genus *Mansonia,* but both genera are now placed in the tribe Mansoniini. The larvae have the air tube pointed, enabling them to pierce the stem of roots of aquatic plants, from which they obtain air and to which they remain attached throughout larval development. Several Asiatic species are vectors of brugian filariasis. Since early references to the isolation of pathogens from *"Mansonia* spp.'' may apply really to *Coquillettidia*, it is best to designate the mosquitoes involved as Mansoniini.

Coquillettidia (formerly *Mansonia*) *perturbans* is widely distributed throughout North America from southern Canada to the Gulf of Mexico; Europe from Sweden and Britain to Palestine. It has severe biting habits and evidently travels some distance from its breeding place. The larvae hibernate in the muddy bottom of aquatic habitats. Bidlingmayer (1968) and Siverly (1974) provide details on the life history of this mosquito in two US locations.

Genera Haemagogus and Sabethes. The genera *Haemagogus* and *Sabethes* are New World tropical forms close to *Aedes*. These are mainly arboreal species that breed in tree holes or broken bamboo internodes, plant containers such as bromeliads and fallen fruits, and occasionally in ground pools or rock holes. They bite man readily and are important sylvan vectors of yellow fever virus. Several other viruses have been isolated from species in these genera. A recent revision of the genus *Haemagogus* includes information on their biology (Arnell, 1973).

Subfamily Anophelinae

Characteristics. The following features characterize the Anophelinae: palpi of both sexes about as long as the proboscis (except in *Bironella*), male palpi paddle-shaped at tip; the scutellum (Figs. 10-3, 10-10) evenly

rounded or straplike (except for slightly trilobed condition in *Chagasia*); legs long and slender (Fig. 10-10) with no distinct tibial bristles and no pulvilli; abdomen with sternites largely lacking scales; wings usually with distinct markings; larvae lacking an air tube, and dorsal surface of body with palmate hairs.

The subfamily Anophelinae is considered by Edwards and other culicidologists to contain three genera: *Chagasia*, with the scutellum slightly trilobed (4 species in tropical America); *Bironella*, with scutellum evenly rounded, wing with stem of median (M) fork wavy (7 species in New Guinea and Melanesia); and *Anopheles*, with scutellum evenly

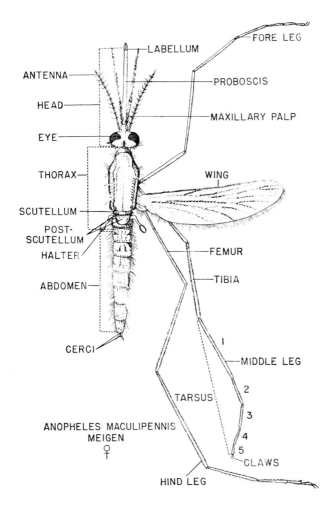

Fig. 10-10 Dorsal view of *Anopheles* mosquito (female) showing certain structural details useful in classification. (After Essig.)

rounded, wing with stem median (M) fork straight (about 390 species and subspecies, about 90 occurring in the Americas and 15 in North America).

The common species of *Anopheles* rest with the proboscis, head, and abdomen nearly in a straight line and when feeding hold the body at an angle from a given surface. In exceptional cases, as in *An. culicifacies* of India, the resting position is *Culex*-like. The hum of anophelines is distinctly low pitched and almost inaudible. As most are not strong fliers, dispersal is usually the result of brief and interrupted flights through vegetation, though high altitude flights may take place. In California, *An. freeborni* engages in an annual dispersal flight of overwintering females about mid-February, during which time the mosquitoes may invade much territory, traveling more than 40 km. *An. pharoensis* appears to fly extensive distances in Palestine.

The usual method of overwintering is as the inseminated adult female, though treehole-infesting species may overwinter as larvae.

Mating and Oviposition. Fertilization of the females takes place soon after emergence. Males emerge first and may be seen hovering over or near the breeding places in small swarms; mating occurs when females dart into the swarms. In temperate regions overwintering females are fertilized by the last brood of males during autumn,

and the females probably lay only a single group of eggs soon after the spring dispersal flight. The female either lays eggs singly while resting on still water or drops them while hovering. Eggs in each batch range from 150 to more than 300. At least three batches of eggs may be laid during the lifetime of a female, with a total of more than a thousand recorded.

Egg Characteristics. The characteristics of anopheline eggs used in classification are presence or absence of floats, position and length of the float, presence or absence of frill, and pattern (Fig. 10-11).

Species Complexes. Many *Anopheles* occur in closely related groups with rather obvious morphological relationships. In other cases separation of species may be extremely difficult, and dependable only by characters of the immature forms, or by genetic or biochemical techniques. See Chapter 4 regarding specialized procedures used in making separations. Certain complexes have come to light during intensive control programs, being noted as greater resistance to insecticides or apparently changed adult habits. Discriminating the members of complexes is very important during control campaigns, for there are often significant differences in the general behavior and vectorial capacity for malaria transmission among females in a complex. Reid (in Ward and Scanlon, 1970) discusses concepts of species complexes and groups in relation to

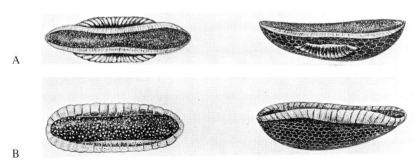

Fig. 10-11 Eggs of certain *Anopheles* mosquitoes. *A. An. punctipennis. B. An. pseudopunctipennis franciscanus,* usual form.

malaria control. A few complexes are discussed here as examples.

The *Anopheles maculipennis* complex of Europe was one of early interest. Hackett and Missiroli (1935) discuss the relationship of different habits to the occurrence of malaria in Europe, Horsfall (1955) gives a comprehensive account of the problem. The complex was initially divided into seven subspecies, most reliably separable by egg character, and by the fact that some strains were zoophilous and exophilous in contrast to others that preferred to feed on man (anthropophilous) and readily entered houses (endophilous). *Anopheles maculipennis* is now recognized as occurring in continental Europe and southwest Asia to the Persian Gulf. *An. messeae* is from the northern Palaearctic region, and *An. melanoon* and its subspecies *subalpinus* are from southern Europe, the Caucasian region of the USSR, and Iran. *Anopheles labranchiae* of Italy, Spain, and North Africa, and *An. atroparvus* occurring somewhat more north in Europe, were both formerly considered subspecies of *maculipennis*. *Anopheles sacharovi* is found in the USSR, Italy, Austria, Greece, Syria, Israel, and Jordan. American representatives of the *maculipennis* complex are *Anopheles quadrimaculatus, An. earlei, An. freeborni, An. occidentalis,* and *An. aztecus.*

Intensive control programs against *Anopheles gambiae* in Africa have provided the stimulus to clarify a species complex. Crosses of dieldrin-resistant and -susceptible strains yielded sterile male progeny with atrophied testes (Davidson and Jackson, 1962). Differences were noted in house-resting or outdoor forms, as well as freshwater and brackish water forms. By combining behavioral differences with crossing experiments as reviewed by Mattingly (1964) and others, and by morphological and cytotaxonomic techniques (cf. White, 1974), it has been possible to distinguish six species in this complex. *Anopheles melas* is a west African saltwater species, and *Anopheles merus*

a saltwater species from east Africa and the larger islands except Zanzibar, and also spreading inland. There are three freshwater species, designated as species, A, B, and C, and mineral-water species D. Species A is designated as true *Anopheles gambiae,* species B is *Anopheles arabiensis* (Mattingly, 1977). Reproductive isolation and genetic barriers exist between all six species of this complex in nature. Species A and B occur together in most areas, extending southwards to subtropical latitudes and eastward to Mauritius, with A in humid zones and B more typically in arid savannas and steppes. Species B is noted for genetic polymorphism and variable adult behavior in regard to entering houses and kinds of mammalian hosts fed upon; species A is more endophilous and easier to control by insecticide deposits on the inside walls of houses. Species A and B have developed physiological resistance to the two main chlorinated hydrocarbon insecticides used to control them. Species C and D have relict distributions, the first of these sometimes surviving at high densities when A and B are controlled.

Anopheles hyrcanus apparently forms a complex across central and southern Asia, the northern Mediterranean, and Libya. The taxonomy of this complex is by no means clear. Published information suggests varieties with widely divergent habits, and therefore probably very different in their capacity to serve as vectors of malaria. The relationships of this complex have been reviewed in southeast Asia and China (Harrison, 1972; Xu and Feng, 1975).

With the development of the Amazon basin, and attendant malaria control problems, it is evident that a species complex in the *Anopheles nuneztovari–An. albitarsus* group is present.

Breeding Habits. The breeding habits of anophelines differ considerably for even very closely related species. The North American *Anopheles freeborni* and *An. quadrimaculatus,* both four-spotted anophelines separable with accuracy as adults only on differences in male terminalia, have very different larval habitats, the former breeding largely in open sunlit shallow seepage water and the latter in impounded water with floating debris and aquatic vegetation. The European members of the *Anopheles maculipennis* complex have characteristic larval sites that, within certain limits, differ from one another.

The common malaria mosquito, *Anopheles quadrimaculatus,* is the chief vector of malaria in the eastern, central, and southern USA. During the period of active malaria transmission in that country, man provided ideal conditions for this mosquito. *An. quadrimaculatus* prefers relatively clean still water, and larvae move between sunshine and shade during development. Ideal breeding places are in freshly impounded water containing some underbrush and having a sparse tree cover, such as in dam reservoirs along wooded valleys, or dense swamp forests opened up by lumbering. Flotage of leaves and twigs provides shade, a food supply of algae and associated microflora, and protection from wave action and natural enemies. Artificial and natural impoundments provide excellent breeding places. Swamps containing mature timber are usually too shady for heavy production of this mosquito. Timber harvest of large trees opens the area to sunshine, but leaves some shade, and resultant flotage blocks natural swamp channels to form a series of ponds ideal for the development of *An. quadrimaculatus.*

The following examples illustrate the importance of knowing anopheline breeding habits in the conduct of malaria-control operations, and the diversity of breeding sites. In Malaya *Anopheles umbrosus* transmits malaria in the coastal plain, breeding in practically stagnant water densely shaded by mangroves. Its production is controlled by clearing swamps and letting in sunshine, or by cutting ditches and confining the water to definite channels. The same procedures practiced on high inland plateaus increases the population of *Anopheles maculatus,* an important vector that prefers the quiet edges of trickling streams in open sunshine. Many *Anopheles* are associated with lake and pond margins, but *An. minimus* of the Philippines breeds along margins of small flowing streams in the foothills. Several species of *Anopheles,* though seldom important as vectors, are tree hole breeders, viz.: *An. plumbeus* (European) and *An. barberi* (American). *An. bellator,* an important malaria vector in the Caribbean, breeds in collections of water among epiphytic bromeliads.

Life History.   Although there is much variation in the life histories of species of *Anopheles,* as well as considerable variation due to temperature and food supply, the length of time required for development from egg to adult is generally longer than in other genera, except for *Aedes* and similar genera where the egg stage may be greatly prolonged.

The usual incubation period is 2 to 6 days. Apparently, the only stimulus needed for hatching is that the egg be floating in water suitable for the development of the larva. The egg shows very little resistance to desiccation. Some tropical species may survive 2 weeks or even longer on a moist surface, but severe drying kills, injures, or retards the embryo (Horsfall, 1955). Larval development takes 2 weeks or longer, and the pupal stage takes about 3 days. Thus from egg to adult takes about 3 weeks to a month. Oviposition often does not commence until about a week after a blood meal.

The life cycle of *Anopheles albimanus,* the important vector of malaria in much of Central America, has been described by Rozeboom (1936). With room temperature from 27 to 32° C and water temperature for larvae 21 to 27°, and eggs and pupae at 27 to 30°, the entire cycle (egg to adult) required 18 to 23 days, an average of 3 weeks. A period of 7 days, or a little over, was necessary for the development of eggs in the ovaries; the incubation period was 40 to 48 hours; the larval stage 6 to 22 days (usually 8 to 13 days in hay infusion water); the pupal stage 30 to 33 hours; the longest observed adult life of a female was 31 days and of a male 27 days. The dynamics of a population of this vector has been described (Weidhaas *et al.*, 1974).

Mosquito Transmission of Pathogens: General Considerations

Mosquitoes are potent vectors of three types of organisms pathogenic to man and animals. These are (1) the plasmodia, causal organisms of the malarias, belonging to the Protozoa; (2) filarial worms of the genera *Wuchereria* and *Brugia*, the causal organisms of human lymphatic filariasis; (3) viruses, especially arboviruses causing such important diseases as yellow fever, dengue, and the American encephalitides. Transmission of the bacterium causing tularemia has been reported, and experimental transmission studies include such pathogens as the bacterial agent causing leprosy.

Plasmodial Infections

HUMAN MALARIA

Disease: Malaria (ague, paludism, marsh fever, intermittent fever, *Wechselfieber*, *Kaltesfieber*).

Pathogen: Protozoa, Sporozoa, *Plasmodium;* four species, with *P. vivax* and *P. falciparum* of greatest importance.

Victims: Humans, almost entirely.

Vectors: *Anopheles* mosquitoes.

Reservoirs: Man, frequently marginally symptomatic. Nonhuman primate reservoirs are known or suspected for two human plasmodia, but are not required to maintain human disease.

Distribution: Mainly humid tropics between 45° N and 40° S latitude; especially Africa, Asia, Central and South America. Formerly widespread in temperate Europe and North America (Fig. 10-12).

Importance: Considered as the principal debilitating parasitic disease of the tropics; currently about 200 million cases worldwide annually. A major deterrent to the economies of underdeveloped countries.

Malaria is essentially a disease of the tropics and subtropics, common among impoverished and low-income groups of people wherever a high rate of biting by *Anopheles* mosquitoes is characteristic. Outside of these circumstances outbreaks develop whenever major displacements of people occur through military or social upheaval, or because of natural disasters. At one time this disease was prevalent to a greater or lesser degree on every inhabited continent and on many islands. There is archeological evidence of human malaria in the eastern Mediterranean region as far back as the beginning of the Neolithic period, and possibly also in Southeast Asia (Thailand) at about the same time (Mattingly, 1973). At present human malaria is mainly a problem in the Asian and African tropics, yet Sweden, Finland, northern Russia, and much of the USA have been subject to serious past epidemics. The Mediterranean region was apparently not seriously affected by malaria in classical times, yet man appears to have changed conditions to make this disease a dominant health problem in the latter days of the Roman empire (Zulueta, 1973).

It seems doubtful that malaria existed in the New World prior to the voyages of Columbus, but Drake lost 500 men in the Caribbean in 1588, apparently mainly from malaria (Gillett, 1972), and this disease was recognized as a factor in colonization on the Massachusetts coast and the Georgia-Carolina coast as early as the middle of the seventeenth century (Boyd, 1941). As recently as the 1930s there were 6 to 7 million cases annually in the continental USA (Russell, 1959), and Schultz (1974) reviews four waves of imported malaria into that country. For the past 75 years, however, malaria has been declining in many parts of the world. In

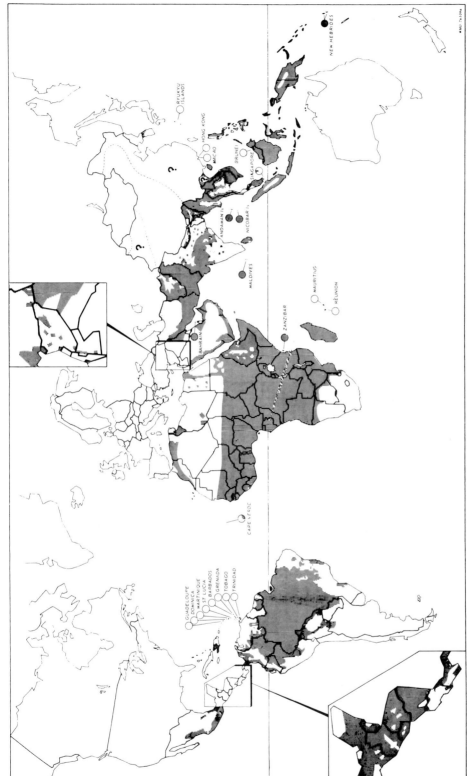

Fig. 10-12 Status of human malaria in 1975. (Courtesy of WHO.)

the USA there have been only two or three proven cases of natural transmission by mosquitoes each year since 1953. In 1967 a resurgence, mainly due to military personnel returning from Viet Nam, peaked in 1970 with the reporting of some 4,059 cases (Schultz, 1974). Several cases were attributed to transmission due to local vectors (AUTOCHTHONOUS TRANSMISSION).

Europe, once widely malarious, is virtually free of the disease, and South America has achieved eradication in great sections. The Netherlands can serve as an example in understanding factors involved in the decline of malaria. In that country, in addition to the introduction of residual insecticide programs in houses, the main vector declined owing to the reduction in numbers of pigsties and stables affording vector blood meals and shelter, insecticide resin strips, pollution of larval breeding sites with anionic detergents, the presence of duckweed (*Lemna*) in larval sites, and weed-free maintenance of ditches (Seventer, 1969). For Europe in general, decline is believed due to improved health conditions since the eighteenth century, and to more recent antimalarial measures, including the introduction of residual insecticides. In the Americas malaria has been eradicated from 12 countries, good progress has been realized in 8 countries, but presently there are poor prospects in 14 countries (Ayalde, 1976). In much of tropical Africa, and the tropics generally, problems continue because of difficulties in preventing vector-human contacts, and technical and administrative shortcomings (cf. Bruce-Chwatt, 1973).

To summarize recent conditions, 1975 WHO estimates state that in Africa south of the Sahara (322 million population) more than one million infants and children died of malaria, and this region is a main source of exportation into malaria-free areas; in Central America (107 million) there have been radical increases since 1974, with 55,000

cases reported from Nicaragua and Honduras; in South America (218 million), Colombia reported nearly 33,000 cases; in Europe including Turkey and the USSR (771 million), Turkey had 10,000 cases and in the whole area there were 2,402 imported cases; in Asia west of the Indian subcontinent (174 million), Pakistan had at least 230,000 cases; on the Indian subcontinent (721 million), the situation regressed into a major public health problem; in eastern Asia and Oceania /minus China/ (512 million), increasing incidence was reported in Sabah (Malaysia), Burma, and Thailand. In this same year in the USA 447 cases were reported (445 imported), an increase of 38 percent over 1974.

Endemic malaria depends upon a complex of environmental factors favoring the development of large numbers of anthropophilous *Anopheles* mosquitoes, the plasmodia causing the disease, and partial immunity of adult humans. Temperature, particularly as it affects the development of the plasmodium in the mosquitoes, and temperature combined with humidity as it affects the longevity of vectors, are critical factors; a mean summer isotherm of 15° C, situated near 45° N and 40° S latitude, in general limits the geographic distribution of malaria. Within this range incidence is dependent upon the availability of water for mosquito breeding, not necessarily heavy rainfall. Naturally arid regions may raise numbers of *Anopheles* in sinkhole water accumulations, open wells, or standing water from irrigation systems. Although lowlands are more likely to be affected, this does not hold as a general rule, because if one or more important factors are lacking in a lowland region, the area is nonmalarious.

High levels of malaria transmission are usually associated with anthropophilous and endophilic vectors. As vector control is achieved, particularly when applied to endophilic vectors, a higher proportion of transmission by exophilic vectors is noted.

This is particularly true among nomadic peoples and other circumstances causing population movements, as in Afghanistan, Iran, and New Guinea (cf. Montabar, 1974). Night exposure of humans protecting crops from pilferage will increase transmission (cf. Ponnampalam, 1975). Military campaigns increase transmission, such as in Viet Nam where total cases in US forces exceeded 10,000 in less than 2 years, mainly owing to exophilic vectors (Holway *et al.*, 1967). A complex of ecological changes has set the stage in modern times for epidemics of malaria, as has been interestingly documented by Desowitz, (1976).

Human malaria is caused by an infection with one or more of four species of blood-inhabiting protozoa (Class Sporozoa) belonging to the genus *Plasmodium*. These closely related parasites have a common means of transmission; nevertheless, they possess individual characteristics discernible on microscopic examination of blood cell preparations.

The plasmodial parasites develop within and destroy red blood corpuscles while undergoing asexual reproduction. This asexual reproduction or sporulation occurs at intervals typical for the species of *Plasmodium*. Symptoms commonly include more or less regularly occurring febrile paroxysms, usually with chills, hyperthermia, and perspiration. But frequently the stages are not well marked and paroxysms do not recur at the same intervals. When paroxysms recur at 24-hour intervals, as is often true in early stages of infection or in multiple infections, the disease is described as QUOTIDIAN; an interval of 48 hours or every third day is TERTIAN; an interval of 72 hours or every fourth day is QUARTAN.

The infection results in (1) changes in organs such as enlargement of the spleen (SPLENOMEGALY) and liver; in fatal cases of *falciparum* malaria, capillaries in the brain and pia are congested or blocked by schizonts and sporulating forms; (2) leukopenia with anemia due to destruction of many red cells by plasmodia and indirect degeneration of others; (3) production of malarial pigment derived from hemoglobin in macrophages in the splenic sinuses and liver; these pigments are produced in red cells infected by plasmodia and released with the rupture of infested red cells; (4) changes in physiology, such as periodic febrile paroxysms, regular in benign *vivax* malaria but often concealed in *falciparum* malaria because of irregular maturing of plasmodia; (5) MALARIA CACHEXIA, a chronic condition characterized as inability to concentrate and tendency to depression following repeated malarial attacks. It has been noted in Africa that malaria can result in low birth weight of children, because of heavy infection of placental tissues (Jelliffe, 1968), and lethal infection may be less common in severely malnourished than in well-nourished children (Hendrickse, 1967). The debilitating effects of malaria profoundly influence economic development (cf. Conly, 1975).

Persons concerned technically with malaria should consult books by Russell and colleagues (1963), Boyd (1950), MacDonald (1957) and Wilcocks and Manson-Bahr (1972). A comprehensive review of malaria parasites and other haemosporidia has been prepared by Garnham (1966), and this source should be referred to for details of life history and characterization of various plasmodia.

Life Cycle of Plasmodia. For purposes of simplicity the cycle in man (Fig. 10-13) illustrates the generalized condition found in all plasmodia and related blood parasites. The main point is that two quite different cycles occur, one in the vertebrate host and the other in the arthropod vector, and both are necessary to complete the life cycle.

Mature sexual forms (microgametocytes and macrogametocytes) are ingested by an *Anopheles* mosquito in the act of blood feed-

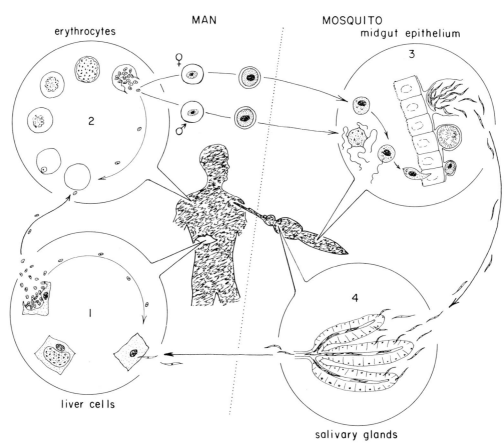

Fig. 10-13 Life cycle of human malaria.

Man Cycle

1. Sporozoites are introduced into man from salivary glands of mosquito in act of biting (arrow from *4* to *1*) and, *1,* sporozoites enter parenchyma cells of liver to establish primary exerythrocytic schizogony. In man there is only this single cycle in *falciparum* malaria; in malaria and various Haemosporidia of man and other animals secondary cycles are characteristic.

2. Merozoites are formed and invade (arrow from *1* to *2*) erythrocytes (red blood cells) where the developing stage is called a trophozoite. The trophozoite grows, destroys its erythrocyte to release merozoites which enter more erythrocytes to continue erythrocytic cycles. Micro (male) and macro (female) gametocytes are also periodically released from erythrocytes and, when ingested by an *Anopheles* mosquito (arrows from *2* to *3*), the microgametocyte divides to form motile microgametes.

Mosquito Cycle

3. A microgamete enters a macrogamete to form a zygote that becomes a motile ookinete. Ookinetes pass through the midgut epithelium to form oocysts that grow, burst, and release sporozoites into the hemocoel (arrow from *3* to *4*).

4. Sporozoites are active, pass freely throughout body cavity, and concentrate in salivary glands where they can be introduced into blood of vertebrate in act of feeding.

192

ing. Asexual parasites predominate in parasitemic blood and are destroyed by digestion, but the gametocytes shed their erythrocyte covering and undergo maturation to form male and female gametes.

Formation of male gametes takes place in a matter of minutes, and is characterized by the formation of a number of nuclei accompanied by exflagellation, each nucleus becoming a flagellated MICROGAMETE. The product in human malaria is generally eight or fewer microgametes that separate and actively swim like spermatozoa. The MACROGAMETOCYTE undergoes less obvious changes, principally an increase in size to form a MACROGAMETE. The nucleus of the macrogamete moves toward the surface, where it is entered by a microgamete, and fusion of the nuclei forms a ZYGOTE.

The newly formed zygote is initially quiescent, but soon elongates and develops into a motile vermicule termed the OOKINETE. The ookinete grows in size, progresses by means of a pseudopod, and internally develops a crystalloid body thought to provide the material that ultimately forms a cyst wall. Electron microscope studies have revealed that the ookinete liquefies the brush border and cell membrane of a midgut epithelial cell in the mosquito. After crossing the epithelial layer the ookinete settles between the outer region of the epithelium and a musculoelastic layer. The ookinete now becomes spherical, secretes a cyst wall, and is known as an OOCYST (Fig. 10-14). It grows from about 6 microns in diameter to as much as 8 microns at maturity, projecting into the mosquito's hemocoel. Meiotic and mitotic divisions follow to form a large number of haploid SPOROZOITES. Sporozoites emerge through the wall of the oocyst, often weakening it so that it bursts. A mature oocyst (probably *P. falciparum*) from *Anopheles funestus* contained some 9,500 sporozoites. The number of oocysts on the midgut of a mosquito may range from a few to more than

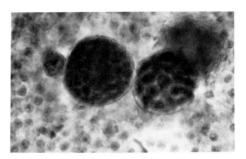

Fig. 10-14 Oocysts of a bird malaria, *Plasmodium gallinaceum*, in midgut wall of *Aedes aegypti*, 200 ×. (Courtesy of R. A. Ward, Smithsonian Institution.)

a hundred, and at optimum temperature maturation of oocysts of the fastest-developing species can be completed in 4 days. Sinden and Garnham (1973) have described the ultrastructure of several species of sporozoites.

Sporozoites invade all parts of the mosquito's body, and most eventually die, but those reaching the salivary gands penetrate and accumulate in the acinal cells (cf. Sterling *et al.*, 1973). When the mosquito feeds, the sporozoites pass into the salivary ducts and thence enter the blood stream of the vertebrate host. In one study naturally infected *An. funestus* and *An. gambiae* had an estimated range of a few hundred to nearly 70,000 sporozoites in the salivary glands. Dissection and microscopic examination of vector *Anopheles* salivary glands is a common epidemiological technique used in determining the proportion of infective vectors during malaria control campaigns.

The sporozoites introduced by a vector are carried by man's circulatory system to the next site of development. One or more cycles termed primary and secondary EX-ERYTHROCYTIC SCHIZOGONY (spore formation) occur in cells of the reticuloendothelial system of a vertebrate host.

The sporozoites that enter reticuloendothelial cells develop into CRYPTOZOITES, assuming a spheroid shape and vastly increasing in numbers through repeated divisions. Finally the parasites, as TROPHO-ZOITES, enter the blood circulation to invade erythrocytes and commence erythrocytic development. Only primary exerythrocytic schizogony occurs in *P. falciparum,* but in the three other human malarias and in most other malarias there is extensive secondary schizogony. The time elapsing from inoculation by the mosquito to appearance of infected erythrocytes varies with the species as well as with the strain and other factors. In *falciparum* malaria the period from blood feeding by an infective mosquito to fever and schizonts in the erythrocytes (the PREPATENT PERIOD), presumably the time for completion of primary schizogony, ranged in the Coker strain from 6 to 25 days, and in the Long strain from 9 to 10 days.

The blood phase is composed primarily of asexual ERYTHROCYTIC SCHIZOGONY. Fully developed erythrocytic schizonts disintegrate the red blood cells to release MEROZOITES that enter parasite-free erythrocytes to continue the blood cell cycle. Through a mechanism not understood, some GAMETOCYTES capable of undergoing sexual development if ingested by a mosquito are also produced. It is the rupture and release of merozoites with accompanying toxins, occurring at periods typical for the species of *Plasmodium,* that causes chills and fever cycles typifying human malarias. Hawking and coworkers (1968) conclude that periodicity is to ensure that mature and viable short-lived gametocytes in the blood will coincide with the activity periods of vector mosquitoes.

Kinds of Plasmodia. Garnham (1966) discusses the taxonomic relationships of *Plasmodium* and near relatives. They produce small numbers of microgametes from each microgametocyte, and each microgamete is like a flagellum in structure, gametocytes of both sexes that are of near equal size, oocysts that expand to form abundant sporozoites, and numerous merozoites that lack dimorphism. Malarial pigments formed from breakdown products of hemoglobin are another characteristic.

If parasites are present in the blood, they should be visible, after proper staining, on careful thin-smear microscopic examination. Pigmented intracorpuscular bodies in the form of signet rings, ameboid forms, segmenting forms, or as crescents are visible, dependent on the *Plasmodium* species. Various books on hematology, parasitology, or tropical disease are available for the microscopic diagnosis of malaria in man. Even in the presence of fever, it may require prolonged searching to find parasitized erythrocytes, and thick-smear preparations may be needed to increase the probability of finding parasites. There are current efforts to improve the search process by utilizing fluorescent antibody staining.

PLASMODIUM FALCIPARUM. This is the causal organism of aestivo-autumnal fever (malignant tertian, subtertian) of the tropics and subtropics; the most severe malaria and often fatal. It rates as the greatest cause of human death by any pathogen over most of Africa and elsewhere in the tropics, though as a result of eradication programs *falciparum* malaria has disappeared from around the Mediterranean, the Balkans, Taiwan, Puerto Rico, and the USA.

Malignant malaria has asserted prime historical importance in several instances. It has been associated with the decline of ancient Greek civilization, the termination of Alexander the Great's eastern expansion, and disruption of the Crusades. During World War II, particularly in Asia but also in the Mediterranean region, there were situations where *falciparum* malaria caused more military casualties than did enemy action, and civilian populations were affected even more severely.

In *falciparum* malaria there is considerable irregularity in the occurrence and duration of the febrile stage owing to irregularity in sporulation of the parasites. Schizogony usually requires about 48 hours, although often less. There is a decided tendency for schizogony to occur in deeper body tissues, mainly among the endothelial cells lining capillaries and sinuses. The assumption is that at a certain stage of growth infected corpuscles become sticky and clump together and adhere to the lining of internal vessels where blood circulation is slow because of toxic vasoconstriction. Merozoites of *falciparum* invade all stages of red blood cells and cause an abundant near-fatal level of parasitemia up to 500,000/mm^3 of blood, whereas *vivax* cannot invade mature erythrocytes and parasitemia is usually less than 20,000/mm^3 of blood.

Man is the only known natural host of *P. falciparum,* and if the human population infected has not been previously exposed to develop a degree of immunity, the parasites develop rapidly and mortality rates are high. Exposure for generations results in a selected human population that survives childhood infection with considerable natural immunity that is often restricted to a particular *falciparum* strain. In naturally resistant children spleen enlargement is less evident than with other malarias. The genetic sickle cell trait makes man a poor host for the parasites; thus, in regions of continuous attack by *falciparum* the heterozygous sickle cell trait is favored. The distribution of sickle cell gene in India, Africa, and southern Europe coincides with present or past high endemicity of *P. falciparum.*

BLACKWATER FEVER is a condition practially restricted to *falciparum* malaria inadequately treated with quinine. Administration of this drug to previously treated susceptibles may precipitate hemolysis that causes hemoglobinuria (hence the name blackwater), which is often fatal. This puzzling complication is virtually nonexistent now that quinine is seldom used.

Malignant malaria consists of a number of strains that vary in virulence and in their degree of adaptation to various *Anopheles.* Thus the Asiatic *An. stephensi* was found to be an efficient vector of a Nigerian strain of *falciparum,* though the European *An. labranchiae,* which readily supported European strains of the parasite, was a very poor carrier of this African strain. There is much other evidence of strain differences in *P. falciparum,* including development of specific types of drug resistance. Instances of collective strain resistance to all the common presently used plasmodicides are documented.

PLASMODIUM VIVAX. This pathogen causes tertian (benign tertian) fever of temperate climates, also abundant in the tropics and subtropics. There are regularly recurring paroxysms every 48 hours. The exerythrocytic stages of this pathogen persist in the liver (*falciparum* does not), and thus relapses occur. Because malaria eradication has been most successful in temperate regions, *vivax* malaria has practically disappeared from them and the highest incidence of this malaria remains in Asia. On the African continent its distribution is a puzzle, being common in North Africa but of low significance in much of tropical Africa. Although *P. vivax* is a characteristic temperate-zone parasite, it is not normally the species causing epidemics in the cold equatorial highlands.

Generally *P. vivax* infections do not cause death, though fatal infections occur in young children when associated with severe anemia. Chronic bad health is more typical, with SPLENOMEGALY (enlarged spleen). Spontaneous immunity that is strain-specific against erythrocytic forms has been known to arise after infection, and to last at least 3 years.

PLASMODIUM MALARIAE. This is the agent of quartan fever, with recurrent paroxysms because of sporulation every 72 hours. This malaria is much less common than, but coincides in distribution with, *falciparum* malaria. It is nearly as common as *falciparum* in parts of tropical Africa, particularly East Africa and Zaire (Congo), and in Burma and parts of India and Sri Lanka (Ceylon). Further eccentricities in its prevalence remain a mystery. It is long-lived, so that it tends to become a dominant surviving species during eradication.

Primary vectors of *P. malariae* are not known, for development has been poor in all *Anopheles* tested. Only a couple of oocysts may develop on mosquito stomachs, in marked contrast to experimental infections with *P. vivax* where hundreds of oocysts may occur. However, though the level is always low, *P. malariae* can develop in a wide range of *Anopheles* species, though only some six species have been found naturally infected (Garnham, 1966). One reason for low levels of mosquito infection appears to be the production of few gametocytes in man. Compared to other human malarias, development in the mosquito is slow, with sporozoites not reaching the salivary glands at 25° C until the sixteenth day.

Blood parasitemia builds slowly, often undetectable until well after quartan fever symptoms have been present for some time. There is prolonged survival of exerythrocytic stages, and recurrence of natural infections after 40 years without the possibility of new transmission suggests that a secondary exerythrocytic cycle has occurred in the liver following prolonged dormancy.

P. malariae has a wild animal reservoir, though it can be maintained in man in the absence of a wild host. Rodhain (1948) succeeded in demonstrating infection by this parasite in chimpanzees, with immunity characteristics identical to those found in humans. Thus the chimpanzee parasite known as *P. rodhaini* is synonymous. Chimpanzees are infected at low levels along the West African coast and into the interior of the Congolese Republic, but the vector to champanzees is unknown and might not bite man readily.

Symptoms of *malariae* malaria are rather distinct in humans. Chronicity with low blood parasitemia may leave the disease undiagnosed, but gradual spleen enlargement results in greater hypertrophy than is found with other malarias. Kidney involvement with observable edema is also frequent. The highest incidence of *P. malariae* infection occurs in children and steadily declines with age accompanied by reduced blood parasitemia, and strong immunity is prevalent in hyperendemic areas.

PLASMODIUM OVALE. This parasite causes a mild tertian fever. It is the rarest human malaria, though found throughout the tropics and subtropics and most prevalent on the west coast of Africa. Coatney (1971) presents evidence that this pathogen is a true zoonosis of nonhuman primates termed *Plasmodium schwetzi*. Prolonged latency is common, with the first attack several months after the feeding of an infectious vector. It may be present in mixed infections, though inapparent until the more virulent species is eliminated by drugs. Immunity in adults is commonplace, with evidence of cross strain resistance but no resistance to other human malarias.

Natural vectors of *P. ovale* have not been clearly incriminated. *Anopheles gambiae sensu lato* and *An. funestus* are probable transmitters in tropical Africa and there is generally poor development in all but African *Anopheles*. Sporozoites do not reach the mosquito salivary glands before 12 days.

Effects of Temperature on Plasmodia in the Mosquito. Despite the appearance of favorable conditions (i.e., presence of numerous anopheline mosquitoes contacting infected humans), malaria may not be transmitted in particular localities. An analysis of such conditions may reveal that the average temperature is low because the nights are cool, or because of prevailing cool fogs. It is generally agreed that malaria gametocytes cannot develop successfully within the mosquito host when the average temperature is below about 15°C. It is nevertheless of interest that malaria parasites have been observed to survive in mosquitoes for a few days at temperatures near −1°C. Knowles and Basu (1943) working with *An. stephensi,* found that the heaviest salivary gland infection of *P. vivax* occurred at 26°C. Using the same *Anopheles,* they found that at a temperature of 38°C no infection with any species of malaria was obtained. From such studies it is apparent that vectors must seek favorable environments. The sporozoite stage of *P. vivax* was reached in the salivary glands of *An. stephensi* in 18 days at 15°C, 15 days at 21°C, 11 days at 26°C, and 9 days at 32°C.

Hibernation of the anopheline host in temperate climates presents the problem of overwintering of the parasites. There is evidence that man and not the mosquito is the main winter season reservoir host, though anophelines overwintering in warm stables and houses may play an important but highly circumscribed role in malaria transmission. The same general circumstances must hold during hot and dry seasons when vector *Anopheles* aestivate.

Qualities of Effective Vectors (Principal Vectors). Conclusive experimental evidence shows that the plasmodia of human malaria and other mammalian malarias do not develop within culicine mosquitoes even though gametocytes are ingested. The reason for such specificity of *Anopheles* as vectors of mammalian malarias is by no means clear, as it has been shown that Malayan *Mansonia uniformis* can host *Plasmodium cynomolgi bastianelli* of monkeys through sporozoite development, but virtually no sporozoites establish in the salivary glands and the infection is not transmitted to uninfected monkeys (Warren *et al.,* 1962). Whether all anopheline species are infectible is unknown; many malariologists believe that laboratory tests suggest a matter of degree only, and no anopheline is completely refractory to the plasmodia of human malaria. Only some 85 of the approximately 400 known species and subspecies of *Anopheles* have been conclusively incriminated as vectors of human malaria. There is variation in infectibility among strains of the species of plasmodia. For example, *An. quadrimaculatus* is susceptible to both *P. vivax* and *P. falciparum; An. punctipennis,* though equally susceptible to *P. vivax,* seems refractory to certain strains but not to others of *P. falciparum.* Several North American *Anopheles* can develop native strains of *P. vivax,* but of these, when exotic *vivax* strains were tested, *An. albimanus* was nearly refractory.

An effective natural vector of human malaria is freely infectible by one or all species of human plasmodia, offers a favorable environment for development to the sporozoite stage in the salivary glands, feeds readily on man and less frequently on other vertebrates, breeds abundantly, is a house invader, and lives long enough to take human blood repeatedly. High populations often occur in response to man's activities, as when *An. albimanus,* a sun-loving species, flourishes where man destroys forests and establishes agriculture. MacDonald (1957) stresses vector longevity as a very significant factor. For example, *An. culicifacies* of India is short-lived and a poor vector, whereas the long-lived *An. gambiae sensu lato* of Africa is a very important vector. The maintenance of close and constant contact between an anthropophilous *Anopheles* and its source of food supply is an important ingredient of endemicity. Species sanitation based on a knowledge of the factors referred to above avoids wasteful control measures against anophelines that may be common, even abundant, yet of little importance as vectors for one or more reasons.

Vectors of Human Plasmodia. The causal plasmodia of human malaria are nearly always transmitted from man to man solely through the agency of *Anopheles* mosquitoes. Exceptions include infectious blood transfusions, blood contamination connected with drug addiction, and rare transplacental infections. Russell and associates (1943) provide identification keys to 240 species and 78 subspecies of anophelines of the world. Habitat for main species may be consulted in Horsfall (1955). Ecology, bionomics, and behavior of vector *Anopheles* have been reviewed for Central and South America (Elliott, 1969), the Pacific and western Pacific (Chow, 1969, 1970), the Oriental region (Scanlon *et al.,* 1968), the eastern Mediterranean region (Zahar, 1974) and Africa (Fritz, 1972). A recent listing of important vectors on a worldwide basis is provided in a pending WHO publication. This listing, based on various authorities, is shown in Table 10-1.

Epidemiology of Malaria. Malaria remains the major debilitating parasitic disease of the tropics. So much has been done to control or eradicate this problem that a specific vocabulary and specific epidemiological theories have been developed. All aspects of the study of malaria, including advanced epidemiological theories and simple terminology, may be included under the term MALARIOLOGY.

A number of books are available on malariological topics (see McDonald, 1957; Pampana, E. 1963; Swaroop, 1959; WHO, 1963). In addition the series of World Health Organization publications under the title World Health Organization, Expert Committee on Malaria, the WHO Tech. Rep, Ser. (numbered) is periodically published to pro-

TABLE 10-1 More Important Anopheles Vectors of Malaria*

1. **North America:**
   a. **Southeastern**—*quadrimaculatus*
   b. **Southwestern**—*freeborni*
   c. **Mexico**—*albimanus, aztecus, pseudopunctipennis.*
2. **Central America and West Indies:** *albimanus, aquasalis, bellator, pseudopunctipennis, punctimacula.*
3. **South America:** *albimanus* (Ecuador, Colombia, Venezuela), *albitarsis, aquasalis, bellator, cruzii, darlingi, nuneztovari* (Northern), *pseudopunctipennis* (Northern and Western), *punctimacula.*
4. **North European and Asiatic:** *atroparvus, messeae, pattoni* (northern China), *sacharovi, sinensis* (southern China).
5. **Mediterranean—Southern Europe; Morocco, Algeria, Tunisia; Through the Levant to the Sea of Aral:** *atroparvus* (Spain, Portugal), *claviger, dthali, labranchiae, messeae, pulcherrimus, sacharovi.*
6. **Desert—North Africa and Arabia:** *hispaniola, multicolor, pharoensis, sergentii.*
7. **Ethiopian:**
   a. **African**—*dthali, funestus, gambiae* (sp. A, B),† *melas* (West Coast), *merus* (East Coast), *moucheti, nili, pharoensis.*
   b. **Yemen**—*culicifacies, gambiae, sergentii.*
8. **Indo-Persian—Iraq, Oman, Persia, Afghanistan, Pakistan, India, Sri Lanka:** *annularis, culicifacies, dthali, fluviatilis, hyrcanus, minimus, philippinensis, pulcherrimus, stephensi, sundaicus, superpictus, varuna.*
9. **Indochinese Hill Zone—Foothills of Himalayas to Hills of S. China, Burma, Thailand, and Indo-China:** *annularis, balabacensis,‡ maculatus, minimus.*
10. **Malaysian—Malaya, Indonesia, Borneo, Philippines, Coastal Plains from S. China to Bengal:** *aconitus, balabacensis,‡ campestris, donaldi, flavirostris* (Philippines), *letifer, leucosphyrus, maculatus, minimus, philippinensis, sinensis, subpictus, sundaicus.*
11. **Chinese—Central China, Korea, Japan:** *lesteri, pattoni, sacharovi, sinensis.*
12. **Australasian:** *bancrofti, farauti, karwari, koliensis, punctulatus, subpictus.*

*Courtesy of L. E. Rozeboom, professor emeritus, Johns Hopkins University.
†*Anopheles gambiae* species A is regarded as true *Anopheles gambiae; Anopheles gambiae* species B as *Anopheles arabiensis* (Mattingly, 1977).
‡For Thailand, and probably most of mainland Southeast Asia north of 8° latitude, the correct name for *balabacensis* is *Anopheles dirus* (Peyton and Harrison, 1979).

vide current viewpoints on the status of malaria. Certain terms recognized by WHO are included here as a minimal background on malariology.

Malaria occurs in relatively stable and unstable conditions. In STABLE MALARIA there is a high rate of transmission to the affected population, no marked fluctuations occur over the years, and a high collective human immunity makes epidemics unlikely. In UNSTABLE MALARIA there are marked seasonal or other fluctuations, and immunity of the population is often low. Where there is some measurable incidence of the disease continuously present in a given area, it is said to be ENDEMIC, and various degrees of endemicity (such as percentage of spleen involvement in children 2 to 9 years old) are recognized by the prefixes hypo, meso, hyper, and holoendemic. An epidemiological investigation of malaria in the broad sense involves the study of environmental, organismal, ecological, and other factors determining its incidence. The objectives may be to ascertain the origin and means of acquisition of current malaria cases; the existence and nature of malaria foci; reasons for seasonal fluctuations in number of cases; evaluate the success of control or eradication programs.

Female *Anopheles* are examined for the presence of malaria parasites; those that have oocysts are termed INFECTED (pathogen present) and those with sporozoites in the salivary glands are classified INFECTIVE (pathogen transmittable at next blood meal).

Newer methods have been employed to detect active or past infections by IMMUNOFLUORESCENCE TESTS (example in Ambroise-Thomas *et al.*, 1974, document WHO/MAL/74.834; Bruce-Chwatt *et al.*, 1973, document WHO/MAL/73.816). SUPERINFECTION refers to more than one infection occurring because of fresh transmission and additional extraerythrocytic development of cryptozoites in liver cells, or to mixed infections with more than one *Plas-*

modium species or distinctly separate strains of the same species.

Vector control and chemotherapy are the main means of reducing transmission. Vector control, of greatest importance to medical entomologists, is dealt with at the end of this chapter.

The presence of *Anopheles* mosquitoes without the presence of malaria, in the absence of concentrated control efforts, is explainable on the basis of (1) lack of suitable vectors, as defined previously; (2) vector density below a level sufficient to maintain the disease; (3) lack of infected humans; (4) climatic conditions unfavorable to the maintenance of the parasite at an infective level (even though a sizable population of vector mosquitoes may be present); (5) lack of contacts between humans and vectors, and (6) a high level of human immunity.

Sir Ronald Ross believed that mathematical analysis of the dynamics of malaria transmission was possible, and he developed differential equations to explain conditions that influenced the proportion of infected people in a population, and conditions describing vector density in a locality. But there were difficulties in applying these equations, and those of other investigators who followed him, to real situations, and so MacDonald (Bruce-Chwatt and Glanville, 1973) contributed further to the theoretical base for the quantitative epidemiology of malaria (Bruce-Chwatt, 1976). MacDonald recognized that high levels of transmission led to partial immunity expressed as lower gametocyte production, particularly in the older segment of a human population, thus leading to equilibrium in which the infective reservoir (man) and the transmission rate fell below potential maxima. Further tests of MacDonald's models in real transmission situations revealed that all factors were still not taken into account. Dietz and fellow workers have expanded these analyses to demonstrate that immunity and superinfections in the human population alter the expected degree of success during control campaigns (cf. Dietz *et al.*, 1974). Further models on transmission dynamics are provided by Pull and Grab (1974), and Rao and colleagues (1974).

The following activities are likely to be engaged in by entomologists during epidemiological studies of malaria: *field investigations* that include biting collections, indoor collections of resting anophelines, trap collections of anophelines exiting from houses, collections of resting anophelines outdoors, and studies on larval bionomics; *laboratory procedures* that include identifying anophelines, dissecting to determine rates of sporozoite infections, determining parous conditions in female anophelines, calculating the age composition of vector populations, testing the level of vector susceptibility to insecticides, and the human blood infective index (Venters *et al.*, 1974).

Malaria Control and Eradication. Prior to the end of World War II malariologists were content to hold onto gains and to reduce malaria, particularly in highly populated areas, to a low level. The appearance of DDT and other potent residual insecticides, new antimalaria drugs such as chloroquine and primaquine, and advances in our knowledge of the bionomics of anophelines led to a concept of eradication. This does not mean eradication of the mosquito vectors. If vectors are reduced below the critical level of density so that transmission of the human parasite is eliminated for a period of around 3 years, the two major malaria pathogens will disappear spontaneously from the human host. If reintroduction can be prevented, an extensive area such as the continental USA or major areas in Europe can be freed completely from the disease. Thus, even though we cannot eradicate the mosquitoes involved in transmission, "anophelism without malaria" may persist.

On a worldwide basis this concept is still far from being realized, for in 1955 there were an estimated 200 million cases of ma-

laria in the world, with 2 million deaths (Russell, 1958), and the situation was little changed in 1975 (WHO sources). Technical difficulties stand in the way of malaria eradication from many parts of Asia, Africa, and South or Central America. Rao (1974) reviews problems in India. For example, nomadism, isolation of populations, frequent replacement of human dwellings in jungle cultures, inflexibility of monetary and technical resources, disruption of control procedures during times of political instability—all present problems for the achievable goal of malaria eradication. Vector response by behavioral changes and insecticide resistance are further serious problems. Technical problems and lack of insufficient resources require that in many areas of the underdeveloped world we must presently be content with holding down active transmission to the lowest level possible (Lepeš, 1974). As Bruce-Chwatt states (1976), "we have come to realize that in the developing countries of the tropics endemic malaria is one of the parts of the whole Gordian Knot of socioeconomic advance in conditions of poverty aggravated by population pressure."

Russell (1958) sums up effectively the steps that permit malaria eradication, symbolically expressed in Figure 10-15, and a planning manual has been prepared for eradication and control programs (WHO, 1972a). First there is a *preparatory phase*, including an initial survey, planning, and preliminary operations. For survey methods consult MacDonald (1957). This preparatory phase is followed by the *attack phase*, with total spraying coverage until malaria transmission has ceased and until the parasite has been virtually eliminated from the reservoir host. For *P. vivax* this means 3 to 5 years, for *P. falciparum* 2.5 to 3 years. The *consolidation phase* begins with the wiping out of any residual pockets that may remain. This is the difficult phase; discovery of isolated cases and the application of antimalarial drugs takes first priority. In the *maintenance phase*

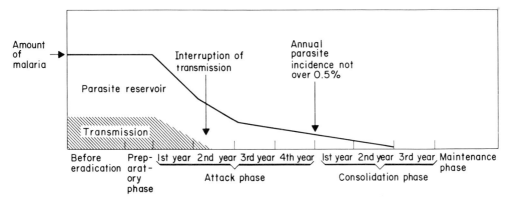

Fig. 10-15 Diagram of malaria eradication. In the attack phase the mosquito population is lowered so that transmission ceases; thereafter annual parasite incidence shows a slow decline. (Courtesy of WHO.)

surveillance is continued over the entire area. Local health agencies are on the alert for the reintroduction of malaria, which has become an exotic disease in temperate areas of the world.

Where malaria was already on the decline in the improving socioeconomic conditions of the developed world, mainly within temperate climates, eradication was spectacularly achieved. An important technique was the residual treatment of walls to decrease longevity of endophilous vectors of malaria. This method also succeeded in many underdeveloped countries in the tropics, but resistance to the main residual insecticides has caused serious problems. The search for substitute residual insecticides has been difficult and costly. Consolidation phase efforts to find active cases and treat them with antimalarial drugs has been expensive, and drug resistance in *P. falciparum* has incurred further problems. It is now apparent that concentrated efforts involving all possible avenues of attack against the vectors and the pathogens will be required to achieve eradication in the remaining malaria-ridden areas of the world. This must employ methods to prevent larval breeding of *Anopheles* (WHO, 1973a), pesticide applications to destroy principal vector adults, prophylaxis and cure with antimalarial drugs. A very promising

additional method that has the potential for a real breakthrough in malaria control involves activating the immune mechanism in man (Maugh, 1977; WHO, 1975e), possibly by the development of satisfactory vaccines. A particularly inclusive account of viewpoints on control and eradication is provided by Brown and associates (1976).

Malaria Chemotherapy. Antimalarial drugs have been developed for (1) curing clinical symptoms of the disease, and (2) reduction of malaria transmission. The first medicine known (about 1633) for curative powers was cinchona, the bark of the cinchona tree indigenous to certain regions of South America. Much later the chief alkaloid of cinchona, *quinine,* was identified and used extensively in malaria therapy. Inaccessibility of quinine sources during World Wars I and II led to an intensified search for synthetic antimalarials; candidate compounds were first tested on avian and rodent malarias. Human malarias have been adapted to certain monkey hosts for chemotherapeutic testing, and *P. falciparum* in the owl monkey (*Aotus*) has been an especially useful system. The achievement of culture techniques for several stages of malarias promises to provide more rapid initial testing of antimalarials. Very effective synthetics have been developed; those chiefly used for malaria therapy are chloroquine, pyrimethamine, primaquine, sulfadiazine and tetracycline. Resistant cases may require quinine or combinations of these synthetics. Drug action may be referred to by the stage of the malaria parasite affected (e.g., sporontocidal, gametocidal, etc.). An overall discussion of malaria chemotherapy may be found in a review by Rollo (in Goodman and Gilman, 1975).

The aim of chemotherapy is to disrupt the malaria cycle within the human host at any of a number of susceptible stages. However, extensive and indiscriminate use of antimalarials in control programs has selected out resistant strains of plasmodia. The World Health Organization indicates that resistance to chloroquine occurs in sections of all the major continents and some island groups where *P. falciparum* occurs. However, an intensive screening program at the Walter Reed Army Institute of Research, US Army research and Development Command, had by 1977 screened some 300,000 chemicals, and a number are promising candidates for clinical trials (Maugh, 1977).

MALARIAS OF APES AND MONKEYS

A number of malarias of higher apes and Old World and New World monkeys can also develop in man. Malaysia contains more species of primate malarias than any other region, suggesting it is the original site of origin for these pathogens (Coatney, 1971). Mattingly (1973) believes that the abundant plasmodia fauna of southern Asia might also represent the appearance of a recent evolutionary expansion, and he cites arguments that African primate stock may have provided the precursor from which the human-infesting *Plasmodium falciparum* developed, while *Plasmodium vivax,* or its precursor, seems likely to have been acquired from simian hosts in Southeast Asia. Garnham (1963) has outlined the possible evolution of primates and their malaria parasites to ac-

count for susceptibility of man to malarias of apes and monkeys. Suffice it to say that intersusceptibility raises questions concerning the possible natural occurrence of simian malaria in man, undetected as unusual because of clinical and morphological similarities to known human malarias.

Coatney (1968, 1971) and Coatney and coauthors (1971) provide a history of the discovery of transmissibility of simian malarias to man, and discuss all primate malarias. The natural *Anopheles* vectors of wild primate malarias are incompletely known, but there is sufficient evidence to show that some of these also serve as natural vectors of human malaria, and experimentally a number of *Anopheles* are capable of transmitting various monkey and ape plasmodia. A worldwide review of simian malaria vectors has been provided by Warren and Wharton (1963), and Bray and Garnham (1964) provide a table of proven and suspected vectors of *Plasmodium* in the wild state and by laboratory transmission.

Five species of simian malaria known to infect humans will be mentioned. The human-infecting species, *P. malariae,* which also occurs in chimpanzees, and zoonotic relationships of *P. ovale,* have been mentioned under the discussion of these pathogens.

Plasmodium inui is the most widely distributed malaria of Old World monkeys, occurring in a number of *Macaca* spp. from Taiwan in the east to possibly India and Pakistan in the west, with island and mainland distribution intervening. Various African monkeys and man are experimentally susceptible.

Plasmodium cynomolgi occurs in *Macaca* monkeys in Malaya, the East Indies, and India, and has been described also from *Presbytis* monkeys. Man is slightly susceptible, with increased virulence after further passage, and though natural infections may occur, these may be diagnosed as *P. vivax.*

Plasmodium knowlesi of monkeys (particularly *Macaca irus*) in Malayan jungle and swamp forests, extends to the Philippines with a subspecies described from Formosa. More widespread natural distribution is suspected. Many species of monkeys can be experimentally infected, and it is generally fatal to the rhesus, *Macaca mulatta*. Humans are susceptible, and this parasite has been used for fever therapy of certain diseases, though continued passage in man increases virulence. A human case of *P. knowlesi* was naturally acquired in Malaya where *Anopheles letifer* and *An. balabacensis introlatus* feed on man and monkeys.

New World primate malarias are thought to be derived from anthroponoses that appeared after 1492 when *P. malariae* and *P. vivax* were introduced from Europe (cf. Coatney, 1971; Wood, 1975). According to this view, the simian *P. brasilianum* was derived from *P. malariae*, and *P. simium* from *P. vivax*. Evidence for earlier lack of malaria in the New World includes the point that there are only these two monkey malarias in this region, and related *Hepatocystis* that are common in Asia and Africa in non-human primates are entirely lacking in the Neotropics.

Plasmodium brasilianum occurs widely in Panama and perhaps most of tropical South America, infesting some 16 species of monkeys in nature. This parasite was easily transmitted to man by the bite of *Anopheles freeborni,* but no adaptation to the human host was observed.

Plasmodium simium is apparently limited to howler monkeys (*Alouatta* spp.) in the forests of southern Brazil. It caused a case of tertian fever in a human, probably transmitted naturally by the bite of *Anopheles cruzii.*

OTHER MALARIAS AND RELATED PATHOGENS

All members of the genus *Plasmodium* are mosquito-transmitted, except for a reptile *Plasmodium* transmitted by phlebotomine flies (cf. Ayala, 1973). Several genera of mosquitoes may be involved, though transmission of mammalian malarias is limited to *Anopheles*. Mammalian hosts include primates, buffaloes, antelopes, bats and other insectivores, and a variety of rodents; many wild and domestic birds serve as hosts, as do reptiles, especially lizards. Closely related blood parasites, not in the genus *Plasmodium,* are also found in many kinds of vertebrates. Frequently the characteristics of these other parasites are known only from studying various stages in the vertebrate host. The vectors, when known, have proved to be bloodsucking flies other than mosquitoes. Clearly much remains to be done in studying these interesting parasites.

Mosquito-Borne Filariases

Infection of vertebrates with filarial nematodes and the transmission of the immature forms by biting arthropods raise many evolutionary questions. The range of relationships includes a simple mechanical transfer on the surface of higher flies (see *Stephanofilaria*) to a more complex developmental cycle within the vector, and release into the host during the act of feeding. Vectors are often mosquitoes and other nematocerous biting flies, but other bloodsucking arthropods, such as brachycerous flies, fleas, ticks and mites, may serve as hosts and transmitters for various filariae. The subject of filarioid nematode transmission is reviewed by Lavoipierre (1958) and Hawking and Worms (1961), with epidemiological aspects covered by Edeson and Wilson (1964).

Human filarial parasites transmitted by mosquitoes are *Wuchereria bancrofti*, known only from man, and *Brugia malayi* (=former *Wuchereria malayi*) of man and other mammals. Strain variants are known, and likely account for records of new filariae (see David and Edeson, 1965). The vertebrate hosts of various mosquito-transmitted filariae include wild and domestic carnivorous and herbivorous mammals, insectivores, birds, monkeys, frogs, and lizards.

W. bancrofti probably originated in Southeast Asia (Hawking, 1974), spreading eastward to adapt in the Pacific Islands to daybiting *Aedes* mosquitoes; northward to Japan and China; westward to India, which it reached more than 2,000 years ago; by dhow traffic to East Africa and adaptation to *Anopheles* vectors; to the Americas by slave trade after 1500. *Brugia malayi* also probably originated in Southeast Asia, but its spread was more limited. *W. bancrofti* established in Charleston, South Carolina, USA, and disappeared around World War II. Current increases in bancroftian filariasis are associated mainly with urbanization, breakdown in sanitary services and attendant increases of *Culex pipiens quinquefasciatus,* movements of people to and from endemic areas, and dam impoundments that increase populations of *Anopheles* vectors (WHO, 1974a).

The development cycle of mosquito-transmitted filariae in the vectors is similar. Microfilariae of human-infesting species, each in a saclike sheath (Fig. 10-16), are ingested from the peripheral blood during the act of feeding, shed their sheaths, and migrate through the walls of the midgut, apparently with the aid of an anterior hook. Migration through the midgut wall can occur in minutes, though live and possibly unsuccessful filariae may be found in the mosquito gut up to 4 days after ingestion. Practically all microfilariae passing through the gut migrate into the thorax within 12 hours. Development of filariae takes place within the large indirect flight muscles of the thorax, where the larvae become slightly shorter and much thicker as a "sausage" form. In *Dirofilaria immitis* of canines the microfilariae undergo development in the Malpighian tubes rather than in the thoracic muscles.

A number of internal changes and two molts occur during development in vector tissues, ultimately resulting in infective third-stage larvae. These changes are not accompanied by an increase in numbers. The infective larvae migrate with little difficulty

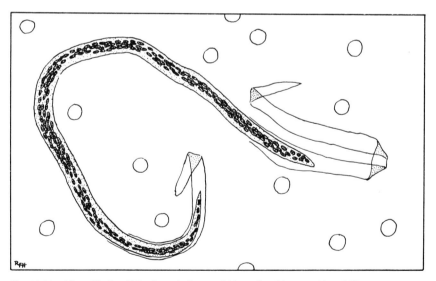

Fig. 10-16 Microfilaria of *Wuchereria bancrofti* in stained human blood film preparation. Sheath has broken and is pulling away from ends. Presence or absence of sheath, and position and numbers of nuclei in the tip of the tail (thinner and lower end), are used in determining the identity of microfilariae.

to the proboscis, and when the mosquito feeds on a vertebrate these larvae emerge through the wall of the mosquito's labellum. Filariae leave during blood feeding and not during sugar feeding, apparently because of pressure of the labium as it is bent back during a blood meal (Zielke, 1976; Ho and Lavoipierrre, 1975). The filariae can only enter host skin where the mouth parts traversed it, and a small amount of fluid (hemolymph?) discharges onto the skin surface with the filariae and prevents their immediate desiccation (Zielke, 1973).

A study in Ceylon indicated *Cx. pipiens quinquefasciatus* can transmit *W. bancrofti* by the fourth blood meal. Infective larvae enter the vertebrate's lymphstream and develop to sexual maturity in the lymphatics to produce new microfilariae that circulate in the peripheral blood. Bloodstream infection of children less than 5 years old is rare in bancroftian filariasis, presumably because there is a long development period before maturity is reached by these worms, and quite common in brugian filariasis.

Lymphatic filariasis is manifested in humans in a number of ways. In endemic areas a considerable proportion of the population is infected, as confirmed by blood examination, but otherwise lacks observable symptoms. An inflammatory or acute phase, characterized by fever and chills as well as swollen lymphatics, may occur months after exposure to infection. The acute phase is believed to represent allergic responses to products of the worms or to an accompanying bacterial infection. In obstructive filariasis (elephantiasis) there will be gross enlargement of scrotum, vulvae, breasts, legs, or arms due to blockage of lymph drainage, a consequence of years of exposure to infection by bancroftian filariae. In brugian filariasis obstructive symptoms are generally only a less obvious swelling of the legs and ankles. Fortunately elephantiasis is the exception in endemic areas.

In canine filariasis infections are often rel-atively asymptomatic and undetected for years. The adult worms live in the right ventricle and pulmonary artery of the dog. The domestic cat, and the fox and coyote have also been found infected, and there is a report of infection in captive California sea lions (Forrester *et al.*, 1973). With an infection of 20 to 50 adult worms in dogs, particularly working dogs, signs of cardiac insufficiency develop, particularly when the animals are exercised.

Periodicity in Filariasis. Manson observed in 1877–1878 that microfilariae are particularly prevalent in the peripheral blood at night. Since his observations it has been found that markedly PERIODIC forms and less periodic, or SUBPERIODIC, forms occur. Samples of peripheral blood should be taken at night in surveying for infestation where nocturnal periodism occurs. Undoubtedly the selective factor dictating periodicity has been the feeding period of vectors. Experimental evidence reveals that the migrating characteristics of microfilariae may be in response to certain daily physiological changes within the vertebrate host or to a self-propagating internal rhythm of the parasite.

Filariae-Mosquito Interactions. Interactions between the microfilariae ingested from a vertebrate, and their mosquito host, cause limitations on both organisms. There is a maximum number of microfilariae that may develop, and filarial development is detrimental to the mosquito. These interactions must be understood as important components of the epidemiology of filariases. The variety of interactions is summarized from observations on human and animal filariases.

Within the mosquito, limitations to filarial development are noted at several levels. The number of final-stage infective larvae that develops is fewer than the number of microfilariae ingested, and reduction occurs mainly at the time of passage through the gut wall, with the viability of microfilariae decreasing as the number ingested increases (Pichon *et al.*, 1974; Bain, 1971). Passage through the mosquito gut includes a period of time when the filariae remain in the epithelium before traversing the resistant basal membrane, and initial damage to the epithelium permits more ready access for other microfilariae (Bain and Brengues, 1972). Heavy buccopharyngeal armature in mosquitoes injures microfilariae during the ingestion of a blood meal (Coluzzi and Trabucchi, 1968), and rapid coagulation of ingested blood in some mosquito species traps microfilariae (Nayar and Sauerman, 1975). Hormonal relationships governing normal growth and reproduction in mosquitoes do not affect filarial development (Gwadz and Spielman, 1974), but as with malaria a low temperature threshold must be exceeded for satisfactory development of the parasites (Kutz and Dobson, 1974). Zielke (1975) has provided a quantified account of the migration of *W. bancrofti* in a mosquito, and stochastic models have indicated that limitations in filarial development occur either through host physiological mechanisms or by mutual inhibition of the microfilariae (Mougey and Bain, 1976).

Vector mosquitoes are harmed when large numbers of microfilariae start development. Damage includes mortality, tissue damage (flight muscles or Malpighian tubes), reduced vector flight ability, and reduced vector fecundity (examples in Husain and Kershaw, 1971; Javadian and MacDonald, 1974).

Even though completely unnatural filaria-arthropod development may be possible, for example, *Loa loa* and *Mansonia africana* (Ogunba, 1972), in natural vectors there is genetic control of suitability or refractoriness. Variation has been observed in natural conditions and can be selected in the laboratory (examples in McGreevy *et al.*, 1974; Paige and Craig, 1975).

MOSQUITO-BORNE HUMAN FILARIASES

Disease: Bancroftian filariasis, brugian filariasis (lymphatic filariasis, obstructive filariasis, elephantiasis).

Pathogen: Filarial nematodes, *Wuchereria bancrofti* and *Brugia malayi*.

Victims: Humans, exclusively (*W.b.*) or mainly (*B.m.*).

Vectors: Several species of night- or day-active mosquitoes; especially *Culex pipiens quinquefasciatus*.

Reservoirs: Bancroftian filariasis, man; Brugian filariasis, man and variety of mammals (subperiodic strain).

Distribution: Humid tropics and around the Mediterranean (Fig. 10-17).

Importance: At least 250 million cases worldwide. Debilitation continuously affects the economies of underdeveloped countries.

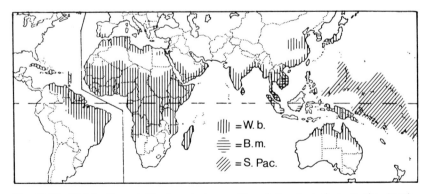

Fig. 10-17 Distribution of *Wuchereria bancrofti* and *Brugia malayi*. In the South Pacific mainly the subperiodic form of *W. bancrofti* is found. (Redrawn from several sources.)

Bancroftian Filariasis. *Wuchereria bancrofti* occurs in periodic and subperiodic forms. The periodic form, with marked nocturnal periodicity, is characteristic of humid tropics of the world and is transmitted by night-biting mosquitoes. This form also exists around the Mediterranean and in Turkey. The other form is diurnally subperiodic, restricted to Polynesia, and transmitted by mainly day-biting *Aedes*.

Bancroftian filariasis is largely an urban and suburban disease owing to the habits of its principal vector *Culex pipiens quinquefasciatus* (=*Cx. p. fatigans*). In rural areas the vectors are *Anopheles* species, or in forests and plantations of Southeast Asia forest-dwelling *Aedes* may be of local importance. Attempts to find wild vertebrate hosts for this filaria, and laboratory attempts to infect various animals, have been unsuccessful.

IMPORTANT VECTORS. Despite the listing of many mosquito species as being infectible, Edeson and Wilson (1964) make the distinction of "infective" (capable of infecting a vertebrate because of the presence of mature infective larvae at time of capture in nature) to derive a listing of vectors of bancroftian filariasis. Their listing may be referred to, but a more current list of important vectors (courtesy of L. E. Rozeboom, professor emeritus, Johns Hopkins University) is provided herewith.

Periodic Form. The vectors in tropico-politan (except Polynesia) and some subtropical areas are *Anopheles gambiae, An. funestus, An. darlingi, An. minimus flavirostris, An. campestris, An. punctulatus* group, *Aedes niveus, Ae. kochi, Ae. poicilius,* and *Culex pipiens quinquefasciatus.*

Subperiodic Form. The vectors in Polynesia and New Caledonia are *Aedes polynesiensis, Ae. tongae, Ae. pseudoscutellaris, Ae. fijiensis,* and *Ae. vigilax.*

Brugian Filariasis. This disease is caused by *Brugia malayi,* predominantly infects rural populations in the Far East between 75 and 140° E longitude, and occurs in small endemic foci. The periodic strain in man is transmitted by night-biting *Mansonia, Anopheles,* and an *Aedes,* and is the main strain of this parasite in man with only rare occurrence in other animals. The subperiodic strain is transmitted by *Mansonia* of swamp forests that feed at any time, with infections common in a variety of mammals. It may be difficult to separate *B. malayi* in mosquitoes from other species of mammal-infesting *Brugia* where these occur.

IMPORTANT VECTORS. Edeson and Wilson (1964) list the vectors of *B. malayi.* According to Rozeboom the more important vectors are:

Periodic Strain. The vectors in Japan, Coastal China, Taiwan, South Korea, Southeast Asia, India, and Indonesia are *Mansonia annulifera, Anopheles campestris,* and *An. donaldi.*

Subperiodic Strain. The vectors in the swamp-forests of Malaya and the Philippines (Palawan, Sulu, Mindanao) are *Mansonia dives, Ma. bonneae, Ma. annulata,* and *Ma. uniformis.*

OTHER BRUGIA FILARIAE. These may be difficult to distinguish. *Brugia pahangi* naturally infests a variety of mammals in various locations in Malaya, and has been experimentally transmitted to man. *B. patei* occurs as a natural infection of domestic dogs and cats in East Africa; also in genet cats (*Genetta tigrina*), bush baby (*Galago*), and probably other mammals (Nelson *et al.,* 1962). *B. buckleyi* of hares and *B. ceylonensis* of dogs occur in Ceylon (Jayawardene, 1963).

Control of Human Filariasis. The control of mosquito-borne filariasis has been studied extensively under the auspices of the World Health Organization and various regional organizations. Natural declines of bancroftian filariasis have been associated with the reduced human-vector contacts, namely, by improved (screened) housing, enclosed water systems, proper sewage disposal, and better drainage of agricultural lands (WHO, 1974). Control may be approached by reducing the vector populations to low levels, and for this purpose control of *Cx. p. quinquefasciatus,* the main urban vector, has received particular attention in bancroftian filariasis. But refer to discussion of this mosquito to note that it has been increasing worldwide. Where *Anopheles* are the main vectors in the Solomon Islands, a theoretical model reveals that the degree of vector control required to stop transmission is less than that needed for similar reductions in malaria (Webber, 1975). In India Bhatia and Wattal (1958) concluded transmission by *Cx. p. quinquefasciatus* does not occur where biting density is below 3.4 per man hour in seasons when climate favors development of the parasite (i.e., temperature over 26°C and high humidity, according to Hawking, 1974). In Calcutta, India, bites by the urban vector were estimated at 115,000 per person annually for exposed people (Gubler and Bhattacharya, 1974).

Webber (1975) discusses the theory behind vector control as an approach to controlling the disease. Chemotherapy may be practiced for radical cure and to reduce the number of microfilariae in peripheral blood to a subinfective level. Reduced numbers of microfilariae also diminish the likelihood of development of obstructive filariasis. Surgery may be resorted to in advanced obstructive filariasis to improve lymphatic drainage.

Chemotherapy is reviewed by Rollo (in Goodman and Gilman, 1975). The drug commonly used is orally administered diethylcarbamazine, which is much more lethal to microfilariae than to adult worms, and can be used for prophylaxis in endemic areas. Resistance to this compound has not appeared because of its low lethality for adult worms. Kessel found that in Tahiti this medication administered for 4 years reduced acute bancroftian filariasis by 84 percent, and in a follow-up study (Kessel, 1967) not only was the proportion of positives reduced, but the microfilariae level in blood became so low

that transmission by vectors seemed much less a possibility. Comparable success has been realized in Samoa (Suzuki and Sone, 1975). A similar program against brugian filariasis in Malaya failed, probably because the areas controlled were too small and there was an untreated reservoir in domestic and forest animals. In parts of South America, where urban filariasis transmission by *Cx. p. quinquefasciatus* occurs, it has been necessary to combine vector control with chemotherapy (Kuhlow, 1971).

DIROFILARIASIS OF DOGS (HEARTWORM DISEASE)

Filariasis of dogs, caused by *Dirofilaria* species, presents a major veterinary problem in much of the world (cf. Bradley, 1972). *Dirofilaria immitis,* a cosmopolitan species of nearly all tropical and subtropical regions, has become prevalent in much of the USA, with more than 40 percent infestation in some midwestern and eastern states (AHS, 1975). This species occurs in dogs, and occasionally in cats and other carnivores. Adult worms reside in the heart and pulmonary arteries of the host where they often form tangled knots that may cause death by embolism, asphyxia, and dilation of the heart. Ludlam and associates (1970) list some 60 anopheline and culicine mosquitoes as potential vectors of this filaria, but as Weinmann and Garcia (1974) note, the vector species in any given location can only be determined by careful analysis of locally acquired cases of the disease, and a knowledge of the distribution and behavior of potential local vectors. *Dirofilaria repens* also affects dogs and is transmitted by *Aedes* and *Anopheles*.

Numbers of microfilariae in the blood do not necessarily relate to severity of the disease. Concentration techniques may be required to verify that *Dirofilaria* are present, and careful microscopic identification of microfilariae is necessary to differentiate from *Dipetalonema reconditum,* a quite similar flea-transmitted filaria that is not considered very pathogenic. In highly endemic areas, keeping dogs indoors, or otherwise protected from mosquito bites, will reduce the incidence of infection. Mosquito control can seldom be applied effectively on individual premises except for those mosquitoes that may be breeding in home-associated water collections. The destruction of adult heartworms by a series of injections of an arsenical compound, and later administration of a compound against microfilariae in the blood, is the usual practice. However, dead parasites can cause occlusion of pulmonary vessels, leading to death or serious complications.

DIROFILARIASIS OF MAN

The filariae of *D. immitis* may reach maturity in humans, to be discovered, at autopsy, in heart and adjacent vessels, or dead and folded in a pulmonary artery when lung sections are removed surgically on suspicion of cancer. They cannot reproduce in this host. This problem is likely more widespread than is realized, particularly in the southeastern USA (Nearfie and Piggott, 1971; see Weinmann and Garcia, 1974). Over 70 cases of such complications have been reported for the USA., and they are also known from other countries. In the southeastern USA there are records of subcutaneous nodules or abscesses from which *Dirofilaria* (probably *D. tenuis* of raccoon) have been removed (Beaver and Orihel, 1965).

DIROFILARIASIS OF OTHER VERTEBRATES

Other mosquito-transmitted *Dirofilaria* are listed by Hawking and Worms (1961) as *D. carynodes* and *D. magnilarvatum* of monkeys, *D. scapiceps* of cottontail rabbits, *D. subdermata* of porcupines (*Erethizon*). Further filariae of warm- and cold-blooded vertebrates, transmitted by mosquitoes, are listed in Hawking and Worms, 1961.

Viral Infections: General Considerations

About 90 viruses have been isolated from mosquitoes in nature, the majority of these from Culicinae. Those demonstrated as causing human infections under natural circumstances, as well as some of those found to be detrimental to animals other than man, are listed here. These have been selected from the 1975 *International Catalogue of Arboviruses Including Certain Other Viruses of Vertebrates* (Berge, 1975). The abbreviations used are from that source, and key references to original literature may also be found there. For the most part space limitations confine these listings to brief inclusion of the probable vectors, general clinical symptoms, and geographic location. More extensive discussion is provided for a few viruses recognized as causing widespread serious and frequent epidemics or epizootics.

Mosquito-Associated Viruses Isolated from Man in Nature: Togaviridae, Alphavirus (= Group A)

CHIKUNGUNYA (CHIK)

Transmitted mainly by *Aedes aegypti* and several other *Aedes;* occasionally *Mansonia* and *Culex* spp. Incubation period 3 to 12 days, high rise in temperature followed by severe pains in limbs and spine, rash. Apparently not dangerous but often associated with dengue and yellow fever (B viruses) because of common urban vector. Reservoir may be chimpanzees and monkeys. Occurs in much of Africa near equatorial belt and south; widespread in Southeast Asia.

EASTERN EQUINE ENCEPHALITIS VIRUS

Disease: Eastern equine encephalitis (EEE)
Pathogen: A virus; Togaviridae, *Alphavirus.*
Victims: Equines, humans, introduced game birds.

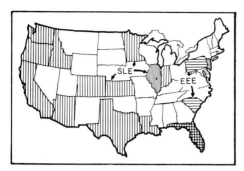

Fig. 10-18 Distribution by states of human cases of St. Louis encephalitis (*SLE*) and Eastern equine encephalitis (*EEE*) reported in the USA during 1968 and 1969. Closer lines indicate states with higher incidence during at least one of these two years. (Redrawn from *MMWR*, annual supplements.)

Vectors: *Culiseta melanura* and other bird-biting mosquitoes in enzootic cycles; *Aedes taeniorhynchus*, *Ae. sollicitans*, *Ae. vexans* mainly in equine and human outbreaks.

Reservoirs: Probably birds, possibly the mosquito *Cs. melanura*

Distribution: Equine and human cases known from eastern seaboard portion of Americas, Canada to Argentina, Caribbean (Fig. 10-18); virus isolated from enzootic situations in Southeast Asia and eastern Europe.

Importance: Irregular outbreaks of several hundred cases in horses, generally fewer than 100 human cases. A few equine and/or human cases usually identified each year.

EEE virus causes a severe and frequently fatal encephalitis in equines and man. First isolated from the brain of horses in 1933, it was identified as the agent causing an epidemic of man in Massachusetts in 1938. The virus has caused equine and human disease from the eastern seaboard of the USA to Argentina, and the Caribbean; it was isolated from sentinel hamsters in the Peruvian Amazon region. The first known Canadian cases involved horses in Quebec during 1972 (Harrison *et al.*, 1975). Isolations reported outside the New World from the Philippines, and especially from eastern Europe, have not been associated with human outbreaks or epizootics. Apparently in Europe this virus cir-

culates in bird-mosquito enzootic cycles involving *Culex pipiens* (Ising, 1975).

Encephalitis due to EEE virus mainly affects infants and children, generally with a sudden onset of high fever, vomiting, drowsiness or coma, twitching, and severe convulsions. Fatal cases usually terminate 3 to 5 days from onset of symptoms, though sometimes later from complications. Survivors under 5 years in age often have mental retardation, convulsions, and paralysis; those over 40 generally recover completely.

In the eastern USA, by far the most isolations have been from *Culiseta melanura* and *Cs. morsitans*, but the virus has also been associated with at least five species of *Culex*, four *Aedes*, an *Anopheles*, a *Mansonia* and other *Culiseta*. Infrequent isolations from biting gnats and poultry ectoparasites are of uncertain epidemiological significance. High density of *Culiseta melanura* populations seems to correlate well with epizootics of pheasants and horses, and antibody levels in swamp birds (Wallis *et al.*, 1974; Williams *et al.*, 1974). Development of EEE virus has been studied in the salivary glands of *Aedes triseriatus* (Fig. 10-19) (Whitfield *et al.*, 1971).

The bulk of developing evidence from investigations in the field suggests that EEE virus is maintained in North America in enzootic bird-mosquito cycles, mainly in swamp environments. The main enzootic vector is *Cs. melanura*, a freshwater swamp mosquito that rarely bites horses or man, and that rarely contacts peridomestic situations. Epidemics in man and equines occur when the amount of virus in swamp environments becomes high, and a high population of mosquitoes carries the virus to birds and mammals outside of the more contained environment. *Aedes taeniorhynchus* and *Ae. sollicitans* on the coast, and *Ae. vexans* inland, are probably the main vectors during outbreaks in horses and man.

Birds of many types frequently have high antibody titers, and yield virus but show no

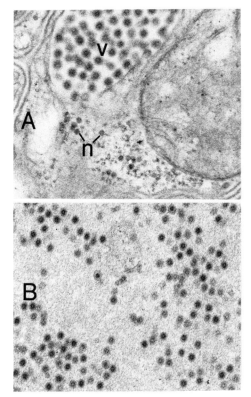

Fig. 10-19 Development of Eastern equine encephalitis virus in the mosquito *Aedes triseriatus.* *A.* Nucleocapsids and enveloped virus particles, 55,000 X. *B.* Large number of enveloped virus particles in lumen of salivary gland at 21 days, 43,000 X. (From Whitfield *et al.,* 1971. *Virology,* **43**:110–22)

clinical signs of disease, yet curiously the Chinese ring-neck pheasant, and probably other introduced game birds (see Ranck *et al.,* 1965) are very susceptible. The pheasant, when penned, may suffer devastating epizootics, apparently disseminated by feather picking, and such epizootics may coincide with encephalitis cases in horses and humans (MMWR, 1973, 22:283).

Major EEE outbreaks in the northeastern USA have been associated with excessive rainfall during summer and the preceding autumn, presumably causing increased production of mosquitoes (Hayes *et al.,* 1962; Hayes and Hess, 1964). Outbreaks in horses coincided in Massachusetts and New Jersey with those in man, were less consistently associated in Delaware and Maryland, and were only occasionally noted as related in southern states. An epizootic affecting horses in Cuba is reported to have been brought under control by combined quarantine, immunization, and mosquito control (Fernandez *et al.,* 1971).

EVERGLADES (EVE)

Vectors are mainly *Culex nigripalpus* and other *Culex* spp. in the subgenus *Melanoconion.* Endemic strain of Venezuelan equine encephalitis from south Florida, USA.

MAYARO (MAY)

Vectors are mainly *Haemagogus* spp. Possibly a variant of Venezuelan equine encephalitis; Bolivia, Brazil, Colombia, Surinam, Trinidad.

MUCAMBO (MUC)

Vectors are mainly *Culex portesi* and other *Culex.* Considered enzootic strain of Venezuelan equine encephalitis; Brazil, Trinidad, Surinam, French Guiana.

O'NYONG/NYONG (ONN)

Vectors are *Anopheles funestus* and *An. gambiae.* Causes fever, severe joint pains, backache and headache, rash, swollen lymphatics; not considered dangerous. Estimated two million persons affected 1959 to 1962 in Uganda, Kenya, and into Congo.

ROSS RIVER (RR)

Vectors are *Aedes vigilax* and *Culex annulirostris.* Caused more than 100 cases of epidemic polyarthritis in Queensland and New South Wales, Australia. A rodent reservoir is suspected (Gard *et al.,* 1973).

SINDBIS (SIN)

Vectors are mainly *Culex* spp. Generally mild human disease; strains from several locations on African continent, India, Philippines, Malaysia, Australia, eastern Europe.

VENEZUELAN EQUINE ENCEPHALITIS VIRUS

Disease: Venezuelan equine encephalitis (VEE)
Pathogen: A virus; Togaviridae, *Alphavirus*. Several epidemic and endemic strains.
Victims: Equines, humans.
Vectors: Many species of mosquitoes.
Reservoirs: Probably mainly rodents.
Distribution: Outbreaks in equines and humans from Peru to Texas; antibodies indicate more extensive range (Fig. 10-20).
Importance: Equine epizootics have caused losses of several thousand horses—up to 20 percent of local stocks—and been a serious problem for agricultural production; human clinical cases may number in thousands, with deaths in hundreds.

This virus causes widespread fatalities in horses, mules, and donkeys, and extensive epidemics in man. Because of the serious threat to the USA of a 1971 outbreak in Texas, a significant number of papers has been published in recent years. Publications by PAHO (1972), and USDA (1972, 1973) provide comprehensive reviews that are mainly utilized in this discussion.

It is theorized that VEE virus originated in humid swamplands of the Amazon and Orinoco rivers. Outbreaks of equine and human disease have occurred from 14° S latitude in Peru to 28° N in Texas, USA; antibodies have been detected at even further extremes. The first recognized outbreaks involved equines in Colombia in 1935. From that year through 1971 some 20 major outbreaks have been recorded, mainly involving horses and related equines. These have occurred in 11 countries, but have especially centered in Colombia and Venezuela. In the latter country during 1962–64 there were more than 23,000 human cases, 960 with neurological involvement, and 156 deaths. A 1969 outbreak in Ecuador was estimated to have caused the death of 27,000 equines, and 31,000 human cases including 310 deaths. Some 20 to 40 percent of equines have died in regions experiencing epizootics, and about 40 to 80 percent fatality has been noted among equines showing clinical symptoms.

In 1969 epizootics and human cases started in Central America. There had been isolated horse and human cases in southern Mexico in the 1962–1969 period, and in 1970 about 10,000 equine deaths occurred in an epizootic that reached the USA border in eastern Texas. Horse cases in Texas, USA, commenced in June, 1971 and resulted in the death of more than 1,500 equines and in 84 nonfatal human cases. Circulating VEE virus had been recognizd by serological studies in the absence of obvious disease in many locations, including high rates in Seminole and Miccosukee Indian residents north of the Florida Everglades in southern Florida. The virus was isolated from a human case in that region during 1968.

Clinical symptoms seen most frequently in horses are depression and stupor, high temperature, impaired vision, and incoordination. The animals do not eat well and a braced stance and nervous movements are typical. In man infection commonly resembles influenza, with severe headache, fever and weakness. Encephalitis followed by death or residual postrecovery effects are most characteristic in children.

VEE virus strains are recognized by serology, geographic location, and effect on equine or human hosts. These appear to be four types and eight subtypes (Fig. 10-20). Subtypes A, B, C of type I are epidemic and highly virulent for equines, also causing human disease. Subtypes D, E of type I are endemic strains that produce disease in humans and some rodents; infection in equines is nonclinical but results in protective antibodies against epidemic strains. Types II and III are distinct and nonpathogenic for equines. Type IV includes two serologically identical strains that are nonpathogenic for equines. The complexities of interactions suggest that epizootic strains invade areas when high vector populations

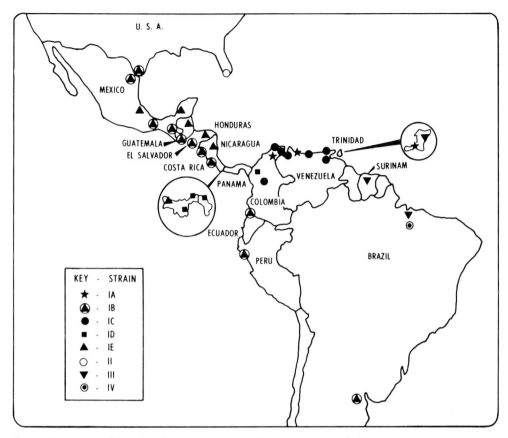

Fig. 10-20 Geographical distribution of Venezuelan equine encephalitis strains. (From USDA, *APHIS*, 1973.)

coincide with low levels of immunity in large susceptible hosts. Equines have a high enough viremia to allow equine-mosquito-equine transfer, and dogs may also have a sufficiently high viremia for this kind of passage (Bivin *et al.*, 1967). Enzootics involve rodents and mosquitoes, but sufficiently high titer is found in equines and some rodents to suggest they may infest their own kind directly through infective oral secretions or urine. During epidemics viruses or antibodies occur in rodents, carnivores, and many other mammals; in interepidemic periods and in endemic areas rodents appear to serve mainly as reservoirs (see Bigler *et al.*, 1974). Birds seldom have a high incidence

of VEE virus or antibodies. It is possible that under certain conditions enzootic strains of type I virus convert to epizootic strains causing outbreaks in equines accompanied by cases in humans.

Some 37 species of mosquitoes have been implicated as vectors of VEE virus; 8 *Aedes*, 4 *Anopheles*, 11 *Culex*, 2 *Haemagogus*, 7 *Psorophora*, and 1 each in *Deinocerites*, *Mansonia*, *Sabethes* and *Wyeomyia*. Isolations of the pathogen from mosquitoes in various regions have most frequently involved *Ae. serratus*, *Ae. taeniorhynchus*, *Ps. confinnis*, and *Ps. ferox*. A study of VEE virus development in *Ae. aegypti* shows many tissues are invaded, but the highest level of infection is in the salivary glands (Larsen and Ashley, 1971). A number of isolations from *Simulium* occurred during a 1967 outbreak in Colombia.

Control of VEE has mainly been attempted through quarantine and immunization of equines during epizootics, sometimes in concert with mosquito control programs. However, high viremia in equines before clinical symptoms are noted, coupled with abundant vector contacts, makes containment very difficult. Comprehensive immunization of equines before an epizootic invades a region should be beneficial. In the 1971 outbreak in Texas, USA, vaccine was administered extensively to horses, quarantines restricted equine movements into or out of the affected area, and some 5.5 million hectares (13.5 million acres) were treated in a short period for mosquito control. Main coverage was achieved by aerially applied ULV malathion or naled. Extensive surveys in the 2 following years for virus in mosquitoes, and VEE virus or antibodies in susceptible sentinel equines and wildlife, suggest that the epidemic form of Venezuelan equine encephalitis did not become established in the USA. This outbreak continued in Mexico during 1972, with sizable numbers of equines and humans affected.

WESTERN EQUINE ENCEPHALITIS VIRUS

Disease: Western equine encephalitis (WEE)
Pathogen: A virus; Togaviridae, *Alphavirus*.
Victims: Equines, humans.
Vectors: *Culex* and *Culiseta* mosquitoes; especially *Cx. tarsalis*
Reservoirs: Nesting birds (spring and summer); possibly reptiles and amphibians (winter).
Distribution: Outbreaks in equines and humans in western USA, north-central USA and adjacent Canada (Fig. 10-21); virus isolations and antibody studies indicate presence on eastern seaboard of North America and South America, eastern Europe.

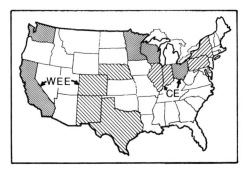

Fig. 10-21 Distribution by states of human cases of Western equine encephalitis (*WEE*) and California-type encephalitides (*CE*) reported in the USA during 1972 and 1973. Closer lines indicate states with higher incidence during at least one of these two years. (Redrawn from *MMWR*, annual supplements.)

Importance: Outbreaks have involved thousands of equines and hundreds of human clinical cases. Some incidence annually in horses and man in western USA.

WEE virus causes encephalitis, generally not fatal in man but causing high equine mortality. The virus was isolated in 1930 from horses with encephalitis, and later from wild mosquitoes by Hammon and associates (1941). Initially thought limited to the western USA, it has also been identified on the Eastern seaboard, and from Canada and eastern South America. Specific antibodies in man have been reported from Mexico and some eastern European locations. An outbreak in 1941 in the north central USA and adjacent Canada involved over 3,000 persons; endemic areas in the western USA have some human cases nearly every year. Specific antibody studies reveal that inapparent infections have occurred in up to 11 per cent of humans in endemic zones. Reeves and coworkers (1962) estimated the ratio of human infections to encephalitis is 58:1 in infants and children, and 1,150:1 in persons older than 15 years.

The active disease includes fever and drowsiness (hence sometimes called sleeping sickness), and convulsions in children.

Symptoms are virtually impossible to distinguish from other arbovirus encephalitides. Small children are most likely to have grave sequelae, such as convulsions and motor and behavioral problems. High postrecovery antibody levels are diagnostic, but virus is generally isolated only from the nervous tissues of fatal cases.

In western endemic areas *Culex tarsalis* has consistently been found infective during epidemic and interepidemic periods. Thomas (1963), studying development of the virus in this mosquito, concludes its importance is due to a high suitability for propagating and retaining WEE virus, and also to many ecological and behavioral factors. In eastern areas where *Cx. tarsalis* is rare, WEE virus has frequently been isolated from *Culiseta melanura* (Saugstad *et al.,* 1972). Isolations have also implicated several other *Culex, Culiseta inornata,* species of *Psorophora,* a number of *Aedes,* and an *Anopheles.* In Saskatchewan, Canada, McLintock and colleagues (1970) conclude that *Cx. tarsalis* is the main vector during WEE epidemics, *Culiseta inornata* may be an important vector to horses during epidemics, and several mosquitoes maintain the virus in an enzootic state.

The reservoir for WEE is of great interest. This virus has frequently been isolated from wild birds, such as sparrows and red-winged blackbird, and from domestic fowl; and birds have a prolonged viremia of high titer that infects feeding mosquitoes. In Hale County, Texas, nestling house sparrows (*Passer domesticus*) appear to be closely associated with epidemics in humans, obtaining virus earlier in the season and at higher levels of incidence during epidemic years (Holden *et al.,* 1973). Mammals, however, develop low titer viremia, so summer maintenance is mainly in *Cx. tarsalis* and birds. The virus has only rarely been isolated from overwintering *Cx. tarsalis* despite numerous attempts. Overwintering in vertebrates in colder northern zones is suggested from field and laboratory studies with cold-blooded vertebrates in Utah and Saskatchewan, where virus has been isolated from garter snakes (*Thamnophis* spp.) and the American leopard frog *Rana pipiens* (Burton *et al.,* 1966; Gebhardt *et al.,* 1973). A close variant of the pathogen has been isolated from *Oeciacus* bugs (Cimicidae) associated with cliff swallow nests.

Extreme water conditions are related to WEE epidemics. Drought can bring vectors and vertebrates into close proximity around existing water, or early spring flooding creates large vector populations and consequent increased vector-vertebrate contacts. The abortion of a potential epidemic in Kern County, California, was attributed to the combined effect of low transmission rates in early season because of a prolonged extrinsic incubation period in *Cx. tarsalis* due to low temperature, and successful mosquito control by midsummer in areas with a susceptible human population.

Several practices have been suggested to reduce the number of human and equine cases when conditions indicate epidemics are likely. Immunization is available for equines, and is a sound routine practice in areas with a history of equine WEE cases. Populations of *Cx. tarsalis* can be monitored by light trap or resting site collections, and if numbers are high and pools of this vector yield high rates of WEE virus, control programs may be considered. If nestling bird or sentinel flock isolations indicate high early-season transmission rates, this also is an indicator of the advisability of mosquito control programs. It has been suggested that selective zooprophylaxis by lowering the incidence of columbiform (pigeons and doves) and passerine (sparrows) birds locally and increasing the availability of attractive dead-end hosts, such as cattle and dogs, would reduce the numbers of mosquitoes biting man (Hess and Hayes, 1970). This type of zooprophylaxis should also reduce the proportion of infective bites received by humans around farm premises.

Mosquito-Associated Viruses Isolated from Man in Nature: Togaviridae, Flavivirus (= Group B)

BANZI (BAN)

Transmitted by *Culex rubinotus,* other *Culex,* and *Mansonia africana.* Fever, not serious; eastern and southern Africa.

BUSSUQUARA (BSQ)

Transmitted mainly by *Culex* spp. Fever, not serious; rodent reservoir suspected. Panama, Colombia, Brazil.

DENGUE (DEN) VIRUS

Disease: Dengue (DEN 1, 2, 3, 4), dengue hemorrhagic fever, dengue shock syndrome, epidemic hemorrhagic fever, breakbone fever.
Pathogen: A virus, Togaviridae, *Flavivirus;* four distinct serotypes.
Victims: Humans. Hemorrhagic and shock manifestations especially serious in infants and children.
Vectors: *Aedes* mosquitoes in subgenus *Stegomyia;* especially *Ae. aegypti.*
Reservoir: Continuous transmission by *Stegomyia* between humans, other mechanisms not substantiated.
Distribution: Coincides with *Aedes aegypti,* within about 40° N and S latitude. Currently mainly in Southeast Asia and Caribbean.
Importance: Outbreaks of debilitating but nonfatal disease may involve a half million to 2 million humans. Large urban centers in Asia, particularly Bangkok, experience frequent hemorrhagic outbreaks with notable mortality, especially in children.

There are in reality four distinct serotypes of DEN virus, infection providing prolonged homologous immunity and transient heterologous immunity. Clinical symptoms do not permit clear separation of dengue from other arboviruses. One characteristic is abnormalities in small blood vessels, with mottling of skin and rash in mild cases, or marked hemorrhagic complications, such as hemorrhagic cutaneous petechiae and bleeding in the lungs or digestive tract, in severe cases. Onset of clinical symptoms occurs after a usual incubation period of 5 to 8 days. In uncomplicated dengue there is fever and severe headaches, backache, and pains in muscles and joints. Weakness and prostration are common. Hemorrhagic manifestations with significant mortality have been especially characteristic in indigenous Asian children 3 to 6 years old since World War II. The severe hemorrhagic manifestations are referred to as dengue hemorrhagic fever, and are believed to represent a hypersensitivity reaction to simultaneous or sequential infection by more than one dengue serotype.

Dengue was first thought of exclusively as a temporarily debilitating but seldom fatal disease. Denguelike disease was reported in the Caribbean in the seventeenth century, further expansion was noted in the eighteenth century, and in the nineteenth century large epidemics occurred throughout the tropics and subtropics. Seaports and coastal regions were particularly affected, though evidence of the disease has occurred 2,000 km up the Amazon river, and inland for 300 km in the southern USA. Halstead (1966) provides an account of hemorrhagic outbreaks through 1964, and Brown (*in* Howe, 1977) thoroughly reviews the historical and current aspects of this disease and its hemorrhagic variant.

Since 1920 outbreaks involving 500,000 to 2 million cases have occurred in the USA, the Caribbean, Australia, Greece and the Mediterranean, and Japan. An epidemic that started around 1920 in the southern USA infected some 600,000 people in the coastal Texas region, and with adjacent areas to the east and northwest, plus the Caribbean, may have involved a total of some 2 million. Dengue is still endemic throughout the Caribbean, with rather sizable outbreaks recurring frequently, but the last continental US epidemic consisted of several hundred cases in Louisiana in 1945 (Hayes *et al.*, 1971). Cases of dengue enter the USA from the Caribbean area (MMWR, 1977, 26:255; 1978, 27:74).

Hemorrhagic outbreaks with mortality, especially of children, have been a prominent feature of recent epidemics in the Philippines, Thailand, Viet Nam, Indonesia, Singapore, and India. The problem has become particularly acute in Thailand, with some 23,000 cases in 1972 and 17,000 cases in 1975. Case histories during hemorrhagic outbreaks indicate a mortality rate up to 7 percent among those hospitalized. The symptoms include lack of usual joint pains associated with dengue, hemorrhagia, and frequent shock syndrome. Prior to World War II classic dengue involved serotypes I and II, but in the most severe hemorrhagic outbreaks either or both serotypes III and IV are also reported, and in the outbreaks in Thailand simultaneous infection with Chikungunya (a painful A group virus disease not considered dangerous) has occurred in 6.3 percent of cases (*Arthropod-borne virus information exchange*, 1974, No. 27).

In tropical areas outbreaks generally coincide with the rainy season and high mosquito populations; in more temperate areas hot weather is associated, probably because of slow development of dengue virus in vectors under cooler conditions. The disease is virtually confined to about 40° on either side of the equator, closely approximating the distribution of *Aedes aegypti* (Wisseman and Sweet, *in* May, 1961).

Dengue viruses have been exclusively associated with *Aedes* mosquitoes in the subgenus *Stegomyia*, particularly *Aedes aegypti*. *Aedes albopictus, Ae. scutellaris,* and *Ae. polynesiensis* have also been implicated, the latter suspected as being responsible for an extensive 1944 epidemic in the Society Islands.

The role of vertebrates in the reservoir is not entirely clear. Man is the most sensitive experimental host, and most ordinary laboratory animals are refractory except by intracerebral inoculation. Although dengue can no doubt be maintained by *Aedes aegypti*–human cycles in continuous transmission where this vector is numerous, Rudnick and associates (1967) conclude from investigations in Malaya that *Ae. albopictus* is involved in rural infections of humans, and a monkey reservoir is maintained in forest environments with unknown vectors. Suggestions of other hosts, such as birds or bats (flying foxes), have not been substantiated.

Epidemic dengue is nearly exclusively associated with *Aedes aegypti*, and can be controlled by reducing populations of this mosquito to low levels. Disruption of transmission has been noted in Puerto Rico when aerial and ground ULV applications of pesticides, and other pesticide applications, temporarily reduced vector populations to a very low level (MMWR, 1976, 25:65–66).

ILHEUS (ILH)

Isolated from *Psorophora* spp, several other genera. Fever, headaches, experimental central nervous system involvement; birds may be reservoir. Panama through much of eastern South America.

JAPANESE ENCEPHALITIS VIRUS

Disease: Japanese encephalitis (JE, JBE).
Pathogen: A virus, Togaviridae, *Flavivirus.*
Victims: Humans.
Vectors: Mosquitoes, *Culex tritaeniorhynchus* in Japan and mainly this species and certain other *Culex* elsewhere.
Reservoirs: Uncertain, may include birds. Pigs are amplifying hosts.
Distribution: From eastern Siberia westward across Asia to India.
Importance: Epidemics irregular, may involve up to around 6,000 clinical cases with 30 percent mortality. Permanent impairment known.

JBE has caused epidemics of several hundred to several thousand cases in Japan, Korea, and Taiwan, and occurs in eastern Siberia, China, Okinawa, Java, Thailand,

Malaya, Singapore, and India. Reports of a disease resembling JBE go back to 1871. The etiologic agent was characterized by Japanese workers during a severe epidemic in 1924. In 1958 there were 5,700 cases with 1,322 deaths reported in Korea, and 1,800 cases with 519 deaths in Japan. A 1961 outbreak in Taiwan involved 704 reported cases. Elsewhere (except perhaps in mainland China) the disease occurs as small outbreaks or sporadic cases.

Inapparent infection and mild systemic illness is common. Cases developing encephalitis have an onset of severe headache and vomiting; high fever, and cerebral and meningeal involvement and transient ocular aberrations are frequent. Fatal cases usually undergo coma and die within 10 days. Convalescence is generally prolonged and accompanied by weakness, tremors, nervousness, and incoordination; permanent mental impairment and personality changes are known. Diagnosis of nonfatal cases is confirmed by a postrecovery rise in specific antibodies.

Japanese encephalitis virus is maintained in mosquitoes and hosts other than man, the latter becoming accidentally involved. In major outbreaks the principal vector is *Culex tritaeniorhynchus,* which feeds mainly on large animals and birds. The disease occurs in warm weather in temperate regions; in the tropics it occurs in any season. In Japan *Cx. tritaeniorhynchus* is the only mosquito consistently infected, a rural species that overwinters as inseminated females and reaches maximum population size by late June. Japanese studies indicate that the number of human cases is directly proportional to the density of this vector when an epizootic in pigs is in progress, and that the virus does not overwinter in this mosquito (Fukumi *et al.,* 1975). There is experimental evidence of transovarial transmission in *Aedes albopictus* and *Ae. togoi* (Rosen *et al.,* 1978). Fluorescent antibody studies in *Cx. tritaeniorhynchus* reveal that primary multiplication of the virus is in the posterior midgut epithelium, fat body cells and several other tissues support further multiplication, and the salivary gland cells become heavily and permanently infected (Doi, 1970). The virus develops well in ovarian cell cultures of this mosquito (Hsu *et al.,* 1975).

Virus is frequently isolated from black-crowned night herons, egrets, and pigs. All these are amplifying hosts that serve to infect more vectors; horse infection is not of epidemiological importance because these animals experience a low and transient blood titer. Bats and *Cx. vishnui* in bat caves were found infected with JBE virus in Taiwan (Cross *et al.,* 1971). Other vectors than *Cx. tritaeniorhynchus* are six common species of *Culex* across the range of distribution of the disease, and also two *Aedes* in the Maritime Province of the Soviet Union. There have been occasional isolations from *Anopheles* mosquitoes.

Where major outbreaks occur, immunization of pigs should reduce amplification of JBE virus and lower the infective level of *Culex tritaeniorhynchus* populations (Ueba *et al.,* 1972). When surveys indicate that virus incidence and antibody conversion rates in pigs are high, and there are large vector populations, mosquito control measures could be applied to reduce the possibility of major epidemics in humans.

MURRAY VALLEY ENCEPHALITIS (MVE) VIRUS

Isolated from *Culex annulirostris, Cx. bitaeniorhynchus,* and *Aedes normanensis.* A few cases of acute encephalitis in man; birds suspected as reservoir. Southeastern Australia and New Guinea. (Doherty *et al.,* 1972).

ROCIO (ROC)

Vectors uncertain. Clinical manifestations include encephalitis similar to those observed in SLE or JE. About 900 human cases observed, with significant incidence of death and residua. Isolated from wild birds. Cases from São Paulo State, Brazil (information provided by R. W. Chamberlain, CDC, for supplement to DHEW, 1975).

SEPIK (SEP)

Isolated from *Ficalbia* spp., *Armigeres* spp., and *Mansonia septempunctata*. Causes fever and headache; similar to Wesselsbron virus. New Guinea and throughout New South Wales, Australia.

SPONDWENI (SPO)

Isolated from *Aedes circumluteolus, Mansonia africana,* and five other genera. Causes mild fever. Southern and western Africa.

ST. LOUIS ENCEPHALITIS VIRUS

Disease: St. Louis encephalitis (SLE).
Pathogen: A virus, Togaviridae, *Flavivirus.*
Victims: Humans, particularly the elderly.
Vectors: *Culex tarsalis* (western USA), *Cx. pipiens* and *Cx. p. quinquefasciatus* (central and eastern USA), *Cx. nigripalpus* (Florida). *Reservoirs:* Probably overwinters in mosquitoes, birds serving as amplifying hosts.
Distribution: Human cases recorded mainly from continental USA (Fig. 10-18), also from Canada. Virus isolations and antibody studies extend distribution to Mexico, the Caribbean, and parts of South America.
Importance: Some human cases each year in USA, major epidemics of up to 2,000 cases occur there in about 10-year cycles.

Basically SLE virus has an active mosquito-bird cycle, with mammals being accidental end points (titer of virus too low to infect feeding mosquitoes) and man being seriously affected on occasion (CDC, 1976). It is a very important mosquito-borne disease in the contiguous continental USA, with several cases occurring every year (see MMWR, 1977, 26:370) and up to 2,000 cases in large epidemics. An epidemic of 1,100 cases and more than 200 deaths occurred in St. Louis in 1933; one of the largest outbreaks occurred during 1954 in the Lower Rio Grande Valley of Texas, with 373 cases

recorded and probably more than 1,000 cases present; in 1975 some 2,000 cases were reported from 30 States, especially in the Mississippi and Ohio River valleys (MMWR, 1976, 25:116). A 1964 outbreak in the New Jersey–Pennsylvania region was unusual because of its occurrence above 40° N latitude, and because of the rarity of this disease east of the Allegheny mountains (except for Florida) in the continental USA. Although antibodies for SLE have been detected in man in Canada, clinical disease was unrecognized until the major North American outbreak of 1975, when by late September there had been some 72 cases in Windsor-Essex County, Ontario (MMWR, 1975, 24:363).

Generally SLE virus causes a clinically inapparent infection of man, serological studies showing prior exposure of 10 to 70 percent of people in endemic areas. Most clinical manifestations consist of a few days of fever and severe headache followed by complete recovery. More grave illness, particularly noted in older people, is characterized by abrupt onset of malaise, chills, and nausea; rapid temperature rise with severe headache, confusion, and drowsiness; nausea and vomiting, convulsions; commonly tremors, speech problems, and visual difficulties. Recovery is often dramatically sudden, generally without complications, though weakness, dulled mentality, and paralysis may follow. Postrecovery demonstration of a high level of specific antibodies, or recovery of the virus (mainly from central nervous system in fatal cases), is required to identify the etiological agent.

Mosquitoes were suspected as vectors of SLE virus in 1933, but this was not verified until isolations from *Culex tarsalis* by Hammon and coworkers (1941) in the Yakima Valley, Washington. *Culex pipiens* and *Cx. p. quinquefasciatus* are main vectors of SLE virus in the central states, and *Cx. nigripalpus* has been the main vector during outbreaks in Florida. In the widespread epidemic of 1975 *Cx. salinarius* and *Cx. restuans* were also involved. All of these *Culex* feed readily on birds and mammals. The virus has been isolated in the Caribbean and Brazil, mainly from *Culex* and *Sabethes* mosquitoes, and serological investigations suggest its presence in Mexico and elsewhere in South America. Many species of birds may become infected during SLE epizootics, but most studies in the USA indicate that house sparrows play the primary role in the infection chain, with pigeons, blue jays and robins also important. Domestic fowl do not appear to be good amplifying hosts.

A temperature dependence of SLE epidemics occurs in the USA, maximal activity following unusually warm spring temperatures (Hess *et al.,* 1963). In *Cx. p. quinquefasciatus* the median incubation time in the range 20 to 30° C, from infectious blood meal to transmission, has been expressed by a temperature-time equation, and the optimum mean is about 30° C, with greatest mosquito survival at that temperature and high humidity (Hurlbut, 1973). The virus in *Cx. pipiens* develops at moderate levels in cells of the gut and several other internal tissues, concentrating heavily in the salivary glands (Fig. 10-22) (Whitfield *et al.,* 1973). Overwintering mechanism of the virus in temperate zones remains uncertain, though it has been isolated from bats in Texas, and from hibernating *Cx. pipiens* females (Bailey *et al.,* 1978).

Since there are different vector species, the epidemiology of SLE is distinct in the western USA from that in the central and eastern states. In the western states the disease is mainly rural with a fairly even distribution among human age groups; in the central states the disease is more urban and suburban with a higher morbidity in older age groups. A case fatality of 20 percent in a 1962 outbreak in the Tampa Bay area, Florida, is attributed to the high incidence in elderly people, and mean age of patients with confirmed or probable SLE in Ohio in

1975 was 51, and that of fatal cases was 71 (MMWR, 1975, 24:363).

It is noteworthy that main outbreaks in the continental USA since 1955 have occurred at about 10-year intervals; 1956 (ca. 550 cases), 1964 (ca. 450), 1975 (ca. 2,000). From the information presently available one might surmise that this is owing to the coincidence of early warm and wet spring conditions favoring the buildup of vector *Culex* populations in or near urban human populations, virus exchange between mosquitoes and virus-amplifying species of birds, high levels of infection in vectors because of favorable temperature and humidity conditions, and a sufficiently long interepidemic period for a significant proportion of older unexposed humans to be present. Movement of the elderly in retirement from northern latitudes where incidence of SLE is low, to southern latitudes where the virus incidence and vector contacts are high, very likely is an important factor increasing the proportion of susceptibles in the human population.

Effective areawide control of mosquitoes by ULV application of insecticides from aircraft shows promise in controlling outbreaks of SLE and other arboviruses. Epidemic areas of SLE in Texas, in August 1966, were treated by aerial ULV malathion to control *Cx. p. quinquefasciatus.* This resulted in a radical reduction in vector population and vector infection rate, and rapid decline of human encephalitis cases approximately 14 days (one incubation period) after spraying was initiated (Kilpatrick and Adams, 1967).

WESSELSBRON (WSL)

Causes moderately severe influenzalike illness in man. See section on animal-associated mosquito-borne viruses.

WEST NILE (WN)

Isolated from many *Culex* spp., some *Anopheles* spp., and a *Mansonia;* also from ticks. Has caused hundreds of cases of encephalitis, myocarditis and respiratory dif-

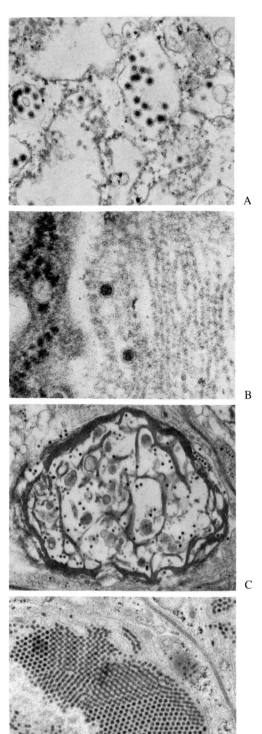

A

B

C

Fig. 10-22 Development of St. Louis encephalitis virus (SLE) in the mosquito *Culex p.
pipiens*, illustrating the "gut barrier" phenomenon. A. Virus particles in distended gut
cell endoplasmic reticulum at 6 days, 55,000 ×.
Only some cells develop virus, and the levels are
low. B. Virus particles between layers of the
complex cellular basement lamina of the midgut
cells at 9 days, 114,000 ×. This region must be
penetrated to enter the hemolymph. C. Dispersed virus particles penetrating through
fenestrations of a chitinous duct connecting to
the main salivary duct, 19 days, 24,600 ×. D. Virus
particles packed in crystalline array in the salivary
gland cells at 25 days, 31,650 ×. (A,B,C from
Whitfield *et al.*, 1973. *Virology*, **56**:70–87; D
courtesy of F. A. Murphy, CDC, US Public Health
D   Service.)

ficulties; especially serious in the elderly. Egypt and several African countries, India and other Asian locations, France and Cyprus. A 1974 epidemic in the ordinarily arid Karoo region of South Africa involved a probable 30,000 cases, and was due to heavy rainfall and consequent large populations of *Culex univittatus* (Arthropod-borne virus information exchange, 1974, No. 27). Many isolations from wild birds, with man a dead end host because of low level viremia.

YELLOW FEVER VIRUS

Disease: Yellow fever (YF), sylvatic and urban.
Pathogen: A virus, Togaviridae, *Flavivirus*.
Victims: Humans, monkeys.
Vectors: Urban form, *Aedes aegypti*. Sylvatic form: Africa-forest canopy and peridomestic *Aedes;* South and Central America-forest canopy *Haemagogus* and *Sabethes*.
Reservoir: Partially resistant species of monkeys. There is continuous transmission by mosquitoes among susceptible species of monkeys.
Distribution: Humid tropics of Africa, Central and South America (Fig. 10-23). Urban form once widespread wherever *Aedes aegypti* occurred, except never established in Asia.
Importance: Formerly a cause of epidemics in urban centers of Africa, the New World, and Europe. Presently about 200 clinical cases of sylvan YF are reported annually from Africa and South America with about 30 to 75 percent mortality, and undoubtedly more cases occur.

Yellow fever consists of a sylvatic form circulating among wild primates in Africa and South and Central America that is transmitted by forest or scrub forms of mosquitoes, and an urban form transmitted by *Aedes aegypti*. The present status of the disease is better understood by reference to its past history, and for this purpose accounts by Gillet (1972) and Brown (*in* Howe, 1977) provide background, the latter source also presenting an excellent review of present circumstances.

Like the main urban vector, *Aedes aegypti,* yellow fever originated in Africa. Along with this vector it was brought to the New World in slave trade ships in the 1500s.

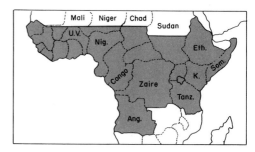

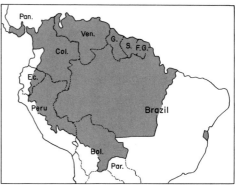

Fig. 10-23 Present-day endemic zones of yellow fever, Africa and the Americas. (Redrawn from USDHEW, *Health Information for International Travel.*)

These ships maintained their own cultures of *Ae. aegypti* in freshwater storage containers, and there are reports of European crew members succumbing to the ravages of YF while their African slave cargo suffered from other diseases but apparently were wholly or partially immune to the more serious effects of YF. The virus and urban vector became established in the Caribbean and east coast of South America, and later throughout Central America and the southern USA. An early epidemic in Barbados in 1647 caused some 6,000 deaths. Similar disasters struck the Caribbean region for the next 2½ centuries, particularly among Europeans who had no immunity to the disease. As examples, the British lost 20,000 of 27,000 men to an epidemic of "black vomit" in 1741 during an ill-fated expedition to conquer Mexico and Peru; British and Spanish forces on Cuba suffered heavy losses from the disease in the 1760s; the French in 1802 lost 29,000 of

33,000 men to yellow fever in attempting to acquire Haiti and the Mississippi Valley, contributing the next year to the sale of the vast area bought by the USA in the Louisiana Purchase. Serious epidemics hit coastal towns of South America, and the USA as far north as New York (1668); sporadic cases extended even further. In cities such as New Orleans, Charleston, and Philadelphia there were heavy losses of life, and great disruption as the populace fled during epidemics. In 1878 yellow fever affected over 100 American towns and killed some 20,000 people. The virus and its urban vector spread to Europe, causing great losses in seaports of Portugal, Spain, France, and Italy, and extending as far north as Dublin, Ireland.

Yellow fever and malaria in Panama were important factors influencing the French to lose interest in completion of the canal. It was not until William Gorgas had cleared Havana of yellow fever in 1901, and applied his comprehensive mosquito-control program to the Canal Zone in 1904, that the canal project was completed.

In 1932 yellow fever was discovered in interior Brazil in a region where at that time *Aedes aegypti* was absent. Waves of the disease passed through jungle areas, and evidence of the pathogen was obtained in resistant *Cebus* monkeys and *Haemagogus* mosquitoes. In 1948 an epizootic causing heavy mortality in *Ateles* and *Alouatta* monkeys was observed in Panama, continuing to the Mexican border in 1956; accompanied by some 98 human deaths. From additional studies in Central and South America, and in Africa, it became apparent that sylvatic yellow fever could cause epizootics in monkeys, and the virus could be maintained in partially resistant species or passed from one susceptible troop to the next via mosquito bites. Urban yellow fever could occur when humans entered savannah or jungle, during hunting or clearing of forests for lumbering or development of crop lands, and were bitten by mosquitoes that ordinarily occur in the forest canopy in close contact with infectious monkeys. Alternatively, infectious monkeys could invade plantation areas and infect mosquitoes that would then bite agricultural workers.

At present sylvatic yellow fever annually causes a few cases in man in South America, centering on the Amazon and Orinoco river basins and occasionally continuing into Central America. In Africa sporadic outbreaks occur near the equatorial belt. In 1960–1962 an outbreak struck southwestern Ethiopia, with an estimated 100,000 cases and 30,000 deaths among the 1 million inhabitants of the affected region. *Aedes aegypti* was not responsible for this outbreak; rather, passage in the forests involved monkeys and *Ae. africanus,* and foraging baboons infected *Ae. simpsoni* breeding in false-banana plantations maintained by humans (Serie *et al.,* 1968).

Reviewing the WHO *Weekly Epidemiological Record* and other reports for the 1970–1973 period, the incidence of yellow fever was as follows: 1970, Africa—322 cases (119 fatal), South America—86 (64 fatal); 1971, Africa—182 (58 fatal), South America—28 (21 fatal); 1972, Africa—6, South America—54; 1973; Africa—8, South America—207, Central America (Panama)—2. Most of these were sylvan YF, though a sizable outbreak in Angola in 1971 involved *Aedes aegypti*. These are summations of reported cases, and doubtless many others were unreported. It is apparent that yellow fever is endemic on both continents and merely awaits suitable conditions for enlarged sylvan outbreaks, or sufficiently high populations of *Aedes aegypti* to permit urban epidemics. Of great interest is the fact that YF does not occur in Asia, though *Aedes aegypti* and other potential vectors are present, and there are large populations of susceptible macaque (*Macacus*) monkeys. The danger for India is discussed by Pandit (1971).

When clinical manifestations of YF occur,

the incubation period is usually 3 to 6 days with abrupt onset of illness. Mild symptoms lasting less than 1 week include fever, headache, generalized aches and pains, and nausea. Severe cases frequently are diphasic; first with rapid onset of fever, headache, dizziness, muscular pain, nausea and vomiting; next with high fever, some jaundice, bradycardia, and various hemorrhagic symptoms. There may be profuse vomiting of dark-brown or black material, collapse and death, or termination with a comatose condition or delirium. Severely ill patients may require prolonged convalescence, but no lasting complications are known.

The vectors of yellow fever virus are urban and wild forms of *Aedes aegypti,* and forest mosquitoes. *Aedes aegypti* can transmit the virus up to 168 days after an infective meal (Philip, 1962). In Uganda a modified sylvatic form involves *Aedes africanus,* which is a forest canopy species associated with primate hosts, and *Aedes simpsoni,* which breeds in vegetation around human dwellings. Other East African vectors implicated are *Ae. vittatus, Ae. metallicus,* and *Ae. taylori.* In Central and West Africa *Ae. simpsoni* and *Ae. aegypti* seem most involved (see Cordellier *et al.,* 1974). Vectors of sylvatic YF in the Americas are mainly *Haemagogus,* particularly *Hg. janthinomys, Hg. spegazzinii* and *Hg. leucocelaenus* and *Sabethes chloropterus.* The last species lives long enough to explain survival through the dry season, but its inability to transmit some strains of yellow fever virus, plus an extrinsic incubation period of 34 to 44 days, suggests it is not an efficient vector (Rodaniche *et al.,* 1959). Smith (1972) discusses epidemiological concepts and information still required to understand yellow fever.

Old World and New World monkeys provide a known reservoir for YF, though antibodies in some rodents and other mammals have brought these under suspicion. In Africa Cercopithecidae are important. The genus *Cercopithecus* contains widespread arboreal monkeys that demonstrate high antibody ratios, and *Erythrocebus* and *Cercocebus* include arboreal-terrestrial forms important in savannas. Baboons are cercopithecids that leave forests to raid plantations, and these monkeys may also have high levels of YF antibodies. Some African colobid monkeys, and lemurs or bush babies (*Galago*), are probably of importance. In the Americas howler (*Alouatta*) and spider (*Ateles*) monkeys are widespread and highly susceptible species that keep YF virus circulating through vector transfer to susceptible members of adjacent troops. Capuchin monkeys (*Cebus*) are important resistant reservoir primates. Other cebids and marmosets may serve as reservoirs.

Control of yellow fever is a continuing concern because of modern rapid air travel, the presence and even increase of *Aedes aegypti* in many urban centers of the world, and absence of the disease in Asia. Where YF in sylvatic form is endemic, immunization may be practiced, particularly in Africa. Epidemics in such areas have been countered by widespread immunization, and in some cases by rapid control of vectors through ULV aerial applications of insecticides. Countries in which *Ae. aegypti* or similar potential vectors exist require travelers from infected areas to show proof of valid vaccination. Airports and seaports attempt to keep *Ae. aegypti* completely controlled or at levels too low to support epidemics. Aircraft leaving an *aegypti*-infested area for an *aegypti*-free area, or aircraft leaving any yellow fever–infested area, must be disinsected by pesticide application.

ZIKA (ZIKA)

Isolated from *Aedes africanus* and other *Aedes.* Caused fever and rash in single case observed. Distribution across equatorial belt of Africa, Malaya.

Mosquito-Associated Viruses Isolated from Man in Nature: BUNYAVIRIDAE, Bunyavirus

BUNYAMWERA GROUP

Bunyamwera (BUN). Isolated from *Aedes circumluteolus* and several other *Aedes,* a *Culex,* and a *Mansonia.* Mild fever, rash, backache, joint pains; sometimes encephalitis. Several African locations.

Calovo (CVO). Transmitted by *Anopheles maculipennis.* Causes fever and headache. Czechoslovakia, Austria, Yugoslavia.

Guaroa (GRO). Transmitted by *Anopheles.* Fever, headache, muscle and joint pains. Colombia, Brazil, Panama.

Ilesha (ILE). Isolated from *Anopheles gambiae.* Mild febrile disease; across equatorial belt of Africa.

Tensaw (TEN). Several isolations from species of *Anopheles, Aedes, Culex,* and *Mansonia.* Symptomless infection in man. Southeastern USA.

Wyeomyia (WYO). Isolated from several *Wyeomyia, Psorophora, Aedes* and *Anopheles* mosquitoes, and other genera. Caused a febrile condition in man. Panama and northeastern South America.

C GROUP

Apeu (APEU), Caraparu (CAR), Itaqui (ITQ), Madrid (MAD), Marituba (MTB), Murutucu (MUR), Oriboca (ORI), Ossa (OSSA), and Restan (RES). All are C group Bunyaviruses. Transmitted by *Culex* mainly and *Aedes;* other genera. These viruses cause febrile illness in man that may include headache and muscle or joint pains. Generally not serious, but Ossa (Panama) and Restan (Trinidad, Surinam) have caused encephalitis. Generally distributed from Panama to Brazil. Reservoir may include *Cebus* monkeys and rodents.

CALIFORNIA GROUP

Viruses in the California group are antigenically similar Bunyaviridae, but represent several separate etiologic agents from the USA and other countries. Review information may be obtained from papers by Sudia and associates (1971) and Parkin and colleagues (1972). Originally found mainly by isolations from pools of biting arthropods, and especially mosquitoes, members of this group are now recognized as pathogens causing human encephalitis. In the USA the California group encephalitides annually cause a number of cases of human encephalitis, second only to St. Louis encephalitis among mosquito-borne viruses (Fig. 10-21). The North American forms are known as **California encephalitis** (CE), **Jamestown Canyon** (JC), **Jerry Slough** (JS), **Keystone** (KEY), **La Crosse** (LAC), **San Angelo** (SA), **snowshoe hare** (SH), **South River** (SR) and **Trivittatus** (TVT). **Melao** strain (MEL) is known from Brazil, and TVT from Panama; **Tahyna** (TAH) from Czechoslovakia, Austria, France, Italy and West Germany; **Inkoo** (INK) from Finland; **Lumbo** (LU) from Mozambique. Serological surveys reveal exposure in a variety of rodents and lagomorphs as well as skunk, opossum, raccoon, coyote, several wild ungulates, reptiles, and a number of species of wild birds. Isolations in nature from arthropods have included some 25 species of *Aedes* mosquitoes, several species of *Anopheles, Culex, Culiseta* and *Psorophora,* and fewer species in other genera; also three tabanid and two hard tick species.

The picture emerging regarding viruses of the California complex is that these are mainly pathogens that amplify in rodents and lagomorphs. They generally cause subclinical or very mild symptoms in humans, but some strains have been associated with encephalitis. Transmission to vertebrates is mainly by tree-hole, floodwater, or snow pool *Aedes,* and in some instances the vectors serve as the reservoir and overwintering mechanism owing to transovarial transmission (LeDuc *et al.,* 1975). Watts and Eldridge (1974) review the subject of such transmission in mosquitoes, with particular reference to California encephalitis group arboviruses. Further discussion will consist of reviewing certain individual members of the complex.

Tahyna Virus (TAH). This virus causes fever with occasional encephalitis of man in Central Europe, and is associated with *Aedes* of forests and meadows. *Aedes vexans* is the main vector, and young hares and suckling pigs are the most important hosts and probable amplifiers of the virus (Vater, 1973; Bádoš, 1975).

La Crosse Virus (LAC). This virus is responsible for human encephalitis in the midwestern USA, especially Ohio, Indiana, Wisconsin, and Minnesota. It has been associated with chipmunks and marmots, and appears to be mainly transmitted by the tree-hole mosquito *Aedes triseriatus* (Gauld *et al.,* 1975; Berry *et al.,* 1975). Transovarial transmission results in infected eggs and postembryonic stages of this mosquito, and no doubt means that this vector is the main reservoir and overwintering mechanism (Watts *et al.,* 1973, 1974). The proportion of infective adults by this mechanism may be quite low, but further distribution of the virus in adult female *Ae. triseriatus* can occur through venereal infection (Thompson and Beaty, 1977). Transovarial infection with LAC virus was possible in an *Aedes* that is not a natural vector, but not in a *Culex* (Tesh and Gubler, 1975).

Inkoo (INK) Strain. This strain of California group virus occurs in Finland and seems to be mainly associated with *Aedes caspius* and other *Aedes;* a few cases of febrile disease have occurred. In Lapland, where mosquitoes are an extreme nuisance, antibodies to this strain at an incidence of 69 percent in man, 88 percent in cattle, and 81 percent in reindeer represent among the highest exposure levels recorded for California group arboviruses (Brummer-Korvenkontio, 1973).

GUAMA GROUP

Catu (CATU). Transmitted by *Culex portesi* mainly, other *Culex,* and an *Anopheles.* Causes fever, headache and muscular aches; rodent reservoir suspected. Brazil, Trinidad, French Guiana.

Guama (GMA). Transmitted by *Culex portesi, Cx. vomerifer, Aedes, Mansonia,* and other genera. Febrile illness and encephalitis; monkey and rodent reservoir suspected. Occurs in Panama to Brazil.

NYANDO (NDO) GROUP

Transmitted by *Anopheles funestus.* One isolation from a human in the Central African Republic with biphasic fever, myalgia, and vomiting; reservoir unknown.

Ungrouped Mosquito-Associated Viruses Isolated from Man

COTIA (COT; POXVIRIDAE)

Infrequent isolations from species of *Culex, Aedes, Psorophora, Mansonia;* sentinel mice. One isolation from symptomless human. Brazil, French Guiana.

TATAGUINE (TAT)

Isolated from *Anopheles gambiae, An. funestus, Coquillettidia aurites.* Several isolations from symptomless humans, reservoir unknown. Africa across equatorial belt.

ZINGA (ZGA)

Isolated from *Aedes palpalis* group, *Mansonia africana.* Two febrile human cases with headache, muscle and joint pains. Antibodies in man, African buffalo, wart hog and other large game animals. Central African Republic.

Mosquito-Associated Viruses, Not Arboviruses, Isolated from Man

HEPATITIS B VIRUS

Australia antigen, the antigen associated with hepatitis B virus of man, can be maintained for long periods in mosquitoes, though it does not become concentrated in the salivary glands. Detection of the antigen in wild mosquitoes from several parts of the world, particularly where hepatitis levels are high, leads to the suspicion that mosquitoes, and possibly other biting arthropods, may mechanically transmit this virus (Blumberg, 1977).

Mosquito-Associated Arboviruses Isolated from Animals

Venezuelan equine encephalitis (VEE) and western equine encephalitis (WEE) affect horses and related equines, and eastern equine encephalitis (EEE) affects equines and some exotic gallinaceous birds; Japanese encephalitis (JBE) usually causes inapparent infections in pigs, horses and wild birds. See previous discussion of these diseases.

RIFT VALLEY FEVER VIRUS (RVF, UNCLASSIFIED)

The disease caused by Rift Valley fever virus is fairly widespread among domestic animals in Kenya, Uganda, Mozambique, Rhodesia, and South Africa. Man may become infected by direct contact with sick animals, and presumably by the bite of an infective vector. This virus causes abortions in sheep, cows, and goats, and heavy mortality of lambs and calves. In man the disease is almost never fatal, but symptoms include fever, headache, muscular pains, liver engorgement; there may be permanent visual impairment. An epizootic in South Africa in 1950–1951 was estimated to cause the deaths of 100,000 sheep and cattle and to have in-

volved 20,000 human cases. An outbreak occurred in 1968–1969, centering on Rhodesia (McIntosh, 1972).

RVF virus appears to exist in enzootic forest foci involving mosquitoes and wild animals, particularly rodents. From such foci epizootics can occur when vector populations and susceptible vertebrates are plentiful. The likely mosquito vectors are *Eretmapodites chrysogaster, Aedes caballus, Ae. circumluteolus* and *Culex theileri*.

WESSELSBRON (WSL, TOGAVIRIDAE, *Flavivirus*

Aedes caballus, Ae. circumluteolus, Ae. lineatopennis, other *Aedes*, species of *Culex, Mansonia*, and *Anopheles* are the mosquito vectors of this virus. In sheep this virus causes abortion and death of lambs. It is associated in man with a moderately severe influenzalike illness, sometimes requiring prolonged convalescence. Nigeria and Uganda through southern Africa, Madagascar; Thailand.

Mosquito-Associated Viruses, not Arboviruses, Isolated from Animals

AVIAN POX (FOWL POX) VIRUS

Avian pox is prevalent wherever poultry is raised. This pathogen causes skin lesions that affect the health of a variety of domestic and wild birds, and that seriously reduce the market value of chickens, turkeys, and squabs. Warner (1968) provides persuasive evidence that the introduction of this pathogen, bird malaria, and a mosquito vector to Hawaii had a profound influence on the decimation of native bird populations. There apparently are strains of the virus that are less infectious and pathogenic on heterologous avian species. Transmission is accomplished mechanically and naturally from injured skin, or through the agency of biting arthropods, particularly mosquitoes. Lesions usually develop 6 to 8 days after an infective mosquito bite, and there is experimental evidence that mosquitoes may remain infective for up to 60 days. *Culex* mosquitoes, mainly because many species feed preferentially on birds, are most frequently implicated as vectors. On a worldwide basis the *Culex pipiens* complex is of great importance, and in the western USA *Cx. tarsalis* is a major vector. Mosquito control, or preventing mosquito bites by rearing poultry in screened quarters, will reduce the incidence of fowl pox; immunizing vaccine is available for chickens and pigeons (Cunningham, in Hofstad, 1972).

ENCEPHALOMYOCARDITIS VIRUS

This virus causes a zoonosis affecting the central nervous system and heart of a number of wild and domestic animals. Several strains have been isolated in many parts of the world from primates, swine, and rodents, the latter frequently being associated with human cases. The etiologic agent was isolated in Uganda from *Mansonia fuscopennata;* study of an epidemic in Para, Brazil, yielded isolations from wild rodents, opossums, horses, birds, sentinel mice, *Aedes* species, two species of *Mansonia* (or *Coquillettidia*) and one *Culex.*

EQUINE INFECTIOUS ANEMIA VIRUS

This virus has been isolated from several biting arthropods, including mosquitoes. It is discussed under Tabanidae.

MYXOMA VIRUS (POXVIRIDAE, LEPORIPOXVIRUS)

Rabbit myxomatosis (big head in rabbits) was observed in a devastating epizootic of domestic European rabbits in Uruguay in 1896. This disease, immediately suspected as caused by a virus, was characterized by numerous mucinous skin tumors, further lesions affecting several other body tissues, and death ensuing in a week or two. The disease affected rabbits, but seriously affected only the European wild rabbit, *Oryctolagus cuniculis,* being mild or asymptomatic in the American *Lepus* and *Sylvilagus.* Long regarded as confined to South America, in 1930 myxomatosis broke out in commercial rabbit farms of southern California. It is now recognized as indigenous to *Sylvilagus braziliensis* in Central and South America, and *S. bachmani* in California. The main means of transmission is by biting arthropods that acquire the virus where it is in sufficiently high titer in the skin on and near tumors, and mechanically transfer the pathogen to the skin of uninfected rabbits during the act of feeding.

Because of its devastating effects on the European rabbit, myxoma virus was introduced to Australia, where this leporid had become a major threat to agriculture. Comprehensive historical and biological accounts of this disease are available (Fenner and Ratcliffe, 1965; Joubert *et al.,* 1973; Marshall, in Gibbs, 1973).

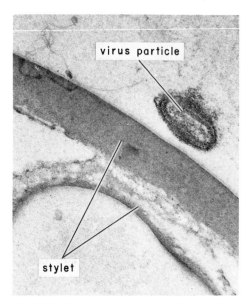

virus particle

stylet

Fig. 10-24 Myxoma virus on maxillary stylet of *Aedes aegypti.* Note that virus particle does not adhere very strongly to stylet. (Courtesy of B. Filshie, CSIRO, Canberra, Australia.)

Mechanical transmission of myxoma virus is accomplished in the laboratory by a wide variety of bloodsucking arthropods including mosquitoes (Fig. 10-24), black flies, fleas, sucking lice, mites, and ticks. In Australia natural transmission has typically spread along river valleys, and has primarily involved the vectors *Culex annulirostris, Anopheles annulipes,* and various *Aedes* species as well as Simuliidae. In France the principal means of spread is thought to be *Anopheles,* but in Great Britain the rabbit flea, *Spilopsyllus cuniculi* is most suspect.

Biological control of rabbits by myxomatosis has largely been successful in Australia, and in Europe the disease threatened the existence of rabbits even where wild species were considered of value. Continued natural exposure of rabbits to the virus, as well as laboratory experimental exposures, have caused the selection of virulent and attenuated virus strains and the development of a certain degree of resistance in rabbit populations.

Control of Mosquitoes

More research has been devoted to mosquito control than to the control of any other group of arthropods affecting man and animals, and manuals on mosquito control have been produced by regional and international agencies (examples are Mulhern 1973; (WHO, 1973a). Since many of the principles of arthropod control are stated in Chapter 5, with illustrations often chosen from mosquito control, that information should be reviewed and understood before putting into practice the techniques suggested here.

Mosquito control is generally a comprehensive communitywide task requiring technical supervision provided by public health agencies or mosquito-abatement districts. However, individuals can protect themselves from mosquitoes on their home premises if the total level of annoyance is not too great, and if removing or modifying mosquito breeding sites around the home or adjacent agricultural lands will significantly reduce local annoyance.

In view of the fact that many mosquito problems are due to the incursion of pest or vector species from extensive breeding areas well removed from home premises, the following methods can be considered to remedy or alleviate only very local problems. (1) Surrounding premises should be inspected thoroughly to see that there are no accumulations of standing water in which mosquitoes are developing. This applies to cans, tires, cisterns, roof drains, and a variety of improper soil drainage situations from marshy home grounds to waterlogged agricultural lands and puddles below outdoor water taps. Water accumulations should be removed by draining or filling, and control in natural or ornamental ponds may be practiced by ensuring that predacious fish are present. (2) Proper screening should be maintained in windows and door entrances, and may be considered for porches. Sleeping under bed nets also provides relief. (3) At times of heaviest attack around homes, or when a person enters mosquito-ridden areas, repellent may be applied to exposed skin surfaces, or a repellent jacket may be worn. (4) Space repellents such as mosquito coils can be used in rooms, or outdoors in limited areas when breezes are not excessive. (5) Insecticides approved for mosquito control in the home may be applied to surfaces likely to be contacted by mosquitoes, that is, wall surfaces of houses, and adjacent shrubbery where mosquitoes might rest; or as an aerosol application in rooms already infested.

In regard to problems of more widespread mosquito control, one usual first requirement is an assessment of the severity of the current problem, and a running account of population levels for possible predictive purposes or to determine the success of control practices. The methods used will depend on objectives of the survey, life history and behavior of the kinds of mosquitoes involved, availability of types of survey equipment, and technical training of persons conducting surveys. A very inclusive account of field sampling methods for mosquitoes is provided by Service (1976).

Surveys of eggs and oviposition sites can provide information regarding levels of mosquito abundance to be anticipated in the near or more distant future. Looking for egg rafts of *Culex* and *Culiseta,* and single floating eggs of *Anopheles* is generally too difficult to be useful. Surveying for eggs of floodwater *Aedes* and *Psorophora* (see Chapter 4) can help to determine the potential population in the next active season, and may be used in planning an effective prehatch application of insecticide. An oviposition trap constructed with a dark jar, a wooden paddle as oviposition site, and a trace of ethyl acetate as attractant, has proved valuable in sampling the presence of *Aedes aegypti,* especially when populations are low and natural breeding sites scarce. Similar containers have been developed for tree-hole-breeding species of *Aedes*.

Larval surveys are used to decide whether control measures should be applied to aquatic sites. A dipper provided with a long handle is the collecting utensil most often used; more quantitative information can be gained by pushing a walled metal cylinder into the muddy pond bottom and removing and examining all the water enclosed. When dipping for mosquito larvae it is important to not overlook obscure larval sites, such as cattle hoofprints in wet pastures or on the edge of water holes and ponds (Fig. 10-25). To examine tree holes and similarly inaccessible cavities a large-capacity rubber suction bulb and flexible extension tube can be used to draw out the water. Determining the presence of *Mansonia* and *Coquillettidia* larvae requires uprooting aquatic plants, to which the larvae attach, and vigorously agitating the plants including the root system in a container of water. As the resulting water is

Fig. 10-25 Hoofprints around puddles, ponds, and streams are good breeding places. The water in the hoofprints shown here produced 60 to 80 larvae, mostly *Culex tarsalis,* per pint; much greater productivity often occurs in similar breeding areas. (Courtesy of H. G. Davis, USDA.)

usually muddy, it should be strained through a fine screen.

Adult survey procedures, further described in Chapter 4, consist of light trapping, aspiration from resting sites, hand catching or trapping mosquitoes attracted to man or animal hosts or carbon dioxide or moving visual patterns, Malaise trapping, and use of moving net devices.

Environmental modification has been a common procedure for mosquito control. Draining or filling of marshy areas is a normal practice in many abatement districts. In salt marsh areas diking and flooding with seawater can destroy mosquitoes, as most salt marsh species cannot withstand so intense a salinity. Diking and impounding of particularly troublesome salt marsh areas will also change the character of marsh vegetation and reduce populations of annoying salt marsh *Aedes* mosquitoes (La Salle and Knight, 1974). Channels from marshy areas into tidal streams will permit free access of predatory fish to water collections where mosquitoes breed. Irrigation canals and drainage ditches must be kept free of vegetation, thus maintaining active flow of water and preventing mosquito breeding, and providing open access for fish and insect predators. Concrete lining of such canals reduces water loss, maintains a steady flow detrimental to mosquitoes, and prevents ground seepage that develops additional breeding sites. Dropping water levels of impoundments in quick steps will prevent the growth of emergent vegetation and thus effectively deter the development of *Anopheles,* but this technique requires careful management to minimize the production of *Aedes.* Some effective control schemes have employed a schedule of periodically flowing rushing water to flush out sluggish small streams and drains, and automatic flushing systems have been designed for this purpose. In Czechoslovakia carp and ducks in rice fields removed water weeds and provided some control of *Anopheles maculipennis* (Trpiš,

1960), and carp have been used in sewage settling basins in California, USA, with some similar success. In Kerala, India, the vector of brugian filariasis, *Mansonia annulifera,* breeds in abundance on roots of the aquatic plant *Pistia stratiotes.* An introduced water fern, *Salvinia auriculata,* crowded out *Pistia* and formed an impervious growth that increased larval mortality, (Joseph *et al.,* 1963)

More pathological microorganisms are known for mosquitoes than for any other group of medically important arthropods. Thus far use of these pathogens has met with only limited success, but read the section on microbial control in Chapter 5 to learn of the possibilities.

Natural predation of larval and pupal mosquitoes by insects may be encouraged. Keeping aquatic vegetation in control allows predators access to their prey. Insect predators include nymphs of dragonflies and damselflies, various predacious aquatic Hemiptera and Coleoptera, and predacious Culicidae in the genus *Toxorhynchites.* The latter have particular possibilities against mosquito larvae developing in tree holes, though the introduction of exotic species into new areas has thus far met with limited success (Trimble and Smith, 1975).

The most effective larval and pupal predators found to the present time are various fishes. These predators are discussed in Chapter 5, particularly the mosquito fish *Gambusia affinis* and the guppy *Poecilia reticulata.* The high reproductive potential of both of these predators makes them especially effective in building up their numbers in the presence of suitable prey, and both species can survive on other kinds of small aquatic invertebrates and other foods in natural environments. They can be transported readily, and in temperate climates they may be distributed early in the spring to start to control mosquito populations and to increase their own numbers as warm temperatures favoring mosquito development appear.

Many predators of adult mosquitoes are known, though their effectiveness is in doubt. Commonly believed to be effective are spiders, dragonflies and damselflies, frogs and toads, lizards, and many birds, such as fly catchers, swallows, martens and warblers, and bats. While all of these predators eat adult mosquitoes, there is no evidence that any of them are specific enough on these insects to provide reliably effective control.

Various plants have been associated with reduced mosquito populations, though aquatic plants usually encourage mosquitoes through blocking access by predators. Some observers suggest that certain algae release toxic substances that inhibit larval growth. Carnivorous plants, especially *Utricularia,* capture mosquito larvae. Other plants, such as the duckweed *Lemna,* can cause such a heavy surface covering on water that they prevent oviposition and larval development.

Repellents have mainly been developed as protectants against mosquito attack. These are applied directly to the skin, or to fabrics, particularly in protective outer garments made of open mesh material. Review the repellent section in Chapter 5.

Insecticides are deposited on potential hatching sites just before flooding occurs, or against mosquito larvae or adults. In addition to conventional toxicants, some IGR (insect growth regulator) compounds have been registered for use against larvae. A number of precautions must be kept in mind when using pesticides, and it would be best to review the general information on insecticide development in Chapter 5 before planning control programs in which insecticides will be used. In all cases, when employing insecticides the status of resistance to the proposed toxicant should be known and overuse should be avoided. Damage to fish and wildlife, and contamination of crop lands must also be avoided.

Direct treatment of larval waters may be practiced. For this purpose boats or amphibious equipment may be used as a base for application equipment, or ordinary terrestrial vehicles may be satisfactory if the body of water is small. For extensive applications it is generally necessary to employ aircraft (Figs. 5-6, 5-9). Treatment is generally with sprays of emulsifiable or oil base insecticides that are appropriately registered for this purpose. In situations where there is heavy vegetative cover it may be advantageous to use granular formulations because of their superior penetration characteristics. There are also special larvicidal oils that cause mortality mainly by blocking the respiratory openings.

Problems peculiar to specific aquatic environments may be encountered when applying insecticides. Control in waters of high organic content can be particularly difficult, owing to toxicant absorption on organic substances, pH extremes hastening toxicant breakdown, and microbial degradation. Log storage ponds, sewage settling basins, holding tank latrines, and food-processing settling ponds may all present special difficulties.

Adult mosquitoes are treated with aerosols, ultra low volume sprays (ULV), residual deposits, or an organophosphorus compound with direct toxicity and fumigant action. Aerosols for control of mosquitoes already in a home are generally dispensed from small pressurized cans and contain compounds with low mammalian toxicity and rapid knockdown characteristics, such as pyrethrins or their synthetic derivatives. Outdoor aerosol applications in the form of thermal fog (Fig. 5-10) or cold fogs (Fig. 5-11) will provide temporary relief against existing populations of adult mosquitoes, and are mainly used when biting rates are causing annoyance. ULV aerial application of suitably registered compounds will rapidly reduce biting attacks by adult mosquitoes over wide areas. This procedure is used to provide relief from annoyance, and has also been practiced during outbreaks of mosquito-associated diseases. A number of residual treatments are satisfactory, such as vegetation coverage with residual toxicants (using a formulation that is not phytotoxic), or spraying walls in dwellings with residual contact insecticides. Fumigant action is a characteristic of some organophosphorus compounds, and these have been sprayed into confined spaces such as storm sewer catch basins, or resin strips have been impregnated with such toxicants and placed in confined spaces to provide long-term control.

Reproductive manipulation still remains largely experimental. Radiation sterilization and chemosterilants have been tested extensively with several species of mosquitoes in laboratory and limited field experiments, but the main present drawbacks to the use of these techniques are human safety, logistical problems in producing sufficient numbers of male mosquitoes for mass releases, and suitable competitiveness of treated males versus normal wild males in the release area. In sterile insect release programs it is essential that mosquito population structure be assessed to evaluate progress, and indeed population models can be utilized in any control program. Simplified population models assessing mosquito control tactics have been developed (Weidhaas, 1974), and the use of models during field operations may be reviewed in analyses of populations of *Anopheles albimanus* during sterile male releases in El Salvador (Weidhaas *et al.,* 1974).

11

HORSE FLIES, DEER FLIES, AND SNIPE FLIES

ORDER DIPTERA; SUBORDER BRACHYCERA

The suborder Brachycera, aside from the Tabanidae, does not contain many flies of medical importance. However, the families Rhagionidae and Athericidae include a number of species of vicious bloodsuckers. This chapter deals only with the bloodsucking Brachycera; the only nonbloodsucker with which we are concerned, the stratiomyid *Hermetia illucens,* is discussed in Chapter 13.

Family Tabanidae

HORSE FLIES AND DEER FLIES

The large and cosmopolitan family Tabanidae includes avidly bloodsucking flies known by a variety of names, the most common being horse flies, deer flies, clegs, and mango flies. In Australia the term "March flies" is used, although in the Northern hemisphere this name applies to an entirely different family of Diptera (Bibionidae). Tabanids are usually moderately large and heavy-bodied, measuring in length from 6 to 10 mm in the smaller species up to 25 mm in the larger ones. They are strong fliers and

notorious pests of horses, cattle, deer, and many other warm-blooded animals; at times these flies, particularly the persistent members of the genus *Chrysops,* annoy man. It has generally been assumed that most species were anautogenous, but recent work indicates that autogeny is not uncommon. Only the females bite. The males feed on vegetable materials and do not bite; in fact, they cannot do so, because they have no mandibles. The eyes are very large and widely separated (DICHOPTIC) in the females; the males are usually HOLOPTIC, the eyes being contiguous. The wing venation is characteristic in that the branches of vein R_{4+5} diverge broadly, thereby enclosing the apex of the wing between them. The mouth parts of the female are bladelike and function as cutting instruments, though the labella are fitted for sponging (Fig. 3-5).

The breeding habits of most species are aquatic or semiaquatic. The eggs are normally deposited in layers on objects over water or situations favorable for the larvae, such as overhanging foliage, projecting rocks, sticks, and emergent aquatic vegetation. The narrow cylindrical eggs, 1.0 to 2.5 mm long, vary in number from 100 to 1,000 and are deposited commonly in layers and

234

Fig. 11-1 *Tabanus punctifer.* Egg mass on willow leaf, larva, pupa, and adult female.

may be covered with a waterproof secretion that also binds the eggs tightly together.

The larva (Figs. 11-1, 11-2) has a slender, cylindrical, contractile body consisting of a small head and twelve additional segments. The head is retractile, with pointed mandibles capable of inflicting a sharp bite; at the posterior end is situated a tracheal siphon that telescopes into the anal segment. The pupa (Figs. 11-1, 11-2), resembling those of naked Lepidoptera, is obtect and abruptly rounded anteriorly, tapering posteriorly, with leg and wing case attached to the body; the abdominal segments are free and about equal in length, segments two to seven each bearing a more or less complete ring of spines near the posterior third. The adult fly emerges from the pupal case through a slit along the dorsum of the thorax, as do the rest of the Brachycera.

Breeding Habits and Life History. The eggs are deposited during the warmer months of the year. The incubation period is greatly influenced by weather conditions, but during midsummer the usual range is from 5 to 7 days. The larvae of the aquatic species, on hatching, fall to the surface of the water or upon the mud or moist earth, in clumps, and quickly drop to the bottom or burrow individually into the wet or damp earth, where they begin to feed on organic matter. Species of *Tabanus* and *Haematopota* are voracious predators on insect larvae, crustaceans, snails, earthworms, and other soft-bodied animals; cannibalism has been observed in several species. *Chrysops*

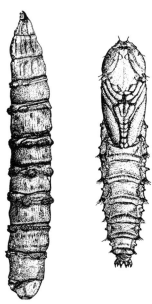

Fig. 11-2 Lateral view of larva (*left*) and ventral view of pupa (*right*) of *Tabanus gilanus.*

larvae, according to Oldroyd, are probably vegetarians. However larvae of both *Chrysops* and *Tabanus* have been reported by Otsuru and Ogawa (cited by Leclercq, 1971) as biting painfully the hands and feet of workers in rice fields in Japan.

The larvae of Tabanidae are commonly encountered buried in wet soil in such places as along the edge of marshy ponds and salt marshes, roadside ditches, and the overflow from rice fields; certain species may be found in moist leaf mold and debris, in rotting logs, or in water. Some inhabit tree holes and brackish water. A few, for example, *Tabanus fairchildi* and *T. dorsifer,* live in rapidly flowing water. Some species have terrestrial larvae. Schomberg and Howell (1955) have shown that *Tabanus abactor* and *T. equalis* breed in soil, in the shade of trees, and under either dry or short sparse grass, where standing water never or seldom occurs, and Jones and Anthony (1964) have cited references to similar habitats for *Tabanus sulcifrons* and *T. quinquevittatus;* the latter species, mistakenly identified as *T. vicarius,* was reported by Logathetis and Schwardt as an open-pasture breeder in New York State.

Most temperate climate species produce one generation a year. In these, the larva grows rapidly during the summer and autumn, and very slowly, if at all during the winter; it attains full growth in the following early spring. Some species, however, such as *Tabanus subsimilis subsimilis,* may have two generations a year, whereas others, such as *Tabanus calens,* require 2 or 3 years to complete their development. Individual larvae of other species may require 2 or 3 years for development if conditions are unfavorable during the first season or if they do not get enough food. At least one species, and possibly more, has been known to survive desiccation by constructing cylinders of mud for protection (Oldroyd, in K. G. V. Smith, 1973).

When the fully grown larva prepares to pupate, it usually moves into drier earth, usually an inch or 2 below the surface, and in a day or 2 the pupal stage is reached. This stage requires from 5 days to 3 weeks, varying with the species. The flies emerge from the pupal case and make their way to the surface; the wings soon unfold, and the insects take refuge among nearby foliage or rest on objects near at hand; in a short time they begin to feed, the females seeking blood and the males feeding on flowers and vegetable juices.

In some species, according to Oldroyd (personal communication to MTJ), mating may take place during the hardening period following a mass emergence of both sexes together. The flies then disperse, and presumably the males soon die. Other species emerge individually, and mating takes place through the agency of male swarms, which are visited by the females. These swarms may be small; in fact, a so-called swarm might even consist of a single male.

In temperate regions females are active only during the warmer months. In the tropics, the time of activity varies with the species; in some, it may extend throughout the year or there may be no clearly defined season; in others, it may be limited to a definite period during either the wet or dry season (Leclercq, 1971). Most tabanids respond clearly to warmer temperatures and are for the most part active during the warmer parts of the day; others show greatest activity at or near the hours of dawn and dusk. Tabanids often follow moving objects.

Tabanids are of medical and veterinary importance in two ways: their bites may cause serious annoyance to man and animals and significant blood loss to the latter, and they may serve as mechanical vectors and biological hosts of human and animal pathogens.

Genera and Species of Tabanidae. The family Tabanidae is a fairly large one of almost worldwide distribution, being absent only in high altitudes and latitudes

A   B

Fig. 11-3 The black horse fly, *Tabanus atratus. A.* Male. *B.* Female. × 1.5. (Photograph by Hine.)

and in such insular areas as Hawaii. There are about 3,000 species distributed through 30 to 80 genera, the number depending upon the interpretation of the authority. Only a few of these concern us, notably *Chrysops* and *Silvius,* which have apical spurs on their hind tibiae, and *Tabanus, Hybomitra,* and *Haematopota,* which lack these spurs. Some of the most vicious-looking species, namely, certain Pangoniinae that have long and erect, but slender, proboscids, are nonbloodsucking.

Tabanus atratus (Fig. 11-3), the widespread black horse fly of eastern North America, is uniformly black, and the thorax and abdomen in well-preserved specimens are thinly covered with a whitish pollen (dust). In certain lights, unless this pollen is rubbed off, the abdomen shows a distinctly bluish cast. Curran believes, probably correctly so, that this is the "blue-tailed fly" of the well-known American ballad.

Among the more serious pests of livestock in the United States are *Tabanus punctifer* (Fig. 11-1) in the West and *T. sulcifrons* in the Midwest and the Great Plains states. *Tabanus quinquevittatus* and *T. nigrovittatus,* the latter a salt marsh breeder, the notorious "greenheads," and *T. lineola* and *T. similis,* the striped horse flies, are very annoying pests in the eastern United States. In the western and northern United States and

in Canada, various species of *Hybomitra* are annoying to man and animals alike.

The genus *Chrysops* (Fig. 11-4) contains about 80 North American species. The term *deer fly* is usually applied to this genus. Leclercq (1960) lists 45, some with several subspecies each, in the Palearctic Region, and Oldroyd (1952–1957) lists 38 Ethiopian species. Many members of this genus attack man readily and persistently. They are, as individuals, smaller than most *Tabanus* and *Hybomitra* species and have pictured wings that are usually definitely cross-banded.

Haematopota americana occurs from central Alaska to New Mexico. It is quite different in appearance from *Tabanus* and

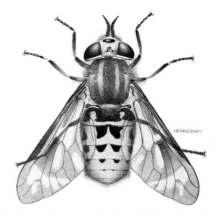

Fig. 11-4 *Chrysops discalis* adult female. (Courtesy of Agriculture Canada.)

Chrysops in that the wings are quite characteristically densely mottled. As was previously indicated, this genus, though poorly represented in the New World, is richly represented in some parts of the Old World, though absent in others; for example, Oriental species are numerous, but there are none in New Guinea or Australia.

Silvius pollinosus and *S. quadrivittatus,* small gray species with distinctly but rather sparsely mottled wings, are viciously annoying to man and animals in the area just east of the Rocky Mountains.

Among the more comprehensive taxonomic works on Tabanidae are Stone (1938) and Brennan (1935) for the Nearctic Region; the general studies of Mackerras (1954–1955) and those of the same author for the Australian Region (Mackerras, 1956, 1960, 1961; 1965–1966); those of Leclercq (1960) and of Chvála, and colleagues (1972) for the Palearctic Region; and that of Oldroyd (1952–1957) for the Ethiopian Region. Many excellent faunistic studies on a more restricted geographical scale exist.

Bites. Female horse flies have broad, bladelike mouth parts (Fig. 3-5) that inflict a deep, painful wound, causing a considerable flow of blood, which they lap up by means of their sponging labella. Man may be seriously annoyed, particularly by *Chrysops, Silvius,* and some species of *Tabanus, Hybomitra, Diachlorus,* and other genera, but, of course, he is better able to protect himself than are wild and domestic animals. Reactions to the bite may vary considerably with the individual. Usually there is no swelling, reddening, or irritation, or at most a moderate reaction that will last from a few minutes to several hours. However, a few persons experience allergic reactions, which, in extreme cases, may be so serious as to require hospitalization. Tabanids may be so annoying as to interfere seriously with the use of recreational areas, or they may cause considerable economic loss in work areas, such as

areas where timber or other crops are being harvested.

Blood loss in livestock may constitute a serious problem. Reductions in beef and milk production in the United States in 1965 were estimated at $30 million and $10 million, respectively, the result of combined losses in energy, blood, and grazing time. Also, heavy infestations can seriously lower butterfat production in dairy animals.

Infestations much heavier than 50 flies per animal frequently occur. Tashiro and Schwardt reported that, in New York State, it was not uncommon for 720 *Tabanus quinquevittatus* or 320 *T. sulcifrons* to feed on one animal for 8 hours, with an estimated blood consumption of .074 and .359 ml per fly for these two species respectively. Leclercq (1971) gives a very graphic account of the harassment of asses and horses that were bound during the day and consequently unable to flee their tormentors; at the end of the day the animals were barely able to stand and the vegetation around them was stained with blood. Heavily infested livestock commonly show hair on infested areas matted with blood that has come from feeding punctures made by tabanids.

Feeding habits vary with the genera and even with the species. Oldroyd (1964) believes that *Haematopota,* which feeds especially on Bovidae, has evolved along with its mammalian hosts. He points out that there are no members of this genus in Australia or South America, where native Bovidae are lacking; that only 5 species occur in North America; but that 50 occur in the Palearctic, 70 in the Oriental, and 180 in the Ethiopian region, areas where cattle and antelopes have likewise undergone great development. *Tabanus* has a much wider host range, some species attacking reptiles as well as mammals. *Chrysops,* which commonly lives in broken woodland areas, commonly attacks members of the deer family, but many species readily and severely attack man. No tabanid is absolutely host-specific, even though definite preferences may be chosen. Mammals other than Bovidae, deer, and camels may be chosen, and man may be attacked in the absence of more preferred hosts of the zoophilous species. Birds have been reported as hosts on several occasions. Amphibians may be attacked; reptilian hosts include crocodiles, lizards, and land and sea turtles.

TABANID TRANSMISSION OF PATHOGENS: GENERAL CONSIDERATIONS

A thorough review of this subject by Krinsky (1976) brings information up-to-date and evaluates the various aspects of it. The overall impression one gets is that, despite the large amount of information available, much of it is circumstantial evidence and one cannot, in many cases, draw firm conclusions without more solid investigation. Nevertheless, the mass of such information is very impressive, and the potential of tabanid transmission of pathogens of man and domestic and wild mammals is high.

Krinsky (1976) discusses five biological adaptations of tabanids that enhance such possibilities and probabilities. (1) ANAUTOGENY, the requirement of a blood meal for maturation of eggs, stimulates the host-seeking behavior, to which the transfer of the pathogen from a carrier to a susceptible host is incidental. (2) TELMOPHAGY, or pool feeding, is the rule; microorganisms from the lacerated superficial tissues, as well as from the peripheral blood, are drawn into the pools from which the fly sucks blood. (3) Tabanids take relatively large blood meals, which consequently can contain a large number of pathogens. (4) Some species have relatively long engorgement times, permitting adequate contact for the transfer of pathogens. (5) The intermittent feeding habit of tabanids increases the likelihood of their being involved in mechanical transmission.

A fly interrupted in or ceasing its feeding activities on one individual may readily move to another, so that sick and healthy individuals may be attacked in succession by the same fly. This behavior is particularly noticeable where animals are bunched together for mutual protection from their tormentors. The pain inflicted by the tabanid bite, which usually is considerable, might be considered a negative adaptation for fly survival. Krinsky points out that it increases the chances of interrupted feeding and consequently of movement from one animal to another.

Tabanids may be involved in the transmission of protozoan, helminthic, bacterial, and viral diseases of man and animals. In most cases transmission is merely mechanical, and it is in relation to such that our conclusions are least secure. However, in at least two protozoans, *Trypanosoma theileri* and *Haemoproteus metchnikovi,* transmission involves multiplication of the pathogens, and at least four nematodes have tabanids as invertebrate hosts. In addition to the actual feeding by tabanids, the flow of blood caused by this activity attracts such facultative blood feeders as *Hippelates* species and certain muscoids. These insects may be secondarily involved in transmission of such pathogens as the virus of equine infectious anemia, *Anaplasma marginale, Bacillus anthracis, Listeria monocytogenes,* and *Trypanosoma evansi.*

PROTOZOAN INFECTIONS

Tabanids have the potential for transmitting a number of pathogenic protozoans to mammals and, to a limited extent, other vertebrates. *Haemoproteus metchnikovi,* the only haemosporidian protozoan known to be associated with tabanids, parasitizes turtles; the fly is an essential host and transmission is cyclopropagative. *Besnoitia besnoiti,* the causative agent of bovine besnoitiosis, is, according to Bigalke's studies in 1968 (cited by Krinsky, 1976) transmitted by tabanids mechanically.

The Trypanosomiases. Tabanids transmit several species of mammalian trypanosomes, either mechanically or biologically. In some instances the process of transmission is of extremely short duration and of little importance, as it is in the almost immediate transfer of a tsetse-harbored trypanosome to a nearby susceptible host. On the other hand, it may involve the mechanical transmission of an important parasite of domestic animals, for example, *Trypanosoma vivax* in Central and South America or *T. evansi,* or the biologically transmitted *T. theileri.* A valuable reference on the trypanosomes of mammals is Hoare (1972).

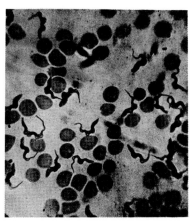

Fig. 11-5 *Trypanosoma evansi,* the causal organism of surra. (After Yutuc.)

Surra, caused by *Trypanosoma evansi,* is a frequently fatal disease of horses, camels, and dogs; it occurs in cattle and buffaloes as usually a nonpathogenic infection, and consequently these animals may become a source of infection for more susceptible hosts. Some wild animals are recorded as hosts, and guinea pigs, white mice, and monkeys are highly susceptible laboratory animals. Surra is widespread in Africa, roughly north of the tsetse zone, in much of southern Asia including the Philippines and Indonesia, and in Central and South America. The disease has been known, according to Hoare, "since time immemorial," but its etiology has been known only since the investigations of Evans and Steel in the 1880's.

Trypanosoma evansi (Fig. 11-5) is very closely related to the tsetse-borne *T. brucei,* from which it may have been derived. Hoare considers *T. hippicum, T. equinum,* and *T. venezuelense* as either outright synonyms or only forms of *T. evansi,* though these specific names have been used extensively in the literature. Among the popular names given to the disease are gufar and debab. The name *mard el debab,* which translated means "sickness (of) from the gadflies," suggests that the Arabs had made some association between the disease and horse flies as early as 1905.

Surra as a disease is subject to many variations, which are discussed by Hoare. The course of infection depends upon the susceptibility of the host, the strain of the parasite, and the severity of the disease. The trypanosomes, after an incubation period of several days, invade the blood, and a series of crises and relapses occurs concurrent with the rise and decrease of the parasitemia. Death may result, or the disease may go into a chronic stage, or recovery may occur. The clinical pattern is similar to that of nagana.

All evidence indicates that transmission, unlike that of nagana, is mechanical. *Stomoxys* species and other bloodsucking flies, including to a limited extent tsetse, have been suspected, but the evidence is that tabanids, particularly species of *Tabanus,* are the chief vectors. *Haematopota* and *Chrysops* species appear to be poor vectors, but extensive studies by Nieschulz, in Indonesia, in 1925–1930, and many others, have incriminated about 25 species of *Tabanus* as at least probable vectors. In the New World, the vampire bat, *Desmodus rotundus,* enters the transmission picture, along with *Tabanus* species. An important aspect of this relationship is that the trypanosome can survive longer in the bat than in the horse fly, but the *Tabanus*-horse/cow cycle and the bat-horse/cow cycle occur together. Dogs and other carnivores can acquire the parasite by eating the fresh flesh of an infected animal. Therefore, though the transmission picture is more complex than one involving only an insect vector, the role of *Tabanus* spp. seems of prime importance.

Trypanosoma theileri is a parasite of domestic cattle and other oxen and antelopes throughout a large part of the world; it is present on all continents except Antarctica. The effect of parasitism by this species on its host is a matter of controversy; the common opinion is that it is harmless, though, in association with pathogens of certain other diseases, such as rinderpest, it may be pathogenic, and there is some evidence that under certain circumstances it may lead to bovine abortion. What was probably this parasite was observed in the gut of a horse fly by Leeuwenhoek in 1680 (Hoare, 1972). Largely because of the work of Nöller in 1925, and as indicated by the fact that the small number of trypanosomes found in ordinary cattle blood could not, without some multiplication process, account for a parasitemia large enough for successful transmission by biting flies, it is generally accepted that a cyclopropagative development takes place in the tabanid host. So, unlike *T. evansi*, *T. theileri* requires a tabanid intermediate host. Tabanid species known to serve as such are *Haematopota pluvialis* and *Tabanus glaucopis*, of Europe, and *Tabanus striatus*, of Java.

Tsetse-borne trypanosomes may in some cases be transmitted mechanically to man or animals within a very brief period and when susceptible individuals are in proximity to infected ones. There is some circumstantial evidence that *Trypanosoma brucei gambiense*, rhodesian strain, may be transmitted in this way. There is evidence that *T. congolense*, which causes a type of nagana and usually develops cyclically in tsetse flies,

may be transmitted by other bloodsucking flies, chiefly tabanids, in tsetse-free areas, either within or outside the tsetse zone, in a purely mechanical fashion. The same is true of *T. simiae*, chiefly a parasite of pigs, despite the implication from its name that it is a monkey parasite; not only the fact that epizootics may occur away from known tsetse infestations, but also the rapidity with which the disease may occur within a herd, suggests mechanical transmission by a bloodsucking fly such as *Haematopota* species or *Stomoxys* species. An important cattle parasite, *T. vivax*, has been introduced into Mauritius and it has been established and has assumed considerable importance in parts of the West Indies and Central and South America, all areas where *Glossina* do not exist and where tabanids appear to be its chief vectors. According to Hoare, the American strain has lost its capability of developing cyclically in *Glossina* species.

Animal trypanosomiases of an unrecognized, benign, or asymptomatic nature and caused by tabanid-transmitted agents may occur outside the areas in which known forms now occur. Krinsky and Pechuman (1975) recovered trypanosomes in 27 percent of the 641 specimens and 69 percent of the 36 species of tabanids trapped by them in central New York State. Though the identities of most of these trypanosomes could not be established, a monogenetic parasite, *Crithidia* sp., and a nonmonogenetic one that the authors considered a *theileri*-like species of *Trypanosoma*, were found or cultured in several instances. There is evidence that the latter was probably acquired from mammalian blood; the tabanids involved with it in this study were *Tabanus quinquevittatus* and *T. superjumentarius*.

HELMINTH INFESTATIONS

Three filarioid nematode parasites of mammals, namely, *Dirofilaria roemeri*, *Loa loa*, and *Elaeophora schneideri*, are known to develop in tabanid invertebrate hosts.

Others may do so, but present evidence does not justify their inclusion. *Onchocerca gibsoni,* once thought to be tabanid-associated, apparently should be removed from that category.

The Australian *Dirofilaria roemeri,* a parasite of the wallaroo, has *Dasybasis hebes* as its tabanid host. There is lack of experimental evidence for including the closely related *D. repens* as a second tabanid-associated member of its genus.

Loiasis. *Loa loa* (Fig. 11-6), an African eye worm that occurs in various parts of the tropical rain forests of western and central Africa, is the best known of the tabanid-associated nematodes of mammals. The adult worms inhabit the superficial subcutaneous connective tissue and are known to move around quite rapidly. The parasites have been observed in many parts of the body, such as the scrotum, penis, breast, eyelid, anterior chamber of the eye, tongue, finger, and back. They may be most readily excised when they travel across the bridge of the nose or the conjunctiva.

Microfilariae of *Loa loa* are found in the peripheral bloodstream in either a diurnal or nocturnal periodicity. In this stage they are ingested by tabanids of the genus *Chrysops,* and they undergo development (Fig. 11-6) similar to that of *Wuchereria bancrofti* in the mosquito. Metamorphosis is completed in 10 to 12 days. When the infected fly bites, the mature infective larvae issue from the proboscis and enter the skin of the host.

A nocturnally periodic, or simian, and a diurnally periodic, or human, loiasis, have been recognized, and it has been thought that the former was transmitted from one sleeping monkey to another by night-biting *Chrysops* species and through the human cycle by day-feeding species. However, Duke (1972), in reviewing several years' work on simian and human loiasis, has noted that there is no epidemiological nor experimental evidence for considering the two strains identical and, on the other hand, there are physiological, ecological, and genetic reasons why exposure of monkeys to the diurnal strain would not result in an infection that could be vectored by night-biting *Chrysops;* he therefore concludes that there are two strains of *Loa loa* that may be in the process of separating specifically.

So far as is known, only *Chrysops* species serve as invertebrate hosts of *Loa loa. C.*

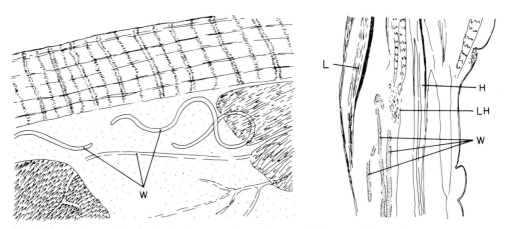

Fig. 11-6 Sections of *Loa loa* in the tissues and mouth parts of *Chrysops silacea.* On the *left* larvae may be seen among muscles of the head region. On the *right* infective larvae are in the process of escaping through the labiohypopharyngeal membrane. *H,* Hypopharynx; *L,* labium; *W,* worms. (Redrawn from photomicrographs by Lavoipierre, 1958, *Ann. Trop. Med. Parasit.,* **52:**103.)

silaceus and *C. dimidiatus,* the mango fly, are the chief vectors of the diurnal form, and *C. centurionis* and *C. langi* are primarily involved in transferring the parasite among monkeys in the forest canopy. *C. atlanticus* of the eastern United States will support the development of *Loa loa* to the infective stage (Orihel and Lowrie, 1975).

Elaeophorosis. The filarial parasite, *Elaeophora schneideri,* also called the arterial worm or blood worm, is found in the arteries of deer, domestic sheep, elk (American), and moose in the western and southwestern United States. The parasites make their way into the arteries supplying the brain, eyes, ears, muzzle, and other parts of the head; the result in elk is damage to the central nervous system, blindness, cropping of the ears, necrosis of the muzzle, and antler deformity. Reinfection of animals that survive may be fatal, though recovery is usually spontaneous in the initial attack, perhaps coinciding with the death of the parasites. The disease is asymptomatic in American deer, probably an indication of a normally developed host-parasite relationship; it is serious in elk, and domestic sheep may develop a severe dermatitis.

Species of *Hybomitra* and *Tabanus* serve as invertebrate hosts (Hibler and Adcock, 1971). In studies by these authors in the Gila National Forest, 183 of 927 tabanids, or 19.7 percent, were infected with 3,189 larvae, an average of 17.4 per fly. Third-stage larvae were recovered from the fly proboscids. Species of *Hybomitra* sp. served as hosts much more frequently than did those of *Tabanus* sp.

BACTERIAL INFECTIONS

Although best documented in respect to anthrax and tularemia, and to a somewhat lesser extent in anaplasmosis, involvement of tabanids as vectors of other pathogenic bacteria is a strong possibility. Experimental transmission of *Pasteurella multocida* (buf-

falo sickness of Asia and Africa), *Brucella* spp. (brucellosis), *Listeria monocytogenes* (listeriosis of man and animals), and *Erysipelothrix rhusiopathiae* (swine erysipelas) has been accomplished, but without support by natural transmission or isolation of the infective agent from tabanids. These disease agents, and some others under suspicion, must merely be considered as possibly tabanid-transmitted.

Anaplasmosis. Anaplasmosis, or infectious anemia, a frequently fatal disease of cattle, is caused by *Anaplasma marginale,* at one time referred to the Protozoa but now considered a bacterium. Ticks transmit this organism biologically, but tabanids and other bloodsucking flies are frequently involved mechanically. Tabanids of the genus *Tabanus* have been shown to be experimental vectors, and natural transmission has been effected. It is possible that these flies may be vectors of some importance in at least some parts of the world, although the lack of epizootological evidence correlating the incidence of the disease in nature with information on naturally infected flies fails to substantiate claims as to the importance of tabanids in the transmission of this disease agent.

Anthrax. Anthrax is caused by *Bacillus anthracis.* Nearly all species of domestic animals and man are susceptible; herbivores and rodents are most likely to become infected. After the inoculation of the organism into the animal, its incubation period is from 3 to 6 days. Entrance to the body is gained in various ways: through local pricks and lesions, including insect bites; through inhalation of the spores; through ingestion of food, as in grazing by livestock in contaminated pastures; through dry insect feces; and through drinking from contaminated streams and ponds.

Ample evidence connects horse flies with the transmission of the anthrax bacillus, and it is possible that an epizootic may be started, or at least to an extent maintained, through the injection of this organism into a susceptible population by tabanid bites. Where tabanids abound, they may be of importance in transmitting the anthrax pathogen through their bites and excreta.

Instances are recorded in which the simple bite of an infected fly was all that was needed to produce malignant pustules in human beings. A notable case, cited by Herms, involved a man in the act of burying a cow that had died of anthrax; he was severely bitten by a horse fly on the back of the neck and in due time developed a malignant pustule at the site of the bite.

Krinsky (1976) points out that, despite the "circumstantial or anecdotal" nature of much of the evidence, the successful transmission studies made by a number of investigators indicates that tabanids have the potential for mechanical transmission of the anthrax bacillus. We lack firm evidence, however, to evaluate the importance of this transmission in natural outbreaks.

Tularemia.　In 1919 a disease of man of hitherto unknown etiology occurring in Utah was reported by Francis as deer fly fever or Pahvant Valley plague. The same author identified it with a disease reported among rodents in Tulare County, California, by McCoy and Chapin in 1912, and gave it the name tularemia. Its causative agent is *Francisella tularensis*. In the western United States, it is a disease of rural populations occurring during the summer months, coinciding with the prevalence of *Chrysops discalis,* a deer fly. The disease occurs also throughout the Northern Hemisphere, between latitudes 30° and 71° (Hopla, 1974).

In its acute form, a primary ulcer (eschar) develops at the site of the inoculation. Pneumonic complications may result. In highly susceptible persons it may assume the form of a septicemia and result in death between the fourth and fourteenth days. In nonfatal cases, convalescence is slow.

In the western United States, rabbits constitute an important reservoir for the bacillus. Tularemia is known to exist in nature among many species of vertebrates, among which are meadow mice, ground squirrels, beavers, coyotes, sheep, and quail and other game birds. Both experimental and natural transmission studies have been extensive. These have been reviewed by Krinsky (1976) and Hopla (1974). *Francisella tularensis* has been isolated from populations of several *Chrysops* species, including, in America, *C. discalis, C. fulvaster,* and *C. aestuans.* In the Utah epidemic of 1971, there was abundant evidence of tabanid transmission of possible importance in consideration of the proximity of infected flies to the site of human cases.

However, the role of tabanids in the transmission of *Francisella tularensis* is uncertain except in those localized endemic regions where rabbits serve as an important reservoir. More than 50 arthropods can harbor the bacterium. There is evidence to indicate that, on the whole, ticks may be more important than flies. There are diverse mechanisms, other than arthropod transmission, by means of which tularemia bacilli can be spread, such as infected animals and animal products and contaminated water.

VIRAL INFECTIONS

Equine infectious anemia (EIA), or swamp fever of horses, is a viral disease that may result on occasion from tabanid transmission, where interrupted feeding may occur between infected horses and noninfected ones. Only a few vector species have been tested in only a few areas, but evidence suggests that tabanids may play a role in maintaining transmission and may increase the size of epidemics. Hog cholera virus (HCV) may also be transmitted in this way; *Tabanus lineola* and *T. quinquevittatus* have been shown experimentally to transmit this virus mechanically.

Families Rhagionidae and Athericidae

SNIPE FLIES AND ATHERICIDS

The Rhagionidae and Athericidae have recently been separated from each other (Stuckenberg, 1973) but it is not practical for our purposes to separate them. Their classification is discussed by Stuckenberg (1973) and by Nagatomi (1977), and the latter author provides a key that will separate them.

Both families, though predominantly nonbloodsucking, include species that suck the blood of man, certain domestic animals, and other vertebrates. Some species of *Symphoromyia,* family Rhagionidae (Fig. 11-7), are without doubt important annoyers of man and animals. In this genus the antenna,

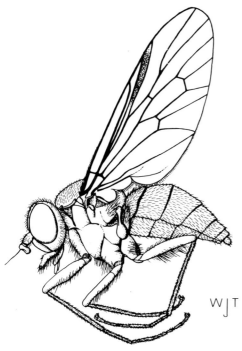

Fig. 11-7 *Symphoromyia atripes,* a common biting snipe fly, family Rhagionidae, from northwestern North America. Body length about 5 mm. (Courtesy of W. J. Turner, Washington State University.)

which is highly variable in the family as a whole, is characteristic, the flagellum being kidney-shaped or nearly so, with a subterminal arista. Another rhagionid genus, the Australian *Spaniopsis,* the species of which also suck blood, has an elongated antennal flagellum that terminates in a style. Three genera of Athericidae include or consist of bloodsucking species: some species of *Suragina* are known to suck the blood of human beings, horses, cattle, and owls; the known species of *Atrichops* attack frogs; and some species of *Dasyomma* are reported to be bloodsuckers (Nagatomi, 1977).

The females of some species of *Symphoromyia* are vicious biters, behaving somewhat as do the tabanid flies of the genus *Chrysops.* They alight on the exposed parts of the body quite silently and singly and often inflict a sudden painful bite before their presence is known. Both the severity of the bite and the accompanying pathologic changes vary with the individual bitten. Man is the most commonly recorded host, but deer are frequently attacked; other hosts include cattle, horses, and dogs. Some species of *Symphoromyia* may be mistaken by the uninformed for stable flies. Among the severe biters are *Symphoromyia atripes, S. hirta, S. limata,* and *S. sackeni.* The genus is best represented in mountain and coastal areas in western North America, although some species are eastern North American or European.

The mouth parts are of the tabanid type, the strongly sclerotized piercing and cutting structures being enclosed in a retractile labial sheath; as in the tabanids, blood is lapped up by the labella.

Little is known about the breeding habits and life histories of *Symphoromyia.* Larvae breed in wet, peaty soil and moss along temporary stream banks and in similar places, a characteristic of the whole family, so far as is known.

The potential medical and veterinary importance of *Symphoromyia* is at the present time undetermined. Shemanchuk and Weintraub (1961) have shown that a fly probes repeatedly in taking a meal. The fact that some species, such as *S. sackeni,* are relatively abundant, may take large volumes of blood, and may feed several times in a lifetime, suggests that such species may serve as vectors of pathogens of deer or other wild animals and may be involved in epizootics or even in the epidemiology of a zoonosis.

The annoyance of the bite and its pathologic consequences, however, may be considerable, even to the extent of interfering with the efficient use of recreational areas. In addition to the pain caused by the bite, some individuals experience an allergic response that, on rare occasions, may be so severe as to necessitate bed rest or hospitalization. Turner (1979) describes in detail a case of violent hypersensitivity so severe that, except for prompt treatment by a physician, it might have terminated fatally. Turner's review of the subject of *Symphoromyia* bites and their effects is a useful general reference.

Control of Horse Flies, Deer Flies, and Snipe Flies

Control of these biting flies can be very difficult because their developmental sites may include extensive marshy or aquatic zones, as well as relatively dry soil environments. Their larval sites are not usually associated closely with human or animal dwellings, and in fact they frequently are not adequately identified. Some species of tabanids develop in salt marshes, in which situations temporary flooding may provide good control (Anderson and Kneen, 1969).

Determining the production of tabanid larvae in marshy or aquatic habitats may be difficult, but Freeman and Hansens (1972) describe a procedure used in salt marshes of New Jersey, USA. Adult tabanids are usually surveyed by counting the numbers attacking cattle, horses, or other large animals. Many tabanids are attracted to large and dark objects, and a number of traps have been developed to take advantage of this behavioral characteristic (Fig. 4-10). In California, Anderson and associates found Malaise traps in combination with the carbon dioxide vaporizing from dry ice to be an effective means of sampling some species of tabanids and for the biting snipe flies, *Symphoromyia*. Attractant traps have been tested as control devices around recreational areas, and though they may literally catch millions of tabanids, there is no adequate indication that this approach reduces the nuisance of their attack (Spencer, 1972).

Jones and Anthony (1964) have presented a review of natural enemies of tabanids. These include hymenopterous egg parasites, dipterous larval parasites, hymenopterous pupal parasites, and predators including insects, spiders, lizards, birds, and fish. Egg masses are often heavily parasitized, wasps in the genus *Telenomus* being quite effective (cf. Dukes and Hays, 1971). A predatory wasp mentioned frequently in the literature is the horse guard, *Bembix carolina* (often referred to the genus *Strictia*).

Personal protection measures seem to be fairly effective, such as use of the repellent deet on humans, and a number of repellents on livestock (cf. Shreck *et al.,* 1976). Control on range animals is a problem because treatment must be renewed once or twice a day. Dairy cattle or horses handled daily may be hand sprayed; cattle in enclosures causing them to go through a chute at least once a day in search of food or water can be automatically treated, using a treadle-step switch or electric eye-operated sprayer (cf. Cheng *et al.,* 1957; Berry and Hoffman, 1963); dusting bag applicators (Fig. 5-7) can also provide some relief.

Chemical control away from the host is difficult. A pesticide applied to vegetation where tabanids congregate can produce complete, though temporary, control (Brown and

Morrison, 1955). Larvae of *Chrysops* were killed by the application of pesticide to breeding sites, and *Tabanus* larvae have been controlled in salt marshes by applying pesticides in granular form (Hansens, 1956; Jamnback and Wall, 1957). However, the generaly utility of such procedures is questionable, because of the extensive areas that must be treated, and the very real danger of undesirable side effects.

12

MUSCOID FLIES AND LOUSE FLIES

ORDER DIPTERA; SUBORDER CYCLORRHAPHA

Characteristics of Adults. In order to understand the descriptive terms used in reference to the higher Diptera, or Cyclorrhapha, certain anatomical structures need to be discussed. The most conspicuous feature of the head (Fig. 12-1) is the pair of large compound eyes, widely separated from each other in the female, but often contiguous or very narrowly separated in the male, at least in the calyptrate muscoids, thereby greatly restricting the areas that lie between them. The ocelli are located on a vertical triangle that is usually clearly demarcated; in the Chloropidae (Fig 2-5), this triangle is very large and may occupy most of the vertex and a considerable area of the frons. Special areas of the head mentioned in the text may be identified by reference to Figure 12-1.

The antenna (Fig. 9-2c) is aristate, with an undivided flagellum (except for the arista, which is actually a part of the flagellum). The second segment is set at a distinct angle to the first, so that the antenna is PENDANT rather than produced horizontally forward

(PORRECT), as in most insects and many Diptera. In the calyptrate muscoids the second segment is partially divided by a seam or suture that runs most of the length of its outer dorsal side; this is not true of the acalyptrate muscoids (Drosophilidae, Chloropidae, Ephydridae, and others). The arista may be bare or hairy; the hairs may be situated above and below, as in the house fly, or above only; if the latter, they may be simple, as in the stable or horn fly (Fig. 12-20), or plumose, as in the tsetse (Fig. 12-12). The house fly antennal type is called plumose, the stable fly or tsetse antennal type, pectinate.

CHAETOTAXY is the arrangement of bristles (MACROCHAETAE); it is shown for the head and thorax in Figures 12-1 and 2. The bristles of the abdomen are of less importance for our purposes.

Certain other thoracic structures should be mentioned. The two THORACIC SPIRACLES have been shown to be the MESOTHORACIC and METATHORACIC, respectively (Fig. 12-2, *mss* and *mts*). The PROPLEURON lies in front of the mesothoracic spiracle; whether its central area is hairy or not may be of taxonomic significance. In the calyptrate muscoids, a

248

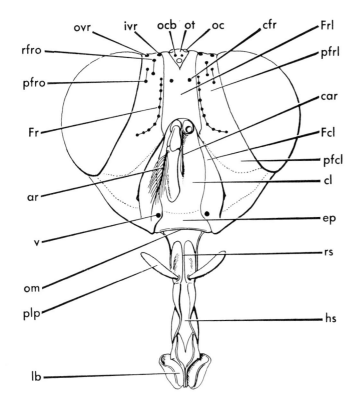

Fig. 12-1 Diagrammatic drawing of the head of a muscoid fly, from front view. Explanation of abbreviations: *ar,* arista; *car,* facial carina; *cfr,* cruciate frontal bristle (scar); *cl,* clypeus; *ep,* epistoma; *fcl,* faciale; *fr,* frontal row (bristle scars); *frl,* frontale (frontal stripe or vitta); *hs,* haustellum; *ivr,* inner vertical bristle (scar); *lb,* labella; *oc,* ocellus; *ocb,* ocellar bristle (scar); *ot,* ocellar triangle; *ovr,* outer vertical bristle (scar); *pfcl,* parafaciale; *pfrl,* parafrontale; *pfro,* proclinate fronto-orbital row (bristle scars); *plp,* palpus; *rfro,* reclinate fronto-orbital row (bristle scars); *rs,* rostrum; *v,* vibrissa. (Drawing by A. Cushman; USDA.)

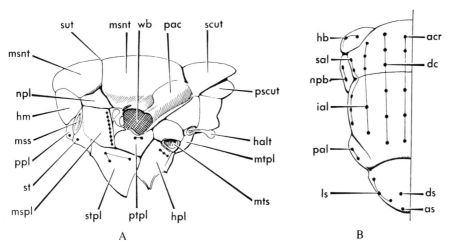

A   B

Fig. 12-2 Diagrammatic drawing of the thorax of a muscoid fly. *A.* Side view. *B.* Left half, dorsal view. Abbreviations of areas and bristle scars: *acr,* acrostichal row; *as,* apical scutellar; *dc,* dorsocentral row; *ds,* discal scutellar; *halt,* halter; *hb,* humeral row; *hm,* humerus; *hpl,* hypopleuron; *ial,* intra-alar row; *ls,* lateral scutellar row; *msnt,* mesonotum; *mspl,* mesopleuron; *mss,* mesothoracic spiracle; *mtpl,* metapleuron; *mts,* metathoracic spiracle; *npb,* notopleural row; *npl,* notopleuron; *pac,* postalar callus; *pal,* postalar row; *pscut,* postscutellum; *ptpl,* pteropleuron; *sal,* supra-alar row; *scut,* scutellum; *st,* stigmatal; *stpl,* sternopleuron; *sut,* mesonotal suture; *wb,* wing base. Bristle scars not labeled are named in accordance with the sclerite on which they are located (propleural, sternopleural, mesopleural, hypopleural, pteropleural). (Drawing by A. Cushman; USDA.)

well-defined TRANSVERSE MESONOTAL SUTURE is present; this is absent or incomplete in the acalyptrates. The area between each wing base and the corresponding side of the scutellum, the POSTALAR CALLUS, is well differentiated. Two pairs of SQUAMAE, one, the ALAR, more closely associated with the wing, and the other, the THORACIC, closer to the wall of the thorax, are present and often large.

Larvae. The typical muscoid larva, called a MAGGOT, is legless, more or less cylindrical but strongly tapering anteriorly and truncate posteriorly. It is distinctly segmented, with 12 segments including the so-called "CEPHALIC" segment (Fig. 12-3) usually clearly visible; however, Zumpt

(1965) has pointed out that the last apparent segment is a composite of 2 actual ones, and sometimes 13 segments may be counted. Some muscoid larvae differ from this general pattern. The cattle grubs and bot flies of sheep and horses, for example, are robust and more oval; so is the tsetse larva during its brief free stage, and it is further peculiar in the presence of two prominent, projecting, POSTERIOR RESPIRATORY LOBES which are also very evident in the pupae (Fig. 12-14). The larva of *Fannia* is flattened, with conspicuous processes extending from the body (Fig. 12-7).

At the somewhat pointed anterior end, the MOUTH HOOKS are prominent, unless the mouth parts are nonfunctional, as in the

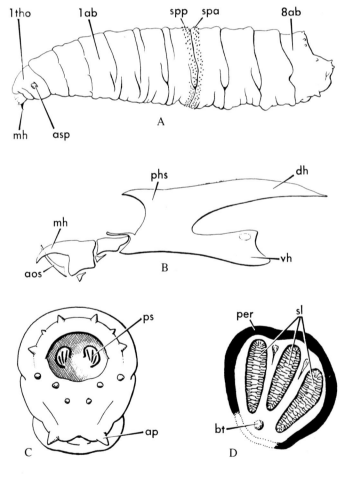

Fig. 12-3 Diagrammatic drawing of the mature larva of a muscoid fly. *A.* Lateral view. *B.* Cephalopharyngeal skeleton. *C.* Posterior view. *D.* A posterior spiracle. Explanation of abbreviations: *ab,* abdomen; *aos,* accessory oral sclerite; *asp,* anterior spiracle; *bt,* button; *dh,* dorsal horn of the pharyngeal sclerite; *mh,* mouth hooks; *per,* peritreme; *phs,* pharyngeal sclerite; *ps,* posterior spiracle; *sl,* slits of the posterior spiracle; *spa,* spines of the anterior margin of the segment; *spp,* spines of the posterior margin of the segment; *tho,* thorax; *vh,* ventral horn of the pharyngeal sclerite. (Drawing by A. Cushman; USDA.)

tsetse or the mature cattle grubs. They move vertically, not horizontally, as in the lower flies, and form a part of the CEPHALO-PHARYNGEAL SKELETON (Fig. 12-3), which assumes a variety of forms, though usually conforming to a general pattern. There are two pairs of SPIRACLES, the ANTERIOR (Fig. 12-3, *asp*) and the POSTERIOR (Fig. 12-3, *ps*), though the former may be lacking; the latter present considerable variation. The posterior spiracular openings usually consist of two or three slits of characteristic form for the species, although sometimes, as in *Hypoderma,* there are instead numerous small openings. The body is more or less covered with spines, which may be fine or coarse, either located in transverse rows or bands or generally distributed over the body. No true appendages occur, though there may be PSEUDOPODS, TRACTION BANDS, or more or less conspicuous body processes.

Pupae. The mature larva ceases to feed and seeks a place for pupation, often entering the ground. The third instar integument then hardens, forming an encasing PU-PARIUM; this process is usually called pupation, though the term PUPARIATION is more apt. The puparium, since it is formed of the last larval integument, presents many of the characters of the mature larva. True pupation takes place within the puparium, and, after the necessary transformation period, the adult by means of a bladderlike, eversible and withdrawable structure called the PTI-LINUM pushes its way through a circular opening thus formed at the anterior end of the puparium.

Literature. The literature on this highly important group of insects is voluminous. References pertinent to our subject will be given in connection with the special groups discussed. A highly important work of general nature is the two-volume monograph of Greenberg (1971, 1973). To make identifications is often the task of the specialist, but the manuals cited at the introduction to the Diptera section will be of general value.

In addition to these, some useful manuals are James (1948), Van Emden (1954), Zumpt (1965), Shtakelberg (1956, in Russian), Greenberg (1971), and K. G. V. Smith (1973). A very useful work that deals with third-stage larvae only is Ishijima (1967). Although some of these works are concerned primarily with a restricted fauna, they assume more general importance than might be expected because of the wide distribution of many synanthropic flies.

Family Chloropidae

HIPPELATES AND SIPHUNCULINA—EYE GNATS

Hippelates. *Hippelates* flies are acalyptrates of the family Chloropidae. Members of this genus are as a rule very small, 1.5 to 2.5 mm in length. They are called "eye flies" and "eye gnats" because they frequently come to the eyes of the victim; they also are attracted to sebaceous secretions, pus, and blood, exposed genital organs of mammals (e.g., *Hippelates pallipes* clustered around a dog's penis), and sores (e.g., *H. flavipes* on yaws sores). They approach their mammalian host quietly, usually alighting some distance from their feeding site; to reach it, they will then crawl over the skin or resort to intermittent flying and alighting, thus adding to the annoyance of the host. They are extraordinarily persistent and if brushed away will quickly return and continue engorging themselves. They do not bite; however, the labella are provided with spines that in some species act as scarifying instruments capable of producing multiple minute incisions, thus aiding in the entrance of pathogenic organisms.

Unlike the gnats discussed in Chapter 9, all of which are Nematocera, the Chloropidae have short aristate antennae and are more like the house fly in form or structure, though much smaller. The pomace or vinegar flies, *Drosophila*, resemble some *Hippelates,* but they have a distinctly feathered arista, whereas that of *Hippelates* is at most pubescent. A distinct feature that will separate the Chloropidae from related and similarly appearing families is the very large vertical triangle (see Fig. 2-5). *Hippelates* may be distinguished from most members of the same family by the presence of a distinct shining black apical or subapical spur on the hind tibia. The larvae of many Chloropidae live in grass or other plants (stem maggots); however, those of *Hippelates* develop in a wide variety of materials, usually involving decaying plant or animal material incorporated into the soil during farming operations.

LIFE HISTORY AND HABITS. These have been summarized very ably by Greenberg (1973). The life history and biology of *Hippelates collusor,* a member of the very troublesome *H. pusio* group, have been studied extensively in southern California. The information detailed below appears to be representative of the group. The fluted, distinctly curved egg (Fig. 2-5), about 0.5 mm in length, is deposited on or below the surface of the soil; the average incubation period under optimum conditions (about 32° C) is about 2 days. The larvae feed on a great variety of decaying organic matter, including excrement, provided the material is rather loose and well aerated. According to Burgess (1951) the larvae will not develop naturally in closely compacted soil or putrid material, nor will they breed in excrement unless it is mixed with loose earth. The larval stage under optimum conditions requires about 7 to 12 days. During the winter months the larval and pupal stages may be prolonged over many weeks.

Pupation takes place close to the surface of the material in which the larvae develop. The pupal stage requires about 6 days and the preovipositional adult about 7, giving a total of about 21 days from egg to adult fly, or about 28 days from egg to egg. Eggs are deposited in batches up to 50, usually followed by a second but smaller batch after about 7 days (Burgess, 1951). Unfavorable breeding media and low temperatures may slow larval development.

The majority of *Hippelates* gnats develop in light, well-drained sandy soils that are freshly plowed (i.e., plowed not more than 3 weeks before) and contain abundant humus or vegetable matter, such as cover crops or manure, and sufficient moisture. In studies in California Mulla and Axelrod (1973) found that leafy plant tops in the soil produced greater numbers of *Hippelates* than did maize stubble and Bermuda grass; steer manure, chicken manure, and blood meal were favorable to the flies' development, but nitrohumus and the chemical fertilizer ammonium sulfate sustained little or no development or emergence. The authors concluded that the two chemicals could be used for soil improvement without affording a breeding ground for *Hippelates*.

Hippelates gnats are generally strong fliers. They fly with and against the general direction of the wind, although wind velocities of 8 km per hour considerably reduce their flight activity. Dow (1959) found *H. pusio* in Georgia to fly as far as 1.6 km from release point. In a more extensive study in the Coachella Valley of California, Mulla and March (1959) found *H. collusor* to become widely distributed within 4 to 6 hours from the time of release and to travel as far as 7.5 km.

Annoyance may occur in a variety of places, such as residential areas, sprinkled lawns, irrigated areas, school yards, golf courses, and dense shrubbery.

CLASSIFICATION. The genus *Hippelates* is strictly American, with a number of species extending from Quebec and the northern United States southward to Argentina and northern Chile. Sabrosky (1941) reviewed the *Hippelates* of the United States and in 1951 clarified the status of certain species. Those of most medical importance belong to the *pusio* group, in which the body is black and shining, at most interrupted by an opaque (pollinose) area at the base of the wing. These are *H. pusio, H. flavipes, H. pallipes, H. bishoppi,* and *H. collusor;* two other species belong to this group. *H. flavipes* and *H. pallipes* have been confused with each other in the literature, but *H. flavipes* is a tropical species, and *H. pallipes* belongs to the temperate faunas. Some members of other groups of *Hippelates* may annoy man, but on the other hand, some species (e.g., *H. hermsi*) are not markedly attracted to humans.

Siphunculina. In southeastern Asia *Siphunculina funicola* plays a role in annoyance and pathogen transmission similar to that of *Hippelates* in America. This species is attracted to blood on horses and cattle that has flowed from wounds made by biting flies, to serous discharges from the eyes and from sores, and to human and animal excrement. It can be an intolerable nuisance to man in parts of India throughout the year during hot weather. Breeding places are varied, but all known ones are characterized by organic pollution. The life cycle according to Syddiq (cited by Greenberg) requires about 10 days from egg to egg.

Medical and Veterinary Importance. *Hippelates* and *Siphunculina* gnats are probably of greater medical importance than is generally realized. Annoyance alone by them is significant. The habit of these flies of freely visiting sources of contamination, plus the scarification of tissues in the feeding process, can readily dispose to mechanical transmission. Even though these flies are not bloodsuckers, they congregate near and lap up blood from wounds created by the feeding of tabanids, *Stomoxys,* and other bloodsuckers. In this way, *Hippelates* species could aid in the transmission of the anaplasmosis organism to cattle (Roberts, 1968), though proof of this is lacking.

Transmission by *Hippelates* species of the agents causing human pinkeye, yaws, and bovine mastitis has been accepted as probable, though not essential. The habits of *Hip-pelates* and *Siphunculina* should be considered also in relation to certain transmissible tropical ulcers. A streptococcal skin infection caused by *Streptococcus pyogenes* in Trinidad implicated *Hippelates peruanus, H. flavidus,* and *H. currani* by epidemiological information and the recovery of the pathogen from flies in the vicinity of infected children. Naturally infected flies retained *S. pyogenes* for 28 hours and visited humans within that time (Bassett, 1970). Staphylococcal infections may also involve *Hippelates flavipes,* as indicated by the studies of Talpin and associates (1967) in Panama.

CONJUNCTIVITIS. *Hippelates* flies have long been looked upon with suspicion in certain parts of the southern United States and Mexico as possible vectors of the bacilli of pinkeye, an acute bacterial conjunctivitis. Since 1912 the correlation between outbreaks of pinkeye and adult *Hippelates* abundance in southern California has been noted; a seasonal conjunctivitis in Florida, Georgia, and other parts of the South seems associated with *Hippelates* abundance; the seasonal prevalence of *Siphunculina funicola* coincides with epidemic conjunctivitis in Assam, India. Yet we lack the critical experimental evidence linking *Hippelates* and *Siphunculina* with transmission of bacteria causing pinkeye, and not a single isolation of an eye pathogen has been reported from these flies (Greenberg, 1973).

The *Hippelates* species most closely associated with human annoyance and probable transmission of the pinkeye agents are *H. collusor* in the southwestern United States and *H. pusio* in the southeast. *Siphunculina funicola* fulfills this role in southeast Asia and other parts of the Oriental Region.

YAWS. As indicated in Chapter 1, flies have been suspected for many years as vectors of the pathogen of yaws (Fig. 12-4). This disease is caused by a spirochete, *Treponema pertenue;* it is widespread in the tropics of both hemispheres and is essentially a rural disease, particularly among peoples

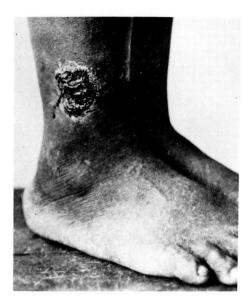

Fig. 12-4 Primary lesion of yaws. (Courtesy of US Armed Forces Institute of Pathology. Negative No. N-39207.)

of the lower social and economic levels. Though positive experimental evidence is lacking, the weight of circumstantial evidence is so strong that there is little doubt but that *Hippelates,* and probably also *Siphunculina,* are involved in the transmission of this pathogen to an important extent. The most convincing evidence was collected by Kumm and his associates in Jamaica in 1935 and 1936 in studies with *Hippelates flavipes* (incorrectly identified as *H. pallipes*). They found it relatively easy to demonstrate motile *Treponema pertenue* in the vomit drops of flies that had fed on infectious lesions of yaws. Of 500 such flies studied, 71 percent were infected. The spirochetes, however, were presumably digested in the midgut and hindgut, as none were seen after an interval of 2 days. It is important to note several epidemiologic facts: the distribution of yaws is much broader than that of its reputed chloropid vectors; transmission is mechanical and can be made other than through the agency of chloropid or other flies; fly transmission is effected through feeding by the in-sect on available primary lesions that exude serum containing large numbers of spirochetes; and positive data as to the extent of fly involvement are lacking. Nevertheless, the case against *Hippelates* in mechanical transmission of yaws remains strong.

BOVINE MASTITIS. Like yaws in man, the pathogens of this important disease of dairy cattle are transmitted in various ways, and the situation is complicated because not one, but a complex of pathogens, is involved. Sanders (1940) reported investigations in Florida that incriminate *Hippelates* (species not given) as well as the house fly as mechanical transmitters of bovine mastitis. *Hippelates* flies were seen to hover around natural openings of calves, yearlings, pregnant heifers, and lactating cows. They fed on lacrimal fluid, fatty body secretions, accidentally spilled milk droplets, and secretions at the tip of the teats of animals in herds where mastitis was prevalent. In exposure tests made with flies feeding alternately on infected material and the teat orifice, mastitis developed in each of the experimental animals.

CONTROL OF EYE GNATS (HIPPELATES)

Hippelates flies are readily sampled by using attractive baits. Simple traps consist of bait covered by a cone leading into a collecting chamber, and rather coarse screen preventing access by large flies. A number of decaying materials are suitable attractants, and Mulla and co-workers (1973) have developed a bait of dried fermented poultry egg product.

Methods for finding natural parasites of *Hippelates* have included exposing larvae and pupae in natural environments, protecting them from predators by screening. By this means Legner and Bay (1964) discovered two species of cynipid wasp parasites in the West Indies; of these *Spalangia drosophilae* proved to be the most efficient species in seeking *Hippelates* pupae.

Insecticides worked into soil may adequately control *Hippelates,* but the expense rarely justifies such a procedure. Ultralow volume applications of pesticides may provide temporary relief from adult annoyance (Axtell, 1972). Attractant baits combined with a pesticide can provide highly effective area control of adult *Hippelates* (Mulla, Axelrod, and Ikeshoji, 1974). Legner (1970) indicates how toxicant baits and soil management can be combined to provide integrated control of *Hippelates collusor.*

Acalyptrates of Minor Medical Importance

A number of small Acalyptratae (Cyclorrhapha) belonging to several families may be nuisances indoors and may even be mechanical transmitters of pathogens. The most familiar of these are the vinegar flies, family Drosophilidae. The numerous species of *Drosophila* breed under a variety of conditions, such as in overripe fruits, decaying potatoes and other vegetables, garbage cans, and milk bottles that are not thoroughly cleaned. Except for the nuisance value, most species are harmless, though some, such as *Drosophila repleta,* breed in human feces and may be mechanical vectors of pathogens.

Several other acalyptrate families may include mechanical vectors. The numerous but less familiar Sphaeroceridae usually do not enter houses, though they may do so at times. They breed in a variety of decaying matter and excrement; they are proven carriers of *Salmonella,* and they frequent outdoor food markets in warmer climates, for example, in Mexico (Greenberg, personal communication to James). The European *Leptocera caenosa* often occurs in houses; its presence there can be traced to defects in the sewage system where it has been breeding. *Teichomyza fusca,* family Ephydridae, has been recorded breeding in large numbers in a faulty toilet. A cosmopolitan species,

Piophila casei, the cheese skipper, family Piophilidae, infests a variety of human foods, not only cheese, but bacon, dried fish and meat, and other high-protein foods.

In addition to the eye gnats, already discussed, the chloropid *Thaumatomyia notata* deserves mention. This fly is harmless, actually a beneficial predator, though at times it may enter houses in such large numbers as to become a nuisance. Several other acalyptrate flies have this same habit; so do certain Muscoidea, but they will be considered later in this chapter.

Some saltwater breeders may occur outdoors in such large numbers, usually because of a paucity of natural enemies, as to become nuisances. The ephydrids *Ephydra cinerea* and *Hydropyrus hians* interfere with the tourist trade around Great Salt Lake (Winget *et al.,* 1969). Coelopids breeding on the seashore may be a nuisance, especially when large numbers of them are blown inland. Other seashore breeders, such as *Fucellia,* a calyptrate fly (Family Anthomyiidae), may annoy vacationers because of their vast numbers.

Some other acalyptrates will be discussed in Chapter 13.

Superfamily Muscoidea

HOUSE FLY AND RELATIVES

House-Invading and Other Man-Associated Flies. Many species of robust flies belonging to several families of Diptera are commonly found indoors. Some of these are actual or potential menaces to human health in that they habitually enter the house and come in contact with human food or drink after breeding or feeding in excrement, dead animal material, or other contaminated media. Other species, which do not necessarily enter houses, may make similar contacts in markets or may be so closely associated with man as to form a definite threat to his health. These same flies, and others that have no or little relationship to the transmission of pathogenic organisms, may also be of importance because of annoyance and interference with human comfort. Finally, such flies may affect the well-being or even existence of man's domestic animals. Though other groups are represented, most flies belonging to this category fall into the families Muscidae, Calliphoridae, and Sarcophagidae, superfamily Muscoidea.

The bionomics of the house fly and related forms of medical and sanitary importance have been treated in a valuable monograph by West (1951), and a condensed account of the relationship to disease, with an extensive list of citations, has been presented by Lindsay and Scudder (1956). An important inclusive bibliographical source for *Musca domestica*, through 1969, is West and Peters (1973); a supplement to this is now in preparation. A very important contribution on flies in relation to disease, including extensive bibliographical references, tables, keys, and illustrations, ecology, and biotic associations, is the two-volume work of Greenberg (1971, 1973), in collaboration with other workers.

Synanthropy in Muscoids. Those flies that have entered the man-dominated ecological community (or human biocoenosis), and consequently coexist with man over an extended period of time, have been referred to by European workers as SYNANTHROPIC species. The chapter by Povolný in Greenberg (1971) gives a very lucid account of synanthropy in flies and can well serve as an introduction to this subject. A significant fact is that a large number of Northern Hemisphere synanthropes are very widespread, often being holarctic or circumpolar. A basic concept is that the synanthropic biocoenosis has developed, with man as its creator and his culture and domestic animals as its products, with the synanthropic animals adapting themselves to becoming its spontaneous, though often unwelcome, members.

Various degrees of synanthropy exist, from a total association with man to one that is quite loose and facultative. Total association involves complete dependence upon the man-controlled environment, including households, food-processing plants, slaughterhouses, and such, for their complete development (the true or EUSYNANTHROPES). Requirements for this development include larval food as well as other biotic needs of the environment. Most synanthropic flies (the HEMISYNANTHROPES) do not depend totally on man's environment, though they readily take advantage of it and under its influence show marked increases in population density; they can exist, however, independently of it. A still looser association involves those species whose contact with man is through his domestic animals, either in stables or barns or in the pastures (SYMBOVINES). This group is linked with the man-dominated biocoenosis through the utilization of the excreta of domestic herbivores, either in pastures or in stables, for breeding purposes.

The chief medical and veterinary significance of synanthropy lies in the potential epidemiological and hygienic implications of the individual requirements of the flies. A classical synanthrope, like the house fly over most of its range, is highly significant medically; its pathogen-transmitting potential, because of its breeding and feeding habits, is great. The degree of synanthropy for many man-associated flies may vary considerably within the species; a closely synanthropic species in temperate areas may be totally disassociated with man, or almost so, in the tropics and subtropics, and even within temperate regions may have strains with a lower degree of synanthropy. The human habitat does, however, provide a breeding and living ground for many species, to their very definite advantage and often with a threat to the human key-occupant of the habitat.

Family Muscidae. This family, to which the house fly and a number of other important synanthropes belong, includes usually dull-colored flies of medium to small size, with well-developed squamae. The hypopleural bristles are absent, whereas in the closely related Calliphoridae and Sarcophagidae they are present; the mouth parts are well developed. The limits of the family are a controversial matter, but they are accepted here in conformity with the usage in the Stone and colleagues (1965) catalog of the Nearctic Diptera.

GENUS MUSCA: THE HOUSE FLY. *Musca domestica* is the most familiar and, in most respects, medically the most important member of the family. It is a gray species, 6 to 9 mm in length, with four dark stripes running lengthwise on the thoracic dorsum. The eyes are separated in both sexes but are much closer in the male than in the female. The central part of the hypopleuron, unlike most species of *Musca,* bears at least a few fine

hairs; however, these might be difficult to see or may be rubbed off.

The prevalence of flies in houses in most parts of the United Sates has been decreasing during the present century. In 1900, Howard found that about 99 percent of all flies in dining rooms were *Musca domestica.* However, there are extensive areas in the United States today where other muscoids, for example, *Fannia canicularis* or certain blow flies, predominate. Such changes as superior sanitation and decreased dependence on horses have greatly altered the composition of the domestic fly fauna since Howard's time. As West (1951) points out, "in many parts of the United States today *Musca domestica* is less a housefly than a 'picnic fly,' 'park fly,' 'dairy fly' or 'stable fly,' but wherever found it is almost certain to be the availability of human food or drink which brings it there."

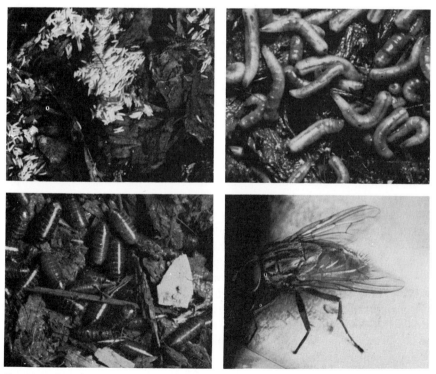

Fig. 12-5 Developmental stages of the house fly, *Musca domestica. (Upper left)* eggs; *(upper right)* larvae; *(lower left)* pupae; *(lower right)* adult. (Photograph of adult by E. S. Ross, others by A. C. Smith; courtesy of California Bureau of Vector Control.)

Musca domestica is almost cosmopolitan, but in some parts of the world its place is taken by other species, particularly *Musca sorbens*.

Life History. The house fly passes through a complex metamorphosis, that is, egg, larva (maggot), pupa, and adult (Fig. 12-5). Under moderately warm summer temperatures, the egg stage requires 8 to 12 hours, the larval stage about 5 days, the pupal 4 to 5 days; a total of about 10 days from egg to adult insect. This allows for the development of 10 to 12 generations in one summer. Under conditions that favor more rapid development the number of generations may be increased. The more usual length of the life cycle, under temperate conditions, is about 3 weeks. The determination of the minimum time may be an important consideration in fly control.

Temperature influences both the survival of the immature stages and the time required for development from egg to adult. The following figures for the duration of this period may be considered as representative: at 16° C, 45 days; at 18° C, 27 days; at 20° C, 20 days; at 25° C, 16 days; at 30° C, 10 days.

The larval stage is the growing period of the fly, and the size of the adult will depend upon the growth that the larva attains. An underfed larva will result in an undersized adult. The mature maggots usually crawl away from their breeding place and burrow into loose ground nearby, where they ultimately pupate; pupae may be found massed together in large numbers near prolific breeding grounds.

As in other muscoids, the adult fly, when transformation is completed, pushes the end of the puparium open by means of the extrusible saclike PTILINUM; then, by alternate expansion and contraction of this organ between the compound eyes the fly opens a passageway through the loose soil and debris to the surface. If the soil is compacted, this may impede the fly's progress or result in its inability to reach the surface.

The house fly is very prolific, and several authors have commented on its fantastic potential for multiplication, if all eggs produced by the fly were to develop adults that, in turn, could reproduce to their full capacity. On the assumption that one adult female deposits 120 to 150 eggs per lot, with at least 6 lots at intervals of 3 or 4 days, C. F. Hodge, in 1911, estimated that "a pair of flies beginning operations in April may be progenitors, if all were to live, of 191,010,000,000,000,000,000 flies in August." The practical value of such a fantasy is, of course, to demonstrate how rapidly a badly depleted population can recover and to show how important control measures early in the season can be.

Breeding Places. Excrement is one material on which *Musca domestica* habitually deposits its eggs, the larvae feeding on this material and the contained microorganisms. Horse manure is a favorite larval food, and at one time it was considered the chief factor in the production of house flies in many rural and village areas of the United States. Cow manure is frequently an important medium for the development of flies. Flies will also breed in hog manure, but swarms of flies around pig pens usually originate from such sources as waste feed and slops. Chicken manure is the most important factor in the breeding of flies in poultry districts. Human excrement, if exposed to flies in open privies, becomes not only a prolific breeding medium but also an important source of pathogens. This consideration emphasizes the need for making privies flyproof or for using other means to prevent flies from breeding in them.

In this day of automobiles and tractors, however, other sources of breeding may be vastly more important than excrement. Smith (1956), studying conditions in dairy barns in California, concludes that "the fly-breeding potential of these enormous amounts of organic waste [chicken and cow manure, garbage, etc.] is far greater than it could possibly have been back in horse-and-buggy days.'' Great swarms of flies are often found around feed troughs; the animals (hogs and cattle) may be literally covered with them. An examination of the waste feed behind or beneath the troughs or in and about the mixing vats will almost invariably reveal numerous maggots. Storage receptacles for slop sometimes present a wriggling mass of maggots. Waste brewer's grain or spent hops, bran mash, and ensilage that is partly consumed by the animals may commonly be a source of enormous numbers of flies (nearly all *Musca domestica*) about dairies where otherwise conditions suggest no apparent reason for swarms of flies. Waste from food-processing plants, when used as feed for beef and dairy cattle, may likewise constitute a major problem.

Garbage heaps, particularly when fermentation and decomposition begin, are commonly sources of many kinds of flies. Heaps of decaying onions, Lima beans, and other vegetables may become infested with maggots. Watermelons decaying in the field may constitute a source. The fly-breeding potential of a garbage can, even an "emtpy" one (Fig. 12-6), under certain circumstances, is amazing; a single can, under experimental conditions, produced more than 20,000 larvae each week, according to Arthur C. Smith (personal communication to James).

In the country, in the absence of septic tanks and sewers, dishwater from the kitchen is frequently piped from the sink to a ditch in the back yard. On occasions, these ditches become clogged and vile smelling, and an

Fig. 12-6 An uncovered, ill-kept garbage can may breed flies even when "empty." An examination of the bottom of this can revealed larvae of the house fly, the lesser house fly, the greenbottle fly, and the drone fly. (Photograph by E. A. Smith; courtesy of California Bureau of Vector Control.)

examination will reveal numerous maggots developing in the muck—a source of flies that is commonly overlooked. Also, maggots may be found in great numbers in the soft sludge mat covering the liquid in defective septic tanks, chiefly those of older construction.

The breeding of fly larvae is not without its beneficial aspects. The house fly, and many other species, help to degrade organic wastes into humus. House fly larvae and pupae also can furnish a supply of food for poultry. As early as 1944 Feldman-Muhsam (cited in West, 1951) suggested that flies be bred intentionally for the latter purpose, and this idea, namely, the combination of degradation of poultry manure for crop fertilization and the utilization of the pupae as a high quality protein source for chicken food, has recently been investigated by several workers (cf. Beard and Sands, 1973).

Range of flight. Ordinarily under city conditions flies, wherever abundant, have bred in the immediate vicinity. The house fly can, however, fly considerable distances. Marking with radioactive isotopes, releasing and recapturing, has indicated that they can fly as far as 32 km from their source and they may disperse as far as 5 or 6 km in large numbers. Their dispersal is usually limited to a distance of 1 to 3 km; consequently, the danger of contamination from fly-borne pathogens attains a strongly local significance near their breeding places. Where houses are situated close together, flies have the opportunity to travel greater distances through a series of short flights, and they are often carried on garbage vehicles and animals.

Longevity. Several factors, such as the availability of food and water, but particularly temperature, influence the longevity of adult flies. Various studies have shown the average life span of the adult to range from 2 weeks to 70 days. As is generally the case, the female lives longer than the male. Where the adult hibernates, it may live over winter.

GENUS MUSCA: OTHER SPECIES. There are 26 species, not including subspecies (according to Greenberg, 1971), belonging to the genus *Musca*. Of these, *M. vicina* should be considered at most a subspecies or even a synonym of *M. domestica,* and *M. nebulo,* a member of the *M. domestica* complex according to general usage, is unidentifiable (Paterson, 1975). On the other hand, some supposed species may actually be complexes of closely related ones.

An important species, *M. sorbens,* is widespread in Africa, the Orient, and the Pacific Islands. In a way, it replaces *M. domestia* over much of its area; it does not enter houses as freely as *M. domestica* does, and it is aptly referred to as the bazaar fly and eye fly. The adult is freely attracted to wounds and ulcers, and it will persistently settle upon the human body, especially around the eyes, nose, and mouth. It tends to lap sores with its mouth parts until serum, on which it feeds, accumulates. It breeds in human and animal excrement, as well as in other suitable media; human excrement is usually preferred. Consequently, it has much the same potential as *M. domestica* for transmitting enteric pathogens, and, in additon, its strong attraction to the eyes and sores gives it added importance in relation to the ophthalmias and yaws. The list of associated pathogens given by Greenberg (1971) includes the causative agent of trachoma and numerous bacteria, protozoa, and helminths.

In Australia, a closely similar species, *M. vetustissima,* is very important because of its nuisance value to man and animals and its possible relationship to pathogen transmission. This species, known as the bush fly, has been considered the same as *M. sorbens* by many authors, although genetic studies (Paterson and Norris, 1970) have supported the usage by Australian workers in a separate specific sense. A valuable review of its biology is that of Norris (1966). This fly has been a pest of man since before the advent of the European inhabitants of the continent, but it has been attracting progressively more attention in recent years, probably because of more human settlement in agricultural areas and of increasing standards of health and comfort. It breeds in the excrement of man, cattle, horses, sheep, dogs, and swine and has been reported in the paunch contents of dead ruminants. Norris, however, believes that human excrement is of importance, at the present time, only under local conditions, where the feces are left exposed without burial; under aboriginal conditions, the situation might have been quite different. Today cattle dung is an important breeding medium. The fly may disperse according to experimental evidence, as far as 5 km from the point of release. There is evidence that seasonal migration may be on a much larger scale; for example, the fly seems unable to survive the winter in southeastern Australia and may there be replaced each spring by warm-wind-borne migrants from the north (Hughes and Nicholas, 1974).

Another species, *Musca crassirostris* (sometimes placed in a separate genus, *Philaematomyia*), is an obligate bloodsucker. Though its proboscis is similar to that of other *Musca* species, it is more bulbular, with strong labellar teeth, by means of which it can scratch the skin or scabs and cause blood to flow. Thus, this fly marks a transition between the house fly type and the bloodsucking *Stomoxys* type. It breeds chiefly in cow dung. It is widespread in the Mediterranean area and in Africa and the Oriental Region. It attacks mainly cattle, causing severe irritation and sometimes weakening animals so much that death may result. Other domestic animals may be attacked; sometimes, though rarely, man is bitten. A useful study of its bionomics is that of Hafez and Gamel-Eddin (1968a).

Musca atumnalis, the face fly, a native of Europe, Asia, and Africa, was introduced into North America probably in 1950 (it was first recorded in 1952). Since then it has spread rapidly, now extending from coast to coast in Canada and in all but the southernmost parts of the United States. It has assumed considerable importance as a nuisance and in pathogen transmission in the New World. A concise discussion of this insect, with an annotated bibliography of the American literature, is given by Smith, Linsdale, and Burdick (1966) and by Smith and Linsdale (1967).

The face fly is a little larger than the house fly; the abdomen of the female is black on the sides, in contrast to the yellowish coloration of that area in the house fly, whereas that of the male is orange or cinnamon-buff laterally. The propleura arc bare; there is a tuft of stiff black hairs at the base of and between the squamae (difficult to see unless the wings are expanded) that is absent in the house fly. The eyes of the male almost touch each other, the frontal stripe consequently being almost interrupted, unlike that of the house fly.

The eggs, which differ from those of the house fly in the possession of a respiratory stalk, are laid just beneath the surface of fresh cow droppings. In about a day they hatch. The larva is yellowish rather than creamy white, but otherwise it looks much like that of the house fly. Its development requires 2½ to 4 days. The puparium is dirty white in distinct contrast to the reddish-brown color of the house fly and most other higher Diptera. The life cycle is completed in about 14 days.

Unmated adults hibernate, often entering houses and massing together in large groups, where their presence can, like that of the cluster fly, cause considerable annoyance. Face flies may be associated with the cluster fly, *Pollenia rudis,* in hibernating groups. During the summer months the adult females feed on secretions around the heads of cattle and other animals, including horses, bison, deer, and American antelopes. In areas where the flies are common, 50 to 100 per animal are not unusual. Males are found on fence posts, tree leaves, and such places, rarely on animals. Flies of both sexes spend the night on vegetation, away from animals.

In addition to the annoyance of man and animals, the face fly plays a role in pathogen transmission. In America it has been shown to be capable of transmitting the bacterium *Moraxella bovis,* the pathogen of keratoconjunctivitis (infectious bovine keratitis, pinkeye of cattle) to cattle; observations have to a limited extent been backed up by experimental evidence (Brown and Adkins, 1972). The face fly, along with the Old World *M. larvipara* and *M. convexifrons,* serve as intermediate hosts of the eye worm *Thelazia rhodesii.*

GENUS FANNIA. About 200 species of this genus are known. *Fannia canicularis* and *F. scalaris* are widespread in both the Eastern and Western hemispheres; these and several others of more limited distribution are of sanitary importance. For a key to some of the more important species and for references to the more important taxonomic works, see Pont, in K. G. V. Smith (1973). Exact identification, however, in most cases will require the help of a specialist.

Males of *Fannia canicularis,* the little house fly, are frequently seen hovering in midair or flying hither and thither in the middle of the room. Whereas the house fly is encountered most abundantly in the kitchen or dining room, particularly on or near food, the little house fly is seen as frequently in one room as in another and seldom on a spread table. Flies of this species often constitute a significantly high percentage of the total fly population in an average house, at times even up to virtually 100 percent.

The little house fly is more slender than the species of *Musca;* it is largely black, but the sides of the abdomen are marked with yellow, more noticeably so in the male. The thorax has three brown longitudinal stripes that remain evident even in some rubbed specimens. The eggs are deposited chiefly on decaying vegetable matter and excrement,

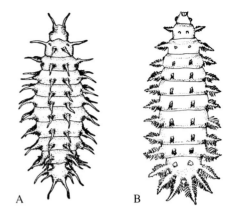

Fig. 12-7 *A.* Larva of *Fannia canicularis. B.* Larva of *Fannia scalaris.* × 6. (After Hewitt.)

particularly of chickens, humans, horses, and cows. The larvae emerge in about 24 hours and may be recognized as flattened organisms, about 6 mm long when fully grown and conspicuously fringed laterally and sometimes dorsally with long projections (Fig. 12-7A). The puparium resembles the larva; the developmental period of the pupa lasts about 7 days under favorable conditions. The complete life cycle requires 15 to 30 days.

Fannia scalaris, the latrine fly, is very similar to the foregoing and of about the same size. It tends to occur away from human habitats, except under primitive conditions where human dwellings are relatively close to latrines and dung pits. The thorax and abdomen are bluish black; the antennae, palpi, and legs are thick. In contrast to the preceding, the middle tibia of the male is provided with a distinct tubercle. The abdomen has a dark median stripe, with segmentally arranged transverse bands, producing a series of dorsal triangular markings.

The eggs of this fly are deposited chiefly in or on semiliquid excrement of humans, horses, cows, and particularly swine. The egg stage lasts about 24 hours, the larval stage about 6 days or more, and the pupal stage about 9 days. The larva of the latrine fly resembles that of the little house fly in general, but its lateral processes are distinctly feathered (Fig. 12-7*B*).

Similar species, with similar breeding habits, that occur in both hemispheres are *F. manicata* and *F. incisurata*. The latter, at least, may have public health implications. *Fannia benjamini*, an American species, not only readily enters houses but may cause annoyance by darting around the eyes, ears, and mouths of persons. It is common in dry chapparal and oak woodland areas throughout California. Though it has been recorded as an intermediate host and vector of the eye worm, *Thelazia californiensis,* a parasite of deer, dogs, rabbits, other mammals, and occasionally man, it has been shown that the species really involved as such is a closely similar one, *Fannia thelaziae* (Weinmann *et al.,* 1974; Turner, 1976).

Species of *Fannia* have also been implicated in human myiasis and pseudomyiasis (see Chapter 13).

GENUS MUSCINA. *Muscina stabulans,* the false stable fly, is larger and more robust than the house fly, 7 to 10 mm in length. The thorax is gray with four longitudinal black lines; the scutellum is orange at the tip. The legs are largely reddish gold or cinnamon in color. The eggs are laid upon decaying organic matter and excrement, in which the larvae develop. The food range is broad and varied, but human excrement is a preferred medium. The first instar is saprophagous, the later ones predacious. The complete life cycle takes from 5 to 6 weeks. *Muscina assimilis,* a similar species with wholly black legs, has much the same habits, except that it is not so domestic.

The sanitary importance of *Muscina* is undetermined. Contact with human food appears not to be common; yet *M. stabulans* may be attracted to and defecate on fruits, and the number of reported instances of association with pathogenic organisms is impressive (Greenberg, 1971). It is possible that these flies transmit enteric bacteria to man and the pathogens of bovine mastitis and brucellosis to cattle. Larvae sometimes inhabit wastes in bird nests where they have been known to attack and kill the nestlings.

GENUS HYDROTAEA. Certain species of the Holarctic genus *Hydrotaea* have for a long time been known to be annoying, sometimes seriously so, to man and animals, but their importance as bloodsuckers and pests to domestic animals has only recently been realized. Makhan'ko (1972a, 1972b) recorded five species in the Soviet Union as facultative or obligatory bloodsuckers; the most important of these were *H. meteorica,* which attacks man both in his living quarters and in the field, and *H. pandellei,* which not merely imbibes blood but actually causes it to flow by rasping the skin.

In the United Kingdom, *H. irritans,* known there as the sheep head fly, is a serious pest to sheep and will attack man, horses, cattle, and deer. A number of recent studies on this fly include those of Tarry and Kirkwood (1974, 1976) and Hunter (1973). The flies are apparently attracted to serous exudates from the eyes and nose; this irritation causes the sheep to produce self-inflicted injuries that, in turn, attract the fly to the freshly flowing blood. Widespread clinical damage may occur, especially in lambs. The larvae occur in soil and litter in pastures beneath thickets and in woodland; they leave the egg as a saprophagous second instar, becoming predatory on other insects as a third instar. There apparently is one generation a year.

Hydrotaea is characteristically a forest or woodland genus. Several species, including *H. meteorica,* occur in America. There is a possible relationship to pathogen transmission, for example, summer mastitis in heifers and dry cows.

OTHER GENERA OF MUSCIDAE. *Atherigona orientalis* is a very abundant species in the tropics of both the Old and

New World. Though essentially a field fly, it readily enters houses where it is plentiful; its breeding media include human excrement, and the adult may visit human food. It may be involved in the transmission of the yaws pathogen. The genus *Ophyra* is represented by several species in both the tropics and warmer temperate regions of both hemispheres. Adults of the widespread *O. chalcogaster* and *O. aenescens* and the Australasian *O. nigra* probably rank next to *Musca* and *Atherigona* as vectors of fecal pathogens in the tropics (Pont, in K. G. V. Smith, 1973). Though essentially field flies, they sometimes enter houses and they may walk over food, utensils, and other objects in the kitchen. Flies of the genus *Morellia* have been recorded as intermediate hosts of the eye worm *Thelazia rhodesii;* they are also suspects in the transmission of the pathogen of brucellosis (Petrova, 1971). These flies breed in animal excrement and occasionally enter houses. They may be very annoying as "sweat" flies when they are attracted to sweat and mucus on man and livestock; sometimes they lap blood flowing from wounds in cattle made by *Stomoxys* or tabanids.

Family Calliphoridae. The blow flies are the common bluebottle, greenbottle, and related forms, including many species of medical and veterinary importance. Most of the more familiar ones are at least in part metallic blue, green, or copper; the common cluster fly and a number of Old World species of importance are nonmetallic. The arista is haired above and below; vein M_2 of the wing bends strongly forward, greatly narrowing but not closing the apical cell.

Several species of bluebottle flies (metallic blue in color) are quite common, notably the widespread *Calliphora vomitoria* (Fig. 12-8) and *C. vicina. Calliphora uralensis* appears to be an important species in the Soviet Union. The eggs of these species hatch in from 6 to 48 hours; the growing larvae feed on flesh from 3 to 9 days, and, after

Fig. 12-8 A common blow fly, *Calliphora vomitoria.*

attaining full growth, leave the food and bury themselves in loose earth and debris. The entire life history requires 16 to 35 days, usually about 22 days. The life span of the adult averages about 35 days. Several nonmetallic species of *Calliphora* are important scavengers in Australia, where they may also be involved in sheep strike (see Chapter 13). The most important of these are *C. augur* and *C. stygia.*

Phaenicia sericata is yellowish green or cupreous green and metallic, with the abdomen varying from metallic green or blue to coppery. It is typically an outdoors scavenger, but it may enter houses. At 27° to 28° C its life cycle, from deposition of the egg to emergence of the fly, requires about 12 days. It is the most abundant species in the genus in the temperate parts of the Holarctic Region and is an important synanthrope, some strains being strongly eusynanthropic. The larvae are essentially carrion breeders, but they are adapted to a wide variety of media, including various types of excrement and garbage containing meat scraps. This species is commonly involved in sheep strike in Great Britain, South Africa, and Australia, and other forms of myiasis (see Chapter 13), and it was at one time used in sterile maggot wound therapy.

Lucilia illustris is a widely distributed holarctic species. It is largely an open woodland and meadow fly; in some areas it will oviposit on carcasses in competition with *Phaenicia sericata.* It has erroneously been recorded in some American literature as *L. caesar;* the latter is not known to occur in the New World. *Lucilia papuensis* is a common and widespread scavenger in Southeastern Asia, New Guinea, and other Pacific Islands.

The black blow fly, *Phormia regina,* is a widely distributed holarctic species; it is a broadly feeding scavenger and is often involved in sheep strike. It is a cool weather fly, occurring more abundantly during the early spring months and becoming less abundant as hot weather approaches. The thorax is black with a metallic bluish-green luster, and the mesothoracic spiracle is orange; there are darker black longitudinal stripes on the dorsum extending somewhat beyond the suture.

The secondary screwworm of the New World, *Cochliomyia macellaria,* and its Old World equivalent, *Chrysomya megacephala,* are important synanthropic flies. They are essentially outdoors flies, but they may make frequent contact with both human food and sources of human pathogens. *C. megacephala* is common and widespread in the Oriental and Australasian Regions, where it often becomes a serious nuisance in marketplaces, fouling with its excrements meats, fish, fruits, and other foodstuffs. It breeds in various substances, such as garbage, carrion, excrements, and slaughterhouse offal. Some other blow flies that deserve mention are discussed by Greenberg (1971).

The cluster fly, *Pollenia rudis,* may be distinguished from other common synanthropic blow flies of North America and Europe by its pollinose, nonmetallic abdomen. It is somewhat larger than the house fly, which it superficially resembles, and it is somewhat more heavily built and slower in its movements. The thorax, particularly on its sides, is clothed with silky, curly, yellow hairs that are clearly visible to the naked eye unless they are abraded.

So far as known, the larvae of this fly parasitize earthworms. Nine species of these annelids are listed from the literature as hosts by Greenberg (1971). The name "cluster fly" is applied to this species because of its habit of entering houses and clustering together in large numbers for purposes of hibernation. Its presence can cause a severe nuisance problem and embarrassment, but the fly is of little medical importance otherwise. Mixed aggregations of this and other species, such as *Musca autumnalis,* can occur.

Family Sarcophagidae. The Sarcophagidae are included by some authors in the Calliphoridae. The abdominal pattern in the more familiar species consists of a TESSELLATED gray and black, that is, a checkerboard pattern in which the spots change from black to gray and back with the light incidence. An exception in a medically important group is the myiasis-producing *Wohlfahrtia,* in which the abdominal pattern, though variegated, is constant regardless of light incidence. A variety of biotic types occurs. The larvae may breed in carrion, excrement, or decaying vegetable matter; some parasitize grasshoppers, Lepidoptera, and other insects, snails, and other invertebrates; a few parasitize vertebrates, including man.

Sarcophaga haemorrhoidalis, the red-tailed flesh fly, may be taken as representative of this group. This is one of many species of flesh flies; it occurs throughout a large part of the tropics and warm-temperate areas of the world. It measures 10 to 14 mm in length; its general color is gray and the prominent terminalia of the male are red. It reminds one somewhat of an oversized house fly. Like most members of the family, it is larviparous. The larvae have a wide range of feeding habits, but they are primarily scavengers, feeding on such materials as dead insects, carrion, and mammalian excrement. The life cycle, in the presence of ample food and warm temperature requires not more than 14 to 18 days. The growth of the larva is very rapid after extrusion when food such as carrion is available. The larval stage may be completed in as few as 4 days. The pupal stage requires 8 to 10 days.

Sarcophagids are not as apt to enter houses as calliphorids are, but many of them, including excrement and carrion feeders in many instances, are definitely synanthropic.

FLIES AND HUMAN WELFARE

Flies as Pests. Lindsay and Scudder (1956) have pointed out the importance of flies as nuisances or pests ''in an age when those who enjoy a high standard of living are spending progressively more for creature comforts.'' There is no reason why we should tolerate a fly nuisance such as that shown in Figure 12-9 any more than we should endure hordes of biting mosquitoes. This is particularly true in an age in which we are recognizing more and more the importance of mental as well as physical health. One can only speculate as to what effect buzzing flies in the home may have on mental health. It is an established fact that some persons have a pathological fear of insects (entomophobia), whether the cause of this fear be real or imagined. The effect on

Fig. 12-9 Such a fly nuisance as this is intolerable. (Photograph by E. A. Smith; courtesy of California Bureau of Vector Control.)

more normal individuals in unknown, but certainly a noisily buzzing blow fly or a swarm of house flies within the house does not lead to mental well-being.

Control of a fly nuisance must, like mosquito control, be a public responsibility, but there is much that the individual can do to relieve the fly nuisance on his own premises, chiefly by keeping them sanitary and forestalling breeding.

The clustering habits of *Pollenia rudis* and *Musca autumnalis* have already been discussed.

Flies as Germ Carriers. On a worldwide basis, the house fly is the chief offender among the filth-breeding and filth-feeding flies, and this section is concerned primarily with it. However, much of this discussion will concern, usually but not always to a lesser degree, many of the other muscoid flies previously discussed. Any of the higher flies are dangerous that breed in or frequent sources of infection and then come in contact with man or his foods, or that have the opportunity to transmit infective materials to his domestic animals. The important consideration is the extent to which these contagious contacts are developed.

The house fly, *Musca domestica,* is by nature of its structure and opportunistic habits an important pathogen-transmitting insect. Its importance can be fully appreciated if one considers the following facts. (1) The house fly is an eusynanthrope over most of its range, at least; it freely enters houses and areas where persons congregate, as well as restaurants, stores, and other places where human food is available; it freely leaves houses to frequent human and animal excrement. (2) It freely feeds on human food and excrement alike. Because the fly can take in only liquid food, it constantly emits vomit spots for the purpose of liquefying solid materials. In the feeding process, droplets of excrement may be deposited. Fecal and vomit spots together may be seen to mottle a surface upon which flies have been resting.

(3) Structurally the fly is well adapted for picking up pathogens. Its proboscis is provided with a profusion of fine hairs and pseudotracheal channels that readily collect contaminants. The foot of the fly, when examined under the microscope, presents a complexity of structure; each of the six feet is fitted with hairy structures and pads that secrete a sticky material, thus adding to its collecting ability. Habit and structure therefore combine to result in a remarkably effective mechanism for mechanical transmission of any pathogen small enough to be carried, particularly microorganisms and helminth ova.

The fly acquires microorganisms by crawling over them, its proboscis, legs, wings, and body thus becoming contaminated. Immense numbers of microorganisms may be found on the exterior surfaces of the fly. In one study in Connecticut, the number of bacteria on a single fly was found to average more than 1¼ million and to run as high as 6 million. A study of about 385,000 flies in Peiping, China, gave an estimate of nearly 4 million bacteria per fly in a slum district and almost 2 million per fly in the cleanest district studied. Although a fly may destroy a large part of this phoretic fauna by preening itself, and though the external environment is rather hostile, one should remember that a very few pathogenic bacteria implanted on or in moist or liquid food can soon multiply to become a dangerous source of infection.

The digestive tract of a feeding fly can also become charged with infective agents that are deposited in fly specks and vomit droplets. The ingested microbe also is subject to certain exigencies in several forms, competition with other microorganisms, for example, or particularly in the case of viruses, which cannot multiply outside a suitable cell. Despite numerous studies that have been made, it is impossible to pinpoint the role of the fly in transmitting pathogenic microorganisms. Nevertheless, the opportunity for flies to become infected is so great in all communities, even the most sanitary, that no fly should be permitted to alight on food prepared for human consumption.

In some cases flies might become infected in the larval stage by developing in infectious matter. The great majority of organisms present in the digestive tract of fly larvae fail to carry over to the adult, and larvae reared in an abundance of them may produce completely bacteria-free flies (cf. Greenberg, 1973). However, some *Salmonella,* including one species that is responsible for food poisoning in man and another that is the agent of the deadly pullorum disease of poultry, can pass from the larva to the adult fly.

The evidence for incriminating flies as mechanical vectors of human pathogens is essentially circumstantial. Flies are known to become contaminated with more than 100 species of pathogenic organisms, including the causative organisms of amebic dysentery, typhoid fever, cholera, shigellosis, salmonellosis, anthrax, leprosy, yaws, trachoma, poliomyelitis, and infectious hepatitis; they can also carry certain helminth eggs, such as those of pinworm, whipworm, hookworm, *Ascaris,* and tapeworm. The student is referred to Greenberg (1971) for detailed documentation. In consideration of this fact, as well as the fly's potential, by habit and structure, of transmitting organisms, plus the epidemiologic circumstances of such diseases as trachoma and shigellosis, the evidence seems so conclusive as to be virtually undeniable. We do, however, lack a clear-cut picture such as the one that exists in relation to the transmission of such pathogens as those of malaria, plague, and Rocky Mountain spotted fever.

Bacterial Enteric Infections. The role of flies in the transmission of pathogenic Enterobacteriaceae is discussed at length by Greenberg (1973). Although evidence is abundant and in overwhelming support of the conclusion that flies are important vectors of organisms of this group, the nature of this evidence and the taxonomic complexity of the pathogens make generalizations difficult at the present time. Nevertheless, it is here that the role of the house fly in pathogen transmission has been most firmly established.

The bacteria involved belong chiefly to three genera. (1) Species of *Shigella,* with about 30 serotypes, occur almost exclusively in man, though animal infections are known rarely to occur. Shigellae are almost worldwide in distribution and are often the most important agents of dysentery and diarrhea (shigellosis). (2) *Salmonella* species are more numerous and more difficult to define, but more than a thousand antigenic types are known. *Salmonella typhi,* the typhoid fever bacillus, is specific for man, and *S. pullorum* and *S. gallinorum* are restricted in general to poultry, but most *Salmonella* have a wide range of hosts, including numerous mammals, birds, reptiles, and perhaps even amphibia and fishes. Salmonellosis is in general a zoonosis. (3) Pathogenic serotypes of *Escherichia coli,* as distinguished from the normal gut types, occur in a number of mammals and birds, particularly the young; these are often associated with infant diarrhea in man and may be involved in "travelers' disease" in adults. As in the salmonelloses, the status as a zoonosis adds to the epidemiological complexity.

The clearest picture linking control of flies with that of enteric diseases may be seen, expectedly, with shigellosis. The work of Watt and Lindsay (1948) in semiarid, subtropical Texas near the mouth of the Rio Grande, where shigellosis is so highly prevalent and epidemic that in some areas a reduction of 50 percent would be readily detectable in a sample of feasible size, has given particularly significant results. Fly control was achieved by spraying five out of nine similar towns with DDT. This resulted in reduction of *Shigella* infections, reported attacks of diarrheal disease, and infant mortality. After 20 months the procedure was reversed, the previously unsprayed towns then being treated instead of the original five. The change of schedule corresponded with the normal seasonal increase in flies in the area. A reversal in trends followed to the extent that the possibility of chance was elimated. The results demonstrated the role of flies in vectoring this disease under the conditions involved in the area studied. This study is the most dramatic of its kind available, but other investigations and analyses of the situation bear out the relationship of fly control to control of shigellosis.

Although flies are not a necessary link in the transmission of the Salmonellas, the review of the subject by Greenberg (1973) will show convincingly that their role is a very important one. Salmonellosis epidemics have often followed military operations and have been associated with the availability to flies of raw sewage and human feces, on the one hand, and the development of large numbers of flies on the other. Numerous studies have implicated the house fly in the transmission of *Salmonella typhi;* indeed, L. O. Howard advocated that the common name "typhoid fly" be used for *Musca domestica*. Studies such as those made in New York City by Hudson in 1915 (cited by Greenberg) have shown that abundance of flies as related to epidemiology may be correlated with sanitary conditions; in sanitary homes investigated a high population of flies did not increase the incidence of dysentery, whereas in unsanitary homes fly abundance had a marked effect on it. Pathogenic salmonellae are often transmitted through food, milk, and water, particularly wash-water, and epidemics occur where there is no evidence of fly involvement. The latter, however, can not be disregarded.

Cholera. Cholera, the causative organism of which is *Vibrio comma,* was among the first diseases in which the house fly was incriminated as a vector. Tizzoni and Cattani in Bologna, Italy, in 1886 isolated cholera vibrios from flies caught in cholera wards.

Subsequent isolation from flies and their feces, along with epidemiological evidence, leave little doubt but that flies play an important role in transmitting the pathogen from human feces to uninfected individuals in endemic situations where water is not the major vehicle.

Musca domestica is the fly that has been chiefly associated with the cholera vibrio, but other such flies include *Calliphora vomitoria, C. vicina, Lucilia caesar,* and *Sarcophaga carnaria*.

Yaws. Yaws (frambesia, frambesia tropica, pian, boubas) is caused by the spirochete *Treponema pertenue*. This disease is widely distributed in the tropics. The spirochetes are found in superficial ulcers on the hands, feet, face, and other parts of the body (Fig. 12-4). Suspicion of fly involvement in the spread of the disease goes back almost four centuries (see Chapter 1). As early as 1907 Castellani demonstrated the presence of the yaws organism on the mouth parts and legs of flies that had been feeding on ulcers in yaws patients; moreover, he demonstrated that flies could transfer the pathogen to scarified areas on the eyelids of monkeys. The work of Satchell and Harrison (1953) in Samoa indicates quite convincingly that wound-feeding flies, particularly *Musca domestica* and *M. sorbens,* are involved in the transmission of the yaws organism in that area. Elsewhere, these and other flies have the necessary habits for disseminating the pathogen, but despite abundant epidemiological evidence, the degree of fly involvement needs to be determined. The role of *Hippelates* and *Siphunculina* has been discussed earlier in this chapter.

Eye Diseases. In addition to *Hippelates* and *Siphunculina, Musca sorbens* and *M. domestica* may be involved in the transmission of bacterial agents of conjunctivitis. In respect to *Musca,* however, the situation is more complicated in that it involves a complex of bacterial infections, which may pose problems in the identification of the etiological agents. Other bacterial infections may occur simultaneously and may include trachoma.

Evidence of fly involvement as vectors of this complex of causative agents is circumstantial, though convincing. Greenberg (1973) has reviewed a series of studies made by various workers in which reduction of *Musca* populations was strongly correlated with significant reductions in trachoma and bacterial conjunctivitis among native populations in Algeria and Morocco. The trachoma pathogen may survive from 4 to 24 hours on the exterior of the fly and as many as 15 days in its gut; the bacteria are more vulnerable to exposure to air. Epidemiological information points strongly to flies as important vectors of these organisms. Nevertheless, despite the preponderance of incriminating evidence, the exact role of these flies in the epidemiology of these diseases is uncertain. Some workers believe that transmission, particularly of trachoma, is predominantly through contact within human families.

Poliomyelitis and Other Enteroviral Infections. Although flies, biting and nonbiting, as well as mosquitoes, have long been under suspicion as vectors of the virus of poliomyelitis, suspicion now rests only on flies that feed and breed in excrement, carrion, and garbage. The virus has been isolated repeatedly from human stools and sewage, and its presence has been demonstrated in flies collected in the field during both urban and rural epidemics by various investigators. The fly species involved most frequently in field isolations are *Musca domestica, Phaenicia sericata,* and *Phormia regina,* although a number of others have been recorded. Positive experimental evidence of potential transmission has been secured by intracerebral and intraperitoneal inoculation of etherized fly extract into *Cynomolgus* monkeys and by the introduction of unetherized material both intranasally and by mouth. Persistence of the virus on or in the fly seems to vary with the strain of the virus, but ranges from about 48 hours to 12 days or more, ample time for the fly to contact the food or person of a susceptible individual.

The conclusion that the house fly and certain flesh flies, blow flies, and muscoids can harbor the poliovirus and, under proper circumstances, can transmit it either by external contamination or contact, or by internal passage following ingestion, seems quite secure. However, any role that flies may play is undoubtedly incidental. The exact mechanism of transmission of the poliovirus, despite all the work that has been done on the subject, is not proven but probably involves intrafamilial means primarily.

We know that man can readily infect flies with the virus, but we do not know to what extent the fly can transmit it to man. Involvement of flies probably does no more than add to the severity of an epidemic. Obviously, epidemics occurring at times other than the fly season must have other causes. Several facts add to the complexity of the picture; for example, poliomyelitis is asymptomatic in the vast majority of cases, and the paralytic disease is rare in tropical areas where sanitary conditions are poor, indicating early exposure and immunity. In less sanitary, warmer regions flies probably play an important role in poliovirus transmission as they do in the transmission of other enteric disease pathogens.

Evidence for the transmission of the Coxsackie and ECHO viruses is of a similar nature, though not so abundant. Viruses of both groups have been isolated from *Musca domestica, Phormia regina,* and *Phaenicia sericata* under conditions that suggest a possible involvement in transmission to man.

Tuberculosis. Though flies have the mechanism and habits for the transmission of the tubercle bacillus, no conclusive work has established the relationship of flies to such transmission. Flies have been observed to feed on the sputum of tuberculosis patients and to pass infective bacilli in their feces.

Lamborn, working with *Musca sorbens* in Nyasaland, found that the tubercle bacillus might remain viable in the body of the fly for a week. Greenberg (1973), after reviewing the literature, concludes that the house fly is probably "a standby vector called to active duty when people are careless or sloppy."

Parasitic Associations. INTESTINAL PROTOZOA. Several of these, notably *Entamoeba histolytica*, the most frequent cause of amebic dysentery in man, but also *Giardia lamblia*, *Chilomastix mesnili*, and others, have often been isolated from flies taken on or around food destined for human consumption.

Roubaud in 1918 found that the cysts of *Entamoeba coli*, *E. histolytica*, and *G. lamblia* passed through the intestine of the fly uninjured, but that free amebae (both *coli* and *histolytica*) when fed to flies were dead in less than an hour. The role of the house fly and other muscoids and calliphorids has been supported by the research of other workers. The evidence, despite the extreme paucity of experimental results, indicates unquestionably that flies can carry pathogenic Protozoa and that they often do so under natural conditions (Greenberg, 1973). Fly involvement is of greatest importance in areas where unsanitary conditions prevail. The chain of infection is from feces to fly to food to host. The pathogens may be transferred from feces to host by careless handling of food or by contaminated water, and the role of flies is probably incidental.

EGGS OF PARASITIC WORMS. Extensive and careful work on the dispersal of eggs of parasitic worms by the house fly has been done by Nicoll (1911). The following is a summary of his investigations. Flies feed readily on excrement in which eggs of parasitic worms occur. Eggs may be conveyed by flies from excrement to human food in two ways, namely, on the external surface of the body and through the digestive tract. Eggs ingested by the house fly must be less than 0.05 mm in diameter; larger flies, such as some blow flies and sarcophagids, can ingest eggs of greater diameter. Eggs also may be conveyed on the external surface; however, these are usually shed or preened off by the fly within a short time, whereas those harbored in the intestine may remain for several days. Material containing eggs of parasites, and in particular tapeworms (*Taenia pisiformis*), remain a source of infection through flies for as many as two weeks.

The eggs of nematodes and cestodes reported as being carried by *Musca domestica* and other calyptrate flies include *Taenia solium*, *T. hydatigena*, *T. pisiformis*, *Hymenolepis nana*, *H. diminuta*, *Dipylidium caninum*, *Diphyllobothrium latum*, *Enterobius vermicularis*, *Trichuris trichura*, *Necator americanus*, *Ancylostoma duodenale*, *Ascaris lumbricoides*, *A. equorum*, *Toxascaris leonina*, and the hydatid *Echinococcus granulosus*. The flies involved in potential helminth egg transmission include a number of synanthropic species in addition to the house fly; mainly other muscids (such as *Fannia* and *Muscina*), blow flies (*Calliphora*, *Chrysomya* especially *megacephala*, and others) and sarcophagids.

The ability of flies to transmit helminth ova, particularly through fecal contamination, is unquestionable. The literature is fairly abundant; it is reviewed by Greenberg (1973). The incidence of contamination in flies is often high in areas where poverty prevails and low sanitation, common exposure of food, and high human infection rates occur.

DISEASES OF DOMESTIC ANIMALS

Domestic, semidomestic, and undoubtedly some wild animals are subject to diseases caused by fly-transmitted pathogens, in much the same way as man is. Our information here is for the most part less secure, resulting as it does from less extensive experimental and epizootiological studies; moreover, in some instances, for example, in respect to mink enteritis, it may be complicated by lack or incomplete understanding of the etiological agents. Undoubtedly, however, the house fly and certain other nonbiting muscoids can transmit certain animal pathogens.

The virus of mink enteritis (MEV), one of several agents of this disease complex, can be transmitted by *Musca domestica,* as shown by Bouillant and associates (1965). Rogoff and associates (1977) have shown that the virus of velogenic viscerotropic Newcastle disease (VVND), exotic Newcastle disease of poultry, may be transmitted by *Fannia canicularis.* Experimental transmission was obtained, wild flies were found infected, and the fly was shown to be capable of retaining the virus for at least six days. Transmission is probably mechanical, though this has yet to be determined. Possibly other muscids, such as *Musca domestica* and other species of *Fannia,* may also be vectors, though this still has to be determined. Involvement of *Fannia* is significant, however, since poultry excrement is a highly important breeding medium for *F. canicularis.*

Pasteurella multocida transmission by *Musca domestica* has been linked to an epizootic of rabbit septicemia in rabbits and, to at least a limited extent, to one of fowl cholera in turkeys; both diseases are highly fatal. The bacterial agents of brucellosis and anthrax can be transmitted by nonbloodsucking as well as by biting flies, and *Clostridium chauvoei,* the causative organism of blackleg, a fatal disease of cattle, can infect *Musca domestica* and might, under certain cirumstances, be transmitted by that fly.

Bovine mastitis is a disease complex, affecting sheep and goats as well as cattle, and resulting from the invasion of the udder by various microbial agents. The role of *Hippelates* spp. has been discussed earlier in this chapter. Although transmission of the pathogens by means other than flies is probably most important, fly involvement almost certainly occurs, though positive proof is lacking. Three bacteria may at times be fly-transmitted, namely, *Streptococcus agalactiae, Corynebacterium pyogenes* (mastitis), and *Staphylococcus aureus.* There is convincing evidence implicating *Hydrotaea irritans* in transmission of *C. pyogenes* and against *Musca domestica* in respect to *S. agalactiae.* Greenberg (1973) lists some symbovines that might be vectors of mastitis agents.

Limberneck, a highly fatal botulism of domestic fowls caused by the bacterium *Clostridium botulinum,* is acquired by poultry through ingesting blow fly maggots that have developed in carcasses in which the pathogen was present in large enough numbers to be infective. Wild pheasants and ducks are also susceptible to the disease. *Lucilia illustris, Phaenicia sericata,* and *Cochliomyia macellaria* larvae are known vehicles for the transfer of the infective agent; other carcass-feeding calliphorids are reasonable suspects. The subject is discussed concisely by Greenberg (1973).

Other microbial pathogens may be transmitted on certain occasions and under certain conditions by nonbloodsucking muscoid flies. The relationship of *Musca autumnalis* to infectious keratoconjunctivitis of cattle has already been discussed.

Larvae of the nematode *Habronema megastoma* may cause persistent ulcerations, or summer sores, on the lower parts of bodies of horses, and also habronemic conjunctivitis, which manifests itself as sores on the eyes. Adults of this and two other species, *H. microstoma* and *H. muscae,* occur in the stomach of the horse, where they cause a condition known as *habronemiasis.* There they lay their eggs, which pass with the feces. The newly hatched larvae find their way into fly larvae in which development begins and continues through the pupal into the adult stage. The infective worm larvae escape from the proboscis of the adult fly into sores or around the lips or nostrils of the horse, where they are then ingested and swallowed. Flies are true biological hosts of the nematodes, *Habronema muscae* and *H. megastoma* developing in *Musca domestica*, and *H. microstoma* in *Stomoxys calcitrans*.

Thelazia rhodesii, an eyeworm of domestic animals, has *Musca larvipara, M. convexifrons*, and *M. autumnalis* as known invertebrate hosts. The larvae of this nematode are taken up by the adult fly during feeding; nematode development involves migration from the gut to the ovarian follicles and finally, in its infective stage, to the proboscis of the fly and transfer to the vertebrate host. It occurs in cattle, sheep, goats, and buffalo in Europe, Asia, and Africa, and it has been reported from California. Parasitism of cattle may result in blindness or severe dysfunction of the eye. A similar role is played by the New World *Thelazia californiensis*, a widespread parasite of wild and domestic mammals, and even man (Weinmann *et al,* 1974) in California and nearby areas. Its invertebrate host is *Fannia thelaziae. Thelazia gulosa* and *T. skrjabini* of Eurasia, also eyeworms, have species of *Musca* and possibly other Muscidae as insect hosts.

Another nematode, *Stephanofilaria assamensis*, produces hump sores (Cascado, equine dhobie, Krian sore) in cattle, horses, goats, and pigs. According to the work of Patnaik (1973) with *Musca conducens*, the worm larva is ingested from sores in the feeding process of the fly, develops in the gut of the latter, and ultimately migrates to the proboscis. The long period of development in the fly, 20 to 25 days, and the small number of parasitized wild flies, fewer than 1 percent, raise questions as to the role of this fly in the transmission of the parasite.

Domestic fowls are commonly infested by tapeworms, several of which may have the house fly or stable fly as invertebrate hosts. The most important of the fowl tapeworms is *Choanotaenia infundibulum*. A parasitized fly larva or adult fly, if ingested, may introduce the infective cestode to the bird, but the evidence that this occurs frequently in nature is not clear and is a subject of dispute.

Flies are suspect in relation to transmission of certain other pathogens of domestic animals, and in some cases there is evidence that should be considered seriously. The works of Greenberg (1971, 1973) should be consulted by those who wish further information on the subject and Stoffolano (1970) has provided a useful review of nematode associations with the genus *Musca*.

CONTROL OF HOUSE FLY AND RELATIVES

Population estimates of these flies (Muscidae, Calliphoridae, Sarcophagidae) are based upon observing the numbers on resting sites (Scudder, 1947), or caught by traps baited with meat or similar attractants, sticky traps (Legner *et al.,* 1973), or by examining larval developmental sites. Numbers of face fly adults can be readily counted on cattle.

Environmental modifications consist of preventing the accumulation of larval development media (cf. Fig. 12-10), feces in the case of the house fly and its close relatives, lesser house fly, and some sarcophagids, and carcasses for calliphorids and other sarcophagids. Sanitation codes generally provide adequate information for the proper disposal of municipal refuse. A Mexican dung beetle, *Copris incertus*, has been successfully introduced to Samoa where it plays a significant part in fly control by removing animal droppings, and an extensive program of dung disposal by introduced dung beetles has been developed in Australia. Swine-waste-disposal methods for fly control have been described (Dobson and Kutz, 1970), and weekly removal and spreading of cattle manure are satisfactory for controlling many dung-breeding flies, but may have to be shortened to a four-day period for *Musca domestica* (Smith, 1969). Habitat modification to minimize annoyance by the sheep head fly, *Hydrotaea irritans*, in Britain consists of housing the animals or clearing ad-

Fig. 12-10 An excellently operated sanitary landfill. (Photograph by A. C. Smith; courtesy of California Bureau of Vector Control.)

jacent scrub vegetation and thickets (Hunter, 1975).

Screening, an effective means of excluding flies from homes and food, is widely used. Curtains of fabric or air jets at doorways will also reduce the entrance of flies into buildings (Morgan *et al.*, 1972; Frazerhurst *et al.*, 1973). Once flies enter homes they may be controlled by sticky tapes, household aerosol sprays, or resin strips that release vapor toxicants.

A number of biocontrol agents of flies are known. Legner, Moore, and Olton (1976) provide keys to common parasitoids of Diptera breeding in animal wastes. Among acarina, members of the genus *Macrocheles* are widely distributed and have received particular attention. Fly eggs and small larvae are eaten by *M. muscaedomesticae,* which is credited with considerable control in field tests. Beetles may also be efficient predators. The histerid *Platylister chinensis* feeds on fly larvae, and upon its introduction to Fiji and Samoa it helped to control flies infesting poultry manure if a sufficient depth of this substrate was maintained (Bills, 1973). Staphylinid beetles of the genus *Aleochara* eat fly eggs and larvae; in the USA these are used to control the face fly and other dung-breeding flies. Inundative releases of para-

sitic wasps have achieved quite effective control of filth-breeding flies (cf. Legner and Dietrick, 1972; Olton and Legner, 1975; Morgan *et al.*, 1975). Parasitic nematodes in the genus *Heterotylenchus* are quite effective in reducing populations of face flies and other pasture flies (Fig. 5-3), with their main effect on reproduction (Nappi, 1973). Hughes and co-authors (1974) discuss the rationale for selection of biocontrol agents effective against *Musca vetustissima* in Australia. Poultry readily eat fly larvae and pupae, and the presence of cockerel chicks under battery-raised chickens in a 1:10 ratio exerts considerable fly control by keeping the feces scattered, and consequently dry, along with the consumption of the developing flies (Rodriguez, 1959). Integration of practices, employing parasites and predators, and confining pesticide control to sprays or baits against adult flies, have proved to be quite effective in poultry houses (Axtell, 1970; Legner, *et al.*, 1975).

Insecticides should be used as a last measure in fly control, not only because sanitation is more permanently effective but also because insecticide resistance develops so readily. Sprays are applied as residual deposits, with particular attention to favored resting sites, such as sunny surfaces of barns, shelters, and fences. Dry sugar or syrup baits with toxicant are very effective. These can be broadcast, sprayed, or painted onto surfaces frequented by filth flies.

A number of chemicals can sterilize both sexes of house flies and their relatives. Use of some of these may be practical particularly because their incorporation into baits would permit self-treatment in field use and limited careful application of such baits should not endanger other forms of life. A chemosterilant-bait test on Grand Turk Island (B.W.I.) yielded a 90 percent reduction of the house fly population in 18 months (Meifert *et al.,* 1967). A female sex attractant is present in house flies, but is not as effective as with many other insects because visual factors also strongly influence mating behavior. There is interest in the biochemical factor present in the reproductive tract of males that, when transferred to females, prevents second matings (Adams and Nelson, 1968).

Face flies on cattle are generally controlled by the use of automatic devices that will apply pesticide to the facial area. For this purpose dust bag application can be quite effective (Fig. 5-7). Experimental approaches have included feed additives that render the fresh feces of cattle toxic to developing face fly larvae. If adult face flies or cluster flies congregate in attics to overwinter, these may be controlled by the use of aerosol sprays, or by hanging up resin strips that slowly release toxicant in the vapor phase.

TSETSE

The genus *Glossina,* family Muscidae, comprises the tsetse of Africa. According to Buxton (1955) the word "tsetse" comes from the Sechuana language of Botswana and means "fly destructive to cattle." The term *tsetse fly,* therefore, is a tautology, though well established in the literature and in common usage. "Tsetse" may be used either as a singular or plural noun. According

to Bequaert, the word was introduced into the English language by R. Gordon Cumming in 1850 in his "Five Years of a Hunter's Life in the Far Interior of South Africa," and David Livingstone in 1857 "focused the attention of the scientific world upon the ravages of the fly."

Evidently tsetse enjoyed a wide distribution in earlier geologic times; four species of fossil *Glossina* have been described from the Oligocene shales of Colorado, USA. Today, so far as is known, tsetse are restricted to Africa south of the Sahara Desert.

The literature on *Glossina,* the trypanosomes that they harbor, and the diseases of man and animals associated with these flies is voluminous, and only a few of the important references can be cited here. A somewhat more extensive list is given in Potts (in K. G. V. Smith, 1973), and still more extensive lists of references may be found in the other comprehensive works cited here. Hoare's (1972) monograph of the mammalian trypanosomes presents much critical information and the reviews of Willett (1963, 1967) will be very useful to the general reader. *Africa's Bane—The Tsetse Fly* by T. A. M. Nash (1969) is an authoritative work designed for the nontechnical reader and, as such, serves a very useful function. Another useful popular treatment by an authority is *Man Against Tsetse—Struggle for Africa* by J. J. McKelvey, Jr. (1973). The encyclopedic work of Buxton (1955) though, of course, not up to date, and the comprehensive studies of Ford (1971) and of Mulligan and Potts (1970) are very important contributions to the literature.

Characteristics. Tsetse (Fig. 12-11) are flies of medium size, ranging from 7.5 to 14 mm in length. They are brownish in color; the body is somewhat wasplike; and the wings, when at rest, are crossed scissorslike and extend well beyond the tip of the abdomen. The wing venation (Fig. 9-1*F*) is characteristic: the discal cell (1st M) is shaped remarkably like a meat cleaver and is

A   B

Fig. 12-11 *A. Glossina palpalis. B. Glossina morsitans.* (After Newstead.)

called the CLEAVER CELL or HATCHET CELL; just beyond its apex the fourth longitudinal vein (M_{1+2}) bends suddenly upward, in line or almost so with the apical margin of the discal cell, which is oblique. The palpi are nearly as long as the proboscis, which points bayonetlike in front of the head. The antennae are highly characteristic (Fig. 12-12) in that the rays of the arista are themselves branched bilaterally. The mouth parts are of the stable fly type described in Chapter 3. A characteristic "onion-shaped" bulb occupies the basal part of the haustellum (Fig. 12-13).

Food Habits. Both sexes feed avidly and exclusively on vertebrate blood. This food source must suffice for all stages of the fly because the larva takes no nourishment other than that furnished by the mother. The flies are attracted to moving objects. They usually feed in broad daylight. Sight apparently plays the larger part in host finding, although the olfactory sense may also be involved, particularly in certain species. As a group, tsetse feed on a wide range of mammals, reptiles, and, rarely, birds. Some species seem to be restricted in their food habits, whereas others are largely opportunistic, though with certain food preferences. Al-

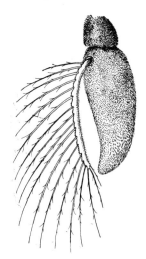

Fig. 12-12 Antenna of *Glossina*, showing arista with branched hairs. (Much enlarged.)

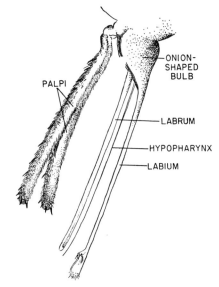

PALPI

ONION-
SHAPED
BULB

LABRUM

HYPOPHARYNX

LABIUM

Fig. 12-13 Mouth parts of *Glossina*. × 17.

though man is freely attacked by some tsetse, he is not considered, in general, a favored host.

Weitz (1963; see also Weitz in Mulligan and Potts, 1970) recognized five general habitat and feeding patterns among tsetse, and Ford (1971; see also Ford in Mulligan and Potts) summarized, rearranged, and modified Weitz's tabulations. The following is based upon this rearrangement.

1. Lacustrine and riverine feeders include the *palpalis* group tsetse: *G. palpalis, G. fuscipes,* and *G. tachinoides.* These feed on most available hosts, including man and livestock, but they are predominantly reptile feeders, with a bovid, the bushbuck, also a favored host.

2. Forest, thicket, or forest-edge feeders include two subgroups: *G. tabaniformis, G. fuscipleuris,* and *G. austeni* feed mainly on suids, notably the bushpig and forest hog, whereas *G. fusca, G. pallidipes,* and *G. longipalpis* are largely bovid feeders, the bushbuck being their favored host.

3. Savanna feeders include *G. morsitans* and its subspecies and *G. swynnertoni,* which attack both bovids and suids.

4. Specialized East African *fusca* group flies, including *G. brevipalpis* and *G. longipennis,* feed mostly on mammals other than pigs and bovids, including elephants, rhinoceros, and hippopotamus; *G. longipennis* is also known to feed significantly on the ostrich.

Tsetse as a whole show a marked attraction to the Suidae, quite out of proportion to the abundance of other available hosts. Antelope and other Bovidae are favored hosts, but most *Glossina* species are highly selective. Removal of selective game from a given area can, consequently, bring about the starvation and disappearance of a target species such as *G. swynnertoni.*

Life History. The female tsetse gives birth to fully grown larvae which are extruded at intervals of 8 to 25 days, depending upon temperature, during the lifetime of the mother. Only one larva is carried by the female at any given time. During the intrauterine stage, which involves the first two larva stages and the greater part of the third, the larva is fed on fluids from special glands commonly known as "MILK GLANDS." The newly extruded, or late-third-stage larva, is creamy white to pale yellow and has a pair of intensely black, shining respiratory lobes at the posterior end. Larval thoracic spiracles are lacking. The female fly requires several regular blood meals at intervals not exceeding four days to complete the development period of each larva. Evidence concerning the number of larvae produced during the lifetime of a female is inconclusive; the average number is probably 8 to 10, the maximum probably about 20.

Tsetse are unique among flies, except the Pupipara, in having virtually no free larval stage. The larvae are usually deposited on moderately dry, loose, friable soil, under fallen trees or decumbent trunks, on sandy beaches, in thickets, and such places. Deposition sites vary with the species, and they may differ in the wet and dry seasons. It has recently become apparent that scattered sites may occur throughout the floor of the woodland through which the females range, and these may be of more importance than has been previously realized (Potts, in K. G. V. Smith, 1973).

The larvae are unable to crawl, as do other muscid larvae, because of the reduced cephalopharyngeal armature and the lack of spinose pads. Instead, they move and burrow by peristaltic movements and longitudinal contractions of the whole body. Coarse, pebbly sand favors the larva in burrowing, although a depth of only a few centimeters is reached. The larva at this time is exposed to possible attack by parasites or predators, so prompt penetration of the soil is necessary to its survival. Hardening of the integument to form the puparium takes place within an hour of larviposition. The puparium rapidly darkens to a blackish brown color; it is

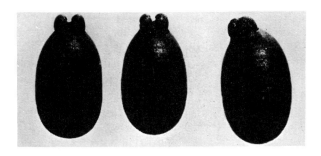

Fig. 12-14 Pupae of *Glossina*. × 4.8.

barrel-shaped, with prominent posterior lobes (Fig. 12-14). A fourth larval stage occurs within the puparium prior to true pupation. The pupal stage lasts two to four weeks or longer, depending on soil temperature and moisture. The fly emerges by breaking loose the end of the puparium through pressure from the ptilinum.

Glossina Species. Twenty-two species of *Glossina* are commonly recognized today; the number may vary slightly, depending on whether certain ones are given specific or subspecific status. They may be indentified by the keys, illustrations, and descriptive material given by Potts in K. G. V. Smith (1973) or in Mulligan and Potts (1970). Probably all of these are capable of serving as hosts of trypanosomes of man or animals; all but three have either been shown to be capable of such transmission or have been found in nature infected with trypanosomes. Six species are of outstanding medical or veterinary importance. Three of these are riverine in habitat and belong to the *palpalis* group: *G. palpalis, G. fuscipes,* and *G. tachinoides*. The other three are savanna or "game" flies, all of the *morsitans* group: *G. morsitans, G. swynnertoni,* and *G. pallidipes*. A number of additional species are of some importance in the transmission of the nagana trypanomoses. Those mentioned or discussed in this account are *G. austeni, G. longipalpis, G. vanhoofi, G. brevipalpis, G. fuscipleuris, G. fusca,* and *G. longipennis*.

Glossina palpalis (Fig. 12-11A), *G. fuscipes,* and *G. tachinoides* are riverine and lacustrine species, that is, they characteristically occur along the shores and banks of the rivers and lakes of forested areas or savanna woodlands. They occur mostly in central and west Africa in river basins that drain into the Atlantic Ocean or Mediterranean Sea. Sometimes they occur well away from the surface water, even in peridomestic situations where at times they may feed on domestic pigs and, of course, the human inhabitants of those areas. These three species, as will be noted, are the important vectors of the Gambian sleeping sickness pathogen. Their combined distribution extends over a large area from eastern Zaire to Senegal and Gambia.

Glossina morsitans (Fig. 12-11B) has a wide distribution in Africa; it is of importance in the Sudan, Zaire, Zambia, Rhodesia, and many other areas. This species, according to Swynnerton, requires savanna "of sufficient shade value, and with sufficient logs, rocks, or tree rot holes to form a good rest-haunt and breeding ground, and relatively open plains in which to hunt for its prey." It is essentially a game fly but, since it attacks human beings readily, it is important from the human standpoint. Three subspecies, with different geographical ranges, are recognized.

Glossina swynnertoni, like *G. morsitans,* is an important vector of trypanosomes of both Rhodesian sleeping sickness and nagana. It is largely confined to the northern part of Tanzania. Swynnerton (1936) describes it as "the fly of the dryest and most open areas and apparently unable to inhabit the more mesophytic savannas. It breeds normally in the thicket, though rock suits it as well. . . . It utilizes open spaces as feeding grounds. . . . It is primarily and essentially a 'game' fly.'' It attacks human beings readily, however. *Glossina pallidipes,* widely distributed through East Africa as far north as Ethiopia, is also an important vector of the Rhodesian parasite.

AFRICAN SLEEPING SICKNESS

Disease: Sleeping sickness (African trypanosomiasis).

Pathogen: Flagellate protozoon, *Trypanosoma brucei gambiense,* including two nosodemes (see definition below), *gambiense* (Gambian disease) and *rhodesiense* (Rhodesian disease).

Victim: Man.

Vectors: Chiefly *Glossina palpalis, G. fuscipes,* and *G. tachinoides* (Gambian); chiefly *G. morsitans,* but also *G. swynnertoni* and *G. pallidipes* (Rhodesian); other *Glossina* and other bloodsucking flies may be involved in a minor way.

Reservoirs: Man (Gambian); man, antelopes, and perhaps other mammals (Rhodesian).

Distribution: Tropical Africa, between 15° N and 29° S (Fig. 12-15).

Importance: A severely disabling and highly fatal disease; along with nagana, it has been and still is a major obstacle to development of Africa between the Sahara and its southern part. It is responsible for 7,000 human deaths each year.

Although human sleeping sickness has been known as a disease since the fourteenth century, its cause and vectors were not known until 1902, when Ford and Dutton discovered the parasites in human blood smears, and 1903 when Bruce and Nabarro proved *Glossina morsitans* to be a vector. Dutton named these parasites *Trypanosoma gambiense.* A second "species" of human parasite, *T. rhodesiense,* was described in 1910.

Sleeping sickness is now widely distributed in Africa, extending along the west coast from Senegal to Angola and eastward to Mozambique and the valley of the upper Nile. Apparently, it was limited to small foci until the late nineteenth century when, as a

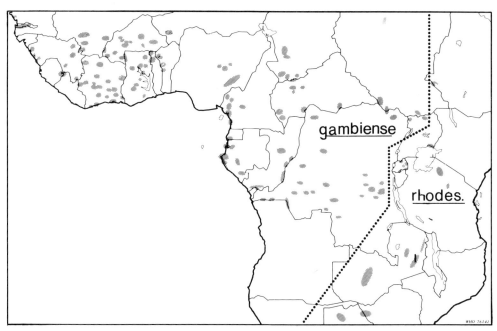

Fig. 12-15 Distribution of human sleeping sickness in Africa. (Courtesy of WHO.)

result of the penetration of Europeans and the lessening in conflict among African tribes, migration in Africa became more prevalent, thus enabling human carriers to spread the infection more widely. Transportation of flies on river steamers and by moving cattle has also helped to spread the flies and other pathogens. Even when the disease was more limited geographically, it made a very definite impact upon the natives. Slave traders in the seventeenth and eighteenth centuries were well aware of its dangers and would not accept slaves that showed the characteristic swellings in the neck now known as Winterbottom's sign (Fig. 12-18). Within the last hundred years, devastating epidemics have occurred. It has been estimated that between 1896 and 1906 from 400,000 to 500,000 Africans perished from this disease. Dutton and Todd found that in some villages from 30 to 50 percent of the population was infected. An epidemic, in Uganda near Lake Victoria over a five-year period, took 200,000 human lives; the population of the area involved was 300,000 before the epidemic, the mortality consequently being 67 percent of the entire population.

Duggan (1962) cites a case where an entire tribe, the Rukuba of northern Nigeria, was virtually brought to extinction following the close of a local war. After the war ended, these people left their overcrowded hilltop fortresses and mingled with returning farmers who harbored the trypanosomes, with the result that transmission from these infected humans via tsetse bites caused the susceptible population to succumb to the infection.

Ford (1971) describes in detail numerous instances of epidemics and epizootics of trypanosomiases in man and cattle and their effects upon the African economy. The problem of diseases caused by tsetse-transmitted pathogens is enormous and very complex. Sleeping sickness has resulted in great loss of human life from the disease itself, in addition to starvation resulting from the removal of cattle by nagana. Even in areas where the human disease has been brought more or less under control, constant vigilance is necessary. Cessation or lessening of control and prophylactic measures in recent years has led to the renewal of epidemics.

The pathogens of human sleeping sickness, as well as of nagana in cattle, are flagellate protozoans of the genus *Trypanosoma,* family Trypanosomatidae. Other pathogenic *Trypanosoma* are considered in Chapters 7 and 11. More than a hundred species of *Trypanosoma* infect mammals; these are dealt with comprehensively by Hoare (1972).

The pathogens of the two types of human sleeping sickness are usually designated as *Trypanosoma gambiense* and *T. rhodesiense,* for Gambian and Rhodesian sleeping sickness, respectively. There has been a growing belief among investigators that these disease producers are not separate species or even subspecies, but that together they form a subspecies of the nagana pathogen *T. brucei,* and should therefore, to use proper taxonomic nomenclature, be designated as *T. brucei gambiense* (Fig. 12-16). This conclusion is based, among other reasons, upon the fact that there is no morphological distinction between the two that is constant enough to separate them. Hoare (1972) discusses the problem extensively and concludes that typical *gambiense* and *rhodesiense* should be considered as NOSO-DEMES, that is, demes or strains that differ in the type of disease that they produce. *T. brucei brucei,* also, does not show consistent morphologic differences from *T. brucei gambiense,* but the host difference and other considerations, according to Hoare, warrant its subspecific separation.

The Disease. Sleeping sickness is essentially a rural disease, and urban dwellers, even in geographic areas where it is rampant, often are not aware of its importance. It occurs in two forms: the Gambian, a chronic form and distinctly human disease, and the

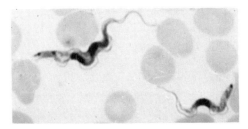

Fig. 12-16 *Trypanosoma brucei gambiense.* (Courtesy of US Armed Forces Institute of Pathology. Negative No. 74-19698.)

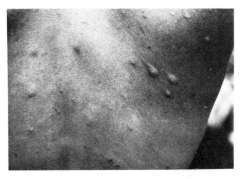

Fig. 12-17 Welts on back from tsetse bites. (Courtesy of US Armed Forces Institute of Pathology. Negative No. 75-5783.)

Rhodesian, an acute form and distinctly a zoonosis. The distinction between the two is not as clear-cut as necessary textbook treatment might indicate, but exceptions (cf. Ford, 1971) do not invalidate the general picture.

Gambian sleeping sickness is caused by the *gambiense* nosodeme, transmitted by flies of the *Glossina palpalis* group, particularly *G. palpalis, G. fuscipes,* and *G. tachinoides.* These are riverine species. This form of the disease is west and central African in distribution (Fig. 12-15). The concentration of trypanosomes in the blood is relatively low, thus requiring for their transmission a situation where constant and frequent contact between man and suitable tsetse vectors occurs. The relatively low parasitemia and the greater possibility of building antigens against the parasites probably are responsible for the milder nature of the disease. Nevertheless, the Gambian as well as the Rhodesian disease may ultimately be highly fatal. Death may result, before the disease has run its course, through susceptibility to other diseases such as dysentery or pneumonia.

Rhodesian sleeping sickness is caused by the *rhodesiense* nosodeme, transmitted by flies of the *Glossina morsitans* group, particularly *G. morsitans, G. swynnertoni,* and *G. pallidipes.* These are savanna or "game" flies. This form of the disease is east African in distribution (Fig. 12-15). A high parasitemia may occur in the bloodstream of antelopes and other mammals, thereby making them available as reservoir hosts. In man the

course of the disease is similar to that of the Gambian form, though more rapid; death usually occurs in three to nine months, compared to two or three years in the Gambian disease.

Certain symptoms may be more severe; others, even the sleeping phase, may fail to appear because of the rapid development.

The following will serve particularly as a description of the Gambian disease. The bite of an infected tsetse (Fig. 12-17) appears to produce more local reaction than that of an uninfected fly. A trypanosome "chancre" is usually found. The usual duration of the incubation period is 5 to 20 days, but clinical symptoms may be delayed for months or even years. In some individuals trypanosomiasis is asymptomatic, and such persons may serve as carriers; the extent to which this condition exists is unknown. During the first phase of the disease, which may continue for many months, the trypanosomes are in the blood; this phase is characterized by an irregular fever, glandular enlargement, debility, and languor. In the second, or sleeping sickness phase, the trypanosomes are consistently found in the cerebrospinal fluid; a constant accompaniment is the enlargement of the posterior cervical lymph nodes, WINTERBOTTOM'S SIGN (Fig. 12-18); there are tremors of the tongue and a speech impairment; there are nervousness, pronounced languor, and drowsiness, which give way to

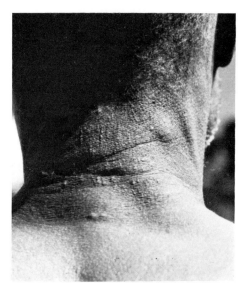

Fig. 12-18 Winterbottom's sign, swollen cervical lymph glands characteristic of human trypanosomiasis. (Courtesy of US Armed Forces Institute of Pathology. Negative No. 74-8337.)

lethargy; finally the victim falls into a comatose state, wasting rapidly, largely as a result of starvation, until death ensues.

Epidemiology and Reservoirs. Very good discussions of this subject are presented in the chapters by D. Scott and F. I. C. Apted in Mulligan and Potts (1970) and by J. R. Baker (1974). These sources have supplied much of the information used in this section.

Man himself is the only known effective reservoir of the Gambian sleeping sickness pathogen; any evidence of a nonhuman reservoir is not convincing. During the earlier stages of the disease, and sometimes for months or even years, the victim is ambulatory and can effectively supply the source of the infection to the fly. Also, healthy human carriers exist. A small colony of tsetse, located where it is restricted to human feeding, can be very effective in producing a localized high-prevalence outbreak; a constantly changing human population can provide new individuals to spread the pathogen more extensively. Contact between fly and man is

usually made along stream sides, water holes, and such places close to the water's edge, which, of course, man will frequent for various purposes, although in the more humid forested and coastal areas, in contrast to the savanna, *Glossina palpalis* is not so closely confined to areas near the water's surface.

The Rhodesian nosodeme is quite another matter. Because of the rapid course of the disease, man is often removed from the reservoir status through early death or because of the severity of the illness, which prevents him from leaving his village or hut. Effective reservoir hosts must be animals that are at least highly resistant to the Rhodesian pathogen but that may have a sufficiently high and long-lasting parasitemia for efficient transmission. These include various antelopes, particularly the bushbuck, which lives in thickets near human habitations, but also the eland, duiker, and impala. It is interesting to note that the first isolation of the Rhodesian nosedeme from a wild animal, the bushbuck, was accomplished as late as 1958 by Heisch and associates.

The movement of man away from villages and the establishment of farms in wooded areas has brought him in contact with the "game" tsetse and made him more vulnerable to attack by the fly and transmission of the pathogen to him. Rhodesian sleeping sickness can also be considered an occupational disease. Fishing, in which contact with the bushbuck environment becomes more significant, and honey and beeswax hunting are of particular importance in this respect.

Transmission. The natives of French Guinea long attributed the transmission of sleeping sickness to flies. In 1909 Kleine demonstrated the development of *Trypanosome brucei* in *Glossina palpalis* and from 1911 to 1914 M. Robertson and Bruce and his associates furnished the basic details of the life cycle of *T. brucei gambiense*.

Tsetse can usually become infected with

brucei parasites only if an infective feed is taken during the first 24 to 48 hours after emergence from the puparium. The trypanosomes are first established in the posterior part of the midgut of the insect, within the endoperitrophic cavity. Here they metamorphose into the long, slender form (trypomastigotes); these multiply enormously and, between the tenth and twentieth days, migrate forward to the proventriculus and ultimately to the salivary glands, where they complete the development to the infective stage. The cycle in the tsetse is much more complicated than this brief account would indicate; a more detailed and technical one is given by Hoare (1972). The parasite becomes infective in the salivary glands 18 to 38 days after the fly has taken the infective meal. Once infective, the tsetse remains so for life. Transmission is through the bite. Of course, many of the flies may not be harboring the parasite, or may be harboring it in a noninfective stage, at the time that a bite is administered. Either sex may transmit the parasite, but the female, being more long-lived and consequently more abundant in populations, is the more apt to do so.

Many aspects of transmission involve the complex ecology of the flies, the trypanosomes, man, nonhuman carriers, and the plant and edaphic environment. The following is given mainly as illustrative of this. Willet (1962),in discussing the relative roles of *Glossina palpalis* and *G. tachinoides* in the transmission of the Gambian parasites, advances the interesting theory that the riverine tsetse are more dangerous when circumstances are such that they have difficulty in surviving at all. Under favorable conditions the fly is more generally distributed over the area that it occupies, but when conditions are marginal for its existence, which in this case means a critical low humidity, it will occur near temporary pools, which it may be unable to leave. Man often visits such pools for water, washing, and such activities, thus bringing victim and fly

together. This theory explains such anomalies as the apparent greater ability of the fly to transmit the parasite near the limits of the tsetse's geographic range. *G. pallidipes* requires moister conditions for its existence, so may show the restricted distributional pattern where the associated *G. tachinoides* may enjoy a general distribution. Consequently, where *G. palpalis* and *G. tachinoides* coexist, the former may be the effective vector; where only *G. tachinoides* can survive, it assumes that role.

After the discovery that the nagana pathogens were transmitted by tsetse, Bruce and his coworkers demonstrated that *Glossina palpalis* could transmit *Trypanosoma brucei gambiense* mechanically for a period of less than 48 hours; the virulence of the organism became more and more attenuated after the fly had bitten the infected individual, and soon the power of infectivity was lost. Thus, the tsetse may be a mechanical vector for a short period of time, during which the trypanosomes are introduced into the wound produced by the bite, before the proboscis is completely cleaned. Interrupted feeding would thus be a factor. The bunching together of people in the presence of tsetse, such as at water holes, is also of importance. Probably cases of purely mechanical transmission are not uncommon, but this is almost impossible to prove.

ANIMAL TRYPANOSOMIASES

About 10 million square kilometers of land in Africa, which could otherwise support about 125 million head of cattle, are severely restricted for such usage because of tsetse. As a result, man has been denied the use of any domestic animals except poultry over an enormous area estimated to be as great as one-fourth of the continent. The result is not only a loss of much-needed protein supply, so essential to the development of an already protein-starved continent; there is also the influence on the mores of African peoples of certain tribes to whom the possession of cattle is an important status symbol and form of currency. Agriculture also has suffered through the lack of cattle manure used as a fertilizer, necessitating a primitive pattern of shifting land usage. On the other hand, tsetse and their trypanosomes might be considered "the most stalwart guardians of the African ecosystem and its magnificent wild fauna" (Desowitz, 1976). But for them, vast areas of Africa would turn into a desert or dust bowl through overgrazing and excessive human agricultural practices. Unwise methods of removing the tsetse scourge and improper subsequent management could result in the replacement of one bad system by another equally or even more vicious one.

What has commonly been called nagana in cattle is actually a complex disease, with three chief pathogens involved: *Trypanosoma brucei brucei,* the classical one; *T. congolense,* probably the most important one; and *T. vivax,* the most widely distributed one and the only tsetse transmitted species to become established without *Glossina* transmission, probably by tabanids and other biting flies, in the New World (See Chapter 11). Strains of various virulence occur within the species, and different hosts may be attacked and affected to varying extents. Mixed infections may occur. *Vivax* nagana is known as *souma* in French-speaking Africa and as *secadera* in America. Though the term *nagana* refers properly to the cattle disease, the same trypanosomes that cause it may cause serious disease, often fatal, in other domestic animals.

In cattle the disease is characterized by anemia, localized edema, and progressive emaciation; the eyes are sunken and the general condition becomes poor. An ill ox is conspicuous in a well-nourished herd but may pass for a while unnoticed in an emaciated one. There is a characteristic fever pattern, with early peaks associated with periodic increases of the parasites in the blood stream. The disease may terminate in death during an acute phase or it may lead to a chronic phase of longer duration; under favorable conditions, involving adequate nutrition and freedom from work, undue exposure to the elements, and absence of complicating diseases, an animal may recover. High or low virulence or strain of parasite may be the determining factor. Antibody build-up may render the animal temporarily resistant to the strain from which it has recovered. A complication of trypanosomiases in domestic animals, however, is its tendency to relapse.

Although *T. brucei brucei* was the species classically involved in the earlier studies of nagana, it is considered to be of much less importance today than the other two species. It is highly pathogenic to horses, camels, donkeys, and dogs. *T. congolense* affects all domestic mammals; it has numerous strains, varying in virulence, host restriction, and resistance to drugs. Apparently it is the last factor that has enabled it to replace *T. vivax* as the dominant species in some areas (Hoare, 1972). *T. vivax,* the dominant species in most of Africa, affects cattle, sheep, and goats; it has both virulent and mild strains. The life cycles of the three trypanosomes differ from one another. That of *T. vivax* is the simplest and shortest; that of *T. congolense* is intermediate, both in complexity and time; that of *T. brucei brucei* is the most complex and longest.

Without doubt, certain game animals, especially ruminants, serve as reservoir hosts. Many of these have lived with tsetse and *Trypanosoma* so long that a parasite-host equilibrium has been established. Other game species, on the other hand, may succumb to the infection. As in the epidemiology of human sleeping sickness, the feeding habits of tsetse species are involved here. Important reservoir hosts, resistant mammals that are commonly fed on by tsetse, include the common duiker, eland, Bohar reedbuck, impala, and spotted hyena (Ford, 1971).

Glossina palpalis, G. fuscipes, G. tachi-

noides, G. morsitans, and *G. pallidipes* serve as vectors of all three species of *Trypanosoma* discussed here. Other recorded vectors are *G. austeni, G. swynnertoni, G. longipalpis,* and *G. vanhoofi* for *T. congolense* and *T. vivax;* and *G. brevipalpis* for *T. congolense* and *T. brucei brucei.*

A tsetse-borne trypanosome of swine, of considerable importance and wide distribution in tropical Africa, is *Trypanosoma simiae.* This is the only trypanosome that is highly pathogenic to domestic pigs. Bruce described the disease as "the lightning destroyer of the domestic pig" (Mulligan and Potts, 1970); its course is rapid, according to Stephen (1966), who summarized the information on pig trypanosomiasis, lasting only an average of 3.6 days from the time of discovery of the parasites in the blood to its fatal termination. Other mammals are affected in various ways; the name "simiae" bears witness that it was originally studied in an infected monkey. The reservoir is apparently the nonsusceptible warthog, *Phacochoerus aethiopicus.* At least 11 species of *Glossina* are known to be vectors. Of these, *G. palpalis, G. fuscipes,* and *G. morsitans* appear to be most important. Janssen and Wijers (1974) found a correlation between the *Glossina* species and the virulence of the trypanosome strain; of the three species studied, it was highest in *G. brevipalpis,* with no survival of an infected pig, intermediate in *G. austeni,* and lowest in *G. pallidipes.*

Fig. 12-19 The stable fly, *Stomoxys calcitrans.* Note slender projecting proboscis. (Courtesy of Oregon State University and Ken Gray.)

STABLE FLY

The genus *Stomoxys* includes the well-known cosmopolitan stable fly *S. calcitrans* (Figs. 12-19, 12-20*A*), the only member of this genus that occurs in the New World, and 17 other known species. All, so far as is known, feed on mammalian blood, and several attack man. Zumpt (1973) has treated the world species of this and other Stomoxyinae (*Haematobia, Haematobosca,* and others), and has included in this work, in addition to the taxonomy, considerable information on their biologies, with valuable bibliographical references.

In addition to the stable fly, two additional species of *Stomoxys* should be mentioned. *S. sitiens* is widespread in Africa and the Oriental Region; it is reported to be common locally, and it can be a pest to donkeys, cattle,

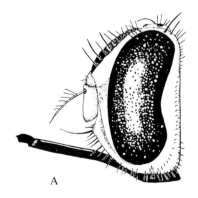

Fig. 12-20 *A.* The stable fly, *Stomoxys calcitrans,* head from side view. *B.* The horn fly, *Haematobia irritans,* same; both drawn to the same scale. (Drawings by B. Eldridge.)

A   B

and other animals. *S. nigra* occurs throughout Africa and is probably the most common species there, according to Zumpt. It attacks chiefly horses, donkeys, and cattle, but its bite is quite painful to man. Around cattle camps in southern Sudan, it occurs in enormous numbers and is the most troublesome of biting flies, according to Reid (cited by Zumpt). It was a major vector of the surra pathogen in Mauritius prior to the eradication of that disease there. Its bite has been responsible for the death of cattle on that island (Monty, 1972) and, when concentrated on or about the eyelids of cattle, it has resulted in ophthalmia and near-blindness (Moutla, cited by Zumpt).

Characteristics. *Stomoxys calcitrans* can be distinguished from other species of the genus by the broader frons of the male (about one-third or more as broad as the eye height) and the color pattern of the abdomen. Owing to similarity in color and size, it is often mistaken for the house fly, *Musca domestica,* though it is more robust, with a broader abdomen, and has piercing rather than sponging mouth parts. In color the stable fly is brownish gray with a greenish yellow sheen; the outer of four thoracic stripes are broken, and the abdomen is more or less checkered. The wings when at rest are widely spread apart at the tips and are distinctly irridescent; the apical cell is open. The proboscis protrudes bayonetlike in front of the head. (For its structure, see Chapter 3.) The antennal arista, unlike that of the house fly, bears hairs on the upper side only; these hairs are simple, however, not plumose as in the tsetse.

Food Habits. Although *Stomoxys calcitrans* is commonly called the "stable fly," in many areas it occurs much less abundantly around stables than does the house fly. It is sometimes called the "biting house fly" because it may occur indoors, especially in the autumn and during rainy weather, and at such times it may bite human beings viciously. It is typically an out-of-doors, day-biting fly and is usually to be found in abundance during summer and autumn where large numbers of domestic animals occur; horses and cattle afford plentiful food supply. Sunny fences, walls, and light-colored surfaces in general, when in the proximity of animals, are abundantly frequented by stable flies.

The stable fly is a vicious biter that draws blood quickly and feeds to full capacity in 3 to 4 minutes if undisturbed; ordinarily, even at such times, it changes position frequently or flies from one animal to another, where the meal is continued. It feeds readily on many species of mammals, such as rats, guinea pigs, rabbits, monkeys, cattle, horses, and man. Both sexes suck blood. The flight of the stable fly is direct, swift, and of long range, the fly sometimes traveling many kilometers.

Breeding Habits and Life History. Although the stable fly can be successfully reared in the manure of horses, cattle, and sheep, it does not commonly breed in excrement under field conditions unless this is mixed with decaying vegetable matter, straw, or hay. Very good breeding places are afforded by leftover soggy hay, alfalfa, or grain in the bottoms of or underneath out-of-doors feed racks in connection with cattle feed lots. This material becomes soggy and ferments, and here virtually pure cultures of stable fly larvae may be found. The materials must be moist, for dryness prevents larval development. They should also be loose and porous, with temperatures ranging between 15° and 30° C.

Old straw piles that remain in the field through the year may produce an abundance of stable flies in the moist fermenting straw near the ground, particularly if cattle have access to it and moisten it with urine. Other fermenting and decaying vegetable matter, such as windrowed seaweed, piles of moist fermented weeds and lawn cuttings, piles of waste vegetables, and fermenting peanut litter, may provide breeding places. In climates where the fly may breed throughout the year, breeding during the colder months may occur in stables and other protected places.

The larvae of the stable fly and of the house fly can readily be differentiated by the form, size, and position of the posterior spiracles; otherwise they resemble each other closely. The posterior spiracles of the stable fly are roughly triangular, widely separated from each other, and situated near the periphery; in the house fly they are elliptical, large, close together, and more central in position. The pattern of the spiracular slits, also, is different.

The eggs of the stable fly are about 1 mm long, curved on one side and straight and grooved on the other. In depositing her eggs, the female fly often crawls far into the loose material, placing them usually in little pockets in small numbers, often in pairs. Depositions range in number from 23 to 100 eggs, usually between 25 and 50; there are ordinarily 4 or 5 layings, but there may be as many as 20. Egg production exceeding 800 has been reported.

The developmental period varies considerably, depending upon temperature and other environmental conditions. The incubation period varies from less than 1 to 5 days at 28° C; under cold conditions it may extend to several weeks. The newly hatched larvae bury themselves in their food at once, thus avoiding desiccation. Larval development requires 6 to 26 days, sometimes up to several months at temperatures near the lower range of development.

Before pupation the larvae usually crawl into the drier parts of the breeding medium, where the chestnut-colored puparia may be found in enormous numbers. The puparia are 6 to 7 mm long and may be recognized by the posterior spiracles, which resemble those of the larvae. The pupal period, including the last larval stage that begins with the formation of the puparium, varies, depending on temperature. At 21° to 26° C it ranges from 5 to 26 days.

The imago emerges with astonishing rapidity, unfolds its wings, and is ready to fly in less than an hour. The fact that the proboscis is temporarily attached beneath the thorax gives the newly emerged adult a very peculiar appearance, and it then may be mistaken for a house fly. The total time for development of the fly, from egg laying to emergence of adults, was determined by Herms to be 33 to 36 days at 21° C. As short a total developmental period as 10 days has been reported, and under adverse conditions this period may extend over 2 or more months.

Longevity. Herms has found that the average length of life of an adult stable fly under favorable laboratory conditions of feeding is about 20 days, with a maximum of 69 days. A maximum of 72 days for the female and 94 days for the male in the Philippine Islands was determined by Mitzmain.

The Stable Fly as a Human Pest. Newsom (1977) has pointed out that although there is relatively little literature on the subject, the stable fly is well recognized as an important human pest. On the Gulf Coast of Florida it can make beach areas untenable for recreational use from mid-August to mid-September, when the fly is breeding in fermented marine vegetation deposited above the normal high tide mark by high storm tides. Newsom cites estimates that these infestations may cost the Florida tourist industry a million dollars a day in lost revenues. The beach pest problem exists much farther north along the Atlantic Coast, as well as on inland beaches, such as along the Tennessee River reservoirs and Lake Superior. Accumulations of dead mayfly bodies added to tidal debris may enhance breeding possibilities, and the flies may move from conventional farm breeding areas to nearby beaches. Similar infestations caused by this and other species of *Stomoxys* and *Haematobia* occur in other parts of the world. Zumpt (1973) cites a report of a serious infestation by *Haematobia minuta,* apparently breeding in marine debris, along the Persian Gulf in Saudi Arabia.

The Stable Fly as a Livestock Pest. This fly is generally recognized as one of the most important sources of annoyance to livestock. Injury by it is caused in various ways: worry resulting from mass attacks of the flies, loss of blood, loss of flesh, lowered milk production and reduced butter fat content, reduced vitality making the animal more susceptible to disease, loss of pasturing time, and actual damage to the victim's tissues and hide caused by the bite itself.

Losses to cattle production in the United States during 1965 were estimated at $142 million, about half of which was in diminished weight gains. Milk production is reduced up to 60 percent according to one estimate, and butter fat content may be lowered for weeks or even months after the removal of the fly menace. Bishopp (1939), in a study that despite its age is frequently cited today, describes the seriousness of the problem. Animals, chiefly horses and cattle, at times of heavy infestation have to wage a constant fight against the pest, and they may become so weakened from the worry and annoyance that they give up and die, if not given attention. The economic threshold for the stable fly on cattle, according to Steelman (1976), is 25 flies per animal per day; this number is often exceeded in nonprotected areas, as many as a thousand flies having been observed on one animal at one time.

Moorhouse (1972) reports that in southeastern Queensland, Australia, in recent years cattle have demonstrated a hypersensitivity to stable fly bites leading to cutaneous lesions formed by the coalescence of intradermal blisters that follow bites. Though uncommon, Moorhouse believes that this condition could become more prevalent in feed lots where cattle are fed on silage, an important breeding medium for *Stomoxys*.

The Stable Fly and Disease. The stable fly is no exception to the rule that any blood-sucking fly must be suspect in the transmission of causative organisms of disease in man and animals. Its characteristic interrupted feeding habits and ready passage from one host to another are of undoubted importance in mechanical transmission. Yet, it is apparent, from a considerable amount of published information, that the true role of this fly in this respect is not adequately known.

Several pathogenic Protozoa can be transmitted for a very limited period of time by the bite of *Stomoxys*. One of these is *Trypanosoma evansi,* the causative organism of surra. This disease is nearly always fatal to horses and mules, and it often affects camels and dogs seriously; it may under certain circumstances be serious in cattle, though in them it is usually mild or asymptomatic, with the result that cattle may serve as carriers. Although the role of the stable fly as a vector of *T. evansi* is purely mechanical, its transmission by this and other bloodsucking flies, particularly *Tabanus,* is quite important. Mechanical transmission of *T. brucei* and its *rhodesiense* and *gambiense* nosodemes also occurs but is of very little importance. *Leishmania tropica* and *L. mexicana* have also been transmitted experimentally through the bite of the stable fly. Transmission of the above-mentioned Protozoa occurs only in areas where the pathogen is maintained by cyclical transmission in the insectan invertebrate hosts. It is possible, however, that *Stomoxys,* as well as tabanids, may be involved in the transmission of *Trypanosoma vivax* in America and elsewhere where the cyclically involved tsetse does not occur.

Occasional transmission of several pathogenic bacteria from animal to animal, animal to man, or man to man has been demonstrated, always mechanically and for a limited period of time. Among these are *Borrelia recurrentis,* the agent of the louse-borne epidemic relapsing fever; *Bacillus anthracis,* which causes anthrax in animals and man; *Dermatophilus congolensis,* the causal agent of cutaneous streptothrichosis in cattle, horses, goats, and man; the brucellosis organisms *Brucella abortus* and *B. melitensis;* and *Erysipelothrix insidiosa,* the cause of swine erysipelas, which affects other animals, including birds, and is transmissible to man.

Among the viral diseases, equine infectious anemia (EIA), also known as swamp fever, has received considerable attention, though as in the case of anthrax, tabanids are probably of more importance than *Stomoxys.* The stable fly seems to play some role in the transmission of the viruses of African horse sickness and fowl pox. The supposed relation to the transmission of the polio virus, which received so much attention during the decade from 1910 to 1920, has been disproven and is only of historic interest.

In addition to its role in mechanical transmission of Protozoa, bacteria, and viruses, *Stomoxys calcitrans* is a proven intermediate host of the nematode *Habronema microstoma,* a stomach parasite of horses. The infective larvae of this worm interfere with the biting process of the fly; Zumpt has suggested that the irritation caused by the excessive probing of the parasitized fly may lead to ingestion of the fly by the vertebrate host and consequently pathogen transmission through this route.

HORN FLY

The horn fly, *Haematobia irritans irritans* (Fig. 12-20*B*), is a well-established pest of cattle throughout most of Europe, North Africa, Asia Minor, and the Americas. It has been referred to in the literature as *Siphona irritans, Lyperosia irritans,* and *Haematobia stimulans.* The true *stimulans* is a different fly, belonging to a different, though closely related genus, *Haematobosca.* For a bibliography of the horn fly and buffalo fly *Haematobia irritans exigua,* see Morgan and Thomas (1974).

Characteristics. The horn fly is about 4 mm long. It has the same general color as the stable fly and in most respects resembles it, though it is a much more slender species. The mouth parts (Fig. 12-20) are like those of the stable fly except that the labium of the horn fly is relatively heavier and the palpi, almost as long as the proboscis, are flattened and loosely ensheath that structure. The wing venation is similar to that of the stable fly.

The season of horn fly abundance varies with climate and latitude. In southern Alberta, Canada, according to K. R. Depner (personal communication), there is one peak of abundance in midsummer, whereas in Texas there are two peaks, in early and late summer respectively. The adult fly remains on the host day and night, the females leaving only briefly to deposit their eggs. In feeding the fly characteristically orients itself with its head downward, toward the ground, in contrast to the stable fly, which usually orients itself with the head directed upward. At least in western North America, the habits of clustering around the horns of cattle, which gives the fly its common name, seems to relate to weather conditions. It often is evident when a storm front is approaching and may be influenced by a change in moisture, pressure, or a combination of the two.

Life History. A comprehensive study of the life history of this fly and its habits was published by McLintock and Depner (1954). The fly deposits its eggs chiefly, if not exclusively, on freshly passed cow manure. Eggs are deposited sometimes singly, more often in groups of 4 to 6, usually under the sides of the cake of dung or in the grass or soil beneath it. A maximum of 20 to 24 eggs may be laid at one time, but a female is capable of producing up to 400 eggs in her lifetime. At temperatures of 24° to 26° C eggs hatch in 24 hours. A relative humidity of close to 100 percent is required for maximum hatching.

The larvae burrow through the droppings, reaching full growth in approximately 4 to 8 days. Temperature has a profound effect upon the developmental period. Depner (1961), for example, found that the development of the larva, from hatching to pupariation, required 10.5, 5.6, and 3.7 days at respective temperatures of 18°, 24°, and 30° C. From 6 to 8 days after puparium formation, under summer conditions, the adult is ready to emerge. In cooler climates overwintering is in the pupal stage in a state of diapause, the potential for which is brought about by the effects of decreasing photoperiod and a corresponding decrease in ultraviolet radiation of the adult female of the preceding generation (Depner, 1962, 1965).

Damage and Relation to Disease. The damage occasioned by the horn fly is chiefly through irritation and annoyance, which in dairy animals results in disturbed feeding and improper digestion, causing loss of flesh and reduced milk production. Cattle heavily attacked by these flies may suffer a loss of 0.5 lb of flesh per animal per day and milk production may be reduced 10 to 20 percent. The estimated loss to cattle production in the United States in 1965 was $115 million in weight loss or reduction of weight gains and $64 million in milk production.

It is not uncommon in areas where the horn fly abounds for from 1,000 to 4,000 flies to be on one animal at any one time. Depner (personal communication) has observed as many as 10,000 flies per animal in southern Alberta, and he states that claims as high as 20,000 per animal have been made. Under such conditions the actual loss of blood must certainly be significant. From 10 to 20 minutes are required for the fly fully to engorge itself; during this time it withdraws and reinserts its proboscis in the same puncture as in a pumping motion. Much digested blood is discharged from the anus of the fly while it is feeding.

As in the case of other bloodsucking pests of man and domestic animals the horn fly becomes suspect as a possible transmitter of pathogens. There is little firm evidence, however. The lesser mobility from host to host, in comparison with the stable fly, may be an important factor in this respect. Stirrat and associates (1955) have shown that though horn flies harbor a number of bacteria, these are normal associates of cattle. The horn fly is the intermediate host of *Stephano-filaria stilesi,* a filarial nematode of cattle that reduces the value of the hides and causes blemishes that interfere with the exhibit of registered animals.

Man is only rarely attacked; the importance of the horn fly is chiefly veterinary rather than medical. A supposed case of human traumatic myiasis is on record, but Zumpt (1973) dismisses it as purely contamination.

Haematobia irritans exigua, commonly known as the buffalo fly, is important to the cattle and dairy industries of Australia and in much of the Oriental Region. Among the animals that it attacks are water buffalo, cattle, horses, dogs, and man. The fly oviposits in fresh dung, particularly of cattle and buffalo. Other species of *Haematobia* may attack man and domestic animals. *H. minuta,* for example, attacks cattle and water buffalo and may, when abundant, be a very unpleasant pest of man (Zumpt, 1973).

CONTROL OF BLOODSUCKING MUSCOIDS: TSETSE, STABLE, AND HORN FLIES

A reasonable degree of personal protection can be achieved against these bloodsucking flies by the use of repellents or fast-acting insecticides. Such methods are not practiced against tsetse because conditions would require nearly continuous application for nomadic or village populations of humans, and for cattle herds in unfenced grazing conditions.

Tsetse (Glossina). Surveying tsetse populations is largely accomplished by human capture and by trapping. One common procedure is to have personnel acting as catchers with nets patrol a standard prescribed route for a standard period of time—

the so-called fly-round (Ford *et al.*, 1959). More recently quite effective traps have been developed, the Morris trap being notably useful with some species of tsetse (Glasgow, 1967). Traps, particularly a biconical design, catch a better proportion of females than the fly-round capture, and are therefore more representative of tsetse populations (Challier and Laveissière, 1973).

Several articles and reviews concentrate on the control of *Glossina* (WHO, 1969; Leach, 1973; Jordan, 1974; Hagen and Wilmhurst, 1975). While there is no question that *Glossina*-associated trypanosomes hinder cattle production and human settlement, the consequences of control may not be entirely beneficial, and may lead to the disruption of stable ecosystems over vast areas of Africa (cf. Ford, 1971).

Personal protection against tsetse bites has been moderately effective with the use of a repellent-treated wide mesh jacket (Sholdt *et al.*, 1975), but there is yet much to be done in testing repellency of compounds against these biting flies.

Environmental modification has been an important tsetse control practice, often in conjunction with other procedures. *Glossina* adults are restricted to high-humidity zones, and adults seldom fly over cleared areas, particularly in drier savannah zones; puparia are found in moist soil conditions, especially associated with rivers and streams. In heavily forested areas reduction of tsetse attack is achieved by clearing vegetation from a wide area around villages. In savannah conditions blocks of land are laid out for control, with natural isolation where brushy vegetation ends, or with cleared zones at boundaries. Thicket vegetation may also be reduced in such zones to provide fewer resting sites for adults. Within control blocks wild mammalian hosts, such as antelopes and pigs, may be destroyed to reduce adult food sources. Reduction of potential blood meals is especially applicable with tsetse because they do not feed on nectar or other sugar sources.

Game destruction alone is often not very effective and has been criticized (Dias and Rosinha, 1973; Wilson, 1975). Areas cleared of tsetse are best maintained by encouraging human settlement (Wooff, 1967; Kavemba, 1975).

Biological control agents of several types have been noted for tsetse (see Buxton, 1955), with some occasionally causing quite high natural mortality but none having been effectively utilized. Insect parasites and predators are perhaps most noteworthy. A eulophid wasp, *Syntomosphyrum albiclavus,* which is a pupal parasite recorded only from *Glossina* in nature, can be cultured on a variety of cyclorraphous Diptera in the laboratory. *S. glossinae* has been noted to parasitize as high as 84 percent of *G. morsitans* pupae in the dry season. Various ants and shrews occasionally exert considerable control as predators of *Glossina* pupae, and adult tsetse may be preyed on extensively by spiders and robber flies (Asilidae). This variety of insect predators suggests that widespread application of insecticides may drastically affect natural biocontrol agents.

Insecticides have been very effective in destroying tsetse in some control projects. Insecticide treatments along 300 miles of river system in Kenya caused *Glossina palpalis fuscipes* to become exceedingly difficult to find within three years, and human cases of Gambian sleeping sickness were reduced from a previous level of 107 to 157 cases per year to 16 (Glover *et al.*, 1960). Because many species of *Glossina* wait on vegetation for hosts along paths and game trails, spraying of resting surfaces in such locations has provided exceptional control, and this more restricted area of pesticide application should have fewer side effects than widespread spraying. Ground and aerial applications of insecticides can provide very effective temporary control of adult tsetse in open savannah or forested riverine zones, but as might be expected there are undesirable side effects (Koeman and Pennings,

1970; Wilson, 1972). Insect growth hormonal analogs are promising as pesticides against tsetse (Denlinger, 1975), and these may prove to be more selective.

Stable Fly (Stomoxys). Survey methods consist largely of observing stable flies on animals or on favored resting sites. A translucent fiberglass panel sticky trap has been found to be highly effective for monitoring *S. calcitrans* (Williams, 1973). Much can be done to reduce populations by controlling the amount of wet and rotting straw and manure around premises. Where development is occurring in such media as beach seaweed, this debris may be hauled away, spread out to dry, or burned.

Biological control agents are generally identical or similar to those attacking house flies, but with the exception of some hymenopterous parasites they seem to exert minimal control. Several parasites and predators have been introduced and established on Mauritius, where they provide fair control (Monty, 1972). An electrocutor grid trap with carbon dioxide attractant has proved highly selective against *S. calcitrans* during winter in Florida, USA (Schreck *et al.*, 1975).

Insecticides and repellents, as listed for horse flies, are satisfactory for direct treatment of cattle. In addition, residual insecticides applied to resting surfaces around farm premises provide good control.

Reproductive manipulation may have distinct possibilities, the more so because stable fly populations tend to be focal in nature and this insect can be reared readily. Chemosterilant treatment in the field would have to rely on general contact, since sweet baits are not eaten. One factor favoring sterile male techniques is the observation that *S. calcitrans* females apparently mate but once.

Horn Fly (Haematobia). Horn fly populations are readily estimated by direct counting of adults, particularly along the back, flanks, and legs of cattle.

Control may be effected by modifying the cattle droppings in which larvae develop. For example, pigs in pastures scatter cattle dung so it dries rapidly, preventing larval development, though this method is impractical for range conditions. Feeding of insecticides, or spores of the bacterium *Bacillus thuringiensis*, to make fresh manure unsuitable for larval development has been one experimental approach. Species of coprophagous insects, such as dung beetles, that compete with horn fly larvae are helpful, and this method has been studied in Texas, USA (Blume *et al.*, 1973) and developed into a program in Australia against this pest and bush fly (Ferrar, 1973). No doubt some of the same egg and larval predators that affect house flies and face flies, such as mites and staphylinid beetles, exert an influence.

Insecticides may be applied to cattle in conjunction with repellents. The automatic sprayer systems discussed for tabanids are useful in dairy herds; more applicable for range cattle are back dusters of pesticide in burlap sacks. Self-applicators are not always satisfactory on range cattle, so systemic insecticides have been tested in feed supplements (cf. Miller and Loomis, 1966).

Two additional approaches are sterile male techniques (Kunz and Eschle, 1971), and use of less attractive Brahman (zebu) or Brahman-cross cattle (Tugwell *et al.*, 1969).

Families Hippoboscidae, Nycteribiidae, and Streblidae

LOUSE FLIES AND BAT FLIES

These constitute four families of curious, sometimes wingless, louselike, ticklike, or spiderlike flies, three of which attack warm-blooded animals: the Hippoboscidae (louse flies), the Nycteribiidae (spiderlike bat flies), and the Streblidae (bat flies). Little is known of the life histories of the streblids and nycteribiids, but they parasitize bats externally. The fact that bats are involved in the epidemiology and epizootiology of rabies is of interest here, because until proven otherwise there exists the possibility that bat parasites may help to maintain the virus in bat populations.

Characteristics of Hippoboscidae. The bloodsucking parasites belonging to the family Hippoboscidae are readily recognized as Diptera when winged. The larvae are retained within the body of the female, where they are nourished by glandular secretions until time for pupation; they are then extruded, and pupation quickly follows. The adult flies are flattened, leathery in appearance, with wings well developed, shortened, or absent; the legs are short and strong and broadly separated. The antennae are inserted in pits or depressions near the border of the mouth; they appear to consist of one segment, which may terminate in a bristle or long hairs.

The adults are all bloodsucking external parasites on birds and mammals. They range in size from 2.5 to 10 mm. The American species have been admirably monographed by Bequaert (1953–1957).

Sheep Ked. The sheep ked, *Melophagus ovinus,* is wingless, reddish brown in color, 5 to 7 mm in length. The head is short and sunken into the thorax, the body is saclike, leathery, and spiny (Fig. 12-21). It is a widely distributed parasite of sheep and goats in the temperate areas of the world; in the tropics it cannot survive except in the cooler mountain areas. Well-nourished animals may acquire a degree of resistance to it (Nelson and Hironaka, 1966).

Development of the larva within the body of the female fly requires about seven days. The extruded larva pupates during the course of a few hours, becoming then chestnut brown in color; the secretion with which it is covered hardens and serves to glue the puparium firmly to the wool of the host. The pupae are commonly found on infested animals in the region of the shoulders, croup, thighs, and belly. Pupae may be found on

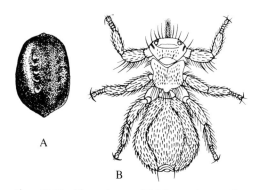

Fig. 12-21 The sheep "tick" or louse fly, *Melophagus ovinus. A.* Pupa. *B.* Adult. (Courtesy of US Public Health Service.)

sheep at all times of the year. The time required for development is about three weeks in the summer but may be twice as long in the winter. Females reach sexual maturity in from 14 to 30 days; they then begin to extrude young at the rate of one about every 7 or 8 days. The average life of the insect is about 4 months; each female produces from 10 to 12 young.

The entire life of the ked is spent on its host; when off the sheep the insects die in from 2 to 8, usually in about 4, days.

DAMAGE DONE. The presence of a few louse flies on the body of a sheep does not materially affect it. In heavy infestations the presence of the insect is indicated by the fact that the animal rubs itself vigorously, bites the wool, and scratches. Badly infested animals show emaciation, anemia, and general unthriftiness. The injury to lambs is especially marked.

A defect of the skin known as sheepskin cockle may result from heavy feeding by this fly. This may seriously reduce the value of the skin; added to the loss in carcass value and wool, the economic loss in maintaining this parasite may be serious (Everett, Roberts and Naghski, 1971), contrary to the statement sometimes made that the cost of control of the sheep ked can not be justified by the results.

Melophagus ovinus is a host of the African

Trypanosoma melophagium; though this parasite is common in that continent, there is no evidence that it has any harmful effect on sheep (Hoare, 1972).

Other Louse Flies of Mammals.
Lipoptena depressa and *Neolipoptena ferrisi* are common parasites of deer in North America. These species are smaller than *Melophagus ovinus* but otherwise resemble it; they are wingless when established on the host, but have well-developed wings when they emerge from the puparium. The parasites have been found in chains of three or four individuals attached to one another, the first drawing blood from the host, the subsequent ones drawing blood from the abdomen of the individual preceding it. *Lipoptena cervi,* the deer ked, is reported to be a common species on deer in Europe. According to Bequaert (1953–1957) it has become naturalized in the northeastern United States on the Virginia white-tailed deer; he also reports it as occurring on wapiti. In Europe, in addition to deer it has been reported as attacking dogs, horses, cattle, and sheep. *Lipoptena mazamae* occurs on deer in South and Central America and in the southeastern United States.

In the genus *Hippobosca* wings are always well developed and functional throughout adult life. With the exception of the ostrich louse fly, *H. struthionis,* the species of this genus are ectoparasitic on mammals. *Hippobosca equina,* known in England as the forest fly, is usually found on horses, mules, and donkeys, sometimes on cattle or other mammals. *H. longipennis* is commonly found on dogs in Asia and the Mediterranean region, and has been introduced, but apparently not established, into several of the United States (Westcott, 1973, Federal State Cooperative and Economic Insect Report, Oregon). *H. variegata* occurs on domestic cattle and equines and is widespread in distribution; *H. camelina* is a parasite of the camel and dromedary. No species of *Hippobosca* seems to be established in America.

Louse Flies of Birds. The pigeon fly, *Pseudolynchia canariensis* is an important parasite of pigeons throughout the tropics and warmer temperate regions of the world. It is found throughout the southern United States and northward along the Atlantic Coast to New England. The dark brown flies have long wings and are able to fly swiftly from the host, but usually alight nearby. They move about quickly among the feathers of the host and bite and suck blood from parts that are not well feathered. The mature larvae, at first yellow and later jet black in color, are deposited on the body of the bird while it is quiet, but they soon roll off and collect in the nests. The pupal stage lasts about 30 days at a temperature of 23° C. Consequently, cleaning the nest at intervals of not more than 25 days is probably the most important step in control.

In addition to being a bloodsucking parasite, the pigeon fly is the vector of a pigeon blood protozoon, *Haemoproteus columbae.* The parasite undergoes sporogeny in the body of the fly and consequently requires it as a link in its life cycle. A similar pathogen of the California valley quail, *Haemoproteus lophortyx,* is transmitted by *Lynchia hirsuta* and *Stilbometopa impressa,* and other bird hippoboscids may be involved in transmission of other species of *Haemoproteus.*

Hippoboscids Attacking Man. This subject has been reviewed critically by Bequaert (1953–57). No known hippoboscid has man as its normal or habitual host, but at least 13 species, including *Hippobosca equina, H. camelina, H. variegata, H. longipennis, Melophagus ovinus, Lipoptena cervi,* and *Pseudolynchia canariensis,* have been authentically reported as biting man. The bite of *H. longipennis* is said to be as painful as the sting of a yellow jacket. The sheep ked may become quite annoying to persons employed in shearing sheep or handling wool. The pigeon fly may readily attack persons who handle squabs and adult birds; the bite is said to be as painful as a bee

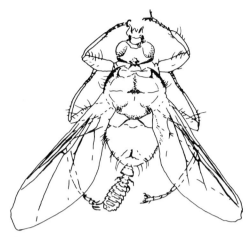

Fig. 12-22 A chewing louse (Mallophaga) attached to a louse fly for phoretic transfer. (Drawn from a photograph in Bequaert, 1953–1957.)

sting, and its effects may persist for five days or more. The deer ked is reported by Ivanov (1975) as attacking man readily and causing dermatitis in Byelorussian forests of western European USSR, where European elk are abundant.

Phoresy. Keirans (1975) has reviewed the subject of phoretic relationships in the role of dispersal of Mallophaga by hippobos-

cids (Fig. 12-22). He cites 405 cases taken from collections of hippoboscids from 22 passerine families and 11 orders of nonpasserine birds and concludes that the relationship is not purely accidental; it may have a distinct survival value to a wingless parasite that must, on the death of its host, find a new one.

CONTROL OF LOUSE FLIES (HIPPOBOSCIDAE)

This discussion is limited to the sheep ked, *Melophagus ovinus*. Close examination is required to find adult keds or puparia in fleece when infestations are low. High levels of infestation are indicated by the animals' engaging in vigorous rubbing, biting of wool, and scratching. Sheep infested by keds for a considerable time develop a certain degree of resistance (Nelson, 1962), suggesting that some form of immunization or breeding of resistant strains of sheep might be possible. Dusting with appropriate pesticides is frequently a practical means of control. Carrying out this control procedure right after shearing helps to achieve better coverage of pesticide over the animals.

13

MYIASIS

GENERAL CONSIDERATIONS

MYIASIS is a term meaning an infestation of the organs and tissues of man or animals by fly larvae that, at least for a period of time, feed upon the living or dead tissues or upon the ingested food of the host. Though the host may, in a broad sense, be any animal, myiasis usually implies the infestation of a vertebrate. Such invasions may be benign in effect or even asymptomatic. On the other hand, they may result in mild to violent disturbances, even death.

Various terms have been used to indicate the localization of the myiasis, such as GASTRIC, INTESTINAL, or ENTERIC (digestive system); RECTAL; URINARY or UROGENITAL (urogenital tract); AURICULAR (ears); OPHTHALMIC (eyes); DERMAL, SUBDERMAL, and CUTANEOUS (skin); or NASOPHARYNGEAL (nose and pharynx). When wounds are involved, the term "TRAUMATIC" may be used; when the lesion produced is boillike, it may be called FURUNCULAR. When larvae burrow in the skin in such a way that progress may be followed as the larva advances, the term "CREEPING MYIASIS" (CREEPING ERUPTION) is applied. Finally, larvae may be INTERMITTENT BLOODSUCKERS; although this is an aberrant type of myiasis, it is recognized as

a form by Zumpt (1965), who calls it "SANGUINIVOROUS MYIASIS."

As in the case of other types of parasitism, myiasis may be OBLIGATORY, when the parasite is dependent on the animal or human host, at least for a certain period of its life, to complete its development; and FACULTATIVE, when the larva, though normally free-living, can under certain circumstances adapt itself to a parasitic existence.

Zumpt (1965) has presented a logical and convincing theory of the origin of myiasis. An unspecialized feeder, such as the false stable fly *Muscina stabulans,* utilizes both decomposing organic matter and dead and living insects for food. It has also been reported as a faculative parasite on numerous occasions in gastrointestinal involvement and traumatic myiasis of several kinds, including sheep strike.

From such a beginning Zumpt hypothesizes evolution along two lines, the saprophagous and the sanguinivorous. The saprophagous root develops through three steps: (1) the carrion- or excrement-feeding larva invades diseased or ill-smelling wounds, usually in a benign fashion; (2) then healthy tissues, usually those in contact with the necrotic ones, are invaded; (3) finally the parasitism becomes obligatory and malign, requiring healthy tissues for its development.

296

The final step is not a large one and in some cases it has not been thoroughly bridged. For example, *Wohlfahrtia vigil,* though an obligatory parasite in its earlier stages and usually throughout its larval development, can survive as a final instar after its host has died.

The sanguinivorous root begins with the larva first piercing the body and sucking the blood of other insects that share its habitat in excrement or carrion. The maggot then proceeds to attack vertebrates as a bloodsucker and finally becomes a warble-forming or body-invading parasite. An interesting transition is found in the blow fly genus *Protocalliphora,* in which the larvae usually attack nestling birds as bloodsuckers but are known in at least one and probably several species to form burrows under the skin and to persist there even into the fledgling stage of the host. The transition from insect feeding to bird feeding is easy to explain here, since bird nests harbor a considerable scavenger fly fauna.

James (1969) has discussed the transition between the saprophagous and the parasitic way of life and has shown that the three obligatory screwworms, *Cochliomyia hominivorax, Chrysomya bezziana,* and *Wohlfahrtia magnifica,* each has a scavenger-facultative counterpart, respectively *Cochliomyia macellaria, Chrysomya megacephala,* and *Wohlfahrtia meigenii.* Races of the common green-bottle fly, *Phaenicia sericata,* include one harmless enough that it could be used safely in maggot therapy of wounds and osteomyelitis, as well as one that has importance as a sheep wool maggot. The higher flies in many cases are very adaptable in their food habits.

Some studies of significance on the subject of myiasis are those of James (1948), Zumpt (1965), and Morikawa (1958). The reviews by Scott (1964) for North America and Lee (1968) for Australia are useful to the general student.

ACCIDENTAL MYIASIS

Accidental Enteric Myiasis. About fifty species of fly larvae have been reported, either positively or questionably, from cases of enteric myiasis in man. These belong largely to the Muscidae and Sarcophagidae, although a number of other families are involved. In most cases, no indication of development within the digestive tract is evident, the larvae merely being carried through passively. Infestation is effected through ingestion of larvae or eggs with the food or by other means, such as the intake or contact with contaminated liquids or other substances.

The reality of enteric involvement as a pathological condition in man has been questioned by some workers. The subject has been discussed in considerable detail by Riley (1939), James (1948), and West (1951), who agree that, despite experimental evidence to the contrary (which these authors do not consider conclusive), "there seems no doubt that genuine enteric myiasis does occur from time to time, when chemical and physical conditions within the patient's alimentary tract are such as to favor survival of the parasites" (West). The evidence for the reality of enteric myiasis in man rests mostly on findings in a number of clinical cases by competent entomologists and physicians. A recent report based on an autopsy (Anon, 1975), in which unidentified but apparently muscoid larvae were found in the lumen of the appendix, indicates that at least an occasional true case may occur. Nevertheless, great care should be exercised in diagnosing human enteric myiasis.

Human enteric myiasis may show clinical symptoms that are sometimes severe. They may depend upon the species of fly larvae, their number, and their location within the digestive tract. No doubt, in many cases larvae are passed without causing any severe disturbances. In severe infestations the patient may be depressed at times and suffer considerable malaise; there may be vomiting, nausea, vertigo, and more or less violent pain in the abdomen; diarrhea with discharge of blood may occur as a result of injury to the intestinal mucosa caused by the larvae. Living and dead larvae may be expelled with either vomit or stool, or both. Repeated vomiting may occur either from multiple infestations or as the result of an accompanying pathological state such as bacillary infection or helminthiasis.

Leclercq (1974) has presented a concise account of the disease and its associated dipteran pathogens.

Zumpt (1965) does not consider the kind of enteric infestation that occurs in man as a true myiasis, and he proposes the term *pseudomyiasis* for it. He believes that the ingested larvae are not normally able to feed and consequently to continue their development in the human digestive tract. This would, of course, remove this type of infestation from that defined at the beginning of this chapter as myiasis. What is involved is merely the ability of the fly maggots to resist an extremely unfavorable environment, rather than adaptation to a facultative type of parasitism. A true and obligatory type of enteric myiasis does, of course, occur in the Gasterophilidae that invade the digestive tract of herbivorous mammals.

ETIOLOGICAL AGENTS. Muscids involved in human enteric myiasis include the house fly, *Musca domestica;* the little house fly, *Fannia canicularis;* the latrine fly, *Fannia scalaris,* and the false stable fly, *Muscina stabulans.* Several species of Calliphoridae and Sarcophagidae have been recorded in this capacity; at least some of these records appear to be authentic, although in dealing with those based on fecal samples care must be taken to eliminate cases of contamination.

Larvae of the cheese skipper, *Piophila casei,* family Piophilidae, are reported frequently in enteric myiasis. The eggs are deposited on cured meats, old cheese, dried bones, smoked fish, and other materials. The larvae often penetrate the food medium quite deeply, particularly meat, and may thus remain unobserved. They are exceptionally hardy; they have been shown experimentally to be capable of passing through the intestines of dogs alive and of producing serious intestinal lesions while doing so. This species is practically cosmopolitan in distribution. It is probably the most important producer of human enteric myiasis.

Several cases of enteric myiasis caused by the larvae of the black soldier fly, *Hermetia illucens,* family Stratiomyidae, are on record. The larvae breed in overripe or decaying fruits and vegetables, excrement, and a wide variety of other animal and vegetable material. The large, active larvae have been reported as causing rather severe gastrointestinal disturbances. Though originally a New World species, it has become widely distributed through commerce in the warmer temperate and tropical areas of the Old World.

The rat-tailed larvae of the drone fly (Fig. 13-1) *Eristalis tenax,* family Syrphidae, have reportedly been passed with stools on a number of occasions. Although fecal contamination, as in the Sarcophagidae, is a possibility with this species, many of the reports appear authentic. Capelle (personal communication, James), has reared them from a privy in Montana, USA, and has noted abundant pupae in and around the privy. Hall and Muir (1913) describe a case involving a five-year-old boy who, after suffering from digestive disturbances for about 5 weeks, discharged rat-tailed larvae upon receiving treatment, after which his health became normal. Like the two preceding species, *E. tenax* larvae are highly resistant to adverse conditions, and they breed in media from which they might readily be ingested, particularly by small children. This species is almost cosmopolitan, though uncommon or absent in many tropical areas.

Rectal Myiasis. Certain fly larvae may invade the intestine after having gained en-

A   B

Fig. 13-1 The drone fly, *Eristalis tenax*. A. Adult. × 3.5. B. The "rat-tailed" larva. × 2.

trance through the anus. Some cases of so-called intestinal myiasis caused by *Eristalis tenax, Fannia scalaris, F. canicularis, Muscina stabulans,* and certain species of *Sarcophaga* may be explained in this way. These larvae are excrement feeders and may complete their immature stages in the rectum or terminal part of the intestine. Parasitism of this type may occur in man, where humans live under filthy conditions, or in domestic animals that are partially paralyzed or otherwise helpless.

FACULTATIVE MYIASIS

Facultative myiasis may occur when a species that is normally a saprophage or carrion-feeder can, and will on occasion, adapt itself successfully to a parasitic existence. Some such fly larvae may assume considerable importance and may, within certain physiological strains, become virtually obligatory parasites. The distinction between the two types can be tenuous. For example, *Phaenicia cuprina,* a benign or harmless feeder in many parts of the world but a vicious wool maggot in Australia, is considered a facultative parasite, whereas *Wohlfahrtia vigil,* which requires living flesh in its earlier stages but can complete the third stage in the body of an animal that has died of the parasitism, is considered obligatory.

Urinary Myiasis. This is, as the name indicates, myiasis of the bladder and urinary passages. Its manifestations depend, in kind and severity, upon the number and kind of larvae involved and on their localization. There may be obstruction and pain; pus, mucus, and blood in the urine; and a frequent desire to urinate. Larvae are ultimately expelled with the urine. The little house fly, *Fannia canicularis,* is the species most frequently involved, although *F. scalaris, Musca domestica, Muscina stabulans,* the ephydrid *Teichomyza fusca,* and other species have been encountered. Albumen, sugar, mucus, and leukocytes in the urine provide food for the larvae; the shortage of oxygen presents their chief difficulty, although oxygen is needed only in limited quantities.

Infestation is probably usually accomplished at night in warm weather when persons may sleep without covering. Oviposition may be stimulated by discharges from diseased organs; the eggs are laid around the orifice, and the larvae, upon hatching, enter it. Use of unsanitary toilets may be another source of infestation.

Facultative Traumatic and Cutaneous Myiasis. Many species of larvae that normally breed in meat or carrion may become involved in traumatic and cutaneous myiasis. Blow flies (Calliphoridae) are most commonly involved; these include several species of *Calliphora,* such as *C. vicina, Phaenicia sericata, P. cuprina, Lucilia illustris, L. caesar, Phormia regina, Cochliomyia macellaria,* and several species of *Chrysomya.* Flies of other families that may occur in this type of myiasis include such sarcophagids as *Sarcophaga haemorrhoidalis,* the house fly *Musca domestica,* and the phorid *Megaselia scalaris.*

Such species as *Phormia regina* and *Phaenicia sericata* may be attracted to neglected, suppurating, malodorous wounds, especially if the patient is to a degree helpless. Considerable pain may accompany the invasion of the maggots and, in heavy infestations, the patient may become delirious. *Phaenicia sericata* is often benign as a human wound parasite, feeding primarily on necrotic tissues; however, healthy tissues may be invaded. Some strains seem to be more malignant than others. Wounds in domestic animals may be invaded by both these species.

The secondary screwworm of the Americas, *Cochliomyia macellaria,* may be a secondary wound invader of some consequence, particularly in domestic animals. However, much of the damage attributed to this fly in the earlier literature was really caused by its close relative, the primary screwworm *C. hominivorax.* A similar situation occurs in the Old World, where *Chrysomya bezziana* was responsible for much of the traumatic myiasis previously attributed to *C. megacephala,* although the latter species as-

sumes considerable importance at some times and places.

Other *Chrysomya* species may be either primary or secondary parasites, but their role is usually a minor one. However, there is evidence that *C. rufifacies,* whose larvae are usually predatory, may become an important pest. In Hawaii, where this species occurs but where *C. bezziana* does not, *C. rufifacies* produces a strange type of myiasis in newborn calves on the island of Maui; the larvae eat the epidermis, causing death by dehydration and possibly by the production of a toxic element. As many as 30 percent of the calves at one ranch have been attacked (Shishido and Hardy, 1969).

Wool Maggots. Though of importance in certain other parts of the world, the wool maggot or fleeceworm problem is most serious in Australia. Froggatt (1922) has suggested that prior to the introduction of cattle and sheep into that country, blow flies existed as simple scavengers, ovipositing on animal carcases that were festering in the sun. The transition from wool of dead sheep to damp or soiled wool of living animals was not a great one. The introduction of rabbits added to the problem; their carcasses, as well as those of slaughtered or poisoned dingoes, hawks, and carrion crows added quantities of decaying flesh in which the scavenger flies could breed.

The next, and perhaps most important, factor in the development of the sheep maggot pest was the work of the sheepbreeders themselves. The bare-bellied, bare-legged type of Merino sheep was replaced in time by the modern type, which produces three times as much wool as its predecessor. A sheep clothed with a mass of thick, close, fine wool, extending over parts of the body that in previously used breeds were bare and increased further in quantity by the wrinkled skin of some breeds, is sure to get more or less damp around the crotch and attract flies. This artificial increase in weight, quantity, and fineness of the wool is accompanied by an increased secretion of yolk, which, rising from the skin and spreading all through the wool fiber, forms an additional attraction for the flies and supplies food for the maggots.

Zumpt (1965) gives a concise account of sheep myiasis. Although exact reasons for primary strike are not thoroughly understood, the main predisposing cause is bacterial activity on the wool resulting from excess water and profuse sweating. Hard driving of the animals, for example, may tend to set up the proper conditions for strike. Larvae may remain on the surface or they may bore inward in susceptible areas or enter previously existing wounds, even small ones. Parasitized sheep may refuse to eat and death may occur, probably the result of a toxemia or even a septicemia.

Losses from sheep maggots in Australia were estimated at $28 million during the season of 1969–1970. The most important species there, responsible for about 90 percent of all cases of sheep myiasis, is *Phaenicia cuprina*. This is also the most important species in South Africa, where sheep strike is likewise an important problem. *Phaenicia sericata* replaces *P. cuprina* in importance in some areas, such as in Scotland. In the United States *P. sericata, Phormia regina* and *Cochliomyia macellaria* are responsible for most of the strike.

Calliphora stygia is an important species in New Zealand, less so in Tasmania and the Australian mainland, where another native species *C. augur,* assumes importance. Other *Calliphora* species and a muscid, *Ophyra rostrata,* sometimes assume secondary importance in South Africa.

Surgical Maggots. Although now mainly of historical interest, the use of sterile maggots in wound therapy was practiced from 1931 until the advent of sulfa drugs and antibiotics a decade later. Healing results from allantoin excreted by the facultative larvae scavenging necrotic tissue. Actually, as early as the sixteenth century Paré observed a remarkable recovery from a maggot-in-fested wound inflicted on the battlefield, but he did not realize that the maggots were responsible for the unexplained healing. The role of maggots in healing was discovered as early as 1799 and was first utilized during the US Civil War. During World War I, W. S. Baer noticed that soldiers brought in from the field with maggot-infested wounds did not develop infections as did the men whose wounds had been treated. Interestingly, maggot therapy has been practiced by at least one aboriginal Australian tribe and by certain isolated hill people of Burma and Yunnan Province, China. The subject of maggot therapy is summarized concisely by Greenberg (1973).

OBLIGATORY MYIASIS

Domestic and wild land vertebrates, particularly mammals but also on occasion birds, reptiles, and amphibians, are subject to attack in varied ways by fly larvae and under circumstances in which the parasite is obligatory, that is, when the fly larva is unable to develop other than at the expense of the living host. Man often falls victim to these same fly larvae, either in the same fashion as the lower vertebrates or through some modification of it; the relationship is therefore a zoonosis. Flies of several families may be involved, mostly Calliphoridae, Sarcophagidae, Oestridae, Gasterophilidae, and Cuterebridae.

Primary Screwworms. The several changes in the scientific name of the primary screwworm since the discovery of its distinctness from the secondary screwworm in 1930 has been confusing to the nontaxonomist. The name *Cochliomyia hominivorax* is now accepted. Unlike the secondary screwworm, *C. macellaria,* which is usually a scavenger fly, *C. hominovorax* is an obligatory parasite and has been responsible for the majority of cases of screwworm infestation in man and animals in the United States

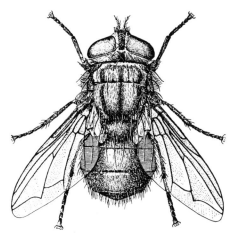

Fig. 13-2 *Cochliomyia hominivorax,* adult female. (Drawing by A. Cushman; USDA.)

and the entire Neotropical Region. It is also known as the cause of nasopharyngeal myiasis in man.

The fly (Fig. 13-2) is strongly attracted to the wounds and sores of animals; even a tick bite or a wound produced by needle grass may be sufficient to produce an infestation. Laake (1936) pointed out that the most common causes of screwworm attack are due to farm practices that can be corrected. He stressed care in shearing, dehorning of cattle, and the removal of old barbed wire from fences; he also advocated that dehorning, castrating, and branding be done so as to expose as little tissue as possible to flies during the season of abundance.

As a human parasite, the primary screwworm has assumed an importance that cannot be disregarded. In 1935, when a serious outbreak in Texas resulted in more than 1.2 million cases in livestock, there were 55 reported human cases and probably twice as many unreported ones. An epidemic of 81 human cases in five provinces of Chile was reported in 1945–1946. Human cases are apt to occur in any area where screwworm infestations are prevalent in livestock.

An early description of a human case was given by Richardson in the Peoria, Illinois, *Medical Monthly* for February, 1883. A traveler in Kansas, in August, while asleep, received a deposit of eggs in the nose. A nasal discharge probably was the attractant to the fly. The first symptoms were those of a severe cold. As the larvae cut away through the tissues of the head, the patient became slightly delirious and complained about the intense misery and annoyance in his nose and head. When the larvae finally cut through the soft palate, his speech was impaired. Despite attempts to remove the larvae (more than 250 in all), the patient, after an apparent trend toward recovery, had a relapse as the eustachian tubes were invaded, and died. The tissue damage was extensive and the head and face showed the characteristic swelling of severe screwworm myiasis.

LIFE HISTORY. Individual females may lay more than 2,800 eggs deposited in characteristic batches of 10 to 400 eggs each; the laying of as many as 300 eggs may be completed in 4 to 6 minutes. The incubation period on wounds of animals ranges from 11 to 21.5 hours under natural conditions. The larval feeding period ranges from 3.5 to 4.5 days or more; the pupal stage, which is spent in the ground, lasts about 7 days. The life history, from egg to egg, requires about 24 days under optimum natural conditions.

The adult fly has a deep greenish blue metallic color with a yellow, orange, or reddish face, and three dark stripes on the dorsal surface of the thorax. Unless one is experienced, it is difficult to separate this species from *C. macellaria* (cf. Fig. 13-3 and James, 1948; for differentiation of larvae, see Fig. 13-4).

The primary screwworm cannot overwinter in cold temperate climates, but migration of adult flies or transportation by moving vehicles and transportation of infested animals may carry flies a considerable distance beyond their winter range. In this way, in the past, temporary breeding sites were established, for example, in parts of Montana and Minnesota, far to the north of the permanent breeding areas. Migration is currently

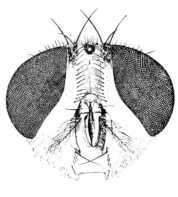
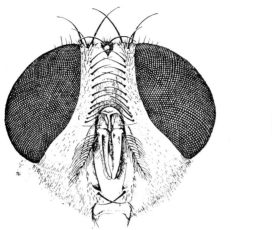

Fig. 13-3 Head, front view, of females of *Cochliomyia hominivorax* (*left*) and *C. macellaria* (*right*). Note small proclinate fronto-orbital bristles of *C. macellaria*, absent in *C. hominivorax*. The hairs of the lower parafrontalia are black in *C. hominivorax*, pale in *C. macellaria*. (Drawings by A. Cushman; USDA.)

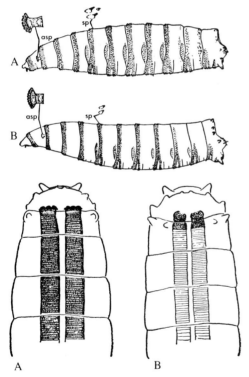

Fig. 13-4 Mature larva of (*A*) *Cochliomyia hominivorax*, (*B*) *C. macellaria*. (*Above*) Larva, side view. (*Below*) Dorsal view of posterior segments, showing pigmented tracheal trunks of *C. hominivorax* and unpigmented tracheal trunks of *C. macellaria*. (After Laake, Cushing, and Parish.)

posing a problem of reinfestation of certain areas from which the pest had previously been exterminated.

The primary screwworm is a true parasite and lives only in the living flesh of warm-blooded animals; it is not found in snakes, lizards, or other cold-blooded vertebrates, nor in carcasses, dead fish, decaying meats, or vegetable matter. Parasitism is apparently due to oviposition habits of the female, as mass rearing experience proves that larvae readily grow in meat and animal products. The maggots are not found in dead animals and similarly appearing flies collected by most methods, other than directly from screwworm infestations, are, in the vast majority of cases, secondary, not primary screwworms. This fact was not realized until 1930, when Cushing and Patton first recognized the distinctness between the two and named the primary screwworm fly *Cochliomyia americana*, new species. The identity of the fly with the previously described *C. hominivorax* was later established.

Today, the primary screwworm exists as an important economic problem and a threat to man throughout the warmer temperate, subtropical parts of the New World, from Mexico to northern Chile and Argentina. It appeared at one time to have been virtually eradicated from the United States through the release of mass-reared sterile flies, but extensive reinfestations have occurred in the southwestern states (Bushland, 1975; Steelman, 1976). These reinfestations are attributed mainly to the relaxation of preventive animal husbandry and two years of ideal overwintering conditions. The eradication program continues with a goal of setting up a permanent barrier across a narrow portion of Mexico or Central America.

Old World Screwworms. In Africa, India, and the nearby island areas of the Pacific and Indian Oceans, including such areas as Indonesia, New Guinea, and the Philippine Islands, another calliphorid fly, *Chrysomya bezziana* (Fig. 13-5) occupies a position similar to that of *Cochliomyia hominivorax* in America. It is also an obligatory parasite, differing in this respect from other members of the genus, including the widespread *C. megacephala,* which it closely resembles.

About 150 to 500 or more eggs are deposited in a batch around the edges of wounds.

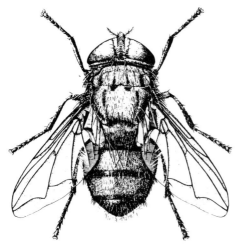

Fig. 13-5 The Old World screwworm fly, *Chrysomya bezziana.* (Drawing by A. Cushman; USDA.)

Larvae develop to the third instar about 2 days after hatching; at this time they are imbedded in the wound to such a depth that only their posterior ends are visible. The entire larval stage lasts 5 to 6 days and the pupal stage 7 to 9 days, under tropical conditions, somewhat longer in cooler areas.

Norris and Murray (1964) have presented an excellent study of the biology of this fly in New Guinea, and Zumpt (1965) has given a very good summary of human and animal myiasis caused by it. Human parasitism is much more common in India and other parts of the Oriental Region than in Africa, although animal cases are apparently as common there as in the rest of the range of the fly. All parts of the body where preexisting wounds or ulcers occur, but particularly areas where the skin is soft or mucous tissue is present, may be affected. The genital openings, eyes, and orifices of the head may be attacked, as by its American counterpart.

The most commonly attacked domestic animals, as in the case of the American screwworm, are cattle. Zumpt cites Cuthbertson as stating that in Rhodesia this species is, next to tsetse, the most important pest of cattle and other domestic animals. It is also a major pest of cattle and other domestic animals in New Guinea. Parasitism may lead to the death of the animal within a short time if the wounds are not treated. Other domesticated hosts of the fly include sheep, goats, buffaloes, horses, pigs, and dogs.

Wohlfahrtia Traumatic Myiasis. A third screwworm fly, belonging to the family Sarcophagidae, is *Wohlfahrtia magnifica.* This fly is widespread over the warmer parts of the Palaearctic Region. The female does not lay eggs; as in most other Sarcophagidae, she gives birth to active first-stage maggots. Skin lesions in prospective hosts, even one as small as a tick bite, may be the site of larval entry, but the mucous membranes of the head openings, the eyes, the ears, and the genital openings may also be used. For larval

Fig. 13-6 The horse bot fly, *Gasterophilus intestinalis,* adult. (Courtesy of Oregon State University and Ken Gray.)

Fig. 13-7 *Gasterophilus* eggs. (Courtesy of Oregon State University and Ken Gray.)

development, 5 to 7 days are required, during which time tissue destruction may be considerable. Fatal cases in man are on record. Domestic animals that are attacked include dogs, horses and donkeys, cattle, goats, pigs, sheep, water buffaloes, camels, and poultry, especially geese.

Gasterophilus Enteric Myiasis. True enteric myiasis occurs in horses, donkeys, mules, zebras, elephants, and rhinoceroses. It occurs in man only as an extreme rarity (James, 1948). Zumpt (1965) lists six species, all in the genus *Gasterophilus,* as attacking domestic Equidae. These are *G. intestinalis, G. haemorrhoidalis, G. nasalis, G. inermis, G. pecorum,* and *G. nigricornis.* All are widely distributed in the Old World, and the first three have been introduced into the Americas; *G. inermis* was also introduced into North America but apparently never established there.

The flies of this genus are about the size of, or slightly smaller than, a honey bee; superficially they are honey-bee-like in appearance (Fig. 13-6). The ovipositor is strong and protuberant. The adults are strong fliers. Larvae are normally restricted to the digestive tracts of Equidae except in their early, migrating first instar. They are not highly host-specific. A taxonomic review of the group has been presented by Zumpt and Paterson (1953), and keys to adults, eggs, and larvae are included in Zumpt (1965).

Gasterophilus intestinalis (Fig. 13-6) is the horse bot fly. Each female deposits about 1,000 light yellow eggs (Fig. 13-7), which are firmly attached to the hairs of the forelegs, belly, flanks, shoulders, and other parts of the body of the horse, but chiefly on the inner surface of the knees, where they are accessible to the tongue, teeth, and lips. The sudden increase in temperature arising from the warmth of the tongue provides the necessary stimulus for the hatching of the eggs. The incubation period is five days, but hatching may be greatly delayed by cool weather so that viable eggs may be found on the horse until late autumn, long after the adult flies have disappeared. The larvae (Fig. 13-8) on hatching are provided with an armature that enables them to excavate galleries in the subepithelial layer of the mucous membrane of the tongue. From the mouth they pass rapidly to the left sac or esophageal portion of the stomach. There, as second and third instars, they remain fixed with little or no change in position (Figs. 13-9, 13-10) until the following spring and early summer,

Fig. 13-8 Newly emerged larva of the horse bot fly, *Gasterophilus intestinalis.*

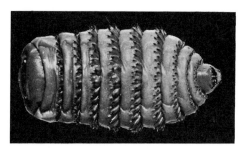

Fig. 13-9 Mature larva of the horse bot fly, *Gasterophilus intestinalis*.

when they detach themselves and pass out with the droppings. Pupation takes place shortly thereafter in loose earth or dry droppings. The usual duration of the pupal stage is three to five weeks, depending on moisture and temperature. The adult bot fly emerges, and egg laying begins in early summer.

Gasterophilus haemorrhoidalis is called the nose bot fly because the female forcibly "strikes" the host in the region of the nose, where she attaches her eggs to the fine hairs of the lips. The fully grown larvae move during the early spring from the stomach to the lower digestive tract near the anus, where

Fig. 13-10 *Gasterophilus* larvae on lining of horse stomach. Lesions in the lining indicate sites where larvae have become detached. (Photograph by R. D. Akre, Washington State University.)

they attach temporarily and finally drop to the ground with the feces.

Gasterophilus nasalis, the throat bot fly, is very annoying to horses because its eggs are attached to hairs under the jaws, and when the fly darts at the throat to deposit eggs it causes the host to throw its head upward as though struck under the chin. Oviposition takes place during late spring and early summer. The larvae hatch in from four to five days without need for heat, moisture, or friction. The newly hatched larvae travel along the jaw and enter the mouth between the lips. They finally reach the pyloric portion of the stomach or the anterior part of the duodenum, where they may be found in dense clusters.

Gasterophilus pecorum is the most common and, apparently the most pathogenic species of this genus, in the Old World. It does not occur in America. The eggs are deposited on grasses and other plants, and they hatch only when they are taken into the mouth of the horse with the food. The larvae mature in the stomach of the host. *Gasterophilus nigricornis* is a strikingly stout species. Eggs are deposited on the cheek or nose of the host, and the act of oviposition is very disturbing to the animal. The eggs of *Gasterophilus inermis* are deposited on hairs on the cheeks of the host; the larvae penetrate the epidermis and work their way under it until the mouth is reached.

Other genera of Gasterophilidae are involved in a similar type of myiasis in two groups of wild mammals: *Gyrostigma* in rhinoceroses and *Cobboldia, Platycobboldia,* and *Rodhainomyia* in elephants.

PATHOGENESIS. Although a moderate infestation of bots will give no outward indications of the parasitism, a heavy infestation will cause digestive disorders (which may, of course, result from other causes).

The discovery of bots in the manure is sufficient evidence. A light infestation is probably of no consequence. On the other hand, infestations of horses by *G. pecorum* may result in such serious injury as to kill the animal or require that it be destroyed. (Zumpt, 1965).

The injuries that bots produce are (1) extraction of nutriment, both from the stomach and its contents; (2) obstruction to the food passing from the stomach to the intestine, particularly when the larvae are in or near the pylorus; (3) irritation, injury, and secondary infections to the mucous membrane of the stomach (Fig. 13-10) caused by the penetration of the oral hooklets; and (4) irritation to the intestine, rectum, and anus caused by the passage of the bot through these areas.

GASTEROPHILUS AS A HUMAN PARASITE. Larvae of the bot flies of horses may burrow freely and cause a form of cutaneous creeping myiasis in humans. Since man is not the common host, the larvae cannot live beyond the first stage in him. The newly hatched larva enters the unbroken skin and begins to burrow, much as it does normally in the tongue, mucous membrane, or the skin of its normal host. The burrow is very tortuous and plainly visible. It may cause severe itching but never serious consequences. The larva, which measures 1 to 2 mm in length, is easily detected a short distance beyond the apparent end of the burrow. Surgical removal is usually a simple matter.

Gasterophilus pecorum, G. haemorrhoidalis, G. inermis, G. nigricornis, and possibly other species may be involved in creeping myiasis. A prerequisite for human parasitism, however, is the ability of the larva to pierce the skin; some species apparently cannot do that. Human infestation usually occurs when there is close association between man and horse; the larva may come in contact with human skin, for example, when one is in the process of grooming the horse.

Cattle Grubs. These are the larvae of *Hypoderma*, the heel flies or ox warble flies,

Fig. 13-11 Adult northern cattle grub, *Hypoderma bovis*. (Courtesy of Pacific Supply Coop and Ken Gray.)

now commonly considered to be Oestridae, although a separate family, the Hypodermatidae, has been recognized for them. Although the normal hosts are cattle and Old World deer, they have been known to parasitize horses and humans. Two well-known species infest cattle: *Hypoderma lineatum,* the cattle grub, widely distributed in the United States, Europe, and Asia; and *H. bovis,* the northern cattle grub, which broadly overlaps the preceding species in distribution but extends farther northward. At least four Old World species attack native deer; one of them, *H. diana,* has been implicated in human myiasis.

Hypoderma bovis (Fig. 13-11) is the larger of the two species that infest cattle, adults measuring about 15 mm in length as compared with 13 mm for *H. lineatum.* Both are bumble-bee-like in appearance, covered with dense hairs, the pattern of light and dark coloration differing in the two species. The fully grown larva of *H. bovis* measures about 27 to 28 mm in length, whereas that of *H. lineatum* (Fig. 13-12) is about 25 mm.

LIFE HISTORY AND HABITS. Females of both species lay their eggs on the hairs of cattle. As many as 800 eggs may be laid by one female. Although no pain is inflicted at the time of oviposition, cattle become terror-

Fig. 13-12 Larva of the common cattle grub, *Hypoderma lineatum.* × 1.3.

stricken when the flies are discovered and gallop madly for water or shade to escape. This ''gadding'' behavior often spreads to the whole herd.

Eggs hatch within a week and tiny armored larvae crawl down the hairs of the host and burrow either directly into the skin or into the hair follicles. In doing so, they cause considerable irritation. The first instar then migrates through the body of the host, finally lodging beneath the skin of the back. Details of this migration vary with the species and some aspects of the migration have not been thoroughly determined. After about four months, *H. bovis* reaches the spinal cord, where it burrows between the periosteum and dura mater for a period of time before continuing its way through muscle and connective tissue to the back. *H. lineatum* rests for a while in the walls of the esophagus, probably without having entered the spinal canal. Soon after reaching the skin, the larva cuts a small opening to the surface, then reverses its position. Here the larva molts, following which its skin is closely set with spines. The body of the host now forms a pocket around it. The growth of the grub from this time on is rapid, and a final molt occurs about 25 days after the preceding one. In the last stage of larval development the color gradually darkens, first becoming yellow, finally almost black. During the entire development in the warble a breathing hole is kept open in the skin of the host and the grub rests with its two posterior spiracles appressed closely to the opening. As growth proceeds the hole is gradually enlarged. At the end of the developmental period in the warble, which requires 5 to 8 weeks for *H. lineatum* and up to 11 weeks for *H. bovis,* the grub works its way out and falls to the ground. There it crawls away, enters loose earth or debris to pupate, and in from 4 to 5 weeks emerges as a fly. The complete life cycle requires about a year.

Seasonally, the occurrence of the two species does not coincide; *H. lineatum* adults occur in the spring about a month earlier than *H. bovis*.

Literature on the cattle grub is voluminous; a good comprehensive work is that of Gebauer (1958).

INJURY AND ECONOMIC LOSS. Injury to the host is severalfold. Irritation caused by the migration of the larva through the body of the host and, later, from its emergence through the skin, is of great importance. Growth may be retarded and the general body condition may deteriorate. Protein sensitization and septicemia may be of some, at times considerable, importance. The escape of the larva from the warble leaves an open running wound that sometimes persists for a long time; it is subject to infection and is attractive to screwworm and other tormenting insects.

According to Steelman (1976), economic losses in the United States were estimated by the USDA in 1965 to amount to $192 million a year. These include (1) reduction in milk production, estimated at $55 million, (2) weight loss, estimated at $115 million, and (3) damage to hides, estimated at $22 million. The animal's response to the fly threat is one factor that contributes strongly to the weight loss. This behavior has been described graphically by Holstein (cited by Herms) as follows: ''A cow quietly grazing will suddenly spring forward, throw up her tail, and make for the nearest water at a headlong gait. Seemingly deprived at the moment of every instinct except the desire to escape, she will rush over a high bluff on the way, often being killed by the fall. This, with miring in water holes and the fact that cattle are prevented from feeding, causes the loss." Gadding may cause a pregnant cow to abort, thus resulting both in the loss of the calf and of a lactation period. Depreciation of the value of the carcass through the burrowing of the larva reduces the value of the meat, and emergence of the larvae from the warbles produces damaging holes in the hide (Fig. 13-13).

Attacks on horses, which are not infrequent, may result in similar warbles on the backs of the animals; these may make it impossible to saddle or harness the animals.

HYPODERMA AS A HUMAN PARASITE. Numerous records of attacks on man by *H. lineatum, H. bovis,* and *H. diana* have been published. Most case histories reveal some association with cattle during the summer or fall preceding recognition of the attack. Children are proportionately more often affected than adults. Unlike *Gasterophilus, Hypoderma* can complete its larval development in man, often with serious consequences.

As in cattle, ingress is probably through the skin. The wandering of the larva may be extensive, often involving the arms and legs; it may be indicated as a swelling, without discoloration, and accompanied by tenderness and soreness. Severe discomfort, itching, pains and cramps, sometimes associated stomach disorders, may develop. When the larva is nearing the end of its developmental stage, it moves upward, as in cattle, but because of man's upright position it usually forms its warble in the upper part of the chest, the neck, or the head. Because of the abnormal host the larva may make several attempts to reach the surface, thus resulting in a dermal creeping myiasis. The pain and discomfort accompanying parasitism may be severe. An apparently increased nocturnal activity of the larva may interfere with sleep. Local paralysis may result from the invasion of the spinal cord; in one case on record a boy parasitized by seven larvae suffered almost complete paralysis of the lower extremities for about a year. There are

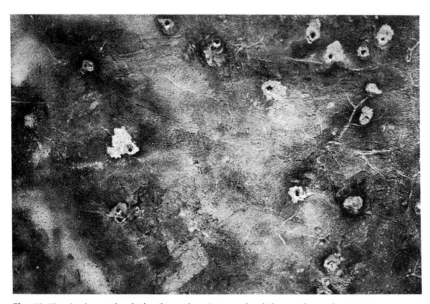

Fig. 13-13 A piece of sole leather, showing work of the cattle grub. × 0.3.

cases on record of an eye being invaded and destroyed. In another recorded case a small boy died after a larva had produced a fetid ulcer under the back teeth of the lower jaw.

The larva may be expressed from the open warble, or surgical removal may be necessary. In the latter case, the exact position of the larva must be determined. In ophthalmomyiasis caused by *Hypoderma,* surgical removal of the parasite may be quite complicated.

A concise account of human hypodermatosis has been given by Leclercq and Letawe-Genin (1976).

Caribou Flies. The caribou or reindeer warble fly, *Oedemagena tarandi,* is most widely distributed over the range of its host in northern Eurasia and northern North America. The fly is yellowish orange and has a beelike appearance. Its life history is similar to that of the warble flies of cattle. This parasite may cause heavy losses in young animals. The warbles tend to form mainly on the rump. Lung involvement may occur, and death may result from strangulation. A certain degree of immunity develops, older animals being more resistant than younger ones, and healthy and well-nourished animals have an immunological advantage. Incidence of infestation may be high, even up to 100 percent of the herd. The fly is of considerable importance in areas where its host is valuable to human economy. In addition to loss and damage to the host, it may be involved in the spread of brucellosis (Vashkevich, 1972). It is unknown as a human parasite.

Warble Flies of Goats and Sheep. In northern Africa, southern Europe, and the warmer parts of temperate Asia, goats and sheep are extensively parasitized by *Przhevalskiana silenus,* also known as *Crivellia silenus,* and perhaps two other species of that genus. The life histories are similar to those of the cattle warble flies. No human cases are known.

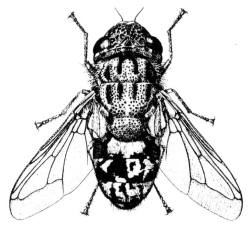

Fig. 13-14 The sheep bot fly, *Oestrus ovis* (Drawing by A. Cushman; USDA.)

Oestrid Head Maggots. SHEEP BOT FLIES. The sheep bot fly, *Oestrus ovis,* is a widely distributed species, a little smaller than the honey bee, which it somewhat resembles (Fig. 13-14). The larvae are often called sheep head maggots or grubs-in-the-head. The female normally deposits active young during early summer to autumn in the nostrils of sheep and goats and related wild hosts. These larvae begin at once to move up the nasal passages, working their way into the nasal and frontal sinuses, often as far as the bases of the horn in rams, and attach themselves to the mucous membranes. Here numbers of these white grubs may be found wedged closely together in various stages of development. The grubs reach full growth by the following spring, a larval period of from 8 to 10 months. At the end of this time they work their way out through the nostrils or are sneezed out by the host, fall to the ground, and pupate in a few hours. The pupal period lasts from 3 to 6 weeks, sometimes much more in areas where low temperatures prevail. Adults may live as long as 28 days. Complete development of the parasitic stage in spring lambs may be shortened to 25 to 35 days. Rarely, two generations may be completed within a year (Breev and Sultanov, 1975).

In the presence of the fly, sheep and goats are greatly excited, shaking their heads, pushing their noses into the dust, snorting, and otherwise indicating attempts to escape something that persists in entering their nostrils. In parasitized animals there is a purulent discharge from the nostrils, vigorous shaking of the head and perhaps the occasional discharge of a maggot, loss of appetite, and grating of the teeth. When the animals walk, their fore feet are lifted in a pawing motion. The majority of cases do not terminate fatally, but death may come in a week or less after the appearance of aggravated symptoms.

HEAD MAGGOTS OF HORSES. *Rhinocephalus purpureus* is an important head maggot of horses in parts of Europe, Asia, and Africa. Its habits are similar to those of *Oestrus ovis,* differing mainly in the details of its life history. The incidence of fatality in parasitized horses in parts of Russia may be as high as 82 percent (cf. Zumpt, 1965).

NASAL BOT FLIES OF ANTELOPES. Two species, *Gedoelstia cristata* and *G. haessleri,* parasitize wildebeest, hartebeest, and other alcelaphine antelopes in Africa. The larva is dropped around the eye orbits of the host, from which it makes its way through a vein, as a first instar, ultimately lodging, in its later instars, in the frontal sinuses. In its native hosts the effects of the parasitism are benign. On occasions, however, generally associated with temporary migrations of wildebeest, it attacks sheep with which the migrating herds come into contact, often in considerable numbers, causing the so-called bulging eye disease or uitpeuloog. The incidence of parasitism may be high, and mortality may reach 75 percent. Encephalitic and cardiac, as well as ophthalmic, involvement may occur. The larva cannot develop beyond the first stage in sheep. Pathogenic symp-

toms, including temporary blindness, have been reported in horses (Wetzel, 1969). Other domestic animals and the native gemsbok are also known as unsuccessful hosts of the larva.

PARASITISM OF MAN BY HEAD MAGGOTS. Ophthalmomyiasis of man is commonly traceable to the larva of *Oestrus ovis* and *Rhinoestrus purpureus.* Because man is not a normal host the larva is unable to progress beyond its first stage, and the infestation is consequently of short duration. The typical patient, who usually has had a close association with sheep or goats, will report being struck in the eye by an insect or small foreign object, with pain and inflammation developing a few hours later. The condition is similar to acute catarrhal conjunctivitis and may be diagnosed as such. It is apparently always benign, though irritating; reports involving destruction of extensive ocular tissue or the entire eye are almost certainly the result of misidentification. This type of myiasis is most common among nomadic shepherds whose food consists to a large extent of goats' milk and cheese. Interestingly, man is most subject to attack in areas where the populations of the normal hosts of the fly are relatively small.

Larvae of *Gedoelstia* may also attack man and produce a similar type of myiasis. It seems that *Gedoelstia* myiasis is usually benign, although a case of intraocular involvement has been reported (Bisley, 1972). In view of the serious development in sheep, like man an abnormal host, the possibility of a malign syndrome in man should be kept in mind.

HEAD MAGGOTS OF DEER. Deer, elk, caribou, and other related animals are commonly infested with head maggots. Among these are the European *Cephenemyia stimulator* in the roe deer, *C. auribarbis* in the red deer, *C. ulrichii* in the European elk, and *C. trompe* in the reindeer. *C. trompe* also occurs in reindeer and caribou in the New

World. Other American species are *C. pho-bifer* from the white-tailed deer, *C. pratti* from the mule deer, *C. jellisoni* from the Pacific black-tailed deer and white-tailed deer, and *C. apicata* primarily in California deer.

Parasitism is known to occur in as high as 75 percent of a herd (Capelle, 1971). Deer are seriously disturbed by the adult fly in the process of larviposition (Anderson, 1975), and the invasion by the larvae, as well as their expulsion when mature, apparently causes the host severe discomfort and pain.

Abnormal invasion of the lungs in reindeer and other host species may have serious consequences, even leading to death. No species of *Cephenemyia* has been known to attack man.

Rodent and Rabbit Bots. Larvae of the species of *Cuterebra,* Family Cuterebridae, are commonly parasitic in rodents and wild and domestic rabbits and hares in the New World. Domestic rabbits (*Oryctolagus*) must be considered an aberrant host, since *Cuterebra* evolution obviously took place in the

Fig. 13-15 The rodent bot fly, *Cuterebra tenebrosa*. *A.* Adult. *B.* Three third-stage larvae *in situ* in warbles of the bushy-tailed wood rat, *Neotoma cinerea*. (Photographs by C. R. Baird, University of Idaho.)

Fig. 13-16 Third-stage larva of *Cuterebra jellisoni,* a parasite of cottontail rabbits and jack rabbits. (Photograph by R. D. Akre, Washington State University.)

American host species. Nevertheless, *Cuterebra* parasitism is of interest in relation to meat animals, pets, and laboratory animals.

Eggs are laid in or near the haunts of the host. The larva (Fig. 13-16) enters natural body openings and the skin after hatching and migrates to its final resting spot, where it forms a warblelike dermal tumor (Fig. 13-15). The host may be heavily infested with these heavily spined, grublike larvae. The adult flies are bumble-bee-like (Fig. 13-15), though as a rule they are much less hairy; usually they may be predominantly blue or black.

Occasionally animals other than rodents or lagomorphs are parasitized by *Cuterebra.* These include dogs, cats, monkeys, and man. Fatal cases have been recorded in a cat and a dog where the larva penetrated the brain (Hatziolos, 1967). Human cases are very rare, only eight being on record up to 1972. Rice and Douglas (1972) reviewed seven of these. Apparently, entry was made through the mucous membranes of the eye, nose, mouth, or anal area, although it has been demonstrated that at least one species can penetrate the unbroken skin. In man, the pathology varies, depending on the tissues attacked; in one known case, serious damage was done to an eye.

Tórsalo. The tórsalo, *Dermatobia hominis* (Fig. 13-17), Family Cuterebridae is common in parts of Mexico and Central and South America. It has been known as the human bot fly; the name berne is often applied to it. The adult superficially resembles a bluebottle fly. It parasitizes a wide range of hosts: cattle, swine, cats, dogs, horses, mules, sheep, goats, monkeys, man, and certain wild mammals; birds, including toucans and ant birds, are known to harbor it. It is a serious pest of cattle in Central America

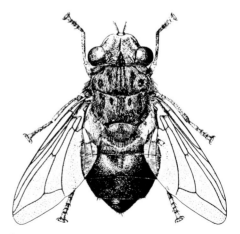

Fig. 13-17 The human bot fly, or tórsalo, *Dermatobia hominis.* (Drawing by A. Cushman; USDA.)

and Brazil, where loss of meat and milk and damage to hides may amount to millions of dollars annually (an estimated $200 million, according to Steelman, 1976). In man the larva has been reported from various parts of the body, mainly the head, arms, back, abdomen, buttocks, thighs, and axilla.

The life history of the fly is extremely interesting. The adult is a forest-inhabiting insect. The female does not deposit eggs directly on the host; rather, she captures another fly, usually a bloodsucker, or, rarely, a tick, and glues the eggs, by use of a quick-drying adhesive, along one side of the carrier's body. The carrier (Fig. 13-18) is an active, day-flying zoophilous species of moderate size. Guimarães and Papavero (1966) list, as known carriers, 48 species of flies and one tick (*Amblyomma cajennense*); the flies include 24 species of mosquitoes as well as black flies, species of *Chrysops* and *Fannia, Musca domestica, Stomoxys calcitrans,* and others.

Eggs are attached to the carrier in such a way that when contact is made with the prospective host the anterior end of the egg is directed downward. This end develops an operculum, through which the larva emerges. The unhatched larva may, however, remain alive and active for as many as 28 days. Upon emergence it penetrates the skin to the subcutaneous tissues (Fig. 13-19).

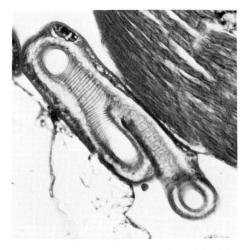

Fig. 13-19 Pathology section of a Colombian woman. Tracheae probably those of *Dermatobia hominis*. × 440. (Courtesy of US Armed Forces Institute of Pathology. Negative No. 72-1874.)

Unlike species of *Hypoderma* and *Cuterebra,* the larva produces a warblelike lesion and swelling at the point at which it enters the skin, without any prior wandering period. Development in the body of the host requires about 6 weeks. The mature larva then drops to the soil and enters it for pupation. The entire life cycle requires 3 to 4 months.

L. H. Dunn has described the life history of the tórsalo on the basis of an infestation that he permitted himself to suffer in the Canal Zone. In his case the fly *Limnophora*

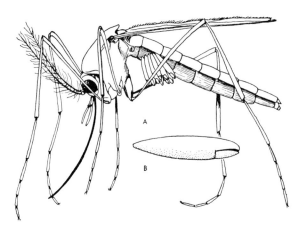

Fig. 13-18 *A. Psorophora* mosquito carrying a load of *Dermatobia* eggs. *B.* Egg enlarged. (Drawing by A. Cushman; USDA.)

sp., not a bloodsucker, was the carrier. Two larvae were observed to enter the skin of his arm. This process required 42 minutes for the first larva and 1 hour and 35 minutes for the second. Dunn experienced "absolutely no sensation caused by the entrance of the first larva after the first 30 minutes. Then, as the posterior end was being drawn inside, a sharp pricking, which lasted for about 2 minutes, was experienced." There was at first a sharp itching by night, and by the end of 2 weeks the lesions had the appearance of small boils; by the end of 3 weeks they were excruciatingly painful. No pain was experienced as the larva emerged from the skin.

Dermatobia hominis myiasis has been diagnosed on several occasions in persons who have acquired the parasite in tropical America and then returned to their homes in North America and Europe before the completion of the development of the parasite.

An important review of the biology, systematics, pathogenesis, economic importance, and control of *D. hominis* is given by Guimarães and Papavero (1966). A very valuable part of this review is its bibliography of 375 titles, a considerable number of which are partially abstracted or more or less extensively quoted.

Cutaneous Wohlfahrtia Myiasis. Traumatic *Wohlfahrtia* myiasis has been discussed earlier in this chapter. A sarcophagid fly of the same genus, *Wohlfahrtia vigil,* causes an entirely different type of myiasis in southern Canada and the northern United States. The injury does not involve invasion of wounds or body openings; rather it is furuncular or boillike, similar to the swollen lesion of the tórsalo or the terminal warble of *Hypoderma* species. In the western part of North America another nominal species, *W. opaca* (Fig. 13-20), is the pathogen; this can best be interpreted, however, as a subspecies of the eastern *W. vigil.* Biologically, the two are very similar, if not identical. A European species, *W. meigenii,* is very similar in appearance to the *opaca* form

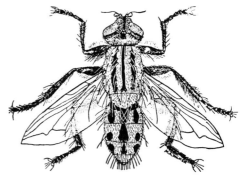

Fig. 13-20 *Wohlfahrtia opaca.* (Drawing by M. A. Palmer.)

of *W. vigil,* but it is a scavenger and is not known to be involved in myiasis of man or other mammals.

The female *W. vigil* deposits active maggots in the neighborhood of a suitable host or directly upon the host itself. The larva penetrates the unbroken skin and usually infests dermal tissue, forming a boil-like swelling with an opening through which the posterior end of the larva may be seen. Here the larva develops to maturity, leaves the swelling, drops to the ground, and pupates. Exceptionally, in small animals, penetration may go deeper than the dermal tissue, even into the coelomic cavity.

W. vigil is a pest in mink and fox ranches in parts of the United States and Canada. The young, often newborn animals, are attacked, usually fatally. Dogs, cats, rodents, rabbits, and humans may be attacked. Human patients are with rare exceptions infants. Cleanliness is no deterrent to the fly; she is not attracted to odors disagreeable to man. Though parasitism of a young child may cause considerable pain and discomfort, the larvae are easily removed.

There is strong evidence that adult rodents and rabbits may be involved in maintaining populations of the fly other than in the whelping season of the carnivore hosts (Eschle and De Foliart, 1965).

Tumbu Flies. The Tumbu fly, an African calliphorid, *Cordylobia anthropophaga,* causes a boillike (furuncular) type of myiasis. This is a heavy-bodied, predominantly yellow fly, in appearance unlike most Calliphoridae of the temperate Northern Hemisphere but of a type commonly encountered in the Old World tropics. Normally, this fly is restricted to Africa south of the Sahara Desert, although there it is widespread. A suspected case of acquisition of the parasite in Spain needs confirmation. Cases have been reported from the United States, Germany, England, the Netherlands, and Italy (Rice and Gleason, 1972), but these were contracted in Africa.

The female deposits her eggs, in batches of 100 to 300, usually in dry sand that has been contaminated by urine or feces. Sometimes soiled diapers, if dry, may stimulate oviposition. Eggs are never deposited on naked skin or attached to hairs. After hatching, a larva may remain alive up to 15 days without feeding. Once it finds a host, it burrows into the subcutaneous tissues, where the characteristic boillike lesion is found. Penetration of the human skin may be accomplished with no more discomfort to the host than a mild itching. As the boillike lesion is enlarged, a serous fluid is exuded, and the surrounding tissues become hardened. Later, febrile reactions and malaise may occur. The patient's sleep may become disturbed. Secondary infection is a possibility.

A wide range of vertebrate hosts has been recorded. Dogs are the domestic animals most commonly affected; wild hosts include native cats, mongoose, monkeys, rats, mice, and other rodents. Rats form the main reservoir.

Another African species of *Cordylobia, C. rodhaini,* may parasitize man in a similar way. Infestation of man by it is relatively rare, but it may be more massive and the consequences more severe than in *C. anthropophaga*. Like that species, it has also been reported as being acquired in Africa and subsequently detected in other parts of the world.

Bloodsucking Maggots. The bloodsucking habit of the immature fly, though not of the adult, can be considered a form of myiasis. If justification for this statement is needed it may be found in the discussion of the sanguinivorous root in the evolution of myiasis near the beginning of this chapter.

CONGO FLOOR MAGGOT. The Congo floor maggot, *Auchmeromyia luteola,* family Calliphoridae, is restricted to Africa south of the Sahara Desert, including the Cape Verde Islands but excluding Madagascar. It is commonly found in and about human habitations. Man has been thought to be the only host; however, Zumpt says this is "certainly not true," though the original animal host is not known.

The eggs are deposited in various situations, such as on sleeping mats on the ground in huts, in dusty crevices, and in dry sand, situations where the larvae, when hatched, may readily find food. The larvae are remarkably resistant to extreme dryness and lack of food. They are nocturnal in their feeding habits, sucking the blood of sleeping persons and producing a wound by means of their powerful mouth hooks and the associated maxillary plates. A maggot will feed for 15 to 20 minutes, then detach and hide in crevices of mats or other protected places during the ensuing day; attacks are repeated almost every night, as long as necessary, if the hosts are available. Persons sleeping on the ground or on low beds are the victims; beds of ordinary height are beyond the reach of the maggots. The duration of the larval stage may be as short as 2 weeks or, with scarcity of food, as long as 3 months. Pupation takes place in protected situations. The pupal period lasts from 9 to 16 days, depending upon temperature.

The bite of the larva is normally felt as a slight prick. As in bed bugs, the effects of their attacks may vary considerably according to the susceptibility of the host. Heavy infestations of huts may cause the natives considerable discomfort. No relationship to any agent of disease is known.

BLOODSUCKING MAGGOTS OF BIRDS. Though the Congo floor maggot is the only dipterous larva known to suck the blood of man, five genera in three families are known to consist of species that suck the blood of nestling birds. Of these the calliphorid genus *Protocalliphora* is widespread in the Northern Hemisphere. Hall (1948) lists more than 50 species of birds, mostly passerines but including hawks and other nonpasserines, that are parasitized by various species of this genus in America, and Zumpt (1965) lists 34 avian hosts in the Old World.

The effects of parasitism on the host are subject to disagreement. According to one report, 5 to 10 percent of the nestlings die from loss of blood, and some of those that become full-fledged are so weakened that they fall easy victims to predators. The presence of the maggots may lead to desertion of the nest by the mother bird. At least one species of the genus, *P. lindneri,* does not suck blood, but rather forms abscesses in the skin of the nestlings. Others enter the nostrils and complete development there.

Species of three genera of Muscidae, namely *Passeromyia* in the Old World and *Philornis* and *Neomusca* in the New, play the same role. These genera are tropical, however, whereas *Protocalliphora* is Holarctic. In Africa *Neottiophilium,* family Neottiophilidae, also has bloodsucking larvae that feed on birds.

CONTROL OF MYIASIS-CAUSING FLIES

Discussion of specific controls for myiasis-causing flies is restricted here to important species having the obligatory habit.

Species with facultative myiasis characteristics may be reduced in numbers by the methods suggested for house flies and their relatives.

Cattle Grubs (Hypoderma spp.). Spectacular success has been realized through the use of animal systemic insecticides for controlling cattle grubs in beef cattle. The same materials cannot be applied to producing dairy animals because milk becomes contaminated with insecticide residues. For dairy animals, wettable powder-water suspensions of approved insecticides must be thoroughly applied to the back. For beef animals the pour-on application procedure, pouring properly diluted approved systemic insecticide along the back of the animals, is convenient and highly effective. A comprehensive and closely supervised program relying mainly on the application of systemics has practically eradicated cattle grubs from Ireland and shows similar promise for Britain and Alberta, Canada (cf. Beesley, 1974a).

Tórsalo (Dermatobia hominis). Control of the torsalo, most serious pest of cattle in the tropics of Mexico and Central and South America, is difficult. Sprays as used for direct application to cattle for horse fly control can be effective if used continuously, killing newly hatched tórsalo larvae and repelling the egg vectors. Systemic insecticides provide control, but cannot be expected to be as effective as against cattle grubs because the tórsalo has a number of animal hosts and development is more continuous. The sterile-male technique and induced cellular immunity are possible future methods of control (Jobsen, 1974; Marín-Rojas, 1975).

Caribou Flies (Oedemagena). Extensive tests of systemics have been carried out in the Soviet Union and Alaska to control bot larvae in reindeer. Pour-on treatments that are effective with cattle do not work properly because of the tendency of reindeer to shed water and water-base liquids. A single intramuscular injection of certain systemic insecticides will provide good to excellent control (Nepoklonov *et al.*, 1973; Ivey *et al.*, 1976).

Horse Stomach Bots (Gasterophilus spp.). Stomach bots of horses were originally treated with gastric application of carbon disulphide, a method often only partially successful. Certain organophosphorous compounds have been highly successful by gastric introduction (Enileeva *et al.*, 1974; Hasslinger and Jonas, 1975).

Primary Screwworms (Cochliomyia hominivorax). The spectacularly successful control of the screwworm in the southern USA by overflooding natural populations with irradiation-sterilized males has progressed to the point where it is planned for extensions southward into a narrow part of Mexico. Excellent reviews have been provided by Baumhover (1966) and Bushland (1975).

Sheep Blow flies. Sheep blow flies, especially *Phaenicia cuprina* in Australia, have proved difficult to control with insecticides because they developed high levels of resistance. The array of potential control options, including genetic methods, has been described by Foster and colleagues (1975). High moisture levels around the crotch of the animals, particularly when associated with staining by urine and feces, predisposes them for sheep strike. Selection of sheep breeds with minimal skin wrinkle and soiling characteristics in the crotch region will reduce strike, or an operation called mulesing (Mule's operation) removes skin strips in the crotch and the resultant wool-free and smooth area remains dry and is much less likely to soil and be fly struck (Anson and Beesley, 1975).

Sheep Bot Fly (Oestrus ovis). The sheep bot fly may be controlled by using a nasal drench of approved insecticide.

14

FLEAS

ORDER SIPHONAPTERA (SUCTORIA, APHANIPTERA)

Fleas are insects, the order Siphonaptera (Suctoria or Aphaniptera of some authors), that are exclusively bloodsucking in the adult stage. They number about 2,000 species and subspecies. Since larvae appear similar to the larvae of nematocerous flies, it has been suggested this group has affinities with the Diptera, but Hinton (1958) argues their origin from a mecopteran stem, believing that siphonapteran larval structures are derived from a *Boreus*-like ancestor. Evidence prevails that they are from a mecopteran stem, based mainly on similar pleural arches and sperm (cf. Rothschild, 1975). Holland (1964) presents the view that although their origins are obscure, their distinctiveness as a group, the existence of small families isolated on ancient continents, and the distribution on hosts of all continents including the Arctic and Antarctic, all suggest a long history of dispersal and evolution. Traub (1972c) points out that the zoogeography of Siphonaptera supports the theory of continental drift. Fossil flea remains are found from the Paleogene to recent times, with two existing genera from western Europe and the Baltic

amber (Rohdendorf, 1962; Rothschild, 1975).

Typical Life History

Fleas are temporary obligate parasites, with the adult stage feeding on hosts and the immature stages almost always developing off hosts. Their eggs are comparatively large (0.5 mm long), glistening white, and rounded at both ends (Figs. 14-1, 14-2). Relatively few, from 3 to 18, are deposited at one laying; however, during the entire lifetime of a female the number may be considerable. Bacot and Martin (1914) record 448 eggs over a period of 196 days from a single female *Pulex irritans*. Fleas often oviposit among the hairs of the host, but in most cases the eggs are dry and do not attach. Oviposition usually occurs in the host's nest where flea excrement and other detritus serve as larval food. Captured fleas readily oviposit in containers. In the case of a heavily infested dog or cat large numbers of eggs may be found on the sleeping mat.

Although most fleas develop their eggs at about any time that temperature, humidity, and food conditions are suitable, those associated with hosts having distinct denning and reproductive periods may synchronize

Fig. 14-1 Life history of the cat flea, *Ctenocephalides felis. A.* Eggs. *B.* Larva. *C.* Pupa, with silken material and outer debris of cocoon surrounding. *D.* Adult feeding. (*A, C,* and *D* courtesy of Oregon State University and Ken Gray; *B* courtesy of Pacific Supply Coop and Ken Gray.)

their reproduction with that of the host. The rabbit flea, *Spilopsyllus cuniculi,* undergoes ovarian maturation only on pregnant does, or experimentally in response to topically applied hydrocortisone or related hormones (Rothschild and Ford, 1964). *Cediopsylla simplex* of American cottontail rabbits (*Sylvilagus*) synchronizes its life cycle on domestic rabbits in a similar manner (Rothschild and Ford, 1972).

Temperatures of 18 to 27° C combined with a humidity of 70 percent and higher appear to favor egg laying. High mean temperatures of 35 to 38° C, in the range of normal body temperature for most mammals, inhibit growth of the developmental stages, which may account for the fact that the eggs do not hatch well on the host. Low temperatures also retard growth of the developmental stages. The incubation period normally varies from 2 to 21 days. Sensitivity of adult and developing fleas to extremes of humidity and temperature may be a principal reason why fleas occur in large numbers on animals that live in burrows or nests; or perhaps their development under conditions of high humidity and relatively stable temperatures,

characteristic of animal lairs, resulted in the loss of ability to withstand environmental extremes. Flea larvae have no mechanism for closing their spiracles, so to prevent water loss they will congregate toward the near saturation end of a humidity gradient (Yinon *et al.,* 1967). A high humidity requirement also explains why, in humid climates that are not excessively cold, developmental stages and adult fleas may be quite abundant outdoors around homes, especially if protein-rich wastes are plentiful in the upper soil layers (cf. Iglisch, 1969).

The flea embryo is provided with a sharp eggburster spine on the head to aid escape through the eggshell. Larvae are very active, slender, 15-segmented, and yellow-white with segmentally arranged bristles (Figs. 14-1, 14-2); the mouth parts are of the chewing type.

Larval food in the common species consists of a variety of materials found in the den and associated areas of the host. It was early noted that droplets of blood and egested blood products are deposited freely by adult fleas in such surroundings, and the key role of blood components as larval food in nature has been confirmed (Strenger, 1973). But in a nutritional study on larvae of the Oriental rat flea, *Xenopsylla cheopis,* it was found that several nutritionally complete rich-protein sources would provide maximum growth. The rather large size of the flea egg, one-twelfth the weight of a fully grown larva in the species studied, suggests that essential nutritional factors are received from the female parent, reducing the need for many specific nutrients essential in the diet of most insects. Some fleas, such as *Stenoponia,* are found primarily in winter, and the gravid female may contain only two huge eggs. Larvae of several species will readily eat flea eggs in their surroundings, even in the presence of adequate proteinaceous nutrients, and this may serve as a population-regulating mechanism (Reitblat and Belokopytova, 1974).

Under favorable conditions the larval period may be but 9 to 15 days; if unfavorable due to low temperatures or inadequate diet it may extend over 200 days. At the end of the active feeding and growth period the larva enters a quiescent stage, spins a cocoon, and pupates. The cocoon is whitish and so loosely spun that one may see the pupa within (Figs. 14-1, 14-2). Debris from the surroundings may adhere to the surface of cocoons.

The pupal period is influenced by temperature and varies from as short a time as 7 days to nearly a year. Accordingly the life cycle (egg to adult) may vary from 18 days to 20 months or more. Figures obtained by

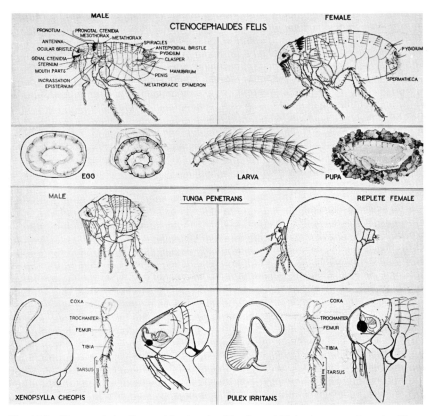

Fig. 14-2 Structural details used in the classification of Siphonaptera; also, the life history.

different workers have varied greatly even within a species, but most fleas require 30 to 75 days for a complete life cycle. Under optimal conditions common rat fleas can produce 5 to 9 generations annually (Nechaeva and Panchenko, 1974). At a temperature of 24° C the life cycle of the cat flea, *Ctenocephalides felis,* has been reported as 20 to 24 days (larval stage 11 to 12 days).

Adult fleas are blood feeders exclusively, generally awaiting the near approach of a host rather than setting out in search of it. Newly developed adults may remain quiescent until vibrations or other stimuli associated with a host's presence cause them to be activated. Adult hen fleas, *Ceratophyllus gallinae,* will leave the host's nest and ascend vegetation to take up a characteristic posture oriented toward light, jumping when the light intensity is suddenly reduced (Humphries, 1968). The rate of host-finding success must be variable, and surely also depends on host behavior, but it can be very fast in rodent fleas, as released deparasitized ground squirrels regained their normal complement of fleas within or before 3 days (Ryckman, 1971).

Longevity of Adult Fleas. In common with other types of lair parasites, adult fleas may survive prolonged periods of starvation while awaiting the return of a host. Bacot states that with nearly saturated air at 7 to 10° C unfed *Pulex irritans* survived for 125 days, *Nosopsyllus fasciatus* for 95 days, *Xenopsylla cheopis* for 38 days, *Ctenocephalides canis* for 58 days, and *Ceratophyllus gallinae* for 127 days. If fed on their natural host, *P. irritans* may live upward of 513 days, *N. fasciatus* for 106 days, and *X. cheopis,* fed on man, 100 days. *Ctenocephalides canis* and *Ceratophyllus gallinae* have lived for periods of 234 and 345 days, respectively, when fed on man. The maximum life cycle of species examined by Bacot ranged from 376 to 966 days. At 7 to 10° C and 100 percent relative humidity

adults of *Ctenophthalmus wladimiri* lived about 150 days under starvation, and a maximum of 1,133 days when fed periodically (Talybov, 1975). Adult longevity in nature permits flea-transmitted pathogens to persist in the absence of vertebrates.

Hosts and Occurrence of Species. The degree of permanence of attachment to the host as adults permits segregation of fleas into four distinct groups (modified from Suter, 1964). (1) The majority of fleas, as represented by *Xenopsylla cheopis,* easily leave their host and may transfer to other hosts of the same or a different species. (2) Females of another group, which includes the sticktight flea, *Echidnophaga gallinacea,* become stationary ectoparasites anchored by their mouth parts to the host's skin. (3) Females of a third group represented by the chigoe, *Tunga penetrans,* develop into proper stationary intracutaneous parasites, though they maintain an opening with the outside. (4) Certain rodent fleas such as species of *Conorhinopsylla* and *Megarthroglossus* are seldom on the host, but occur in abundance in the host's nest and are structurally modified accordingly. From the standpoint of pathogen transmission, attached species probably play no important role. Rate of development and optimal ecological conditions are similar for the first three types of fleas.

About 100 species of fleas are parasites of birds; the remainder are mammalian associates (Holland, 1964). Ordinarily a certain species of flea predominates on a given host, for example, *Nosopsyllus fasciatus* on the rat in Europe and the USA, *Xenopsylla cheopis* on *Arvicanthis* rats in northern Africa, the European mouse flea, *Leptopsylla segnis* on commensal *Rattus, Opisocrostis* on North American ground squirrels. But host specificity is not characteristic of many fleas, and a lack of absolute specificity increases the potential acquisition and transmission of pathogens.

Certain fleas, particularly those of medical

or veterinary importance, readily pass from one host to another. The human flea, *Pulex irritans,* attacks not only man: Hopkins and Rothschild include such hosts as pigs, dogs, cats, goats, domestic rats, and such wild animals as skunks, coyotes, and badgers; it has even been recorded from the echidna (*Tachyglossus aculeatus*), mallard duck (*Anas platyrhynchos*), and short-eared owl (*Asio flammeus*). It may breed in the litter inside pig shelters. *Ctenocephalides canis* and *C. felis* freely attack both dogs and cats, and they readily bite man. Rodent fleas may divide their attentions among various rodent species.

An extended study on host relationships of fleas has been made by Hubbard (1947) and Holland (1949), who include from several sources the following fleas as attacking man in the USA: *Cediopsylla simplex, Ceratcphyllus garei, C. gallinae, C. niger, Ctenocephalides canis, C. felis, Dasypsyllus gallinulae perpinnatus, Diamanus montanus, Hoplopsyllus affinis, Hystrichopsylla* sp., *Monopsyllus ciliatus protinus, M. eumolpi eumolpi, M. vison, M. wagneri wagneri, Nosopsyllus fasciatus, Orchopeas howardii, Pulex irritans, Xenopsylla cheopis. Hoplopsyllus anomalus,* a common parasite of ground squirrels in California and a proven vector of plague, is also known to bite man.

Morphology

External and Sclerous Structures of Fleas. Many of the external features of fleas are unique among the Insecta. As Snodgrass (1946) so aptly states, ''No part of the external anatomy of an adult flea could possibly be mistaken for that of any other insect. The head, the mouth parts, the thorax, the legs, the abdomen, the external genitalia, all present features that are not elsewhere duplicated among the hexapods.'' Specialized external structures in the form of spines and combs, and other unique external features, reflect the ways in which fleas have adapted to the vesture and habits of their hosts (cf. Traub, 1972b), yet internally the flea is a fairly generalized insect.

Adult fleas are laterally compressed, wingless, and generally 1.5 to 4 mm long. The wingless condition appears to be a secondary adaptation to obligate parasitism in adults, and Rothschild (1975) summarizes the evidence that a rubbery protein in conjunction with a pleural arch adapted from flight structures powers the jumping legs. The posterior pair of legs is strikingly adapted for jumping, though some fleas attaching to birds and squirrels in the nest have secondarily become crawlers (Holland, 1964). Males as a rule are smaller than females.

The head is a highly specialized cranial capsule set closely against the pronotum and having limited movement. On its sides are grooves in which are held the tiny knobbed and segmented antennae; in front of the antennae are located inconspicuous eyes, if these are present (Fig. 14-5). The ORAL or GENAL CTENIDIUM (Figs. 14-2, 14-4), a conspicuous comb of heavy spines located just about the mouth parts, is a useful feature in classification. Mouth parts of the adult flea are of the piercing-sucking type (Fig. 3-12).

The flea thorax is compact, the pronotum immediately behind the head with a comb of spinelike processes known as the PRONOTAL CTENIDIUM in many species, useful in group classification. The mesonotum is a simple arched plate. The metathorax is highly developed to sustain the jumping mechanism (cf. Rothschild, 1975). The chaetotaxy of the thoracic sclerites is of some systematic importance. Bristle numbers and arrangement on leg segments, and particularly the tarsi, are likewise of importance on the generic level.

The abdomen comprises 10 obvious segments, which like the thoracic segments are made up of sclerotic plates, but the pleurites are concealed. Numerous backward-pointing bristles aid forward locomotion through the host's pelage (Fig. 14-5). On the apical edge of the seventh tergite are the ANTEPYGIDIAL BRISTLES; the ninth tergite consists of a peculiar pincushionlike structure known as the PYGIDIUM and probably sensory in function.

Male terminalia are particularly important in classification. Among the parts to be observed are the CLASPERS, movable and nonmovable portions, and the MANUBRIUM (Fig. 14-2). In cleared specimens, springlike PENIS RODS may be seen coiled in the region of the fifth and sixth segments. In copulation this coil projects from the upper and lower claspers (cf. Rothschild, 1965, 1975). Females possess a sacculated SPERMATHECA (Figs. 14-2) situated internally in the region of the eighth or ninth segment and easily visible in cleared specimens. The shape of this organ is unique for many species and genera, and therefore an important taxonomic character. Pigmentation of the spermatheca may be used to distinguish mature from young females in *Xenopsylla* fleas (Gerasimova, 1971).

Digestive Tract. As soon as blood flows from the feeding wound it is drawn up into the pharynx by the action of both cibarial and pharyngeal pumps. The blood is carried to the long narrow esophagus that begins in the region of the brain and passes through the circumesophageal ring. The esophagus opens into the stomach through the bulbous PROVENTRICULUS, which is provided internally with radially arranged rows or proventricular spines. When the bands of muscle encircling the proventriculus contract, these spines meet as a valve that prevents regurgitation from the stomach. The stomach (midgut) is highly distensible and nearly as long as the abdomen, emptying into the short hindgut. Where the stomach joins the hindgut, four filamentous Malpighian tubes attach.

Classification

About 2,000 species and subspecies of fleas are known, and it is likely about 3,000 comprise the total existing fauna. Terms of structures used in classification are provided in a glossary by Rothschild and Traub (1971). The most complete catalog is that of Hopkins and Rothschild (1953, 1956, 1962, 1966, 1971). For America north of Mexico (including Greenland), Jellison and colleagues (1953) list 72 genera with 243 species and 55 subspecies. Holland (1964) discusses the difficulties encountered in classifying Siphonaptera. These difficulties occur largely because many groups of mammalian hosts and their associated fleas have become extinct and there are gaps in apparent phylogenetic relationships. In addition, flea taxonomy is based mainly on external structures for which there has been much convergent development.

Most flea identification requires considerable familiarity with group characteristics before one achieves accuracy. For public health or veterinary purposes, keys have been devised to identify common fleas known to attack man, species found on commensal rodents, rodent fleas recognized as involved in the transmission of plague, or fleas attacking livestock and pets (see USD-HEW, 1966, pictorial keys). Hopkins and Rothschild as well as Smit (in Smith, 1973) provide much valuable taxonomic information. Indexes to the literature of Siphonaptera of North America, through 1960, have been prepared by Jellison and coauthors.

COMMON SPECIES OF FLEAS

Frequently encountered species of fleas are discussed by families, with brief comments on their identifying features and biology, their ability to serve as causal agents of illness, and the role of fleas as vectors of pathogens and as intermediate hosts of helminths. While biting, fleas inject salivary secretions to which humans and animals frequently develop allergic sensitization. These reactions are discussed in Chapter 17.

Family Ceratophyllidae. These are fleas that are mainly associated with rodents.

Nosopsyllus fasciatus, the northern rat flea, is widespread over Europe and North America and less common in other parts of the world. It has been recorded on rats, house mice, pocket gophers, skunk, man, and many other hosts. It has only a pronotal ctenidium with a total of 18 to 20 spines. There are three to four hairs on the inner surface of the hind femur. *N. fasciatus* is regarded as unimportant in the causation of natural outbreaks of plague.

The genus *Nosopsyllus* may be distinguished from the genus *Diamanus* by the fact that in *Diamanus* there are long, thin bristles on the inside of the mid and hind coxae from the base to the apex; in *Nosopsyllus* such bristles occur at most in the apical half.

Diamanus montanus is a common ground squirrel flea abundant in much of western North America including Mexico.

Ceratophyllus niger, the western chicken flea, was originally described from specimens taken from man and from *Rattus norvegicus.* This flea is considerably larger than the sticktight flea of poultry (*Echidnophaga gallinacea*), and unlike that latter species does not attach permanently to the host. Additional hosts include cats and dogs. It breeds primarily in fowl droppings. *Ceratophyllus gallinae* is commonly known as the European chicken flea, although it has a wide range of hosts. Cotton (1970), in a study of the development of this flea, notes that most emerging adults overwinter in cocoons and are stimulated to emerge by a rise in ambient temperature in the following spring, completing the life cycle within the total period of nest occupation by most passerine birds. Flea infestation of persons in certain households, due to nearby bird nests or nest boxes infested with this flea, is a problem (Wolff, 1975). The large number of spines on the pronotal ctenidium furnishes a striking characteristic for distinguishing these bird fleas from others discussed here.

Family Leptopsyllidae. *Leptopsylla segnis* is the cosmopolitan European mouse flea, also common on rats. It bites man reluc-

tantly and is regarded as a weak vector of plague. In human plague outbreaks the role of this flea is considered negligible.

Family Pulicidae. This family includes a number of significant pests of man, domestic fowl and pets, as well as important vectors of the plague pathogen and putative vectors of murine typhus to humans.

Pulex irritans, the human flea, is cosmopolitan in distribution and occurs on a surprisingly wide range of hosts including domesticated animals, particularly swine. Zolotova and Yakunin (1973) provide an account of the development of this flea under various temperature and humidity conditions. There are neither oral nor pronotal ctenidia; metacoxae have a row or patch of short spinelets on the inner side. The maxillary laciniae extend about halfway down on the fore coxae, distinguishing this species from *Pulex simulans* (laciniae at least three-fourths the length of the fore coxae). *Pulex irritans* transmits the plague pathogen under laboratory conditions, and transmission between humans via this vector has been suspected in major epidemics. It may also be the chief vector of two unusual types of plague: *viruola pestosa,* a vesicular form, and *angina pestosa,* a tonsillar form, found in Ecuador. *P. irritans* comprised 81 percent of the fleas found in a sampling of dogs in Georgia, USA, and the public health dangers of this relationship have been pointed out (Kalkofen, 1974). Undoubtedly some of the published New World information relative to *P. irritans* belongs properly to *P. simulans.*

Echidnophaga gallinacea, the sticktight flea of poultry, resembles *Tunga penetrans* in the great reduction of the thoracic segments; it differs in having the angles of the head acutely produced. This flea is a serious poultry pest (Fig. 14-3) in many parts of subtropical America, attacking all kinds of domestic fowl and also cats, dogs, rabbits, horses, and man.

Before copulation both sexes are active, hopping about much as do other species of fleas. Shortly after feeding, the females at-

Fig. 14-3 Sticktight fleas, *Echidnophaga gallinacea*, clustered on the head of a young chicken. (After Suter.)

tach firmly by their mouth parts to the skin of the host and copulation takes place. The females deposit eggs in the ulcers produced by the infestation. Eggs are also deposited in the dust or dry droppings of poultry, or in old nests. The usual incubation period is 6 to 8 days at a temperature averaging 25° C. If the eggs are deposited in the ulcer the larvae crawl out and drop to the ground, grow rapidly if conditions are favorable, and feed on nitrogenous matter, dry droppings, and the like. After about a 2-week growth phase the larva spins a cocoon, pupates, and in about 2 weeks emerges as a fully developed flea. The life history requires 30 to 60 days.

Sticktight fleas are most likely to attack the skin around the eyes, the wattles and comb, and the anus or other bare spots. Ulceration and wartlike elevations around the eyes often become so aggravated that blindness occurs and the blinded host starves to death. Because this flea may live on several hosts, suitable precautions should be taken to exclude other animals from chicken pens. Infestations on dogs can be persistent and difficult to control (Kalkofen and Greenberg, 1974).

Ctenocephalides canis and *Ctenocephalides felis* are the dog and cat flea, respectively. Both species attack dogs and cats as well as man, and in households having these pets human annoyance is commonplace. *C. felis* is much more frequent on dogs in North America than is *C. canis*. Both species have the genal ctenidium consisting of 7 to 8 sharp black teeth, a character that distinguishes them from other fleas. These are ordinarily parasites of mammals, but a heavy infestation of chickens with *C. felis* has been reported (Jagannath *et al.,* 1972). *C. felis* (Fig. 14-4) may be separated (from *C. canis*) by observing that in the female the head is fully twice as long as high, and pointed (less than twice as high, and rounded); first and second genal spines of approximately equal length (first spine shorter than second); pronotal ctenidium with about 16 teeth (about 18); 2 or 3 bristles on metathoracic episternum (3 or 4 bristles); bristles on metathoracic epimeron, first row 4 to 8 (7 to 11), second row 5 to 7 (7 to 9); seven to 10 bristles on inner side of hind femur (10 to 13).

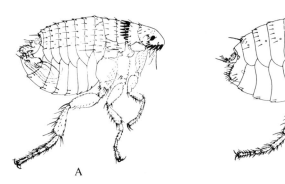

Fig. 14-4 *Ctenocephalides felis,* the cat flea. *A.* Male. *B.* Female. × 17.

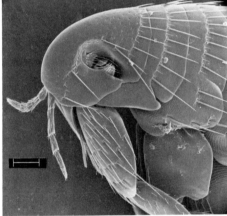

Fig. 14-5 The Oriental rat flea, *Xenopsylla cheopis*, female. Scale lines are 0.1 mm long. Note streamlined shape with setae facing backward. At higher magnification the antenna is seen in a pit, with the eye just below and in front. (Scanning electron micrographs courtesy of A. R. Crooker, Electron Microscope Center, Washington State University.)

Xenopsylla cheopis (Figure 14-5), the so-called Oriental rat flea, is largely cosmopolitan virtually wherever *Rattus rattus* is found, though scarce or absent from northern areas. It habitually occurs in buildings and bites man freely. It resembles *Pulex irritans* in that both oral and pronotal ctenidia are absent; the maxillary laciniae reach nearly to the end of the anterior coxae.

Mellanby demonstrated that *X. cheopis* completes its life history between 18 to 35°C in moist air; between 18 to 29° C, 40 percent relative himidity is unfavorable, but with 60 percent rh pupation takes place successfully. The developmental temperature threshold for pupation is about 15° C. This flea is frequently reared for a number of experimental purposes, and a simple means of mass production has been described (Cole, *et al.*, 1972).

Xenopsylla brasiliensis, an African species, is the predominant rat flea in Uganda, Kenya, and Nigeria. It has spread to South America and certain areas in India. Because it is associated with village huts it is regarded as a more important vector of plague bacteria than *X. cheopis* in Kenya and Uganda; *cheopis* is more urban, infesting rats in stone or brick buildings.

Xenopsylla astia is common on gerbils on the Indo-Pakistan subcontinent, and on rats in seaports of the same area and Burma. It may be the vector of plague in certain circumscribed and isolated outbreaks.

The Australian rat flea, *Xenopsylla vexabilis*, is commonly on the Hawaiian field rat, *Rattus hawaiiensis*. It is rarely found on rats caught in buildings.

Several additional species of *Xenopsylla* occur on *Rattus* in drier areas throughout much of the world. These are often proven or suspect enzootic vectors of plague pathogen among rodents.

Family Tungidae. These fleas are especially adapted for intracutaneous permanent attachment on hosts.

Tunga penetrans, the chigoe, is also known as the "jigger," "chigger," "chique," or "sand flea." The head of this flea is definitely angular and larger proportionately than the head of most other fleas. There are no ctenidia, and the mouth parts are conspicuous, with palpi four-segmented.

The chigoe is a tiny "burrowing" species of the tropical and subtropical regions of the Americas, also of the West Indies and Africa. According to Hoeppli (1963) the first reference to the chigoe was from tropical America in 1526, and it was reported from Africa in 1732. Very probably *T. penetrans* was introduced in the seventeenth century from tropical America into Africa; reintroduced in 1872 with a rapid spread along the west coast, and disseminated by expeditions across the tropical part of the whole African continent. Indian laborers returning to their homeland from Africa carried the parasite to Bombay and later to Karachi.

The chigoe is reddish brown and measures about 1 mm in length, though the impregnated female may become as large as a pea. The adults are intermittent feeders but adhere closely to the host. The female when impregnated "burrows" into the skin of the host (Fig. 14-6), frequently between the toes or under the toe nails and into the soles of the feet, though no body part is exempt from attack. The embedded female causes nodular swellings that ulcerate. Actually, there is no obvious way for the insect to burrow, yet somehow the skin develops over the female which becomes enclosed in a sinus, except for a small opening to the outside. The larvae that emerge in a few days from the eggs are typical flea larvae. Some hatch within the sinus; these usually drop to the ground to develop under conditions similar to those having hatched on the ground. Faust and Maxwell

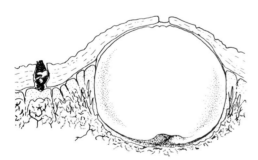

Fig. 14-6 Females of the chigoe, *Tunga penetrans*, embedded in the host's skin. On the *left* the abdominal segments are just starting to swell; on the *right* is the embedded and rounded gravid female, with an opening to the outside maintained at the end of the abdomen. (Redrawn from Grzimek [Ed.], 1975, *Grzimek's Animal Life Encyclopedia*, Vol. 2, Kindler Verlag, Munich; Van Nostrand Reinhold, New York.)

report an exceptional case in which the eggs hatched in or on the body around the sites of the burrows of the gravid females, and the larvae thrived and grew there. The larval period under favorable conditions probably requires not more than 10 to 14 days, and the cocoon or pupal period about the same time.

The attached female chigoe causes extreme irritation, and the elevated area surrounding it becomes charged with pus; ulcerations due to the presence of numerous chigoes become confluent. The commonly observed autoamputation of toes of indigents in Angola is attributed to the work of chigoes and attendant secondary infection. Tetanus and gangrene frequently result, and the latter may require amputation of affected structures. Chigoe lesions are still a very important site of development for the tetanus pathogen in parts of Africa (cf. Ancelle *et al.*, 1974). This flea is rare presently in parts of Costa Rica where it was once abundant, possibly owing to the use of persistent insecticides (communication from R. Zeledon to RFH).

Although the chigoe is primarily feared as a pest of man, it will develop in other animals and has been noted in particular to infest swine on the feet, snout, and scrotum, or the teats of sows, obstructing milk production to result in the death of piglets (Cooper, 1976; Verhulst, 1976). There are other species of *Tunga* that normally occur on wild hosts, such as *Tunga monositus*, which parasitizes rodents on the Pacific coast of Baja California, Mexico (Barnes and Radovsky, 1969).

Where the chigoe flea occurs, walking in bare feet should be avoided. Parts of the body attacked by the fleas should receive immediate attention. The insect can be removed quite easily by means of a sterile needle or fine-pointed knife blade. Wounds caused by this treatment must be dressed carefully to heal properly.

Flea Transmission of Pathogens: General Considerations

The following account concentrates on main flea-pathogen relationships. In addition to these there are reports of fleas as vectors of the etiological agents of Q fever, tularemia, listeriosis, salmonellosis, and *Trypanosoma lewisi* of the rat (cf. Bibikova, 1977). The same review points out that all known instances of fleas serving as vectors involve mechanical transmission, or pathogens only within the digestive tract. That is, no vertebrate pathogens develop beyond the gut to infect salivary glands or other tissues.

Viral Infections

MYXOMATOSIS

The myxoma virus causes a disease of rabbits that is mechanically transmitted by a variety of bloodsucking arthropods, and especially by mosquitoes (see Chapter 10). However, in England there is abundant evidence that fleas play the major role as vectors. The rabbit flea of England, *Spilopsyllus cuniculi,* is primarily involved, readily transferring between rabbits (*Oryctolagus*) but seldom feeding on hares (*Lepus*) (Mead-Briggs, 1964).

There is no evidence that the myxoma virus multiplies in the flea, or harms the flea directly, but fleas feeding on sick rabbits undergo maturation of the ovaries and have internal organs often characteristic of a spent or aged flea. Rothschild (1965) feels this condition may be due to a sudden rise in the temperature of affected rabbits, which causes increased defecation and feeding of fleas. The increased passage of blood through the flea automatically increases the amount of corticosteroid hormones ingested, and these hormones in pregnant rabbits cause this flea to reproduce (Rothschild and Ford, 1964). Virus survival on the fleas in southern England, under starvation conditions in artificial burrows, was demonstrated to occur for as long as 105 days (Chapple and Lewis, 1965).

Spilopsyllus cuniculi has been introduced into Australia where its transmission of myxoma virus is effective in preventing the normal steep rise in the rabbit population during the breeding season (cf. Sobey and Conolly, 1971; Williams *et al.,* 1973).

Bacterial Infections

PLAGUE

Disease: Plague, pest.
Pathogen: A bacterium, *Yersinia pestis.*
Victims: Humans. This is essentially a disease of rodents, which may be decimated by the pathogen. There are reports of infections in cats and dogs.
Vectors: Fleas, especially *Xenopsylla* to man and to rodents; other genera also between rodents.
Reservoir: Partially resistant field rodents and their fleas.
Distribution: Much of the world (Fig. 14-7). Presently for the most part in the field or campestral form associated with wild rodents, formerly the urban form associated with commensal rats was associated with major outbreaks.
Importance: Past pandemics killed millions and profoundly affected human civilization; currently WHO sources list between 1,500 and 5,000 cases worldwide annually.

Plague is caused by the bacterium *Yersinia pestis* (formerly *Pasteurella pestis*) and has been known from ancient times. This disease has had a profound influence on the course of history, being a decisive factor affecting military campaigns, and weakening besieged cities or their attackers. More significantly plague has been characterized by epidemics that have decimated human populations of entire countries or even continents (PANDEMICS). Some early historic highlights of this disease are covered in Chapter 1. This account concentrates on events from the time of the last major pandemic, which started in

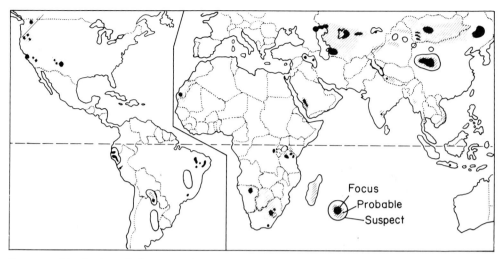

Fig. 14-7 Distribution of plague. (Redrawn from WHO, 1973, *Wkly. Epid. Record,* **48**:160.)

the interior of north China toward the end of the nineteenth century.

The Disease in Man. In addition to the work of Wu and coauthors (1936), two valuable general references are the well-documented monograph of Pollitzer (1954) and Hirst's interesting book, *The Conquest of Plague* (1953). Anisomov and associates (1968) cite some 5,000 Russian references to plague.

For the bubonic type of plague Wu and colleagues give the period of incubation as 2 to 10 days; the onset usually occurring in 3 to 4 days. Fox (1925) describes the disease:

It develops suddenly with a rapid rise of temperature, reaching [ca. 40.5° C] in two or three days, after which it is more or less irregular. There is headache, the eyes are injected and the facies are characteristic of extreme illness. Prostration is profound and comes on early. Delirium also appears early. The characteristic lesion of the disease, the bubo, usually is sufficiently pronounced by the second day to be readily detected. The most common site for the bubo is the femoral or inguino-femoral region, then the axillary region, cervical, iliac and popliteal. Over the enlarged glands oedema appears and pressure elicits great tenderness. The individual lymph nodes cannot be palpated. This swelling forms the primary bubo. Secondary bubos may appear in other parts of the body. In these, the glands are not matted together as in the primary bubo. Four forms of skin eruption may be described—a petechial eruption, ecchymoses, a subcuticular mottling, and the so-called plague pustule . . . a bulbous-like formation containing thin, turbid material teeming with plague bacilli. It is believed to indicate the original point of inoculation, the flea bite. Extending from this to the nearest lymphatic glands faint red lines indicating lymphangitis may be observed. A secondary pneumonia due to the deposit of plague bacilli in the pulmonary tissues may occur. In about a week if the patient survives, the bubo breaks down leaving an ulcer which heals slowly.

Plague is essentially a disease of rodents transmitted by rodent fleas, but it may under certain conditions cause serious outbreaks among humans. The term *bubonic plague* applies when inflammation of lymph glands results from the infection, and buboes are formed (Fig. 14-8); these first foci may remain so localized and cause little discomfort. The pathogenesis of plague infection follows a standard course; from the lymphatics and lymph nodes, to the blood stream, to the liver and spleen. When rapidity of the infection or other reasons prevent the liver and

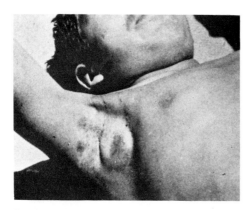

Fig. 14-8 Plague bubo in right axilla of a human case. (Courtesy of US Armed Forces Institute of Pathology. Negative No. ACC 219900-7-B.)

spleen from coping with the invaders, the infection massively invades the bloodstream and becomes *septicemic plague*. As Meyer points out, septicemic plague is really plague in which the buboes are inconspicuous; it is better to distinguish only two forms of human plague: the *primary bubonic* or zootic form, and the *primary pneumonic* (pulmonary) or demic form. The latter is especially dangerous and is transmitted from human to human by infective droplets from the respiratory system of a plague patient.

Several preventive and therapeutic practices greatly reduce the dangers of human plague. Immunization can be accomplished with killed organisms or avirulent cultures, and antiplague serum (see *J. Infect. Dis.* 1974, Supplement on plague immunization, **129**:S1-S120). The sulfonamides and tetracyclines are useful in treating the disease once contracted.

Fleas as Vectors. Ogata (1897) concluded on epidemiological grounds that fleas are the agents of transmission, noting that fleas leave a rat as it cools after death, and so may transmit the pathogen directly to man. He found that fleas ingest plague bacilli while feeding, and he produced plague in mice by injecting a suspension of crushed fleas taken from infected rats. Simond in 1898 succeeded in transmitting plague from

a sick rat to a healthy rat through the bite of infected fleas. His work, initially discredited, was successfully repeated in 1903.

Liston (1905), working in Bombay in 1904, concluded (1) one flea, *Xenopsylla cheopis*, infested rats in India far more commonly than did any other; (2) fleas acquired plague bacilli by feeding on a plague rat, and these bacilli multiplied in them; (3) infested fleas were at large during the incidence of plague fatalities; and (4) after an epizootic of rat plague, man acquired rat fleas, and might become infected as had guinea pigs used in his experiments.

Among other findings, the Indian Plague Commission organized in 1905 found that (1) plague was transferred from rats to guinea pigs, and between the latter, only through the agency of infected fleas; (2) guinea pigs free of fleas introduced into rooms in which persons had died from plague, or from which plague-infected rats had been taken, attracted fleas, and many died under subsequent ordinary confinement; (3) of pairs of susceptible animals confined in plague-associated rooms without access to soil, with one of each pair in a flea-proof cage, only (some) animals accessible to fleas developed plague; (4) among fleas trapped in plague-associated rooms, no cat fleas, less than 0.5 percent of human fleas, and nearly 30 percent of rat fleas harbored plague organisms.

Further observations relating dead rats to human acquisition of plague were reported from San Francisco, USA, in 1906 by Blue (1910), and from Manila, the Philippines (*Public Health Rep.*, Nov. 7, 1913, p. 2356).

Role of the Flea in Plague Bacillus Transmission. The Indian Plague Commission showed that the average capacity of *Xenopsylla cheopis* for blood is 0.5 mm^3, and that it might receive as many as 5,000 plague organisms from an infected rat. The Commission also found that the bacillus multiplies in the stomach of the flea and the percentage of fleas with bacilli varies seasonally. In the epidemic season (dry season) the percentage of infected fleas and duration of infection was greater than in the nonepidemic season. Both male and female fleas could transmit the bacilli. Bacilli were found in the stomach and rectum only, never in salivary glands or body cavity, and rarely in the esophagus if the flea was killed immediately after feeding. After digestion, the blood in the stomach passes into the rectum and is ejected as a dark-red tarry droplet containing plague bacilli that can infect an animal if rubbed into recent flea bites. The actual inoculation was therefore believed accomplished indirectly by the flea-bitten person scratching or rubbing the site of the bite after the infected flea had discharged fecal material upon the skin.

The most frequent and normal mechanism of infection was demonstrated by Bacot and Martin (1914), based on observations by Swellengrebel in 1914 that *X. cheopis* seldom defecates when feeding. They showed infection resulted when the flea's only contact with the experimental animal was by means of the proboscis, that is, the infection is introduced with the bite directly. This mode of infection is due to regurgitation caused by a temporary obstruction at the entrance to the stomach. The bacilli cause a gelatinous mass to develop among the spines in the proventriculus (Fig. 14-9), and when the flea tries repeatedly to feed, back pressure caused by elastic recoil of the esophageal wall causes some of the infectious blood to be driven back into the feeding wound. Infected fleas lived as long as 50 days at 10 to 15° C and 23 days at 27° C, and died infected. Working with the rat fleas *Xenopsylla cheopis* and *Nosopsylla fasciatus,* fed on septicemic blood, they concluded that these species can transmit plague organisms during sucking and that fleas with a temporary obstruction at the entrance to the stomach caused most if not all infections. In the course of time the plague culture forming

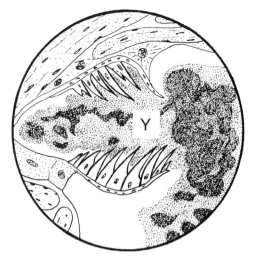

Fig. 14-9 A gelatinous mass of *Yersinia pestis* (*Y*) plugging the digestive tract of a flea in the region of the proventricular spines. (Redrawn from Pollitzer, 1954, *Plague*, WHO, Geneva.)

the proventricular plug dissipates and the normal passage of blood resumes.

In describing the mechanism of plague transmission by fleas, Eskey and Haas (1940) show numerous photomicrographs of blood-distended stomachs of fleas after feeding. The elapsed interval between an infective blood meal and an infective bite for *Xenopsylla cheopis* averaged about 21 days (5 to 31 range); for *Diamanus montanus* 53 days. The average length of life of plague-infected fleas was 17 days (maximum 44) for *X. cheopis* and 47 days (max. 85) for *D. montanus.*

Bibikova and Klasskovskiǐ (1974) have reviewed the literature on the mechanism of transmission of plague bacilli by fleas. Essential components influencing blocking and transmission include variants of the pathogen (cf. Kartman and Quan, 1964), the nature of the alimentary system in fleas, and factors affecting the development of blockage. Bibikova, and colleagues (1975) found no change in blocking frequency of *X. cheopis* after 19 generations of maintenance on plague-infected mice, but mean monthly temperatures below 27.5° C are needed to prevent a trypsinlike enzyme in the gut of this flea from digesting the plug (Cavanaugh, 1971). A minimum number of the pathogen must pass into the flea gut for blocking to occur, erythrocytes in the blood meal are needed for sufficient multiplication of the plague bacillus, the kind of rodent blood affects the incidence of blocking, and the proventricular block usually has a core of coagulated plasma protein. The various factors favoring blockage coincide with the dry season, when *X. cheopis* is most abundant, and this is reflected in the seasonal case-rate of humans (cf. review by Traub, 1972a).

Plague organisms are more constantly infective in the feces of some species of fleas than in others. In one set of experiments plague followed every inoculation of feces deposited by infected *Diamanus montanus,* and less than one-third of fecal inoculations of *Nosopsyllus fasciatus.* Feces of *Xenopsylla cheopis* gave positive reactions, but these fleas did not survive long enough for it to be determined whether or not the results would be constant for any length of time. There seems to be danger of infection from the feces of all fleas infected with the plague pathogen.

Another mode of transmission occurs when a host crushes infected fleas with its teeth. Infection through the mucosa of the buccal cavity will cause lymph node involvement in the region of the neck. This type of transmission applies usually to rodents only, though tonsillar plague among the Indians of Ecuador seems to be caused in this manner (Pollitzer, 1954).

Efficiency of Vectors in Experimental Studies of Plague. Kartman (1957) and associates have ascertained those factors that make for efficiency in flea vectors of the plague pathogen. The terms used in their investigations could often be applied to other arthropods and to their ability to transmit any vertebrate pathogen. VECTOR EFFICIENCY is the percentage of fleas becoming infected after feeding on an infected host. In practical usage this is derived simply by taking the percentage of plague transmissions accomplished within the group of fleas tested. The VECTOR OR BLOCKING POTENTIAL is the percentage of fleas with a blocked proventriculus. The TRANSMISSION POTENTIAL is the ratio of the number of infected rodents to the number of transmitting fleas fed daily until the death of individual rodents.

Field data should include the natural infection rates of fleas, their prevalence and length of life, their transfer potential between wild and commensal rodents, their ability to retain plague bacilli and to become infective (capable of transmission) under given conditions of temperature and humidity, and their association with known rodent reservoirs of plague. Also of importance are quantitative data on flea habits and movements, and habits and prevalence of the principal rodents concerned. As an example, the fleas *Hystrichopsylla linsdalei* and *Malaraeus telchinum* were both associated with a plague epizootic in the San Francisco Bay region. *H. linsdalei* had an experimental blocking index, field infection index, field prevalence index, and vector potential respectively of 0.02, 0.11, 0.08, and 0.02; these same factors for *M. telchinum* were 0.0, 0.04, 0.43, and 0.0. Vector potential on an individual basis showed *M. telchinum* to be inefficient, yet this flea apparently transmitted plague by sheer numbers, and its known prevalence and infection rate in the field suggested it to be important during the epizootic.

Urban and Campestral Plague. The epidemiology of plague in urban conditions involves commensal rodents, particularly *Rattus rattus,* the Oriental rat flea, *Xenopsylla cheopis,* and man. Typical urban plague can occur wherever poor sanitation places large numbers of rats and their fleas in close association with man.

Because rodents other than rats harbor large numbers of fleas, and rodents are variably susceptible to the plague pathogen, the question naturally arises as to whether plague is maintained in other than urban situations. The existence of a plague epizootic in wild rodents in California was suspected as early as 1903, and plague was demonstrated in the ground squirrel, *Citellus beecheyi beecheyi,* under natural conditions in 1908 by McCoy. According to that investigator about a dozen persons contracted the disease under circumstances that pointed conclusively to ground squirrels as the source. The two species of fleas commonly infesting the ground squirrel in California are *Diamanus montanus* and *Hoplopsyllus anomalus,* of which the former is far more numerous. McCoy proved the first of these capable of transmitting the plague bacillus from an infected to a healthy ground squirrel.

The designation SYLVATIC (SELVATIC) PLAGUE was proposed by Ricardo Jorge in 1928 for the plague of field rodents. The designation CAMPESTRAL PLAGUE, related to fields or open country, is more apt, since the term sylvatic implies woodland, which is not typical plague habitat. Fleas play an important role in transmission from rodent to rodent and consequently in the endemicity of the disease.

On a worldwide basis at least 220 species of rodents have been shown to harbor *Yersinia pestis* (Dubos and Hirsch, 1965). Wherever plague is characterized by occasional human cases due to man's coming in close association with wild rodents, or wherever the initial stages of urban outbreaks cannot be accounted for by spread from commensal rodents, there is a strong likelihood that campestral plague is present. A number of studies have shown that the maintenance of plague in focal rural areas is associated with resistant species of wild rodents that survive epizootics, yet experience a prolonged bacteremia. Lengthy bacteremia in wild rodents, and the ability of some rodent fleas to survive for long periods, even under starvation conditions, both help to maintain a plague reservoir.

Pattern of Plague in Various Parts of the World. The last pandemic of plague originated at the close of the nineteenth century in northern China. It soon reached Hong Kong via routes of commerce and was transferred to other continents by way of rats on steamships. Subsequent devastating epidemics in India served as the impetus for studies of the Indian Plague Commission. Northern China has been regarded as the cradle of plague, with permanent foci in wild rodents, and transfer to man and commensal rodents causing passage along ancient land trade routes and hostelries to European urban centers, resulting in the severe pandemics of the past. With the last pandemic and transfer along steamship routes, epidemics occurred in seaports of India (Bombay), the USA (San Francisco), South Africa, the Far East, and the Near East. The general opinion prevails that plague was introduced to these seaport areas, that native rodents became infected and epizootics resulted, and that infection of urban areas may occur through proximity of man or commensal *Rattus* with wild rodents.

The contemporary situation involves substantial numbers of human cases, but nothing approaching peak numbers during pandemics. Velimirovic (1974b) provides a world review of incidence for 1968–1972 that is adequately representative. For this period Asia varied between 1,300 to 5,200 cases reported annually, Africa between 30 to 170, and the Americas between 200 to 400. The actual numbers occurring versus those reported varies according to the adequacy of the diagnostic and reporting systems. The number of countries reporting plague cases annually fluctuated between 9 and 14. Enzootic foci in the Americas are known for Bolivia, Brazil, Ecuador, Peru, and the USA; wild rodent foci occur in five African countries, with no human incidence reported during the survey period from

Kenya, but a new focus discovered in Libya; six countries in Asia reported plague during this five-year period. South Vietnam listed 1,203 cases in 1972 versus about 4,000 cases annually in the preceding years. A world total of 2,737 cases was reported in 1974, and 1,478 in 1975 (MMWR, 1976, 25:291).

Even though plague has been declining on a worldwide basis (Dubos and Hirsch, 1965), the fact remains that permanent foci smolder with varying intensity among rodents and their fleas, and only disruptions of general sanitation and health services associated with earthquakes, famines, floods, wars, or similar catastrophes are required to trigger the recurrence of epidemics. Illustrative of this situation was a jump of plague cases reported in Viet Nam, from 353 in 1966 to some 4,700 in 1967 (*WHO Epidem. Vital Statist. Rep.* 20:380).

The historical and ecological relationships of plague in the USA from the time of introduction in 1899 to establishment of campestral plague have been reviewed by Kartman (1970). Campestral plague occurs presently only in certain western states and particularly in New Mexico. Flea-host relationships of the San Francisco Bay area of California have been described. It was shown that the mouse flea *Malaraeus telchinum* transferred from the field vole *Microtus californicus* to *Rattus norvegicus* and their nests under certain conditions. Further studies indicate that the direction of transfer of fleas occurs principally from wild rodents to commensal rats (Stark and Miles, 1962; Hudson and Quan, 1975).

Under campestral conditions typical of the western USA humans rarely acquire the plague pathogen even though they may be in rather frequent contact with wild rodents. Nonetheless, there is occasional evidence of the disease in such conditions. Lechleitner and others (1962) found that plague exterminated an isolated colony of prairie dogs (*Cynomys*) in Colorado. In this epizootic the

fleas *Opisocrostis labis* and *O. tuberculatus cynomuris* were found infected with the pathogen, the latter species being more numerous and having a higher infection rate. These authors concluded that the fleas acted as reservoirs and, while infected, remained alive in prairie dog burrows for at least one year after the epizootic ceased.

Highest current incidence of plague is in New Mexico, and there is evidence that Indians, particularly Navajos and especially children, contract plague from prairie dogs or their fleas (Pub. Hlth. Rep., 1965, 82:1077–1099). Plague-positive *Nosopsyllus fasciatus* in 1971 from the Norway rat in Tacoma, Washington, represents an instance where the pathogen has transferred from the wild to the urban environment through wild and commensal rodent contacts (MMWR, 1971, 20:190).

There is no evidence that rabbits constitute basic reservoirs of plague in the USA, yet in New Mexico isolated cases of the disease have occurred in men who shot and skinned rabbits and the pathogen was isolated from the cottontail rabbit (*Sylvilagus*) and two species of fleas. Domestic cats in contact with campestral rodents are susceptible to plague, and four cases of feline plague were reported in the USA in 1977, with two human cases suspected as being due to close association with infected cats (MMWR, 1977, 26:362). Dogs seem to be less susceptible to the pathogen, but fleas from dogs that have been in wild areas can be a problem, as indicated in the case of a woman in California who had been in contact with a dog in a campground (Coop. Plant Pest Rept., Calif., week ending October 7, 1977). It is interesting to note that reported cases of plague have been increasing in recent years in the USA (WMMR, annual summary, 1976), and that peaks have occurred at about five-year intervals (Fig. 14-10). The reasons for such cycling of human incidence are not clear, but cycling of rodent populations in response to

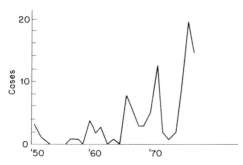

Fig. 14-10 Annual occurrence of reported human cases of plague in the USA, 1950–1976. (After *MMWR*, 1976 annual summary.)

plague epizootics might relate to this phenomenon.

In South Africa, according to Davis (1964) and De Meillon and associates (1961), plague-infected rats left ships following the great pandemic spreading from Hong Kong. Between 1899 and 1905 outbreaks occurred in the major port cities, as well as at inland centers such as Johannesburg. These outbreaks were confined to house rat areas, the last epidemic of this phase taking place in Durban in 1912. During the urban phase of the disease in South Africa wild rodents appear to have become infected, probably at fodder and remount depots established on main communication lines from ports to inland areas during the Anglo-Boer war. Gerbils (*Tatera* and *Desmodillus*) with the fleas *Xenopsylla philoxera* and *X. piriei*, respectively, are the most likely primary reservoir, with many other indigenous rodents and hares implicated. In human surroundings *X. brasiliensis* transmits the pathogen among commensal rodents and man. The first plague outbreak recorded for Rhodesia involved 22 human cases in 1974 (Pugh and Parker, 1975).

In India both urban and campestral plague occur. An analysis of major urban centers shows *Rattus rattus* to be most important, and *R. norvegicus* to be of minor significance. In Calcutta *Xenopsylla astia* is the predominant rat flea, whereas in Bombay *X.*

cheopis predominates. However, in Calcutta, wards with plague have rats more heavily infested with *X. cheopis* (Seal and Bhattacharji, 1961). In Uttar Pradesh (north central India) it was thought that only *R. rattus* and its fleas were implicated in the genesis of rural plague, but a reservoir was found in a relatively resistant and abundant field rodent, the gerbil *Tatera indica*. A similar relationship is found in South India (Chandrahas *et al.*, 1974). The domestic rat acts as a liaison agent between wild rodents and man, with the flea *Nosopsyllus punjabensis* abundant on wild and commensal rodents, along with *Xenopsylla astia* on the latter.

The propagation and persistence of plague in Java occur under conditions much as described for rural India. *Rattus exulans* is the predominant wild rodent that is important for long-term maintenance because of its high resistance to *Y. pestis,* and *R. rattus* acts as the liaison rodent between wild rodents and man. *Xenopsylla cheopis* predominates on *R. rattus,* and the flea *Stivalius cognatus* predominates on rodents in mountain regions although it is also found in domestic areas. Turner and colleagues (1974) provide a comprehensive account of the dynamics of plague transmission and ecology of potential vector fleas in central Java.

In southern Viet Nam plague was first recognized in 1898. Velimirovic (1974a) presents a historical account of the disease and its epidemiology through 1972. The main mammals concerned are *Rattus norvegicus, R. rattus, R. exulans,* and *Suncus murinus. Xenopsylla cheopis* is the principal vector, but *X. astia* may also be involved.

There are some interesting characteristics of human plague in the Kurdistan ethnogeographical region (parts of Persia, Iraq, Turkey, Syria), a plague focus considered united with the southwestern portion of the Soviet Union. Commensal rodents are a rarity in much of this region and even appear to be lacking in some villages, yet rare human cases are contracted in the field, apparently from fleas normally biting plague-resistant or susceptible species of gerbils (*Meriones*). Epidemics spreading the disease from man to man are then by way of the human flea, *Pulex irritans* (Baltazard *et al.*, 1960).

Pollitzer (1966) has reviewed the subject of plague and plague control in the Soviet Union. Plague regions in that country have been divided into four general foci: (1) the Caspian focus, a region surrounding the northern part of the Caspian Sea; (2) the Central Asian focus, comprised of plains and mountain areas; (3) the Transcaucasian focus, lying predominantly west of the mid to southern portion of the Caspian Sea; and (4) the Transbaikalian focus, encompassing Mongolia and surrounding areas. These are all campestral foci, as plague never became established in commensal rat populations. It was probably from the Transbaikalian focus that the last great pandemic of plague originated in men engaged in trapping tarabagans (*Marmota sibirica*) for their valuable fur. From humans the pathogen was transferred to the parasites of commensal rats along trade routes and resulted in the Hong Kong epidemic of 1894 that was spread by shipping to much of the globe. An epidemic of pneumonic plague in Manchuria, resulting in 60,000 deaths in 1910–1911, was also a consequence of trapping tarabagans. Apparently the low body temperature of the tarabagan in hibernation enables this animal to survive infection and thus to maintain plague bacilli from one season to the next. The flea, *Oropsylla silantiewi,* and possibly other blood sucking ectoparasites, transmits the infection from animal to animal. An outbreak in Nepal represents a southerly extension of the Central Asian focus (La Force *et al.*, 1971).

Wherever they occur in sufficient numbers, the following groups of rodents and their associated fleas seem primarily involved in campestral plague in the Soviet Union, with occasional outbreaks in man: marmots (*Marmota* spp.), ground squirrels or susliks (*Citellus* spp.), gerbils (species of *Rhombomys* and *Meriones*).

The World Health Organization has provided a technical guide for a system of plague surveillance (WHO, 1973C). In addition to suggesting how the incidence of human cases, rodents, and vectors should be studied and recorded, this guide explains the importance of classifying strains of *Yersinia pestis*. Strain identification provides clues to the reservoirs of infection and the character of natural foci.

Rickettsial Infections

MURINE TYPHUS (Endemic Typhus, Shop Typhus, Flea Typhus, Rat Typhus)

Sporadic cases of typhus in Europe, Australia, the USA, and Mexico were suspected as being different from epidemic louse-borne typhus. The agent was found to be distinct from that of epidemic typhus and suspected rodent-flea relationships were confirmed, the disease being called murine typhus to signify it is a natural infection of rats. The causative agent is *Rickettsia typhi* (formerly *R. mooseri*) and the disease is found in port areas of many parts of the world. An important review of this disease is presented by Traub and colleagues (1978).

Incubation of *R. typhi* in man is 6 to 14 days before the onset of clinical symptoms. The disease is relatively mild, with negligible mortality except in persons over 50. In comparison with epidemic typhus the rash is shorter in duration, skin lesions are less numerous, involvement of nervous system and kidneys is less severe, and serious complications are infrequent. In the USA from 1931 to 1946 approximately 42,000 cases were reported. Since that time there has been a dramatic reduction in incidence, with about 50 cases recognized annually, the majority of these from Texas. Most present cases in the USA occur along the southeastern or Gulf

Coast area and up the Mississippi River drainage.

The normal biocenose of murine typhus involves rats and the Oriental rat flea *Xenopsylla cheopis;* the fleas *Nosopsyllus fasciatus* and *Leptopsylla segnis* are suspect vectors. Transmission from rat to rat also occurs by way of the spined rat louse *Polyplax spinulosa,* and the tropical rat mite *Ornithonyssus bacoti* may perhaps be similarly involved. In the USA rats on scattered farms are believed to be the most important reservoirs of endemic murine typhus, as since 1943 most human cases have occurred in rural regions. In that country the monthly incidence of *R. typhi* antibodies in the commensal rats *Rattus norvegicus* and *R. rattus* correlated strongly with the monthly abundance of the flea *X. cheopis,* and showed little relationship to the abundance of *P. spinulosa* and *O. bacoti* (Smith, 1957). However, epidemiological studies in Texas during 1969 suggested that

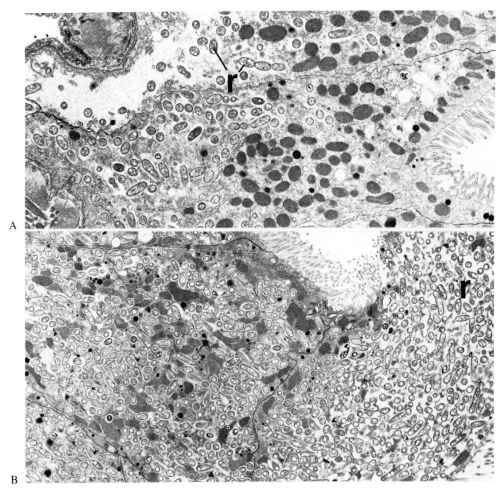

Fig. 14-11 Proliferation and release of *Rickettsia typhi* (*r*) in epithelial cells of the flea midgut. Base of the cells to the *left,* and the cell brush border at lumen surface of gut to the *right.* In micrograph *A* the number of rickettsiae is low and toward the base of the cells; in micrograph *B* the cell is bursting and releasing large numbers of rickettsiae into the gut. (From Ito *et al.,* 1975, *Ann. N.Y. Acad. Sci.,* **266**:35–60.)

most transmission involved fleas other than *X. cheopis* (Older, 1970). Adams and colleagues (1970) present the view that the opossum, *Didelphis marsupialis,* may maintain the pathogen in the Los Angeles region, with the cat flea, *Ctenocephalides felis,* serving as the agent of transfer to man, but significant data are lacking. In Kenya *Xenopsylla cheopis* and *X. brasiliensis* appear to be involved with human infections. Suspect vertebrate-vector relationships for human infections in western Yugoslavia involve the Norway rat and *Nosopsyllus fasciatus* (Urlic *et al.,* 1971).

The pathogen proliferates in great abundance within the epithelial cells of the flea midgut (Fig. 14-11), and when packed these cells burst to release free rickettsiae into the lumen in limited numbers. None of the normal cytoplasmic organelles of the vector are disrupted, and infection does not shorten the life span of the flea; infection of fleas can occur only during the parasitic adult stage (Ito *et al.,* 1975). Transmission from flea to man is accomplished when infective feces are scratched into the skin or transferred to conjunctiva or mucous membranes; respiratory acquisition via flea feces in dust may be important.

Control of this disease consists of rodent control, and reduction of flea numbers by dusting rat runs and harborages with insecticides. Once contracted, the disease is readily cured with antibiotics.

Helminth Infestations

Fleas play a more significant role as an intermediate host of the double-pored dog tapeworm, *Dipylidium caninum,* than the biting louse of the dog, *Trichodectes canis.* Most important are the cat and dog fleas, *Ctenocephalides felis* and *C. canis.* This tapeworm normally develops in the digestive tract of dogs, cats, and certain wild carnivores, but also occurs in man and particularly in young children. The embryonated eggs of the tapeworm are discharged in the feces of the vertebrate host and are ingested by the larval flea, developing into cysticercoids in the body cavity of the insect. Thus the mature flea, which cannot ingest tapeworm eggs because of its piercing-sucking mouth parts, becomes infected. When infected fleas are ingested by a cat, dog, or human the cysticercoids are liberated and develop into tapeworms in the digestive tract. The cat flea can serve as an experimental intermediate host of several cestodes, many causing some mortality in the flea's larval stage, but *D. caninum* causing appreciable mortality in the flea's pupal stage (Marshall, 1967).

A common tapeworm of rats and mice, *Hymenolepis diminuta,* has numerous intermediate arthropod hosts, among them *Nosopsyllus fasciatus* and *Xenopsylla cheopis.* A related rodent tapeworm, *H. nana,* has among its intermediate hosts the fleas *X. cheopis, Ctenocephalides canis,* and *Pulex irritans.* Both cestodes frequently infest children.

Fleas also serve as the main invertebrate hosts and vectors of *Dipetalonema reconditum,* a benign filarial worm commonly occurring in dogs. Nearly 23 percent of stray dogs in Okinawa were infected with this filaria (Pennington and Phelps, 1969). The same condition has been noted in foxes in southwestern France (Bain and Beaucornu, 1974).

A natural infection of encysted *Trichinella spiralis,* ordinarily occurring in the muscles of rats and swine, has been found in *Xenopsylla cheopis* on infected rats and bandicoots in India (Ranade and Bhalchandra, 1976).

Control of Fleas

Flea populations are generally analyzed by examining living or freshly killed hosts. Their presence may be suspected by the host's scratching or by typical welts resulting from bites. A carbon dioxide attractant trap may be placed in rodent burrows to sample (ticks or) adult fleas (Miles, 1968).

Because flea larvae feed on debris found at ground level, beneath old carpets, or in floor cracks, keeping such debris to a minimum indoors helps to reduce their numbers. Rat fleas are controlled, particularly when plague or murine typhus threaten, by controlling their rodent hosts. Personal protection may be achieved by applying repellents to the skin or to clothing.

Fleas may be routinely examined for the presence of parasites and pathogens (Beaucornu and Deunff, 1976). Flea pathogens include the protozoan *Nosema pulicis,* taken up by larvae and causing very high mortality on occasion. The pteromalid wasp, *Bairamlia fuscipes,* may cause heavy parasitization of flea cocoons. Ant predation on fleas in rat nests is suggested as an important factor in the reduction of murine typhus in Puerto Rico (Fox and Garcia-Moll, 1961), and substantial predation of flea eggs and larvae in rodent burrows and nests by gamasid mites has been noted in the Soviet Union (Reitblat *et al.,* 1974).

Insecticides are effective against fleas, but these insects have developed resistance to various groups of insecticides where comprehensive control programs have been conducted (cf. WHO, 1970; Velimirovic, 1974a). Flea powders containing botanical insecticides such as pyrethrins or their synthetic derivatives, plus synergist, or low mammalian toxicity synthetic organic compounds, are safe for use on pets. Special caution must be used in treating cats because of the tendency of these animals to lick themselves in grooming. Flea collars that slowly release toxicant in vapor phase are effective for dogs and cats (but see Bell *et al.,* 1975, regarding problems with cats). Topical or oral treatment of dogs and cats with approved systemic insecticides is effective against fleas. To gain rapid control of fleas around infested premises, in addition to cleaning up debris in which they are developing, it is useful to spray a suitable toxicant at floor and ground level. Insecticide dust applied along rodent runs and around garbage areas and warehouses will control fleas on rats. The toxicant is picked up on the body of rodents and carried back to their nests where it may be very effective against larvae.

In plague and murine typhus control campaigns it is often necessary to control rodents. WHO (1974b) reviews the ecology and control of rodents of public health importance. Some suggestions for commensal rodent control include: (1) elimination of rat and mouse harborages, (2) elimination of exposed foodstuffs, (3) rodentproofing of buildings, and (4) destruction of rodents.

The commensal rodents usually encountered are the house mouse, *Mus musculus,* the brown or Norway rat *Rattus norvegicus,* the black rat, *Rattus rattus,* and the Alexandrine or roof rat, *Rattus alexandrinus*. If poison baits are used, special precautions are required to keep them away from children, pets, and farm animals. Because rats are quite suspicious, prebaiting with unpoisoned bait is commonly practiced. As baits various cereals, bread, raw meat, or fish are satisfactory; containers of drinking water may be used. Rodenticides in current usage are anticoagulins and red squill, a safe but far less effective botanical derivative. Most poison baits should be applied only by licensed operators.

To reduce campestral plague foci before or during outbreaks, rodent control or the control of rodent fleas has been attempted. The latter practice is considered preferable because of the dependence of predatory wildlife on wild rodent populations. Direct treatment of burrows with insecticides is effective (cf. Barnes *et al.,* 1972; Bennett *et al.,* 1975), but requires much labor. A faster approach is to apply baits that rodents will remove to their dens, and systemic or vapor toxicants combined with baits have been successful (cf. Clark *et al.,* 1971; Cole *et al.,* 1976). Scattering of baits, for example, by aircraft, is quicker and may be more effective than the use of discrete bait stations, because territoriality of rodents may prevent adequate distribution of baits from limited locations (Nelson and Smith, 1974).

15

MITES AND MITE-BORNE DISEASES

CLASS ARACHNIDA; SUBCLASS ACARI

Mites are small (usually less than 1 mm long) to microscopic members of the class Arachnida, subclass Acari. There are many free-living and predacious forms, but some families and larger groups are exclusively parasites. Skin-infesting forms may cause hide and wool damage to livestock in the USA estimated in 1965 as $3 million for cattle, and $2 million for sheep (Steelman, 1976).

Characteristics

In mites the abdomen is broadly joined to the cephalothorax with little or no evidence of segmentation (Fig. 15-1). All but a few are minute, that is barely visible to the unaided eye, and some that live in tissues are essentially microscopic. Size is a factor that could favor endoparasitism by mites. Species such as scabies mites and follicular mites, living within or beneath the skin of their host, could have initially developed such a habit with little structural modification if they were derived from free-living ancestors small enough to fit within so limited

an environment. Even less in the way of structural modifications would be required of mites living in respiratory passageways and ear canals.

Mites, like other arachnids, have with the exception of certain animal and plant parasites where the number may be reduced, four pairs of legs as nymphs and adults but only three pairs in the larval stage. The mouth parts are quite varied, but along with those of ticks they follow a general pattern (Fig. 3-14). A hypostome is lacking in many mites, but in Mesostigmata it is well developed (Fig. 15-1), though not as a holdfast organ with rows of teeth as in the ticks. The CHELICERAE of the parasitic species are tearing or piercing structures. Eyes are absent, or one or more pairs of SIMPLE EYES may occur. The respiratory system is tracheal in most species, though some absorb oxygen through the soft general body surface. Nearly all species deposit eggs; however, a few are ovoviviparous, for example, *Pyemotes ventricosus*. From the egg there emerges the HEXAPOD LARVA, which usually molts soon, becoming the nymph and acquiring its fourth pair of legs. The life cycle of many species requires less than four weeks; in some it is as short as eight days.

342

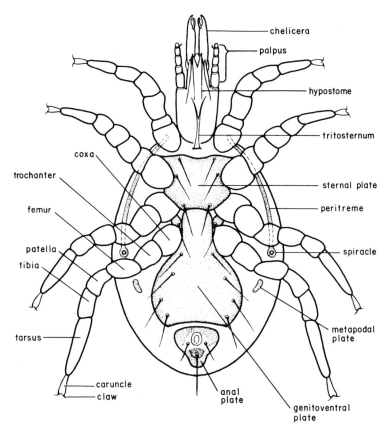

Fig. 15-1 General structure of a mite, ventral view. (Courtesy of US Public Health Service.)

Rohdendorf (1962) provides a review of the origin of parasitic Acarina, namely, the ticks and some mites. These appear to have had a common ancestry, with ticks separating from the mesostigmatid mites at the beginning of the Permian. Parasitic mites are believed to have developed during the Devonian. There are a number of opinions on the phylogeny of mites, and Wooley (1961) provides a review of the subject.

Mites affect the health of man or animals in four general ways: (1) by causing dermatitis or other tissue damage; (2) through loss of blood or other tissue fluids; (3) by transfer of pathogenic agents either as vectors or as developmental hosts; (4) by causing strong allergic reactions in man, pets, and livestock. An infestation of mites is called

ACARIASIS. Although mites are commonly thought of as external parasites, some species infest the inner and middle ear, the respiratory passages and lungs, nasal passages, and lymphatic tissues.

Classification: General Considerations

Baker and Wharton (1952) recognized more than 200 families of mites, and several families have been added since that publication, but only a few contain species that affect man and domestic animals. The identification of mites, even the few species affecting health, is difficult. The student is referred to Baker and Wharton's general introduction to the group (1952); the guide to families by Baker and coworkers (1958) and the manual by Krantz (1978); the manual of parasitic mites by Baker and others (1956), and the Pratt (1963) publication on mites of public health importance. Strandtmann and Wharton (1958) provide a manual of parasitic mesostigmatid mites and Radovsky (1969) outlines relationships in this group. General coverage of chiggers is found in Wharton and Fuller (1952) and Brennan and Jones (1959). The list of genera and species of parasitic mites by Radford (1943) gives the authority, year of validity, and where possible the original host. Host and distribution lists of North American mites parasitic and phoretic in the hair of wild animals, prepared by Whitaker and Wilson (1974), provide brief descriptions of family characteristics for 10 families excluding burrowing species and trombiculids. Oudemans's (1937) monumental work is a standard reference source for acarologists, and Vitzthum (1943) is an important foundation work. Further general information may be obtained from the publications by Evans and coauthors (1961), Hyland (1963), and Audy and Lavoipierre (1964). Quite comprehensive accounts of mites affecting wild animals and laboratory animals may be found in Davis and Anderson (1971) and Flynn (1973).

Earlier classifications placed mites and ticks in the order Acarina of the class Arachnida (see Baker and Wharton, 1952), but acarologists now tend to place these in the subclass Acari. Many specialists unite the Mesostigmata and Ixodides on morphological grounds into the suborder Parasitiformes. We are using the approach presented in Krantz (1970), although ticks in the suborder Metastigmata are dealt with separately in Chapter 16. See Evans and coauthors (1961) and Johnston (1968) for different divisions of the order Acari, and note that Krantz (1978) has made a number of changes.

In identifying mites it is essential to realize that some structures, particularly the spiracles and associated peritremes, are difficult or impossible to see unless the specimen has been properly cleared and mounted. In many cases the use of a phase contrast compound microscope and properly prepared specimens on slide mounts are requirements for discriminating fine details, especially when species determinations are attempted. Figure 15-1 illustrates various features used in identification.

Order Parasitiformes; Suborder Tetrastigmata

HOLOTHYRID MITES

Tetrastigmata are heavily sclerotized nonsegmented mites without propodosomal ocelli, very large and somewhat rounded.

The family Holothyridae includes the largest of known mites, which are free-living, and some of which reach a length of 7 mm. On the island of Mauritius *Holothyrus coccinella* is said to cause the death of ducks and chickens that swallow it, due to a toxic secretion. Children also suffer ill effects through handling the mites and then touching their mouths with their fingers (cf. Southcott, 1976).

Order Parasitiformes; Suborder Mesostigmata

MESOSTIGMATIC MITES

Mesostigmata possess a pair of respiratory spiracles or STIGMATA usually located behind and laterad of the third coxa. The stigmata are usually associated with elongated PERITREMES (Fig. 15-1), shallow surface grooves of uncertain function. Coxae of the pedipalps are fused dorsally so that the base of the GNATHOSOMA (fused ventral parts of three segments enclosing the mouth and bearing chelicerae and pedipalps) forms a tube that encloses the mouth parts. Barker (1968) has reviewed the effect of food quality on reproduction of these mites. Most mesostigmatids affecting man and animals belong to the families Dermanyssidae and Macronyssidae. A few others, however, are included in this discussion.

Family Laelapidae. In this family the subfamily Haemogamasinae are medium-sized oval mites, heavily clothed with setae that give the body on upper and lower surfaces a furry appearance. They are widespread parasites on small mammals. *Haemogamasus pontiger* was suspected of causing dermatitis of soldiers who slept on straw-filled mattresses in England during World War II (Baker *et al.,* 1956). *Haemogamasus ambulans,* the laelapid *Androlaelaps fahrenholzi,* and the laelapid *Hirstionyssus isabellinus* (=*arvicolae*), taken from small mammal nests and burrows, were shown to harbor the etiological agent of hemorrhagic nephrosonephritis (epidemic hemorrhagic fever) of the Far East (Chumakov, 1957). The close association of haemogamasid mites with small mammals suggests that they may play a role in the maintenance of plague, typhus, tularemia, and perhaps other diseases (Baker and Wharton, 1952).

Mites occur as "normal" inhabitants of the respiratory passages of dogs, monkeys, numerous marine and other mammals, and birds and reptiles. Human infestation is accidental and due to species that are ordinarily free-living.

Family Rhinonyssidae. The canary lung mite, *Sternostoma tracheacolum* invades the tracheae, air sacs, bronchi, and parenchyma of canaries and Gouldian finches (*Poephila gouldiae*) and may cause illness and death. This mite appears to be larviparous, with immature forms in the lungs and adults in the tracheae and nasal cavities. Dietary administrations of an appropriate pesticide or a suitable pesticide in aerosol exposure (Jolivet, 1975) appear to provide relief.

Family Halarachnidae. *Pneumonyssoides caninum* inhabits the sinuses and nasal passages of dogs, usually in benign association, though Christensson and Reh-

binder (1971) believe they may cause sinusitis leading to symptoms of central nervous system disorders. A related mite, *Pneumonyssus simicola,* and other species of *Pneumonyssus* infest the lungs of monkeys and baboons. *P. simicola* is apparently tolerated well in a natural state, but causes a catarrhal enteritis that may lead to as high as 70 percent mortality during the stress of captivity (Testi and DeMichelis, 1972). Furman and associates (1974) provide information on diagnostic techniques, and behavior of this mite in macaques, and Kim (1974) further describes the biology and pathogenesis. An electron microscope study by Ogata and colleagues (1971) revealed that lesions due to this mite are characterized by slight fibrosis, inflammation of the bronchi, and pneumonitis. Baboons (*Papio* sp.) in the wild state and in captivity harbor at least three species of *Pneumonyssus* that can cause pneumonia and diffuse pleuritis (McConnell *et al.,* 1974; Kim and Kalter, 1975).

Raillietia auris (family Halarachnidae, subfamily Raillietinae), the cattle ear mite, is often common and has been reported from several continents. It is believed to feed on ear wax and sloughed epidermal cells.

Families Dermanyssidae and Macronyssidae. An important group of mesostigmatids includes very bothersome ectoparasites of poultry, wild birds, and rodents. They may also attack humans and cause discomfort and skin disorders, though they cannot permanently establish on this host.

Baker and Wharton include these mites collectively in the family Dermanyssidae, but Radovsky (1967) separated them into the family Macronyssidae composed of *Ornithonyssus* and its relatives, and the family Dermanyssidae, which includes *Dermanyssus* and *Liponyssoides* (=*Allodermanyssus*). These are discussed together as a matter of convenience.

The genus *Ornithonyssus* (=*Liponyssus* of Authors, *Bdellonysus*) includes a homoge-

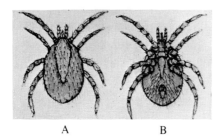

A   B

Fig. 15-2 The tropical rat mite, *Ornithonyssus bacoti*. A. Dorsal view. B. Ventral view. (After Dove and Shelmire.)

neous group of mostly tropical species according to Strandtmann and Wharton (1958), and further characterized by Furman and Radovsky (1963).

The tropical rat mite, *Ornithonyssus bacoti* (Fig. 15-2), was first recorded from the Norway rat in Egypt by Hirst and described in the genus *Leiognathus*. This mite can cause significant debilitation, retarded growth, and high mortality in research colonies of mice, particularly affecting infant animals (Schoenbaum and Rauchbach, 1975). It occurs on all continents in both temperate and tropical regions, particularly where the roof rat is present. When rats abandon their nests or are killed the mites may migrate considerable distances in search of a blood meal, entering homes from rat-infested attics, down chimneys, or through crevices. They accumulate around a source of heat such as hot water pipes or stoves (Larsen, 1973). Humans who are bitten experience a sharp itching pain at the time of attack, and sensitive individuals may develop severe dermatitis.

Strandtmann and Wharton (1958) summarized the life cycle of the tropical rat mite. The developmental stages are adult male and female, egg, nonfeeding larva, bloodsucking protonymph, and nonfeeding deutonymph. Unfertilized females produce males only. From egg to adult requires 7 to 16 days at room temperatures with sufficient food, with a minimum of 13 days from egg to egg. Unfed protonymphs have survived 43 days,

and adult females live an average of 61 days and produce eggs after each feeding for a total of about 100.

O. bacoti has not been incriminated in the natural transmission of any human disease, though there are reports of experimental transmission of the pathogens causing murine typhus, plague, and rickettsialpox; the tularemia bacterium was passed between growth stages and to progeny of the mite, but not to vertebrates by its bite. It acts as an intermediate host of a filariid worm, *Litomosoides carinii,* a parasite of the cotton rat, *Sigmodon hispidus.* This host-filaria-vector system may be used in testing for antifilarial drugs.

Ornithonyssus bursa, the tropical fowl mite, is a poultry ectoparasite in tropical and subtropical areas on all continents. It also appears to be a widespread parasite of the English sparrow, *Passer domesticus.* In Israel it infests wild birds and turkeys, causing these poultry to be listless and to develop poorly (Hadani *et al.,* 1975). Although man is frequently bitten by this mite, only a slight irritation is reported, and this tends to be temporary because the mite cannot exist for more than 10 days apart from an avian host.

Ornithonyssus sylviarum, the northern fowl mite, closely resembles the tropical fowl mite. It is a widespread parasite of poultry in north temperate regions and has been reported from New Zealand and Australia. This mite will at times cause itching of man by its bite and by crawling over the skin, being especially bothersome in the absence of bird hosts, or severely affecting persons collecting and processing eggs from heavily infested flocks. Survival up to six weeks in the absence of avian hosts is known, though most records indicate a starvation survival time of about three weeks. There is some uncertainty over the amount of harm northern fowl mites do to chickens. Infestation tends to be concentrated around the vent, though some birds suffer from a general distribution of mites over the body. Infestation appears to be higher on birds with genetically low plasma corticosterone levels, and initial high levels of infestation on young birds are reduced naturally owing to the development of antibodies. Mite levels and their effect on egg production seem related to dietary protein levels and strain of birds (Hall and Gross, 1975; Matthyse *et al.,* 1974).

Macronyssid mites in the genus *Ophionyssus* infest snakes. Their numbers appear to increase on captive snakes, causing anemia or death (Baker *et al.,* 1956). Infestation of man by these snake mites is known (Schultz, 1974).

Dermanyssus gallinae, the chicken mite (red chicken mite or roost mite), is a pest of chickens throughout the world (Fig. 15-3). Other poultry such as turkeys, as well as pigeons, English sparrows, and other birds may be infested, though some bird infestations attributed to this mite have no doubt been due to other species of *Dermanyssus* (Moss *et al.,* 1970). In daytime the mites hide in crevices of poultry houses and roosts, under boards and debris. In these hiding places they deposit their eggs. At night the

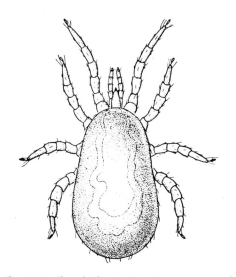

Fig. 15-3 The chicken mite, *Dermanyssus gallinae.*

mites leave their hiding places and attack birds on the roosts. Damage is considerable and may be summarized in heavy infestations as follows: Egg production is greatly reduced, setting hens may be forced to leave their nests or perish, newly hatched chicks die, chickens are unthrifty and unprofitable for marketing, and loss of blood and reduced vitality cause increased susceptibility to disease. The chicken mite survives starvation for as long as 34 weeks.

Chicken mite infestation of pigeons is quite common, and when these birds nest near human dwellings attacks on humans may follow. Abundant mites leave nests to enter by way of windows, attacking man and sometimes causing pruritus (Sexton and Haynes, 1975).

Bird and poultry mites have been found with the viruses of western equine and St. Louis encephalitides. Laboratory and epidemiological studies suggest that mites play no important part in the maintenance of these pathogens, and field isolations probably mean that these ectoparasites had fed recently on viremic birds.

The house mouse mite, *Liponyssoides* (=*Allodermanyssus*) *sanguineus* (Fig. 15-4), is known to occur in northern Africa, Asia, Europe, and the USA. Ewing (1929) provides a description of the species. The house mouse, *Mus musculus,* is the preferred host, but the mite will feed on rats and other rodents and readily attacks humans. This mite is important because of its transmission of the rickettsialpox pathogen to man.

A good summary of the life history of the house mouse mite is given by Baker and co-authors (1956). There are five developmental stages, but unlike the tropical rat mite, both nymphal instars feed on blood. Adult females feed often, each feeding followed by oviposition. From deposition of the egg to emergence of the adult takes 17 to 23 days. Unfed females have lived as long as 51 days, and a female that fed and oviposited twice lived 9 weeks. Engorged nymphs and adults

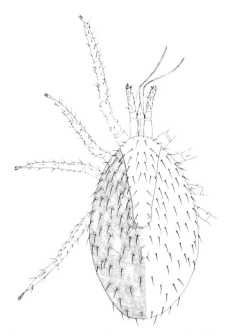

Fig. 15-4 *Liponyssoides sanguineus,* the house mouse mite. Dorsal view of male. (From Baker *et al.,* 1956, *A Manual of Parasitic Mites of Medical or Economic Importance;* courtesy of National Pest Control Association.)

Order Acariformes;
Suborder Prostigmata

PROSTIGMATID MITES

This is a very diverse group, suggesting that in reality it consists of an amalgamation of several subordinal entities. Many prostigmatids have a pair of stigmata at the bases of the chelicerae. They are typically weakly sclerotized in comparison with parasitiform mites. Sexual dimorphism is often obscure.

Family Pyemotidae. Pyemotids are soft-bodied mites with greatly reduced mouth parts; chelicerae tiny, styletlike, and palpi reduced, lying close to the rostrum; third and fourth pair of legs separated from the first and second by a long interspace; sexual dimorphism usually marked. There are several species of *Pyemotes* mites (Fig. 15-5), but the one most often causing dermatitis in man is the straw itch mite, *Pyemotes tritici* (Moser, 1975). This mite commonly attacks a variety of stored grain insects and is highly

may occur in great numbers in buildings that are near rodent nests and runways.

IDENTIFICATION OF DERMANYSSID MITES. To identify dermanyssid mites the student is referred to Baker and associates (1956), Strandtmann and Wharton (1958), Evans and Till (1966), and Radovsky (1967). Moss (1968) provides an illustrated key to females of 13 species of *Dermansyssus.*

Family Macrochelidae. *Macrocheles muscaedomesticae,* the house fly mite, is of interest because the nymphs and adults prey upon the eggs and first-stage larvae of the house fly. There may be appreciable natural control of fly populations by this predator, and there have been a number of attempts to utilize it in fly control by management of the environment in which flies are developing, or by releases of the mite.

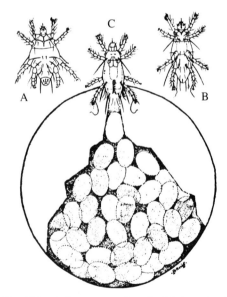

Fig. 15-5 A *Pyemotes* mite. *A.* Adult male. *B.* Adult female. *C.* Gravid female, showing developing eggs which will hatch within the abdomen.

toxic to man, whereas *Pyemotes ventricosus,* which is commonly believed to affect man, is associated with the furniture beetle *Anobium punctatum* and causes little or no human irritation (but see Scott and Fine, 1967). Other *Pyemotes* attacking bark beetles cause no observable symptoms in man. *Pyemotes boylei* parasitizes termites and other crop and stored product pests in Hawaii, causing dermatitis in people handling products infested by its insect hosts, or in buildings fumigated to destroy termites (Vaivanijkul and Haramoto, 1968).

Dermatitis associated with *P. tritici* is known as straw, hay, or grain itch (Fig. 15-6). The male mite is very tiny, barely visible to the unaided eye; the female when gravid becomes enormously swollen, mea-

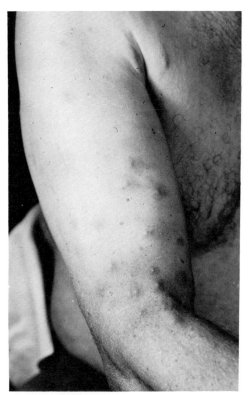

Fig. 15-6 Dermatitis caused by *Pyemotes.* (Courtesy of R. V. Southcott and *Records of the Adelaide Children's Hospital.*)

suring nearly a millimeter in length, the abdomen presenting a globular appearance and resembling a tiny pearl. Within the enlarged abdomen of the female may be found rather large eggs that hatch internally, and the young mites develop to maturity within the body of the mother before being extruded. The number of offspring per female is 200 to 300. Males emerge first and remain clustered around the genital opening of the mother, fertilizing young females as they emerge.

Many epidemics of dermatitis have been traced to these mites, and because the irritation is temporary many additional cases are undoubtedly not reported. Infestation occurs after sleeping on straw mattresses, laboring in grain fields at harvest time, or otherwise coming in contact with grains, straw, hay, grasses, beans, peas, cowpeas, cottonseed, or other materials infested with insect larvae that these mites attack. As many as a thousand wheals appear 12 to 16 hours after attack, though the inflammation caused by these varies with individuals. Each wheal surrounds a vesicle marking the site of puncture, and if the vesicle is ruptured, secondary infection may arise. Headache, anorexia, nausea, vomiting, mild diarrhea, and joint pains may occur in severe cases.

Family Tetranychidae. These are the spider mites or "red spiders," web-spinning mites most commonly infesting vegetation and destructive to various crops. Persons employed in picking hops and harvesting almonds, and other crops, often complain of itching caused by the "red spider," but this itching soon disappears.

In the fall with the advent of cold weather the clover mite, *Bryobia praetiosa* complex, will invade homes, sometimes in great numbers. There may be considerable annoyance, and there are records of this species parasitizing man in Argentina (cf. Ebeling, 1975).

Family Cheyletidae. These are primarily predators, but included are several species that parasitize mammals and wild birds; gnathosoma usually conspicuous, palpi usually well developed; stigma opening at base of chelicerae (Fig. 15-7).

Dermatitis in rabbits, cats, dogs, and people associated with these animals has been attributed to *Cheyletiella parasitivorax* (Fig. 15-8). It was originally believed that this mite is a predator on other mites in fur, and it has been noted to occasionally be attached to the rabbit flea *Spilopsyllus cuniculae,* suggesting it may use an ectoparasite as a means of transfer between hosts (Rothschild, 1970). Reports of human dermatitis caused by *C. parasitivorax* associated with cats and dogs are probably erroneous. It now appears that the *Cheyletiella* of dogs is *Cheyletiella yasguri* and that of cats is *Cheyletiella blakei,* and both species cause a scurfy pruriginous dermatitis (mange) on their normal hosts as well as an itching dermatitis on humans who handle these pets; *C. parasitivorax* is associated with rabbits (Guilhon *et al.,* 1973). Although infestation of humans

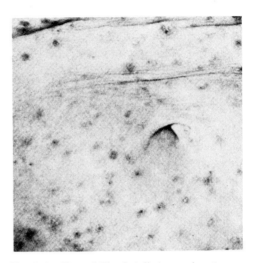

Fig. 15-8 Bites of *Cheyletiella* in navel region on abdomen of a person who acquired a new puppy. (Courtesy of R. V. Southcott and *Records of the Adelaide Children's Hospital.*)

is believed to be temporary, and only when humans are in contact with infested pets, there is a report of a persistent human anal pruritus caused by larval *Cheyletiella,* and chronic urticaria of a woman who had no pets yet harbored *Cheyletiella* and other mites (Hewitt and Turk, 1974; Mougeot and Poirot, 1975).

Family Psorergatidae. These mites infest the skin surface of hosts, causing considerable irritation. *Psorergates ovis,* the itch mite of sheep, is a serious pest in the USA and Australia, and occurs in many major sheep-raising countries. The adult is the only mobile stage and is responsible for spread of the infestation, chiefly from sheep to sheep. Spread is generally slow. The mite is very susceptible to desiccation, which explains the decline reported after shearing and in summer. Characteristic fleece derangement due to this mite in Australian and New Zealand flocks varied from 1 to 10 percent of the sheep (Sinclair, 1975). A related species, *Psorergates simplex,* causes skin injury to laboratory mice. *Psorergates bos* has been described as a parasite of cattle in the state of New Mexico, USA (Johnston, 1964).

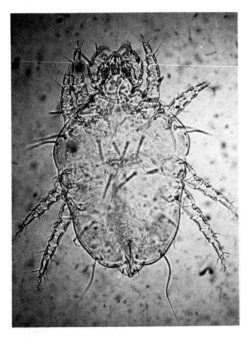

Fig. 15-7 A *Cheyletiella* mite. (Courtesy of R. V. Southcott and *Records of the Adelaide Children's Hospital.*)

Family Demodicidae. Follicle mites; very minute (0.1 to 0.4 mm long) with an elongated transversely striated abdomen and four pairs of stubby, five-segmented legs in a group on the anterior body; parasitize the skin of mammals. All stages of the life history are often present together in a follicle. Smith (1961) has reviewed the morphology, bionomics, world distribution, prevalence, and pathogenicity of *Demodex* infesting large domestic animals. Nutting (1976) reviews gross signs and histological details of *Demodex* infestations in mammalian hosts, noting that these mites are highly host specific, commonly with two species in close association but in different tissues. These mites may penetrate beneath the skin and infest various organs, and subdermal tissues where there is a granulomatous response.

Follicular mites of man, once believed to be a single species, are now known to consist of an elongate species, the follicle mite, *Demodex folliculorum*, that lives in hair follicles, and a stubby species, *Demodex brevis*, that inhabits the sebaceous glands (Akbulatova, 1970; Desch and Nutting, 1972). Human follicle mites are mainly in the region of eyelids, nose, and other facial areas, and the proportion of infested persons varies from about 20 percent of persons age 10 to 20 to nearly 100 percent in the aged. Though infestation is usually benign, there may be blepharitis (loss of eye lashes) and granulomatous acne, in the latter case particularly when mites leave their usual sites in follicles and sebaceous glands to penetrate the dermis (Grosshans *et al.,* 1974).

Demodectic mite infestations may be quite serious, even fatal, in dogs. The species involved in red mange or canine demodectic mange is the dog follicle mite, *Demodex canis* (Fig. 15-9), typically found in association with *Staphylococcus pyogenes albus* or some allied bacterium, which seems to be the real cause of hair loss. The lesions occur chiefly around the muzzle and eyes and on the forefeet. A high proportion of dogs har-

Fig. 15-9 *Demodex canis.* Ventral view of female. Very similar in general appearance to the human follicle mite, *Demodex folliculorum.* (From Baker *et al.,* 1956, *A Manual of Parasitic Mites of Medical or Economic Importance;* courtesy of National Pest Control Association.)

bor demodectic mite with no symptoms, and it appears to be more prevalent in older animals and those with long hair. Mange reactions may be due to a lowering of the immune response in animals (Owen, 1972). Diagnosis may be easier by finding the mites through fecal flotation than by the usual method of making skin scrapings (Derwelis, 1967).

Demodex bovis, the cattle follicle mite, parasitizes cattle. Swellings are produced that may be as large as a hen's egg and that are filled with a cheesy or fluid substance containing the mites. *D. bovis* is transferred from the cow to her calf within the first few days following birth. The mite-filled nodules may produce holes in hides, thus lessening their value. Nodular damage to hides may be as high as 20 percent in African cattle herds, and a study in a midwestern USA abbatoir found 82 to 93 percent of hides to have some level of infestation.

Demodex phylloides, the hog follicle mite, produces nodular lesions in swine; the goat follicle mite, *D. caprae,* may form pustules in goats; the horse follicle mite, *D. equi,* causes a mild mange of horses, as does the cat follicle mite, *D. cati,* of cats. *D. cervi* infests the hair follicles of deer. *D. criceti* and *D. aurati* infest hamsters and may cause mange in colonies of experimental animals; there are other *Demodex* species infesting various rodents.

Family Erythraeidae. These are red or reddish-brown mites that are predators of small arthropods. Members in the genus *Balaustium* are sometimes involved in annoying biting attacks on man (Newell, 1963).

Family Trombiculidae. Chigger mites are unique among acarines affecting man and animals in that the larval stage is the only

Fig. 15-10 *Trombicula autumnalis.* Dorsal and ventral view of larva; details of chelicera and palp, and scutum. (From Baker *et al.,* 1956, *A Manual of Parasitic Mites of Medical or Economic Importance;* courtesy of National Pest Control Association.)

one feeding on vertebrates; consequently it is this stage that is normally seen. The larval characteristics can mostly be seen in Figure 15-10. The larval body is more or less rounded. Genera of medical and veterinary significance have a single dorsal plate or scutum that bears a pair of sensillae and four to six setae. The palpal tibia has three setae and a terminal claw; palpal tarsus usually bears four to seven setae. The palpal tarsus articulates and opposes the tibial claw in a thumb-like fashion. The dorsal body setae posterior to the scutum are usually plumose; similar setae are found on the venter between and posterior to the coxae. Each coxa and trochanter bears at least one plumose seta. Nearly 3,000 species have been described.

Adults are about 1 mm long, oval or more usually figure-eight-shaped (Fig. 15-11), clothed with filiform, densely pilose setae, giving a velvety appearance. The color is often bright red.

GENERALIZED CHIGGER LIFE CYCLE. The development cycle of chiggers (Fig. 15-11) has been reviewed by Sasa (1961). Eggs of chiggers are globular; in about a week the egg shell splits, exposing the maturing larva or DEUTOVUM. The larval stage is six-legged and is the only parasitic stage. Larvae may attach to any of a great variety of vertebrates, or those of some species are very specific (a few exceptional species feed on arthropods or other invertebrates). The engorged larva leaves the host and passes a quiescent NYMPHOCHRYSALIS or prenymph stage. Nymphs and adults are eight-legged and free-living. A nymph passes a quiescent IMAGOCHRYSALIS or preadult stage before molting to the adult. Some chigger nymphs and adults have been reared on eggs of mosquitoes or Collembola, and it is likely their normal food consists of insect eggs or small soft and inactive soil invertebrates. Females are inseminated by means of stalked spermatophores deposited by males on the substrate; there is no evidence of parthenogenesis.

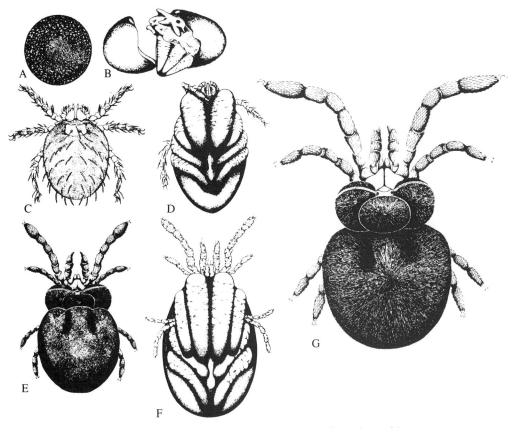

Fig. 15-11 Life cycle of scrub typhus chigger mite, *Leptotrombidium akamushi*. *A.* Egg. *B.* Deutovum with ruptured shell exposing developing larva. *C.* Larva, the stage parasitic on vertebrates. *D.* Prenymph, ventral view, showing remnants of larval legs. *E.* Nymph. *F.* Preadult, ventral view, showing remnants of nymphal appendages and developing legs of adult. *G.* Adult. (From Neal and Barnett, 1961, *Ann. Entomol. Soc. Am.,* **54**:196–203.)

Developmental time is greatly affected by temperature and food supply. Under laboratory conditions at 28° C *Leptotrombidium akamushi* takes about 75 days to complete the life cycle; at "favorable" laboratory conditions *Euschoengastia indica* takes about 60 days, *Leptotrombidium deliense* 40 days; the common North American chiggers require 50 to 70 days.

The number of generations per year also varies with species and local climate. Most temperate zone chiggers have one annual generation. *Trombicula alfreddugesi* in North America has one or two generations in

Ohio, three generations in North Carolina, and continuous development in Florida.

Typical host-seeking behavior for several species of larval chiggers consists of congregating in a shaded niche near the top of an object in close contact with the earth, such as a blade of grass or a fallen leaf. Here they remain quiescent until activated by an air current bearing carbon dioxide, as occurs with the approach of a vertebrate host. Some species may remain on the ground surface, and a few in rodent burrows, until a host arrives.

Wharton and Fuller (1952) divide the

Trombiculidae into four subfamilies, of which most chiggers belong to the Trombiculinae; this is still the most commonly accepted division. This subfamily consists of trombiculid larvae with a median scutal seta; no submedian scutal setae; seven segments in all legs (but note that Brennan and Goff, 1977, are more inclusive regarding number of leg segments); at least four sternal setae, no median, anterior projection on the scutum and no stigmata or trachael trunks. The taxonomy of chiggers presents some difficult problems (cf. Fuller, 1956). Chief among these is the fact that many chigger mites have been described from the larval stage only, and adults and nymphs are either unknown or uncorrelated with larvae.

Literature pertaining to the Trombiculidae, their taxonomy and medical importance, has grown enormously in recent years. For identification and taxonomic studies the student should start with a manual by Wharton and Fuller (1952) that carries these mites through keys to genera and subgenera, but includes detailed bibliographical, distributional, and host information for species. Brennan and Goff (1977) provide keys to genera in the Western Hemisphere. Fuller (1956) cites a number of references to taxonomic papers in the Pacific area and Asia. Vercammen-Grandjean (1963) reviews taxonomic characters of the Trombiculidae and includes correlations between larvae and nymphs. For determination to species the student is referred to keys by Brennan and Jones (1959), to the studies of Jenkins (1949) and Brennan and Wharton (1950), respectively, for the North American species of the subgenera *Eutrombicula* and *Neotrombicula* of the medically important genus *Trombicula,* and to Farrell (1956) for the genus *Euschoengastia*. Representative regional treatments include the chiggers of Kansas (Loomis, 1956) and California (Gould, 1956).

CHIGGER DERMATITIS. It is important to note that not all chiggers cause itch reactions, and indeed those serving as vectors of chigger-borne rickettsiosis (=scrub typhus) are not associated with noticeable itching or skin reactions (cf. Traub and Wisseman, 1974). However, in many parts of the world, particularly in warmer regions and during late summer months in temperate climates, persons who have been outdoors in vegetated areas suffer an intolerable itching, beginning 3 to 6 hours after exposure and followed by a severe dermatitis consisting of pustules and wheals. A careful examination of itching areas of the skin after a trip through tall weeds, grass, or berry brambles will reveal minute red mites, either traveling fast or attached to the skin.

Chigger larvae do not burrow into the skin as commonly believed, nor do they primarily feed on blood. There are reports of some species' partaking of blood, but mainly their food consists of serous elements of tissue, though a few blood cells may be ingested during feeding. When firmly attached, chiggers inject a digestive fluid that causes disintegration of cellular contents and yields disorganized cytoplasm and fragmented nuclei (Jones, 1950). The resulting material is utilized as food. The skin of the host becomes hardened, and a tube (STYLOSTOME) forms in which the mouth parts remain until the mite is replete; then it loosens and drops to the ground. It is presumably the action of the digestive fluid that causes the attachment site in some species to itch after a few hours. Histological preparations of chigger bites on rabbit ears show the epidermis is completely penetrated (Williams, 1946). A tube lined with stratum germinativum forms and extends to the derma and subcutis. This tube represents a combined reaction of the host to the secretion of the chigger, and the secretory material itself (see Chapter 3); its inner layer of cells is necrotic and shows digestive action.

IMPORTANT DERMATITIS-CAUSING CHIGGERS. Among the chigger mites that cause severe dermatitis is the European species

Trombicula (Neotrombicula) autumnalis (Fig. 15-10) known as the harvest mite, *aoutat,* or *lepte automnal.* It differs from American chiggers of the subgenus *Eutrombicula* in that its larvae have trifid rather than bifid claws.

Trombicula (Eutrombicula) alfreddugesi is the common chigger of the continental USA, ranging from the Gulf states and similar ecological sites in New England and eastern Canada to Nebraska, west to California; Texas and south into Mexico, Central America, and South America. It is known by a wide range of names including *tlalzahuatl, bicho colorado,* and *bete rouge.* It is abundant in second-growth cut-over areas, especially wild blackberry patches, forest edges, and river valleys. Its hosts include man, a wide range of domestic and wild animals, birds, reptiles, and even a few amphibians.

Trombicula (Eutrombicula) splendens is distributed in the USA along the Atlantic Coast from Florida to Massachusetts, and the Gulf Coast to Texas, then northward to westward in suitable habitats to Ontario, Michigan, and Minnesota. It is the most abundant species in Florida and parts of Georgia, occurring in moister habitats than does *alfreddugesi,* such as swamps, bogs, rotten logs, and stumps. Although *splendens* and *alfreddugesi* may occur in the same region, seasonal incidence of the two are independent.

Trombicula (Eutrombicula) batatas is tropical and found in Florida, Georgia, Alabama, Kansas, and California. It seldom attacks man in the USA but commonly does so in Panama and other tropical areas, and it may cause intense itching on chickens and turkeys (Confalonieri and Carvalho, 1973).

Wharton and Fuller (1952) report additional species attacking man and producing dermatitis as chiggers in the genera *Trombicula, Euschoengastia, Schoengastia, Apolonia,* and *Acomatacarus* from the Americas, Pacific Islands, Australia, and Indonesia.

Chiggers affect vertebrates other than man. Species such as the turkey chigger, *Neoschoengastia americana,* cause losses to turkey poults and quail, and probably other birds as well. This mite causes downgrading of market turkeys due to lesions in the skin and adjacent tissues, particularly in the southeastern USA. Everett and associates (1973) have reared this mite and provide a complete account of its biology and life cycle. In Texas, USA, daily variations in rainfall and temperature correlated with the number of skin lesions greater than 3 mm diameter on turkeys, and the number of lesions per bird could be suppressed by about 50 percent through the sprinkling of water on the rearing area (Cunningham *et al.,* 1976).

Euschoengastia latchmani caused mange-like dermatitis on horses recently imported into the state of California from Washington, and lesions on hares (*Lepus californicus*) and a goldencrowned sparrow (*Zonotrichia coronata*) (Brennan and Yunker, 1964). Another *Euschoengastia* species has caused occasional lesions on people and horses in Oregon, USA (Easton and Krantz, 1973).

IDENTITY AND BIOLOGY OF SCRUB TYPHUS VECTORS. Information regarding all aspects of chigger-borne rickettsiosis or typhus (=scrub typhus) is found in the comprehensive review of this disease by Traub and Wisseman (1974), and this discussion is derived from that source. Wherever outbreaks of the disease occur, species in the genus *Leptotrombidium* are the vectors. (These were formerly called the subgenus *Leptotrombidium* in the genus *Trombicula*). Within this genus the probable vectors to man are all in what is referred to as the *Leptotrombidium deliense*-group. Other genera may be involved in transmission to rodents because of physiological and behavioral restriction to these mammals, but prevailing admittedly incomplete evidence for enzootic transmission also involves species of *Leptotrombidium*.

Leptotrombidium akamushi is the vector

in classical areas of this disease on northwest Honshu, Japan. *Leptotrombidium deliense* is the main or an important vector from New Guinea and the coastal fringe of Queensland, Australia, the Philippines and China, westward through Southeast Asia to West Pakistan (Fig. 15-12). *L. fletcheri,* erroneously considered *L. akamushi* in previous literature, is an important vector in Malaya, Borneo, New Guinea, and the Philippines. *L. arenicola* (=*Neotrombicula arenicola*) is a vector along sandy beaches in Malaya. *L. pallidum* is a vector in limited locations in certain parts of Japan, Korea, and the Primorye Region, USSR. *L. pavlovskyi* is considered the main vector in Siberia and the Primorye Region, USSR. *L. scutellare* is regarded as a fall and winter vector in the Mt. Fuji area, Japan, and on certain Japanese islands. The species *Leptotrombidium palpale, tosa, orientale,* and *fuji* have also been mentioned as possible vectors, but the evidence is not as strong as that for species listed above. There is little sound evidence

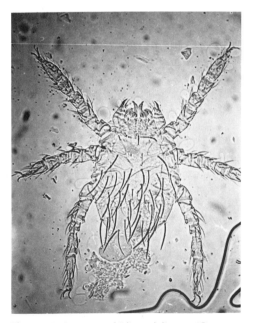

Fig. 15-12 *Leptotrombidium deliense.* (Courtesy of R. V. Southcott and *Records of the Adelaide Children's Hospital.*)

that other genera of Trombiculidae are natural vectors.

Leptotrombidium akamushi inhabits partially cultivated land that becomes inundated by spring and early summer floods, the mite reaching a peak during July and August. *L. deliense* is more characteristically associated with forests.

The adult mites are 1 to 2 mm long and generally reddish in color. In cultures they feed readily on eggs of Collembola and mosquitoes. The winter is spent as adults, except in equatorial climates where there is no hibernation. Eggs are deposited on the soil under leaves or in damp places. The hexapod larvae, at first about 0.22 mm long, emerge in 10 to 12 days at 30° C and wait the coming of a suitable mammal or bird (man is not a usual host). They attack in clusters, often packed in the ears of a field rodent. Numerous hosts are available, the chiggers being habitat- rather than host-specific (Audy, 1958). Rats in the genus *Rattus* are frequent hosts, though this does not include commensal species or forms that are closely associated with human habitations.

Within a generally suitable habitat there will be specific zones (''islands'') from a square meter to fairly extensive areas where chiggers, and particularly infective vector chiggers, are found. General habitat consists of ecologically disturbed vegetation, whether created by human activities such as forest and brush clearing during lumbering or farming, or by natural events such as flooding, landslides, or fire. The vegetation in such zones is transitional, and accordingly the presence of rodents and chiggers is unstable and often changes appreciably in a few seasons; there may also be marked seasonal fluctuations in chigger distribution.

The larval mite (''*akamushi*'') measures about 0.55 mm when fully engorged. After a feeding period of 1 to 10 days it drops to the ground. Some time is spent seeking suitable shelter, and after passing through the inactive nymphochrysalis stage, the predacious nymphal stage, and the inactive imagochrysalis stage, the adult emerges. The life cycle in tropical species takes 2 to 3 months; montane or cold weather species may take 8 months or more. Adults lay eggs singly on the substrate, such as on the soil surface. About 1 to 5 eggs are laid daily for 6 to 12 weeks; there is then no oviposition for a similar period, after which oviposition is resumed for some weeks. Females deposit about 400 eggs over a 3- to 5-month period, and adults can live 15 months or longer.

The occurrence of *Leptotrombidium* chiggers coincides with high humidity. In semi-arid regions these mites are limited to specific canal zones or river bottoms, and in montane zones they are found along stream borders or marshy sites. Clusters of larvae on leaves or twigs, awaiting a vertebrate host, are often in shade and in close association with high humidity at the soil level.

Order Acariformes; Suborder Astigmata

ASTIGMATID MITES

Astigmatids are slow-moving and weakly sclerotized mites. They lack true claws but may have suckerlike or clawlike structures. Respiration is integumental.

Families Acaridae, Glycyphagidae, and Carpoglyphidae. Mites in these families, along with some members of the family Pyroglyphidae, are stored product pests that cause skin reactions and lung or intestinal infections in man and animals. They are very tiny mites, ordinarily about 0.5 mm or less in length. Food habits can generally be described as scavenging, involving grain, flour, meal, dried meat, hams, dried fruits, dried insects, and similar products. Development is so rapid that literally millions of these mites may appear in a stored product in a few days. Persons handling infested products may experience a severe patchy or coalescing dermatitis. It is uncertain whether

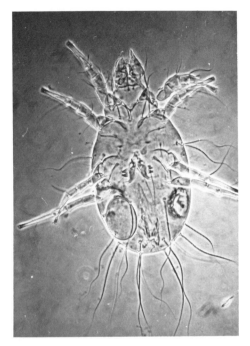

Fig. 15-13 Grocer's itch mite, *Tyrophagus putrescentiae*, female with egg. (Courtesy of R. V. Southcott and *Records of the Adelaide Children's Hospital*.)

the reaction is caused by mite bites or by simple contact allergy, though there is little doubt some of these mites actually bite.

This type of itch is known as OCCUPATIONAL ACARINE DERMATITIS (cf. Southcott, 1976; Sheals, in K. G. V. Smith, 1973). Grocer's itch or copra itch is caused by the mold mite, *Tyrophagus putrescentiae* (=*Tyroglyphus longior* var. *castellani*) (family Acaridae), which also affects dock workers who handle cheese (Fig. 15-13); grocer's itch by the house mite, *Glycyphagus domesticus* (fam. Glycyphagidae); baker's itch by the grain mite, *Acarus siro* (=*Tyrophagus farinae*) (fam. Acaridae) (Fig. 15-14), also "vanillism" among vanilla pod handlers; wheat pollard itch by the scaly grain mite, *Suidasia nesbitti* (fam. Acaridae), and dried fruit dermatitis by the dried fruit mite, *Carpoglyphus lactis* (fam. Carpoglyphidae). Dermatitis is also reported

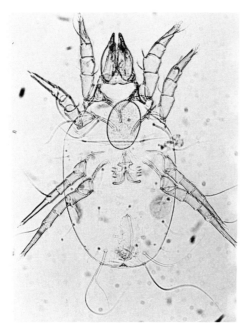

Fig. 15-14 *Acarus siro,* causal agent of baker's itch. Female with egg visible internally. (Courtesy of R. V. Southcott and *Records of the Adelaide Children's Hospital.*)

from *Rhizoglyphus hyacinthi* (fam. Acaridae) (probably referring to the bulb mite, *R. echinopus*), a mite associated with certain flower and onion bulbs.

The metamorphosis of this group may involve a peculiar stage known as the HYPOPUS, appearing in a number of species between the two nymphal stages, very unlike either of these and very different from the adult. There are two types of hypopi; the active type is phoretic and adapted for clinging to arthropods or mammals, the inactive type (DAUERNYMPH) relies on air currents for dispersal or passively awaits more favorable conditions.

Baker and coauthors (1956) and Southcott (1976) have summarized reports of mites involved in human pulmonary, urinary, and intestinal acariasis, and a review by Morini and Roveda (1974) includes effects on animals. *Tyrophagus longior,* a free-living form, has been reported from both the digestive and urinary tract, and *T. putrescentiae* from the urinary tract. Free-living astigma-

tids have been recovered from the sputum of patients suffering from lung disorders. Infestations by these free-living mites appear for the most part to be accidental and temporary in nature. Feed heavily infested with *Acarus siro* and *Tyrophagus putrescentiae* did not appear to adversely affect mice, pigs, or calves (Gesztessy and Nemeseri, 1970).

Family Epidermoptidae. Very small mites, 0.17 to 0.39 mm long, oval to nearly circular shape and flattened; integument soft and striated; tarsi with caruncles. The family has been reviewed by Fain (1966). *Epidermoptes bilobatus* causes a generalized dermatitis in fowl.

Family Laminosioptidae. *Laminosioptes cysticola* is found in North America, South America, Europe, and Australia. This mite causes tiny white caseocalcareous nodules in the skin following death, and this condition results in downgrading of poultry carcasses; heavy infestations may cause death of the host. The lesions have been reported from chickens, turkeys, pheasants, geese, and pigeons, and one study in midwestern states, USA, found up to 0.5 percent of chickens affected (Cassidy and Ketter, 1975).

Family Psoroptidae. Mites of the family Psoroptidae attack domestic animals in a variety of ways. They do not burrow like sarcoptids; rather they live at the base of host hairs, piercing the skin and causing inflammation. Psoroptids are distinguished from sarcoptids by propodosoma lacking vertical setae, though a dorsal shield may be present. The important genus *Psoroptes* can be distinguished from *Sarcoptes* and from other genera of Psoroptidae by tarsal caruncles with long segmented stalks. Legs of Psoroptidae are longer than those of the Sarcoptidae.

Mites of the genus *Chorioptes* produce chorioptic mange, which is restricted to certain parts of the body, such as the feet, tail, and neck. The question has arisen as to whether a number of separate species of *Chorioptes* are present, or whether there is mere physiological adaptation to specific hosts. In a careful study, in which *Chorioptes* were raised on epidermal scrapings of various hosts, Sweatman (1957) concluded that practically all previously described species are one and the same. Domestic cattle appear to be the host species responsible for its cosmopolitan distribution. The accepted name is *Chorioptes bovis* (Fig. 15-15); synonyms are *C. equi* of horses, *C. caprae* of goats, *C. ovis* of sheep, and *C. cuniculi* of rabbits. In vial rearings this mite fed on debris and not hair; the presence of eggs from one female attracted others to lay so that communal groups were formed. Important sites of infestation are the feet and lower hind legs of horses and cattle, and high levels have been associated with a severe decrease in milk production (Diplock and Hyne, 1975). A survey of 40 flocks of sheep in eight widely separated states in the USA found *C. bovis* present in all, and in more than half the flocks it was the commonest arthropod parasite, yet mangy skin lesions were evident in only two animals. Scrotal mange of rams, due to *C. bovis,* could on occasion cause seminal degeneration (Rhodes, 1975).

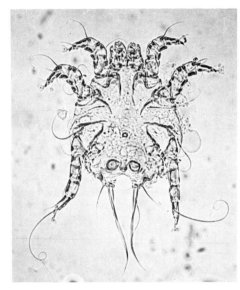

Fig. 15-15 *Chorioptes bovis,* causal agent of chorioptic mange affecting cattle and other livestock and rabbits. Male; note suckers. (Courtesy of R. V. Southcott and *Records of the Adelaide Children's Hospital.*)

Psoroptic scab mites, genus *Psoroptes* (Fig. 15-16), differ from *Sarcoptes* in that all four pairs of legs are long and slender and extend beyond the margin of an elongate

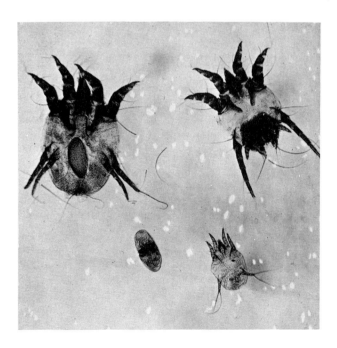

Fig. 15-16 Showing life history and general characteristics of the scab mite, *Psoroptes equi.* (*Lower left*) Egg; (*lower right*) larva; (*upper left*) adult female; (*upper right*) adult male. × 85.

body; pedicel of the caruncles segmented; chelicerae styliform, serrate near tip. They pierce the skin and cause inflammation that results in exudation that partially hardens to form a scab. As mites multiply, bites and itching increase; more serum oozes to form a loose humid crust. Mites are most numerous under the outer edges of the scabs. Owing to the looseness of the scabs and hardiness of the mites, this form of acariasis is highly contagious by contact and by rubbing against fences, trees, and the like. Outbreaks of psoroptic mange in cattle or sheep are often quarantined, shipping restrictions are applied, and stringent control programs are necessary to prevent further outbreaks.

In a comprehensive review Sweatman (1958) concludes that *Psoroptes caprae* of goats, and *P. hippotis* are synonyms of *P. cuniculi* of the rabbit, and that *P. bovis* is a synonym of the sheep scab mite, *P. ovis*, a cosmopolitan body mite of sheep, cattle, and horses. *P. natalensis* occurs on domestic cattle, zebu, Indian water buffalo, and horses from South Africa, Uruguay, Brazil, New Zealand, and probably France; the scab mite, *P. equi*, is a body mite of the horse and probably other equines, arbitrarily restricted to specimens from England. Although Sweatman states that *P. cervinus* is distinctly an ear mite of bighorn sheep (*Ovis canadensis*), it was later found in scrapings from the skin of elk (*Cervus canadensis*), and when in abundance on that host may be responsible for psoroptic mange believed to contribute to the death of the animals. White-tailed deer (*Odocoileus virginianus*) can also serve as a host for *Psoroptes* mites (Strickland *et al.*, 1970).

Unlike the sarcoptic mange mite, scab mites on sheep infest parts of the body most thickly covered with wool. Psoroptic scab is indicated by a "tagging" of the wool; the coat becomes rough, ragged, and matted at the points affected (Fig. 15-17). Tags of wool are torn away by the sheep or rubbed off at posts and other objects, and the ani-

Fig. 15-17 *Psoroptes* scab mange of sheep. (USDA photograph; courtesy of W. P. Meleney.)

mals display signs of intense itching. Contact with infested sheep spreads the infestation. Numbers of sheep scab mites are generally highest in winter, but a basic population may be maintained in summer anywhere on the body surface of the animals. There appear to be strains of this mite on sheep and cattle that vary in aggressiveness and in susceptibility to acaricides (Roberts *et al.*, 1971).

A fairly common infestation of cats, dogs, ferrets, and foxes, known as otoacariasis or *parasitic otitis*, is caused by *Otodectes cynotis*, which resembles *Psoroptes* very closely (Fig. 15-18). These mites literally swarm in the ears of the host, causing tenderness of the ears, auricular catarrh, wasting, twisting of the neck, and "fits." Infestations may involve additional areas on the head, including the area around the eyes (Baker, 1974), and other parts of the body are sometimes infested. Varietal names of this species have no taxonomic status. Infestation is significantly correlated with semierect ear conformation in dogs, and the mites are not found where ear inflammation has become highly suppurative (Grono, 1969).

Family Pyroglyphidae. *Dermatophagoides* mites (Fig. 17-2) were placed in the Psoroptidae (Fain, 1967), but are now considered in the family Pyroglyphidae. These mites do not feed directly on living tis-

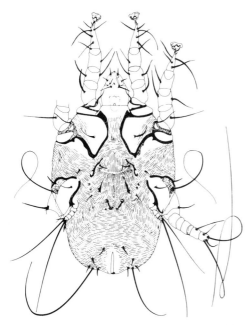

Fig. 15-18 *Otodectes cynotis,* ventral view of female. (From Baker *et al.,* 1956, *A Manual of Parasitic Mites of Medical or Economic Importance;* courtesy of National Pest Control Association.)

sues, but they consume skin scales and similar detritus and may therefore be in close association with man and other animals. *Dermatophagoides saitoi* has been associated with lung disorders. *Dermatophagoides scheremetewskyi* is in close association with bats, rodents, and sparrows in the eastern USA. Human cases of infestation are sometimes accompanied by severe and persistent dermatitis, as summarized by Baker, and colleagues (1956) for the period prior to 1956. Several *Dermatophagoides* and related pyroglyphids are associated with house dust allergy, a condition discussed in Chapter 17.

Family Sarcoptidae. Sarcoptidae are known as the sarcoptic itch or scabies mites. They are at the limit of human visibility, somewhat hemispherical, whitish, and skin parasites of mammals. The PROPODOSOMA (part of body bearing the two fore pairs of legs) is not separated from the HYSTEROSOMA

(posterior body) by a suture; they frequently bear a propodosomal shield and always bear a pair of vertical setae. In other areas the skin has fine striae that may be interrupted by scaly areas or surfaces bearing small points or spines. Legs are very short, claws or caruncles may be present or absent.

The Family Sarcoptidae includes the genera, *Sarcoptes, Notoedres,* and *Trixacarus,* each producing a particular type of dermatosis. Several other genera lack medical or veterinary importance.

Sarcoptes mange or itch mites have very short legs, the posterior pair not extending beyond the margin of the nearly circular body (Fig. 15-19); caruncles with nonsegmented pedicels on first and second pairs of legs. Sarcoptic mites form definite burrows in the skin in which the females deposit eggs. The adults hold onto skin by means of suckers on the anterior legs, raise the hind end of the body on bristles of the posterior legs until nearly perpendicular, cut in and completely disappear from the surface in 2.5 minutes. A fertilized female burrows into the epidermis and deposits eggs in the tunnel behind her. The eggs hatch in 3 to 8 days, and the larvae migrate to the skin surface and

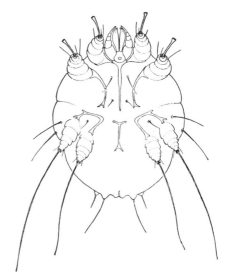

Fig. 15-19 The itch mite, *Sarcoptes scabiei.*

molt twice, forming two nymphal stages and finally adults. The adult stage is reached 4 to 6 days after egg hatch and and the entire life cycle takes 10 to 14 days. Mating occurs on the skin surface; eggs and debris are left within the burrows.

There has been considerable speculation as to whether the scabies mite of man, and scabies mites found in a variety of other animals, are several species or variants of a single species. Some investigators have believed that the various forms are referable to separate species (cf. Kutzer, 1966), but after a detailed review of the variability of morphological characters Fain (1968) concluded that all forms from a variety of mammalian hosts over widely separated parts of the world are referable to the itch mite, *Sarcoptes scabiei*. Nonetheless there are host-adapted varieties, for even though the forms from mammals such as dogs, pigs, and horses will cause itching on man, their infestation is temporary and can be distinguished from the human form by the relatively low established population on humans and an absence of tunnels in human skin. But these temporary infections, termed pseudoscabies by Kutzer and Grünberg (1969), may be intensely irritating owing to secondary bacterial infections or allergic reactions. Host strains have in the literature been designated by varietal names, for example, the swine form is *S. scabiei* var. *suis,* equine is var. *equi,* and canine is var. *canis;* however, these have no taxonomic status.

HUMAN SCABIES OR ITCH. *Sarcoptes scabiei* causes scabies, "seven-year itch," or "Norwegian itch" of man. The mite is universal in distribution. The female is 330 to 450 microns long and 250 to 350 microns wide; the male slightly more than half as large. Skin between the fingers, the bend of the knee and elbow, the penis, the breasts, and the shoulder blades are most often attacked. Apparently newly infested persons do not experience itching, and infestations may progress extensively before noticed. A rash and intense itching appears in about a month in the neighborhood of the burrows. The itching, caused by toxic secretions and excretions, is associated directly with the burrowing. The sinuous burrows of the mite in the epidermis may reach 3 cm in length, and tiny vesicles and papules form on the surface. Scratching may cause secondary infection and result in suppuration and bleeding.

Infestation is generally effected by the adult fertilized female mites through intimate personal contact, usually sleeping in the same bed with an infested person. Apparently bed linen does not serve as a means of transfer (Samšiňsák *et al.,* 1974). Immunity responses seem to develop, for long-standing chronic cases show few parasites.

The gravid female mite deposits rather large oval eggs (150 to 100 microns) at intervals in the tortuous tunnel she makes in the epidermis. Usually she remains in the burrow for her lifetime of about 2 months, depositing eggs at 2- to 3-day intervals. Larvae move freely over the skin, and they and nymphs are frequently in hair follicles. Within 4 to 6 days after egg hatch development is completed into a male (rarely seen) or an immature female; the female makes a temporary gallery in the skin before mating.

Human scabies is variable in infestation rates, appearance, and complications. It has been noted that increased case incidence follows 15- to 20-year cycles, and a worldwide resurgence in the 1970s is generally thought to be due to large population movements, tourism, and previous insufficient recognition of cases by the medical profession. Immunological studies suggest that cycling is mainly due to changing levels of immunity in the human population (Sönnichsen and Barthelmes, 1976). Infestation levels are highest in lower socioeconomic groups, and it is here that the mites are endemic. Infestations in school children during epidemics can reach 5 percent, even in so-called advanced cultures, and can require school closures and remedial treatments. These outbreaks are more related to intimate family contacts than to classroom exposures (cf. WMMR, 1975, 24:123). In situations of poor hygiene and overcrowding, as in certain villages of southern Tanzania, the incidence in preschool children may be as high as 31 percent (Masawe and Nsanzumuhire, 1975). In addition to typical rashlike scabies, the associated skin vesicles may be even more extreme in the form of bullous lesions (Bean, 1974), or there may be thick crusts over the skin, accompanied by abundant mites but only slight itching—a form called Norwegian scabies. Acute glomerulonephritis may develop if scabies lesions are complicated by the presence of beta-hemolytic streptococci (Syartman *et al.,* 1972).

SARCOPTIC MANGE OF DOMESTIC OR WILD ANIMALS. Mange of swine commonly occurs about the top of the neck, shoulders, ears, withers, and along the back to the root of the tail. A microscopic examination of deeper tissues from beneath scabs usually reveals the mites. Suckling pigs and young shoats suffer most. Animals that scratch and rub vigorously, and that have the skin cracked and thickly encrusted with heavy scabs, should be examined for scab mites. Histological examinations suggest reactions are caused by chronic irritation and by allergic responses (Bindseil, 1974). Control of sarcoptic mange in pigs can result in significant weight-gain improvement (Sheahan *et al.,* 1974).

Sarcoptic acariasis in horses, mules, and asses is caused by *S. scabiei* var. *equi*. Sarcoptic mange of cattle caused by *S. scabiei* var. *bovis* (also called the cattle itch mite, *Sarcoptes bovis*) is not so common as the psoroptic form (*Psoroptes* mites), but is far more difficult to cure. It usually occurs where hair is short, namely on the brisket and around the base of the tail.

Common mange of dogs closely resembles that of swine. Dog mange usually appears first on the muzzle, around the eyes, on ears, breast, and later spreads to back, abdomen, and elsewhere. Pruritic dermatitis of humans is readily contracted by exposure to infested dogs (cf. Charlesworth and Johnson, 1974).

Sarcoptes scabiei in sheep occurs primarily on the face and causes "black muzzle." In more severe cases the limbs and rarely the body, but not the woolly parts, may be affected.

Sarcoptic mange has been noted on a wide variety of mammals in zoos. In nature this condition may also be prevalent and serious, for example, severely affecting lion cubs and a variety of large bovids in Kruger National Park, South Africa (Young, 1975).

NOTOEDRIC MANGE. Mange of cats is caused by *Notoedres cati,* smaller and more circular than *Sarcoptes* but otherwise quite similar. It affects dogs and certain rodents, but apparently not man. Notoedric mange of cats begins at the tips of the ears and gradually spreads over the face and head. A form in rabbits causes severe mange, beginning at the muzzle and in serious cases spreading over the whole body.

Trixacarus caviae is a sarcoptid found to cause mange of laboratory guinea pigs in England (Fain *et al.,* 1972).

Family Knemidokoptidae. *Knemidokoptes* (also seen spelled as *Cnemidocoptes* or *Knemidocoptes*) mites cause a variety of mange manifestations in birds. The family Knemidokoptidae has been reviewed by Fain and Elsen (1967). Legs of domestic chickens, turkeys, pheasants, and other gallinaceous birds are frequently attacked by a microscopic burrowing mite, the scalyleg mite, *Knemidokoptes mutans* (Fig. 15-20), which causes a lifting of the scales and a swollen condition of the shank with deformity and encrustation. The mites burrow and live in the skin, depositing their eggs in channels as do sarcoptic mange mites. Infestation occasionally results in death of the birds, and involves the visceral organs as a result of sensitization. Scaly leg is easily transmitted

A   B

Fig. 15-20 *A.* Normal lower leg of a fowl. *B.* A leg affected with the scaly leg mite, *Knemidokoptes mutans,* causing scaly leg.

from fowl to fowl, so segregation is important in control.

The depluming mite, *Knemidokoptes gallinae,* closely related to the scaly leg mite, attacks the skin near the bases of feathers. Intense itching associated with the mites impels the host to pluck its feathers.

Knemidokoptes jamaicensis infests canaries. *Knemidokoptes pilae* causes mange of budgerigars (parakeets, *Melopsittacus*) in the USA. Affected areas are generally the face and base of beak and legs; it is not highly contagious. Rickards (1975) gives information on these mites, and on chemical control.

Family Analgidae (Feather Mites). *Megninia* species have been reported to infest domestic fowl, pigeons, and parakeets (Baker *et al.,* 1956). *Megninia ginglymura* (Fig. 15-21) is suspected of causing feather pickling and plumage loss in pullets. The feather mite, *M. cubitalis,* causes a depluming itch of poultry.

Family Cytoditidae. *Cytodites nudus,* found in the bronchiae of poultry and canaries, is widespread in the Americas, Australia and New Zealand, and India. It may cause tissue reactions consisting of an encapsulated

granuloma containing giant cells (Kaliner, 1970). *Speleognathopsis galli* (Family Ereynetidae, suborder Prostigmata) infests the anterior respiratory passages of domestic fowl and guinea fowl in Rwanda, and domestic fowl in New Guinea (Talbot, 1968).

Family Myocoptidae. *Myocoptes musculinus* (and another astigmatid, *Radfordia affinis,* as well as a prostigmatid, *Myobia musculi*) commonly infests the pelage of mice in laboratory colonies. When in low numbers they are usually overlooked, but when abundant they cause a bothersome dermatitis on these rodents (Flynn, 1973).

Family Dermoglyphidae (Near Psoroptids). This family includes *Falculifer rostratus,* a parasite of pigeons that is thought to be the cause of human discomfort associated with proximity to these birds (Heath, *et al.,* 1971).

Family Audycoptidae. This family has been described by Lavoipierre (1964) from specimens in the hair follicles of primates. The paper with these descriptions is of additional interest because it reviews the literature on other mite parasites or primates.

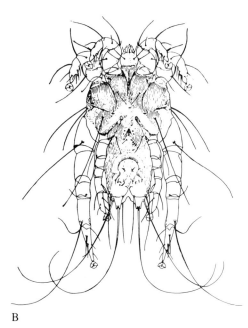

A

B

Fig. 15-21 *Megninia ginglymura. A.* Ventral view of female. *B.* Ventral view of male. (From Baker *et al.*, 1956, *A Manual of Parasitic Mites of Medical or Economic Importance;* courtesy of National Pest Control Association.)

Suborder Cryptostigmata

CRYPTOSTIGMATID MITES

These are termed *beetle mites,* a large group of mainly slow-moving and strongly sclerotized mites. Respiration is tracheal, with openings hidden. They are often saprophytic, and common inhabitants of humus and soil.

Oribatid Mites. This is a large assemblage called the oribatei. Oribatei may serve as developmental hosts of tapeworms including *Moniezia expansa,* a common parasite of ruminants (Allred, 1954). Oribatids are hosts of other species of *Moniezia* and other cestodes, in fact, this appears to be a widespread phenomenon. Further details on desert pasture relationships and oribatid hosts may be referred to for Uzbekistan, USSR, and Chad, Africa, in Nazarova (1971) and Graber and Gruvel (1970). The vertebrate ingests free-living oribatids that are common in pastures and may harbor tapeworm cysticercoids. It has been suggested that some reduction of tapeworm acquisition could be achieved by restricting the period of grazing to times when the host mites are less numerous in vegetation, but a study in the state of Kentucky, USA, showed one oribatid host to be common in vegetation

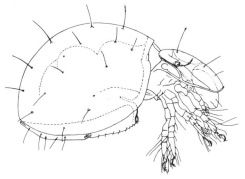

Fig. 15-22 An oribatid mite, *Pseudotritia ardua,* lateral view of female. (From Baker *et al.*, 1956, *A Manual of Parasitic Mites of Medical or Economic Importance;* courtesy of National Pest Control Association.)

at a steady density throughout the day (Wallwork and Rodriguez, 1961).

The integument of oribatids is leathery or strongly sclerotized; tarsi have one to three claws but lack delicate transparent suckerlike structures between the claws called caruncles; the sexes are similar; the respiratory system is a complex of tracheae opening through stigmata or porous areas in many regions of the body (Fig. 15-22).

Mite Transmission of Pathogens

FOWL POX

This virus of poultry is believed to be normally transmitted by mosquitoes (see Chapter 10), but natural and experimental transmission of the pathogen, as well as transovarian transmission, is reported for the chicken mite, *Dermanyssus gallinae,* in Azerbaijan, USSR (Shirinov *et al.,* 1972).

RICKETTSIALPOX

The etiological agent of this disease is *Rickettsia akari,* transmitted by the house mouse mite *Liponyssoides sanguineus.* This mite normally parasitizes the house mouse, *Mus musculus,* and transmission to man is by more or less accidental biting. The disease was first observed in New York City in the summer of 1946 when the causal organism was isolated by Huebner and associates (1946).

Rickettsialpox is usually a mild febrile condition commencing 7 to 10 days after the bite of the mite, with a vesicular rash appearing some 3 to 4 days after the onset of fever. There is an initial lesion at the site of the bite, and a black scab develops; healing is slow. Fever is accompanied by chills, sweating, backache, and muscle pains. Untreated cases recover in 1 to 2 weeks, and fatalities are unknown.

Larvae of the vector mite do not feed, so transmission is by feeding nymphs or adults of both sexes. Transovarian passage of the

rickettsia has been shown (Kiselev and Volchanetskaya, cited in Burgdorfer and Varma, 1967), thereby making the vector an important part of the reservoir.

The epidemiology of rickettsialpox indicates an urban pattern in the northeastern USA. A similar relationship has been observed in the USSR, but there wild commensal rodents have also been found to harbor the mite. A disease clinically and serologically resembling rickettsialpox has been described from Africa under conditions suggesting a different cycle. Evidence of independence from man is provided by the isolation of the pathogen from a wild Korean field mouse (*Microtus*).

SCRUB TYPHUS

Disease: Scrub typhus or chigger-borne rickettsiosis, tstutsugamushi disease, mite-borne typhus, Japanese river fever, tropical or rural typhus.

Pathogen: *Rickettsia tsutsugamushi* (*R. orientalis* of some authors).

Victims: Humans exposed outdoors to bites of certain chigger mites.

Vectors: Mites in the family Trombiculidae, genus *Leptotrombidium* (formerly a subgenus in the genus *Trombicula*).

Reservoir: *Leptotrombidium* mites through transovarial transmission; infection of rodents is transitory.

Distribution: Southeastern Asia and adjacent islands of Indian Ocean and southwest Pacific; coastal north Queensland, Australia (Fig. 15-23).

Importance: High incidence among humans exposed outdoors in certain disturbed environments such as river bottoms, plantations, cutover forests. Mortality of untreated cases may be high, but early treatment with appropriate broad-spectrum antibiotics is very successful.

This is an ancient disease (Blake *et al.,* 1945) first described from Japan. It was long said that the infection is transferred to man by the "akamushi" (Japanese for "dangerous bug") from rodent reservoirs. It was not, however, until 1916 that positive experimental evidence was secured by Miyajima

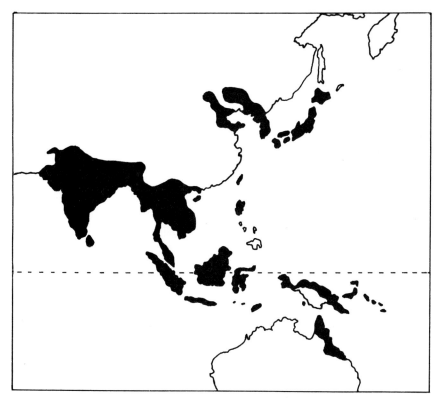

Fig. 15-23 Areas in southeastern Asia and the western Pacific region where cases of scrub typhus occur. Distribution of *Leptotrombidium* chiggers, and the etiological agent, is somewhat more extensive. (Redrawn from several sources.)

and Okumura in tests with monkeys. The causal organism is *Rickettsia tsutsugamushi* (*=R. orientalis*).

The disease has an incubation period of 6 to 21 days, usually 10 to 12. During the first 5 to 7 days of clinical symptoms it is characterized by headache, apathy and general malaise, fever, relative bradycardia, anorexia, conjunctival congestion, lymphadenitis, often regional.

An eschar or primary lesion originates at the point of infectious chigger attack in the majority of cases (Fig. 15-24). It is a painless papule at first, usually unnoticed by the patient and usually absent in Asians, slowly enlarging to a diameter of 8 to 12 mm, the center becoming very dark and necrotic. A shallow ulcer may result, leaving a

scar. Between the fifth and eighth days, in nearly all cases, a dull red macular or maculopapular rash appears on the trunk and may

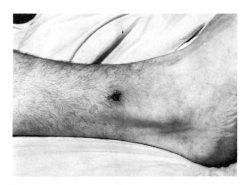

Fig. 15-24 Primary eschar at site of chigger bite transmitting *Rickettsia tsutsugamushi*. (Courtesy of US Armed Forces Institute of Pathology. Negative No. Det. 4451.)

spread to the extremities, persisting for several days or disappearing within a few hours. Enlargement of the spleen, nervous disturbances, delirium, and prostration are common; in many cases there is deafness. The majority of patients recover in 3 to 5 weeks.

Mortality rates in epidemics have varied from 6 to 35.3 percent, in some instances as high as 60 percent depending on localities and human populations. Death results approximately equally from secondary bacterial pneumonia, encephalitis, or circulatory failure at about the end of the second week.

Although most cases are found at low elevations, infections occur up to 900 m and, in Taiwan, as high as 2,000 m. Natural infections of rodents and chiggers have been located as high as 3,150 m in Pakistan (Traub et al., 1967). Chigger-borne rickettsiosis is endemic in many parts of southeastern Asia and adjacent islands in the Indian Ocean and southwest Pacific, and the coastal area of north Queensland, Australia.

The importance of scrub typhus as a cause of casualties in some Asiatic-Pacific military operations in 1941–1945, according to Philip (1948) was second only to malaria and more dreaded. Philip states, on the Assam-Burma front:

. . . 18 percent of a single battalion got scrub typhus in two months and in that time 5 percent of the total strength had died of it.

American Task Force operations in the Schouten Islands resulted in a thousand cases in the first two months on Owi and Biak, reaching a total of 1,469 casualties in six months time, while at Sansapor beach head the curve for weekly admissions on a thousand-per-year basis shot up to over 900 at the end of the second week, a rate higher for an individual episode than any yearly rate for all causes in the entire American Army in all theatres. . . . These two disasters alone provided a potential estimated loss of over 150,000 man days to the American Sixth Army.

The epidemiology of scrub typhus has been further clarified in more recent investigations, and the review of Traub and Wisseman (1974) provides a particularly comprehensive account. The most fundamentally important fact is that only larval chiggers feed on vertebrates; therefore, those infective to man or rodent hosts have received the pathogen transovarially from the female parent. Limited foci of the disease, though they differ markedly in superficial appearance, are characterized by a suitable rodent population, adequate ground moisture for the mite vectors, and the etiological agent in the mites. Traub and Wisseman term the necessary factors as the "zoonotic tetrad," consisting of a conjunction of (1) vector *Leptotrombidium* mites, especially the *L. deliense*-group, (2) wild rats in the genus *Rattus*, (3) the pathogen, *Rickettsia tsutsugamushi*, and (4) disturbed environments with transitional vegetation. Larger mammals and birds help to disperse infected mites. Occurrence of the disease in man varies by region, being year-round in Malaya but often restricted to warm rainy periods elsewhere.

The vector *Leptotrombidium* mites are listed in the earlier discussion on chigger mites, Trombiculidae. Other kinds of chiggers seem to be of minimal importance though species of *Neotrombicula* and *Gahrliepia* may be enzootic vectors among rodents.

Experimental transmission studies by chigger bite indicate that uninfected chiggers readily acquire the pathogen from infected rodents, but very rarely pass this infection transovarially to offspring. Nonetheless certain familial lines of chiggers transmit the pathogen to virtually all of their offspring (Rapmund, et al., 1969). This further explains the highly focal characteristic of transmission to man in nature—that is, not only are vector chiggers in restricted locations, but also only certain lines of these are infected. However, the pathogen is widespread in rodents and there may be enormous numbers of chiggers, so that even though the rate of permanent infection (transovarial) is extremely low when uninfected chiggers

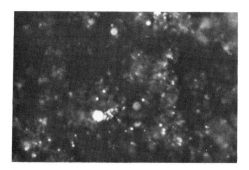

Fig. 15-25 Fluorescent antibody staining of *Rickettsia tsutsugamushi* (small, bright punctations) in tissues of vector chigger. (Courtesy of US Army Medical Research Unit, Kuala Lumpur, Malaya.)

feed on an infected rodent, this still must occur in nature and such infections are magnified by transmission to offspring. Microdissection and direct fluorescent antibody staining techniques have made it possible to observe *R. tsutsugamushi* in all stages of the mites (Roberts *et al.*, 1975). (Fig. 15-25). All organs are positive at some time during the life cycle but tissue infection may be transitory. The high level of rickettsiae in the salivary glands of unengorged larvae, together with the low incidence in engorged larvae, suggest that infectious salivary fluid is the source of infection for mammalian hosts.

Q FEVER

Common gamasid mites from natural foci of Q fever in the Soviet Union have been found to harbor *Coxiella burneti* (Zemskaya and Pchelkina, 1967). Experimental transmission was also obtained with *Dermanyssus gallinae, Liponyssoides sanguineus,* and *Ornithonyssus bacoti,* and the latter species transmitted the pathogen transovarially. This appears to be an enzootic relationship and unimportant for transmission to humans, in consideration of the many other possible routes of infection.

FOWL SPIROCHAETOSIS

This disease, caused by *Borrelia anserina,* has been noted in poultry in the So-

viet Union at times when the usual vectors, soft ticks in the *Argas persicus* complex, were not present. Under these circumstances the chicken mite, *Dermanyssus gallinae,* was abundant. Experiments showed that *D. gallinae* could readily transmit the pathogen mechanically (Reshetnikov, 1967; Ciolca *et al.*, 1968).

Control of Mites

Survey procedures for mites vary considerably, depending on the habits of these parasites. For ectoparasites on small hosts such as rodents, washing with water and detergent is quite effective (Henry and McKeever, 1971). Intradermal forms are removed in skin scrapings. Chigger larvae are collected on hosts, or they may be surveyed for by placing black plastic discs on the ground, or a fabric drag is used much as in flagging for ticks (Daniel, 1969).

Methods chosen for mite control are based on whether they are aimed at parasites that are intradermal or at even deeper levels, ectoparasitic forms on the skin surface, roost- or lair-associated species that are only on the host to feed, or chigger mites that occur in the outdoor environment. Other modifying factors concern the vertebrate host and its susceptibility to pesticides; song birds, reptiles, and cats being notably sensitive to toxicants. Whenever a pesticide is mentioned in the following discussion, regulations must be consulted regarding legality of a suggested chemical or practice, since some of the information is from experimental approaches, and in other instances an application legal in one country may not be so in another. When treating animals used for food, in addition to direct toxicity one must be concerned with pesticide residues in meat, milk, and eggs.

Intradermal mites include those causing sarcoptic or demodectic mange of mammals, and knemidocoptic skin afflictions of birds. Treatments must generally be applied by a physician or veterinarian. First it is necessary to identify the problem, and this is usually accomplished by applying mineral oil to the affected skin surface, scraping the skin vigorously, and examining the scrapings microscopically. For human scabies repeated treatments with oitments containing sulfur, benzyl benzoate, thiabendazole, or an approved insecticide such as lindane have been successful (Shinskiĭ et al., 1973; Villalobos and Neuman, 1975). On animals such as pigs or foxes, treatment of sarcoptic mange requires repeated thorough spraying or dipping with appropriate insecticides (Maĭorov, 1969; Keller et al., 1972). Demodectic mange often involves subcutaneous complications, and treatment of dogs may require subcutaneous injection and vigorous surface rubbing-in of a pesticide (Stoenescu et al., 1972). Thorough repeated spraying with pesticide can control demodectic mange of goats (Das and Misra, 1972).

Psoroptic mange on cattle and sheep often requires quarantines and shipping restrictions while the problem is treated. Related conditions in sheep, dogs and cats, and mouse colonies also require persistent and thorough application of pesticides. *Psoroptes* of cattle and sheep are generally treated by whole body immersion in dipping vats, and Hourrigan (1968) describes an eradication program devised for sheep in the USA. For itch mite of sheep, *Psorergates ovis,* it is suggested that treatment is best applied following shearing (Sinclair and Gibson, 1975). *Otodectes cynotis* in the ears and facial region of dogs and cats may be controlled with a solution of benzyl benzoate or other pesticide (Bollweg, 1975). Acariasis in mouse colonies has been controlled with sulfur dressings, pesticide dips, or the use of pelleted toxicants in nest boxes that control mites by the release of toxic vapors.

Mites on poultry are controlled with sprays or dusts of appropriate pesticides, or to control chicken mite roosts and crevices are thoroughly sprayed or painted with toxicant suspensions or solutions. The deplum-

ing mite *Knemidokoptes gallinae* can be controlled by a thorough dusting or water dip in sulfur. *Knemidokoptes* infestations of parrots or budgerigars may be treated by repeated applications of appropriate pesticides (Bendheim, 1966; Levi et al., 1974). Sulphaquinoxaline, used in low doses for common poultry diseases, reduces infestations of northern fowl mite *Ornithonyssus sylviarum*. When given continuously in feed at rates recommended for control of coccidiosis this compound controlled northern fowl mite infestations within five weeks (Furman and Stratton, 1963). Similar results are claimed for the use of a combined sulfonamide feed additive (Goldhaft, 1969). Pesticides in dust bath boxes also control this mite, but the pest has become notably resistant to a number of acaricides.

Attack of trombiculid or chigger mites can be almost completely prevented by using repellents. Best control is achieved by application to socks and to the bottom of pant legs, though skin may also be treated. Satisfactory repellents are deet, dimethyl phthalate, dimethyl carbate, and ethyl hexanediol. Impregnation of clothing with repellents is useful, and benzoyl benzoate in particular can withstand some washing and rinsing in water. Controlling habitat and hosts of trombiculid mites can also reduce their numbers. Sprays or dusts of approved acaricides when applied to vegetation around premises will control chiggers (cf. Traub and Wisseman, 1968). Consult USDA (1976) for control of chiggers causing dermatitis in man. To this may be added the use of ULV applications of pesticides (Mount et al., 1975).

When humans are annoyed in the home by fowl mites or rodent mites, control of the normal vertebrate hosts is recommended. This consists of destroying bird nests under eaves and cleaning out rodent harborage. A pesticide may have to be applied to surroundings to bring the nuisance under more rapid control.

16

TICKS AND TICK-ASSOCIATED DISEASES

ORDER ACARINA; SUPERFAMILY IXODOIDEA

Most terrestrial vertebrates are subject to attack by ticks, but particularly vulnerable are mammals, whose warmth and odor are highly attractive to these parasites. The food of both sexes of adults and all active immature stages is blood, other tissue fluids, and cellular debris. Hunters have long observed heavy infestations on wild animals; stockmen suffer enormous losses due to ticks on cattle, horses, and other stock; poultry is often severely infested in the subtropics.

Evidence shows that tick feeding alone, without the transmission of pathogens, can harm wild vertebrates and livestock. Heavy populations of immature *Ixodes* ticks greatly affected voles in Siberia (Okulova and Aristova, 1974). Hunter and Hooker (1907) reported that as many as 200 pounds of blood may be withdrawn from a large host animal by ticks in a single season. Anemia is frequent in heavily infested hosts. Woodward and Turner (1915), using the cattle tick, *Boophilus annulatus,* found infested cows under experimental conditions gave only 75.8 percent as much milk as tick-free

cows. Furthermore, tick-free cows gained 6.1 percent in body weight during the experiment while the infested animals gained but 3.6 percent. Host nutrition influences whether ticks feed successfully, as it has been noted that steers on low protein and fat diet produce more ticks than animals on proper diets, and cattle losing weight produce more ticks than those gaining weight (Gladney *et al.,* 1973). Beef rather than dairy animals are mainly affected, the losses for Australia being recently calculated as $25 million annually (Fig. 16-1). In 1965 the US Department of Agriculture estimated tick-associated losses annually for cattle production at $60 million, and sheep at $4.7 million (Steelman, 1976).

All blood losses are probably accompanied by some toxicosis caused by tick salivary secretions. Toxicosis and paralysis from tick bite are two serious conditions not always clearly separable. These topics are discussed in Chapter 17.

There are two basic types of ticks (Fig. 16-2). The IXODIDAE, or hard ticks, attach for prolonged periods to hosts and are therefore familiar to most people who have seen these on themselves, pets, or livestock. The

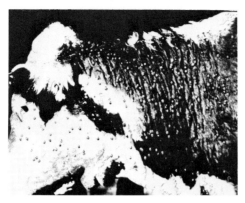

Fig. 16-1 Head and neck region of a European breed of cattle heavily infested with ticks; of common occurrence on rangeland stock in some areas of Australia. (Courtesy of R. H. Wharton, CSIRO, Australia.)

ARGASIDAE, or soft ticks, mainly feed for very brief periods and at night; consequently most people have never seen these.

The mechanism for penetration and anchoring of tick mouth parts is discussed in Chapter 3. One remarkable feature, considering the extensive laceration of tissues during the act of feeding, is that in most cases, particularly with ixodid or hard ticks, the bite is not felt at the time of feeding. However, persistent lesions following tick bite are frequent even if the tick is not forcibly removed, leaving pieces of mouth parts in tissue as sources of infection. The bites are characterized by vascular trauma and edematous swelling, and accumulation of host neutrophils at the attachment site. Neutrophils and other inflammatory responses can be suppressed by nitrogen mustard, with the result that tick feeding lesions are insignificant (Tatchell and Moorhouse, 1970). Neutrophil infiltration is believed due to cleavage of a component of serum complement by tick saliva (Berenberg *et al.*, 1972).

Once attached, great care should be taken in removing a tick, for the mouth parts are easily left in the flesh as an irritant and source of secondary infection. Several investigators have studied chemicals singly and in combination as treatments to cause tick detachment, and more recent findings indicate that formamidine derivatives are effective for this purpose (Gladney *et al.*, 1974; Stone *et al.*, 1974).

The length of time a tick remains attached depends on the family, the species, whether mating has taken place, and the stage of development. As a rule ixodids stay attached to a host for a prolonged time; many argasids are parasites associated with the lairs, roosts,

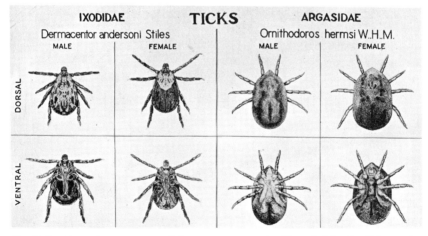

Fig. 16-2 General appearance of the two common families of ticks, Ixodidae and Argasidae.

or dens of their hosts, attaching only briefly to the host and remaining at other times in or near its habitation. As examples, nymphs and adults of the ixodid cattle tick *Boophilus annulatus* feed 6 to 8 days to become fully engorged; common argasid poultry ticks, *Argas persicus* and related species, feed nightly and intermittently. Both male and female ticks suck blood; great enlargement occurs in female Ixodidae only, but in both sexes of Argasidae.

Among publications on the general biology of ticks, that of Arthur (1962) includes their relation to disease, Theiler (1964) discusses ecogeographical aspects of their distribution, Balashov (1972) provides a comprehensive review on many aspects of tick biology, morphology, physiology, and reproduction, and Hoogstraal (*in* Gibbs, 1973) reviews the biology of tick families and genera. Bibliographies on ticks and tick-associated diseases have been prepared by Doss and coauthors (1974-continuing) and Hoogstraal (1970-continuing) with completion of five volumes and two additional volumes anticipated.

Many disorders and diseases of man and animals are traceable to ticks; among these are (1) *dermatosis,* inflammation, itching, swelling, and ulceration at the site of the bite, or skin ulceration and lesions resulting from improper or partial removal of tick mouth parts; (2) *exsanguination,* a serious condition in which a heavily infested animal develops anemia; (3) *otoacariasis,* infestation of the auditory canal by ticks, with possible serious secondary infection; (4) *infections* transmitted by ticks including *piroplasmoses (babesioses),* for example Texas cattle fever; *rickettsioses,* for example, Rocky Mountain spotted fever; *viruses,* for example, Colorado tick fever; *spirochetoses,* for example, tick-borne relapsing fever; *bacterioses,* for example, spread of tularemia by ticks. Transmission of a filaria, and isolation from ticks of a spiroplasma, each affecting rodents, also occur but seem exceptional.

High Vector Potential of Ticks

Factors accounting for the potency of ticks in the spread of diseases of man and animals are the following: they are first of all (1) *persistent bloodsuckers*—they attach firmly while feeding and do not dislodge easily; their (2) *slow feeding,* especially characteristic of ixodid ticks, permits wide dispersion while attached to a bird or large mammal and allows ample time for the transfer of pathogens to a host; a (3) *wide host range* in some species insures more certain sources of blood and opportunities to acquire and transmit pathogens; (4) *longevity,* particularly among argasids, enhances the chances of acquiring and transmitting pathogens; and (5) *transovarial transmission* of some pathogens ensures infectivity in some members of the next generation, a matter of greatest importance in one-host ticks. To these main factors might be added (6) *relative freedom from natural enemies;* the nymphs and adults are (7) *highly sclerotized* and hence quite resistant to environmental stresses, and (8) the *reproductive potential* of ticks is great, as some species may deposit as many as 18,000 eggs and parthenogenesis occurs sometimes. The last three factors simply help to provide large populations of ticks, so that these acarines are widespread and vector-host contacts are high.

Ticks are generally parasites of wild animals in nature. They parasitize man and his domesticated mammals fortuitously, though the latter may be heavily attacked owing to conditions caused by man. Hoogstraal (1972) points out that about 10 percent of the world tick species infest domestic mammals and birds; clearly a result of man introducing these hosts into former wild tick habitats and certain ticks' adapting from wild hosts to domestic relatives so that in some cases only relict wild host relationships remain. Travel on early caravan routes and more recent shipping of domestic animals has resulted in continental and intercontinental extension of the range of ticks infesting cattle, horses, sheep, pigs, dogs, and poultry. Wildlife parks have introduced exotic tick species to the USA, though thus far these have been controlled. Changes in habitat in the eastern USA from farmland reverting to scrub growth, and movement of human populations into adjacent suburban communities, have increased cases of Rocky Mountain spotted fever in recent years. Increased human participation in outdoor recreation has enlarged the risk of acquiring tick-associated diseases. Certain tick-borne infections in wild animals have persisted for so long that a host-parasite accommodation has resulted in benign infections; when man and domestic animals intervene as fortuitous hosts they may be severely affected.

Structural Characteristics

Ticks are easily distinguished from insects as the body is not definitely divided and has a saclike leathery appearance (Fig. 16-2). A distinct head is lacking, but the mouth parts, together with the *basis capituli* in many species, form a headlike structure known as the CAPITULUM (Fig. 16-3). Like most other Arachnida, mature ticks and nymphs have four pairs of legs, the larvae three pairs.

All adult and nymphal ticks have a pair of spiracles situated lateroventrally on the abdomen, one on each side near the third and fourth leg bases. In ixodids the dorsum of the adult male is largely or wholly covered by a plate called the SCUTUM. In immature and female ixodids the scutum covers only the anterior part of the dorsum behind the capitulum. The scutum is ORNATE if it has a pattern of gray or white imposed upon a dark background (Fig. 16-10), INORNATE without such a pattern.

A pair of simple eyes may be located on the lateral margins or submargins of the scutum in hard ticks, or along the submargins in certain soft ticks. Many species are eyeless, in which case there are photosensitive areas in the locations where eyes are found in eyed species; such ticks react positively to low light intensities and negatively to high intensities as do eyed species (Panfilova, 1976). These photosensitive areas lie beneath the integument and are primitive light-concentrating structures with nerves connecting to the central nerve mass or "brain" (Binnington, 1972). Light regulates many biological processes in various ticks including questing for hosts, metamorphosis, time of detachment from hosts, and diapause induction or termination.

In certain hard ticks more or less rectangular areas, along the posterior submarginal area of the dorsum separated from adjacent ones by grooves, are called FESTOONS. Festoons may be present in both sexes, though not evident in engorged females. Ventrally (Fig. 16-4), an ANAL GROOVE may set off a plate or area on which the anus is located; the anal groove may be evident only in front of

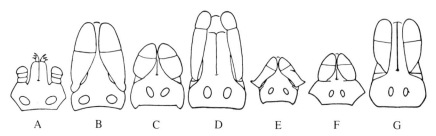

Fig. 16-3 Characteristic capituli of several genera of ixodid ticks. *A. Boophilus. B. Ixodes. C. Dermacentor. D. Amblyomma. E. Haemaphysalis. F. Rhipicephalus. G. Hyalomma.* (After Cooley.)

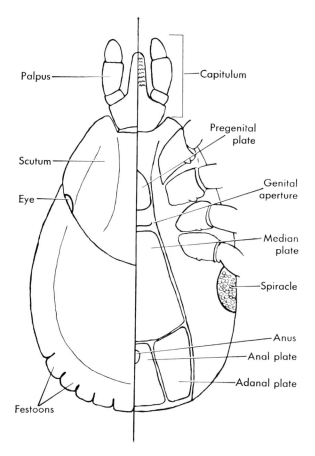

Fig. 16-4 Diagrammatic illustration of the anatomy of an ixodid tick. (*Left*) Dorsal view; (*right*) ventral view. (Redrawn and adapted as a composite of two illustrations from Gregson, 1956.)

the anus, or it may occur only beside and behind the anus. Other plates of the ventral surface may be of some taxonomic significance, such as ADANAL PLATES (to each side of the anal plate) and ACCESSORY PLATES (anterior to the adanal plates).

Ticks vary considerably in size according to species, but even fully engorged females rarely exceed 20 mm in length. Females can become greatly distended and, when fully engorged, are beanlike in form.

Life History

Under natural conditions a few species of ticks show a rather marked host specificity, for example, *Boophilus annulatus* on cattle, and *Dermacentor parumapertus* on jack rabbits. However, other species have a fairly

wide range of hosts. The life histories of ticks vary among the many species; hence it is quite impossible to generalize except that it may be said all species pass through four stages (EGG, LARVA, NYMPH, and ADULT) in from six weeks to three years. Fully engorged females usually deposit their eggs on the ground, the number varying from 100 in lair- or den-associated species up to 18,000 in some free-ranging species. The larvae are HEXAPOD (six-legged) and remain in this condition until after the first molt. The nymph emerging from the first molt has four pairs of legs and remains in this stage until transformation to the sexually mature adult. Ixodid ticks have one nymphal stage, but there may be as many as five nymphal molts in argasids. In argasids a small blood meal causes molting and the presence of addi-

Fig. 16-5 A female hard tick laying a group of eggs at the front of the body. (Courtesy of Pacific Supply Coop and Ken Gray.)

tional instars, resulting in adults with abnormal setal patterns. Copulation takes place after the last molt; the females then engorge and deposit eggs (Fig. 16-5). The majority of tick species drop off the host animal to molt, but in several species, notably *Dermacentor albipictus* and members of the genus *Boophilus,* molting takes place on the host, an adaptation to infesting large wandering mammalian hosts.

In addition to questing for hosts at relatively fixed positions by ixodids, both argasids and ixodids may use host-associated stimuli in orientation. Carbon dioxide from host respiration is readily perceived and responded to at some distance, and this may be taken advantage of in sampling traps utilizing dry ice (cf. Howell, 1975; Semtner and Hair, 1975). Heat from the host seems to be attractive only at short range.

According to Theiler (1964) there are records of tick eggs hatching as long as 80 days after submersion in water, and even adults have been submerged 3 weeks without ill effects. Eggs of *Rhipicephalus appendiculatus* and *Amblyomma variegatum* can hatch under water, and a good proportion of the larvae climb up onto emergent vegetation, and larvae and nymphs are quite resistant to submersion (M. W. Smith, 1973); unfed nymphs of *Ixodes ricinus* survived submergence of 160 days in clear water at 9° C (Honzáková, 1971). The eggs are coated with a waxy secretion from a struc-

ture called Gene's organ (Lees and Beament, 1948). When Gene's organ is everted to wax the eggs it covers porose areas that have been demonstrated to be a site of secretion of an antioxidant that inhibits oxidation of steroids and other compounds in egg waxes (Atkinson and Binnington, 1973). To conserve on water, guanine is the principal end product of tick nitrogenous excretion, and embryonic development can be monitored in the eggs by spectrophotometric measurement of the accumulation of this substance (Londt, 1975).

Even though all stages of ticks, but particularly immature stages, are subject to desiccation, these pests have developed behavioral and physiological adaptations to prevent excessive water loss. Behaviorally, the active stages leave ground level (to search for hosts) when they are water saturated, returning to humid ground level when dehydrated. Water may be recovered by imbibing accumulations such as dew (Wilkinson, 1953), or by absorbing water in the vapor phase above a certain equilibrium humidity. Absorption above equilibrium humidities is accomplished by secretion of a hygroscopic crystalline solid at the mouth parts (Rudolph and Knulle, 1974). Excretion of excess water, and salt balance, is accomplished by the coxal glands in argasids and by the salivary glands in ixodids (Kaufman and Phillips, 1973). Argasids generally withstand lower equilibrium humidities than ixodids and consequently live in drier environments. Different equilibrium humidities among species of ticks are believed to be important among the several factors that dictate which habitats they can successfully live in (Hair *et al.*, 1975).

Larval ticks emerging from eggs on the ground commonly climb grasses and other low vegetation to come within easy reach of grazing or passing animals. The height of larval ascent varies markedly among species, and probably is an important factor involved in selecting the size of host that will be infested. Questing is characteristic of all postembryonic stages of ixodid ticks that leave a host to molt (Fig. 4-7); in some instances an adventitious aggregation assembles on the ground and then climbs vegetation stems to quest together (Browning, 1976). *Dermacentor variabilis* adults reach saturated weight at ground level and respond to a photostimulus to commence questing, resulting in a diurnal questing pattern (McEnroe and McEnroe, 1973).

Upon reaching the body of a host a larval tick follows a sequence of feeding and molting until maturity is reached. When feeding is completed on one animal, as in the case of *Boophilus annulatus,* the species is said to be a ONE-HOST TICK. Hoogstraal (1972) points out limited host relationships are atypical and coincide with infestation of large wandering mammalian hosts, and in more recent times with man's introduction of domesticated herd animals.

When two hosts are involved, as with the African red tick, *Rhipicephalus evertsi,* the species is called a TWO-HOST TICK. The larva of this species hatches on the ground like other ticks, then attaches to the inner surface of the ear of the host animal where it engorges. The engorged nymph drops off to molt, then the adult must find a second host and engorge. Replete adults drop to the ground, where the females lay eggs.

When a tick requires three different hosts to complete its cycle it is called a THREE-HOST TICK, as for example *Dermacentor andersoni,* the Rocky Mountain wood tick. In this tick the larva engorges on smaller mammals, such as ground squirrels, then drops to the ground to molt. The nymph also feeds on one of the smaller mammals and then drops to the ground, molts, and becomes adult. The adult finds a host (usually one of the larger game or livestock mammals, or sometimes man) upon which it feeds, then drops to the ground where the female lays eggs.

In ticks such as the argasid *Ornithodoros hermsi* of several western states, USA, a number of individual host animals are utilized; such species are known as MANY-HOST TICKS. There are usually five molts in this species, each completed off the host; hence, at least five host animals are needed to complete the cycle. Moreover, the adult may feed intermittently on additional hosts.

Obviously the variety of hosts utilized by a tick species determines its potential sources of pathogens and opportunities for transmission. One curious phenomenon is the feeding on female ticks by males (cf. Moorhouse and Heath, 1975), and general obtaining of a blood meal by hungry immature ticks feeding on recently blood-fed members of the same species, thus obtaining pathogens (Votava *et al.,* 1974).

Chemical communication via pheromones affecting assembly, mating, and host finding has been demonstrated in ticks. Pheromone produced by several species of female ixodids to attract males has been reported, and identified as 2,6-dichlorophenol (cf. Berger, 1972; Chow *et al.,* 1975); males of *Amblyomma maculatum* produce a pheromone that attracts females (Gladney *et al.,* 1974). Assembly pheromones causing concentrations of ticks on hosts have been noted in *Amblyomma* ticks of two species (cf. Rechav *et al.,* 1976), and there appears to be a non-species-specific assembly pheromone in several soft ticks (Leahy *et al.,* 1975).

Studies on tick reproduction reveal a number of adaptations. During copulation a spermatophore containing sperm is transferred to females by males. Parthenogenesis occurs in some species, and parthenogenetic races are known in several instances where bisexual reproduction is normal (Oliver, 1971). A blood meal is normally required for reproduction, but autogenous strains of ticks are known (Feldman-Muhsam and Havivi, 1973). Oliver (1977) has reviewed the cytogenetics of mites and ticks.

The longevity and hardiness of ticks are truly remarkable and must be considered when control measures are applied or when the persistence of tick-borne diseases is considered. Particularly noteworthy are argasids, with survival under starvation conditions for as long as 16 years in some species. Gregson (1949) recorded an adult life span of 7 years for the ixodid *Ixodes texanus,* and the possibility of a total life cycle of more than 21 years.

Classification: General Considerations

Although ticks were referred to as "disgusting parasitic animals" by Aristotle in the fourth century B.C., the orderly classification of these parasites dates from Linnaeus in 1746 (*Fauna Suecica*), where they were placed among the Acari in the genus *Acarus* (according to Nuttall and Warburton, *Monograph of the Ixodoidae,* Cambridge University Press, 1911). In 1844 Koch separated ticks from the Acari, which included both ticks and mites, and in 1896 Neumann placed ticks in the order Acarina and divided them into the two subfamilies Argasinae and Ixodinae, subsequently raised in 1901 by Salmon and Stiles to family rank in the superfamily Ixodoidae established by Banks in 1894. Alternative higher classifications are provided by Baker and Wharton (1952) and Krantz (1970).

The ticks and parasitic mites appear to have had a common origin, with ticks separating from the mesostigmatid mites at the beginning of the Permian (Rohdendorf, 1962). Parasitic mites are believed to have developed during the Devonian. The present ixodid genera *Ixodes* and *Dermacentor* are known from the Oligocene of North America.

Ticks are considered by some authors to form a separate suborder of Acarina, the Ixodides, and by others to form, along with a number of mite families, the suborder Mesostigmata (they have also been placed in Metastigmata, or in the order Parasitiformes, suborder Ixodida according to Krantz, 1970; Krantz, 1978). They constitute the super-family Ixodoidea, which includes about 800 species divided into two major families: (1) Ixodidae or hard ticks; scutate ticks with a terminal capitulum, sexual dimorphism obvious, scutum of males covering the dorsum, males incapable of great expansion, scutum of females a small shield immediately behind the capitulum, females capable of enormous expansion. (2) Argasidae or soft ticks, nonscutate, sexual dimorphism not obvious; capitulum ventral and palpi leglike, eyes when present lateral and on the supracoxal folds, spiracles very small. The family Nutalliellidae is represented by a single rare African species associated with rodents and small burrowing carnivores.

The literature on tick taxonomy is rather widely scattered and still unsettled. For the ticks found on livestock in the USA, including species that are commonly pests on other animals, refer to Strickland and associates (1976 revision). Cooley and Kohls (1944) deal with the Argasidae, and this family has been more recently reviewed in a number of papers by Clifford and associates. An alternative treatment of argasids, presenting the eastern European viewpoint, has been provided by Pospelova-Shtrom (1969).

Hoogstraal (*in* Gibbs, 1973) provides an excellent review of tick families and genera, and refers to the main taxonomic literature for genera. The discussion of families and genera presented here is mainly from the previous edition of this book, Hoogstraal's review, and recent literature sources. Discussion is limited to genera of medical and veterinary importance that contain those ticks most frequently encountered.

Family Ixodidae

HARD TICKS

There are about 660 species of hard ticks in some 14 genera, only 10 of which are considered here. In some classifications the Ixodidae are divided into subfamilies: Ixodinae with the genus *Ixodes* (ca. 250 species);

Amblyominae with the genera *Hae-maphysalis, Aponomma, Amblyomma,* and *Dermacentor* (ca. 307 species); and Rhipicephalinae containing the remaining genera (ca. 101 species). There is a tendency for ixodids to be three-host ticks, with females leaving the third host after feeding, then ovipositing and dying. Size and kind of host infested by immature and adult stages is usually quite different, and tick population density frequently decreases significantly at each stage, likely because of difficulties in finding and attaching to suitable hosts. Most ixodids also have a rather narrow host range, but there are notable exceptions. Species that will feed on larger animals as larvae may also infest man, and these are very bothersome and referred to as "seed ticks," causing notable annoyance of humans outdoors in such areas as the southcentral USA, Central America, and East Africa. Seed ticks are particularly prevalent in the genera *Amblyomma* and *Hyalomma*.

Genus Ixodes. This is the largest ixodid genus, worldwide in distribution, with about 40 North American species and almost 250 species total. The scanning electron microscope has been utilized to identify *Ixodes* of the USA (Keirans and Clifford, 1978). These are three-host ticks.

In Europe *Ixodes ricinus* and *I. persulcatus* play an important role in the transmission of viruses affecting man, and the protozoa of bovine piroplasmosis. The literature on these two species is scattered and conflicting, and a review by Arthur (1966) is the primary source for this account. Both species utilize an extraordinary range of hosts that includes birds, small rodents, insectivores, and intermediate-sized and large mammals. *Ixodes ricinus,* the European castor bean tick, is more western in distribution, extending eastward in places to about 55° longitude; westward along the western margins of the British Isles and Norway to about 65° N latitude; southward to about 35° N latitude in Iran through the mountains to Turkey,

Bulgaria, Italy, and the Pyrenees. Aeschlimann (1972) provides a detailed life history of this species in Switzerland, Uspenskaya (1974) records differences in populations from four diverse areas of the Soviet Union, and overwintering activities in Czechoslovakia are described by Dusbábek and colleagues (1971) and Dyk and Zavadil (1971).

Ixodes persulcatus is more Eurasian in distribution, below 65° N to as far south as the island of Kyushu, Japan, and westward into north Germany. *I. persulcatus* is more tolerant of temperature extremes and more cold hardy than *I. ricinus*. Where the two species overlap there are microclimatic conditions separating their distribution; *I. ricinus* occurs mainly in cut-over, secondary small-leaved forests of alder, aspen, and birch intermingled with shrubby undergrowth and pastures and *I. persulcatus* inhabits small-leaved forests near primary coniferous forests, such as spruce-basswood combinations. This latter condition is commonly referred to as TAIGA, and *I. persulcatus* is known as the taiga tick. A related species, *I. pavlovskyi* is more abundant in the western part of the distribution range in tick-borne encephalitis foci, and Filippova (1971) indicates that this species is a relic of the Pliocene fauna.

An American species close to these European *Ixodes* is *I. scapularis* (Fig. 16-6), the blacklegged tick, widespread in the south-

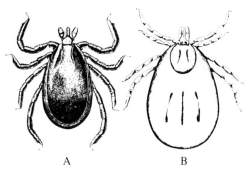

A   B

Fig. 16-6 *Ixodes scapularis.* A. Male. B. Engorged female (not to same scale). (USDA drawing.)

eastern to south central USA and along the east coast. It congregates along paths, trails, and roadways and may inflict a painful bite. *Ixodes pacificus* occurs along the Pacific coastal margins of the USA. It is common on deer in California but flourishes on cattle as well, and bites humans freely to often cause generalized discomfort.

Severe systemic disturbance may be characteristic of the bite of many *Ixodes*. In Israel a 4-year-old girl suffered fever, vomiting, and severe local pain from the attachment of a female of *I. redikorzevi* (Boger *et al.*, 1964); fever and headache may follow the attachment of *I. pacificus* in the state of Washington, USA (specimens and reports received by R.F.H.); in Australia *I. holocyclus* is the major cause of tick paralysis (see Chapter 17), and populations are related to numbers of bandicoots, which are increasing near urban areas owing to control campaigns against dingoes and foxes (Bagnall and Doube, 1975). *I. rubicundus* is called the Karoo paralysis tick in South Africa; its microclimatological requirements in relation to distribution have been studied in depth (Stampa, 1971).

Genus Haemaphysalis. There are about 150 species, worldwide in distribution with two species in North America; they are usually small, with both sexes similar. This genus probably originated as relatively large ticks in humid and tropical southern Asia, but when mammals and birds evolved, smaller representatives of this genus also developed, with structural adaptations for locomotion on hairy and feathered hosts. Few *Haemaphysalis* spend their entire active life on birds, though several species sometimes infest these hosts. Most species specialize on mammals, and adults of a few species are adapted to domestic cattle, sheep, and goats.

Haemaphysalis leporispalustris, the rabbit tick, is widely distributed in the New World from Alaska and Canada to Argentina (Fig. 16-7). Although commonly known as the rabbit tick, it has been taken on a number of

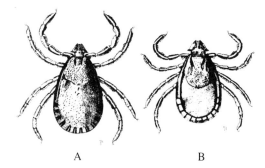

Fig. 16-7 *Haemaphysalis leporispalustris*. *A. Male. B. Female.* (USDA drawing.)

species of birds and rarely on domestic animals such as horses, cats, and dogs. It infrequently bites man but is important in spreading the pathogens of Rocky Mountain spotted fever and tularemia among wild animals.

Haemaphysalis chordeilis, the bird tick, occurs commonly on upland game birds in North America and rarely attacks cattle. This tick is of importance as a parasite of turkeys, and as a vector of pathogens of wildlife.

Haemaphysalis leachii, the yellow dog tick of Africa, is common in parts of Asia and Africa; usually on wild and domestic carnivores, frequently on small rodents and rarely on cattle; it is a vector of the protozoan pathogen of malignant jaundice of dogs. *Haemaphysalis bispinosa* causes severe irritation of cattle and other farm animals. In New Zealand a basic population of a related tick, *H. longicornis*, is maintained on the red deer, *Cervus elaphus*. It has been reported as reproducing by obligatory parthenogenesis in Queensland, Australia, and in Japan a related species on deer, *H. mageshimaensis*, has bisexual and parthenogenetic reproduction (Saito and Hoogstraal, 1973). The taxonomy and biology of all stages of *H. bispinosa*, a reservoir of Kysanur Forest disease virus in Mysore state, India, has been thoroughly detailed (Ghalsasi and Dhanda, 1975).

Genus Amblyomma. About 100 species of mostly large, highly ornamented,

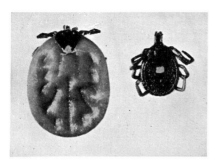

Fig. 16-8 The lone star tick, *Amblyomma americanum.* × 3.5.

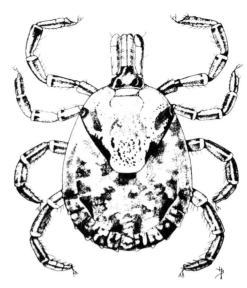

Fig. 16-9 *Amblyomma cajennense,* female. (USDA drawing.)

tropical ticks with long mouth parts; possessing eyes and festoons. The few southern USA species are the only ticks of this genus from temperate regions, though nymphs of some species are on migratory birds from Africa to Europe and southwestern Asia. All *Amblyomma* appear to have a three-host life cycle. Immatures of many species will infest humans and are known as seed ticks. There is an exceptional host range among immatures and adults in this genus, including birds, reptiles, amphibians (rarely), rodents, and large mammals that are hairy or relatively hairless.

Amblyomma americanum is the lone star tick of the southern USA (Fig. 16-8), with range extending considerably northward and southerly into Mexico; it has a wide variety of hosts, including wild and domestic animals, birds, and man. This three-host tick attacks man in all its active stages. It is very pestiferous and a vector of Rocky Mountain spotted fever and tularemia bacterial pathogens. Studies on the ecology and behavior of this tick in various habitats have been conducted recently in Virginia and Oklahoma, USA (Sonenshine and Levy, 1971; Semtner and Hair, 1973b).

Amblyomma cajennense, the Cayenne tick, occurs in Texas, Mexico, Central and South America, and the West Indies (Fig. 16-9). All active stages commonly attack man, domestic and many other animals. This species is abundant in a few counties in

Texas and is considered a vector of Rocky Mountain spotted fever rickettsia in Mexico, Panama, Colombia, and Brazil.

The Gulf Coast tick, *Amblyomma maculatum,* can be quite troublesome on cattle. Its distribution, seasonal abundance, hosts, and habitat relationships have been studied in Oklahoma, USA (Semtner and Hair, 1973a).

The tropical bont tick, *Amblyomma variegatum,* is common throughout much of the tropics and has had five viruses isolated from it, including Crimean-Congo hemorrhagic fever virus. This tick was found on cattle in St. Croix, US Virgin Islands, in 1967, and although an eradication program was initiated, this species still persists there.

Genus Dermacentor. Ornate ticks with eyes, moderate to large and of generally similar appearance; there are 31 species. Most are three-host ticks, but a few one-host species occur. Immatures of three-host species infest rodents, insectivores, and lagomorphs; adults infest many types of large mammals, including humans. The distribution of three common North American *Dermacentor* has been described for Canada in relation to bioclimatic zones (Wilkinson, 1967), and hybridization studies show separate species but close relationships between *D. andersoni, variabilis,* and *occidentalis* (Oliver *et al.,* 1972).

Dermacentor variabilis, the American dog tick, is the principal if not only vector of Rocky Mountain spotted fever pathogen to humans in the central and eastern portion of the USA (Fig. 16-10); it is also an important vector of the bacterium causing tularemia. It may cause canine paralysis and is a common pest of dogs, which are preferred hosts of adults; it freely attacks horses and other animals, including man. Though this is a typically ornamented *Dermacentor,* unornamented dwarf forms are known and are due to an inadequate nymphal blood meal (Homsher and Sonenshine, 1973). Immature stages feed almost exclusively on small rodents, with a decided preference for meadow mice (*Microtus pennsylvanicus pennsylvanicus* in the Atlantic states). This tick commonly congregates along paths and roadways where humans are apt to be attacked. The American dog tick has shown an increasing distribution in Nova Scotia, Canada, especially in the western region (Dodds *et al.,* 1969).

A fully engorged female *D. variabilis* drops from a host and in 4 to 10 days lays 4,000 to 6,500 eggs on the ground. Incubation during summer is about 35 days, but temperature influences this stage greatly. Larvae remain on the ground or ascend low-growing vegetation while awaiting a host. Larvae engorge for 3 to 12 days (ave. ca. 4) then drop to the ground to molt in about a week. Nymphs usually engorge on a *Microtus* or *Peromyscus* mouse for 3 to 11 days (ave. ca. 6) and leave to molt in from 3 weeks to several months. Unfed adults may live more than 2 years, but this instar usually soon attacks dogs and other large hosts. Females engorge in 6 to 13 days. Mating occurs on the host as the attached female feeds. Like unengorged adults, immatures have considerable longevity in the absence of suitable hosts, which may prolong the life cycle to two or more years. Under favorable conditions, from egg to adult may require but three months. In Virginia, USA, bimodal seasonal larval activity is attributed to a spring larval population of survivors from the previous season, and late summer larvae emerging from eggs of that spring and summer. Larval and adult spring activity occurs when a minimum daily average solar radiation is reached or exceeded.

D. variabilis continues to expand its range and to be a significant pest of man and a potent vector of the Rocky Mountain spotted fever rickettsia. Studies in Virginia include ecological investigations by radioecological methods, and distribution in relation to vegetation zones (Sonenshine, 1972; Sonenshine *et al.,* 1972). Population spread and characteristics have changed in Massachusetts, USA, where winter mean temperatures around the 0° C level were shown to affect winter survival of adults significantly (McEnroe, 1975). This tick has been introduced over the years into the northwest USA, where a distinct local population has become firmly established in Washington, Idaho, and Oregon, mainly along certain river valleys (Stout *et al.,* 1971).

Dermacentor andersoni, the Rocky Mountain wood tick, is widely distributed and common throughout western North America from British Columbia near 53° N and eastward to 105° latitude in Saskatche-

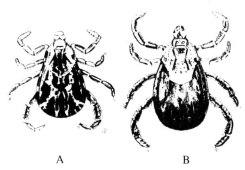

Fig. 16-10 *Dermacentor variabilis.* A. Male. B. Female. (USDA drawing.)

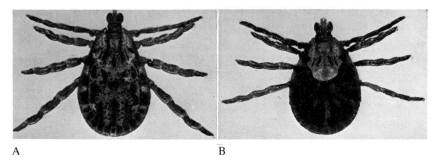

Fig. 16-11 The Rocky Mountain spotted fever tick. *Dermacentor andersoni.* A. Male. B. Unengorged female. × 3.5.

wan and North Dakota southward to New Mexico, Arizona, and California, It is especially prevalent where the predominant vegetation is brushy, with good protection for small mammalian hosts of the larvae and nymphs and with sufficient forage to attract large hosts required by the adults. Adults (Fig. 16-11) feed mostly on large mammals such as horses, cattle, sheep, deer, bears, and coyotes; larvae and nymphs on small mammals such as rabbits, ground squirrels, pine squirrels, woodchucks, and chipmunks; all three stages may feed on intermediate-sized mammals such as jack rabbits and porcupines. Wilkinson (1972) notes that adult Canadian montane forms tend to settle on the upper body of cattle around head, neck, and shoulders, and prairie forms on the lower body. Larval numbers in a study area in western Montana, USA, were estimated by radiolabeling techniques to be around 247,000 to 494,000 per hectare in two seasons (Sonenshine *et al.,* 1976).

The life cycle of *D. andersoni* is fully described by Cooley (1932) with further details provided by Wilkinson (1967, 1968). Copulation occurs on the host and the fully fed and greatly distended female drops to the ground. After a preoviposition period of about a week, egg-laying of a mass at the front of the female and averaging some 6,400 eggs continues for about 3 weeks. The eggs incubate about 35 days, then the larvae hatch, find suitable hosts, feed for 3 to 5 days, drop off, molt in 6 to 21 days, and emerge as eight-legged nymphs. Nymphs that find a host feed, molt, and overwinter as adults, a 1 year life cycle being common where rodents are plentiful. If they do not feed, nymphs hibernate and seek hosts the next spring. Thus a seasonal bimodal nymphal feeding period can be found with spring-feeding nymphs on the two-year cycle and late summer nymphs on the one-year cycle. Fully engorged nymphs drop to the ground, molt in 12 to 16 or more days, and transform into adults. The cycle can therefore be either one or two years, and a three-year cycle is believed to be possible at high altitudes and at the northern limits of the range. Larvae feed throughout the summer, and adults commonly disappear by about July 1, but nymphs continue in diminishing numbers until late summer. Since man is usually bitten only by adult *D. andersoni,* danger from this source exists from early spring to about the first of July, or well into August at high elevations.

Like other ticks, *D. andersoni* is very resistant to starvation. Hunter and Bishopp recorded more than 317 days between the beginning of hatching and the death of the last larva. Unfed nymphs have survived about a year, and adults collected on vegetation in spring survived 413 days without food.

Dermacentor albipictus, the winter tick (or elk or horse tick), is widely distributed in

North America. It occurs in Canada from the east to the west coast and north to nearly 60° (Wilkinson, 1967). This one-host species does not feed in summer. The eggs are laid in spring and hatch in three to six weeks. Larvae bunch tightly together, remaining in torpor until the first cold weather in autumn, when they actively seek hosts. Molting occurs on the original host animal. Females reach maturity with the final molt and engorge, usually in about six weeks after larval attachment. Although females drop off the host after final engorgement, egg-laying is delayed until spring, often after several months. In British Columbia, Canada, seed ticks are active from October to April, and fed females occur in late March (Gregson, communication to R.F.H.).

Heavy infestations on horses, moose, elk, or deer can cause death; cattle are seldom attacked. Man may be infested through transfer of the tick by direct contact, as when skinning and dressing elk or deer. The species named *D. nigrolineatus* from the southeastern USA will experimentally interbreed freely with *D. albipictus* and is consequently believed to be the same (Ernst and Gladney, 1975).

Dermacentor occidentalis, the Pacific Coast tick, is distributed in Oregon and California. Adults of this tick have been taken from many large domestic animals, deer, and man; immature stages occur on many kinds of smaller mammals.

Genus Anocentor. Inornate ticks, eyes present but obsolescent. The only known species, the tropical horse tick, *Anocentor nitens,* chiefly attacks horses. It is known from Florida, Georgia, and extreme southern Texas, USA, to Brazil (Fig. 16-12). In some classifications this tick is designated as a species of *Otocentor* or *Dermacentor*.

Genus Hyalomma. Usually large ticks; no body ornamentation; eyes present; festoons coalesced. The genus consists of 21 species, probably originating in semidesert

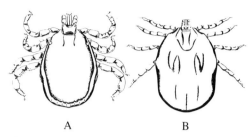

Fig. 16-12 *Anocentor nitens. A.* Male. *B.* Female (not same scale). (USDA drawing.)

or steppe lowlands of Central Asia. It is presently distributed from the Indian subcontinent through much of the Soviet Union, the Middle East, Arabia, North Africa into southern Europe, and in southern Africa. Adults of many species parasitize all domestic and some wild mammals, waiting for hosts in the shade of rodent burrows or plants and moving toward hosts when these are sensed. Larvae and nymphs of some species also attack domestic animals; others parasitize small wild mammals, birds, and/or reptiles. Life cycles are the one-, two-, or three-host type and often variable within a single species.

Among the hardiest of ticks, many species exist under extreme conditions of cold, heat, and aridity. They are variable in appearance and often difficult to identify. The number and variety of mechanical injuries caused, and human and animal pathogens harbored and transmitted by *Hyalomma* ticks are exceptional. In a study of the movement of ticks on birds migrating between Europe and Asia into Africa, *Hyalomma marginatum rufipes* proved to be the commonest species, and cattle grazing and birds associated with agriculture appear to increase the ability of *H. marginatum* to vector the virus of Crimean-Congo hemorrhagic fever.

Genus Nosomma. This genus is represented by a single species, *Nosomma monstrosum* from India and Southeast Asia. The adults mainly parasitize cattle and buffaloes, but also humans, wild boar, bear, horses, and dogs; larvae and nymphs of this three-host tick infest rodents (Avsatthi and Hiregoudar, 1971).

Genus Rhipicephalus. About 63 species and subspecies make up this genus, all Old World and mainly Ethiopian. The brown dog tick, *R. sanguineus,* is now cosmopolitan. *Rhipicephalus* ticks infest a variety of mammals but seldom birds or reptiles. Most African species are three-host ticks with different kinds and sizes of hosts for immatures and adults.

Rhipicephalus sanguineus, the brown dog tick (Fig. 16-13), has the dog as a principal host, although it attacks numerous other animals, and man infrequently (Nelson, 1969). It may be a significant pest of large mammals in zoos. This is probably the most widely distributed of all ticks, found in practically all countries between 50° N and 35° S, including most of the USA and parts of southeastern Canada. It is a vector of the protozoan pathogen causing malignant jaundice of dogs and is considered important in spreading the rickettsia of boutonneuse fever, which man may acquire by crushing

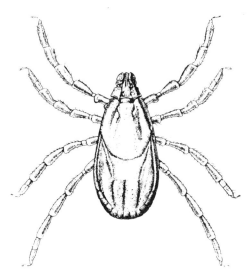

Fig. 16-13 *Rhipicephalus sanguineus,* female. (USDA drawing.)

the ticks. It is a vector of Rocky Mountain spotted fever in parts of Mexico.

Adults are found most often in the ears and between the toes of dogs, larvae and nymphs in the long hair at the back of the neck. Eggs are deposited in cracks and crevices of the kennel or other quarters frequented by dogs. This tick has a strong tendency to crawl upward; hence, it often hides in cracks in kennel roofs or in ceilings of porches. The eggs hatch in 20 to 30 days or more, depending on temperature. The life cycle corresponds to that of three-host ticks and has been described in detail (Sardey and Rao, 1973). Morel and Vassiliades (1962) conclude that ticks identified as this species from Africa, including the Mediterranean area, actually comprise five species separable by morphology, bionomics, and distribution. Three distinct populations in this complex occur in Israel (Paperna and Giladi, 1974). The brown dog tick is frequently introduced into temperate climates but only persists in rather critical temperature and humidity limits provided within kennels and homes.

Genus Boophilus. Five species are recognized; no festoons or ornamentations, eyes present. Unfed *Boophilus* are small and easily overlooked on hosts, making it possible for these ticks to pass quarantine inspections and to be shipped on livestock into uninfested areas. Some question has remained as to whether the nominal species *Boophilus annulatus* and *Boophilus microplus* are truly distinct, since both are widely scattered throughout the tropics and experimental hybridization is known. Studies on hybrids show F_1 females to be essentially normal, but the testes of F_1 males are absent or vestigial, leading Newton, and colleagues (1972) to conclude that these are separate species.

The cattle tick of the western hemisphere, *Boophilus annulatus,* is found in the southern USA and Mexico. A tick known as *B.*

calcaratus, from the Mediterranean basin and the Near East, is believed to be the same, and the origin of *B. annulatus* is thought to be southwestern Asia. It was transported to the New World, probably on Brahmin (zebu) cattle, and became firmly established in certain Mexican states with low daytime relative humidity. Populations apparently referable to this species also occur in central Africa and certain other parts of the world. Although typically infesting cattle, this tick occurs at times on horses, donkeys, sheep, goats, and other animals. In Texas, USA, wild animals in general are poor hosts, and are very adept at removing ticks in all stages. Deer are physiologically suitable, and wild hosts, especially deer, could serve as carriers into tick-free areas but would not be important in maintaining populations (Graham *et al.,* 1972). This tick and the southern cattle tick, *B. microplus,* of similar habits and economic importance, have been eradicated from the USA except for a small, narrow quarantine zone along the Texas-Mexico border, where reintroduction and eradication occur at times. A cooperative USA-Mexico program has been initiated to eradicate this tick from Mexico. In 1906 it was estimated that the economic effect to the southern USA from all losses directly and indirectly occasioned by the cattle tick amounted to $130.5 million. From 1906 to 1918, through cooperative tick-eradication procedures, a total of nearly 1.17 million square kilometers of territory was released from quarantine against Texas cattle fever.

LIFE HISTORY. The life cycle of this one-host tick is in two phases: (1) the *parasitic phase* in which the tick is attached to the host until the adult drops to the ground after fertilization; (2) the *nonparasitic phase* when the tick is on the ground and when eggs and free larvae are subject to severe environmental stresses. After the mature female drops, oviposition usually begins in about 72 hours and continues for 8 to 9 days, but this may be prolonged at adverse temperatures. About 2,000 to 4,000 eggs are usual. Incubation ranges seasonally in Texas, USA, from 19 to 180 days, averaging about 43 days for April, 20 to 26 days from May through August, and 40 days for September. Larvae are active immediately, ascending blades of grass or other objects, where they cluster to attach on a suitable passing host. Longevity of newly hatched larvae varies with temperature: about 39 days in July and 167 days in October. The life cycle may be completed in about 40 days under most favorable conditions but is usually nearer 60. Cattle must be kept on tick-free fields about 65 days to become free of this tick.

Boophilus microplus, the southern cattle tick (Fig. 16-14), seriously affects cattle production in Australia, parts of Mexico, Central and South America, parts of southern Africa, Madagascar, and Taiwan. In Australia strict control schemes may be practiced, including restriction of stock movements, spraying or dipping of cattle and horses, spelling of pastures, and use of resistant zebu-European cross cattle.

McCulloch and Lewis (1968) performed ecological investigations in New South Wales, Australia, that indicated nonparasitic stages could survive no longer than 7.5 months, despite previous studies that concluded longer survival was possible. In open

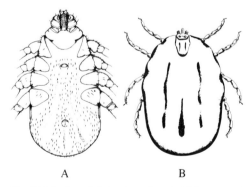

Fig. 16-14 *Boophilus microplus. A.* Male, ventral aspect. *B.* Engorged female, dorsal aspect (not to same scale). (USDA drawing.)

rangeland larvae of this tick may be wind distributed 30 to 80 m, and casual passage on birds and other unsuitable hosts may range from 30 m to 0.8 km (Lewis, 1970). Even in drought periods larvae can take up water from a humid atmosphere prior to dawn, and dew can also be drunk directly (Wilkinson and Wilson, 1959). Survival of low populations is possible through a fairly high incidence of parthenogenesis (Stone, 1963). A mathematical model has been developed to study the population dynamics of this tick, taking into account the effects of climatic conditions on all stages, tick-host contacts, mortality on the host (host resistance), and productivity of engorged female ticks (Sutherst and Wharton, 1971).

Genus Margaropus. These are relict boophilids, with two species infesting giraffes in Central Sudan and East Africa. *Margaropus winthemi*, probably originally from the giraffe or zebra, is found on domestic horses in winter at higher altitudes in southern Africa.

Family Argasidae

SOFT TICKS

The family Argasidae includes about 140 species and four genera of so-called soft or nonscutate ticks in which sexual dimorphism is slight. The integument of all stages except larvae is leathery, wrinkled, granulated, mammillated, or with tubercles (Fig. 16-16). The capitulum is either subterminal or distant from the anterior margin and in adults and nymphs lies in a more or less marked depression, the CAMEROSTOME (Fig. 16-15). The anterior part of the integument extending above the capitulum and forming part of the camerostome wall is called the HOOD. The articulations of the palpi of all stages are free; porose areas at the base of the capitulum are absent in both sexes. Cooley and Kohls (1944) provide a general account of this family, recognizing the four genera *Argas, Ornithodoros, Otobius,* and An-

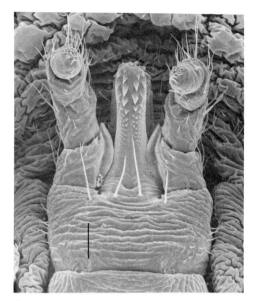

Fig. 16-15 Capitulum on undersurface of *Argas persicus*, lying in a depression, the camerostome. Indicator line is 0.1 mm long. (Scanning electron micrograph by D. Corwin and C. Clifford, Rocky Mountain Laboratory, NIH.)

tricola. Antricola infest cave-dwelling bats in Central and North America.

Argasidae live in deserts, or under dry conditions in wet climates, hiding in crevices or burrowing into loose soil, or climbing the walls of caves and stables inhabited by hosts. As an adaptation to hosts such as birds, bats, or desert vertebrates that shift habitations, these ticks can drastically reduce their metabolic rate and undergo a torpor that permits survival under actual- or near-starvation conditions, often for many years. They are mainly also adapted for feeding rapidly and leaving promptly so that they are seldom collected on hosts (but see *Otobius*). Mating occurs off hosts. Larvae of a few species do not feed before molting, and there are 2 to 8 nymphal molts with the first nymphal instar not feeding in a few species.

Females feed and oviposit several times. It is likely that studies on egg production in *Argas arboreus* are representative of argasids in general, namely that a blood meal provides nutrient for egg development and also stimulates production of a hormone in the central ganglion that is released to the hemolymph to cause egg development in the ovaries (Shanbaky and Khalil, 1975). The life cycle may be completed in as little as 2.5 months or may take several years.

Genus Argas. Members of this genus are distinctly flattened, with margins quite evident even when the tick is fully engorged; integument is leathery, minutely wrinkled in folds, often intermingled with small rounded "buttons," each with a pit on top and often bearing a hair in the pit; eyes are absent. Hosts are exclusively birds, bats, limbed reptiles, or small insectivores.

Argas persicus, the fowl tick, is one of the most important cosmopolitan poultry parasites (Fig. 16-16). It is necessary to note, however, that in the New World what is termed the fowl tick is in reality *A. persicus* (rarely) plus *Argas radiatus, Argas sanchezi,* and *Argas miniatus,* some of which also infest wild species of birds (Kohls *et al.,* 1970). *Argas persicus* is prevalent in the Old World, and there are representatives of this *Persicargas* subgenus of ticks that include wild bird and domestic poultry parasites, including *Argas walkerae* of wild birds and poultry in southern Africa (Kaiser and Hoogstraal, 1969). These species designations were initially based on morphological grounds, but biological validation of species in some New World and Old World *Persicargas* has been provided (Medley and Ahrens, 1970; Gothe and Koop, 1974).

In addition to "fowl tick," these ectoparasites are commonly called "adobetick," "tampan," or "blue bug." They vary from light reddish brown to a dark brown, depending on the stage of engorgement. The ovoid, flattened adults average about 8.5 by 5.5 mm (females), and 6.5 by 4.5 mm (males). The margin of the body is composed of irregular quadrangular plates of cells, often with one or more circular pits (Fig. 16-16).

Nymphs and adults of *Argas persicus*

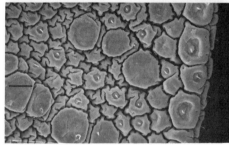

Fig. 16-16 The fowl tick, *Argas persicus.* Lower picture shows upper surface of lower right quadrant, with evident margin and wrinkled folding. Upper indicator line is 1.0 mm long; lower line, 0.1 mm long. (Scanning electron micrograph by D. Corwin and C. Clifford, Rocky Mountain Laboratory, NIH.)

sensu lato are strikingly active at night, traveling some distance to their host and back to hiding places where they stay inactively during the day. Females deposit reddish-brown eggs in crevices occupied during the day. Eggs are laid in masses of 25 to 100 or more, and there are usually several layings, each preceded by a meal of blood, with a total of about 700 eggs per female. In 10 to 28 days eggs hatch; the hexapod larvae apparently attach as readily by day as by night.

Once attached, larvae feed about 5 days,

and when fully engorged they resemble little reddish globules. At the end of this feeding they detach and hide in a convenient crevice, where they molt in about a week, acquiring a fourth pair of legs. Short nocturnal feeding now occurs; there is a molt to the second nymphal stage and one more feed and molt that results in the adult stage. Under favorable conditions the adult stage is reached in about 30 days, but absence of hosts may greatly prolong the developmental time. Adults engorge in 20 to 45 minutes. Instances are recorded in which transient laborers occupying long-vacated but renovated poultry houses have been bitten severely.

Argas reflexus, commonly known as the "pigeon tick," is a European bird argasid that differs from the poultry tick in that the body often narrows rather suddenly toward the anterior end and the thin margin is flexed upward; the body margin is composed of irregular striations. This species and *Ornithodoros coniceps* infest pigeons in Spain and elsewhere in southern Europe, occasionally entering houses and biting man. References to the occurrence of *A. reflexus* in the USA prior to 1960, and mainly in association with cliff swallows, are actually *Argas cooleyi.* There are other species of bird-infesting *Argas* in what is called the reflexus group, and interested students should refer to publications of Hoogstraal and coworkers.

Other *Argas* of note are the following: *A. brumpti,* the largest known species, measuring 15 to 20 by 10 mm; it feeds on a variety of hosts in Africa and will attack man; *A. verspertilionis,* a bat-infesting tick of wide distribution in the Old World, occasionally attacks man.

Genus Ornithodoros. In this genus the capitulum is either subterminal or distant from the anterior margin; the hypostome well developed. The body is more or less flattened but strongly convex dorsally when distended. The integument pattern is continuous over the sides from dorsal to ventral

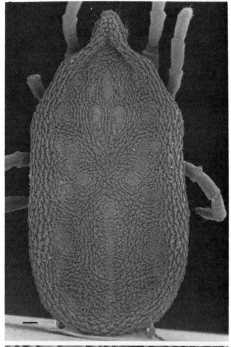

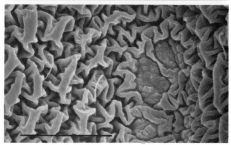

Fig. 16-17 *Ornithodoros hermsi.* Lower picture shows upper surface of lower right quadrant with characteristic wrinkling. (Scanning electron micrograph by D. Corwin and C. Clifford, Rocky Mountain Laboratory, NIH.)

surfaces (Cooley and Kohls, 1944) (Fig. 16-17). The genus includes about 90 species in seven subgenera; some 20 species occur in North America, Central America, and Cuba. Balashov (1970) has reviewed experimental interspecific hybridization in 3 species of *Ornithodoros* ticks. There are bird- and mammal-infesting representatives.

THE ORNITHODOROS MOUBATA COMPLEX. What was originally regarded as a single widespread species of eastern, central, and southern Africa, with several varieties, has been separated on the basis of crossing and morphological studies into four distinct species and a subspecies by Walton (1962). Because this species is the only known vector of relapsing fever spirochetes in those areas, the habits and ecology of each member of this mainly night-feeding group are matters of great importance. According to earlier published information on *O. moubata sensu lato* (Fig. 16-18), eggs are deposited in batches of 35 to 340, with a total maximum of 1,217 per female. The larvae are completely quiescent and do not feed, molting in a few hours within the split eggshell to the first nymphal stage. About five feeding periods and molts follow to reach the adult stage. The list of possible hosts is believed to be very long.

According to Walton's classification the *moubata* complex is divided as follows: (1) *Ornithodoros moubata*, a domestic and sometimes wild species from Angola, southwest Africa, Bechuanaland, Mozambique, Tanzania; a wild species from Kenya and Rhodesia. This is apparently a tick of arid conditions that may be widely distributed in the wild state. (2) *O. compactus* from tortoises in Cape Province, and in an area bounded in the north by the Zambezi river. (3) *O. apertus,* a large and rare species from the burrows of porcupines (*Hystrix*) in the Kenya highlands and probably also from Ghana and Bechuanaland. (4) *O. porcinus,* common and widely distributed in large burrows and animal lairs, and hollow baobab trees. The warthog (*Phacochoerus*) is its main host.

O. porcinus is known from east, central, and southern Africa; also from Madagascar. A study of the distribution of this species in animal burrows in East Africa showed that populations ranged from a few to 250,000 per burrow, that optimum conditions were found at about 1,000 to 1,600 m altitude; the warthog was the favorite among four hosts and from this host it circulates African swine fever virus. Possibly high populations of this tick contribute to the mortality of young warthogs (Peirce, 1974). *O. porcinus domesticus* is common in native dwellings at all altitudes in East Africa and northern Mozambique, and overlaps *O. moubata* in southern Tanzania and areas of southwest Africa. *O. p. domesticus* occurs in three races (Walton, 1964) typified by their occurrence (1) in damp cool conditions at high altitude, feeding chiefly on man; (2) in hot climates with long dry periods, feeding on man and domestic fowls; (3) in hot moist climates, feeding almost exclusively on fowls. Typical *O. porcinus* in human dwellings can survive starvation at least 5 years.

Ornithodoros erraticus occurs in Spain, Portugal, and northern Africa. It is an important vector of relapsing fever pathogen in northern Africa. This tick is particularly sedentary and adapted to burrows, and because hybridization attempts among strains yielded no normally fertile offspring, a species complex is suspected.

Ornithodoros talaje is a South and Central American (south to Argentina) and Mexican species occurring also in Florida, Texas, Arizona, Nevada, Kansas, and California, USA. It feeds on wild rodents, swine, cattle, horses, man, and other animals, inflicting a very painful bite. It is a vector of the relapsing fever spirochete in Guatemala, Panama, and Colombia.

Ornithodoros rudis (=*venzuelensis*) is a Central and South American species, considered the most important vector of relapsing

Fig. 16-18 African relapsing fever tick, *Ornithodoros moubata.* × 3.

fever pathogens in Panama, Colombia, Venezuela, and Ecuador. It appears to be essentially a parasite of man, but feeds on other animals as well.

Ornithodoros tholozani is a vector of *Borrelia* in central Asia, and several other species in Asia, Africa, and Europe are proven or suspected vectors.

Ornithodoros hermsi is a rodent parasite and proven vector of relapsing fever spirochetes. It is widespread in the Rocky Mountain and Pacific Coast states, USA. Herms and Wheeler (1936) described its life history. Tiny amber-colored eggs are deposited at intervals in batches of 12 to 140 from May to October and range over a total of 200 per female. In natural conditions the eggs are deposited in hiding places of the ticks; in summer cabins in such corners and crevices as afford protection of adult ticks. The incubation period at 24° C and 90 percent humidity ranges from 15 to 21 days. The larvae remain attached to the host for about 12 to 15 minutes; in later stages attachment may be a half hour to an hour.

Molting occurs about 15 days after feeding, with four pairs of legs appearing in the first nymphal instar. There follow two additional blood meals and nymphal instars, another blood meal, and the resultant molt to the adult. Egg laying begins about 30 days after last molt, fertilization and blood feeding taking place a few days after maturity. The cycle from egg to egg under laboratory conditions is about 4 months, but may be greatly prolonged in the absence of food. Larvae live up to 95 days without food, unfed first-stage nymphs up to 154 days, unfed second-stage nymphs to 79, and third-stage nymphs to 109 days, adults well over 7 months. Adults have been kept alive for 4 years with occasional feedings.

Ornithodoros coriaceus, known as the *tlalaja* in Mexico and the *pajaroello* in California, is a large argasid associated with deer and cattle bedding areas, where these ticks emerge from soil to attack hosts. They may be collected in such sites by attracting them with carbon dioxide emanating from blocks of dry ice. The bite is reported to be painful, with serious venomous aftereffects, though this is often exaggerated (see Chapter 17). A survey by Loomis and colleagues (1974) shows this tick is widespread in mountainous areas throughout the length of California, and it has been reported from adjacent counties in Oregon (USDA, 1975. *Coop. Econ. Insect Rept.,* **25**:821).

A number of *Ornithodoros* that normally infest freshwater or marine birds transmit viruses of uncertain medical significance. These ticks will bite humans when they are in close association (Hoogstraal, *in* Gibbs, 1973).

Genus Otobius. Cooley and Kohls (1944) recognize two species, *O. megnini* and *O. lagophilus,* the latter a parasite on the face of lagomorphs in the western USA and Canada. In adults the integument is granulated, nymphs have spines; sexes similar; capitulum distant from anterior margin in adults and near margin in nymphs; hypostome well developed in nymphs but vestigial in adults.

The ear tick, *Otobius megnini,* is widely distributed in warmer parts of the USA and in British Columbia, Canada. It has been transported to other parts of the world and is a serious problem in South America, South Africa (Theiler and Salisbury, 1958) and India (Chellappa, 1973). It receives its common name from the spiny covering of the nymphs (Fig. 16-19), which along with larvae invade the ears of cattle, horses, mules, sheep, cats, dogs, and other domesticated animals as well as deer, coyotes, rabbits, and other wild animals. In British Columbia this tick has been found in the ears of mountain sheep, and deaths of cattle are attributed to it. There are several records of nymphs in the ears of man.

Rather large dark eggs are deposited by this species on the ground; under laboratory conditions at about 21° C the incubation period is 18 to 23 days. In the field newly

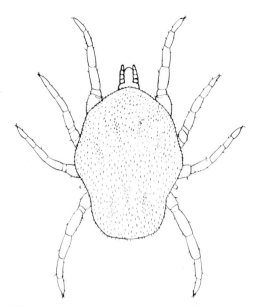

Fig. 16-19 *Otobius megnini* nymph. (USDA drawing.)

emerged larvae ascend weeds and other vegetation, contact suitable hosts, and gradually work their way to the shoulders, neck, and head, and thence to deeper inner folds of the outer ear where they engorge and assume a peculiar saclike form. In British Columbia larvae were abundant in a cave where wild sheep sheltered (Rich and Gregson, 1968). After molting in the ear, nymphs attach and remain for as long as 121 days; the second nymphal stage is the one in which this species is most easily distinguished from other ticks. On detaching they drop to the ground and molt to adults. Herms noted that copulation takes place within a day or 2 and oviposition occurs in 14 to 22 days, with a maximum oviposition period of 155 days and a total of 562 eggs produced. Unfed larvae at room temperature lived 19 to 63 days (ave. 44).

Tick Transmission of Arboviruses

The arboviruses associated with ticks are Togaviridae, *Flavivirus,* in B group; Reoviridae, *Orbivirus;* or ungrouped viruses.

Thus far they have proved to be transmitted to vertebrates in tick salivary secretions; transstadial passage is often essential for maintenance, and transovarial and sexual transmission in ticks are known (cf. Il'enko *et al.*, 1970; Plowright *et al.*, 1974). With minor exceptions, all those affecting humans and livestock are transmitted by ixodid ticks. Reliance has been placed on Horsfall and Tamm (1965), Hoogstraal (1966), recent original papers, and particularly Hoogstraal (*in* Gibbs, 1973) and Berge (1975) in preparing this account.

Hoogstraal points out some 68 viruses (about 21 known to affect man) have been identified from 60 ixodid and 20 argasid species of ticks. A complete picture of these biocenoses has often not been developed because many virus isolations have been the result of casual tick collecting. It is doubtful that the few tick-transmitted viruses identified from South America and China truly represent conditions in those vast areas, and great portions of the southern and eastern Pacific regions have been inadequately investigated. While it would seem that relatively sedentary parasites like ticks would be vectors in rather stable virus-vertebrate host associations, ticks are distributed great distances on hosts, and the ecology of regions continues to change through man's activities; therefore, formerly unnoticed viruses circulating in wild hosts may suddenly affect man and domestic animals. Although tick vectors may remain in some of these areas, the intricate total relationships of vertebrates responding pathologically to viruses, as well as nonsymptomatic vertebrate portions of the reservoirs, may cause marked shifts in pathogen circulation. The population dynamics of vectors of tick-borne arboviruses has been reviewed by Sonenshine (1974).

Tick-vertebrate-virus habitats are extraordinarily inclusive: continental areas, isolated islands, primary and secondary forests, wide altitudinal ranges, savannas, steppes and plains. Hoogstraal points out that there are two basic patterns of habitat-tick-vertebrate relationships that influence virus circulation: (1) restricted vertebrate habitats and shelter-seeking ticks that feed at all active stages only on sedentary nesters in burrows or trees, or ground-level colonies; and (2) more generalized habitats where adult ticks feed on wandering hosts that are usually larger and ecologically distinct from small or medium-sized hosts utilized by immature ticks.

The discussion that follows concentrates on viruses that affect humans and domestic animals. For this reason the interesting findings on a large number of viruses isolated from birds, and especially freshwater and marine bird colonies, pigeon flocks, heron and stork rookeries, swallow and martin colonies, and vulture nests will not be included except as they involve viruses of man and domestic animals. However, it should be noted that viruses of marine birds may cause mortality in these hosts and could be of importance in decreasing their populations (see Hoogstraal *et al.*, 1976). Refer to Hoogstraal (in Gibbs, 1973) for an inclusive discussion of these relationships.

TOGAVIRIDAE, FLAVIVIRUS (GROUP B)

Viruses in this complex cause serious diseases in humans, and though some of these pathogens have caused hundreds of cases, others appear to circulate in nature in quite isolated zoonoses.

Russian Spring-Summer Encephalitis Virus (RSSE). Russian spring-summer encephalitis is in a complex that includes a number of closely related entities with clinical and antigenic similarities, and that may also have considerable geographical overlap. The disease is associated with taiga forest of the Soviet Union and with the tick *Ixodes persulcatus;* it has been recorded from Prussia and East Germany. Incidence of this disease has increased with man's further development of Siberian and Far Eastern forest zones.

The disease caused by RSSE virus is characterized by sudden onset of violent head-ache, rapidly rising fever, nausea, and vomiting. The incubation period is 8 to 14 days. At the height of illness delirium and coma may be present, as well as convulsions and paralysis. Bulbar center involvement and ascending paralysis occur, and residual paralysis is common. The mortality rate may be 25 to 30 percent.

The severe paralytic form of RSSE is apparently associated with *I. persulcatus*. A possible variant from northern China and the Khabarovsk region of the Soviet Union near the Sea of Japan is associated with the same tick, and with *Haemaphysalis concinna* and *H. japonica douglasi;* it is clinically more severe than tickborne encephalitis (TBE), but like that disease may be acquired from infected milk. Over 50 percent of the permanent human residents of forest zones may have RSSE antibodies and no history of symptoms; clinical manifestations are more frequent in newcomers. The focal nature of this virus has made it possible in some areas to relate highest incidence of cases to coniferous foothill taiga forest seen in aerial photographs.

Ixodes persulcatus is associated with a wide variety of small forest mammals and birds in immature stages, and with larger wild and domestic animals as the adult. *Haemaphysalis concinna* is found extensively in Europe, northern Asia, and western Siberia and may be the principal RSSE virus vector in regions where it outnumbers *I. persulcatus;* it feeds in immature stages within cut-over forest areas on small mammals and birds, and often as the adult on cattle. *H. japonica douglasi* is an important vector of RSSE in the southern portion of the far eastern USSR; it feeds on small and large vertebrates, and all stages attack humans. *Dermacentor silvarum* is naturally infected in the easternmost USSR, but seems of less epidemiological importance than the other ticks mentioned. Other vectors of RSSE include *Ixodes lividus, Haemaphysalis longicornis, Dermacentor reticulatus,* and *D. marginatus.*

Lack of obvious tick relationship in the occurrence of some RSSE cases has raised the question of the role of other arthropods and other means of reservoir maintenance.

Tickborne Encephalitis Virus (TBE). This virus is listed in Hoogstraal (*in* Gibbs, 1973), and as Hypr virus (HYPR) in Berge (1975). It causes a disease that is difficult to separate clinically from RSSE. TBE infection in humans is often a consequence of drinking infected milk. The virus reaches a high level in mammary glands, and at low temperature remains active for days in raw milk and uncooked milk products. By this mechanism, though TBE virus is in low incidence in ticks and wild vertebrates in nature, milk-producing animals serve as a common infection route for man.

The tick *Ixodes ricinus* has been implicated as the main vector, but other European ticks, *Haemaphysalis inermis, H. punctata, H. concinna, Ixodes hexagonus, Dermacentor marginatus,* and *D. reticulatus* are also involved exceptionally (Blaškovič and Nosek, 1972). *Ixodes ricinus* feeds on a wide variety of vertebrates, and TBE virus has been isolated from rodents (*Apodemus, Clethrionomys*) and the European mole *Talpa europaea.* Antibodies to this virus have been found in a wide variety of wild vertebrates, including shrews, hares, bats, and several species of birds. Two hosts of *I. ricinus,* the dormouse *Glis* and hedgehog *Erinaceus,* can experimentally maintain the virus through torpor or hibernation.

This virus may be very focal in nature. In Austria it is associated with high rodent populations in dense, moist, young mixed forests at least 10,000 m² in area, and in quite similar conditions in Switzerland (Radda *et al.,* 1974). Planted protective forest belts of the Western Siberian Railway have created conditions of TBE foci affecting railway workers (Dorontsova and Chudinov, 1971). Though often considered to be associated with heavy coniferous forest, it will also extend into the transitional zone (Vasenin *et al.,* 1975). Finnish strains are

termed Kumlinge virus (KUM) (Berge, 1975). Birds serve as a host for *I. ricinus,* which could explain how isolated but otherwise suitable ecological pockets acquire the virus.

Louping Ill Virus (LI). Primarily involving sheep, this disease has received extensive review by Varma (1964). LI virus causes an encephalomyelitis of high mortality in sheep and can affect cattle and humans. Serious involvement in sheep is indicated by a jumping vertigo, or loup, known from Scotland for over a century and a half. The disease occurs in hills and moors of Scotland and northern England, and is widespread in Ireland. Walton (1967) discusses the ecology of this disease.

Cases in humans usually occur through close contact with sheep by laboratory workers, butchers, veterinarians, and sheep farmers; infection by tick bite is apparently also possible. In humans the disease resembles tickborne encephalitis, involving influenzalike symptoms; complications include temporary paralysis, and one death is known.

This virus seems to circulate chiefly among sheep and *Ixodes ricinus,* though because this tick feeds on a wide variety of animals many vertebrates have been examined. Isolations of LI virus have been made from red grouse (*Lagopus*), wood mouse (*Apodemus*), and shrew (*Sorex*); antibody surveys suggest the involvement of other mammals. Transstadial transmission of LI virus has been demonstrated in *I. ricinus* and transovarial transmission is inadequately proved.

Negishi Encephalitis Virus (NEGE). This virus was isolated from a human corpse in Japan. Because of antigenic relationships to RSSE complex viruses, it is assumed to be tick transmitted.

Omsk Hemorrhagic Fever Virus (OHF). This pathogen is the causal agent of an acute febrile disease of man. It was first isolated in 1947 from human cases in the Omsk area of southwestern Siberia; further distribution in

Central Asia is not clear, but a disease termed "Bukovinian hemorrhagic fever" from Bukovina, northern Romania, and a similar syndrome in the Ukraine may be identical.

The disease has an incubation period of 3 to 7 days with an acute onset of headache and fever, often biphasic. There is frequently atypical bronchopneumonia, hemorrhagic rash, and extensive internal hemorrhage; the mortality rate is 0.5 to 3 percent.

The ticks *Dermacentor pictus, D. marginatus, Ixodes persulcatus,* and *I. apronophorus,* a parasite in rodent burrows, appear to constitute the reservoir. Long-term existence of natural foci is possible through transstadial transmission and long maintenance of the virus in ticks with a 3-year development cycle (Ravdonikas *et al.,* 1971). Birds may circulate this virus in the western part of its distribution (Ravkin *et al.,* 1973). Introduced muskrats (*Ondatra*) along streams in the Omsk focus died in large numbers between 1960 and 1962, and gamasid mites were implicated in passing the virus between these mammals and voles (*Arvicola*); a number of cases of Omsk hemorrhagic fever occurred among persons handling muskrat carcasses and pelts. The Bukovinian agent is associated with *Ixodes ricinus* infestation of people entering forests.

Kyasanur Forest Disease Virus (KFD). Kyasanur forest disease is a well-documented, seemingly new disease caused by an RSSE complex arbovirus. A number of excellent reviews include that of Work (1962). Discovered in 1957 as a zoonosis killing monkeys, and first suspected to be sylvan yellow fever, it was the first demonstration in Asia of a RSSE-like agent. As an active clinical entity KFD is thus far known only from the 1,500-square kilometer Shimoga District of Mysore State, India. Serological studies indicate the virus is widespread elsewhere in that country. Curiously these other foci include arid regions sharply differing from the forest environment where human disease occurs. Since 1951 the hu-

man population of the affected area has more than doubled, and alterations of the ecosystem have favored expression of a former hidden enzootic process (Boshell, 1969).

Clinical manifestations of the disease are sudden onset of fever, headache, and severe muscle pains; prostration is common. In a significant number of cases a diphasic course is seen, with mild meningoencephalitis in the second phase. Bronchiolar involvement with a persistent cough and gastrointestinal disturbance are usual. Convalescence is prolonged, but lasting damage is not seen; about a 5 percent fatality occurs.

Human KFD has coincided with epizootics of local monkeys. Langur monkeys (*Presbytis entellus*) and bonnet macaques (*Macaca radiata*) are often fatally affected, and examination of these and humans in the disease area reveals the presence of ticks, especially of the genus *Haemaphysalis.*

Virus isolations from a number of pools of ticks have implicated *H. spinigera* in particular and *H. turturis,* in which the virus persists in nymphs throughout the year; six additional species of *Haemaphysalis* are involved, and there have been isolations from *Ixodes ceylonicus, I. petauristae, Dermacentor auratus,* and *Rhipicephalus turanicus.* Isolations from *Ornithodoros chiropterphila,* an argasid parasite of insectivorous bats, and from bats, suggest a mechanism for widespread circulation of KFD virus.

A variety of small rodents, squirrels, shrews, and porcupine (*Hystrix*) that serve as hosts of immature ticks in the Kyasanur Forest area have yielded KFD virus. Monkeys of the region are infested by *Haemaphysalis* ticks, and peak infestation, predominantly by larvae, occurs in the postmonsoon dry month of November. During dry months humans of the region turn from agricultural pursuits to wood gathering in forests, thereby entering infectious foci at a time of high virus transmission.

Haemaphysalis spinigera is very inclusive in its host range. Small mammals are frequent hosts of the immature stages; birds are also attacked, and from these KFD antibodies have been demonstrated. The adults prefer large mammals, either wild, or domestic forms such as cattle. Cattle grazing within forests therefore also spread the principal vector tick and bring it into close association with villages; restricting roaming of cattle in forests, and tick control on village cattle, may be useful for KFD prevention.

Langat Encephalitis Virus (LANE). Isolated from ticks and rats in Malaysia, this virus is experimentally capable of causing human encephalitis transmissible by several species of ticks. *Ixodes granulatus* of tropical Asian forest rodent and shrew burrows has been found infected in nature, and LANE virus seems to occur in restricted habitats.

Royal Farm Virus (RF). This pathogen has been isolated from *Argas hermanni* in pigeon cotes in Afghanistan. Human infections are presumed to be mild.

Powassan Encephalitis Virus (POWE). This is the only known American representative of the tick-borne RSSE complex. It was first isolated from a fatal encephalitis case in northern Ontario, Canada, in 1958. Widely distributed foci are recognized in nature, but clinical cases have been restricted to Ontario and Quebec, Canada (Harrison *et al.,* 1975), and New York State, USA (MMWR, 1972, 21:207–208; 1975, 24:375). A strain of this agent has been isolated in the southern Maritime Province, Soviet Union, from *Haemaphysalis longicornis* (L'vov *et al., 1974).*

Natural isolations have involved *Dermacentor andersoni* in Colorado and South Dakota, and *Ixodes spinipalpis* from *Peromyscus* mice in Connecticut and South Dakota. The virus has been followed transstadially in *D. andersoni* by organ titration and immunofluorescence (Chernesky and McLean, 1969). Serological evidence from New York State implies exposure of racoons and foxes, and man only minimally. In northern Ontario isolates of virus along with

seasonal serological sampling implicate the groundhog *Marmota monax* and the tick *Ixodes cookei,* also the red squirrel *Tamiasciurus hudsonicus* and the tick *Ixodes marxi.* Evidence from other serological studies reinforces the conclusion that wild rodents and hares, and their tick parasites, constitute the reservoir; man seldom enters the picture because the ticks normally involved rarely bite humans.

MINOR B GROUP ARBOVIRUSES

Viruses Isolated from Ticks and Humans. These include Absettarov virus (ABS) from *Ixodes persulcatus* and *I. ricinus,* central Europe; Hanzolova virus (HAN) from *Ixodes ricinus,* Czechoslovakia; West Nile virus (WN), generally mosquito-borne, from *Argas hermanni* (Egypt), and *Hyalomma m. marginatum,* southwestern USSR.

Viruses Not Known to Cause Human Disease. These include Kadam virus (KAD) from *Rhipicephalus pravus* in Uganda, Tyuleniy virus (TYU) from *Ixodes uriae* in northern USSR and Oregon, USA.

TICKBORNE ARBOVIRUS GROUPS OTHER THAN GROUP B

Kemerovo Group, Kemerovo Subgroup (Reoviridae, *Orbivirus***).** There have been six isolates from *Ixodes* ticks in eastern Europe; marine bird colonies in Oregon and Alaska, USA, and Newfoundland, Canada.

KEMEROVO VIRUS (KEM). This pathogen has been isolated from *Ixodes persulcatus* and humans in Siberia. Clinical symptoms are like those of tickborne encephalitis, with febrile, benign, and nonparalytic characteristics.

TRIBEC VIRUS (TRB). This virus has been isolated from *Ixodes ricinus, Haemaphysalis punctata,* rodents, and humans in eastern Europe. It is not considered dangerous.

Kemerovo Group, Wad Medani Subgroup (Reoviridae, *Orbivirus*). WAD MEDANI VIRUS (WM). This virus has been isolated from ixodid ticks and livestock, particularly cattle, goats and sheep, and may be quite pathogenic in these last two hosts. Tick isolations have included *Hyalomma marginatum isaaci* in India, *H. a. anatolicum* in Pakistan, *Amblyomma cajennense* in Jamaica, *Rhipicephalus sanguineus* in Sudan and Egypt, *Boophilus microplus* in Singapore and Malaya.

Uukuniemi Group (Bunyavirus-Like). This group includes five entities from *Argas, Rhipicephalus,* and *Ixodes ricinus* in Eurasia, parasitizing pigeons, goats, and a variety of small and large hosts.

Crimean-Congo Hemorrhagic Fever Group (Bunyavirus-Like). CRIMEAN-CONGO HEMORRHAGIC FEVER VIRUS (CCHF). This virus has been isolated mainly from *Hyalomma m. marginatum* from southwestern USSR and southern Europe on birds and domestic animals in rural settings. A translation of an extensive Russian review of this disease is available (Hoogstraal, 1974). Isolations have also occurred from ticks mainly associated with livestock, especially cattle in Africa—*Dermacentor marginatus, Rhipicephalus rossicus, Boophilus annulatus, B. decoloratus, Amblyomma variegatum, Hyalomma marginatum rufipes, H. truncatum, H. impeltatum, H. a. anatolicum*—this last species also from Central Asia (USSR) and Pakistan.

CCHF was first observed in 1944–1945 among presumably nonimmune Russian troops assisting farmers in the war-devastated Crimea. Later cases were noted from the Rostov and Astrakhan regions. Hemorrhagic symptoms occur along with seriously acute febrile conditions. Human cases originated from the wide Crimean steppe, not from cities, forests, or mountains. Birulya and colleagues (1971) point out that foci of human cases from Bulgaria to Central Asia are areas with 2,800 to 5,000 day-degrees above a 10° C threshold in transitional zones between forest-steppe and desert, namely, land suited only for pasture and on loess plains in large river valleys. *H. marginatum* occurs in a widespread complex, members of which infest birds readily and may be transported to nearby localities and from Africa to Asia and Europe. Human cases occur possibly from crushing ticks with the fingers as well as by tick bite.

Kaisodi Group (Bunyavirus-Like). This group consists of three isolates, mainly from *Haemaphysalis* ticks infesting small and large mammals in Malaya, India, Canada, and the USA.

Ganjam Group (Bunyavirus-Like). GANJAM VIRUS (GAN). This pathogen has been isolated from *Haemaphysalis intermedia* and *H. wellingtoni* infesting small and large vertebrates on the Indian subcontinent. It has also been isolated from mosquitoes and causes fever in humans.

DUGBE VIRUS (DUG). This virus is associated with *Hyalomma marginatum rufipes, Amblyomma variegatum, A. lepidum* and *Boophilus decoloratus* in Nigeria. It has been isolated from cattle and humans, causing moderately severe fever in the latter.

Hughes Group (Unclassified). These are mainly marine bird-associated viruses, generally from *Ornithodoros* ticks in Ethiopia, the Arabian Gulf, and the Americas.

HUGHES, PUNTA SALINAS, AND ZIRQA VIRUSES. Five isolates, including Hughes (HUG) and Punta Salinas (PUNS) viruses, may infect humans; Zirqa virus (ZIR) apparently causes fever, headache, itching, and erythema in affected humans.

Quaranfil Group (Unclassified). QUARANFIL VIRUS (QUA). This virus has caused human febrile illness and has been isolated from *Argas* ticks with herons in rural surroundings on the African continent, and from rural pigeon houses in Afghanistan and Nepal. In this group isolates from *Ornithodoros* ticks have been associated with Austrialian and mid-Pacific marine bird colonies.

Dera Ghazi Khan Group (Unclassified). This group contains five viruses from various bird-infesting *Argas* ticks in Egypt and southern to southwestern Asia, and *Hyalomma dromedarii* from camels in Pakistan.

UNGROUPED TICK-BORNE VIRUSES

There are about 20 entities that are ungrouped, and the number will continue to grow except as some of these can be placed into groups based on antigenic or viroid similarities. Only those presently known to affect humans or seriously affect livestock are mentioned.

African Swine Fever Virus (ASF). This virus causes an acute, contagious disease with high mortality in domestic pigs. This disease and its tick relationships are reviewed by Plowright and associates (1970, 1974). In Africa domestic swine usually acquire the virus by association with wild pigs, especially warthogs (*Phacochoerus aethiopicus*), and it is maintained in their burrows by *Ornithodoros porcinus*. This tick transmits ASF virus sexually; it persisted in the vector for 12 to 15 months, and its distribution in the systems of this acarine was studied by immunofluorescence (Greig, 1972). The disease spread to Portugal in 1957 and subsequently to Spain, Italy, and France, causing great economic losses. In Spain *O. erraticus* can apparently maintain and transmit the virus. Wild specimens of *O. porcinus* have rather low rates of infection, and the virus level in infected warthogs, which are relatively resistant, is quite low. Epizootics among domestic pigs involve highly virulent strains, and transmission is accomplished by tick bite or by contact between pig.

Colorado Tick Fever Virus (CTF). The features of disease caused by this virus are summarized below.
Disease: Colorado tick fever (CTF).
Pathogen: An ungrouped arbovirus.
Victims: Humans.
Vectors: Mainly ixodid ticks, especially *Dermacentor andersoni*.
Reservoir: Ticks; transstadial passage occurs and the virus persists in overwintering nymphs.
Distribution: USA and Canada; all the Rocky Mountain states, Black Hills of South Dakota, British Columbia in western Canada.
Importance: 312 cases reported in USA during 1976, 220 of these from Colorado. Fatalities not common, but encephalitis and bleeding may occur in children.

In 1885 Ewing called what was apparently CTF "mountain fever," and the disease was noted soon thereafter in the Bitter Root Valley of Montana, USA. The etiological agent was subsequently shown to be a virus transmitted by ticks (Fig. 16-20), now recognized as the only tick-transmitted viral disease common in North America. Only subclinical human infections are known from Canada (McLean, 1975).

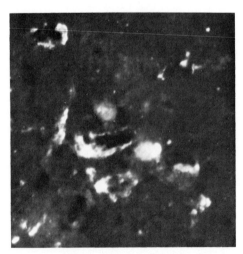

Fig. 16-20 Reaction between Colorado tick fever virus and fluoroscein isothiocyanate-labeled antibody globulin in mouse brain, approximately 2,250 ×. (Photograph by W. Burgdorfer, Rocky Mountain Laboratory, NIH.)

Frequently the disease is diphasic, with denguelike symptoms and a remission period lasting 2 to 3 days. Incubation period, following the tick bite, is 3 to 6 days. Rash occurs but is not usual. Almost exclusively in children, there may be encephalitis and severe bleeding. No lasting complications are reported.

Human cases of this disease are closely associated with the tick *Dermacentor andersoni; D. occidentalis* has been found infected in Oregon and California. The cycle in nature has also been associated with the rabbit and hare ticks *D. parumapertus, Otobius lagophilus,* and *Haemaphysalis leporispalustris*. Rodents having a prolonged viremia are important in maintaining the pathogen; in western Montana the golden-mantled ground squirrel, *Citellus lateralis tescorum,* is considered the main vertebrate host (Burgdorfer and Eklund, 1959), and several other small rodents and porcupine (*Erethizon*) are hosts. Transstadial passage of CTF virus occurs in *D. andersoni,* but transovarial passage has not been demonstrated; unfed nymphs overwinter the virus. This virus has been grown in tissue cultures of *D. andersoni* cells, with no differences observed in the growth of virus-infected and uninfected cultures.

Bhanja Virus (BHA). This virus has been isolated from *Haemaphysalis intermedia* (India), *H. punctata* (Italy), *Hyalomma truncatum, Amblyomma variegatum,* and *Boophilus decoloratus* (Nigeria), the last species also in Cameroon (Vinograd *et al.,* 1975); all species infest domestic animals, and birds and rodents in one case. It may affect goats and sheep in India, cattle, goats and sheep elsewhere, and antibodies have been demonstrated in these animals, humans, rodents, and birds in Italy.

Thogoto Virus (THO). This virus has been isolated from *Hyalomma a. anatolicum* (Egypt), *Rhipicephalus* sp., *Boophilus decoloratus* (Kenya); this last species and *H. truncatum* and *Amblyomma variegatum*

(Nigeria); *R. bursa* (Sicily). It is associated with camels and cattle, and rarely causes meningoencephalitis in humans.

Nairobi Sheep Disease Virus (NSD). This pathogen occurs in *Rhipicephalus appendiculatus* and causes considerable economic losses to sheep breeders in East Africa. The agent passes transovarially to progeny, which can transmit it more than 100 days after hatching. Surveys show up to 20 percent of humans with NSD antibodies near Entebbe, Uganda.

OTHER POSSIBLE TICK INVOLVEMENT WITH VIRUSES

There are other situations where ticks have been demonstrated to be potential vectors or reservoirs of arboviruses, or viruses not ordinarily associated with arthropods. The more important and interesting of these possibilities include the following:

Lymphocytic Choriomeningitis. The causative virus has been isolated in Ethiopia from the blood of human patients and from *Amblyomma variegatum* and *Rhipicephalus sanguineus;* from *Dermacentor andersoni* in Canada.

West Nile Virus. Experimental work has suggested the potential of ticks in maintaining this mosquito-transmitted virus, and the agent has been isolated in midwinter from *Argas reflexus hermanni* in Egyptian pigeon cotes.

Other Viruses. Other viruses definitely or possibly transmitted by ticks include tick-borne meningopolyneuritis virus in Cologne, Germany (Hörstrup and Ackermann, 1973); enzootic abortion agent of sheep and *Ornithodoros lahorensis* (Terskikh *et al.,* 1973); canine hepatitis virus and *Rhipicephalus sanguineus* (Corrado and Mantovani, 1966); a swine-pox virus and *Hyalomma a. anatolicum* in India (Tewari *et al.,* 1974); fowlpox virus and *Argas persicus* (Shirinov *et al.,* 1969); and the virus of Newcastle disease and *Argas persicus* (Petrov, 1972).

Tick Transmission of Bacteria

The compilations by Dubos and Hirsch (1965) and Buchanan and Gibbs (1974) provide a basic background on these diseases, as well as reviews by Varma (1962), and Burgdorfer and Varma (1967).

RELAPSING FEVER

Disease: Relapsing fever, tick-borne spirochetosis, endemic relapsing fever; louse-borne relapsing fever, epidemic relapsing fever.

Pathogen: Spirochete bacterium, *Borrelia recurrentis* (louse-borne) and tick-adapted strains

Victims: Humans, rodents.

Vectors: Argasid ticks in genus *Ornithodoros;* body louse of humans (see Chapter 8).

Reservoir: *Ornithodoros* ticks, with transovarian transmission; rodents.

Distribution: Africa, Near East, central Asia, eastern Europe and Mediterranean, western USA and Canada to Central and South America.

Importance: Several hundred cases annually worldwide, 15 cases reported in USA during 1976; morbidity high, mortality low when treated with appropriate antibiotics.

Taxonomic status of the relapsing fever pathogen is in question, the literature suggesting a great number of species but some authorities believing that these are variants of the single species *Borrelia recurrentis* (Fig. 16-21). Characteristically these variants are specific in their ability to develop and to be transmitted by species of argasid ticks in the genus *Ornithodoros* and by lice. Apparently, because of the isolation caused by burrow- or den-infesting proclivities of argasid ticks, a single pathogen has developed strains specifically adapted to single or limited groups of arthropod vectors. Some tick-borne and louse-borne strains can develop in the complementary vectors, and in turn are transmissible to mammals. Weyer (1968) found that tick strains can be maintained in human body lice by intracoelomic injection of the spirochetes, and further preserved for long periods by freezing infected

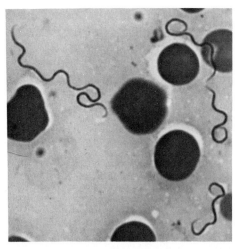

Fig. 16-21 Relapsing fever spirochetes, *Borrelia hermsii*, in smear of mouse blood, Giemsa stain, × 1,950. (Photograph by W. Burgdorfer, Rocky Mountain Laboratory, NIH.)

lice. A "species" designation such as is often made in the literature is used here after the listing in Buchanan and Gibbs (1974). Volzhinskii and coworkers (1974) also provide a list of tick and louse vectors, named strains of spirochetes, and the world distribution of 14 types of tick-borne relapsing fever spirochetes and their *Ornithodoros* vectors and rodent hosts.

The important tick-borne "species" of *Borrelia* are the following: *Borrelia duttoni* from *Ornithodoros moubata* in central and southern Africa; *B. hispanica* from *O. erraticus* in North Africa; *B. persica* and *O. tholozani* from Uzbekistan and Kashmir to Cyprus and Tripoli; *B. crocidurae* from *O. erraticus sonrai* in Africa, Near East, and Central Asia; *B. caucasica* from *O. verrucosus* in the Caucasus mountains; *B. hermsii* from *O. hermsi* at high elevations in the western USA and Canada; *B. turicata* from the relapsing fever tick, *O. turicata,* in the western USA, Kansas to Mexico; *B. parkeri* from *O. parkeri* in western USA; *B. venezuelensis* from *O. rudis* in Central and northern South America; *B. mazzottii* and *O. talaje* in Mexico and Guatemala. There are additional strains from rodents, monkeys, and livestock, but the strains listed above have been found in man. Some strains are highly enzootic, and human infection is rare; there is great variation in strain pathogenicity.

The disease in man is characterized by an acute onset of fever 3 to 10 days (typically 6 to 7) after the bite of an infectious tick. Since *Ornithodoros* ticks feed for a brief period, usually at night, the victim is usually unaware of recent tick exposure. The first attack of fever lasts about 4 days, followed by a similar afebrile period, and so on for 3 to 10 fever episodes. The spirochetes are only readily found in blood and other fluids during attacks of fever, particularly during the first attack. Mortality in endemic situations is 2 to 5 percent, but may be 50 percent or more in epidemics. Penicillin and the tetracyclines provide effective treatment.

Human relapsing fever infections may be transmitted by arthropod vectors from man to man, from animal to animal, and from animal to man. Ticks can serve as a long-term reservoir, with transovarian passage for many generations. The pathogen may be maintained naturally in some species of ticks for at least 13 years (Gugushvili, 1973). The percentage of transovarian transmission varies greatly among tick species. Rodents may serve as natural sources of infection for ticks. Transmission is by the bite of ticks of either sex in all active stages. Fluids of the coxal glands play an important role, inasmuch as many species produce copious infectious coxal fluids from which spirochetes are introduced into the bite wound or penetrate unbroken skin.

Relapsing fever is essentially worldwide in distribution except for Australia, New Zealand, and Oceania. Epidemics of the disease are normally louse-borne; endemicity is characteristic of tick-borne cases. Tick-borne endemicity is prevalent in central and south Africa, much of Asia, and the Americas.

Development of *Borrelia* in tick hosts has been well studied, though not with all tick species involved. The early studies were with *Ornithodoros moubata* (or ticks in that complex) and *Borrelia duttoni*. The findings of Dutton and Todd, and Koch, in 1905, confirmed the beliefs of natives in many parts of Africa, who are reported to have dreaded tick bites. According to information cited by Walton (1962), some African tribes carried ticks on their journeys in order to retain immunity. Burgdorfer (1951) reviewed further studies on development, and through his own investigations provided a detailed picture of the progressive development of the spirochetes in the tick (Fig. 16-22). The pathogens concentrated along the gut wall after ingestion, and were in the hemocoel in as early as 24 hours. Multiplication followed, with invasion of various tissues. As early as the third day following ingestion spirochetes were in salivary glands, coxal glands, and central ganglion, and by the fourth day in the malpighian tubes. Nymphs showed a prolonged heavy infection of salivary glands but in adults heavy infection of these glands was temporary. Tests demonstrated that the central ganglion and coxal glands seemed most attractive to spirochetes. Spirochetes invaded the ovaries to infect eggs prior to the time they had developed an eggs shell (cf. Aeschlimann, 1958).

Further studies indicate that infectivity of *Borrelia* for man and other mammals may best be maintained under natural conditions when transmission occurs alternately between ticks and mammals. Experimentally, prolonged serial passage through one type of host seems to modify infectivity, increasing it in some studies and decreasing it in others. Foci of tick-borne spirochetosis in uninhabited parts of Turkmenia can exist 16 to 30 years, involving *Ornithodoros tholozani, O. tartakovskyi,* and *O. nereensis* (Petrischeva, 1961).

Tick-Borne Relapsing Fever in the USA. The earliest known focus of endemic relapsing fever in the USA is believed to have been reported by Meador in 1915 from mountainous areas near Denver, Colorado. Prospectors and others who worked in the Sierra Nevada of California at altitudes of 1,600 m and higher frequently suffered from a malarialike disease they called "squirrel fever," which was probably relapsing fever due to contact with the blood of spirochete-infected squirrels, or due to unsuspected tick bites. Proof that *Ornithodoros hermsi* transmits the infection to humans was secured by Wheeler. From 1921 to 1944 inclusive, 283

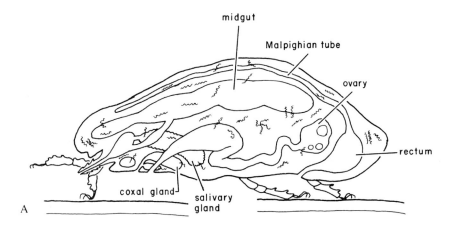

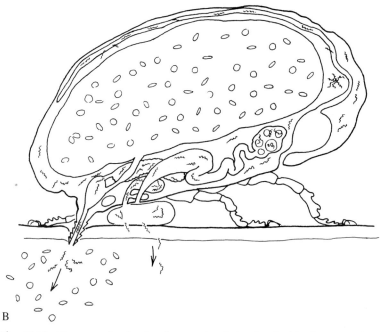

Fig. 16-22 Diagram of relapsing fever spirochetes in *Ornithodoros moubata*. *A.* Within four days after infectious blood meal the spirochetes have extensively invaded body tissues. *B.* Tick introduces spirochetes into host by salivary secretions and by infectious coxal fluid. (After Burgdorfer.)

cases of relapsing fever were reported in California, with epidemiological evidence pointing to their origin in the Sierra Nevada. *O. hermsi* has since been taken in all endemic areas of California, as high as 3,300 m and as far north as Kamloops, British Columbia, Canada.

Tick-borne relapsing fever in the USA generally occurs as single cases, but there may be focal outbreaks that usually involve rodent-infested cabins. The disease has been recorded from twelve western states. In 1968 an outbreak involved 11 of 42 boy scouts and scoutmasters camping near Spokane, Washington, all but one of the cases limited to those who slept in cabins infested by rodents and *O. hermsi* (Thompson *et al.,* 1969). In 1973 an outbreak among visitors to rustic cabins on the north rim of the Grand Canyon, Arizona, resulted in 62 cases, 16 confirmed and 46 clinically diagnosed. Cases from this focus occurred in people who traveled to midwestern and southern states, and as far as West Germany (Pratt and Darsie, 1975). A cluster of 6 cases occurred in July 1976 among 11 persons sharing a rodent-infested cabin in San Bernardino County, California (*Coop. Plant Pest Rept.,* Calif., week ending Oct. 8, 1976).

Ornithodoros hermsi (Fig. 16-17) transmits the infection by the *bite* of both sexes in all stages. The strain of *Borrelia hermsii* is specific for this tick. Weller and Graham (1930) reported infected *O. turicata* from a cave in central Texas that was frequented by a variety of mammals and in which these ticks were abundant.

The control of tick-borne relapsing fever is best accomplished by avoiding caves and other natural sites where *Ornithodoros* ticks are likely to occur, and by controlling rodent hosts around cabins and campsites.

AVIAN SPIROCHETOSIS (FOWL SPIROCHETOSIS)

This is a very destructive disease of poultry. The causal agent is *Borrelia anserina,* and the disease is found in India, Australia, Brazil, Egypt, South Africa, Persia, and no doubt elsewhere. *B. anserina* is highly pathogenic to chickens, geese, ducks, turkeys, pheasants, grouse, and canaries, and less so to guinea fowl and pigeons.

Argas persicus (or ticks in the *Persicargas* subgenus) was proved to be a vector of this spirochete by Marchoux and Salimbeni, Balfour, Nuttal, and others. The pathogen is transmitted by contamination from infectious tick feces; the tick is said to be infective 6 months or more. The infection is carried over from one generation of ticks to the next through the eggs. The incubation period in the fowl is 4 to 9 days. Recovery from the disease is followed by immunity. A study in the Transvaal, South Africa, showed the pathogen to be prevalent in ticks in the subgenus *Persicargus* from 6 of 11 populations collected in chicken houses, and similar ticks from a white egret colony harbored *Borrelia* pathogenic to fowl (Gothe and Schrecke, 1972).

TULAREMIA

Disease: Tularemia, rabbit fever.
Pathogen: Bacterium, *Francisella tularensis.*
Victims: Humans; lagomorphs and some rodents.
Vectors: Several ixodid ticks, tabanid flies (mechanically), possibly other bloodsucking arthropods; contact with infected rabbits and rodents, or infectious water.
Reservoir: Ticks, with transstadial and transovarial transmission; continuous infection in lagomorphs and rodents.
Distribution: Extensive in New World, much of Europe; known from Japan, Israel, Africa.
Importance: Ranged from 129 to 186 cases reported annually in contiguous USA during 1967–1976; high morbidity, low mortality, particularly when treated with antibiotics.

Tularemia in man follows a history of skinning rabbits or rodents, or the bite of a tick or deer fly. Symptoms include an influenzalike attack with initial severe fever, temporary remission, and a further fever period of 2 weeks followed by a local lesion, possible conjunctivitis, and enlarged and tender lymph nodes. Epidemiological circumstances and clinical symptoms may initially suggest campestral plague. Streptomycin and broad-spectrum antibiotics provide effective treatment.

Ixodid ticks have been shown to be infected with tularemia in nature. Tularemia infection in ticks was suspected in the USA in numerous instances during the seasons of 1922 and 1923 because of characteristic lesions at death in guinea pigs into which such ticks had been injected, and this was later confirmed by cultivation of the pathogen. *Dermacentor andersoni* collected from nature proved infective, and in experiments the pathogen was transmitted transstadially. Transovarial transmission was demonstrated later, and is believed to be the first demonstration of transovarial transmission of a known bacterial infection.

Several species of ticks can transmit the pathogen; among those in the USA are *Dermacentor andersoni, D. occidentalis, D. variabilis, Rhipicephalus sanguineus, Amblyomma americanum,* and *Haemaphysalis leporispalustris. Ixodes pacificus* may also transmit the pathogen to humans. The current incidence of tularemia in the USA tends to be during the fall and winter in the east and midwest because of the shooting and skinning of rabbits, and in summer in the southwest and west as a result of tick bites.

Tularemia is known from Alaska, Canada, Venezuela, Mexico, and Japan; since 1926 epizootics have spread from Siberia to Turkey, Iran, Israel, and over most of Europe, with cases also reported from Africa (Gelman, 1961). In western Siberia epizootics of tularemia and Omsk hemorrhagic fever affect populations of the muskrat, *Ondatra zibethica,* up to 68° N latitude. *Ixodes apronophorus* circulates the pathogen between muskrats and the water vole *Arvicola terrestris* (Korsh *et al.,* 1975).

OTHER BACTERIAL INFECTIONS

Ticks are known or suspected to be reservoirs or vectors of a number of bacterial infections that mainly affect domestic animals. In Israel *Hyalomma anatolicum excavatum* and *Rhipicephalus sanguineus* are suspect in maintaining and occasionally transmitting *Leptospira* organisms to cattle and goats (Hoeden, 1968). Ticks may play a role in maintenance and transmission of the *Mycobacterium* species causing tuberculosis in poultry and cattle (Blagodarnyi *et al.,* 1971). *Salmonella pullorum-gallorum* that causes so-called pullorosis-typhus of poultry can be transstadially maintained and transmitted by the bite of *Argas persicus* (Glukhov, 1970).

Caldwell and Belden (1973) provide field and laboratory evidence that suggests nymphs of the Pacific Coast tick, *Dermacentor occidentalis,* may transmit chlamydiae causing bovine chlamydial abortion in the foothills and mountain ranges of California, USA, and sporadically elsewhere. A pathogenic mycoplasma has been isolated from ticks by Tully and associates (1977).

Tick Transmission of Rickettsiae

Ticks are vectors of a number of important rickettsiae affecting man and animals. As background for this discussion Weyer (1964), Horsfall and Tamm (1965), and Hoogstraal (1967) provided important reviews. In most studies the rickettsiae transmitted by ticks do not occur for a prolonged period at high titer in the peripheral circulation of vertebrates, thereby making ticks the reservoir of infection.

ROCKY MOUNTAIN SPOTTED FEVER

Disease: Rocky Mountain spotted fever (RMSF, tick-borne typhus, Mexican spotted fever, Tobia fever, São Paulo fever).

Pathogen: Bacterium, rickettsia, *Rickettsia rickettsii.*

Victims: Humans.

Vectors: Ticks, especially ixodids in the genus *Dermacentor* (*D. andersoni, D. variabilis*).

Reservoir: Primarily ticks; transovarial transmission occurs commonly.

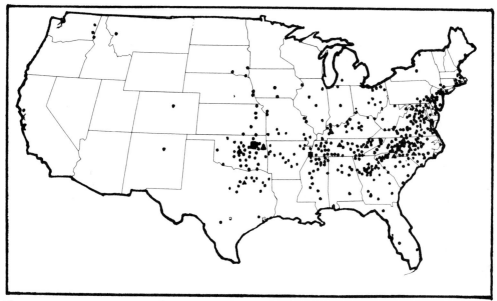

Fig. 16-23 Reported cases of Rocky Mountain spotted fever in the USA during 1976. Triangular and square marks indicate clustered concentrations of cases. (*MMWR,* 1976 annual summary.)

Distribution: Canada to South America; in USA now most prevalent in east (Fig. 16-23).

Importance: In the USA this disease declined to 298 reported cases in 1968, but has been increasing, and 937 cases were reported in 1976. Mortality formerly high, presently infrequent with early treatment employing antibiotics.

Rocky Mountain spotted fever has been known in the Bitter Root Valley of western Montana, USA, since 1872. The most characteristic and constant symptom is a rash that appears about the second to the fifth day after onset of symptoms on the wrists, ankles, and less commonly on the back, later spreading to all parts of the body (Fig. 16-24). The commonest initial complaints are frontal and occipital headaches, intense aching in the lumbar region, and marked malaise. The incubation period is 2 to 5 days in severe infections and 3 to 14 in milder cases. Temperature rises in more virulent infections to 40° C or higher; in fatal infections death usually occurs 9 to 15 days after onset of symptoms. Treatment employs broad-spectrum antibiotics such as tetracyclines. Commercial vac-

cines are available. Strains of spotted fever of different virulence occur, sometimes in the same locality, which caused many points of dispute among early investigators.

RMSF is endemic in the USA, some parts of Canada, Mexico, and parts of South America. The greatest number of cases occur

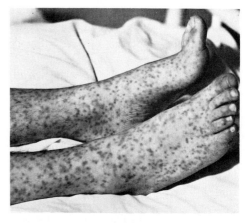

Fig. 16-24 Extensive rash characteristic of Rocky Mountain spotted fever. (Courtesy of US Armed Forces Institute of Pathology. Negative No. N-67987.)

in persons exposed outdoors, principally through agricultural pursuits and outdoor recreation. In the western USA most cases involve men; in the eastern USA more women and children contract the disease (but see Rothenberg and Sonenshine, 1970), probably because the eastern vector is *Dermacentor variabilis,* which infests the dog, a household animal; the principal western vector is *D. andersoni,* the adults of which are associated with large game animals and cattle.

In 1902 Wilson and Chowning advanced the theory that a "wood tick" acts as the natural transmitter of Rocky Mountain spotted fever. According to Ricketts, as recorded by Hunger and Bishopp (1911):

. . . the first experiments which resulted in the proof of the transmission of spotted fever by a tick were conducted by Drs. McCalla and Brereton of Boise, Idaho, in 1905. . . . A tick which was found attached to a spotted fever patient was removed and allowed to bite a healthy person. In eight days this person developed a typical case of spotted fever. . . . (the same tick bit a second person) . . . again a typical case of spotted fever resulted.

The famous experiments of Ricketts have been collected in a memorial volume (Ricketts, 1911): First it was shown that the pathogen could be transmitted to guinea pigs by direct inoculation, and this animal was chosen for further experimentation. A female tick was fed on the ear of an intraperitoneally infected guinea pig, and later this tick infected two healthy guinea pigs. Guinea pigs in association with infected animals showed no symptoms. Undoubtedly the tick Ricketts used was *D. andersoni.* Additional findings included the observation that the disease can follow the bite of a male tick and that one disease episode establishes rather high immunity to subsequent inoculations. Field-collected ticks transmitted the disease to a guinea pig in the laboratory. It was also ascertained that any active stage of the tick could transmit the pathogen (transovarial and transstadial transfer) and transmission is probably through tick saliva.

The Infection in Nature. Parker thought, from field observations in 1916–1917, that some agent other than *D. andersoni* was involved in the natural maintenance of the pathogen, and in 1923 established that the rabbit tick, *Haemaphysalis leporispalustris,* can transmit the infection between rabbits, and infected rabbit ticks occur in nature; also the infection is transmitted transovarially in this tick. The rabbit tick rarely bites man but it is important in maintaining the pathogen and is the only known vector that occurs in all parts of the USA.

Small rodents and rabbits are possible reservoirs of the pathogen, though it has not been proved that they have a prolonged rickettsemia. The agent has been isolated from meadow mouse (*Microtus*), cottontail rabbits (*Sylvilagus*), white-footed mouse (*Peromyscus*), pine vole (*Pitymys*), cotton rat (*Sigmodon*), and opossum (*Didelphis*). Burgdorfer and associates (1962) isolated the rickettsia in western Montana from the blood of snowshoe hare, and spleen tissue of chipmunks (*Eutamias*) and golden-mantled ground squirrel (*Citellus*).

The American dog tick, *Dermacentor variabilis,* was proved a carrier of RMSF rickettsia in the eastern USA through experiments using larvae bred from eggs. *Amblyomma americanum* has been demonstrated as a vector of spotted fever in Oklahoma and Texas, and the Cayenne tick, *A. cajennense,* is a vector in Brazil and Colombia. Transmission has been demonstrated with *D. occidentalis, D. parumapertus,* and *Rhipicephalus sanguineus.*

Burgdorfer (1969, 1975) has reviewed the ecology of tick vectors of RMSF, and the changing status of the disease in the USA.

Transmission of *Rickettsia rickettsii* by other ticks is a distinct possibility. *D. albipictus* may, under certain circumstances, leave its single host and transmit the pathogen to humans. *D. occidentalis* and *R. sanguineus* are possible present or future vectors, and *Anocentor nitens,* species of *Amblyomma* other than the two mentioned, and even species of the argasid genera *Ornithodoros* and *Otobius* may at times be involved. The presence of the rickettsia in *D. parumapertus* feeding on native rabbits has been demonstrated, so this species may be important in maintaining the natural reservoir.

Mechanism of Infection. The pathogen is acquired from an infected animal by a feeding tick in any stage of its life history, and females pass it on transovarially. The great majority of persons with spotted fever give a definite history of tick bite 2 to 12 days before onset of symptoms. Tick tissues are highly infectious when in contact with abraded skin, so crushing ticks with the fingernails can be a dangerous practice. It may be a matter of some hours and not infrequently a day or more after a tick attaches before reactivation of the pathogen occurs in overwintered ticks. If a tick that is attached is removed within a few hours the danger of infection is reduced materially.

The proportion of ticks containing the infectious agent is reported as small; it may be less than 1 percent and is seldom as high as 5 percent. Burgdorfer and Brinton (1975) have reviewed findings on transovarial infection, using conventional staining, electron microscopy, and immunofluorescence. Transovarial infection occurred to all progeny in naturally infected *D. andersoni, D. variabilis,* and *H. leporispalustris,* in one instance through twelve generations. There are strain differences in *R. rickettsii* and in the degree of success in transovarial transmission. Invasion of ovarian germinal cells begins during the feeding of nymphal females, and there are massive infections of the cytoplasm in each oocyte with minor infection of nuclei (Fig. 16-25). Intensive infection of many tissues in ticks is in some cases fatal for engorged females or adversely affects ovi-

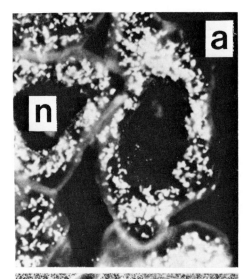

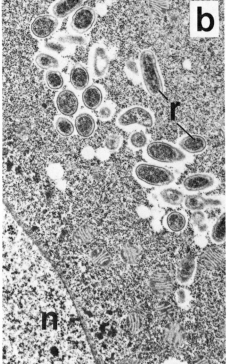

Fig. 16-25 Development of *Rickettsia rickettsii* in tick oocytes. *a.* Fluorescent antibody staining of abundant rickettsii in cytoplasm. *b.* Electron micrograph of similar tissues. *n,* nucleus; *r,* rickettsii. (From Burgdorfer and Brinton, 1975, *Ann. N.Y. Acad. Sci.,* **266**:61–72.)

position and egg development. The low level of infection of ticks in nature seems due to poor survival to the adult stage, and since most tick host animals experience a rather short-lived and mild infection, the rate of acquisition by ticks feeding in nature must be low.

A significant factor in the epidemiology of RMSF is the so-called interference phenomenon. Guinea pigs infected intraperitoneally with a strain of *Rickettsia rickettsii* of low virulence were protected from the highly virulent strain, provided that the concentration of the former was 10 to 30 times that of the latter. Infection with the rickettsiae of Q fever, scrub typhus, endemic typhus, and epidemic typhus gave the same type of protection under the same conditions. Considering that a number of mild or nonpathogenic spotted fever–related rickettsiae also circulate through tick bites in RMSF areas, the interference phenomenon may help to explain why in specific localities Rocky Mountain spotted fever of low and high virulence persists year after year.

Because of the present highest incidence of RMSF in the eastern USA there have been a number of studies reviewing the natural history of the disease in that region. In Virginia *Amblyomma americanum* may be playing an increasingly important role, and incidence of the disease appears to fall in endemic suburban areas. In relating human cases to vegetation types it was found that highest numbers of cases are in areas in which potential dominants are mesic eastern deciduous forest, particularly oak-hickory-pine and northeastern coastal oak-pine (Sonenshine *et al.,* 1972).

TICK TYPHUS

Siberian Tick Typhus (North Asian Tick Typhus). First recognized in the 1930s when virgin steppes came under extensive utilization, *Rickettsia sibirica* was later established as present elsewhere in the USSR; foci occur on islands in the Sea of Japan, the Pacific Far East (Maritime Territory), and northern and southern Siberia.

Symptoms resemble moderately severe Rocky Mountain spotted fever cases with fever, headache, and extensive rash. Survival is the rule, and treatment is much like that for RMSF.

Several species of ixodid ticks are implicated as vectors or reservoirs. These include four species of *Dermacentor,* three of *Haemaphysalis, Rhipicephalus sanguineus* from sheep and dogs in Armenia, and *Hyalomma asiaticum* from rodents (*Meriones*) in Kirgizia. Other arthropods implicated, but of uncertain epidemiological significance, are various mites and fleas. Several species of rodents and a hare have yielded strains of the pathogen. Ground squirrels may become infected while still in burrows in the spring (Merinov, 1962). Ticks removed from cattle, sheep, dogs, and birds have been found infected. The role of mammals as reservoir is uncertain; if infection is short-lived, as with RMSF, then ticks likely serve as the main reservoir.

Transstadial and transovarial survival of *R. sibirica* has been demonstrated through at least 5 years in *D. marginatus,* and development of the pathogen has been followed under experimental conditions with this species and *Ornithodoros lahorensis* (Kryuchechnikov and Sidorov, 1969; Kokorin *et al.,* 1969). This type of survival has also been demonstrated in a number of other ticks. For main vectors implicated, Hoogstraal (1967) has summarized information on distribution and typical habitat.

Boutonneuse Fever (South African Tick Typhus, Kenya Tick Typhus, Crimean Tick Typhus, Marseilles Fever, Indian Tick Typhus). This disease, caused by *Rickettsia conori,* is widespread on the African continent, and has also been reported from European Mediterranean regions, Israel, Turkey, Crimea, and much of Southeast Asia. Variations in virulence, along with epidemiology within diverse areas have led investigators to suspect more than one species of pathogen is present, but most workers agree a single etiological agent is involved.

An extensive rash occurs in human cases but is not easily detected in Africans; mild cases are the rule in Asia. In serious cases there may be severe and persistent headache and mild delirium; full recovery may require a prolonged period. At the site of tick attachment a black buttonlike (hence *boutonneuse*) lesion develops with a central dark necrotic area; local swelling of the lymph glands draining the lesion is common. In addition to tick bite, infection may occur by contact of hands with skin or eyes after crushing ticks removed from dogs.

The reservoir of *R. conori* appears to involve primarily a variety of ticks and rodents. In the Mediterranean area lagomorphs are perhaps the most important reservoir. Birds can be in the reservoir and can also carry ticks to other areas; dogs seemingly do not sustain infection for long but are frequently significant because they bring ticks into close contact with humans. *Rhipicephalus sanguineus* is an important vector in South Africa, East Africa, Ethiopia, and around the Mediterranean and Black Seas. Other tick vectors are in the genera *Haemaphysalis, Hyalomma, Amblyomma, Boophilus, Rhipicephalus, Ixodes,* and *Dermacentor;* in fact the general impression is that virtually any ixodid tick can serve as a host (see Hoogstraal, 1967; Camicas, 1975). Wild rodents of many species harbor the rickettsia. From a Malayan primary forest cycle wild species of rats exchanged ectoparasites with man; also dogs and people passing through forests may acquire infective vectors. Kokorin and coworkers (1969) describe development of the pathogen in a tick.

Queensland Tick Typhus. The agent causing this disease, *Rickettsia australis,* is close to the rickettsialpox agent, and clinical symptomatology resembles the latter disease. Queensland tick typhus has been recognized over much of coastal Queensland, particularly from grassy savanna or secondary scrub areas. On circumstantial evidence *Ixodes holocyclus,* a tick that feeds on a great variety of warm-blooded vertebrates, has been implicated as the main vector, and the pathogen has been isolated from nymphs and adults of this species and *I. tasmani* (Campbell and Domrow, 1974). Serological evidence suggests a number of small marsupial mammals are part of the biocenose.

Epidemic (Louse-Borne) Typhus. The pathogen, *Rickettsia prowazekii,* was believed to be maintained in large domestic animals and ticks in Ethiopia (Hoogstraal, 1967), but further laboratory and field evaluations do not confirm this (Burgdorfer *et al.,* 1972).

OTHER TICK-BORNE RICKETTSIAE OF MINOR IMPORTANCE

Two nonpathogenic rickettsiae have been described from Montana, USA (Lackman *et al.,* 1965). Eastern Montana agent, *Rickettsia montana,* has been isolated from *Dermacentor variabilis* and *D. andersoni.* An unnamed western Montana agent has been recovered from *D. andersoni.* Another agent close to *R. montana* has been isolated from *D. variabilis* in Massachusetts, USA. Two spotted fever group rickettsiae of low pathogenicity have been isolated from *Rhipicephalus sanguineus* in Mississippi, USA (Burgdorfer *et al.,* 1975), and *Ixodes pacificus* in Oregon, USA (Hughes *et al.,* 1976). *Rickettsia canada* is a member of the typhus group of rickettsiae isolated from *Haemaphysalis leporispalustris* ticks in Canada (McKiel *et al.,* 1967). A disease among troops in Texas during 1942–1943, called Bullis fever and associated with the bite of *Amblyomma americanum,* is suggested as having been caused by a rickettsia and provided the name *R. texiana* (Anigstein and Anigstein, 1975).

Rickettsia phagocytophila, an agent causing fever in sheep of Great Britain, can be transmitted transstadially by *Ixodes ricinus* (Foggie, 1951). *Rickettsia parkeri,* or maculatum agent, has been isolated from *Amblyomma maculatum* on cattle in three southern states, USA. Isolations have also been made from ticks on sheep. More widespread distribution probably includes the range of this tick, which encompasses coastal areas of the southeastern USA, Mexico, and to northern South America.

Q FEVER

The infectious agent of Q fever, *Coxiella burneti,* is widely distributed and readily transferred to humans by the simplest contaminative routes, such as milk or inhalation of infectious dust. The disease is generally mild, often without obvious symptoms, but persistent and severe cases can occur. Easy contaminative transfer does not require ticks; nonetheless, Philip and Burgdorfer (1961) note tick maintenance is significant in natural environments. In reviewing world distribution of Q fever, and particularly its occurrence in eastern and southeastern Europe, Thiel (1974) notes that it is found in both hemispheres within the 10° C annual isotherms, especially where climate is warm and dry as in the Mediterranean and Black Sea areas, the steppes of central Asia, African savannas and grasslands of North America and Australia. Where conditions are favorable there are natural foci in wild mammals, birds, and ticks. Balashov and associates (1972) observed that in *Hyalomma asiaticum* this rickettsia was most abundant in ectoderm derivatives and gut, and was passed through saliva and the excreta.

RICKETTSIAL DISEASES OF VETERINARY IMPORTANCE

Ehrlichia bovis, E. canis, and *E. ovina* of cattle, dogs, and sheep in Africa, are transmitted respectively by *Hyalomma* spp., *Rhipicephalus sanguineus,* and *R. bursa* (Philip and Burgdorfer, 1961; Burgdorfer and Varma, 1967). *E. canis* causes canine ehrlichiosis or tropical canine pancytopenia, which was associated with an epizootic in military working dogs in Thailand. There is transstadial transmission in *R. sanguineus,* but transovarian transmission does not occur (Smith *et. al.,* 1976).

Cowdria ruminantium causes heartwater fever of ruminant stock in Africa, and can be transstadially transmitted by two species of *Amblyomma.* This disease is recognized as enzootic in sheep, goats, and possibly camels in a large area of the Sudan, with *Amblyomma lepidum* the vector. Several African *Amblyomma* ticks serve as vectors, and the pathogen will develop in tissue cultures derived from ticks of this genus (Andreasen, 1974).

Bovine anaplasmosis is an important and virtually worldwide infection of cattle caused by minute punctiform blood parasites, described by Theiler in 1910 as *Anaplasma marginale* with the organism at or near the periphery of the red cells, and *A. centrale,* a somewhat smaller body, located approximately in the center of the infected corpuscle. The latter species is relatively benign.

Anaplasmosis is an acute, subacute, or chronic febrile infectious disease. The average mortality ranges from 30 to 50 percent in affected animals. Mechanical transmission of the infection by several species of tabanid flies is known, and Stiles (1939) records 17 species of ticks incriminated by various investigators; among them *Boophilus annulatus, B. decoloratus, B. microplus, Rhipicephalus simus, R. bursa, R. sanguineus, Ixodes ricinus, Hyalomma lusitanicum, Dermacentor variabilis, D. andersoni,* and *D. occidentalis.* Transovarial passage is reported, though it was not seen in the one-host tick *Boophilus microplus* in Australia (Connell, 1974). *Anaplasma marginale* has been followed with some difficulty by fluorescent antibody techniques in nymphs of *D. andersoni* (Bram and Romanowski, 1970).

Aegyptianella pullorum is the etiological agent of a rather widespread disease of poultry. Geese and fowl of different ages are susceptible, but the infection is more severe in young chickens. The tick *Argas persicus* has been demonstrated as a vector, though this probably refers to more than one *Argas* in the complex of poultry ectoparasites. There is transstadial and transovarial transmission (Hadani and Dinur, 1968).

Rhipicephalus sanguineus is suspect as the vector of *Haemobartonella canis,* a usually nonpathogenic agent that parasitizes the erythrocytes of dogs in many parts of the world (Seneviratna *et al.,* 1973).

Tick Transmission of Fungi

In Tadzhikistan, USSR, a disease in sheep and certain small wild animals causing multiple abscesses on and under the skin and several internal tissue lesions is due to *Nocardia asteroides* and *Dermatophilus dermatonomus.* Both fungi were recovered from and transmitted by *Hyalomma asiaticum* (Kusel'tan, 1967). A skin condition of man and animals is caused by *Dermatophilus congolensis* and is called "lumpy wool" of sheep in South Africa and Britain, and by a variety of other names as it affects herbivorous domestic animals in these countries, Africa, and Australia. The primary lesions occur at the favored sites of attachment of ticks, especially *Amblyomma variegatum* (Stewart, 1972). *Nocardia farcinica,* the causal fungus of bovine farcy, was transmitted transstadially and by bite of *Amblyomma variegatum* (Al-Janabi *et al.,* 1975).

Tick Transmission of Protozoa

Ticks are the vectors of several important protozoan blood parasites of domestic and wild animals. The life cycle and identity of several of these is still not completely settled, nor is the identity of the tick vectors certain in all instances. Recent studies on several species of *Babesia* blood parasites in a number of tick vectors indicate that there is no sexual cycle involving gametes. For the role of ticks in these diseases, reviews by Riek (1966) and Arthur (1966) are useful. For an understanding of the developmental cycle of *Babesia* and *Theileria* in the vertebrate host refer to Riek (1966).

Particularly from the historical viewpoint, the elucidation of the life cycle of *Babesia bigemina,* the causal organism of Texas cattle fever, serves as a suitable introduction to other babesioses, to piroplasms in general, and to other protozoa transmitted by ticks. *Babesia* species infect herbivorous livestock and wild ungulates, canines, and several other kinds of mammals including rodents. The rodent-infecting species have caused disease in humans.

BABESIOSES

Texas Cattle Fever. Bovine piroplasmosis (babesiosis, splenic fever, bloody murrain, Mexican fever, redwater, and so forth) is a widely distributed disease of cattle, endemic in southern Europe, Central and South America, parts of Africa, the Philippines, Mexico and formerly in the southern USA where it was known for more than a century, having probably been introduced into that country from Europe. Eradication of the tick vectors of the causal organism, *Babesia bigemina,* a haemosporidian protozoon, eliminated the disease from the USA; constant vigilance prevents its reintroduction from countries where it is still prevalent. No bovine babesiosis has become established in the USA since 1939, though one of the tick vectors occurs and is eradicated from time to time in a few Texas counties bordering on Mexico. There is a current cooperative program between the USA and Mexico to eradicate the disease in the latter country also.

The disease was named Texas cattle fever in the USA because large herds driven northward from Texas passed a certain disease in some mysterious manner to northern cattle. In 1889 Smith made his epoch-making discovery of the intracorpuscular protozoan parasite inhabiting the red blood cells of the diseased cattle. Smith and Kilbourne, on suggestion of Salmon, who studied the disease earlier, proved the disease to be tick-associated. Previously infection was variously attributed to infectious cattle saliva, urine, or feces. The work of Smith and Kilbourne (1893) marks a most important milestone in demonstrating the complete cycle of an arthropod-borne pathogen of vertebrates, in studying a protozoan disease, in demonstrating transovarial transmission of a pathogen by an arthropod, and in the history of preventive medicine. It made possible the elimination of Texas cattle fever from the USA.

The disease may be acute during the summer months and chronic during autumn and early winter. Vast numbers of red corpuscles are destroyed by the parasites, accounting for the reddish color of the urine through the elimination of blood pigments. Mortality in the acute form of the disease ranges from 50 to 75 percent; the chronic form shows milder symptoms resembling the acute type and is accompanied by considerable weight loss.

The tick responsible for transmission of the causal protozoan in the USA was the cattle tick, *Boophilus annulatus.* Other tick vectors are *B. microplus* and *B. decoloratus* within their range—for the former, Mexico, Central America, southern Florida, South America, the Oriental region, Australia and parts of Africa; for the latter, Africa. *Haemaphysalis punctata* is the chief vector in Europe, and species of *Rhipicephalus* are listed as vectors in Africa. Purnell and colleagues (1970) believe, based on their experiments, that *Rhipicephalus appendiculatus* and *R. evertsi,* African species, do not transmit this parasite in the field as is generally held.

There appear to be strain differences in *Boophilus* ticks and *B. bigemina* affecting whether development of the pathogen in these arthropods and subsequent transmission will occur. Heavy infections of *B. bigemina* in *Boophilus microplus* may cause mortality or reduce egg production of the females by about 40 percent, and there are differences in egg infection rates dependent on when the eggs are laid after an infectious feed (Hoffman, 1971; Muangyai, 1974).

Other Bovine Babesioses. There are a number of additional bovine babesioses, some quite pathogenic while others are rather benign. *Babesia argentina,* transmitted by *Boophilus microplus,* is reported in Argentina and Australia, causing mortality of cattle though these hosts may develop immunity naturally or experimentally. *Babesia bovis* is transmitted by *Ixodes ricinus* in Europe, causing a disease called redwater in Britain, and an ill-defined syndrome involving reduced milk production in the Rhone River plain of Switzerland (Morisod *et al.,* 1972). *Babesia major* appears to be benign in Britain and Switzerland where it is transmitted by *Haemaphysalis punctata,* and in parts of the Soviet Union where *Boophilus calcaratus* and *Rhipicephalus bursa* are implicated. *Ixodes ricinus* is the apparent vector of *Babesia divergens* in Britain and France (cf. Donnelly and Peirce, 1975).

Babesioses of Sheep, Goats, and Deer. *Babesia ovis* affects sheep in parts of Europe and Turkey, and its development in the vector, *Rhipicephalus bursa* has been thoroughly studied, with no sexual processes evident (D'yaknov, 1970). *Babesia motasi* may cause heavy losses of sheep and goats in India (Jagannath *et al.,* 1974). Mortality of roe deer (*Capreolus capreolus*) in the Soviet Union is caused by infection with *Babesia capreoli* transmitted by *Ixodes ricinus* (Nikol'skii and Pozov, 1972).

Equine Babesioses. At least two types of piroplasmosis are found in horses, mules, and donkeys, namely, true equine piroplasmosis due to *Babesia caballi,* occurring in Africa, Spain, USA, and the Soviet Union, and a similar though distinct disease due to *Babesia equi,* occurring in Transcaucasia, Italy, Spain, Africa, India, USA, and South America (Brazil). *B. caballi* is transmitted by three species of *Dermacentor,* one *Anocentor,* four *Hyalomma,* and two *Rhipicephalus; B. equi* by two species of *Dermacentor,* one *Anocentor,* four *Hyalomma,* and three *Rhipicephalus.*

Anocentor nitens, the tropical horse tick, was first noted in the USA from southern Texas in about 1908. Surveys in 1960–1963 showed it well established in the southern half of Florida, and it was also collected in Georgia. In a two-year period ending in September, 1963, 141 cases of equine piroplasmosis were registered from Florida. By fluorescent antibody techniques it was established that both equine babesioses were present in Florida (Ristic *et al.,* 1964), and transovarial and transstadial transmission of *B. caballi* by *A. nitens* was demonstrated. The pathogen reduces the reproductive potential and the life span of engorged females of this tick (Anthony *et al.,* 1970).

Canine Babesioses. *Babesia canis* causes a disease known as malignant jaundice of dogs that is prevalent in southern Europe, Asia, South Africa, and the USA. *Rhipicephalus sanguineus* is a vector in many parts of the world; *Hyalomma marginatum* and *H. plumbeum* in Russia; *Haemaphysalis leachii* in South Africa; *Dermacentor reticulatus, D. marginatus,* and *Ixodes ricinus* in southern Europe. *D. andersoni* can carry the disease experimentally. The infection is transovarial in the tick, but transmission to the dog is due to the bite of adults and not immature forms. The incubation period is 10 to 20 days. Shortt (1973) reviews this disease, and describes the life cycle of *B. canis* in *R. sanguineus* and *H. leachii.* Generally young dogs are most susceptible, and no differences in susceptibility were noted in dogs from Europe, Africa, or Asia. Depending on the strain of the pathogen, and susceptibility of the dog, death follows infection, or

a chronic condition with weight loss persists for more than two years.

Another canine babesiosis is caused by *Babesia gibsoni,* for which several tick vectors have been identified. In Japan *Haemaphysalis longicornis* is the vector, with transstadial and transovarial transmission present (Otsuka, 1974).

Babesioses and Human Infections. Human infection with babesiosis organisms, mainly rodent babesia, has been known in splenectomized persons but has only recently occurred in apparently normal humans. The situation is reviewed by Spielman (1976) and Spielman and coauthors (1978). Cases of spleen-intact individuals with malarialike symptoms have involved some 17 persons and seem limited to islands near Cape Cod, Massachusetts, USA, and particularly Nantucket Island. The pathogen is *Babesia microti,* a cosmopolitan parasite of small rodents, and found mainly on Nantucket Island in the white-footed mouse (*Peromyscus*), less often in the meadow vole (*Microtus*), and rarely in rabbits. The vector is a new species of *Ixodes,* near *I. scapularis,* that appears to be abundant only where deer that are principal hosts of the adult ticks are found. Apparently the present limited focus is related to an abundance of deer, increasing contact of humans outdoors with brushy vegetation harboring rodents and ticks, and a mild oceanic climate favoring large populations of ticks.

THEILERIAL DISEASES OF LIVESTOCK

East Coast Fever. This is a highly pathogenic disease of cattle and domestic buffaloes in eastern, central, and southern Africa. Mortality may run over 90 percent and may, in endemic areas, take 80 percent of the cattle crop annually. The disease is caused by the protozoan *Theileria parva,* family Theileridae. Unlike redwater, it is not readily transmitted by means of blood inoculations, nor accompanied by jaundice or hemoglobinuria. A characteristic symptom is swelling of the superficial lymphatic glands.

The disease is transmitted by the bite of several species of ticks as reported by Lounsbury as early as 1906. The adult brown tick, *Rhipicephalus appendiculatus,* is the most important vector, but the disease is also transmitted by the Cape brown tick, *R. capensis,* and the red tick, *R. evertsi.* The cyclic development of *Theileria parva* has been followed in the salivary glands of *R. appendiculatus* (Purnell *et al.,* 1971). Infection in the salivary glands may be determined by sectioning and histological examination or by fluorescent antibody techniques (Kimber *et al.,* 1973). An electron microscope study by Mehlhorn and Schein (1976) indicated stages similar to sexual forms in Haemosporidia are found in the gut of ticks. East Coast fever pathogen is transmitted transstadially and not transovarially in ticks. A single tick can transmit the infection only once, and that during the stage following the one in which it had the infectious meal.

Other Theilerial Diseases. *Theileria annulata* is another cattle parasite transmitted by *Hyalomma anatolicum* in Eurasia. There has been some controversy as to whether sexual stages occur in this protozoan, but more recent investigations, including electron microscope studies, indicate that gametes are present (cf. Mehlhorn *et al.,* 1975).

Theileria mutans, which is generally non-pathogenic, has been associated with cattle in the USA. It is transmitted to cattle by *Haemaphysalis bispinosa* in Hokkaido, Japan, by *Amblyomma variegatum* in East Africa, and by *Haemaphysalis punctata* in Britain.

Theileria lawrencei is found in wild African buffaloes (*Syncerus caffer*) and is transmitted by *Rhipicephalus appendiculatus.* This parasite is highly pathogenic to cattle, and problems occur when these livestock are on land inhabited by wild buffaloes (Burridge, 1975).

Theilerias of sheep and goats include *Theileria hirci* in Iraq, transmitted by *Hyalomma a. anatolicum* (Hooshmand-Rad and Hawa, 1973). *Theileria ovis* causes a benign disease of sheep and goats in Africa, Asia, and Europe; Neitz (1972) lists two species of *Rhipicephalus* as vectors in Africa, and the ticks are known as vectors in the Soviet Union.

Theilerias occur in deer, though the taxonomic status of the pathogens involved is uncertain. The same is true for theilerias of African antelopes, which apparently will not infect cattle. *Amblyomma americanum* transmits a *Theileria* to white-tailed deer fawns that causes anemia and mortality when abundant (Barker, *et al.,* 1973).

TRANSMISSION OF TRYPANOSOMES

Burgdorfer and co-workers (1973) noted that *Rhipicephalus pulchellus* and *Boophilus decoloratus* collected from cattle in Ethiopia harbored all developmental stages of *Trypanosoma theileri,* suggesting they can transmit this protozoan while feeding.

Tick Transmission of Filariae

An argasid tick, *Ornithodoros tartakovskyi,* can develop and transmit *Dipetalonema vitaea* to gerbils (*Meriones*) (Londono, 1973). *Ixodes ricinus* is believed to be the vector of a subcutaneous filaria (*Wehrdikmansia rugosicauda*) of roe deer in the south of West Germany (Schulz-Key, 1975).

Control of Ticks

Tick surveys consist of close observation of hosts, premises, or other habitat. Dragging a white flannel cloth over ground and low vegetation, termed FLAGGING, and examining it frequently for attached ticks will provide an index of ixodid ticks in range and pasture areas and along game trails. For some species the use of dry ice carbon dioxide attractant can effectively indicate numbers (Grothaus et al., 1976). Most argasid ticks are infrequently found on the host; fowl ticks, *Argas* species, are observed in cracks around roosts in chicken houses, and other argasids are in hiding places around the habitations of their hosts.

Biological control of ticks has been attempted, and though pathogens, parasites, and predators are known, they have not proved to be very successful when applied or introduced. African tick birds, *Buphagus* spp., feed heavily on ticks upon large mammalian hosts, and ants have been noted as fairly important natural predators in Australia, the Soviet Union, and elsewhere.

Protection against ticks can be achieved for humans by the use of repellents. Repellents may be applied to skin, but impregnated clothing, particularly the lower garments, provides best control. Several repellents have been registered for use against ticks. Butyl, propyl, and isopropyl acetanilide are very effective but not accepted for general use in the USA. Laboratory tests against *Ornithodoros tholozani* show pyrethrum to be very repellent and to possess considerable tick toxicity (Bar-Zeev and Gothilf, 1973), and further tests should be extended with various synthetic pyrethroids and ixodid ticks. Tick repellents have not been thought practical for livestock, at least on rangeland, but they are marketed for use on pets.

Brush piles, stumps, or other harborages of wild rodents should be removed from around dwellings to deplete hosts for populations of argasids associated with the transmission of relapsing fever.

Controlling ticks on livestock has long been a problem. The economics of control in Australia is analyzed by Johnston (1975). For one-host ticks on cattle, keeping the hosts off tick-infested pasture, termed PASTURE SPELLING, for a period sufficient to cause high mortality of larval ticks, can be very effective and is consistent with the sustained production of adequate range forage. Such a method has been practical against *Boophilus microplus* in Australia. In Queensland, alternating pasture every 2 to 3 months minimizes the amount of insecticidal control necessary, yet leaves a low level of ticks so that cattle do not lose immunity to piroplasms. In that country innate and acquired resistance of cattle to ticks has been studied most thoroughly; consult *New South Wales, Sci. Bull.* No. 78, 1961, for an overall discussion. Crosses of zebu and British breeds have been developed for tick resistance and are used (cf. Wagland, 1975). European breeds of cattle may be killed by heavy tick infestations, but zebu crosses are sufficiently resistant to make tick control on these animals for purposes of increased weight gains of doubtful economic value (Gee et al., 1971; Bainbridge, 1973).

Sprays, dusts, or dips of appropriately registered acaricides can be used most effectively when moving livestock into tick-free paddocks. Pesticide use should be minimal, and greater emphasis placed on environmental manipulation, because of the widespread occurrence of pesticide resistance wherever pesticides have been used heavily. WHO (1975d) has provided instructions for determining resistance in ticks, and Harrison and colleagues (1973) review resistance problems. In some ticks, such as *Otobius megnini* and *Anocentor nitens,* it may be necessary to apply pesticides within the ears and adjacent head areas to achieve satisfactory control. Fumigants can control brown dog ticks in kennels, and ticks on hides in quarantine stations. Area control by the application of pesticides may be worthwhile in recreation areas and military bases, particularly along pathways (Mount et al., 1976).

The most effective control of ticks affecting livestock on rangeland undoubtedly involves the integration of several approaches. This can consist of combining pasture spelling, vegetation modification to increase tick mortality through environmental exposure and in some instances to reduce the small hosts required by some immature ticks, pesticides to reduce the movement of ticks on livestock from tick-infested to tick-free paddocks, and the use of tick-resistant breeds of cattle. The possibilities of integrated control approaches may be reviewed in papers by Harley and Wilkinson (1971), Hoch and co-workers (1971), Morel (1974), Stampa (1969), and Wharton and colleagues (1969).

17

VENOMS, DEFENSE SECRETIONS, AND ALLERGENS OF ARTHROPODS

GENERAL CONSIDERATIONS

The importance of envenomization by arthropods has been, at least in temperate climates, exaggerated in the public mind. Many persons are unjustifiably afraid of spiders, centipedes, millipedes, and in fact, many arthropods that are completely powerless to harm them. The extreme manifestation of such fear is termed ENTOMOPHOBIA, a hallucination that the body is infested with insects, which is mentioned briefly in Chapter 1. However, the danger from venomous arthropods is greater than most entomologists have realized. A report based on examination of death certificates from all parts of the USA (Parrish, 1963) showed that deaths resulting from arthropod bites and stings, covering the period 1950–1959, far outnumbered those resulting from snake bite. Of 460 deaths caused by venomous animals, 30 percent were attributed to venomous snakes; 1.7 percent to scorpions; 14 percent to spiders; and 50 percent to hymenopterous insects, especially bees and wasps. Responses of 135 physicians in Mississippi,

USA (about 15 percent of state total), to a questionnaire concerning venomous bites and stings revealed that in 1971 they had been consulted on 122 snake bite cases, 499 spider bites, 2,381 cases of wasp, bee, or ant sting, and 387 miscellaneous and usually unidentified bites or stings (Keegan, 1972). It seems likely that arthropod-associated deaths are underreported, inasmuch as the bites of some spiders are hardly noticed initially and later effects generally resemble heart attack. In addition, if a person is alone at the time of attack, it may prove very difficult to connect cause of death with a venomous arthropod. In lists of human deaths caused by venomous arthropods many more males than females are accounted, a reflection of the greater outdoor exposure of men and boys. Children are also more subject to serious complications from venoms, because their natural curiosity without appropriate caution makes them more subject to attack, and their small size results in a higher proportion of venom than adults on a comparative weight basis.

Accurate statistics are for the most part lacking for the tropics, though the impression is that all types of envenomization are more commonplace than in temperate regions. In 1941 deaths in Mexico for all venomous animals were listed as snakes 219, scorpions 1,802, spiders 19, and others 170 (Tay and Biagi, 1961). Prior to 1958 more than 1,000 scorpion-associated deaths occurred annually in Mexico, especially in the arid west central part of the country (Minton, 1974), and in Belo Horizonte, Brazil, between 1938 and 1945 an average of 843 scorpion sting incidents and 16 deaths was reported yearly (Bücherl and Buckley, 1971).

Chemicals possessed by arthropods may cause a variety of medical complications of vertebrates. These chemicals are usually multicomponent systems, variable with respect to amounts introduced and route of entry into vertebrates. Chemical variability, coupled with variability in the sensitivity of species and even individual vertebrates, make for differences in pharmacological effects and severity of response.

Chemicals may be classified on the basis of function into (1) venoms that cause immediate responses, and that are introduced beneath the skin of vertebrates by mouth parts or stinging apparatus; (2) irritating defense secretions that are received by a vertebrate on contact, or that may be projected with some accuracy by an arthropod before bodily contact occurs; (3) allergens to which prior exposure and sensitization may result in anaphylaxis, or milder allergic reactions such as eczema, hives, or asthma. Clear-cut classification of venom action may be impossible, because allergic reactions frequently accompany envenomization and contact with defense secretions.

Arthropod Venoms

Venoms and the apparatus for their disbursement may perform dual functions, namely the subduing of prey and self-defense. One characteristic of most of these venoms is the relative rapidity with which

they act. There is usually a painful immediate response to bites or stings that may be followed by complications, though responses to some spider bites may be somewhat delayed. Beard (1963) has reviewed the characteristics of insect venoms, and books that include the subject are by Bücherl and Buckley (1971), Minton (1974), and Tu (1977). Regional coverage of venomous animals includes that of Keegan and Macfarlane (1963) for the Pacific area, and Freyvogel (1972) for East Africa.

For accurate chemical, pharmacological, and allergenic characterization it is necessary to obtain venom that is as pure as possible. A method of electric shocking to cause reflexive release of venom has proven suitable for individual arthropods of several kinds, and this has been modified to collect large amounts of venom from honey bees (Benton et al., 1963). For some arthropods it may be necessary to dissect out the venom reservoir of the stinging or biting apparatus.

How Venoms, Vesicants, and Allergens Contact Vertebrates

Injurious chemicals produced by arthropods are introduced into or onto the body of man and other animals by (1) BITE, the thrust of the piercing mouth parts and introduction of venomous salivary secretions as in conenose bugs, penetration of the chelicerae of spiders and ticks, or the maxillipeds of centipedes; (2) STING, as in aculeate Hymenoptera (ants, bees, wasps) and scorpions; (3) CONTACTS of an essentially passive nature, for example, by means of urticarial hairs of certain caterpillars or adult moths, producing a condition similar to nettling, or with vesicating fluids of certain beetles and millipedes, or through wind-blown allergens, such as cast mayfly skins; (4) ACTIVE PROJECTION, as occurs in arthropods that use secretions as a means of defense; found in some predatory Hemiptera, some ants, some millipedes and in whipscorpions.

Arthropod Defense Secretions

Defense secretions of arthropods are unpleasant fluids or gases that may have a strong smell and may be irritating or toxic to vertebrates on contact. These compounds are poured onto the body surface, or they may be forcefully ejected. Characteristically, defensive fluids are produced by glandular cells, stored in reservoirs, and have a specific apparatus for their expulsion. They are ejected by blood pressure, by tracheal air pressure, pressure from chemical reactions, or through muscular contraction of the reservoir.

Defense secretions are not normally of major medical consequence, for the most part being disagreeable but temporary in their effects unless they contact the eyes or open wounds. The subject of chemical defenses of arthropods has been reviewed by Roth and Eisner (1962), who provide a tabular listing of chemicals found and discuss the associated glands, discharge mechanisms, and medical complications.

Arthropod Allergens

The body constituents of arthropods cause allergic reactions of humans and other higher vertebrates. Insect allergens and human allergic responses have been reviewed by Perlman (1967), Shulman (1967), Feingold and associates (1968), and Frazier (1969). The gravity of allergic reactions depends on individual susceptibility (including hereditary predisposition), amount of previous exposure, and method of exposure. Deaths are generally due to anaphylaxis that is usually associated with prior rapid exposure to allergens by sting. Less severe hypersensitivity may be indicated by hives, eczema, or asthma, and result from sting or by exposure to arthropod particles. Chronic allergic reactions, such as rhinitis, urticaria, and eczema can be caused by stings but are more characteristic of respiratory or skin contact with arthropod parts.

There is considerable controversy over whether venom allergy is due to venom components alone, or whether other arthropod body fractions are also involved. When seeking to desensitize patients, some allergists prefer to use antigens prepared from specific arthropods whereas others incline toward antigen preparation from a group of related arthropods; for instance, preparing antigenic extracts from honey bees exclusively, or making a mixed extract including several species of stinging Hymenoptera. There does seem to be cross reactivity, particularly among fairly closely related groups of arthropods.

Testing for sensitivity and standardization of hyposensitizing doses also present problems. Skin scratch and patch tests obtain reactions from substances to which patients do not necessarily show any sensitivity under normal exposure situations. Skin tests (Fig. 17-1) appear to be of value when used in conjunction with detailed case histories in evaluating insect sting allergy (Barr, 1972). It has been noted that the measurement of IgE (immunoglobulin E) to insects such as honey bees can be accomplished using RAST discs (radioallergoabsorbent test discs), avoiding the nonspecific wheal and flare found with skin tests (Arbesman *et al.*, 1975). Measuring the amount of histamine released in sensitized monkey lung tissue challenged with allergens is another indirect technique that tends to avoid nonspecific responses, and provides a clearer indication of respiratory reactions (cf. Perlman *et al.*, 1976).

Contact or respiratory allergies have been reported as due to a variety of insects and mites situated outdoors or in household environments. Particles from mass emergences of aquatic insects, particularly mayflies and caddis flies, have been well documented as

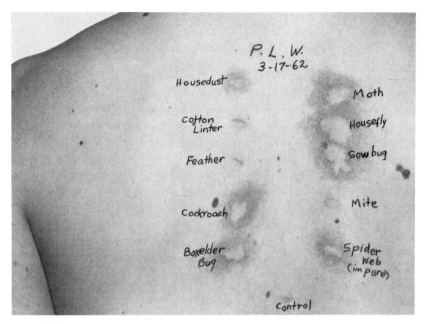

Fig. 17-1 Eruptions on skin of a patient showing strong reactions to scratch tests of several arthropod extracts. (Courtesy of F. Perlman, Allergy Clinic, Portland, Oregon.)

causing seasonal allergies. There have been similar reactions to stored products pests, namely to the mite *Acarus siro* among cheese makers (Molina *et al.*, 1975), and contact with or inhalation of setae or scales of stored grain beetles, particularly Dermestidae (Théodoridès, 1950). Skin irritation has been recorded from the spider mite, *Tetranychus moutensis,* infesting flax when workers are harvesting this plant in New Zealand (Manson, 1970). The Locustidae (Orthoptera) cause allergic reactions, as it has been noted that people who work with locusts in research laboratories and in schools gradually develop what are usually asthmatic reactions (COPR, 1973), and a case of generalized blotchy hives during exposure outdoors was caused by sensitivity to grasshoppers (Schnitzker, 1974). Worth further review are the allergic responses associated with cockroaches and house dust mites.

Cockroach Allergy. This subject has been rather completely reviewed by Ebeling (1975), who states, in part:

In one investigation among 253 normal persons, 7.5% showed positive skin tests with extracts of cockroaches (*Periplaneta americana* and *Blatta orientalis*) compared with 28% of an unselected group of 114 allergic patients. Skin-sensitizing antibodies were present in the blood sera of positive reactors. The allergin is thermostable. . . . In an investigation to test the age of onset of skin reactivity, 38 out of 102 children ranging from infants to 12 years old, gave positive cutaneous reactions to body extracts of the German cockroach, *Blattella germanica,* compared with only 5 out of 100 nonallergic children. . . . An allergen in the feces of *B. germanica* acts as an ingestant when it contaminates food and as an inhalant when dried fecal particles become incorporated with house dust.

The situation described occurs in temperate climates, but cockroaches are particularly abundant in the moist tropics. In Bangkok, Thailand, 10 species of domiciliary cockroaches infest congested and unsanitary housing areas. Constant exposure to house dust contaminated with cockroach allergens is unavoidable in such environments. Of 458 allergic patients tested, 76.4 percent reacted positively in cutaneous tests to housedust extract and 53.7 percent to an extract of the bodies of *Blattella germanica* and *Periplaneta americana,* with incidence higher in younger than in older people and in asthmatic than in nonasthmatic patients (Choovivathanavanich, 1974). Cockroach allergy may be alleviated through hyposensitization using cockroach extracts, and through better sanitation and control measures to reduce or eliminate these pests (see Chapter 6).

House Dust Mite Allergy. Mites in the family Pyroglyphidae commonly infest stored food products and can also be found in debris-accumulating situations that provide a variety of organic materials, such as skin dander or insect parts. They have been reared on mixtures of human dandruff and yeast, or on dog meal. These mites are extremely small and easily overlooked, but are prevalent in such substrates as house dust and bird nests. Mites in this family, particularly in the genus *Dermatophagoides* (Fig. 17-2), have been associated with house dust allergy and climate allergy, and a quite voluminous literature has been developed on the subject, as indicated by a bibliography covering literature through 1974 prepared by Lang and coauthors (1976).

The mite *Dermatophagoides pteronys-*

Fig. 17-2 Scanning electron micrograph of protonymph of mite, *Dermatophagoides farinae,* an allergenic component of house dust. (Scanning electron micrograph from Wharton, 1970, *Science,* **167**:1382–83; copyright 1970 by the American Association for the Advancement of Science.)

sinus was implicated as an allergenic component of house dust by Fain (1967) and by Spieksma and Spieksma-Boezeman (1967). Since then a number of additional pyroglyphids have been found; commonly the American house dust mite, *Dermatophagoides farinae, Euroglyphus maynei,* and *Glycyphagus destructor,* though the European house dust mite, *D. pteronyssinus,* appears to be the species most frequently present and to which most asthmatics are sensitive and that elicits the strongest responses (Billioti *et al.,* 1972; Araujo-Fontaine *et al.,* 1974, Wharton, 1976). These mites are suspect as one possible cause of infant cot death in Australia (Mulvey, 1972; Turner *et al.,* 1975). Lecks (1973) and Voorhorst (1972) provide rather comprehensive reviews on house dust mites and allergy, pointing out that these mites are essentially worldwide in occurrence, various species being associated with household furniture and floor debris, nests of birds and mammals, stored products, and facultatively on human skin and the skin of birds and mammals. The excreta, exoskeleton, and scales of mites have a common antigen. Mites are most abundant wherever their microclimate provides favorable temperature and humidity conditions, 25° C being optimum. The main evidence that house dust mites contain allergens is their worldwide presence in house dust samples; allergic responses to house dust and house dust mite extracts (measured by skin tests and bronchoprovocation) in large numbers of individuals; good correlation of the allergenic activity of a particular dust extract with the proportion of mites present.

In the USA, Wharton found up to 69 *D. farinae* and 9 *D. pteronyssinus* per gram in commercial extracts of house dust, and in England sampling of room air during and just after bedmaking produced from 0.04 to 0.34 mites per cubic meter of air, and 44 to 136 mites per gram of airborne dust.

Control of house dust mite allergy is achieved by desensitization of patients by the use of mite extracts, covering mattresses with plastic, generally keeping mattresses free of dust since mattress dust is a favored site of mite development, keeping airborne dust at low levels, especially in bedrooms, and efficient and frequent housecleaning to reduce the amounts of organic debris on which these mites feed. Exposing mattresses and bedding to winter freezing temperatures has been suggested. Pesticides tested have been ineffective, particularly because large numbers of *D. pteronyssinus* occur in crevices where pesticides do not readily penetrate.

Stinging Insects

Stinging insects of the order Hymenoptera include the ants, bees, velvet ants, and solitary and social wasps in which females have a specialized ovipositor known as a sting, well adapted for piercing the skin of higher animals or other arthropods. Parasitic Hymenoptera also sting, but infrequently bother humans. The sting is used either for defense or offense; in the latter case to procure food for the young. Structure of the venom apparatus of aculeate (stinging) Hymenoptera, such as bees, wasps, and hornets does not vary greatly (Fig. 17-3). Minton (1974) has reviewed what is known about this apparatus. There are one or two tubular poison glands that empty into the venom reservoir. Dufour's gland also empties secretions into the base of the sting, its duct usually fusing with that of the venom reservoir. Contributions of the various secretory cells to the complex of substances in Hymenoptera venoms is incompletely known; Dufour's gland may be the site of synthesis of low-molecular-weight pharmacologically active compounds such as serotonin, histamine, and acetylcholine.

In larger Hymenoptera up to 0.1 ml of venom is introduced when the entire contents of the venom reservoir are emptied. In the case of the honey bee complete emptying is

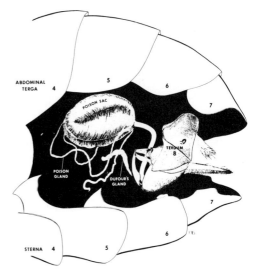

Fig. 17-3 Sting apparatus of yellowjacket wasp. (Courtesy of A. Greene.)

frequent because the sting is markedly barbed (Fig. 17-4), and the sting and venom apparatus tear easily from the end of the insect's abdomen. There is reflexive emptying of the venom reservoir, governed by the fifth abdominal ganglion and its radiating nerves.

General reviews of Hymenoptera sting occurrence, allergic reactions, and the treatment of serious complications are provided by Barr (1971), Cheminat and coauthors (1972), Feingold (1973), Hunt and McLean (1970), Leclercq and Lecomte (1975), and Pursley (1973).

The principal aculeate Hymenoptera, according to the classification followed by Borror and associates (1976), are divided into the following seven superfamilies, all but the first containing representatives that are known to be dangerous or painful: (1) Chrysidoidea, the cuckoo wasps and their relatives; (2) Bethyloidea, the bethylid wasps; (3) Scolioidea, the scoliid wasps, velvet ants, and their relatives; (4) Formicoidea, the ants; (5) Vespoidea, the hornets, yellow jackets, spider wasps, and their relatives; (6) Sphecoidea, the sphecoid or mud dauber wasps; (7) Apoidea, the bees.

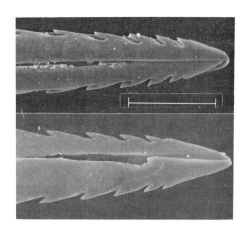

Fig. 17-4 Tip of sting in the honey bee, *Apis mellifera* (*upper* illustration), and a yellowjacket wasp, *Vespula pensylvanica* (*lower* illustration). Index line is 0.1 mm long. The honey bee sting is slightly more barbed, and the venom apparatus rips readily out of the bee's abdomen. (Scanning electron micrographs courtesy of A. R. Crooker, Electron Microscope Center, Washington State University.)

Honey bees and bumble bees (Apoidea), and wasps and hornets (Vespoidea), are frequently misidentified by the public, and doubtless statistics regarding allergies and deaths by these two groups are in considerable error. Bees are readily distinguished from wasps by the presence, especially on the bee thorax, of hairs, which under the microscope are distinctly plumose. Various bees live in solitary or subsocial existence, that is, each female develops her own nest, lays eggs, and provides for her brood without the aid of a worker caste. Many such bees possess stings, but serious cases of bee envenomization are usually due to bumble bees or honey bees.

Families Bombidae and Apidae

BEES

Both bumble bees, *Bombus* spp., and honey bees, *Apis* spp., are social insects, with a worker caste in addition to the fertile males and females. Bumble bees form tem-porary colonies with only the fertilized young queens surviving the winter. In the spring these queens search for a place to found their colony. They are generally ground nesters, often nesting in a deserted rodent burrow, but infrequently in buildings or other structures with suitable soft materials. A person working or hiking in fields where bumble bees are nesting may unwarily step into a nest and be stung severely.

Bees of the genus *Apis*, known as honey bees, form permanent colonies that may survive indefinitely. There are five well-known species: *Apis mellifera*, the cosmopolitan honey bee of commerce that is the cause of most bee stings and many deaths; the giant bee, *A. dorsata*, of Asia, vicious when aroused and with painful stings; the Indian bee, *A. indica*, smaller and more gentle than *mellifera* and with less painful stings; the little bee, *A. florea*, with sting comparable to a simple "needle-prick"; the Japanese or Chinese bee, *A. cerana*, somewhat less aggressive than *mellifera*.

The common honey bee *Apis mellifera* is itself separated into about a dozen named races in European, Asian, and African groups, characterized by differences in aggressiveness. The four best-known economic races are the dark bee, *A. m. mellifera*, generally nervous and often aggressive; the Italian bee, *A. m. ligustica*, the major commercial stock and variably aggressive; the Carniolan bee, *A. m. carnica*, commercial bee of much of eastern Europe and the most gentle; the Caucasian bee, *A. m. caucasica*, usually considered the most gentle in the USA. Two African races are noted for their viciousness, a coastal variety, *A. m. litorea*, and plateau variety, *A. m. adansonii*, generally referred to in the literature as the African bee.

The African bee was introduced into apiaries in Brazil in 1956 to develop a strain of bees adapted to the tropics by crossing them with European strains (Michener, 1973). A number of queens and workers escaped in 1957, and since then the African bee and its hybrid, known as the Brazilian honey bee or africanized bee, have spread into the warmer parts of all adjacent countries to the south and southwest, and northward at a rate of some 200 miles per year. Where they occur they have eliminated the European strains of honey bees through hybridization and competition. There is no obvious barrier to prevent them from reaching the continental USA in the 1980s, though they should be limited to the southern portions of that country because of a lack of winter hardiness. There is some evidence that hybridization of the African stock with European strains is resulting in less aggressiveness, but where africanized strains are presently found in Brazil there are several reports of mass stingings and associated deaths of man and domestic animals. Suggested means of control for North America include setting up a genetic barrier consisting of high populations of European strains at a narrow portion of Central America to prevent further spread northward, and the possibility of breeding and liberating relatively nonaggressive strains of honey bees that could compete effectively with African and Brazilian honey bees.

The cause of attack by many members of a honey bee colony is of interest. There are sting gland alarm pheromones that incite further attack after the first sting. Morse and coworkers (1967) have reviewed alarm substances in the genus *Apis*.

Family Vespidae; Superfamily Vespoidea

VENOMOUS WASPS (PAPER WASPS, HORNETS, AND YELLOW JACKETS)

The stinging wasps consist of solitary, but mostly social, wasps of the family Vespidae. Of prime concern are so-called yellow jackets and hornets in the subfamily Ve-

spinae. The literature on this group has been summarized in a comprehensive bibliography by Akre and colleagues (1974), and Akre and associates (1979) have reviewed the yellow jackets of North America north of Mexico. An analysis of the biology and pest status of the Vespidae has been developed by Akre and Davis (1978), and the information on biology, pest status, and distribution presented here is mainly from that review.

There are about 15,000 species of stinging wasps worldwide, but about 95 percent of these are solitary and nonagressive, possessing venom that subdues prey. If they do happen to sting humans the venom of solitary wasps generally causes only slight and temporary pain. Most stings with complications are by social Vespidae, which use their sting primarily in defense of the colony. There are about 800 species of Vespidae. Colonies vary in size from 30 to several thousand individuals; consequently quite massive stinging attacks may occur. The prey of Vespidae generally consists of live arthropods in order to provision their nests in rearing young, but some species will scavenge protein in the form of animal flesh, and it is these vespids that are main pests because their scavenging habit places them in close contact with humans.

In the Vespidae the tribe Polistini in the subfamily Polistinae are found worldwide except in the coldest regions. The genus *Polistes* or paper wasps comprises about 150 species of concern. Commonly *Polistes* construct a single horizontal naked comb suspended by a short stalk (Fig. 17-5). There are usually 200 or less individuals per colony. There is continuous colony development in the tropics, but in temperate zones fertilized queens overwinter and start new colonies in the spring. Problem species tend to be those that build their nests under the eaves of homes, or choose attics as an overwintering site for queens.

Basically *Polistes* are beneficial, because of a strong tendency to provision their nests

Fig. 17-5 A paper wasp, *Polistes* sp., on the stalked uncovered comb of a newly started nest. Two of the cells may be seen to contain an egg. (Photograph by R. D. Akre, Washington State University.)

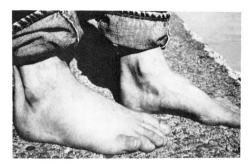

Fig. 17-6 Uncomplicated swelling of the right foot, due to a yellowjacket wasp sting. (Photograph by R. D. Akre, Washington State University.)

with caterpillars, yet when numerous they can be hazardous to humans, particularly in agricultural areas. In Mexico groves of citrus have been abandoned because of the abundance of these wasps, and in Arizona, USA, laborers have refused to pick fruit when large numbers of *P. exclamans* are present. *P. apachus* has injured fig harvesters in the state of California, USA.

The main pest vespids are in the subfamily Vespinae, with the four genera *Provespa* (tropical), *Vespa* (tropical and temperate), and *Vespula* and *Dolichovespula* (temperate).

Vespids in the genus *Vespa* are commonly called hornets; they are large (ca. 20 mm long) and there are about 20 species. In temperate areas colonies are developed annually by overwintering fertilized queens. Various species nest on trees and bushes, under house eaves, in attics, or underground. Fully developed nests contain 600 to 13,000 cells. Hornets hunt insects, and several species are extremely detrimental to colonies of honey bees. They do not ordinarily feed on nectar, but may be attracted to ripening fruit. Some species are easily aroused to sting in great

numbers. *Vespa orientalis* in the Middle East is feared, and causes damage to honey bee colonies and fruit crops; *V. mandarinia* is dangerous around the nest; *V. velutinana* is the probable cause of several reported human and animal deaths, and it has a very painful sting. The only species in the USA is *V. crabro*, the European hornet, introduced into the east around 1840–1860. As a group *Vespa* damage many kinds of ripe fruits, and tea crops have remained unpicked because of nests in the bushes. They have a tendency to feed on meat and sweets, which puts them in unfortunately close contact with man.

Vespula and *Dolichovespula* are called yellow jackets and comprise about 26 species (Fig. 17-7). These are north temperate wasps of Europe, North Africa, Asia, and North America. Colonies are initiated by single overwintered queens. *Vespula* nests are concealed, usually underground but occasionally in hollow logs, trees, or attics; *Dolichovespula* nests are usually aerial. Fully developed nests vary from 300 to 120,000 cells, and the prey is usually many kinds of insects and spiders. *Dolichovespula* are generally not scavenger species, bothering man only when their nests are disturbed. However, some species that tend to nest aerially around house eaves are a fairly frequent problem.

Vespula in general should be considered beneficial, since they feed on many insects,

Fig. 17-7 (*Top, left*) Aerial nest of yellow jacket; (*top, center, and right*) side and top view of the insect, *Vespula diabolica;* and (*bottom*) longitudinal section of nest.

including pest species. However, some species in what is termed the *V. vulgaris* species group may be a serious problem because their proteinaceous foraging habits make them a nuisance around picnics, garbage, and the like. Brief comment on six common pest species is in order. *Vespula germanica,* the German yellow jacket, is a highly pestiferous species native to Europe but introduced into New Zealand, Tasmania, Australia, South Africa, Chile, and the eastern USA where it has spread as far west as Indiana; nests usually subterranean but may be aerial, nearly always in structures in USA; colonies usually annual, but perennial colonies in New Zealand, Tasmania, and Australia reach about 1 million cells; a problem to beekeepers, a general nuisance in markets, butcher shops, and parks; in New Zealand several schools have closed when this pest

was abundant; it is becoming more of a nuisance in the eastern USA wherever it is firmly established. *Vespula maculifrons,* the eastern yellow jacket, is an eastern US species from Minnesota to Texas and eastward, including Canada; nests usually subterranean with 1,500 to 12,000 cells; one of the most troublesome yellow jackets in the East and Southeast USA in parks, at picnics, and in yards of homes. *Vespula lewisi* is a Japanese pest species; nests usually subterranean with up to 16,800 cells; forages for raw and cooked meat, enters houses but not as aggressive as some pest species. *Vespula pensylvanica* of western North America, also found in Mexico and Hawaii, is commonly called the western yellow jacket (Fig. 17-9); nests usually underground in old rodent burrows, with 4,000 to 10,000 cells; a notorious hazard at picnics, in yards, playgrounds,

Fig. 17-8 The bald-faced hornet, *Dolichovespula maculata*. (Courtesy of US Public Health Service.)

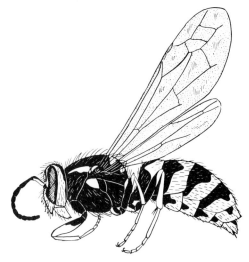

Fig. 17-9 A common yellow jacket, *Vespula pensylvanica*. (Courtesy of US Public Health Service.)

canneries, and picnic sites; causes economic losses due to disuse of camping facilities and resorts, disrupts fruit harvesting, and hampers forest workers. *Vespula vulgaris,* the common yellow jacket, is Holarctic, distributed across North America including Mexico, also Europe and Asia; nests subterranean or in walls of houses or rotten logs, comparable in size to *V. pensylvanica;* nuisance in food-dispensing facilities in Germany, most annoying species to humans in Norway, second only to *V. pensylvanica* in western North America (involved in major yellow jacket outbreak in the Pacific Northwest, USA, in 1973, and a main pest species in a similar outbreak in Alaska in 1974). *Vespula squamosa,* the southern yellow jacket, occurs in the USA from Wisconsin to Texas and throughout the East and Southeast, extends into Central America; nests generally subterranean; it is one of the most troublesome species in the Southeast.

Humans are rarely stung by other vespoid wasps, though Bromley (1933) describes pain and swelling of the hand and forearm from a sting by the tarantula wasp, *Pepsis formosa;* Pompiliidae.

Sphecoid wasps, so-called mud daubers or thread-waisted wasps such as *Chalybion californicum* (=*C. caeruleum*), make mud nests, usually quite small, and provision them with insects and spiders as larval food. These nests are frequently in attics and outbuildings so that the wasps are seen flying about and are feared by the uninitiated, but sphecoids are seldom aggressive, and generally their sting is not very painful and does not cause serious complications.

Control of Bees and Wasps

Several methods have been developed to control stinging insects (cf. Putnam, 1977, concerning campsites). Honey bees and bumble bees are beneficial pollinators that ordinarily should not be controlled, but when they nest too close to human habitations, especially where persons sensitized to their stings reside, control may be necessary. Most municipalities in the USA have ordinances that do not permit the keeping of bees within city limits, but honey bees will forage considerable distances, and this fact coupled with the point that there may be wild swarms around makes these insects nearly ubiquitous wherever blooming plants, including weeds, are found. When honey bee colonies become overcrowded, new off-shoots are developed by a queen leaving with a considerable number of workers in what is termed a swarm. Swarms may settle temporarily in the open while searching for new enclosed quarters, and these settled swarms may be in trees, bushes, or on the sides of houses, presenting some hazard to nearby humans. The most prudent solution is to contact an apiarist who may want the new colony nucleus, or who is willing to remove it in any case because he possesses suitable equipment to prevent stinging. If one decides to personally attempt control it helps to apply an insecticide at night. Swarms recently treated with pesticides should be avoided, because the workers are highly irritable. If a colony has established in a shed or between the outer and inner walls of a home, there are additional problems; the entrance to the colony can be the route of treatment with a suitable pesticide, but it may be necessary to partially dismantle the structure to remove dead bees, brood, and honey stores because these residual materials may cause strong odors as they decompose.

Hornets and yellow jackets may be controlled by a variety of techniques. Aerial nests, particularly if small, can be knocked down and the residue of brood and nest burned. Larger aerial nests and subterranean nests, once the entrance is discovered, can be thoroughly saturated with one of several pesticides. Whatever treatment is employed to destroy a colony, it is better to work after dark and to be thoroughly covered with protective insect-proof clothing. Commercial pressurized containers are available that will treat nests with a stream of pesticide at some 4 m away.

Where yellow jackets are a nuisance around foodstuffs, and colonies have not been located, more indirect methods can be employed. Particular care should be taken to cover garbage and food that is highly attrac-tive. Repellent sprays are sold commercially to control these pests around picnic areas, but these tend to provide only temporary relief and are not sufficient in seasons when pest yellow jacket numbers are high. Poison baits using ground meat or fish with a pesticide can be extremely effective, though it may take up to a week for much noticeable relief and precautions must be taken to prevent pets or children from contacting the baits. The main principle in using pesticide-poisoned protein baits is to choose a pesticide and formulation that has minimal repellency to worker yellow jackets, and to treat the bait with low levels of toxicant so that workers will return with bait to the nest. The poisoned foodstuff will be distributed within the colony and fed to brood and adults. Pesticides for direct treatment of colonies should be fast-acting in order to minimize the time irritable wasps are around. Information on toxicity of pesticides against yellow jackets may be consulted in Johansen and Davis (1972), and Iglisch (1970). It has been found that encapsulated formulations of pesticides help to mask repellency in poisoned baits (Ennik, 1973).

Certain esters of organic acids are potent attractants for yellow jackets in the western USA. These can be used to trap out the bulk of local populations in a matter of a few days. The early standard attractant compounds developed were 2,4-hexadienyl butyrate and heptyl butyrate (Davis *et al.*, 1969), but it was later discovered that octyl butyrate is more attractive, less expensive, and volatilizes more slowly (Davis *et al.*, 1972). Using perimeter traps baited with these attractants, it has been possible to bring considerable relief during yellow jacket season to large outdoor gatherings and to workers in fruit orchards (Davis *et al.*, 1973). Unfortunately these organic esters do not appear to have the same degree of attractancy for yellow jackets in moister areas of the West and in the eastern USA and elsewhere (Grothaus *et al.*, 1973). This attractant method is indiscriminate, affecting yellow jackets that are not serious pests and that provision their nests with insects; therefore it should be reserved for special circumstances, and timed correctly. In Germany it has been possible to gain slow control of the pest yellow jackets in food shops by applying pyrethrins to the inside of display windows and frames, or by using traps with an ultraviolet light attractant source (Coch, 1970).

Nature, Consequence, and Treatment of Bee, Wasp, and Hornet Stings

The general nature of bee, wasp, and hornet venoms has been reviewed by Bücherl and Buckely (1971), Habermann (1972), Minton (1974), and Tu (1977). Among low-molecular-weight components histamine predominates, being present in the amount of 0.1 to 1.5 percent of total components; dopamine and noradrenaline are also present. The most potent agents appear to be polypeptides. Melittin, a polypeptide, is the main component of honey bee venom, comprising 50 percent of the dry weight. It indirectly affects the release in animal tissue of other pharmacological components, produces pain, and affects the blood vessels and musculature; it can cause anaphylaxis reactions in mice.

Two other polypeptides identified are apamin and MCD-peptide. Apamin is the smallest neurotoxic polypeptide known. It interacts with the spinal cord and not at the neuromuscular junctions. MCD-peptide releases histamine from mast cells and causes capillary permeability.

Other components present are hyaluronidase (spreading factor) at about 2 to 3 percent of dry components; phospholipase A and phospholipase B that alter tissue structures by enzymatic hydrolysis; procamine, a histamine-terminal peptide that has been prepared synthetically.

There is a fairly widespread belief that bee stings can help prevent arthritis, and in experimenting with rats Zurier and co-workers (1973) found this to be so with whole bee venom but not with the individual components. In studying the protein bands of bee venoms by electrophoresis, it has been found that *Apis m. adansonii* and *A. m. ligustica* possess 10 similar bands, but the bands of *Bombus atratus* are different. This suggests that the protein components of bee venom causing allergic reactions are alike in various strains of the common honey bee, *A. mellifera*, but that there is no great amount of cross reactivity between the venom of this bee and bumble bees.

Social wasp and hornet venoms do not possess polypeptides similar in action to melittin, apamin, or MCD-peptide of the honey bee, but a number of pharmacologically active compounds have been identified. Among biogenic amines, serotonin and histamine are present in *Vespula vulgaris* and *Vespa crabro* venom, and the latter possesses about 5 percent dry weight of a third amine that is the most concentrated biological source of acetylcholine known. Each of these constituents produces pain. From *Polistes* a kinin has been isolated in purified state that affects smooth musculature and circulation like bradykinin. Compounds similar to these probably occur in other social wasps, but solitary wasp venoms are quite different. Social wasp and hornet venoms have been shown to have hemolytic and anticoagulant properties, and to cause edema and hemorrhaging.

Various authors estimate that on the basis of toxicity alone, without allergic complications, it would require at least 500 honey bee stings to cause mortality in a healthy adult human. Wasp and hornet stings generally elicit a more severe response and it presumably would take fewer stings to produce the same effect. Tolerance to bee stings appears to be developed in beekeepers long exposed to stings; however, this tolerance seems to be lost when exposure to stings is discontinued. In nonsensitized persons a sting produces local pain, swelling, and redness, which passes harmlessly in a few hours. In susceptible individuals there is a gradation in complications: (1) In slight reactors there is general urticaria, malaise, and anxiety developing within an hour after the sting. (2) In somewhat more serious reactions, in addition to the above symptoms, generalized edema, chest constriction, wheezing, abdominal pain, nausea, vomiting, and dizziness may occur. (3) In severe cases there are in addition such symptoms as dyspnea, dysphagia, hoarseness and thickened speech, confusion and a feeling of impending disaster. (4) In shock reaction cases add at least two of the following: cyanosis, drop in blood pressure, collapse, incontinence, unconsciousness. The onset of serious symptoms is more rapid with the increasing severity of the case, only some 2 to 15 minutes in cases involving shock.

Various methods have been devised to protect against or alleviate the effects of Hymenoptera stings. Clothing has some effect on whether one is likely to be stung, and protective clothing has been developed and analyzed (Wagner and Reierson, 1975). For low to moderate sensitivity the use of antihistamine compounds is helpful, and physicians may provide their patients with emergency kits that provide antihistamines and epinephrine to be used immediately after stinging to minimize effects before medical help is available. For the person with a high level of sensitivity to bee and wasp stings, with allergic manifestations such as hives, asthmatic responses, or shock reactions, it is vitally important that a physician, preferably an allergist, be contacted and a course of desensitization be devised. One study shows that the great majority of sensitive patients have allergic responses caused by the release of histamine from the leucocytes, mediated by IgE-type antibodies (Sobotka et al., 1974).

For desensitization most physicians use whole body Hymenoptera extracts made from bees, wasps, and hornets, so-called polyvalent extracts, because there is difficulty in knowing what species of stinging insect has sensitized the patient. There is some cross antigenicity developed between various stinging insects, and whole body extracts are easily obtained. A 1969 study by a special committee of the American Academy of Allergy reported that of 547 patients who were taking desensitizing injections and had then been stung again, 95 percent were less sensitive, 3 percent the same, 2 percent suffered worse reactions; of 275 patients who had stopped desensitizing injections and were stung again, 86 percent experienced less severe reactions, 8 percent the same, and 6 percent were worse (Barr, 1974). Where it is known that sensitivity has been caused by a particular species of stinging insect, and that future stings will likely be from the same source, as in the case of beekeepers and their families, there seem to be advantages in using pure venom from that source for desensitization (Lichtenstein et al., 1974). Hoigné and associates (1974) believe that the usual practice of discontinuing maintenance hyposensitization after three years is not advisable.

Family Formicidae

ANTS

General discussions on ants and other social Hymenoptera, along with excellent illustrations, may be seen in Wilson (1971); reviews of stinging ants, particularly the fire ants *Solenopsis,* are included in Bücherl and Buckley (1971), and Minton (1974). Much of the information summarized here on ant sting venoms has been gleaned from these sources.

Ants belong to the family Formicidae, superfamily Formicoidea. Many ants cannot sting, but some in the subfamilies Ponerinae, Dorylinae, and Myrmicinae sting viciously.

Parrish (1963) lists four human deaths in the USA, during the period 1950–1959, caused by ant stings, and isolated instances of ant-associated death continue to occur. As with bee, wasp, and hornet stings, ant stings are very painful, but since workers in the nests cannot fly and pursue offenders, mass stinging is less likely. Very large numbers of stings on humans by ants have generally involved infants, or handicapped or intoxicated adults.

The more primitive subfamilies of ants show affiliations with other stinging Hymenoptera in possessing a stinging apparatus and venom; Hermann and colleagues (1975) illustrate the histology and function of the venom gland system in formicine ants. Venoms of more primitive myrmicine, ponerine, pseudomyrmicine, and doryline ants are proteinaceous and rather like the venoms of social bees and wasps. Detailed studies of myrmicine ant venoms have been undertaken with Australian bulldog ants of the genus *Myrmecia,* large ants, up to 2 cm long, that excavate nests in dry stony ground in areas of tree roots and large rocks. Venom of the red bulldog ant, *M. gulosa,* has histamine, hyaluronidase (spreading factor), kininlike activity, and hemolytic protein; in *M. pyriformis* fractionation of the venom

yielded an enzyme-free protein fraction that possesses smooth muscle stimulant, and red-cell lysing and histamine-releasing activities resembling the main peptide melittin of honey bees (Wanstall and Lande, 1974).

The subfamily Myrmicinae includes several pugnacious stinging groups found in tropical areas and to some extent in warmer portions of temperate zones. Some of the better-studied dangerous species are the fire ants and harvester ants (Fig. 17-10).

Fire Ants. Fire ants, genus *Solenopsis,* are so named because of the sharp, fiery pain of their sting. In the USA what was originally named the imported fire ant, *Solenopsis saevissima richteri,* is now recognized as a complex from South America of at least two and possibly four species (Lofgren *et al.,* 1975). *S. richteri* is known as the black imported fire ant, imported as early as 1918 near the Mobile, Alabama, area and only moderately successful in its spread. The red imported fire ant, *S. invicta,* entered the same region in the 1940s and has spread spectacularly. The overall spread of imported fire ants encompasses parts of all the coastal southeastern states from North Carolina in the East to Texas in the Southwest. It is believed that their continuing northward and westward spread will ultimately extend

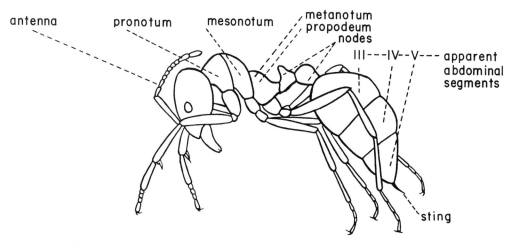

Fig. 17-10 Diagram of a fire ant. (Courtesy of US Public Health Service.)

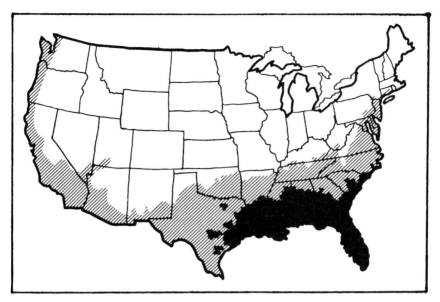

Fig. 17-11 Quarantine areas as of August, 1976 (*solid black*), and potential climatically suitable infestation zones (*shaded*) of imported fire ants in the United States. (From USDA, APHIS, PPQ sources.)

their range to all of the southern USA and strips of infestation along extensive portions of the east and west coasts (Fig. 17-11). They build large hard-crusted earthen mounds (Fig. 17-12).

Fire ants sink their powerful mandibles into the flesh for leverage and then drive the sting into the victim. They attack crops and kill and devour newly hatched quail and poultry, or enter pipped eggs to reach unhatched chicks; they will also attack young pigs and newborn calves. The death of large numbers of bluegill sunfish (*Lepomus* sp.) in Mississippi and Alabama has been attributed to the ingestion of fire ants, though these fish will not readily eat them in the laboratory. Patterns of fire ant invasion and dispersion into new areas suggest that automotive vehicles play an important role in their spread.

Following the initial burning sensation of the fire ant sting, a wheal appears at the site of venom entry, and a few hours later a clear vesicle containing fluid appears; within 24 hours the fluid becomes purulent and the

Fig. 17-12 Mounds of the imported fire ant. (USDA photograph.)

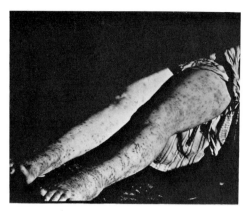

Fig. 17-13 A case of extensive and severe stinging by fire ants. (Courtesy of V. J. Derbes, Tulane University School of Medicine.)

pustule may persist for some time if unbroken, but will eventually leave a crust that may scar. Occasionally there are systemic reactions, particularly if the victim is stung extensively (Fig. 17-13), including nausea, vomiting, dizziness, perspiration, cyanosis, asthma, and other symptoms typical of severe allergic reactions. In such cases, without appropriate medical treatment, victims have died. Immunotherapy of individuals who are highly sensitive to fire ant stings has met with rather good success (Triplett, 1973; Rhoades *et al.*, 1975). It should be mentioned that marked allergic reactions can occur in response to other species of ants, as in the case of a woman from South Carolina who suffered an acute systemic anaphylactic reaction from the sting of an ant in the genus *Tetramorium* (Majeski *et al.*, 1974).

Fire ant venoms are unique among stinging arthropods because they are without protein or polypeptide. Active components are alkaloids distinctive for each of the US forms. *S. saevissima* venom contains 5 alkaloids: three piperidine derivatives and two unsaturated analogs.

Other fire ants of some importance in the USA are the common fire ant, *Solenopsis geminata*, and the southern fire ant, *S. xy-*loni, the latter having an amine in the venom.

Harvester Ants. Harvester ants of the Americas, *Pogonomyrmex* spp., have a vicious sting and readily attack man and animals. They are of concern to agriculture because their low, bare mounds and their destruction of adjacent vegetation can reduce forage for livestock. The red harvester ant, *P. barbatus,* occurs at lower altitudes from Kansas and Louisiana to Utah and California, USA, and into Mexico; the California harvester ant, *P. californicus,* occurs from Texas and Utah to California and Mexico; the western harvester ant, *P. occidentalis,* is widespread from North Dakota, USA, and British Columbia, Canada, southward to Arizona and Oklahoma; the Florida harvester ant, *P. badius,* the only harvester ant east of the Mississippi River, occurs in the southeastern states. These ants readily attack man and animals; young pigs may be killed by their stings. *P. barbatus* has venom with strong cholinergic properties, causing hair erection and sweating.

Ants can disseminate pathogens contaminatively, and serve as intermediate hosts of helminths (see Chapter 6).

Other Hymenoptera

Some solitary Hymenoptera are of occasional medical importance. Of general interest is the isolation in pure form of a toxin in larvae of a sawfly (Tenthredinidae: *Lophyrotoma interrupta*) responsible for poisoning large numbers of sheep and cattle in Queensland, Australia, when ingested (Leonard, 1972). It may well be that the insect sequesters toxic plant components as occurs in the defensive reactions of a number of phytophagous insects. Most medical complications from solitary Hymenoptera are due to stinging in response to forceful contact.

Fig. 17-14 A velvet ant, family Mutillidae. (Photograph by R. D. Akre, Washington State University.)

Mutillid Wasps (Velvet Ants, Woolly Ants, Cow Killers, Mule Killers). Among the less-known stinging insects are wasps belonging to the family Mutillidae, superfamily Scolioidea. Most mutillids are covered with a velvety pubescence (Fig. 17-14), often in bright orange, red, or yellow warning coloration. The females are without wings, are good runners, and can inflict a painful sting. Their larvae are parasites of bees and other wasps. There are many species, some of the commoner forms from 0.75 to 2.5 cm in length. Velvet ants are often associated with dry and sandy environments, and species associated with swimming beaches can cause much distress when stepped on by barefoot bathers.

Bethylid Wasps (Superfamily Bethyloidea). These wasps are parasitic on hymenopterous, lepidopterous, and coleopterous larvae, although comparatively little is known of their bionomics. Species of *Scleroderma* have rather frequently been reported as stinging humans. A woman in Italy, stung in many places on her body by *S. domesticum,* showed localized reactions and general symptoms of the anaphylactic type. The genus *Cephalonomia* has also been reported

as a cause of aggravation in Italy, and *C. gallicola* has often been reported stinging persons in North America. *Epyris californicus* has been associated with quite bothersome stings with complications in California, and a species of *Laelius* stings man in Rumania.

BITING (PIERCING) INSECTS

Insects that pierce the skin with their mouth parts are normally bloodsuckers, and the act of biting or piercing is simply their means of obtaining food. Noteworthy exceptions include a few plant feeders and predacious insects. The pain caused by the mechanical insertion of such tiny mouth parts would be relatively benign, but in perhaps every instance irritating salivary fluid is introduced. This fluid differs among various insects, as shown by the resulting reactions, local and systemic, that are generally specific enough so that one who is experienced may discern whether the offender was a bed bug, flea, mosquito, black fly, or other. Reactions may occur on livestock as well as humans (cf. Pascoe, 1973). To understand the operation of the bloodsucking apparatus of various insects consult Chapter 3.

Some persons suffer no ill effects from insect bites, not even a swelling at the site of the bite; others react violently to even one bite. These differences in tolerance to a given species are not fully understood, but doubtless allergic responses play a role. Hyposensitization to a variety of biting insects, using injected extracts, has yielded good to excellent results in 82 percent of a sample group of patients, although an unfortunate few experience greater sensitivity (Frazier, 1974).

Some allergic responses to the bites of arthropod vectors are thought to be caused by sensitivity to the pathogens they introduce. *Chrysops* deer flies transmitting infective larvae of *Loa* of human or simian origin may cause severe immediate reactions after previous sensitization (Crewe and Gordon, 1959). On the other hand, experiments with various triatomid bugs show that bilateral or unilateral palpebral edema, considered an early symptom of Chagas' disease, is caused by uninfected bugs as well in sensitized persons (Lumbreras *et al.*, 1959).

Order Diptera; Family Culicidae

MOSQUITOES

Several investigators have noted that sensitivity to mosquito bites consists of two stages: an immediate whitish wheal and a response that may occur about 24 hours later consisting of an inflamed reddish swollen area. This general type of response has been noted in humans and guinea pigs. The oral secretions of the mosquito *Aedes aegypti* caused allergic reactions when injected into guinea pigs. Animals bitten for the first time did not react, but an allergic response followed on the next challenge (McKiel, 1959). Hudson and associates (1960) succeeded in cutting the salivary duct of mosquitoes, demonstrating that such insects caused no wheal on humans but their bite was more painful. Strophuluslike skin eruptions in children have been shown to be due to the bites of *Culex pipiens,* and in most cases these eruptions could be reduced or eliminated by a series of desensitizing injections using commercial mosquito extract (Tager *et al.*, 1969). Consult Chapter 10 regarding personal protection from mosquitoes, and means of mosquito control.

Order Hemiptera

BUGS

Anthocoridae, very small predators about 2 to 3 mm long, have a surprisingly vicious bite considering their size. The assassin bugs, aside from the Triatominae, attack species of insects, particularly soft-bodied forms from which they suck body fluids. Among well-known North American forms are the wheel bug *Arilus cristatus,* and corsairs, *Rasahus* spp. (Fig. 7-5). The wheel bug is common from New Mexico through the southern and parts of the eastern USA; corsairs are also widespread, with species extending from the midwestern USA to South America. Attacks on humans are principally, if not wholly, in self-defense. Persons picking up objects may accidentally put pressure on one of these insects; likewise, in handling a plant the fingers may close upon the insect, and a very painful bite may be suffered. The predacious reduviid, *Platymeris rhadamanthus,* from East Africa and Zanzibar, injects its salivary secretions into prey, but also ejects saliva forcibly as a defense against vertebrates. The saliva retains toxic characteristics for at least 3 years when dried and contains six or more proteins; it has trypsinlike activity, strong hyaluronidase (spreading factor), and weak phosopholipase activity; it causes intense local pain, vasodilation, and edema around eye and nose membranes (Edwards, 1961). A case of anaphylactic reaction is reported from the bite of an Asiatic triatomine, *Triatoma rubrofasciata* (Teo and Cheah, 1973), and similar strong reactions are sometimes caused by bites of New World triatomines.

A number of aquatic Hemiptera will bite man, including members of the families Belostomatidae and Notonectidae. The giant water bugs *Lethocerus, Belostoma,* and *Benacus,* family Belostomatidae, are among the largest bugs, measuring up to 6 cm or more in length and possessing formidable beaks. They feed on other aquatic insects, also young frogs and fish. Because they are winged and fly readily to light, they are commonly known as electric light bugs; they are called *Wasserbienen* (water bees) in Germany. The bite of *Benacus griseus* has been described as causing a burning sensation with some swelling, but no very lasting effects. The predacious family Notonectidae, so-called back swimmers, which orient ventral surface uppermost, may also inflict a bite that may be nearly as severe as a bee sting.

Numerous instances of bloodsucking among phytophagous Hemiptera and Homoptera have been reported. Information concerning these cases has been assembled by Usinger (1934). Among species exhibiting this fortuitous bloodsucking are members of the families Membracidae (treehoppers), Cicadellidae (leafhoppers), Miridae (plant bugs), Coreidae (leaf footed bugs), and Tingidae (lace bugs).

Order Siphonaptera

FLEAS

In studying the response of guinea pigs to flea bites it was found that substances causing hypersensitivity were haptenic in nature, possibly needing to conjugate with skin collagen to induce hypersensitivity. Gel filtration showed allergenic activity in a fraction of high molecular weight (4,000 to 10,000) as well as a fluorescing aromatic fraction of below 1,000 mw. Sensitive humans may be desensitized by using commercially available flea antigens in a series of injections to gradually build up their tolerance. Pets can develop eczematous sensitivity to flea bites, known as miliary dermatitis and eczema in cats, and canine summer dermatitis in dogs. Desensitization of these animals by using whole flea extracts or a haptenic fraction of flea saliva is possible, generally requiring booster injections at intervals (Michaeli and Goldfarb, 1968; Reedy, 1975). In cases of sensitivity flea control around premises is advisable (see Chapter 14).

Order Thysanoptera

THRIPS

Thrips are minute, mostly plant-sapsucking insects; however, there have been numerous reports of their attacking man and of their ability to suck blood, mainly a fortuitous response when their plant hosts are maturing and drying up (Bailey, 1936). Reports of annoyance have come from many areas of Europe and the USA. Populated areas in the path of thrips migrating from drying cereals may suffer serious annoyance, the insects being so small they pass through ordinary window screen. Reports include the onion thrips, *Thrips tabaci;* the pear thrips, *Taeniothrips inconsequens;* cereal and grass feeders, *Chirothrips aculeatus* and *Limothrips cerealium; Heliothrips* sp. in the Sudan; *Thrips imaginis* from Australia; *Limothrips denticornis* from Germany; the Cuban laurel thrips, *Gynaitkothrips ficorum,* from Algiers.

URTICATING LEPIDOPTERA

A number of lepidopterous families have species whose larvae possess STINGING HAIRS and SPINES that serve as effective armaments of self-defense. Other types of medical complications with Lepidoptera are possible, and will be mentioned briefly. Bücherl and Buckley (1971) and Minton (1974) provide quite comprehensive reviews on urtication, and the comments that follow are from these sources, the previous edition of this text, or other sources as credited.

Urticating structures are on the integument of Lepidoptera larvae, and in a few cases on adults as well. They consist of flexible hairs that may be long or microscopic, or rigid bristles and spinules that may adorn the whole body surface or occur on tubercles. The urticating setae can be tubular or porous, capable of causing mechanical irritation as with glass fibers but more commonly filled with poison derived from a unicellular gland at the setal base (Figs. 17-15). Usually these toxic hairs cause burning and inflammation on the skin, but they can be much more serious when in contact with mucous membrane, the upper respiratory tract, or the

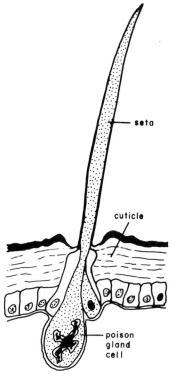

Fig. 17-15 Urticating hair of a caterpillar showing sharp seta and associated poison glands. (From Gilmer, 1925, *Ann. Entomol. Soc. Am.,* **18**:203–39.)

eyes. An incident of eye loss in a child from the entrance of urticating hairs has been recorded.

Some Lepidoptera have toxic substances that may be irritating on contact, or toxic upon ingestion. Kearby (1975) mentions being burned, with subsequent blisters, upon handling larvae of the variable oakleaf caterpillar, which secrete a mist smelling like formic acid. Larvae of *Zygaena* have special stink glands, and have been noted as a seasonal pest in Liguria, Italy. In the swallowtail butterflies, Papilionidae, larvae have eversible glands termed OSMETERIA that protrude from the prothorax and emit a repellent odor when they are annoyed; wood-boring larvae in the family Cossidae produce an emission of fluid around the mandibles that will cause a burning sensation on human skin; several Notodontidae larvae forcibly emit irritating fluid from the body surface. Larvae of the monarch butterfly, and relatives, sequester cardiac glycosides from various milkweeds, and as larvae and adults they may be highly toxic if ingested by predators such as birds.

There are doubtless many instances of unrecorded urticating Lepidoptera. The main families known to contain at least some urticating species are Arctiidae, Bombycidae, Eucleidae (Limacodidae, Cochlidiidae), Lasiocampidae, Lithosiidae, Lymantriidae (Liparidae), Megalopygidae, Morphoidae, Noctuidae, Notodontidae, Nymphalidae, Saturniidae, and Sphingidae. In Bücherl and Buckley there is a list of over 100 species in 41 genera. Some of the better known species will be discussed further.

The suggestion has been made, and it seems valid, that a distinction should be made between urtication by Lepidoptera larvae, termed ERUCISM, and adults, termed LEPIDOPTERISM. Likewise in both kinds of affliction there are DIRECT EFFECTS caused by contact with the insect, and INDIRECT EFFECTS due to contact with airborne irritating hairs and spines.

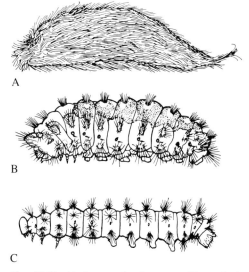

Fig. 17-16 Various urticating caterpillars. A. A flannel moth, *Megalopyge*. B. *Norape crenata*. C. Io moth, *Automeris io*. (Courtesy of US Public Health Service.)

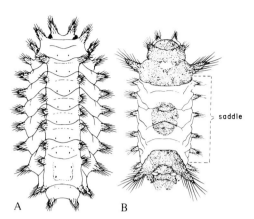

Fig. 17-17 A. A slug caterpillar, *Euclea chloris*. B. The saddleback caterpillar, *Sibine stimulea*. (Courtesy of US Public Health Service.)

Fig. 17-18 A species of *Hemileuca* larva. (Photograph by R. L. Furniss; courtesy of US Forest Service.)

Direct erucism has been extensively documented. Flannel moths, *Megalopyge* species (Fig. 17-16*A*), are known in the southern USA and much of tropical America. The saddleback caterpillar, *Sibine stimulea* (Fig. 17-17*B*), slug caterpillars (Fig. 17-17*A*) such as the tree asp, *Euclea delphinii*, and the Io moth caterpillar, *Automeris io* (Fig. 17-16*C*), are well known in the eastern USA. Pine processionary caterpillar, *Thaumetopoea pinivora*, and relatives are important in Europe and the Mediterranean region. Larvae of the browntail moth, *Euproctis chrysorrhoea*, are troublesome in Europe and in the eastern USA where it is introduced. In southeast Asia and the Far East species of *Euproctis* are most frequently a problem.

Two general effects of direct erucism are evident: immediate and highly painful reactions from some larvae and more delayed and rashlike manifestations caused by contact with other species. Species of *Megalopyge, Dirphia, Automeris,* and *Hemileuca* (Fig. 17-18) cause violent pain on contact, and commonly subsequent headache, nausea, vomiting, and lymphadenitis; shock and convulsions have been reported. Problems have been especially noted in small agricultural villages surrounded by forest in the South American tropics. In a 1958 study of 433 cases, 66 percent involved people harvesting and hoeing, 16 percent lumbering, and 11 percent carrying loads. In species causing delayed effects an itching, papular dermatitis or persistent generalized urticarial eruption are usual (cf. Fig. 17-20). Dias and Azevedo (1973) describe an occupational disease of rubber workers in northern Brazil, caused by contact with *Premolis semirufa* caterpillars, that in chronic form involves long-term disability of the fingers due to lesions in the peripheral connective tissues.

Indirect erucism may be difficult to trace to its source because of the airborne nature of the offending materials, and because the irritants may last for some time after the living source is no longer present. This kind of affliction generally involves regions where there are sizable human populations near the offending irritants, and has characteristically included portions of Europe, Korea, Japan, and the USA. The Bois de Boulogne, Paris, Baltic seaside resorts, schools and colleges in Texas, USA, and several important agricultural centers have been involved. Causal Lepidoptera may also be associated with direct erucism. Larvae in pupating may cover the silken cocoon with irritant hairs, or adults may cover egg masses with these structures, and later these irritants may be disseminated aerially. The peculiarity of circumstances surrounding urticaria can be puzzling. In Israel 600 of 3,000 soldiers encamped in a pine tree grove developed rash and severe irritation. No living larvae were seen, but dead larvae and old cocoons of *Thaumetopoea wilkinsoni* were present, and these caused typical reactions on volunteers' arms.

Lepidopterism as an important problem seems limited to South America, where moths of the genus *Hylesia* occur. Individual instances are quite widely scattered; small outbreaks have occurred in Argentina and the Caribbean region, several hundred cases have occurred during outbreaks in Brazil, and several thousand cases in Peru. *Hylesia* belongs in the family Saturniidae, subfamily Hemileucinae, in which *Hemileuca* has been noted as a cause of highly irritating direct erucism. *Hylesia* urticating setae are barbed or spiny and are called "fleshettes"; they occur on female moths only. Outbreaks are seasonal and most prevalent when large numbers of these moths take to flight after rains. Direct lepidopterism through contact with the moth occurs, but by far the greater number of cases involve urticating rash and upper respiratory tract complications after the moths are attracted to lights, where they flutter around and lose their spines. There is a report of an instance where a tanker ship moored on the Orinoco river at Caripito, Venezuela, in December was invaded by large numbers of *Hylesia canitia* attracted to lights illuminating the deck. Most of the crew developed severe dermatitis, which persisted because the irritants were distributed throughout the ship by the ventilation system.

Other milder instances of lepidopterism are known. Rothschild and associates (1970) list the genera *Epanaphe, Anaphe, Gazalina,* and *Epicoma* in the Notodontidae as having females with barbed setae in the anal tufts. In Japan a lymantriid moth, *Euproctis flava,* is regarded as venomous. Actually a number of moths cover the egg masses with abdominal hairs, and in addition their larvae possess venomous setae, so lepidopterism is not always readily separated from erucism. Perlman and coworkers (1976) review reactions they term "tussockosis" due to the Douglas fir tussock moth, *Orgyia pseudotsugata,* in

A

B

C

Fig. 17-19 *A.* Nearly full-grown larva of the Douglas fir tussock moth, *Orgyia pseudotsugata.* Note dorsal tufts of irritating setae. Scanning electron micrograph of dorsal setal tuft (*B*), and at greater magnification (*C*). (*A* courtesy of US Forest Service; *B* and *C* courtesy of E. Press, Oregon Department of Human Resources, and courtesy of Department of Entomology, Oregon State University.)

the Pacific Northwest, USA. In this insect the larvae possess dorsal tufts of irritating setae (Fig. 17-19), and the females cover the egg masses in body hair. Forest workers coming in contact with these irritants may develop an urticarial rash (Fig. 17-20) and eye and respiratory symptoms.

Lepidoptera may also affect livestock and pets. Adults of *Zygaena,* family Zygaenidae, are zootoxic when ingested. European ducks have been reported to undergo severe enteritis from consuming unspecified caterpillars. In South Africa fatal enteritis has occurred in pigs ingesting *Nudaurelia cytherea* caterpillars, and death of cattle from similar causes has been reported from the Bavenda region, just west of Kruger National Park, South Africa. Severe stomatitis occurs in dogs ingesting grasses contaminated with the setae of *Thaumetopoea pinivora.* The range caterpillar, *Hemileuca oliviae,* is destructive on rangeland grasses in parts of New Mexico and Colorado, USA, in some years infesting thousands of hectares. While its main effect is to consume valuable forage, it will cause blistering of the mouth in cattle on contact.

The chemistry and pharmacology of urticating venoms has been difficult to ascertain because they cannot be obtained in satisfactory amounts and purity. Skin symptoms from the irritating hairs and spines of various Lepidoptera include congestion, edema, infiltration of eosinophilic cells, hemorrhage, bulbous vesicles, and necrosis. In the blood there may be hemolysis. In the ocular conjunctiva granulomas of the foreign body type may develop. One study of victims of the highly painful urtication of *Megalopyge* showed malaise in 27 percent of cases, sensation of high temperature in 29 percent, chills in 14 percent, and histamine-type reactions in 10 percent. Extracts of the setae of these caterpillars yielded typical albuminoid

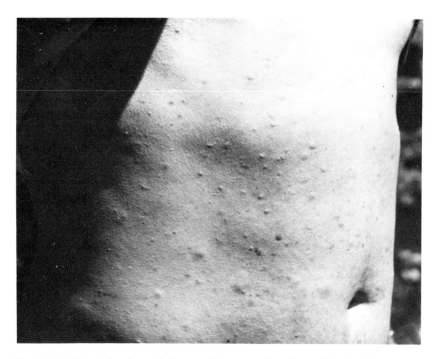

Fig. 17-20 Urticarial rash on abdomen of forest worker, due to irritation from setae of Douglas fir tussock moth. (Courtesy of E. Press and J. A. Googins, Oregon Department of Human Resources.)

reactions, histamine-type responses, and hemolysis, but no acetyl choline was demonstrated. The urticating venom of *Automeris io* has a proteolytic enzyme. *Dirphia* spines contain histamine, and apparently more is caused to be released by host tissues. The toxin of *Hylesia urticans* setae is rather water soluble and alcohol and ether insoluble. The urticating venom of *Lonomia* sp., a saturniid, will lyse fibrin protein and activate human plasminogen. Perlman and coworkers discovered that the irritants of the Douglas fir tussock moth, *Orgyia pseudotsugata,* would cause histamine release from sensitized tissues, but not from unsensitized tissues.

There are a few suggestions, in addition to avoiding contact, for reducing the effects of urtication. The immediate application and removal of adhesive tape, or other sticky tape, will lift off some of the irritants and diminish full effects. The application of analgesics, creams, and lotions with steroids, and intravenous calcium gluconate have helped in many cases. Where moths possess the irritants, as in *Hylesia,* it may be necessary to turn off exposed lights during their most intense flight periods, and bedding should be kept under cover until used, or laundered if exposed.

VESICATING COLEOPTERA

Several instances of beetles containing vesicating substances are known, and while the incidents of human problems are normally quite minimal, special circumstances will bring out these insects in large numbers. If this occurs in heavily populated areas, the contacts between humans and these annoying beetles can cause a number of serious blistering problems. The role of beetles as intermediate hosts of helminths is discussed in Chapter 6. The medical and veterinary importance of Coleoptera has been reviewed extensively by Théodoridès in a series of papers (1950 and several more recent papers). The toxic nature of some beetles is known to primitive hunting groups, as Jolivet (1967) and Freyvogel (1972) refer to toxic hemolymph from the larvae of chrysomelid and carabid beetles used by Kalahari bushmen as sources of arrow poison.

Family Meloidae

Members of this family are known as blister beetles. The application of their pulverized bodies, or contact with the living insects, may cause blistering of the skin. The family is comprised of medium to large beetles with comparatively soft body; head broad, held vertically, and abruptly narrowed to a neck; prothorax narrower than the soft and flexible wing covers (Fig. 17-21).

Adult blister beetles are plant feeders that deposit their eggs on the ground. These beetles often complete development and emerge as adults in large numbers, appearing in gardens and crops very suddenly and leading people to believe they have migrated some distance. Meloid larvae feed on the eggs of grasshoppers and in nests of solitary bees, others are predacious on various soil invertebrates. They undergo a multiplicity of larval stages that differ structurally and in habits from one another, an unusually complex development termed HYPERMETA-MORPHOSIS.

Fig. 17-21 Common vesicating meloid beetles from the central to southeastern-eastern areas of the United States. On the *left* is *Epicauta fabricii,* the ash-gray blister beetle; on the *right* is *E. vittata,* the striped blister beetle. Slightly larger than life size.

The Spanish fly, *Lytta vesicatoria,* is a European meloid found most abundantly during early summer in Spain, southern France, and other parts of Europe. It is golden green or bluish in color, from 1.25 to 1.9 cm in length. This meloid appears suddenly in early summer and may be collected by the hundreds, clinging principally to such vegetation as ash, privet, and lilac.

Cantharadin, a crystalline anhydrid of cantharidic acid, was first isolated from the Spanish fly. The biosynthesis of cantharadin has been studied (Guenther *et al.,* 1969). This substance penetrates the epidermis quite readily and produces violent superficial irritation, even in amounts as small as 0.1 mg, resulting in blistering in a few hours. Former pharmacological uses, including use as an aphrodisiac, have been abandoned because its effect can be so dangerous to life. Equines have developed colic symptoms and died from ingesting the striped blister beetle, *Epicauta vittata* (Bahme, 1968).

All members of Meloidae that have been studied contain cantharadin, though sometimes in very low amounts. Blistering occurs from contact with the dried pulverized body, by the blood of the beetle released when the insect is crushed on the skin, or by blood coming from the leg and body joints through reflexive bleeding. Several North American meloids found in gardens and croplands may cause blistering, the clematis blister beetle, *Epicauta cinerea,* being a common problem species in the eastern USA. Meloids causing severe vesicular dermatitis in Africa are *Zonabris nubica, Epicauta tomentosa,* and *E. sappharina.*

Family Oedemeridae

At least two oedemerids, *Sessinia collaris* and *S. decolor,* cause severe blistering on some of the mid-Pacific islands, where they are called coconut beetles (Herms, 1925). These beetles fairly swarm about the newly opened male flowers of the coconut, where

they feed on pollen. They are readily attracted to light. Crushing on skin and resultant contact with their hemolymph causes a sharp momentary pain; a large blister causing little pain forms in a few hours.

In Puerto Rico a 1.0 to 1.5 cm brown and yellow oedemerid, *Oxacis* (=*Oxycopa*) *vittata,* is a cause of vesication (Fleisher and Fox, 1970). In one location in 1975 an extensive outbreak of dermatitis occurred from this species, affected premises having hundreds of beetles apparently attracted to lights (C. G. Moore, communication to R.F.H.). The larvae are believed to be associated with the littoral zone along rivers and lakes, in decaying wood such as driftwood. The beetle must be crushed while on the skin, resulting in tingling and burning after 1 hour and vesication after 9 hours. Vesicles are replaced by crust, and the lesions heal without scarring. In 1957 large numbers of *Oxycopis suturalis* in homes and tourist attractions on Plantation Key and Islamorada, Florida, USA, caused several instances of human blistering when crushed on the skin (1957 summary, Florida Plant Pest Survey).

Family Staphylinidae

Called rove beetles, these Coleoptera are most readily characterized as having very short elytra. At least 30 species in the staphylinid genus *Paederus* (Fig. 17-22), small beetles less than 1.0 cm long, produce dermatitis and ophthalmic lesions in animals (Pavan, 1959). The beetles breed in damp leaf litter and humus, and rains will bring them out in great abundance, at which time adults are attracted to lights, land on humans, and are brushed off. The toxicant is in the hemolymph, and since these beetles do not bleed reflexively, they must be crushed against the skin to cause vesication. According to Kurosa (1958) the larvae also contain vesicant. The crystalline toxin *pederin,* as well as derivatives called *pseudopederin* and

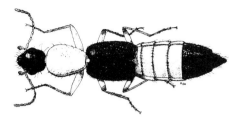

Fig. 17-22 A rove beetle, *Paederus fuscipes* (family Staphylinidae) from Thailand. Actual body length is 5 mm.

pederone, have been chemically characterized, and the biosynthesis of pederin investigated (Cardani *et al.,* 1973).

Agricultural workers in Paraguay suffer from vesication by these beetles. *Paederus fuscipes* is a problem in Europe, Japan, Formosa, and much of southeast Asia and the islands to Australia. In Thailand on one occasion this species caused numerous cases of dermatitis and ocular involvement in the Bangkok area (Papasarathorn *et al.,* 1961); a 1966 outbreak on Okinawa resulted in 2,000 people seeking treatment, and probably an equal number did not seek medical aid (Armstrong and Winfield, 1968). *Paederus ilsae* has caused burning eruptions of the skin in Israel (Adar et al., 1972). In southern Africa a widespread outbreak of vesicular dermatitis was reported with essentially painless lesions caused by contact with crushed adults of *Paederus sabaeus*. Outbreaks in Uganda caused by this species and *Pachypaederus puncticollis,* as well as extraction and bioassy of the toxin, have been reviewed by McCrae and Visser (1975). The term "Nairobi eye" applies to conjunctivitis caused when juices of crushed beetles are rubbed in the eye. In India *Paederus* vesicular dermatitis is called "spider-lick"; frequently the brushing activity to remove the beetles from the skin results in a smear lesion referred to as "whiplash dermatitis." *Paederus* are practically worldwide in distribution, though their numbers are low in temperate zones. Vesicating lesions in many parts of the world, especially if unusual,

could logically include these beetles as a suspect source, since it seems likely that vesicants are present in all *Paederus.*

Family Tenebrionidae

Tenebrionid or darkling beetles of the genus *Blaps* produce a fluid that causes blisters, and that probably contains quinones (Roth and Eisner, 1962). *B. judaeorum* is common in certain dune areas of Palestine, and caustic effects of the secretion of *B. nitens* have been described in detail. Adults of several other tenebrionids produce fluid with quinones that will cause a durable darkening of the skin in light-skinned people.

SPIDERS

Class Arachnida; Order Araneida

Spiders are arachnids in which the cephalothorax is uniform, bearing not more than eight eyes, and a constriction where this section joins the abdomen (Fig. 2-7). The abdomen is usually unsegmented and bears not more than four, usually three, pairs of spinnerets; there is no telson. The chelicerae are two-segmented, moderately large and unchelate, and contain (or are connected with) a poison gland. In most spiders VENOM GLANDS, which may include accessory glands, are situated in the anterior cephalothorax; in tarantulas and bird spiders these are in the basal segments of the chelicerae. The pedipalps are six-segmented, leglike, and tactile in function. In some forms they are so large they might be mistaken for an extra pair of legs. The pedipalps of the male are modified as intromittent organs (see Savory, 1935). The eight legs consist each of seven segments; tarsi have two or three claws. Respiration is by book lungs or tracheae, and normally by both. A good account of spider anatomy is given by Snodgrass (1952).

A number of otherwise unattributed statements that follow are from reviews by Bücherl and Buckely (1971), Minton (1974), and Southcott (1976), as well as the previous edition of this book.

Spiders are universally feared, mainly owing to superstition and their repulsive appearance, but also because of the knowledge that they kill insects and other small animals by introducing a venom with the bite. Nevertheless, of the more than 100,000 species in over 2,000 genera, and in excess of 60 families, only a few species are known to be dangerous to man. But it is well to bear in mind that the chelicerae (poison fangs) and associated venom are developed for the purpose of killing prey, and many species of spiders large enough to penetrate human skin can inflict bites that have local or systemic effects. The venom has toxins to subdue prey, and may have digestive functions to liquefy insect tissues that can then be sucked up, leaving the empty cuticular shell. Various venomous spiders produce 0.2 to 5.0 mg dry weight of venom, with a mouse LD_{50} in mg/Kg of body weight ranging from 0.34 intravenously to 62.5 subcutaneously. Fortunately, the vast majority of spiders are nonaggressive to man and animals, and most bite when pressure is applied unwittingly to them as they are hidden in clothing, lumber piles, or similar places of concealment. Russell and coauthors (in Conn, 1974) state that physicians in southern California, USA, treat about 400 spider bites annually, though some of these cases are likely due to other causes.

TARANTULAS

The term "tarantula" was first applied to a European species, *Lycosa tarentula,* a member of the family Lycosidae, the wolf spiders. It is of interest that the American tarantulas belong to an entirely different group of spiders, and that the arachnid that goes by the generic name *Tarantula* is a tailless whipscorpion, Order Phrynichida. In Italy in the vicinity of Taranto, there occurred a spider scare during the seventeenth century, which gave rise to a condition known as "tarantism," resulting from the bite of the European *L. tarentula.* To rid the body of the venom those bitten, or imagining they had been bitten, engaged in a frenzied dance known as the tarantella (Thorp and Woodson, 1945). Tarantulas, bird spiders, trapdoor spiders, and funnel-web spiders are found in the Americas, Australia, New Zealand, Asia, and Africa, These are mainly large ground-associated and rather slow moving arachnids, placed in the Mygalomorphae by arachnologists.

In the USA the term tarantula is applied to very large mygalomorphs belonging to the family Theraphosidae (Fig. 17-23), also called "bird spiders." Many of the Central and South American forms measure up to 17.5 cm in the spread of legs. About 30 species live within the limits of the continental USA, mostly in the Southwest (Gertsch, 1949), and many others inhabit the New World tropics. They are greatly feared because of their large size and hairiness and are erroneously supposed to have prodigious jumping power.

In a very readable little book, Baerg (1958) explored the supposedly venomous nature of American tarantulas, concluding that most species from Mexico, Central

Fig. 17-23 A Mexican tarantula, *Aphonopelma emilia,* approximately 40 percent natural size. (Photograph by R. D. Akre, Washington State University.)

America, and Trinidad, as well as all from the USA, so far as he determined, are harmless to man. The bite ''is as painful as a couple of pin stabs and has essentially the same effect.''

On the other hand, a few tropical tarantulas, such as the southern Mexican *Aphonopelma emilia,* may be quite poisonous to man, with venom biological and electrophoretic properties similar to *Centruroides* scorpion venom. The large black tarantula of Panama, *Sericopelma communis,* is definitely poisonous to man, but the effects of its bite though severe are local, causing pain and subsequent numbness in the afflicted area. In the venom of the South American tarantula *Pamphobeteus roseus* there is a highly active proteolytic enzyme with a molecular weight of 10,840. The venoms of *Pteronchilus* from Tanzania and *Hapalopus* from Peru cause little local reaction in laboratory animals, through initial irritability may be followed by weakness, paralysis, and death; degenerative lesions are found in the renal tubules and liver. *Aphonopelma* from the southwestern USA has similar effects, but elicits more excitability. The bite of a large Malayan so-called bird-eating spider, *Lampropelma violaceopedes,* in the presence of medical treatment caused human symptoms that included headache and a feeling of chest constriction, subsiding without complications in 72 hours (Lim and Davie, 1970).

Urticating hairs are borne on the dorsal aspect of the abdomen of many New World tarantulas. Contact with these may cause intense pruritus and urticarial lesions. Experience with laboratory animals demonstrates that these hairs, when air-borne, can cause acute respiratory symptoms. Cook and co-workers (1973) have shown that four types of urticating hairs exist, but that only one type of hair is present on tarantulas occurring within the USA. The barbed hairs are rubbed off the back by the rapid, brushing action of the hind legs of the tarantula when it is irritated. Indeed, this, rather than the bite, seems to be the primary line of defense of these spiders. The majority of tarantulas sold in pet shops in the USA are species of *Aphonopelma* with this characteristic.

WIDOW SPIDERS (LATRODECTUS)

Dangerous spiders belonging to the genus *Latrodectus* are found throughout the world. The genus belongs to the family Theridiidae, the combfooted spiders. They have eight eyes and three tarsal claws; the hind pair of legs is comb-footed. They spin irregular webs in which the female spiders hang. Smith and Russell (in Russell and Saunders, 1967) studied the venom gland structure of *L. mactans.*

Arachnologists do not agree on the classification of spiders in this genus. Levi (1958, 1959) recognizes six species in the world: *Latrodectus mactans,* the black widow, present through much of the warmer parts of the Americas and similar climatic zones of all continents; *L. curacaviensis,* widespread from southern Canada and the northern USA through the lesser Antilles southward in America to Chile and Argentina; *L. geometricus,* the brown widow, widespread in the tropics, particularly in Africa but occurring in the USA only in Florida; *L. pallidus,* found from the Turkmen Soviet Republic through Iran and Asia Minor to Libya and Tripolitania in North Africa; *L. hystrix,* known only from Yemen; *L. dahli,* from Iran and the island of Sokotora. Bücherl (in Bücherl and Buckley, 1971) lists only three species, *L. mactans* with five subspecies over much of the world, *L. pallidus* and *L. curacaviensis* with the same distribution as the equivalent named species of Levi. Maretić (1965) provides a map showing the distribution of species and subspecies of *Latrodectus* and states that *L. m. tredecimguttatus* is the only dangerous spider in Europe. It may never be possible to settle *Latrodectus* relationships, at least from the standpoint of geographic distribution, because there are several documented instances where these spiders have been transported great distances in packing crates and have established in new locations (cf. Southcott, 1976, regarding incidents in Australia).

In the USA *L. mactans* is commonly called the black widow, but the names hourglass spider and the shoe-button spider have also been used. In the common US black widow(s) the female is glossy black to sepia and densely clothed with short, almost microscopic hairs, giving it a naked appearance. It is usually wholly black dorsally, although an irregular red (or rarely white) stripe or pattern is sometimes present. The crimson marking on the underside of the abdomen (Fig. 17-24), rarely altogether absent, varies among individuals from the common hourglass marking to a design with two or more distinct triangles or blotches or sometimes only an irregular longitudinal area. The abdomen is globose and often likened to an old-fashioned shoe button. The abdomen of females is 6 mm wide and 9 to 13 mm long, the overall length with legs extended about 40 mm. Males are thinner and about 0.75 as long as females. The color pattern of the adult male (Fig. 17-25) exhibits considerable variation but approaches that of the immature female. Occasionally mature males are almost black but retain some of the abdominal markings of the immature form. Each palpus is shaped like a knob at the front of the head and contains the ejaculatory sexual apparatus, a portion of which resembles a coiled watch spring. The color of immatures varies considerably, with variable patterns of white and yellowish predominating in early growth and tending toward the black and red of adults in later growth stages.

L. mactans occurs in virtually every state in the continental USA except Alaska, and in southern Canada. It has been recorded at an altitude of 2,600 m in Colorado. The female spider remains with her web and egg sacs in

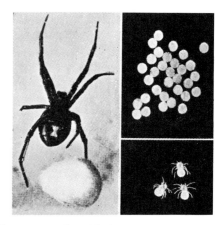

Fig. 17-24 The black widow, *Latrodectus mactans;* (*left*) mature female with egg sac; (*right*) eggs and first instars.

protected places, such as in vacant rodent burrows, under stones, logs, and long grass, in hollow stumps, and in brush piles. Convenient abode is found in dark corners of barns, stables, privies, pump houses, garages, fruit-drying sheds, piles of boxes and crates, wood piles, and stone piles.

Males do not bite. As a rule the females are not aggressive unless agitated or hungry; when guarding the egg sac the female is particularly prone to bite. Once a web is established in a suitable location she spends the rest of her life feeding on the prey ensnared in this crude but effective web, and guarding her eggs.

Herms and associates (1935) indicate that

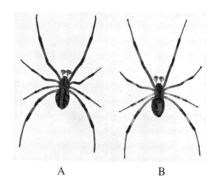

A   B

Fig. 17-25 Male black widow spider, *Latrodectus mactans*. A. Dorsal view. B. Ventral view.

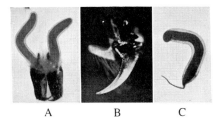

Fig. 17-26 *A.* Chelicerae and venom glands of female black widow spider, *Latrodectus mactans. B.* Separated chelicera. *C.* A freed gland. (Photographs by C. Ladenheim.)

the adult female hibernates, but Baerg cites F. R. Smith as saying that in Pennsylvania the black widow winters in immature stages also. The life history from egg to maturity requires about four months under laboratory conditions with ample food. The average number of skins cast by growing males is five; females take longer to mature with an average of seven molts. The life cycle extends over a year, with late-developing females overwintering.

The black widow bite, similar to a pinprick, is not always felt, and often there is little evidence of a lesion. However, a slight local swelling and two tiny red spots may occur, with local redness usually evident at the point of attack. In the USA *Latrodectus* bites occur frequently in outbuildings; in Italy, Yugoslavia, parts of the Soviet Union and South Africa these incidents are generally associated with hand harvesting of crops.

Case records published by many authors are characterized by severe muscular pain, a rigid abdomen, tightness in the chest, difficulty in breathing and speech, and nausea generally accompanied by profuse sweating. The condition is self-limiting, and in most cases symptoms wane without treatment after 2 or 3 days. Deaths are more prevalent in the very young, the aged, or those with hypertension. Complications include generalized injuries to the liver, kidneys, and spleen. The venom is neurotoxic and causes an enormous increase in frequency of spon-

taneous miniature end plate potentials at the neuromuscular junctions, and depletion of cholinergic vesicles from the presynaptic nerve terminals. Nerve effects consist of irreversible action within membranes to produce a rise in membrane conductance (Finkelstein *et al.*, 1976).

Thorp and Woodson (1945) list 1,291 cases of black widow spider bite in the USA from 1726 to 1943, with 55 fatalities (4.25 percent); these figures have led several authors to conclude that the death rate of 4 to 5 percent is too high on the assumption that all deaths, but not all cases, of arachnidism have been diagnosed, and fatal incidents are undoubtedly higher than official figures indicate. Between 1948 and 1965 there were 176 human cases of latrodectism recorded in Yugoslavia (Maretić, 1965).

Even though latrodectism is usually self-limiting, the intense suffering of the victim justifies consideration of therapy. The patient should be treated with local antiseptic at the point of injury to prevent secondary bacterial infection and should be kept as quiet as possible; a physician should be contacted at once. Victims may go into shock, have visual difficulties, and show electrocardiograph changes. Intravenous calcium gluconate has proved effective. *Latrodectus* antisera produced in the USA and several other countries is quite available and generally effective irrespective of which (species of) *Latrodectus* caused the bite. See Russell and associates (in Conn, 1974) regarding treatment.

VIOLIN SPIDERS (LOXOSCELES)

The genus *Loxosceles,* family Scytodidae (sometimes placed in Loxoscelidae), contains spiders with body 10 to 15 mm in length; tawny, light brown or grayish on the cephalothorax, olive-tannish on the abdomen. The carapace is flattened, with six eyes (most spiders have eight) in a strongly curved row; legs are long and lack unpaired

Fig. 17-27 Female brown recluse spider, *Loxosceles reclusa*. Note dark violin pattern on cephalothorax. (Courtesy of C. W. Wingo, University of Missouri.)

claws on all tarsi. The genus in North, Central, and South America and in the West Indies has been reviewed by Gertsch (1958, 1967). Darker and more heavily sclerotized areas on the cephalothorax dorsum include a violin-shaped anterior central region. (Fig. 17-27), hence the name violin or fiddleback spiders. Violin spiders are nocturnal and may be found under bark, lumber piles,

stored bricks, and the like, and indoors behind pictures or in closets. They are never aggressive but will bite when pressure is placed on them. The two commonly known dangerous species, *Loxosceles reclusa* and *L. laeta,* occur in North, Central, and South America. Less dangerous *Loxosceles* bites have been recorded in the Old World from southern Europe and the Middle East (Efrati, 1969) and South Africa (Newlands, 1975). It is likely that widespread distribution of some species has come about by their dissemination in items of commerce.

Loxosceles reclusa (Brown Recluse Spider). This spider has been responsible for a number of cases of necrotic spider poisoning. According to Wingo (1964) this spider is commonest in Missouri, Arkansas, and eastern Kansas, and is found in the southern USA from Texas to northwestern Georgia and north to Indiana and southern Illinois (see also Gorham, 1968). It has also been reported from Tucson, Arizona, and Los Angeles County, California (Fig. 17-28). The brown recluse occurs outdoors under

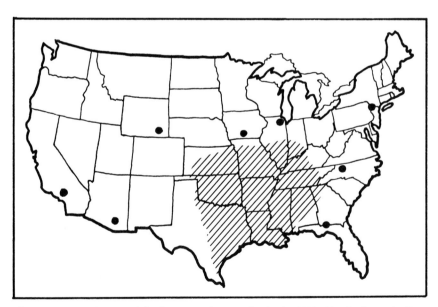

Fig. 17-28 Distribution of the brown recluse spider in the USA. Hatched area indicates regular occurrence; point locations are occasional reports. (Redrawn from Negative No. 1601, US Armed Forces Institute of Pathology.)

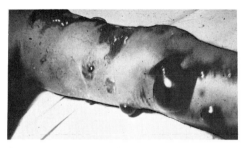

Fig. 17-29 Extensive hemolytic condition of arm due to bite of brown recluse spider. (Courtesy of US Armed Forces Institute of Pathology. Negative No. 75-5876-1.)

Loxosceles laeta. This spider known as the *araña de los rincones, araña de detras de cuadros* ("spider of the corners," "spider behind the pictures"), is a South American species responsible for numerous cases with gangrenous cutaneous involvement and considerable mortality (Fig. 17-30). Reports are prevalent from Peru, Chile, Argentina, Brazil, and Uruguay. This spider is similar to *L. reclusa* in size and appearance and is found on walls, especially in the corners, behind pictures, in cracks, and sometimes in clothing that has been hung on walls. Most cases of bite occur in homes while the victims are sleeping or dressing themselves. In central Chile, Schenone and coworkers (1970) found 40.6 percent of urban and 24.4 percent of rural houses infested with *L. laeta,* with a mean number of 11.9 spiders per rural house and 3.9 per urban house; yet despite this high frequency and density the retiring nature of this spider reduces the hazard of bite incidents. Case histories indicate a much higher rate of spider bite among women than among men. *L. laeta* was found infesting a museum in Massachusetts, northeastern USA (Levi and Spielman, 1964), possibly introduced in shipments; there is an apparently long-standing establishment of this South American spider in a section of Los Angeles County, southern California, USA (Waldron *et al.,* 1975).

rocks and rubble in its southern range, and predominantly in and around homes in its northern range. Hite and associates (1966) indicate that within homes this spider is most frequently in boxes, and outdoors it is usually under rocks; Butz (1971) and Cross (1972) note that it is often in association with fabrics and other fibrous materials, thus in clothing and bedding. The bite of *L. reclusa* is usually localized, but it produces considerable necrosis that may ultimately leave an unsightly scar (Fig. 17-29). Necrotic damage may be so extensive that skin grafting is required. It should be noted, however, that effects may be confined to a comparatively mild skin reaction (Berger, 1973). If systemic complications occur, they are serious, and several deaths have been reported in the USA.

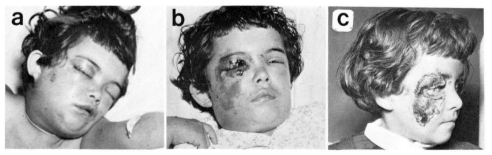

Fig. 17-30 Reaction of seven-year-old girl to bite of *Loxosceles laeta. a.* Twenty hours after bite, severe systemic reactions and local edema. *b.* Four days after bite. *c.* Forty days after bite. Skin grafting was ultimately required. (Courtesy of H. Schenone, University of Chile.)

Other Loxosceles. A number of other species of *Loxosceles* are known. Although laboratory tests have shown some of these to have necrotic effects on laboratory animals and they can cause necrosis in humans, they are not known to be life threatening. *L. rufescens* was introduced and became widely established along the East Coast and Gulf states of the USA (Gertsch). This same species is listed from California, and *L. rufipes* (=*unicolor*) and *L. arizonica* from California and Arizona (Keh, 1967; Russell, *et al.*, 1969). *L. rufipes* is known from coastal Peru (Delgado, 1968) and New South Wales, Australia; *L. rufescens* from Paraguay (Canese, 1972), Israel, and South Australia. Newlands (1975) lists from southern Africa *L. spinulosa,* found under rubble and shaded logs and suspected of causing injury; and two cave-dwelling species.

Loxosceles Venoms and Treatment. The information herein refers to *L. reclusa* and *L. laeta.* The bite is often not noticed initially; there is a bluish-white halo of vasoconstriction in the first few hours, which then blisters and is surrounded by a painful reddish margin. Chills, malaise, and generalized rash may accompany this state. A hemorrhagic zone may spread for up to six days and become necrotic with destruction of skin and subcutaneous fatty tissue; healing is slow and leaves an extensive scar. Life-threatening complications are usually confined to children, with massive intravascular hemolysis, hemoglobinuria, jaundice, high fever, and shock; renal shutdown and pulmonary edema also accompany *L. laeta* bites with complications. Schenone and associates (1975), studying *L. laeta,* note that there is great regional variation in the toxic activity of strains from different countries and districts. Among laboratory animals the rabbit reproduces very closely the cutaneous and visceral lesions found in man. There are two components in *L. reclusa* venom, both with a molecular weight of approximately 24,000, that cause lesions and death in test animals (Geren *et al.*, 1973). Notable tissue changes include vasculitis, endothelial damage, and thrombosis of blood vessels.

The most effective treatment is surgical excision of the bite area as rapidly as feasible, and preferably within the first day of trauma. A sensitive and specific passive hemagglutination-inhibition test to identify *L. reclusa* bite can provide early diagnosis if the spider is not captured and brought in for identification (Finke *et al.*, 1974), and a test is available to determine a person's sensitivity to the bite (Berger *et al.*, 1973). *Loxosceles* antivenom is produced by the Instituto Butantan, São Paulo, Brazil, but it has not received extensive clinical evaluation.

DANGEROUS AGGRESSIVE SPIDERS (GENERA PHONEUTRIA, ATRAX, HARPACTIRELLA)

Spiders in these genera, as well as some tarantulas, are aggressive by nature, assuming defensive and offensive attitudes when threatened. Human deaths have been attributed to members of all three genera.

Phoneutria species, family Ctenidae, are called "wandering spiders" and "bird spiders" and are found in South and Central America; they are discussed thoroughly in Bücherl and Buckley (1971). *Phoneutria nigriventer* is the largest and most aggressive true spider (not tarantula) in South America, being prevalent around beach areas in southern Brazil. Several hundred incidents involving this species occur annually in the state of São Paulo, with occasional fatalities most frequent in children. This species weaves no webs, wanders about hunting very actively in shade, and in the evening, morning, and night. It often enters human dwellings, hiding in clothes and shoes during the day. When threatened it is very aggressive and may bite several times, yielding up to 1.25 mg of dry venom, in the range of small tarantulas. A case of *Phoneutria* bite during the handling of bananas is recorded from Costa Rica (Trejos *et al.*, 1971).

P. nigriventer venom is one of the most pharmacologically active of spider venoms tested in laboratory animals. In man it causes intense pain, neurotoxic effects, sweating, priapism, respiratory paralysis and spasm, and histaminelike effects. It has a high concentration of serotonin. Symptomatic treatment with analgesics and antihistamines may be sufficient if general symptoms do not occur in the first 3 to 12 hours. An antiserum is available from the Butantan Institute, São Paulo, Brazil, and is especially recommended in treating children; it should also be used in any cases of disturbed cardiac output or affected vision and respiration.

Atrax formidabilis and *A. robustus* are cited by Thorp and Woodson (1945) and Southcott (1976) as dangerous mygalomorph spiders in Australia. They are fairly large funnel-web spiders up to 3.75 cm in body length; males are particularly aggressive, and all fatalities are attributed to bites by this sex. The fore part of the body is glossy ebony black, the black abdomen is covered with a velvety pile and the undersurface bears tufts and brushes of red hair (Fig. 17-31). *A. robustus* is called the "Sidney funnel-web spider"; it tends to roam through dwellings and to be found in clothing and shoes. Sutherland (1972a, 1972b) states at least 11 fatalities from bites of this species

Fig. 17-31 A female Sidney funnel-web spider, *Atrax robustus*. (Courtesy of R. V. Southcott and *Records of the Adelaide Children's Hospital.*)

Fig. 17-32 A spider in the genus *Chiracanthium*. These have been reported as causing severe bite reactions. (Courtesy of R. V. Southcott and *Records of the Adelaide Children's Hospital.*)

have occurred in New South Wales since 1927, and further discusses the classical syndrome, studies on crude venom, and recommended treatment. The venom is nonantigenic, and no adequate pharmacological antagonists are known. Treatment is symptomatic and may require the use of a respirator.

Harpactirella spiders are known to be aggressive, and to have caused fatalities in South Africa (Bücherl and Buckley, 1971). The same volume cites human death as possible from bites of spiders in the genus *Trechona*, and possibly from *Mastophora, Chiracanthium* (Fig. 17-32) (see Spielman and Levi, 1970; Ori, 1975, regarding Europe, USA, Japan), and *Lithyphantes*.

OTHER INSTANCES OF ARACHNIDISM

Spiders in the genus *Lycosa*, family Lycosidae, are of definite medical importance in South America, and antivenin for their bites is produced by the Butantan Institute. Listed as causing deep and large wounds (Bücherl, and Buckley, 1971), in addition to *Loxosceles*, are spiders in the genera *Acanthoscurria, Megaphobema, Xenesthis, Theraphosa;* some *Avicularia, Phormictopus,* and *Pamphobeteus*. Moderately severe reactions

Fig. 17-33 A spider in the genus *Miturga*. These have been reported as causing moderately severe bite reactions. (Courtesy of R. V. Southcott and *Records of the Adelaide Children's Hospital*.)

have been recorded in the USA from bites of *Herpyllus ecclesiasticus:* Gnathophosidae (Majeski and Durst, 1975), *Peucetia viridens* (Hall and Madon, 1973), *Trachelas tranquillus:* Clubionidae (Uetz, 1973) and *Lycosa antelucana:* Lycosidae (Redman, 1974); in New Zealand for *Miturga* sp. (Fig. 17-33): Clubionidae (Watt, 1971). Consult Russell and coauthors (in Conn, 1974) for a listing of species affecting humans in the USA, and Southcott (1976) and Main (1976) for Australian spiders causing medical problems.

Eye inflammation related to spiders or their webs, and thought due to quinones, is reported from Japan (Kawashima, 1961).

CONTROL OF SPIDERS

Avoiding bites by highly toxic spiders is largely a matter of educating the public about the habits of species typical of a region, without sensationalizing the problem. Precautions such as shaking out clothing and shoes before dressing, and turning back and examining bedding before retiring, are prudent. Proper house construction, including use of door and window screens and closers to reduce the possibility of entry by outdoors species, is sensible. Removal of debris and rubble such as piles of building materials from premises can reduce sites of concealment and breeding near people. Most spiders are quite sensitive to a number of pesticides, but these should be applied only where there is a clear problem from dangerous species and then only to likely hiding places, bearing in mind that the great majority of spiders are beneficial insectivores. Recent interest in the brown recluse spider in the USA has led to laboratory tests of pesticides for its control (Sterling *et al.*, 1972; Gladney and Dawkins, 1972).

SCORPIONS

Class Arachnida; Order Scorpionida

CHARACTERISTICS

Scorpions are easily recognized by their more-or-less flattened lobsterlike appearance, but particularly by the long fleshy five-segmented taillike postabdomen with a terminal bulbous sac and prominent STING (Fig. 17-34). The pedipalps are greatly enlarged, with distal strong chelae or pincers. The true jaws, chelicerae, are small and partly concealed by the front edge of the carapace. There are four pairs of terminally clawed legs.

The cephalothorax usually bears a pair of conspicuous eyes near the middorsal line (MEDIAN EYES) and several smaller ocelli in groups of from two to five on the lateral margins (LATERAL EYES); some species are eyeless. Scorpions breathe by means of BOOK LUNGS. They are ovoviviparous, and young are carried on top of the body of the mother. Although the sexes are very similar in appearance, the males have a longer cauda and broader chelae. An account of the morphology as well as characters used in classification of scorpions is given by Moreno (1940), and Bücherl (in Bücherl and Buck-

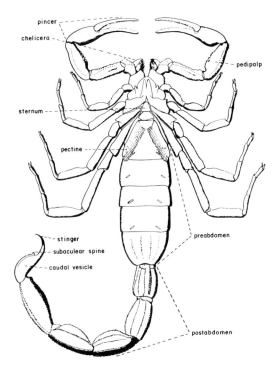

pincer

chelicera

pedipalp

sternum

pectine

stinger

subaculear spine

caudal vesicle

preabdomen

postabdomen

Fig. 17-34 Anatomy of a scorpion, ventral view. (Courtesy of US Public Health Service.)

ley, 1971) reviews scorpion classification and biology and the distribution of dangerous species.

Scorpions are most abundant in warmer climates, yet they are surprisingly widespread, occurring in the southern part of western Canada to Patagonia in the New World, and in the Old World from southern Germany and Mongolia through most of Africa, Asia, and Australia. They are active nocturnally, seeking prey in the open and hiding during the day beneath loose stones, loose bark of fallen trees, boards, piles of lumber, floors of outbuildings, dried cattle droppings, and debris; some enter sand or loose earth. Although generally associated with ground level, some species will climb into trees and shrubby vegetation during night hunting activities. They feed upon insects, spiders, millipedes, and even small rodents, seizing prey with their chelae and striking with the powerful sting thrust forward along one side or in a characteristic fashion over the scorpion's head.

SCORPION STING

The ACULEUS or sting of the scorpion is situated terminally and contains a pair of VENOM GLANDS, which are separated by a muscular septum. From the glands are given off fine ducts opening at the apex of the sting. The development of scorpion poison glands has been studied in the lesser brown scorpion, *Isometrus maculatus* (Probst, 1972), and the ultrastructure of *Centruroides* venom glands and their secretory epithelium has also been investigated (Keegan and Lockwood, 1971; Mazurkiewicz and Bertke, 1972). The sting curves downward when the "tail" is extended behind, but upward when the scorpion poises for attack or defense, the entire postabdomen being curved upward and forward. The victim is struck quickly and repeatedly.

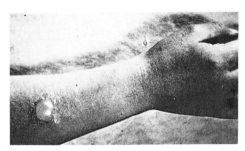

Fig. 17-35 Blistering at the site of a scorpion sting. This response is not present in many cases. (Courtesy of US Armed Forces Institute of Pathology. Negative No. 75-58-76-2.)

The toxicity of scorpion venoms is highly variable. Even among species in the same genus there are differences as to whether or not the sting is painful, whether only localized or dangerously systemic in effect, and whether or not there is much swelling (Fig. 17-35). Many highly toxic species produce intense local pain, insignificant swelling, and principally systemic effects. Mortality is especially high in young children. In Brazil stings by the highly venomous *Tityus serralutus* cause mortality of 0.8 to 1.4 percent in adults, 3 to 5 percent in school-age children, and 15 to 20 percent in very young children. For all of Mexico, Mazzotti and Bravo-Becherelle (1963) state that annual deaths from scorpion stings range from 1,100 to 1,900. According to Balozet (1956), scorpions account for many more deaths in North Africa than are caused by snake bites. Minton (1974) states that, based on antiserum usage, there are presently about 40 to 60 serious cases annually of *Centruroides* stings in Arizona, USA. The most extensive region of morbidity and mortality encompasses arid regions of North Africa and extends through Pakistan and India.

The symptoms produced by the sting of the Durango scorpion, *Centruroides suffusus,* are described by Baerg as sharp pain followed immediately by a feeling of numbness or drowsiness, then an itching sensation in the nose, mouth, and throat that makes the victim distort the face, rub nose and mouth, and sneeze. An excessive production of saliva induces the victim to swallow rapidly. The tongue becomes sluggish and communication difficult, the muscles of the jaw contracted. Disordered movements of arms and legs follow, and the temperature rises rapidly to 40.5° C; salivary secretion diminishes, and there is a scarcity of urine. The senses of touch and sight are affected, a veil seems interposed between the eyes and objects, strong light is unpleasant, and luminous objects are surrounded by a circle. Hemorrhage of stomach, intestine, and lungs may occur; convulsions come in waves and increase in severity up to 2 hours, or until a fatal termination. A patient surviving for 3 hours is usually out of danger, yet death may occur 6 to 8 hours after the sting.

Stahnke in characterizing the effects of the sting of the dangerous species of Arizona scorpion, *Centruroides sculpturatus,* emphasizes that the venom does not produce swelling or discoloration at the sting site; on the other hand the sting of less venomous *Centruroides* and some species of other genera may produce local swellings, with or without discoloration. Bücherl notes that *Tityus* stings cause severe pain for several hours, sometimes edema of the affected area, rhinorrhea, salivation, pallor, muscle twitchings, hypertension, tachycardia, and convulsions. Bartholomew (1970) reports on 30 human cases of acute pancreatitis caused by the sting of *Tityus trinitatis* in Trinidad. In discussing Old World scorpionism, Balozet (in Bücherl and Buckley, 1971) states that the first symptoms in humans, other than pain, occur in 20 minutes to 4 hours; death in 2 to 20 hours. He emphasizes that close surveillance of the victim must continue for at least 12 hours, otherwise general irritability, respiratory problems, and death can ensue.

Treatment of Scorpion Sting. In the absence of immediately available appropriate medical assistance, analgesics and antihistamines can be used to reduce symptoms. Scorpion antisera are generally available in regions where dangerous species are known to occur, but they are quite specific, and if the identity of the offending scorpion is not known, polyvalent antiserum is advised. Antiserum must be administered within 2 hours of the sting to realize full effectiveness. Bücherl advises administering a total dose sufficient to neutralize at least 2 mg of dry scorpion venom, half intravenously and the remainder subcutaneously or intramuscularly.

Nature of Scorpion Venoms. Tu (1977) provides a detailed review of the chemistry of scorpion venoms. *Tityus* venoms, according to Bücherl, have mainly protein components, 2 of 7 of these being toxic and one of the toxins mimicking whole venom symptoms in mice. Hypertensive effects of these venoms appear to be by indirect action through the release of catecholamine (Corrado *et al.*, 1974). In describing Old World scorpion venoms, Balozet speaks of two protein toxins called scorpamines as the main components, with a molecular weight between 10,000 to 18,000; the presence of 5-hydroxytryptamine causes violent local pain. Some venoms are strongly hemolytic, and hyaluronidase has been identified in some, as is the case with the south Indian *Heterometrus scaber* (Nair and Kurup, 1975). Minton's review (1974) of scorpion venoms states that North African and Middle East scorpion venoms are neurotoxic low molecular weight proteins with a high content of sulfur and basic amino acids. Watt and associates (1974) review protein neurotoxins in scorpion venoms.

DISTRIBUTION OF DANGEROUS SCORPIONS

The Order Scorpionida is divided into the families Scorpionidae, Buthidae, Vejovidae, Chactidae, Bothriuridae, and Diplocentridae. Stahnke (1967) has provided a simple key with illustrations for identifying families. Of about 650 species in some 70 genera, approximately 40 species are found in the USA, and only one, *Centruroides sculpturatus,* is highly dangerous to man.

Half of all species and all the most important venomous scorpions of the world belong to the family Buthidae. The dangerous scorpions of Mexico and southwestern USA belong to the genus *Centruroides* (Fig. 17-36). The Durango scorpion, *C. suffusus,* is the common scorpion of the state of Durango, Mexico; it is frequently fatal to children, particularly those under 7 years of age. At least four other Mexican species are dangerous, including *C. limpidus* and *C. norius;* the latter reported to be about six times as venomous as *C. suffusus.*

Centruroides sculpturatus, believed to be confined to Arizona, is a small species about 6.25 cm in length, generally of a solid straw-yellow color. It is dangerously virulent and locally abundant. A dangerous scorpion of Arizona, originally identified as *C. gertschi,* is apparently only one of several color phases of *C. sculpturatus* (Stahnke, 1971). This color phase has two irregular black stripes down the entire back and is basic yellow in color. *Centruroides vittatus,* a common striped scorpion of the USA, is widely distributed—reported from Georgia, Florida, Kansas, Texas, Arkansas, Louisiana, New Mexico and South Carolina. Its sting causes a sharp pain followed by a wheal that soon disappears without complications.

Bücherl (in Bücherl and Buckley, 1971) provides a world map of the distribution of dangerous scorpions. These are species of the following genera:

New World. *Tityus*—Mexico, West Indies, South America; *Centruroides*—USA (Ariz.), Mexico, Central America, Cuba, Haiti, Barbados, Curacao.

Old World. *Heterometrus*—Indian subcontinent; *Pandinus*—Africa and Arabia; *Opisthophthalmus*—South Africa; *Scorpio*—North Africa; *Hadogenes*—South Africa

Fig. 17-36 *Centruroides* scorpions. On the left is the typical color form of *C. sculpturatus*, next to it is the striped form of this species (formerly known as *C. gertschi*); on right is *C. vittatus*. (From Stahnke, 1956, *Scorpions*, Fig. 6, p. 16; courtesy of H. L. Stahnke and the Poisonous Animals Research Laboratory, Tempe, Arizona.)

and Madagascar; *Androctonus*—India and Persia to Atlantic coast of Morocco and eastern Mediterranean to Senegal and upper Egypt; *Leiurus*—Syria, Palestine, Egypt, Yemen; *Buthacus*—Atlantic coast to Palestine, Senegal, Tunis; *Buthotus*—Palestine, Syria, Algeria, Morocco; *Buthus*—North Africa, Egypt, Ethiopia, Somalia, parts of Palestine, Spain, France; *Parabuthus*—South Africa to Sudan.

CONTROL OF SCORPIONS

Seasonal awareness against stings is helpful, the highest rate generally coinciding with the onset of the rainy season in various parts of the world. Clothing and shoes should be shaken out before wearing in scorpion-prone areas, particularly if one sleeps out-of-doors. Human dwellings of adobe and rough stone construction are likely to harbor scorpions and provide shelter for their insect prey, such as cockroaches. Piles of lumber and firewood should be avoided around home premises. Ennik (1972) suggests the floor level of houses should be at least 20 cm above ground, with a space of at least 6 cm between the entry steps and the house wall. The ground-associated surfaces should be made difficult to climb, such as by facing the front of each outside step and outside walls at ground level with a row of glazed ceramic tiles. Plants should not be grown against the house. Within the home use ceramic baseboards on interior walls and furniture with plain legs or legs in glass cups, and a ceiling cloth to prevent scorpions from falling out of thatch or tile roofs. Ducks will prey on scorpions but may receive lethal stings in doing so. Laboratory tests indicate most insecticides with long-lasting deposits, such as organochlorine compounds, are effective but not favored for use in homes. An application of a protective pesticide at ground level around the outside base of the home, and in selected areas of harborage, may be justified.

WHIPSCORPIONS

Class Arachnida; Order Pedipalpida

The Pedipalpida (also known as the Uropygi or the Thelyphonida) are widespread through the tropics and subtropics, although very unevenly distributed. They are scorpionlike in appearance but differ in the form of the pedipalps, the antennalike first legs, and the whip-bearing abdomen. Whipscorpions feed chiefly on insects, worms, and slugs, which they seize quickly with their sharp pedipalps. They are nocturnal, and in the daytime they may be found under stones, in burrows and other protected places. They defend themselves from predators by means of their pedipalps and an acid secretion that they eject from the tail.

The giant whipscorpion, *Mastigoproctus giganteus* (Fig. 17-37) (known as the vinegaroon, grampus, or mule killer), occurs commonly in Florida, Texas, other parts of the South, and westward to California, USA. The term vinegaroon is apt, inasmuch as its secretions contain acetic acid as well as caprylic acid. Forcible ejection of the spray may extend to 80 cm, though discharge is in response to direct body stimulation only (Roth and Eisner, 1962). This whipscorpion feeds on larger insects and other arthropods, if not too hard or active; it has been known to kill small frogs or toads (Cloudsley-Thompson, 1958). Although it is greatly feared, there is little justification for this.

Flower (cited by Cloudsley-Thompson) described the scorpionlike behavior of *Thelyphonus skimkewitchii* in Thailand and the resulting "sting" that hurt for several hours and that required medication. If the irritating fluid secreted by the whipscorpion should come into contact with a scratch on a person's hand, possibly one made by the cheli-

Fig. 17-37 The giant whipscorpion or vinegaroon, *Mastigoproctus giganteus*. (Courtesy of C. S. Crawford, University of New Mexico.)

cerae or pedipalps of the arachnid, it is quite likely that a smarting sensation might be mistaken for the result of a sting. Cloudsley-Thompson cites one case of a blacksmith who experienced blisters as the result of a vinegaroon that was crushed on his chest.

SOLPUGIDS

Class Arachnida; Order Solpugida

The solpugids (Fig. 17-38), commonly known as "sun spiders" and "wind scorpions," are spiderlike, although there is no

Fig. 17-38 A solpugid or sun spider. (Courtesy of C. R. Baird, University of Idaho.)

pedicel, and rather hairy. They occur mainly in desert, tropical, and subtropical regions. The chelicerae are large, powerful, and two-segmented. The second segment is movable and articulates in a more or less vertical plane. Food is crushed to a pulp, the fluid is swallowed, and the hard parts are usually ejected. The first pair of legs is long and rather feeble and used as tactile organs. Respiration is tracheate.

Solpugids are commonly but erroneously regarded as exceedingly venomous. There is not the slightest foundation for the suggestion that any animal drinking from a water trough in which a solpugid was present would die as a result. The question of the venomous nature of these animals cannot be dismissed summarily, however. Cloudsley-Thompson (1958) gives a concise account of the subject. No poison glands associated with the jaws have been found, but it has been suggested that poisoning might result from toxic excretions through the setal pores that could be traced along the tips of the chelicerae. Apparently authentic cases of aftereffects resulting from a solpugid bite

have been recorded but were most probably due to infection of the wound.

The solpugids of the USA have been monographed by Muma (1951), who recognizes about 75 species in 10 genera (in addition to several of uncertain status). Most species are found in the southwest from Texas to California, but *Ammotrechella stimpsoni* inhabits Florida, and *Eremobates pallipes* is widespread throughout the western half of the USA.

CENTIPEDES

Class Chilopoda

Centipedes are wormlike in form, with a distinct head that possesses a pair of antennae and with many fairly similar body segments, each with one pair of segmented appendages (Fig. 17-39). Like the insects they are tracheate and for the most part terrestrial. The individual body segments are somewhat flattened so that a cross section through one of them is oval. The legs are at least moderately conspicuous, in number from 15 to 100 or more pairs. Notwithstanding the confusing number of walking appendages, centipedes crawl very rapidly.

Centipedes feed mainly on insects. The POISON CLAWS, or MAXILLIPEDS, are located ventrad of the mouth and connected via a

Fig. 17-39 A large centipede from New Mexico, USA, *Scolopendra polymorpha*. (Courtesy of C. S. Crawford, University of New Mexico.)

hollow tube to large POISON GLANDS. Insects are killed rapidly when the poison claws of a centipede close upon them. It is mainly large specimens that are reported as causing painful bites, and some tropical species may attain a length of 30 cm or more.

Bücherl (in Bücherl and Buckley, 1971) and Minton (1974) review literature on the biology of centipedes, and the former provides good anatomical descriptions and keys to major groups.

Remington (1950) states that "the soundest conclusion from the published records appears to be that no centipede bites are deadly to man and that the immediate pain diminishes rapidly, much like the sting of a honey bee." Bücherl tested the venom of five larger common Brazilian scolopendromorph centipedes against mice. The venom of *Scolopendra viridicornis* injected intravenously at 0.03 gland equivalent killed half the mice tested, and 0.25 gland equivalent intramuscularly was the median lethal dose. The venom appeared to be neurotoxic, but when this species bit Bücherl it caused pain for about 8 hours and no general effects, leaving a small superficial necrosis at 36 hours and healing completely in 12 days. Reports of human envenomization are from the Philippines, Malaysia, New Guinea, the Crimea, Brazil, the southwestern USA, and Hawaii. Bites are rare because centipedes are essentially nocturnal. General symptoms are reputed to be mental anxiety, vomiting, irregular pulse, dizziness, and headache. Treatment consists of preventing secondary infection and providing relief from pain.

MILLIPEDES

Class Diplopoda

Millipedes differ rather obviously from centipedes in that most apparent body segments possess two pairs of appendages instead of one (Fig. 17-40). Millipedes are

Fig. 17-40 A large millipede, *Polyconoceras alokistus*, from New Guinea. Secretions from this diplopod cause a burning sensation on the skin. (Courtesy of R. V. Southcott and *Records of the Adelaide Children's Hospital*.)

vegetarians and lack poison fangs characteristic of centipedes. In most species the body is cylindrical, and the numerous legs, as well as the antennae, are relatively short and inconspicuous.

The Diplopoda are commonly separated into two groups depending upon the presence or absence of repugnatorial glands that discharge through quite readily discernible lateral openings on most abdominal segments from the sixth backward. In the Chilognatha, to which all North American genera except *Polyxenus* belong, these glands are present and can produce irritating effects. The irritant may seep to the surface, or in some species the fluid is squirted; the Haitian species *Rhinochrichus latespargor* discharges its secretion up to a distance of 82 cm.

Radford (1975) reviews millipede burns in man, and provides a world map showing where injurious effects from millipedes are well known, and where there are occasional reports. Burns are generally associated with removing crawling centipedes from the body during sleep, the lesions usually involving

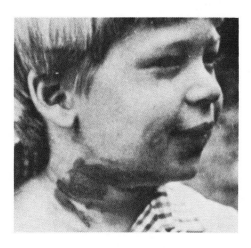

Fig. 17-41 Long-lasting discoloration of skin on the neck of a child, caused by caustic secretions from a New Guinea millipede. (From Radford, 1975, *Papua New Guinea Med. J.*, **18**:138–41.)

the face, eyes, or nose, and occasionally the mouth. Children are most affected, and there are eye injuries to poultry and young dogs. Eye injury consists of a burning sensation, intense lacrimation, pain up to 2 days, conjunctivitis and even ulceration. On skin there is a prickling or burning sensation as with any corrosive fluid. An initial yellowish-brown tanning turns to deep mahogany or purple-brown in a few hours. Blistering follows in a day or two, exfoliating to expose a raw surface with clear serous exudate. In darker-skinned people hypopigmentation remains for many weeks; the site of the lesion in light-skinned people is noticeable for up to 14 months (Fig. 17-41). Malayan and other villagers are known to use hydrogen cyanide from crushed millipedes mixed with certain plants to poison arrow tips.

Treatment, to be applied as soon as possible after contact, consists of washing skin with copious amounts of water to remove the secretions, then applying antiseptic. Eyes exposed should be thoroughly irrigated with warm water and local anesthetic applied to relieve pain.

TICK PARALYSIS AND TOXICOSIS

Ticks cause a paralysis of animals that is reversible when they are removed. The manifestations of paralytic conditions are virtually worldwide (Fig. 17-42). Paralytic responses are sufficiently different, and there are instances where localized toxic reactions occur, to make it impossible to clearly separate paralytic and toxic manifestations. Therefore this continuum of conditions is discussed as one general entity. Main reliance for this discussion is placed on the excellent review by Gregson (1973), which was in part based on Neitz (1962).

The earliest reference to tick paralysis occurred in Australia in 1824 when a diary entry by William Howell referred to a tick ''which buries itself in the flesh and would in the end destroy either man or beast if not removed in time.'' In 1894 hundreds of cases of paralysis in Australian dogs were referred to, but the first syndrome in man from that country was described in 1912. From South Africa, Cape of Good Hope, sheep farmers in 1890 complained of paralysis among their stock. The earliest North American case was noted in 1898. Tick paralysis was definitely recognized in 1912 in North America when a survey of physicians brought out instances of paralysis in British Columbia, Canada; in the same year 12 human cases were listed in Oregon and Idaho.

Tick paralysis is caused by a number of species of ticks in several genera. *Ixodes holocyclus* is responsible for Australian cases of paralysis, exceptions being two cats paralyzed by a female of *Ixodes hirsti,* and paralysis of dogs and cats by *Ixodes cornuatus* in Tasmania. Dogs are the main victims, but a 1968 report noted that 170 sheep died of paralysis in a flock of 404. Paralysis on that continent has also been recorded in horses, goats, pigs, calves, poultry, and a crow.

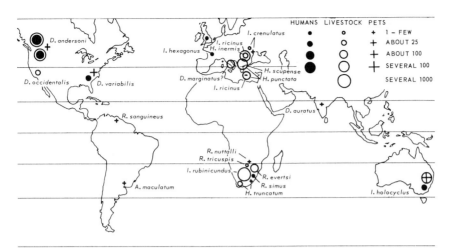

Fig. 17-42 World occurrence of tick paralysis showing approximate numbers of cases, hosts affected, and ticks mainly responsible. (From Gregson, 1973, *Tick Paralysis: An Appraisal of Natural and Experimental Data,* Agriculture Canada, Ottawa [as modified from Neitz, 1962].)

In Africa, so-called Karoo tick paralysis caused by *Ixodes rubicundus* is characterized by all limbs' being affected simultaneously, the main animals involved being sheep, goats, cattle, and rarely dogs and jackals. Spring lamb paralysis is an ascending paralysis due to *Rhiphicephalus evertsi.* Paralysis of fowl is associated with *Argas* ticks in the subgenus *Persicargas* (Gothe and Verhalen, 1975) and appears to be more of a toxicosis. Paralysis of man is rare and regional in nature, attributed to *Rhipicephalus simus, Hyalomma truncatum,* and *I. rubicundus.*

In Eurasia tick paralysis cases are widely scattered. On Crete sheep, dogs, and cats are paralyzed by *Ixodes ricinus* and *Haemaphysalis punctata;* in Yugoslavia *Ixodes ricinus* paralyzes cattle, sheep, and goats; in Macedonia and Bulgaria *Haemaphysalis punctata* and *H. inermis* cause paralysis of chickens; *Dermacentor* ticks in Italy paralyze cattle, goats, and sheep; in England and France three human cases were due to *Ixodes hexagonus;* in India *Haemaphysalis kutchensis* has paralyzed rabbits and *Dermacentor auratus* has caused a limping condition of domestic animals.

In the Americas, Mexico and South America seem largely free of tick paralysis; *Amblyomma maculatum* caused paralysis of dogs in Uruguay, and *Rhipicephalus sanguineus* affected these hosts in Venezuela. The greatest number of tick paralysis cases have occurred in North America and are mainly due to *Dermacentor andersoni,* with highest incidence near the border of the province of British Columbia, Canada, with the states of Montana, Idaho, and Washington, USA (Fig. 17-43). In the eastern USA *D. variabilis* has caused human paralysis

Fig. 17-43 Cattle in Idaho, USA, paralyzed by the tick *Dermacentor andersoni.* (Courtesy of R. L. Kambitsch, University of Idaho.)

from seaboard portions of Virginia, North and South Carolina, and Georgia. There have also been reports from Kentucky and Tennessee, and Mississippi and Oklahoma (*MMWR*, 1973, **22**:263). No human cases have been attributed to *D. occidentalis,* which has paralyzed wildlife and livestock.

Livestock paralysis in North America is limited to the Pacific Coast and the Northwest. In British Columbia, Canada, since 1900 over 3,800 sheep and cattle have been stricken in outbreaks caused by *D. andersoni* and involving up to 320 animals per incident. In the USA livestock paralysis has been less frequent, and that caused by *D. andersoni* extends from the Canadian border to southern limits in eastern Montana across into central Oregon. *D. occidentalis* in California has paralyzed cattle, ponies, blacktail and mule deer (*Odocoileus*). Reports of paralysis of pets in the USA involve dogs and cats and are mainly due to *D. andersoni* and *D. variabilis; Ixodes pacificus* caused paralysis of a dog in California. Occasional or experimental paralysis in British Columbia and Montana involve *D. andersoni* and mule deer, bear cub, foxes, skunk, marmots, and Columbian ground squirrels. Birds have been reported to be paralyzed by *Ixodes brunneus* in Atlanta, Georgia.

The circumstances and symptoms of human paralysis in North America and Australia are quite different. In both cases female ticks are usually involved, but male ticks occasionally cause limited regional paralysis. In North America the victim is often a young female who visited a tick-infested area 4 to 6 days before the first symptoms appeared. Paralysis is ascending with difficulty first in walking, inability to walk, limb numbness, complete locomotory paralysis within 24 hours, difficulties in speech, respiratory paralysis, and death. Since the tick is frequently hidden in the hair, it may be overlooked and neural involvement such as poliomyelitis may be suspected. Removal of the tick results in rapid and complete recovery, the rate of recovery dependent on the degree of paralysis. Extensive paralysis cases may take 1, 2, or even 6 weeks for full recovery. If deep paralysis has occurred, it is well to have pulmonary assistance on standby. In Australia *Ixodes holocyclus* paralysis of humans commonly involves vomiting, acute illness, peak paralytic development about 48 hours *after* the tick is removed, and a slow recovery that may take several weeks.

Paralysis affects the myoneural junctions, and particularly the conduction rate of slower conducting terminal fibers of smaller diameter (McLennan and Oikawa, 1972); central nervous system depression is also noted. No drugs have been conclusively shown to reverse the effects of tick paralysis. Kaire (1966) obtained a protein fraction from *I. holocyclus* that caused a paralysis reversible by antiserum. In Australia the serum from dogs challenged sufficiently by bites of this tick develops a level of immunity that can be used to aid recovery in cases of paralysis, but this procedure is not effective with *Dermacentor andersoni* paralysis.

It is generally believed that tick paralysis is due to a toxin in the salivary glands, but there is no evidence for a common tick toxin. The natural and often rapid recovery noted on removal of a causal tick suggests toxin is excreted quite rapidly or metabolized. One theory suggests a host reaction to tick saliva produces a metabolic toxemia; another hypothesis that could account for geographic variations in virulence suggests that an organism in the tick (possibly symbiotic *Wolbachia rickettsiae* in the salivary glands) produces toxin while the tick is feeding. There are many puzzles remaining, including the observation that although there are regions of high incidence for tick paralysis of humans and other animals, under experimental conditions only some female ticks cause paralysis.

Tick toxicosis is another phenomenon associated with tick bite, and here there may also be paralysis. The reactions caused by argasid ticks are believed to be different from ixodid paralysis, and are mostly produced by early stages rather than adults. *Argas brumpti* and *O. moubata* cause bite reactions that are painful and accompanied by inflammation, and *O. turicata* bites have been reported to kill people. The pajaroello, *O. coriaceus,* is reputed to cause very irritating bites on man, though this is not always so (Failing *et al.,* 1972). The pigeon-infesting *Argas reflexus* of Europe may cause severe reactions including loss of consciousness when it bites humans (Coudert, *et al.,* 1972; Grzywacz and Kuzmicki, 1975). Bovines may be killed overnight by attacks of *O. savignyi,* and this tick possesses proteinlike toxins in the saliva (Howell *et al.,* 1975). *O. lahorensis* on several occasions caused extensive losses of sheep in the southern Soviet Union (Kusov, 1955).

Management of tick paralysis and toxicosis consists mainly of awareness of those conditions associated with incidents and avoidance of the problem. Since ticks must be attached for some time before paralysis occurs, humans and pets that have been in areas noted for paralysis should be examined frequently and thoroughly and these ectoparasites removed. For livestock in Australia, Doube (1975) suggests ecological manipulation of habitat to reduce the density of bandicoots (*Perameles nasuta* and *Isoodon macrourus*) that are associated with high populations of *Ixodes holocyclus*. Since paralysis chiefly affects young animals, calving time may be altered from spring to winter and calving confined to tick-free paddocks. Rich (1973) notes that calves in British Columbia, Canada, are seldom paralyzed, and his experiments demonstrate that grooming by the mothers removes ticks, so young animals should be kept with their herds. On range in British Columbia, loss of animals is reduced by herdsmen, who observe the animals frequently and remove ticks at the first indication of paralysis. Management practices that reduce tick populations, such as pasture spelling and treatment of livestock with pesticides, will lower the incidence of tick paralysis.

CITED REFERENCES

Abonnenc, E. 1972. *The Phlebotomi of the Ethiopian Region (Diptera, Psychodidae).* Cah. Off. Rech. Sci. Tech. Outre-Mer, 55, 289 pp.

Adams, T. S., and D. R. Nelson. 1968. Bioassay of crude extracts for the factor that prevents second matings in female *Musca domestica. Ann. Entomol. Soc. Am.,* **61:**112–16.

Adams, W. H.; R. W. Emmons; and J. E. Brooks. 1970. The changing ecology of murine (endemic) typhus in southern California. *Am. J. Trop. Med. Hyg.,* **19:**311–18.

Adar, H.; S. Bitnun; S. Ben-Meir; G. Barta; J. Herman; and Y. Kersh. 1972. (Blister beetles in the Beth Shean area.) *Harefuah,* **82:**447–48.

Adler, S., and O. Theodor. 1957. Transmission of disease agents by phlebotomine sand flies. *Annu. Rev. Entomol.,* **2:**203–26.

Aeschlimann, A. 1958. Développement embryonnaire d'*Ornithodorus moubata* (Murray) et transmission transovarienne de *Borrelia duttoni. Acta Trop.,* **15:**15–64.

Aeschlimann, A. 1972. (*Ixodes ricinus,* Linne, 1758) (Ixodoidea; Ixodidae). (First integrated account of the biology of this species in Switzerland.) *Acta Trop.,* **29:**321–40.

AHS 1975. *Proceedings of the Heartworm Symposium '74.* American Heartworm Society, VM Publishing, Bonner Springs, Kansas, 161 pp.

Akbulatova, L. K. 1970. (Two forms of the mite *Demodex folliculorum,* found in man.) *Meditsin. Parazitol.* (Mosk.), **39:** 700–704.

Akesson, N., and W. E. Yates. 1974. *The Use of Aircraft in Agriculture.* FAO Ag. Dev. Paper No. 94, 217 pp.

Akre, R. D., and H. G. Davis. 1978. Biology and pest-status of venomous wasps. *Annu. Rev. Entomol.,* **23:**215–38.

Akre, R. D.; A. Greene; J. F. MacDonald; P. J. Landolt; and H. G. Davis. 1979. *The Yellowjackets of America North of Mexico.* USDA, Ag. Handbook, No. 552, in press.

Akre, R. D.; J. F. MacDonald; and W. B. Hill. 1974. Yellowjacket literature (Hymenoptera: Vespidae). *Melanderia,* **18:**67–93.

Alger, N. E., and E. J. Cabrera. 1972. An increase in death rate of *Anopheles stephensi* fed on rabbits immunized with mosquito antigen. *J. Econ. Entomol.,* **65:**165–68.

Al-Janabi, B. M.; D. Branagan; and D. Danskin. 1975. The trans-stadial transmission of the bovine farcy organism, *Nocardia farcinica,* by the Ixodid *Amblyomma variegatum* (Fabricius, 1794). *Trop. Anim. Health. Prod.,* **7:**205–209.

Allred, D. M. 1954. Mites as intermediate hosts of tapeworms. *Proc. Utah Arts, Sci., Lett.,* **31:**44–51.

Altman, R. M.; C. M. Keenan; and M. M. Boreham. 1970. An outbreak of *Culicoides guyanensis* in the Canal Zone. *Mosq. News,* **31:** 231–35.

AMCA. 1961. *Organization for Mosquito Control.* American Mosquito Control Association, Bull. No. 4, 54 pp.

Ancelle, J. P.; P. Hagbe; J. Noutat; and R. Befidi. 1974. (Tetanus in Yaoundé.) *Med. Afr. Noire,* **21:**953–57.

Anderson, D. M. 1975. *Common Names of Insects (1975 Revision).* Entomological Society of America, Special Publ. 75-1, 37 pp.

Anderson, J. F., and F. R. Kneen. 1969. The temporary impoundment of salt marshes for the control of coastal deer flies. *Mosq. News,* **29:**239–43.

Anderson, J. R. 1975. The behavior of nose bot flies (*Cephenemyia apicata* and *C. jellisoni*) when attacking black-tailed deer (*Odocoileus hemionus columbianus*) and the resulting reactions of the deer. *Can. J. Zool.*, **53:**977–92.

Anderson, L. D.; E. C. Bay; and A. A. Ingram. 1964. Studies of chironomid midge control in water-spreading basins near Montebello, California. *Calif. Vector Views*, **11:**13–20.

Andreasen, M. P. 1974. Multiplication of *Cowdria ruminantium* in monolayer of tick cells. *Acta Pathol. Microbiol. Scand.* [B], **82:** 455–56.

Anigstein, L., and D. Anigstein. 1975. A review of the evidence in retrospect for a rickettsial etiology in Bullis fever. *Tex. Rep. Biol. Med.*, **33:**201–11.

Anisimov, P. I.; T. I. Anisimova; and Z. A. Koneva. 1968. (Plague bibliography—Russian literature, 1740–1964.) Pt. I:1–269; Pt. II:270–539. Translated by U.S. Joint Publications Research Service. JPRS 66304-1 & 2. 5 Dec. 1975.

Anonymous. 1975. Rare case of parasitic infection by maggots uncovered at autopsy. *Infect. Dis.*, **12**, Dec.:10–11, 14.

Anson, R. J., and P. S. Beasley. 1975. Radical mulesing pays. *Queensland Ag. J.*, **101:** 299–302.

Anthony, D. W.; A. J. Johnson; and A. A. Holbrook. 1970. Some effects of parasitism by *Babesia caballi* on the tropical horse tick, *Dermacentor* (= *Anocentor*) *nitens*. *J. Invertebr. Pathol.*, **15:**113–17.

Araujo-Fontaine, A.; F. Miltgen; H. Rombourg; B. Molet; G. Pauli; and A. Basset. 1974. (Contribution to the study of the allergenic role of dust mites. Immunological study of the sera of patients and the sera of hyperimmunised rabbits.) *Rev. Fr. Allerg. Immunol. Clin.*, **14:**91–96.

Arbesman, C. E.; J. I. Wypych; and R. E. Reisman. 1975. Standardization of stinging insect extracts. *In Developments in Biological Standardization*. S. Karger, Basel, Vol. 29, pp. 249–57.

Armstrong, R. K., and J. L. Winfield. 1968. Staphylinidae dermatitis on Okinawa. *J. Med. Entomol.*, **5:**362.

Arnell, J. H. 1973. Mosquito studies (Diptera, Culicidae) XXXII. A revision of the genus *Haemagogus*. *Contrib. Am. Entomol. Inst.*, **10**, No. 2, 174 pp.

Arthur, D. R. 1962. *Ticks and Disease*. Pergamon, New York, 445 pp.

Arthur, D. R. 1966. The ecology of ticks with reference to the transmission of Protozoa. In Soulsby, E. J. L. (ed.): *Biology of Parasites*. Academic Press, New York, pp. 61–84.

Ashbury, P. M., and C. F. Craig. 1907. Experimental investigations regarding the etiology of dengue fever. *J. Infect. Dis.*, **4:**440–75.

Atchley, W. R., and W. W. Wirth. 1975. Two new western *Culicoides* (Diptera; Ceratopogonidae) which are vectors of filaria in the California valley quail. *Can. J. Zool.*, **53:**1421–23.

Atchley, W. R.; W. W. Wirth; and C. T. Gaskins. 1975. *A Bibliography and a Keyword-incontext Index of the Ceratopogonidae (Diptera) from 1758 to 1973*. Texas Technical Press, Lubbock, 300 pp.

Atkinson, P. W., and K. C. Binnington. 1973. New evidence on the function of the porose areas of Ixodid ticks. *Experientia*, **29:**799–800.

Audy, J. R. 1958 (1956). The role of mite vectors in the natural history of scrub typhus. *Proc. 10th Internat. Cong. Entomol.*, **3:**639–49.

Audy, J. R., and M. M. J. Lavoipierre. 1964. The laboratory rearing of parasitic Acarina. A general review. *Bull. WHO*, **31:**583–86.

Avsatthi, B. L., and L. S. Hiregoudar. 1971. Life cycle of *Nosomma monstrosum* (Nuttall and Warburton, 1908), a tick of buffaloes and cattle in Gujarat State, India. *Gujvet, Anand,* **5:**43–46.

Axtell, R. C. 1970. Fly control in caged-poultry houses: comparison of larviciding and integrated control programs. *J. Econ. Entomol.*, **63:**1734–37.

Axtell, R. C. 1972. *Hippelates pusio* eye gnat control by ultralow volume aerial sprays. *J. Georgia Entomol. Soc.*, 7:119–27.

Axtell, R. C. (ed.). 1974. *Training Manual for Mosquito and Biting Fly Control in Coastal Areas*. Pub. UNC-SG-74-08, Ag. Exp. Sta., North Carolina State University, Raleigh, 249 pp.

Ayala, S. C. 1973. The Phlebotomine sandfly—Protozoan parasite community of central California grasslands. *Am. Midland Nat.*, **89:** 266–80.

Ayalde, J. 1976. (Analysis of strategy in the carrying out of malaria-eradication programmes in the Americas.) *Bol. Of. Sanit. Panam.*, **80:**45–51.

Azeez, S. A. 1972. Mass invasion of *Carbula pedalis* Bergr. (Pentatomidae) and its control in Bulassa, N.W. State. *Nigerian Entomol. Mag.*, **2:**125–26.

Azevedo, J. F. de; J. Tendeiro; L. T. de A. Franco; M. da C. Mourão; and J. M. de C. Salazar. 1956. Notica sobre a tsé-tsé de Ilha do Principe. *Garcia de Orta*, **4:**507–22.

Bacon, P. R. 1970. The natural enemies of the Ceratopogonidae—a review. *Tech. Bull. Com-*

monwealth Inst. Biol. Control, No. 13, pp. 71–82.

Bacot, A. W., and C. J. Martin. 1914. Observations on the mechanism of the transmission of plague by fleas. J. Hyg., 13(Plague Supp. 3):423–39.

Baerg, W. J. 1958. The Tarantula. University of Kansas Press, Lawrence, 88 pp.

Bagnall, B. G., and B. M. Doube. 1975. The Australian paralysis tick Ixodes holocyclus. Aust. Vet. J., 51:159–60.

Bahme, A. J. 1968. Cantharides toxicosis in the equine. Southwest. Vet., 21:147–48.

Bailey, C. L.; B. F. Eldridge; D. E. Hayes; D. M. Watts; R. F. Tammariello; and J. M. Dalrymple. 1978. Isolation of St. Louis encephalitis virus from overwintering Culex pipiens mosquitoes. Science, 199:1346–49.

Bailey, S. F. 1936. Thrips attacking man. Can. Entolmol., 68:95–98.

Bain, O. 1971. (Transmission off filarial infections. Limitation of passage of ingested microfilariae towards the haemocoel of the vector: interpretation.) Ann. Parasitol. Hum. Comp. 46:613–31.

Bain, O., and J. C. Beaucournu. 1974. (Infective larvae of Dipetalonema sp. in fleas collected on foxes in south-western France.) Ann. Parasitol. Hum. Comp. 49:123–25.

Bain, O., and J. Brengues. 1972. (Transmission of wuchereriasis and bovine setariasis. Histological study of the crossing of the stomach wall of Anopheles gambiae A and Aedes aegypti by microfilariae.) Ann. Parasitol. Hum. Comp. 47:399–412.

Bainbridge, M. H. 1973. (I) Cattle tick. (II) The effects of tick on cattle. Turnoff, Darwin, Australia, 4(2):9–13, 14–18.

Baker, E. 1974. Ectopic ear mite infestation in the dog. Am. Vet. Med. Assoc., 164:1125–26.

Baker, E. W.; J. H. Camin; F. Cunliffe; T. A. Woolley; and C. E. Yunker. 1958. Guide to the Families of Mites. Institute of Acarology, College Park, Md., Contrib. No. 3, 242 pp.

Baker, E. W.; T. M. Evans; D. J. Gould; W. B. Hull; and H. L. Keegan. 1956. A manual of Parasitic Mites of Medical or Economic Importance. National Pest Control Ass., New York, 170 pp.

Baker, E. W., and G. W. Wharton. 1952. An Introduction to Acarology. Macmillan Publishing Co., Inc., New York, 465 pp.

Baker, J. R. 1974. Epidemiology of African Sleeping Sickness. In CIBA Foundation Symposium 20 (n.s.). Trypanosomiasis and Leishmaniasis, with Special Reference to Chagas'

Disease. Elsevier, Amsterdam, 353 pp. (pp. 29–50).

Balashov, Y. S. 1970. (Experimental interspecific hybridization of the argasid ticks Ornithodoros papillipes, O. tartakovsky and O. verrucosus [Argasidae, lxodiodea]). Parazitologiia, 4:274–82.

Balashov, Y. S. 1972. Bloodsucking ticks (Ixodoidea) vectors of diseases of man and animals, 319 pp. (English translation of 1968 Russian publication by Med. Zool. Dept., U.S. NAMRU-3, Cairo, Egypt.) Miscellaneous Publications, Entomological Society of America, 8:161–376.

Balashov, Y. S.; A. B. Daiter; and T. N. Khavkin. 1972. (The distribution of Burnet's rickettsiae in the tick Hyalomma asiaticum [immunofluorescent and histological investigation].) Parazitologiia, 6:22–25.

Balozet, L. 1956. Scorpion venoms and antiscorpion serum. In Buckley, E. E. and N. Porges, Venoms. American Association for the Advancement of Science publ., 44:141–44.

Baltazard, M., and others. 1960. (Articles reviewing the status of plague on a worldwide basis.) Bull. WHO, 23:135–418.

Bänziger, H. 1970. The piercing mechanism of the fruit-piercing moth Calpe (Calyptra) thalictri Bkh. (Noctuidae) with reference to the skin-piercing blood-sucking moth C. eustrigata Hmps. Acta Trop., 27:54–88.

Bänziger, H. 1971. Bloodsucking moths of Malaya. Fauna, 1:5–16.

Bänziger, H. 1975. Skin-piercing blood-sucking moths I: ecological and ethological studies on Calpe eustrigata (Lepid. Noctuidae). Acta Trop., 32:125–44.

Bárdoš, V. 1975. The role of mammals in the circulation of Tahyna virus. Folia Parasitol. 22:257–64.

Barker, P. S. 1968. Effect of food quality on the reproduction of Mesostigmata: a review. Manitoba Entomol., 2:46–48.

Barker, R. W.; A. L. Hoch; R. G. Buckner; and J. A. Hair. 1973. Hematological changes in white-tailed deer fawns, Odocoileus virginianus, infested with Theileria-infected lone star ticks. J. Parasitol., 59:1091–98.

Barnes, A. M.; L. J. Ogden; and E. G. Campos. 1972. Control of the plague vector, Opisocrostis hirsutus, by treatment of prairie dog (Cynomys ludovicianus) burrows with 2% carbaryl dust. J. Med. Entomol., 9:330–33.

Barnes, A. M., and F. J. Radovsky. 1969. A new Tunga (Siphonaptera) from the nearctic region with descriptions of all life stages. J. Med. Entomol., 6:19–36.

Barnett, H. C. 1960. The incrimination of arthropods as vectors of disease. *Proc. 11th Inter. Cong. Entomol.*, **2:**341–45.

Barr, A. R. 1967. Occurrence and distribution of the *Culex pipiens* complex. *Bull. WHO*, **37:**293–96.

Barr, A. R. 1974a. Reproduction in Diptera of medical importance with special reference to mosquitoes. *J. Med. Entomol.*, **11:**35–40.

Barr, A. R. 1974b. New concepts of mosquito taxonomy. *Mosq. Syst.*, **6:**134–36.

Barr, S. E. 1971. Allergy to Hymenoptera stings—review of world literature 1953–1970. *Ann. Allergy*, **29:**49–66.

Barr, S. E. 1972. Skin test reactivity to the stinging insects. *Ann. Allergy*, **30:**282–87.

Barr, S. E. 1974. Allergy to Hymenoptera stings. *J.A.M.A.*, **228:**718–20.

Barrera, M. 1973. Fauna del Nordoeste argentino. Observaciones biologicas sobre *Antiteuchus variolosus* Westwood (Hemiptera-Pentatomidae). *Acta Zool. Lilloana*, **30:**141–62.

Barrow, P. M.; S. B. McIver; and K. A. Wright. 1975. Salivary glands of female *Culex pipiens:* morphological changes associated with maturation and blood-feeding. *Can. Entomol.*, **107:**1153–60.

Bartholomew, C. 1970. Acute scorpion pancreatitis in Trinidad. *Br. Med. J.*, **1:**666–68.

Bar-Zeev, M., and S. Gothilf. 1973. Laboratory evaluation of tick repellents. *J. Med. Entomol.*, **10:**71–74.

Bar-Zeev, M., and S. Sternberg. 1970. Testing space repellents against mosquitoes. *Mosq. News*, **30:**27–29.

Bassett, D. C. J. 1970. *Hippelates* flies and streptococcal skin infection in Trinidad. *Trans. R. Soc. Trop. Med. Hyg.*, **64:**138–47.

Bates, M. 1949. *The Natural History of Mosquitoes.* Macmillan Publishing Co., Inc., New York, 379 pp.

Bauer, S. 1974. Helicopter operations in the control of *Simulium damnosum* in West Africa. *Ag. Aviation*, **16:**65–70.

Baumhover, A. H. 1966. Eradication of the screwworm fly. *J.A.M.A.*, **196:**240–48.

Bay, E. C., and L. D. Anderson, 1965. Chironomid control by carp and goldfish. *Mosq. News*, **25:**310–16.

Bay, E. C., and L. S. Self. 1972. Observations of the guppy, *Poecilia reticulata* Peters, in *Culex pipiens fatigans* breeding sites in Bangkok, Rangoon, and Taipei. *Bull. WHO*, **46:**407–16.

Bean, S. F. 1974. Bullous scabies. *J.A.M.A.*, **230:**878.

Beard, R. L. 1963. Insect toxins and venoms. *Annu. Rev. Entomol.*, **8:**1–18.

Beard, R. L., and D. C. Sands. 1973. Factors affecting degradation of poultry manure by flies. *Environ. Entomol.*, **2:**801–806.

Beatson, S. H. 1971. Control of the martin bug, *Oeciacus hirundinis. Environ. Health*, **79:** 283–85.

Beatson, S. H. 1972. Pharaoh's ants as pathogen vectors in hospitals. *Lancet*, **1:**425–27.

Beaucournu, J. C., and J. Deunff. 1976. (Value of extemporaneous examination of fleas for the study of their parasites.) *Ann. Parasitol. Hum. Comp.* **50:**831–35.

Beaver, P. C., and T. C. Orihel. 1965. Human infection with filariae of animals in the United States. *Am. J. Trop. Med. Hyg.*, **14:**1010–29.

Beck, E. C., and W. M. Beck, Jr. 1969. The Chironomidae of Florida. II. The nuisance species. *Florida Entomol.*, **52:**1–11.

Beesley, W. N. 1973. Control of arthropods of medical and veterinary importance. *Adv. Parasitol.*, **11:**115–92.

Beesley, W. N. 1974a. Economics and progress of warble fly eradication in Britain. *Vet. Med. Rev.* No. 4, 334–47.

Beesley, W. N. 1974b. Recent developments in the control of ectoparasites of animals. *Proc. 7th British Insecticide and Fungicide Conference*, **3:**865–75. Brit. Crop. Prot. Council, London, UK.

Beklemishev, V. H. 1958. (*The Identification of Arthropods Prejudicial to the Health of Man.*) Gosudarstv. Izdatelestvo Medits. Lit. Medgiz, Moscow. 422 pp.

Bell, T. D.; R. K. Farrell; G. A. Padgett; and L. W. Leendertsen. 1975. Ataxia, depression, and dermatitis associated with the use of dichlorvos-impregnated collars in the laboratory cat. *J. Am. Vet. Med. Assoc.*, **167:**579–86.

Bellamy, R. E., and G. K. Bracken. 1971. Quantitative aspects of ovarian development in mosquitos. *Can. Entomol.*, **103:**763–73.

Bellamy, R. E., and P. S. Corbet. 1974. Occurrence of ovariolar dilatations in nulliparous mosquitoes. *Mosq. News*, **34:**334.

Bendheim, U. 1966. Cnemidocoptic mange of budgerigars in Israel. *Refuah Vet.*, **23:**54–57.

Bennet-Clark, H. C. 1963. Negative pressure produced in the pharyngeal pump of the blood-sucking bug, *Rhodius prolixus. J. Exp. Biol.*, **40:**223–29.

Bennett, W. C.; G. N. Graves; J. R. Wheeler; and B. E. Miller. 1975. Field evaluations of dichlorvos as a vapor toxicant for control of prairie dog fleas. *J. Med. Entomol.*, **12:**354–58.

Benton, A. W.; R. A. Morse; and J. D. Stewart.

1963. Venom collection from honey bees. *Science,* **142:**228–30.

Bequaert, J. C. 1953–1957. The Hippoboscidae or louse-flies (Diptera) of mammals and birds. Part I. Structure, physiology, and natural history. Part II. Taxonomy, evolution, and revision of American genera and species. *Entomol. Am.,* **32**(n.s.):1–209; **33**(n.s.):211–422; **34** (n.s.):1–232; **35**(n.s.):233–416; **36**(n.s.): 417–611.

Berenberg, J. L.; P. A. Ward; and D. E. Sonenshine. 1972. Tick-bite injury; mediation by a complement-derived chemotactic factor. *Immunology,* **109:**451–56.

Berg, C. O. 1973. Biological control of snail-borne diseases [with Sciomyzids]: a review. *Exp. Parasitol.,* **33:**318–30.

Berge, T. O. (ed.). 1975. International catalogue of arboviruses including certain other viruses of vertebrates. U.S. Dept. Health, Education, and Welfare, Publication No. (CDC) 75-8301.

Berger, R. S. 1972. 2,6-Dichlorophenol, sex pheromone of the lone star tick. *Science,* **177:**704–705.

Berger, R. S. 1973. The unremarkable brown recluse spider bite. *J.A.M.A.,* **225:**1109–11.

Berger, R. S.; L. E. Millikan; and F. Conway. 1973. An *in vitro* test for *Loxosceles reclusa* spider bites. *Toxicon,* **11:**465–70.

Berry, I. L., and R. A. Hoffman. 1963. Use of step-on switches for control of automatic sprayers. *J. Econ. Entomol.,* **56:**888–90.

Berry, R. L.; M. A. Parson; B. J. LaLonde; H. W. Stegmiller; J. Lebio; M. Jalil; and R. A. Masterson. 1975. Studies on the epidemiology of California encephalitis in an endemic area in Ohio in 1971. *Am. J. Trop. Med. Hyg.,* **24:**992–98.

Bhatia, M. L., and B. L. Wattal. 1958. Tolerated density of *Culex fatigans* Wied. in transmission of bancroftian filaria. *Bull. Nat. Soc. India Malar.,* **6:**117–22.

Bhuangprakone, S., and S. Areekul. 1973. Biology and food habits of the snail-killing fly, *Sepedon plumbella* Wiedemann (Sciomyzidae:Diptera). *Southeast Asian J. Trop. Med. Public Health,* **4:**387–94.

Bibikova, V. A. 1977. Contemporary views on the interrelationships between fleas and the pathogens of human and animal diseases. *Annu. Rev. Entomol.,* **22:**1–22.

Bibikova, V. A., and L. N. Klassovskiĭ. 1974. (*The Transmission of Plague by Fleas.*) Moscow, USSR; Izdatel'stvo "Meditsina", 188 pp.

Bibikova, V. A.; L. N. Klassovskiĭ; and N. M. Khrustselevskaya. 1975. (The effect of pro-longed contact between the flea population and plague bacteria on the infecting activity of vectors.) *Parazitologiia,* **9:**515–17.

Bidlingmayer, W. L. 1961. Field activity studies of adult *Culicoides furens. Ann. Entomol. Soc. Am.,* **54:**149–56.

Bidlingmayer, W. L. 1968. Larval development of *Mansonia* mosquitoes in central Florida. *Mosq. News,* **28:**52–57.

Bigler, W. J.; A. K. Ventura; A. L. Lewis; F. M. Wellings; and N. J. Ehrenkranz. 1974. Venezuelan equine encephalomyelitis in Florida: endemic virus circulation in native rodent populations of Everglades hammocks. *Am. J. Trop. Med. Hyg.,* **23:**513–21.

Biliotti, G.; A. Passaleva; S. Romagnani; and M. Ricci. 1972. Mites and house dust allergy. I. Comparison between house dust and mite (*Dermatophagoides pteronyssinus* and *D. farinae*) skin reactivity. *Clin. Allergy,* **2:**109–13.

Bills, G. T. 1973. Biological fly control in deep-pit poultry houses. *Br. Poult. Sci.,* **14:**209–12.

Bindseil, S. 1974. Histopathological studies of sarcoptic mange in pigs with particular reference to immunological characteristics of the lesions. *Dan. Dyrlaegefor. Medlemsblad.,* **57:**470–75.

Binnington, K. C. 1972. The distribution and morphology of probable photoreceptors in eight species of ticks (Ixodoidea). *Z. Parasitenkd.,* **40:**321–32.

Birulya, N. B.; L. I. Zalutskaya; and V. D. Perelatov. 1971. (Distribution area of natural foci of Crimean hemorrhagic fever.) Trudy Inst. Poliomelita: Virusnykh Entsefalitov, Akad. Meditsinskh Nauk SSSR 19, pp. 180–85.

Bishopp, F. C. 1939. The stable fly; how to prevent its annoyance and its losses to livestock. *USDA Farmer's Bull.,* 1097, 18 pp. (revised).

Bisley, G. G. 1972. A case of intraocular myiasis in man due to the first stage larva of the Oestrid fly *Gedoelstia* sp. *East Afr. Med. J.,* **49:**768–71.

Bivin, W. S.; C. Barry; A. L. Hogge; and E. C. Corristan. 1967. Mosquito-induced infection with equine encephalomyelitis virus in dogs. *Am. J. Trop. Med. Hyg.,* **16:**544–47.

Blagodarnyi, Y. A., and others. 1971. (The role of ticks in the transmission of the mycobacteria of tuberculosis.) *Veterinariia,* No. 7, 48–49.

Blake, F. G.; K. F. Maxcy; J. F. Sadusk, Jr.; G. M. Kohls; and E. J. Bell. 1945. Studies on tsutsugamushi disease (scrub typhus, mite-borne typhus) in New Guinea and adjacent islands. Epidemiological, clinical observations and etiology in the Dobadura area. *Am. J. Hyg.*, **41**:243–373.

Blaškovič, D., and J. Nosek. 1972. The ecological approach to the study of tick-borne encephalitis. *Prog. Med. Virol.*, **14**:275–320.

Blue, S. 1910. *Rodents in Relation to the Transmission of Bubonic Plague. The Rat and its Relation to the Public Health.* Washington, D.C., U.S. Public Health and Marine Hospital Serv., 254 pp.

Blumberg, B. S. 1977. Australia antigen and the biology of hepatitis B. *Science,* **197**:17–25.

Blume, R. R.; J. J. Matter; and J. L. Eschle. 1973. *Onthophagus gazella:* effect on survival of horn flies in the laboratory. *Environ. Entomol.*, **2**:811–13.

Boger, N.; B. Rightblat; R. Cwilich; and A. Adani. 1964. A case of *Ixodes* tick in man. *Refuah Vet.*, **21**:30–39.

Bollweg, G. 1975. (On the treatment of otitis externa parasitaria in dogs and cats.) *Tierärzt. Umschau.*, **30**:140, 142.

Boorman, J. 1975. Semi-automatic device for inoculation of small insects with viruses. *Lab. Pract.* **24**:90.

Boorman, J. P. T. 1961. Observations on the habits of moquitoes of Plateau Province, Northern Nigeria, with particular reference to *Aedes* [*Stegomyia vittatus* (Bigot)]. *Bull. Entomol. Res.*, **52**:709–25.

Boreham, P. F. L. 1975. Some applications of bloodmeal identification in relation to the epidemiology of vector-borne tropical diseases. *J. Trop. Med. Hyg.*, **78**:83–91.

Bořkovec, A. B. 1966. *Insect Chemosterilants.* Adv. Pest Control Res., 7, Wiley, New York, 143 pp.

Borror, D. J.; D. M. DeLong; and C. A. Triplehorn. 1976. *An Introduction to the Study of Insects,* 4th ed. Holt, Rinehart & Winston, New York, 852 pp.

Borror, D. J., and R. E. White. 1970. *A Field Guide to the Insects of America North of Mexico.* Houghton Mifflin, New York, 404 pp.

Boshell, M. J. 1969. Kyasanur Forest disease: ecologic considerations. *Am. J. Trop. Med. Hyg.*, **18**:67–80.

Bouillant, A. M.; V. H. Lee; and R. P. Hanson. 1965. Epizootiology of mink virus: II, *Musca domestica* as a possible vector of virus. *Can. J. Comp. Med. Vet. Sci.*, **29**:148–52.

Boyd, M. F. 1941. An historical sketch of the prevalence of malaria in North America. *Am. J. Trop. Med.*, **21**:223–44.

Boyd, M. F. 1950. *Malariology.* Saunders, Philadelphia, 2 vols., 1,643 pp.

Bradley, R. E. (ed.). 1972. *Canine Heartworm Disease, the Current Knowledge.* State of Florida, Internal Improvement Trust Fund, 148 pp.

Bram, R. A., and R. D. Romanowski. 1970. Recognition of *Anaplasma marginale* Theiler in *Dermacentor andersoni* Stiles (= *D. venustus* Marx) by the fluorescent antibody method. I. Smears of nymphal organs. *J. Parasitol.*, **56**:32–38.

Bray, R. S. 1974. Leishmania. *Annu. Rev. Microbiol.*, **28**:189–211.

Bray, R. S.; R. W. Ashford; and M. A. Bray. 1973. The parasite causing leishmaniasis in Ethiopia. *Trans. R. Soc. Trop. Med. Hyg.*, **67**:345–48.

Bray, R. S., and P. C. C. Garnham. 1964. *Anopheles* as vectors of animal malaria parasites. *Bull. WHO,* **31**:143–47.

Breeland, S. G., and E. Pickard. 1965. The Malaise trap: an efficient and unbiased mosquito collecting device. *Mosq. News,* **25**:19–21.

Breeland, S. G., and E. Pickard. 1967. Field observations on twenty-eight broods of floodwater mosquitoes resulting from controlled floodings of a natural habitat in the valley. *Mosq. News,* **27**:343–58.

Breev, K. A., and F. R. Sultanov. 1975. (On some peculiarities in the development of larvae of the sheep nostril fly *Oestrus ovis* L. [Diptera, Oestridae].) *Parazitologiia,* **9**:47–56.

Brennan, J. M. 1935. The Pangoniinae of Nearctic America, Diptera: Tabanidae. *Univ. Kansas Sci. Bull.*, **22**:249–402.

Brennan, J. M., and M. L. Goff. 1977. Keys to the genera of chiggers of the western hemisphere (Acarina: Trombiculidae). *J. Parasitol.*, **63**:554–66.

Brennan, J. M., and E. K. Jones. 1959. Keys to the chiggers of North America with synonymic notes and descriptions of two new genera (Acarina; Trombiculidae). *Ann. Entomol. Soc. Am.*, **52**:7–16.

Brennan, J. M., and G. W. Wharton. 1950. Studies on North American chiggers No. 3. The subgenus Neotrombicula. *Am. Midland Naturalist,* **44**:153–97.

Brennan, J. M., and C. E. Yunker. 1964. A new species of *Euschoengastia* of potential veterinary importance (Acarina: Trombiculidae). *J. Parasitol.*, **50**:311–12.

Bromley, S. W. 1933. The sting of a tarantula wasp. *Bull. Brooklyn Entomol. Soc.*, **28**:192.

Brown, A. W. A. 1974. The safety of biological agents for arthropod control. *WHO Chron.* **28**:261–64.

Brown, A. W. A.; J. Haworth; and A. R. Zahar. 1976. Malaria eradication and control from a global standpoint. *J. Med. Entomol.*, 13:1–25.

Brown, A. W. A., and P. E. Morrison. 1955. Control of adult tabanids by aerial spraying. *J. Econ. Entomol.*, **48:**125–129.

Brown, A. W. A., and R. Pal. 1971. *Insecticide Resistance in Arthropods.* Monograph Ser. WHO No. 38, 2nd ed., 491 pp.

Brown, J. F., and T. R. Adkins, Jr., 1972. Relationship of feeding activity of face fly (*Musca autumnalis* De Geer) to production of keratoconjunctivitis in calves. *Am. J. Vet. Res.*, **33:**2551–55.

Browning, T. O. 1976. The aggregation of questing ticks, *Rhipicephalus pulchellus,* on grass stems, with observations on *R. appendiculatus. Physiol. Entomol.*, **1:**107–14.

Bruce, D., and D. Nabarro. 1903. *Progress Report on Sleeping Sickness in Uganda.* Report of Sleeping Sickness Comm., R. Soc., London, No. 1.

Bruce-Chwatt, L. J. 1973. (Malaria eradication in Africa.) *Ann. Parasitol. Hum. Comp.* **48:**221–29.

Bruce-Chwatt, L. J. 1976. Mathematical models in the epidemiology and control of malaria. *Trop. Geogr. Med.*, **28:**1–8.

Bruce-Chwatt, L. J., and V. J. Glanville (eds.). 1973. *Macdonald G., Dynamics of Tropical Disease. A Selection of Papers with Biographical Introduction and Bibliography.* Oxford University Press, London, 310 pp.

Brummer-Korvenkontio, M. 1973. Arboviruses in Finland. V. Serological survey of antibodies against Inkoo virus (California group) in human, cow, reindeer, and wildlife sera. *Am. J. Trop. Med. Hyg.*, 22:654–61.

Bryan, R. P. 1973. The effects of dung beetle activity on the numbers of parasitic gastrointestinal helminth larvae recovered from pasture samples. *Aust. J. Agr. Res.*, **24:**161–68.

Buchanan, R. E., and N. E. Gibbs (eds.). 1974. *Bergey's Manual of Determinative Bacteriology,* 8th ed. Williams & Wilkins, Baltimore, 1,246 pp.

Bücherl, W., and E. Buckley (eds.). 1971. *Venomous Animals and Their Venoms,* Vol. III. *Venomous Invertebrates.* Academic Press, New York, 537 pp.

Buck, A. A.; D. H. Connor; and M. M. da Silva (Program committee). 1974. *Research and Control of Onchocerciasis in the Western Hemisphere.* Pan American Health Org. Sci. Pub. No. 298, 154 pp.

Bullini, L., and M. Coluzzi. 1974. *Electrophoretic Studies on Gene-Enzyme Systems in Mosquitoes (Diptera, Culicidae).* WHO/VBC/74.483, 21 pp.

Burden, G. S. 1972. Gas-propelled aerosols and micronized dusts for control of insects in aircraft. 6. Insects of medical importance. *J. Econ. Entomol.*, **65:**1458–62.

Burden, G. S. 1976. Feeding activity of cockroaches on ceiling boards. *Pest Control,* **44:**16.

Burgdorfer, W. 1951. Analyse des Infektionsverlaufes bei *Ornithodorus moubata* (Murray) und der naturlichen Übertragung von *Spirochaeta duttoni. Acta Trop.,* **8:**193–262.

Burgdorfer, W. 1957. Artificial feeding of ixodid ticks for studies on the transmission of disease agents. *J. Infect. Dis.,* **100:**212–14.

Burgdorfer, W. 1969. Ecology of tick vectors of American spotted fever. *Bull. WHO,* **40:** 375–81.

Burgdorfer, W. 1970. Hemolymph test. A technique for detection of rickettsiae in ticks. *Am. J. Trop. Med. Hyg.,* **19:**1010–14.

Burgdorfer, W. 1975. A review of Rocky Mountain spotted fever (tick-borne typhus), its agent, and its tick vectors in the United States. *J. Med. Entomol.,* **12:**269–78.

Burgdorfer, W., and L. P . Brinton. 1975. Mechanisms of transovarial infection of spotted fever rickettsiae in ticks. *Ann. N.Y. Acad. Sci.,* **266:**61–72.

Burgdorfer, W., and C. M. Eklund. 1959. Studies on the ecology of Colorado tick fever virus in western Montana. *Am. J. Hyg.,* **69:**127–37.

Burgdorfer, W.; V. F. Newhouse; E. G. Pickens; and D. B. Lackman. 1962. Ecology of Rocky Mountain spotted fever in western Montana. I. Isolation of *Rickettsia rickettsii* from wild animals. *Am. J. Hyg.,* **76:**293–301.

Burgdorfer, W.; R. A. Ormsbee; and H. Hoogstraal. 1972. Ticks as vectors of *Rickettsia prowazekii*—a controversial issue. *Am. J. Trop. Med. Hyg.,* **21:**989–98.

Burgdorfer, W.; M. L. Schmidt; and H. Hoogstraal. 1973. Detection of *Trypanosoma theileri* in Ethiopian cattle ticks. *Acta Trop.,* 30:340–46.

Burgdorfer, W.; D. J. Sexton; R. K. Gerloff; R. L. Anacker; R. N. Philip; and L. A. Thomas. 1975. *Rhipicephalus sanguineus:* vector of a new spotted fever group rickettsia in the United States. *Infect. Immun.,* **12:**205–10.

Burgdorfer, W., and M. G. R. Varma. 1967. Transstadial and transovarial development of disease agents in arthropods. *Annu. Rev. Entomol.,* **12:**347–76.

Burges, H. D., and N. W. Hussey (eds.). 1971. *Microbial Control of Insects and Mites*. Academic Press, London and New York, 861 pp.

Burgess, R. W. 1951. The life history and breeding habits of the eye gnat, *Hippelates pusio* Loew, in the Coachella Valley, Riverside County, California. *Am. J. Hyg.*, **53:**164–77.

Burridge, M. J. 1975. The role of wild mammals in the epidemiology of bovine theilerioses in East Africa. *J. Wildlife Dis.*, **11:**68–75.

Burton, A. N.; J. McLintock; and J. G. Rempel. 1966. Western equine encephalitis in Saskatchewan garter snakes and leopard frogs. *Science*, **154:**1029–31.

Burton, G. J. 1963. Bedbugs in relation to transmission of human disease. *Public Health* Rep., **78:**513–24.

Bushland, R. C. 1975. Screwworm research and eradication. *Bull. Entomol. Soc. Am.*, **21:**23–26.

Bushland, R. C.; R. D. Radeleff; and R. O. Drummond. 1963. Development of systemic insecticides for pests of animals in the United States. *Annu. Rev. Entomol.*, **8:**215–38.

Busvine, J. R. 1976. *Insects, Hygiene and History*. Athlone Press, University of London, 262 pp.

Büttiker, W. 1958. Observations on physiology of adult anophelines in Asia. *Bull. WHO*, **19:**1063–71.

Büttiker, W. 1959. Observations on feeding habits of adult Westermanniinae (Lepid., Noctuidae) in Cambodia. *Acta Trop.*, **16:**356–61.

Büttiker, W. 1962. Notes on two species of Westermanniinae (Lepidoptera: Noctuidae) from Cambodia. *Proc. R. Entomol. Soc. Lond.* (B), **31:**73–76.

Büttiker, W. 1964. New observations on eye-frequenting Lepidoptera from S.E. Asia. *Verh. Natur. Ges. Basel*, **75:**231–36.

Butz, W. C. 1971. Envenomation by the brown recluse spider (Aranae, Scytodidae) and related species. A public health problem in the United States. *Clin. Toxicol.*, **4:**515–24.

Buxton, P. A. 1955. *The Natural History of Tsetse Flies*. H. K. Lewis, London, 816 pp.

Caldwell, H. D., and E. L. Belden. 1973. Studies of the role of *Dermacentor occidentalis* in the transmission of bovine chlamydial abortion. *Infect. Immun.*, **7:**147–51.

Calisher, C. H.; R. G. McLean; G. C. Smith; D. M. Samyd; D. J. Muth; and J. S. Lazuick. 1977. Rio Grande—a new phlebotomus fever group virus from south Texas. *Am. J. Trop. Med. Hyg.*, **26:**997–1002.

Callahan, J. T., and R. L. Bailey. 1974. The effects of flooding on moth fly larvae and zooglea in a sewage trickling filter system. *J. Environ. Qual.*, **3:**24–25.

Camicas, J. L. 1975. (Present concepts on the epidemiology of boutonneuse fever in the Ethiopian Region and the European Mediterranean Sub-region.) *Cah. ORSTOM, Sér. Entomol. Méd. Parasitol.*, **13:**229–32.

Campbell, R. W., and R. Domrow. 1974. Rickettsioses in Australia: isolation of *Rickettsia tsutsugamushi* and *R. australis* from naturally infected arthropods. *Trans. R. Soc. Trop. Med. Hyg.*, **68:**397–402.

Campion, D. G. 1972. Insect chemosterilants: a review. *Bull. Entomol. Res.*, **61:**577–635.

Canese, A. 1972. (*Loxosceles rufescens* [Dufour 1820] in the Isla Pucú area of the Cordillera Department [Paraguay].) *Revta. Paraguaya de Microbiología*, **7:**83–85.

Capelle, K. 1971. Myiasis. In Davis, J. W., and R. C. Anderson, *Parasitic Diseases of Wild Mammals*. Iowa State University Press, Ames, 364 pp.

Cardani, C.; C. Fuganti; D. Ghiringhelli; P. Grasselli; M. Pavan; and M. D. Valcurone. 1973. Biosynthesis of pederine. *Tetrahedron Lett.*, **30:**2815–18.

Carmichael, G. T. 1972. Anopheline control through water management. *Am. J. Trop. Med. Hyg.*, **21:**782–86.

Carpenter, S. J. 1968. Review of recent literature on mosquitoes of North America. *Calif. Vector Views*, **15:**71–98.

Carpenter, S. J., and W. J. LaCasse. 1955. *Mosquitoes of North America (North of Mexico)*. University of California Press, Berkeley, 360 pp.

Cassidy, D. R., and W. E. Ketler. 1975. The subcutaneous mite of chickens: an incidence report. *Avian Dis.*, **9:**78–81.

Cavallo-Serra, R. J. 1973. (*Arthropods of Medical and Veterinary Interest*.) Institut Universitaire de Microbiologie, Lausanne, Switzerland, 283 pp.

Cavanaugh, D. C. 1971. Specific effect of temperature upon transmission of the plague bacillus by the Oriental rat flea, *Xenopsylla cheopis. Am. J. Trop. Med. Hyg.*, **20:**264–73.

CDC. 1976. *Vector Topics No. 1, Control of St. Louis Encephalitis*. U.S. Dept. Health, Education, and Welfare, Center for Disease Control, 35 pp.

Cerf, D. C., and G. P. Georghiou. 1974. Cross resistance to juvenile hormone analogues in insecticide-resistant strains of *Musca domestica* L. *Pestic. Sci.*, **5:**759–67.

Challet, G. L.; G. T. Reynolds; and D. L. Rohe. 1974. A pilot program for the intensive culture of *Gambusia affinis* (Baird and Girard) and *Tilapia zillii* (Gervais). Part II: initial operation and production efficiencies. *Proceedings and Papers,* 30th Annual Meeting American Mosquito Control Association, pp. 55–58.

Challier, A., and C. Laveissière. 1973. (A new trap for catching tsetse flies [*Glossina:* Diptera, Muscidae]: description and field trials.) *Cah. ORSTOM, Sér. Entomol. Méd. Parasitol.,* **11:**251–62.

Chamberlain, R. W., and W. D. Sudia. 1961. Mechanism of transmission of viruses by mosquitoes. *Annu. Rev. Entomol.,* **6:**371–90.

Chamberlain, W. F. 1975. Insect growth regulating agents for control of arthropods of medical and veterinary importance. *J. Med. Entomol.,* **12:**395–400.

Chance, M. M. 1970. The functional morphology of the mouthparts of blackfly larvae (Diptera: Simuliidae). *Quaest. Entomol.* **6:**245–84.

Chandrahas, R. K.; A. K. Krishnaswami; and C. K. Rao. 1974. Studies on the epidemiology of plague in a South India plague focus. *Indian J. Med. Res.,* **62:**1080–1103.

Chaniotis, B. N., and M. A. Corrêa. 1974. Comparative flying and biting activity of Panamanian Phlebotomine sandflies in a mature forest and adjacent open space. *J. Med. Entomol.,* **11:**115–16.

Chapman, H. C. 1974. Biological control of mosquito larvae. *Annu. Rev. Entomol.,* **19:**33–59.

Chapple, P. J., and N. D. Lewis. 1965. Myxomatosis and the rabbit flea. *Nature,* **207:**388–89.

Charlesworth, E. N., and J. L. Johnson. 1974. An epidemic of canine scabies in man. *Arch. Dermatol.,* **110:**573–74.

Chellappa, D. J. 1973. Note on spinose ear tick infestation in man and domestic animals in India and its control. *Madras Agr. J.,* **60:** 656–58.

Cheminat, J.; J. Brun; C. Grouffal; R. Petit; and C. Molina. 1972. (Allergy to Hymenoptera.) *Rev. Fr. Allerg.,* **12:**239–50.

Cheng, T., and E. M. Kesler. 1961. A three-year study on the effect of fly control on milk production by selected and randomized dairy herds. *J. Econ. Entomol.,* **54:**752–57.

Cheng, T.; R. E. Patterson; B. W. Avery; and J. P. Vandenberg. 1957. *An Electric-Eye-Controlled Sprayer for Application of Insecticides to Livestock.* Pennsylvania Agriculture Experiment Station Bull. 626, 14 pp.

Cherrett, J. M.; J. B. Ford; I. V. Herbert; and A. J. Probert. 1971. *The Control of Injurious Animals.* English University Press, London, 210 pp.

Chiang, C., and W. C. Reeves. 1962. Statistical estimation of virus infection rates in mosquito vector populations. *Am. J. Hyg.,* **75:**377–91.

Chinchilla, M., and A. Ruiz. 1976. Cockroaches as possible transport hosts of *Toxoplasma gondii* in Costa Rica. *J. Parasitol.,* **62:**140–42.

Chinery, W. A. 1973. The nature and origin of the "cement" substance at the site of attachment and feeding of adult *Haemaphysalis spinigera* (Ixodidae). *J. Med. Entomol.,* **10:**355–62.

Choovivathanavanich, P. 1974. Insect allergy: antigenicity of cockroach and its excrement. *J. Med. Assoc. Thai.* **57:**237–41.

Chow, C. Y. 1969. Ecology of malaria vectors in the Pacific. *Cah. ORSTOM, Sér. Entomol. Méd. Parasitol.,* **7:**93–97.

Chow, C. Y., 1970. Bionomics of malaria vectors in the western Pacific region. *Southeast Asian J. Trop. Med. Public Health,* **1:**40–56.

Chow, Y. S.; C. B. Wang; and L. C. Lin. 1975. Identification of a sex pheromone of the female brown dog tick, *Rhipicephalus sanguineus. Ann. Entomol. Soc. Am.,* **68:**485–88.

Christensen, H. A., and A. Herrer. 1975. Predation of adult sand flies, (Diptera: Psychodidae). *Mosq. News,* **35:**233.

Christensson, D., and C. Rehbinder. 1971. (*Pneumonyssus caninum,* a mite in the nasal cavities and sinuses of the dog.) *Nord. Vet. Med.* **23:**499–505.

Christophers, S. E. 1960. *Aedes aegypti (L.) the Yellow Fever Mosquito: its Life History, Bionomics and Structure.* Cambridge University Press, London, 739 pp.

Chumakov, M. P. 1957. *Etiology, Epidemiology* and *Prophylaxis of Hemorrhagic Fevers.* Public Health Service: Publ. Hlth. Mon. No. 50, pp. 19–25.

Chvála, M.; L. Lyneborg; and J. Moucha. 1972–1977. *The Horse Flies of Europe.* Entomological Society, Copenhagen, Denmark, 500 pp.

Ciolca, A.; I. Tănase; and I. May. 1968. (Rôle of the poultry mite, *Dermanyssus gallinae,* in the transmission of spirochaetosis.) *Arch. Vet.,* **5:**207–15.

Clark, P. H.; M. M. Cole; D. L. Forcum; J. R. Wheeler; K. W. Weeks; and B. E. Miller. 1971. Preliminary evaluation of three systemic insecticides in baits for control of fleas of wild rats and rabbits. *J. Econ. Entomol.,* **64:**1190–93.

Clarke, J. L., and F. C. Wray. 1967. Predicting

influxes of *Aedes vexans* into urban areas. *Mosq. News,* **27:**156–65.

Clay, T. 1970. The Amblycera (Phthiraptera: Insecta). *Bull. Br. Mus. Nat. Hist. (Entomol.),* **25:**73–98.

Clements, A. N. 1963. *The Physiology of Mosquitoes.* Pergamon Press, Oxford, 393 pp.

Cloudesley-Thompson, J. L. 1958. *Spiders, Scorpions, Centipedes and Mites.* Pergamon Press, New York, 228 pp.

Coatney, G. R. 1968. Simian malarias in man: facts, implications, and predictions. *Am. J. Trop. Med. Hyg.,* **17:**147–55.

Coatney, G. R. 1971. The simian malarias: zoonoses, anthroponoses, or both? *Am. J. Trop. Med. Hyg.,* **20:**795–803.

Coatney, G. R.; W. E. Collins; McW. Warren; and P. G. Contacos. 1971. *The Primate Malarias.* Superintendent of Documents, Washington, D.C.

Coch, F. 1970. (The control of wasps in food shops.) *Angew. Parasitol.,* **11:**225–31.

Cochran, D. G.; J. M. Grayson; and A. B. Gurney. 1975. Cockroaches—biology and control. WHO/VBC/75.576, 48 pp.

Cole, F. 1969. *The Flies of Western North America.* University of California Press, Berkeley and Los Angeles, 693 pp.

Cole, M. M.; W. C. Bennett; G. N. Graves; J. R. Wheeler; B. E. Miller; and P. H. Clark. 1976. Dichlorvos bait for control of fleas on wild rodents. J. Med. Entomol., **12:**625–30.

Cole, M. M.; D. L. VanNatta; W. Ellerbe; and F. Washington. 1972. Rearing the oriental rat flea. *J. Econ. Entomol.,* **65:**1495–96.

Coluzzi, M., and R. Trabucchi. 1968. (Importance of the buccopharyngeal armature in *Anopheles* and *Culex* in relation to infection with *Dirofilaria.*) *Parassitologia,* **10:**47–59.

Confalonieri, U. E. C., and L. P. de Carvalho. 1973. (Occurrence of *Trombicula* [*Eutrombicula*] *batatas* [L.] on *Gallus gallus domesticus* L. in the State of Rio de Janeiro [Acarina, Trombiculidae].) *Rev. Bras. Biol.,* **33:**7–10.

Conly, G. N. 1975. *The Impact of Malaria on Economic Development: a Case Study.* Sci. Pub. 297, Pan American Health Organization, 117 p.

Conn, H. F. (ed.). 1974. *Current Therapy.* Saunders, Philadelphia, 914 pp.

Connell, M. L. 1974. Transmission of *Anaplasma marginale* by the cattle tick *Boophilus microplus. Queensland Ag. Anim. Sci.,* **31:**185–93.

Conway, G. R.; M. Trpis; and G. A. H. McClelland. 1974. Population parameters of the mos-

quito *Aedes aegypti* (L.) estimated by mark-release-recapture in a suburban habitat in Tanzania. *J. Anim. Ecol.,* **43:**289–304.

Cook, J. A. L.; F. H. Miller; R. W. Grover; and J. L. Duffy. 1973. Urticaria caused by tarantula hairs. *Am. J. Trop. Med. Hyg.* **22:**130–33.

Cook, S. F., Jr., 1967. The increasing chaoborid midge problem in California. *Calif. Vector Views,* **14:**39–44.

Cooley, R. A. 1932. *The Rocky Mountain Wood Tick.* Bozeman, Montana State College, Ag. Exp. Sta. Bull. 268, 58 pp.

Cooley, R. A., and G. M. Kohls. 1944. *The Argasidae of North America, Central America and Cuba.* Am. Midland Naturalist Monogr. 1, 152 pp.

Cooper, J. E. 1976. *Tunga penetrans* infestation in pigs. *Vet. Rec.,* **98:**472.

COPR. 1973. *Note on Allergy to Locusts.* Centre for Overseas Pest Research, London, UK, 3 pp.

Corbet, P. S. 1960. Recognition of nulliparous mosquitoes without dissection. *Nature,* **187:**525–26.

Corbet, P. S. 1967. Facultative autogeny in Arctic mosquitoes. *Nature,* **215:**662–63.

Corbet, P. S., and H. V. Danks. 1975. Egg-laying habits of mosquitoes in the high Arctic. *Mosq. News,* **35:**8–14.

Corbet, P. S., and A. E. R. Downe. 1966. Natural hosts of mosquitoes in northern Ellesmere Island. *Arctic,* **19:**153–61.

Cordellier, R.; M. Germain; and J. Mouchet. 1974. (The vectors of yellow fever in Africa.) *Cah. ORSTOM, Sér. Entomol. Méd. Parasitol.,* **12:**57–75.

Cornwell, P. B. 1968. *The Cockroach.* Volume 1. *A Laboratory Insect and an Industrial Pest.* Hutchinson, London, 391 pp.

Corrado, A., and A. Mantovani. 1966. (Transmission of canine hepatitis virus by *Rhipicephalus sanguineus.*) *Atti Soc. Ital. Sci. Vet.,* **20:**772–76.

Corrado, A. P.; F. R. Neto; and A. Antonio. 1974. The mechanism of the hypertensive effect of Brazilian scorpion venom (*Tityus serrulatus* Lutz e Mello). *Toxicon,* **12:**145–50.

Cotton, M. J. 1970. The life history of the hen flea, *Ceratophyllus gallinae* (Schrank) (Siphonaptera, Ceratophyllidae). *Entomologist,* **103:**45–48.

Coudert, J.; M. R. Battesti; and J. Despeignes. 1972. (A case of allergy to the bites of *Argas reflexus.*) *Bull. Soc. Pathol. Éxot.,* **65:**884–89.

Craig, C. B., Jr. 1967. Mosquitoes: female monogamy induced by male accessory gland substance. *Science,* **150:**1499–1501.

Crans, W. J. 1969. An agar gel diffusion method for the identification of mosquite blood-meals. *Mosq. News,* **29:**563–66.

Crewe, W., and R. M. Gordon. 1959. The immediate reaction of the mammalian host to the bite of uninfected *Chrysops* and of *Chrysops* infected with human and monkey Loa. *Ann. Trop. Med. Parasitol.,* **53:**334–40.

Cross, H. F. 1964. Observations on the formation of the feeding tube by *Trombicula splendens* larvae. *Acarologia,* **6:**255–61.

Cross, H. F. 1971. Survey of an old Mississippi homesite for brown recluse spiders. *J. Miss. Acad. Sci.,* **17:**49–51.

Cross, J. H.; J. C. Lien; W. C. Huang; S. C. Lien; S. F. Chiu; J. Kuo; H. H. Chu; and Y. C. Chang. 1971. Japanese encephalitis virus surveillance in Taiwan. II. Isolations from mosquitoes and bats in Taipei area 1969–1970. *J. Formosan Med. Assoc.,* **70:**681–86.

Crosskey, R. 1980. [A catalogue of the Diptera of the Ethiopian Region.] In preparation.

Crosskey, R. W. 1969. A re-classification of the Simuliidae (Diptera) of Africa and its islands (including Madagascar). *Bull. Br. Mus. Nat. Hist. (Entomol.),* suppl. 14, 195 pp.

Cuellar, C. B. 1973. Spatial units of transmission and the theory of control of vector transmitted diseases at the primary level. In Ninth Int. Congr. Med. Malaria, Athens, pp. 300–301.

Cunningham, J. R.; S. E. Kunz; and M. A. Price. 1976. Effects of selected environmental factors on damaging populations of *Neoschongastia americana. J. Econ. Entomol.,* **69:**161–64.

Curtin, T. J. 1967. Status of *Aedes aegypti* in the Eastern Mediterranean. *J. Med. Entomol.,* **4:**48–50.

Cushing, E. C. 1957. *History of Entomology in World War II.* Smithsonian Institute Publ. 4294, vi+ 117 pp.

Čuturić, S., and E. Topolnik. 1975. (Bread beetle [*Stegobium paniceum* L.] as *Salmonella* vector in fodder and fodder mixture.) *Zentralbl. Bakteriol.* (Orig. A), **232:**545–48.

Dadd, R. H. 1970. Relationship between filtering activity and ingestion of solids by larvae of the mosquito *Culex pipiens:* a method for assessing phagostimulant factors. *J. Med. Entomol.,* **7:**708–12.

Dadd, R. H., and J. E. Kleinjan. 1974. Autophagostimulant from *Culex pipiens* larvae: distinction from other mosquito larval factors. *Env. Entomol.,* **3:**21–28.

Dalmat, H. T. 1955. *The Black Flies (Diptera,*

Simuliidae) of Guatemala and Their Role as Vectors of Onchocerciasis. Smithsonian Misc. Coll. 125, No. 1, 425 pp.

Daniel, M. 1969. (Methods of collecting Trombiculid mites [*Trombicula autumnalis*].) *Angew. Parasitol.,* **10:**224–28.

Darsie, R. F. 1973. A record of changes in mosquito taxonomy in the United States of America 1955–1972. *Mosq. Syst.,* **5:**187–93.

Das, D. N., and S. C. Misra. 1972. Studies on caprine demodectic mange with institution of effective therapeutic measures. *Indian Vet. J.,* **49:**96–101.

Das, Y. T., and A. P. Gupta. 1974. Effects of three juvenile hormone analogs on the female German cockroach, *Blattella germanica* (L.) (Dictyoptera: Blattellidae). *Experientia,* **30:** 109–15.

David, H. L., and J. F. B. Edeson. 1965. Filariasis in Portuguese Timor, with observations on a new microfilaria found in man. *Ann. Trop. Med. Parasitol.,* **59:**193–203.

Davidson, G., and R. H. Hunt. 1973. The crossing and chromosome characteristics of a new, sixth species in the *Anopheles gambiae* complex. *Parassitologia,* **15:**121–28.

Davidson, G., and C. E. Jackson. 1962. Incipient speciation in *Anopheles gambiae* Giles. *Bull. WHO,* **27:**303–305.

David-West, T. S. 1974. Propagation and plaquing of Dugbe virus (an ungrouped Nigerian arbovirus) in various mammalian and arthropod cell lines. *Arch. Gesamte Virusforsch.,* **44:**330–36.

Davies, D. M.; B. V. Peterson; and D. M. Wood. 1962. The black flies (Diptera; Simuliidae) of Ontario: Part I. Adult identification and distribution, with descriptions of six new species. *Proc. Entomol. Soc. Ontario,* **92:**70–154.

Davis, D. H. S. 1964. Ecology of wild rodent plague. In Davis, D. H. S. (ed.) *Ecological Studies in South Africa.* W. Junk, The Hague, pp. 301–14.

Davis, H. G.; G. W. Eddy; T. P. McGovern; and M. Beroza. 1969. Heptyl butyrate, a new synthetic attractant for yellow jackets. *J. Econ. Entomol.,* **62:**1245.

Davis, H. G.; R. J. Peterson; W. M. Rogoff; T. P. McGovern; and M. Beroza. 1972. Oxtyl butyrate, an effective attractant for the yellowjacket. *Env. Entomol.,* **1:**673–74.

Davis, H. G.; R. W. Zwick; W. M. Rogoff; T. P. McGovern; and M. Beroza. 1973. Perimeter traps baited with synthetic lures for suppression of yellowjackets in fruit orchards. *Environ. Entomol.,* **2:**569–71.

Davis, J. W., and R. C. Anderson. 1971. *Parasitic Diseases of Wild Mammals.* Iowa State University Press, Ames, 364 pp.

Davis, S. 1961. Soil, water and crop factors that indicate mosquito production. *Mosq. News,* **21:**44–47.

DeBach, P. 1974. *Biological Control by Natural Enemies.* Cambridge University Press, New York, 323 pp.

De Guisti, D. L. 1971. Acanthocephala. In Davis, J. W., and R. C. Anderson (eds.), *Parasitic Diseases of Wild Mammals.* Iowa State University Press, Ames, Iowa. x + 364 pp. (pp. 140–157).

Delfinado, M. D., and D. E. Hardy. 1973–1977. *A Catalog of the Diptera of the Oriental Region.* Volume 1, 1973, 539 pp; volume II, 1975, 459 pp; volume III, 1977, 854 pp. University of Hawaii Press, Honolulu.

Delgado, A. 1968. (Biocoenotic aspects of loxoscelism.) *Bol. Chil. Parasitol.,* **23:**68–74.

De Meillon, B.; D. H. S. Davis; and F. Hardy. 1961. *Plague in Southern Africa.* I. *The Siphonaptera (Excluding Ischnopsyllidae).* Pretoria, Government Printer, 280 pp.

Denlinger, D. L. 1975. Insect hormones as tsetse abortifacients. *Nature,* **253:**347–48.

Deoras, P. J., and R. S. Prasad. 1967. Feeding mechanism of Indian fleas. *X. cheopis* (Roths.) and *X. astia* (Roths.). *Indian J. Med. Res.,* **55:**1041–50.

Depner, K. R. 1961. The effect of temperature on development and diapause of the horn fly, *Siphona irritans* (L.)(Diptera: Muscidae). *Can. Entomol.,* **93:**855–59.

Depner, K. R. 1962. Continuous propagation of the horn fly, *Haematobia irritans* (L.)(Diptera: Muscidae). *Can. Entomol.,* **94:**893–95.

Depner, K. R. 1965. Ultraviolet irradiation of cattle in relation to diapause in the horn fly, *Haematobia irritans* (L.) Diptera: Muscidae. *Int. J. Biometereol.,* **9:**167–70.

Derwelis, S. K. 1967. Presence of *Demodex canis* in dog skin and feces. *Ill. Vet.,* **10:**10–11.

Desch, C., and W. B. Nutting. 1972. *Demodex folliculorum* (Simon) and *D. brevis* Akbulatova of man: redescription and reevaluation. *J. Parasitol.,* **58:**169–77.

Desowitz, R. S. 1976. How the wise men brought malaria to Africa. *Natural History,* **85**(8):36–44.

Desowitz, R. S. 1977. The fly that would be king. *Natural History,* **86**(2):76–83.

Detinova, T. S. 1962. *Age-Grouping Methods in Diptera of Medical Importance with Special Reference to Some Vectors of Malaria.* WHO Mon. Ser. No. 47, 216 pp.

Detinova, T. S. 1968. Age structure of insect populations of medical importance. *Annu. Rev. Entomol.,* **13:**427–50.

Dias, J. A. T. S., and A. J. Rosinha. 1973. (Is indiscriminate slaughter of game as a tsetse-control measure justified? [An analysis of the case of Mozambique].) *An. Serv. Vet.,* **1969/71:**25–53.

Dias, L. B., and M. C. de Azevedo. 1973. (Pararama, a disease caused by moth larvae; experimental findings.) *Bol. Of. Sanit. Panam.,* **75:**197–203.

Diaz Nájera, A., and L. Vargas. 1973. (Mexican mosquitos. Present geographical distribution.) *Rev. Invest. Salud Pública.,* **33:**111–25.

Dickerson, G., and M. M. J. Lavoipierre. 1959. Studies on the methods of feeding of blood-sucking arthropods: II. The method of feeding adopted by the bed-bug (*Cimex lectularius*) when obtaining a blood-meal from the mammalian host. *Ann. Trop. Med. Parasitol.,* **53:**347–57.

Dietz, K.; L. Molineaux; and A. Thomas. 1974. A malaria model tested in the African savannah. *Bull. WHO,* **50:**347–57.

Dinulescu, G. 1966. *Fauna Republicii Socialiste România. Insecta.* Vol. VI., fasc. 8. *Diptera, Fam. Simuliidae (Mustele columbace).* Editura Acad. Rep. Socialiste România, Bucuresti, 600 pp.

Diplock, P. T., and R. H. J. Hyne. 1975. Chorioptic mange in cattle associated with a severe fall in milk production. *New South Wales Vet. Proc.,* **11:**31–33.

Dixon, R. D., and R. A. Brust. 1971. Predation of mosquito larvae by the fathead minnow *Pimephales promelas* Raf. *Manitoba Entomol.,* **5:**68–70.

Dobson, R. C., and F. W. Kutz. 1970. Control of house flies (*Musca domestica* L.) in swine-finishing units (in Indiana) by improved methods of waste disposal. *J. Econ. Entomol.,* **63:**171–74.

Dodds, D. G.; A. M. Martell; and R. E. Yescott. 1969. Ecology of the American dog tick, *Dermacentor variabilis* (Say), in Nova Scotia. *Can. J. Zool.,* **47:**171–81.

Doherty, R. L.; J. G. Carley; M. R. Cremer; J. T. Rendle-Short; I. J. Hopkins; D. H. Herbert; A. J. Caro; and W. B. Stephens. 1972. Murray Valley encephalitis in eastern Australia, 1971. *Med. J. Aust.,* **2:**1170–73.

Doi, R. 1970. Studies on the mode of development of Japanese encephalitis virus in some groups of mosquitoes by the fluorescent antibody technique. *Jap. J. Exp. Med.,* **40:** 101–15.

Donnelly, J., and M. A. Peirce. 1975. Experiments on the transmission of *Babesia divergens* to cattle by the tick *Ixodes ricinus. Int. J. Parasitol.,* **5:**363–67.

Dorontsova, V. A., and P. I. Chudinov. 1971. (The role of artificial protective forest belts of the West Siberian railroad right of way in the natural development of foci of tick-borne encephalitis.) *Med. Parazitol. (Mosk.),* **40:** 283–86.

Doss, M. A.; M. M. Farr; K. F. Roach; and G. Anastos. 1974. *Ticks and Tickborne Diseases.* I. *Genera and Species of Ticks.* Part 1. Genera A–G. Index-Catalogue of Medical and Veterinary Zoology. Special Publication No. 3. I. Part 1, 429 pp.; part 2, 593 pp.; part 3, 329 pp.

Doube, B. M. 1975. Cattle and the paralysis tick *Ixodes holocyclus. Aust. Vet. J.,* **51:**511–15.

Dove, R. F., and A. B. McKague. 1975. Effects of insect developmental inhibitors on adult emergence of black flies (Diptera: Simuliidae). II. *Can. Entomol.,* **107:**1211–13.

Dow, R. P. 1959. A dispersal of adult *Hippelates pusio,* the eye gnat. *Ann. Entomol. Soc. Am.,* **52:**372–81.

Downe, A. E. R. 1975. Internal regulation of rate of digestion of blood meals in the mosquito, *Aedes aegypti. J. Ins. Phys.,* **21:**1835–39.

Downes, J. A. 1965. Adaptations of insects in the Arctic. *Annu. Rev. Entomol.,* **10:**257–74.

Dremova, V. P.; V. V. Markina; S. N. Smirnova; and T. V. Volkova. 1973. (Effectiveness of application of repellents in the tundra of the Krasnoyarsk territory.) *Med. Parazitol. (Mosk.),* **42:**21–27.

Dubos, R. J., and J. G. Hirsch. 1965. *Bacterial and Mycotic Infections of Man.* Lippincott, Philadelphia, 1,025 pp.

Dubrovsky, Y. A. 1975. Ecological causes of predominance of some animals as reservoirs of *Leishmania tropica major* in Turanian deserts. *Folia Parasitol., (Praha)* **22:**163–69.

Duggan, A. J. 1962. The occurrence of human trypanosomiasis among the Rukuba tribe of Northern Nigeria. *J. Trop. Med. Hyg.,* **65:**151–63.

Duke, B. O. L. 1971. The ecology of onchocerciasis in man and animals. In Fallis, A. M., *Ecology and Physiology of Parasites.* University of Toronto Press, 255 pp. (pp. 213–22).

Duke, B. O. L. 1972. Behavioral aspects of the life cycle of *Loa.* In Canning, E. U., and C. A. Wright (eds.), *Behavioural Aspects of Parasite*

Transmission. Academic Press, New York, pp. 97–108.

Dukes, J. C., and K. L. Hays. 1971. Seasonal distribution and parasitism of eggs of Tabanidae in Alabama. *J. Econ. Entomol.,* **64:** 886–89.

Dusbábek, F.; M. Daniel; and V. Černý. 1971. Stratification of engorged *Ixodes ricinus* larvae overwintering in soil. *Folia Parasitol., (Praha),* **18:**261–66.

D'Yaknov, L. P. 1972. (Development of *Babesia ovis* in ticks and in vertebrate hosts.) *Tr. Vsesoyuznogo Inst. Eksp. Vet.,* **38:**27–35.

Dyk, V., and R. Zavadil. 1971. Winter movement, food, and multiplying activity of the common tick. *Acta Vet. Czech.,* **40:**439–44.

Easton, E. R., and G. W. Krantz. 1973. A *Euschoengastia* species (Acari: Trombiculidae) of possible medical and veterinary importance in Oregon. *J. Med. Entomol.,* **10:**225–26.

Ebeling, W. 1971. Sorptive dusts for pest control. *Annu. Rev. Entomol.,* **16:**123–56.

Ebeling, W. 1975. *Urban Entomology.* University of California Division of Life Sciences, 695 pp.

Ebeling, W.; D. A. Reierson; R. J. Pence; and M. S. Viray. 1975. Silica aerogel and boric acid against cockroaches: external and internal action. *Pestic. Biochem. Phys.,* **5:**81–89.

Edeson, J. G. B., and T. Wilson. 1964. The epidemiology of filariasis due to *Wuchereria bancrofti* and *Brugia malayi. Annu. Rev. Entomol.,* **9:**245–68.

Edeson, J. G. B.; T. Wilson; R. H. Wharton; and A. B. G. Laing. 1960. Experimental transmission of *Brugia malayi* and *B. pahangi* to man. *Trans. R. Soc. Trop. Med. Hyg.,* **54:**229–34.

Edman, J. D., and W. L. Bidlingmayer. 1969. Flight capacity of blood-engorged mosquitoes. *Mosq. News,* **29:**386–92.

Edman, J. D., and H. W. Kale. 1971. Host behavior: its influence on the feeding success of mosquitoes. *Ann. Entomol. Soc. Am.,* **64:** 513–16.

Edwards, J. P. 1975. The effects of a juvenile hormone analogue on laboratory colonies of pharaoh's ant, *Monomorium pharaonis* (L.) (Hymenoptera, Formicidae). *Bull. Entomol. Res.,* **65:**75–80.

Edwards, J. S. 1961. The action and composition of the saliva of an assassin bug *Platymeris rhadamanthus* Gaerst. (Hemiptera, Reduviidae). *J. Exp. Biol.,* **38:**61–77.

Efrati, P. 1969. Bites by *Loxosceles* spiders in Israel. *Toxicon,* **6:**239–41.

Eichler, D. A. 1971. Studies on *Onchocerca gut-*

turosa (Neumann, 1910) and its development in *Simulium ornatum* (Meigen, 1818). II. Behaviour of *S. ornatum* in relation to the transmission of *O. gutturosa*. *J. Helminthol.*, **45:** 259–70.

Eichler, D. A. 1973. Studies on *Onchocerca gutturosa* (Neumann, 1910) and its development in *Simulium ornatum* (Meigen, 1818). 3. Factors affecting the development of the parasite in its vector. *J. Helminthol.*, **47:**73–88.

Eichler, D. A., and G. S. Nelson. 1971. Studies on *Onchocerca gutturosa* (Neumann, 1910) and its development in *Simulium ornatum* (Meigen, 1818). I. Observations in *O. gutturosa* in cattle in south-east England. *J. Helminthol.*, **45:**245–58.

Eldridge, B. F., 1968. The effect of temperature and photoperiod on blood-feeding and ovarian development in mosquitoes of the *Culex pipiens* complex. *Am. J. Trop. Med. Hyg.*, **17:**133–40.

Eldridge, B. F. 1974. The value of mosquito taxonomy to the study of mosquito-borne diseases and their control. *Mosq. Syst.*, **6:**125–29.

Elliott, R. 1969. Ecology and behaviour of malaria vectors in the American region. *Cah. ORSTOM, Sér. Entomol. Méd.*, **7:**29–33.

Emerson, K. C. 1956. Mallophaga (chewing lice) occurring on the domestic chicken. *J. Kans. Entomol. Soc.*, **29:**63–79.

Emerson, K. C. 1962a. Mallophaga (chewing lice) occurring on the turkey. *J. Kans. Entomol. Soc.*, **35:**196–201.

Emerson, K. C. 1962b. *A Tentative List of Mallophaga for North American Birds (North of Mexico)*. Dugway, Utah, Dugway Proving Grounds, 217 pp.

Emerson, K. C. 1962c. *A Tentative List of Mallophaga for North American Mammals (North of Mexico)*. Dugway, Utah, Dugway Proving Grounds, 20 pp.

Emerson, K. C. 1964. *Checklist of the Mallophaga of America (North of Mexico)*. Part I, Suborder Ischnocera; Part II, Suborder Amblycera. Dugway, Utah, Dugway Proving Grounds, pp. 104, 171.

Enileeva, N. K.; U. Y. Uzakov; and B. Ishmirzaev. 1974. (The development of control measures against stomach bots of horses in the Kashkadar'inskaya region of Uzbekistan.) *Tr. Uzbek. Nauch. Vet. Inst.*, **22:**28–33.

Ennik, F. 1972. A short review of scorpion biology, management of stings, and control. *Calif. Vector Views*, **19:**69–80.

Ennik, F. 1973. Abatement of yellowjackets using encapsulated formulations of diazinon and Rabon. *J. Econ. Entomol.*, **66:**1097–98.

Erasmus, B. J. 1975. The control of bluetongue in an enzootic situation. *Aust. Vet. J.*, **51:**209–10.

Ernst, S. E., and W. J. Gladney. 1975. *Dermacentor albipictus:* hybridization of the two forms of the winter tick. *Ann. Entomol. Soc. Am.*, **68:**63–67.

Eschle, J. L., and G. R. DeFoliart. 1965. Rearing and biology of *Wohlfahrtia vigil* (Diptera; Sarcophagidae). *Ann. Entomol. Soc. Am.*, **58:**849–55.

Eskey, C. R., and V. H. Haas. 1940. *Plague in the Western Part of the United States*. U.S. Public Health Bull. 254, 83 pp.

Evans, G. O.; J. G. Sheals; and D. Macfarlane. 1961. *The Terrestrial Acari of the British Isles. An Introduction to Their Morphology, Biology and Classification*. Vol. I. *Introduction and Biology*. British Museum (Natural History), London, 219 pp.

Evans, G. O., and W. M. Till. 1966. Studies on the British Dermanyssidae (Acari: Mesostigmata): Part II. Classification. *Bull. Br. Mus. (Nat. Hist.) Zool.*, **14:**109–370.

Everett, A. L.; I. H. Roberts; and J. Naghski. 1971. Reduction in leather value and yields of meat and wool from sheep infested with keds. *J. Am. Leather Chem. Assoc.*, **66:**118–30.

Everett, R. E.; M. A. Price; and S. E. Kunz. 1973. Biology of the chigger *Neoschöngastia americana* (Acarina: Trombiculidae). *Ann. Entomol. Soc. Am.*, **66:**429–35.

Ewing, H. E. 1929. *A Manual of External Parasites*. Thomas, Springfield, Ill., 225 pp.

Failing, R. M.; C. B. Lyon; and J. E. McKittrick. 1972. The pajaroello tick bite. The frightening folklore and the mild disease. *Calif. Med.*, **116:**16–19.

Fain, A. 1966. *A Review of the Family Epidermoptidae Trouessart Parasitic on the Skin of Birds (Acarina: Sarcoptiformes)*. Verhand. Vlaamse Acad. Wetenschappen, 27 (84), Part I, 176 pp; Part II, 144 pp.

Fain, A. 1967. Le genre *Dermatophagoides* Bogdanov 1864, son importance dans les allergies respiratoires et cutanées chez l'homme (Psoroptidae: Sarcoptiformes). *Acarologia*, **9:**179–225.

Fain, A. 1968. (Study of the variations of *S. Scabiei* with a revision of the Sarcoptidae.) *Acta Zool. Pathol. Antverp.*, No. 47, 196 pp.

Fain, A., and P. Elsen. 1967. Mites of the family Knemidokoptidae that cause mange in birds [a revision with notes].) *Acta Zool. Pathol. Antverp.*, No. 45, 142 pp.

Fain, A.; G. J. R. Hovell; and K. H. Hyatt. 1972. A new sarcoptid mite producing mange in albino guinea-pigs. *Acta Zool. Pathol. Antverp.,* No. 56, 73–82.

Falcon, L. A. 1976. Problems associated with the use of arthropod viruses in pest control. *Annu. Rev. Entomol.,* **21:**305–24.

Fallis, A. M. (ed.). 1971. *Ecology and Physiology of Parasites.* University of Toronto Press, Toronto, 258 pp.

Fallis, A. M.; S. S. Desser; and R. A. Khan. 1974. On species of *Leucocytozoon.* In Dawes, B. (ed.). *Adv. Parasitol.,* **12:**1–67.

FAO. 1974. Recommended methods for the detection and measurement of resistance of agricultural pests to pesticides. Tentative method for larvae and adult of sheep blowflies, *Lucilia* sp.—FAO Method No. 14. *Plant Pro. Bull., FAO,* **22:**122–26.

Farrell, C. E. 1956. Chiggers of the genus *Euschongastia* (Acarina: Trombiculidae) in North America. *Proc. U.S. Nat. Mus.,* **106:**85–235.

Faucheux, M. J. 1975. (Wounding and sensory organs of the proboscis of various tabanids: sexual dimorphism.) *Ann. Soc. Entomol. Fr.,* **11:**709–22.

Feingold, B. F. 1973. *Introduction to Clinical Allergy.* Thomas, Springfield, Ill., 380 pp.

Feingold, B. F.; E. Benjamini; and D. Michaeli. 1968. The allergic responses to insect bites. *Annu. Rev. Entomol.,* **13:**137–58.

Feldman-Muhsam, B., and Y. Havivi. 1973. Autogeny in the tick *Ornithodoros tholozani* (Ixodoidea, Argasidae). *J. Med. Entomol.,* **10:**185–89.

Fenner, F. 1976. Classification and nomenclature of viruses. *Intervirology,* **7:**4–105.

Fenner, F., and F. N. Ratcliffe. 1965. *Myxomatosis.* Cambridge University Press, Cambridge, 379 pp.

Fernandez, H.; P. Chavez; and R. Polanco. 1971. (Report on an outbreak of eastern equine encephalitis. Control measures.) *Rev. Cubana Cien. Vet.,* **2:**83–85.

Ferrar, P. 1973. The CSIRO dung beetle project. *Wool Tech. Sheep Breeding,* **20:**73–75.

Ferris, G. F. 1951. *The Sucking Lice.* Pacific Coast Entomol. Soc. Mem. 1, 320 pp.

Filippova, N. A. 1971. (On species of the group of *Ixodes persulcatus* [Parasitiformes, Ixodidae]. VI. Peculiarities of the distribution of *I. pavlovskyi* Pom. and *I. persulcatus* Schulze in relation to their palaeogenesis.) *Parazitologiia,* **5:**385–91.

Fincher, G. T. 1973. Dung beetles as biological control agents for gastrointestinal parasites of livestock. *J. Parasitol.,* **59:**396–99.

Fine, P. E. M. 1975. Vectors and vertical trans-mission: an epidemiologic perspective. *Ann. N.Y. Acad. Sci.,* **266:**173–94.

Finke, J. H.; B. J. Campbell; and J. T. Barrett. 1974. Serodiagnostic test for *Loxosceles reclusa* bites. *Clin. Toxicol.,* **7:**375–82.

Finkelstein, A.; L. L. Rubin; and M. Tseng. 1976. Black widow spider venom: effect of purified toxin on lipid bilayer membranes. *Science,* **193:**1009–11.

Fleisher, T. L., and I. Fox. 1970. Oedemerid beetle dermatitis. *Arch. Dermatol.,* **101:**601–605.

Flynn, R. J. 1973. *Parasites of Laboratory Animals.* Iowa State University Press, Ames, 884 pp.

Foggie, A. 1951. Studies on the infectious agent of tick-borne fever in sheep. *J. Pathol. Bact.,* **63:**1–15.

Foot, M. A. 1970. An agar-gel immunodiffusion method for blood meal identification in *Aedes (Finlaya) notoscriptus* Skuse (Diptera: Culicidae). *N.Z. Entomol.,* **4:**95–98.

Forattini, O. P. 1971. (On the classification of the subfamily Phlebotominae in the Americas [Diptera: Psychodidae].) *Pap. Avul. Zool. S. Paulo,* **24:**93–111.

Forattini, O. P. 1973. *Entomologia Medica. IV. Psychodidae. Phlebetominae Leishmanioses. Bartonelose.* Edgard Blucher, São Paulo, 658 pp.

Ford, J. 1971. *The Role of the Trypanosomiases in African Ecology. A Study of the Tsetse Fly Problem.* Clarendon, Oxford, 568 pp.

Ford, J. 1973. The consequences of control of African trypanosomiasis. In *Computer Models and Application of the Sterile-Male Technique.* International Atomic Energy Agency, Vienna, pp. 141–44.

Ford, J.; J. P. Glasgow; D. L. Johns; and J. R. Welsh. 1959. Transect flyrounds in field studies of *Glossina. Bull. Entomol. Res.,* **50:**275–85.

Forrester, D. J.; R. F. Jackson; J. M. Miller; and B. C. Townsend. 1973. Heartworms in captive California sea lions. *J. Am. Vet. Med. Assoc.,* **163:**568–70.

Foster, G. G.; K. L. Kitching; W. G. Vogt; and M. J. Whitten. 1975. Sheep blowfly and its control in the pastoral ecosystem in Australia. In J. Kikkawa and H. A. Nix (eds.). Managing terrestrial ecosystems. *Symposium, Brisbane Proc. Ecol. Soc. Aust.,* **9:**213–29.

Fox, C. 1925. *Insects and Diseases of Man.* Blakiston, Philadelphia, 349 pp.

Fox, I., and I. Garcia-Moll. 1961. Ants attacking fleas in Puerto Rico. *J. Econ. Entomol.,* **54:**1065–66.

Fraiha, H.; J. J. Shaw; and R. Lainson. 1971. Phlebotominae Brasileiros—II. *Psychodopygus wellcomei,* nova espécies anthropófila de flebótomo do grupo *squamiventris,* do sul do estado do Para,́ Brasil (Diptera, Psychodidae). *Mem. Inst. Oswaldo Cruz,* **69:**489–500.

Francis, E., and B. Mayne. 1921. Experimental transmission of tularemia by flies of the species *Chrysops discalis. Public Health Rep.* **36:** 1738–46.

Fraser, D., and M. S. Waddell. 1974. The importance of social and self-grooming for the control of ectoparasitic mites on normal and dystrophic laboratory mice. *Lab. Pract.,* **23:** 58–59.

Frazerhurst, L. F.; L. G. Wyborn; and J. H. Elliott. 1973. Air curtains for fly control. *N.Z. J. Sci.,* **16:**763–67.

Frazier, C. A. 1969. *Insect Allergy: Allergic and Toxic Reactions to Insects and Other Arthropods.* W. H. Green, St. Louis, 493 pp.

Frazier, C. A. 1974. Biting insect survey: a statistical report. *Ann. Allergy,* **32:**200–204.

Fredeen, F. J. H. 1969. Outbreaks of the black fly *Simulium arcticum* Malloch in Alberta. *Quaest. Entomol.,* **5:**341–72.

Fredeen, F. J. H. 1970. A constant-rate liquid dispenser for use in blackfly larviciding. *Mosq. News,* **30:**402–405.

Fredeen, F. J. H. 1975. Effects of a single injection of methoxychlor blackfly larvicide on insect larvae in a 161-km (100-mile) section of the North Saskatchewan River. *Can. Entomol.,* **107:**807–17.

Fredeen, F. J. H., and J. A. Shemanchuk. 1960. Black flies (Diptera: Simuliidae) of irrigation systems in Saskatchewan and Alberta. *Can. J. Zool.,* **38:**723–35.

Freeman, J. C. 1973. The penetration of the peritrophic membrane of the tsetse flies by trypanosomes. *Acta Trop.,* **30:**347–55.

Freeman, J. V., and E. J. Hansens. 1972. Collecting larvae of the salt marsh greenhead *Tabanus nigrovittatus* and related species in New Jersey: comparison of methods. *Environ. Entomol.,* **1:**653–58.

Freyvogel, T. A. 1972. Poisonous and venomous animals in East Africa. *Acta Trop.,* **29:**401–51.

Friend, W. G., and J. J. B. Smith. 1977. Factors affecting feeding by bloodsucking insects. *Annu. Rev. Entomol.,* **22:**309–32.

Fritz, R. F. 1972. *References and Contents Index to Some of the Literature Concerning Malaria Vectors in Africa.* WHO/VBC/72.402, 64 pp.

Frogatt, W. W. 1922. *Sheep-Maggot Flies.* New South Wales Department of Agriculture, *Farmers Bull.* 144, 32 pp.

Frommer, R. L.; R. R. Carestia; and R. W. Vavra. 1975. Field evaluation of deet-treated mesh jacket against black flies (Simuliidae). *J. Med. Entomol.,* **12:**558–61.

Fukuda, T., and D. B. Woodard. 1974. Hybridization of *Aedes taeniorhynchus* (Wiedmann) with *Aedes nigromaculis* (Ludlow) or *Aedes sollicitans* (Walker) by induced copulation. *Mosq. News,* **34:**431–35.

Fukumi, H.; K. Hayashi; K. Mifune; A. Shichijo; S. Matsuo; N. Omori; Y. Wada; T. Oda; M. Mogi; and A. Mori. 1975. Ecology of Japanese encephalitis virus in Japan. I. Mosquito and pig infection with the virus in relation to human incidences. *Trop. Med.,* **17:**97–110.

Fuller, H. S. 1956. Veterinary and medical acarology. *Annu. Rev. Entomol.,* **1:**347–66.

Furman, D. P.; H. Bonasch; R. Springsteen; D. Stiller; and D. F. Rahlmann. 1974. Studies on the biology of the lung mite, *Pneumonyssus simicola* Banks (Acarina: Halarachnidae) and diagnosis of infestation in macaques. *Lab. Anim. Sci.,* **24:**622–29.

Furman, D. P., and F. J. Radovsky. 1963. A new species of *Ornithonyssus* from the white-tailed antelope squirrel, with a rediagnosis of the genus *Ornithonyssus* (Acarina: Dermanyssidae). *Pan-Pac. Entomol.,* **39:**75–79.

Furman, D. P., and V. S. Stratton. 1963. Control of northern fowl mites, *Ornithonyssus sylviarum,* with sulfaquinoxaline. *J. Econ. Entomol.,* **56:**904–905.

Furman, D. P.; R. D. Young; and E. P. Catts. 1969. *Hermetia illucens* as a factor in the natural control of *Musca domestica* Linnaeus. *J. Econ. Entomol.,* **52:**917–21.

Fye, R. L., and G. C. LaBrecque. 1976. *Bibliography of Arthropod Chemosterilants.* USDA, ARS, ARS-S-93, 54 pp.

Gaĭdamovich, S. Y.; N. V. Khutoretskaya; A. I. L'vova; and N. A. Sveshnikova. 1974. (Detection of group A arboviruses in the salivary glands of mosquitoes by the method of immunofluorescence.) *Vopr. Virusol.,* No. 1, 13–18.

Galley, R. A. E. 1973. The cost of vector-borne diseases and of their control. *Span,* **16:**56–58.

Gamboa, C. J., 1974. (Ecology of American trypanosomiasis [Chagas' disease] in Venezuela.) *Bol. Dirección Malariol. Saneamiento Ambiental,* **14:**3–20.

Garcia, R., and G. Ponting. 1972. Studies on the ecology of the treehole mosquito *Aedes sierrensis.* Ludlow. Calif. Mosq. Control Assoc. Proceedings and papers, 40th Annual Conference, pp. 63–65.

Gard, G.; I. D. Marshall; and G. M. Woodroofe.

1973. Annually recurrent epidemic polyarthritis and Ross River virus activity in a coastal area of New South Wales. II. Mosquitoes, viruses and wildlife. *Am. J. Trop. Med. Hyg.,* **22:**551–60.

Gardner, D. R., and J. R. Bailey. 1975. *Methoxychlor: Its Effects on Environmental Quality.* National Research Council of Canada, Ottawa, Pub. No. NRCC 14102, 164 pp.

Garnham, P. C. C. 1963. Distribution of simian malaria parasites in various hosts. *J. Parasitol.,* **49:**905–11.

Garnham, P. C. C. 1966. *Malaria Parasites and Other Haemosporidia.* Blackwell, Philadelphia, 1,132 pp.

Garnham, P. C. C. 1971. *Progress in Parasitology.* Athlone, London, 224 pp.

Garret-Jones, C., and B. Grab. 1964. The assessment of insecticidal impact on the malaria mosquito's vectorial capacity, from data on the proportion of parous females. *Bull. WHO,* **31:**71–86.

Gatehouse, A. G. 1970. Interactions between stimuli in the induction of probing by *Stomoxys calcitrans. J. Ins. Phys.,* 16:991–1000.

Gauld, L. W.; T. M. Yuill; R. P. Hanson; and S. K. Sinha. 1975. Isolation of La Crosse virus (California encephalitis group) from the chipmunk (*Tamias striatus*), an amplifier host. *Am. J. Trop. Med. Hyg.,* **24:**999–1005.

Gebauer, O. 1958. *Die Dasselfliegen des Rindes und ihre Bekampfung.* Parasitol. Schriftenreihe. Heft. 9., 97 pp.

Gebhardt, L. P.; S. C. St. Jeor; G. J. Stanton; and D. A. Stringfellow. 1973. Ecology of western encephalitis virus. *Proc. Soc. Exp. Biol. Med.,* **142:**731–33.

Gee, R. W.; M. H. Bainbridge; and J. Y. Haslam. 1971. The effect of cattle tick (*Boophilus microplus*) on beef production in the Northern Territory. *Aust. Vet. J.,* **47:**257–63.

Gelman, A. C. 1961. The ecology of tularemia. In May, J. M. (ed.;, *Studies in Disease Ecology.* Hafner, New York, 89 pp.

Gerasimova, N. G. 1971. (The maturation rates of adult females of *Xenopsylla skrjabini* and *X. nuttalli.*) *Parazitologiia,* **5:**137–39.

Gerberg, E. J. 1970. *Manual for Mosquito Rearing and Experimental Techniques.* Bull. Am. Mosq. Control Assoc. No. 5, 109 pp.

Gerberich, J. B., and M. Laird. 1968. Bibliography of papers related to the control of mosquitoes by the use of fish. An annotated bibliography for the years 1901–1966. *F.A.O. Fish. Tech. Pap. No. 75,* 70 pp.

Geren, C. R.; T. K. Chan; D. E. Howell; and G. V. Odell. 1975. Partial characterization of the low molecular weight fractions of the extract of the venom apparatus of the brown recluse spider and of its hemolymph. *Toxicon,* 13:233–38.

Geren, C. R.; T. K. Chan; B. C. Ward; D. E. Howell; K. Pinkston; and G. V. Odell. 1973. Composition and properties of extract of fiddleback (*Loxosceles reclusa*) spider venom apparatus. *Toxicon,* **11:**471–79.

Gertsch, W. J. 1949. *American Spiders.* Van Nostrand, New York, 285 pp.

Gertsch, W. J. 1958. The spider genus *Loxosceles* in North America, Central America, and the West Indies. *Am. Mus. Novitates,* **1907:**1–46.

Gertsch, W. J. 1967. The spider genus *Loxosceles* in South America (Araneae, Scytodidae). *Am. Mus. Nat. Hist. Bull.,* **136:**117–74.

Gesztessy, T., and L. Nemeseri. 1970. Veterinary hygienic aspects of mite-infested feed. II. Experiments on albino mice, swine and cattle. *Acta Physiol. Acad. Sci. Hung.,* **20:**401–403.

Ghalsasi, G. R., and V. Dhanda. 1975. Taxonomy and biology of *Haemaphysalis (Kaiseriana) spinigera* (Acarina: Ixodidae). *Oriental Insects,* **8:**505–20.

Gibbs, A. J. (ed.). 1973. *Viruses and Invertebrates.* American Elsevier, New York, 673 pp.

Gillett, J. D. 1971. *Mosquitos.* Weidenfeld & Nicolson, London, 274 pp.

Gillett, J. D. 1972. *The Mosquito: its Life, Activities, and Impact on Human Affairs.* Doubleday, Garden City. N.Y., 358 pp.

Gillies, M. T. 1974. Methods for assessing the density and survival of blood-sucking Diptera. *Annu. Rev. Entomol.,* **19:**345–62.

Gjullin, C. M., and G. W. Eddy. 1972. *The Mosquitoes of the Northwestern United States.* Tech. Bull. Ag. Res. Serv., USDA, No. 1447, 111 pp.

Gjullin, C. M.; T. L. Whitfield; and J. F. Buckley. 1967. Male pheromones of *Culex quinquefasciatus, C. tarsalis* and *C. pipiens* that attract females of these species. *Mosq. News,* **27:** 382–87.

Gjullin, C. M.; W. W. Yates; and H. H. Stage. 1950. Studies on *Aedes vexans* (Meig.) and *Aedes sticticus* (Meig.), floodwater mosquitoes, in the Lower Columbia River Valley. *Ann. Entomol. Soc. Am.,* **43:**262–75.

Gladney, W. J., and C. C. Dawkins. 1972. Insecticidal tests against the brown recluse spider. *J. Econ. Entomol.,* **65:**1491–93.

Gladney, W. J.; S. E. Ernst; and R. O. Drummond. 1974. Chlordimeform: a detachment-stimulating chemical for three-host ticks. *J. Med. Entomol.,* **11:**569–72.

Gladney, W. J.; R. R. Grabbe; S. E. Ernst; and D. D. Oehler. 1974. The Gulf Coast tick: evidence of a pheromone produced by males. *J. Med. Entomol.*, **11**:303–306.

Gladney, W. J.; O. H. Graham; J. L. Trevino; and S. E. Ernst. 1973. *Boophilus annulatus:* effect of host nutrition on development of female ticks. *J. Med. Entomol.*, **10**:123–30.

Glasgow, J. P. 1967. Recent fundamental work on tsetse flies. *Annu. Rev. Entomol.*, **12**: 421–38.

Glover, P. E.; J. G. LeRoux; and D. F. Parker. 1960. The extermination of *Glossina palpalis* on the Kuja-Migori river systems with the use of insecticides. In *International Scientific Committee for Trypanosomiasis Research.* Seventh Meeting, Brussels, 1958. Publ. Comm. Tech. Co-op. Afr. S. Sahara No. 41, pp. 331–42.

Glukhov, V. F. 1970. (On the transmission of the causal agent of pullorosis-typhus of birds by *Argas persicus.*) *Veterinariia*, **47**:60–61.

Goldhaft, T. M. 1969. The use of a combined sulfonamide feed additive for the control of a fowl feather mite (*Ornithonyssus sylviarum*). *Refuah Vet.*, **26**:167–70.

Gómez-Núñez, J. C. 1971. *Tapinoma melanocephalum* as an inhibitor of *Rhodnius prolixus* populations. *J. Med. Entomol.*, **8**:735–37.

Gooding, R. H. 1972. Digestive processes of haematophagous insects. I. A literature review. *Quaest. Entomol.*, **8**:5–60.

Goodman, L. S., and A. Gilman (eds.). 1975. *The Pharmacological Basis of Therapeutics*, 5th ed. Macmillan Publishing Co., Inc., New York, 1,704 pp.

Gordon, R., and J. M. Webster. 1974. Biological control of insects by nematodes. *Helminth. Abst., A*, **43**:327–439.

Gordon, R. M., and W. H. R. Lumsden. 1939. A study of the behaviour of the mouth-parts of mosquitoes when taking up blood from living tissues; together with some observations on the ingestion of microfilariae. *Ann. Trop. Med. Parasitol.*, **33**:259–78.

Gorham, J. R. 1968. The geographic distribution of the brown recluse spider *Loxosceles reclusa* (Araneae, Scytodidae) and related species in the United States. *Coop. Econ. Insect. Rpt. USDA*, **18**:171–75.

Gothe, R., and E. Koop. 1974. (On the biological evaluation of the validity of *Argas [Persicargas] persicus* [Oken, 1818], *Argas [Persicargas] arboreus* Kaiser, Hoogstraal and Kohls, 1964 and *Argas [Persicargas] walkerae* Kaiser and Hoogstraal, 1969. II. Cross-breeding experiments.) *Z. Parasitenkd.*, **44**:319–28.

Gothe, R., and W. Schrecke. 1972. (On the epizootiological significance of *Persicargas* ticks of hens in the Transvaal.) *Berl. Munch. Tieraerztl. Wochenschr.*, **85**:9–11.

Gothe, R., and K. H. Verhalen. 1975. (The capacity of various species and populations of *Persicargas* to induce paralysis in fowls.) *Zentralbl. Veterinaermed. (B)*, **22**:98–112.

Gouck, H. K.; D. R. Godwin; K. Posey; C. E. Schreck; and D. E. Weidhaas. 1971. Resistance to aging and rain of repellent-treated netting used against saltmarsh mosquitoes in the field. *Mosq. News*, **31**:95–99.

Gould, D. J. 1956. The larval trombiculid mites of California. *U. Calif. Pub. Entomol.*, **11**: 1–116.

Gould, G. E., and H. O. Deay. 1940. *The Biology of Six Species of Cockroaches which Inhabit Buildings*. Purdue U. Agr. Exp. Sta. Bull. 451, 31 pp.

Graber, M., and J. Gruvel. 1970. (Oribatid vectors of *M. expansa* of sheep in the Fort-Lamy region.) *Revue Méd. Vét. Phys. Trop.*, **22**:521–27.

Graham, H. 1902. Dengue: a study of its mode of propagation and pathology. *Med. Record*, **61**:204–207.

Graham, O. H.; W. J. Gladney; and J. L. Trevino. 1972. Some non-bovine host relationships of *Boophilis annulatus*. *Folia Entomologica Mexicana*, No. 23/24, pp. 89–90.

Granett, P., and E. J. Hansens. 1957. Further observations on the effect of biting fly control on milk production of cattle. *J. Econ. Entomol.*, **50**:332–36.

Greenberg, B. 1971. *Flies and Disease*. Vol. 1. *Ecology, Classification, and Biotic Associations*. Princeton University Press, Princeton, N.J., 856 pp.

Greenberg, B. 1973. *Flies and Disease*. Vol. II. *Biology and Disease Transmission*. Princeton University Press, Princeton, N.J., 447 pp.

Gregson, J. D. 1949. Note on the longevity of certain ticks (Ixodidae). *Proc. Entomol. Soc. Br. Col.* **45**:14.

Gregson, J. D. 1960. Morphology and functioning of the mouthparts of *Dermacentor andersoni* Stiles. *Acta Trop.*, **17**:48–79.

Gregson, J. D. 1967. Observations on the movement of fluids in the vicinity of the mouthparts of naturally feeding *Dermacentor andersoni* Stiles. *Parasitology*, **57**:1–8.

Gregson, J. D. 1973. *Tick paralysis. An Appraisal of Natural and Experimental Data.* Can. Dept. Agr. Monograph No. 9, 109 pp.

Greig, A. 1972. The localization of African swine

fever virus in the tick *Ornithodoros moubata porcinus*. *Arch. Gesamte Virusforsch.*, **39**: 240–47.

Griffiths, H. J. 1971. Some common parasites of small laboratory animals. *Lab. Anim.*, **5**:123–35.

Grinnell, M. E., and I. L. Hawes. 1943, *Bibliography on Lice and Man, With Particular Reference to Wartime Conditions*. USDA Washington, D.C., Bibliographical Bull. No. 1.

Grodhaus, G. 1963. Chironomid midges as a nuisance. *Calif. Vector Views*, **10**:19–24, 27–37.

Grodhaus, G. 1975. Bibliography of chironomid midge nuisance and control. *Calif. Vector Views*, **22**:71–81.

Grono, L. R. 1969. Studies of the ear mite, *Otodectes cynotis*. *Vet. Rec.*, **85**:6–8.

Grosshans, E. M.; M. Kremer; and J. Maleville. 1974. (*Demodex folliculorum* and the histogenesis of granulomatous acne rosacea.) *Hautarzt*, **25**:166–77.

Grothaus, R. H.; H. G. Davis; W. M. Rogoff; J. A. Fluno; and J. M. Hirst. 1973. Baits and attractants for East Coast yellowjackets, *Vespula* spp. *Environ. Entomol.*, **2**:717–18.

Grothaus, R. H.; J. R. Haskins; and J. T. Reed. 1976. A simplified carbon dioxide collection technique for thje recovery of live ticks (Acarina). *J. Med. Entomol.*, **12**:702.

Grothaus, R. H.; J. R. Haskins; C. E. Schreck; and J. K. Gouck. 1976. Insect repellent jacket: status, value and potential. *Mosq. News*, **36**:11–18.

Grzywacz, M., and R. Kuzmicki. 1975. (A case of *Argas reflexus* [Fabricius 1974] infestation in man.) *Wiad. Lek.*, **28**:1571–77.

Gubler, D. J., and N. C. Bhattacharya. 1974. A quantitative approach to the study of bancroftian filariasis. *Am. J. Trop. Med. Hyg.*, **23**:1027–36.

Guenther, H.; E. Ramstad; and H. G. Floss. 1969. On the biosynthesis of cantharadin. *J. Pharm. Sci.*, **58**:1274.

Gugushvili, G. K. 1973. (The length of time for which *Borrelia* persists in *Ornithodoros verrucosus* Olen., Zas. & Fen. and *O. alactagalis* Isaakyan.) *Bull. Acad. Sci. Georgian SSR*, **69**:185–88.

Guilbride, P. D. L.; L. Barber; and A. M. G. Kalikwant. 1959. Bovine infectious keratitis. Suspected moth-borne outbreak in Uganda. *Bull. Epizoot. Dis. Afr.*, **7**:149–54.

Guilhon, J.; A. Marchand; and G. Jolivet. 1973. (Two new species of skin mites in France, responsible for a scurfy pruriginous dermatitis

of domestic carnivores.) *Bull. Acad. Vét. France*, **46**:399–407.

Guimarães, J. H., and N. Papavero. 1966. A tentative annotated bibliography of *Dermatobia hominis* (Linnaeus Jr., 1781) (Diptera, Cuterebridae). *Arq. Zool.*, **14**:223–94.

Guthrie, D. M., and A. R. Tindall. 1968. *The Biology of the Cockroach*. Edward Arnold, London, 405 pp.

Gwadz, R. W. 1969. Regulation of blood meal size in the mosquito. *J. Ins. Phys.*, **15**: 2039–44.

Gwadz, R. W. 1972. Neuro-hormonal regulation of sexual receptivity in female *Aedes aegypti*. *J. Ins. Phys.*, **18**:259–66.

Gwadz, R. W., and A. Spielman. 1974. Development of the filarial nematode, *Brugia pahangi* in *Aedes aegypti* mosquitoes: nondependence upon host hormones. *J. Parasitol.*, **60**:134–37.

Habermann, E. 1972. Bee and wasp venoms. *Science*, **177**:314–22.

Hacker, C. S., and W. L. Kilama. 1974. The relationship between *Plasmodium gallinaceum* density and the fecundity of *Aedes aegypti*. *J. Invertebr. Pathol.*, **23**:101–105.

Hackett, L. W., and A. Missiroli. 1935. The varieties of *Anopheles maculipennis* and their relation to the distribution of malaria in Europe. *Riv. Malar.*, **14**:45–109.

Hackman, R. H. 1975. Expanding abdominal cuticle in the bug *Rhodnius* and the tick *Boophilus*. *J. Ins. Phys.*, **21**:1613–23.

Hadani, A., and Y. Dinur. 1968. *Studies on the Transmission* of *Aegyptianella pullorum by the Tick Argas persicus*. *J. Protozool.*, **15**(Suppl.), 45 pp.

Hadani, A.; K. Rauchbach; Y. Weissman; and R. Bock. 1975. The occurrence of the tropical fowl mite—*Ornithonyssus bursa* (Berlese, 1888), Dermanyssidae, on turkeys in Israel. *Refuah Vet.*, **32**:111–13.

Hafez, M., and F. M. Gamal-Eddin. 1968a. On the bionomics of *Musca crassirostris* Stein, in Egypt (Diptera: Muscidae). *Bull. Soc. Entomol. Egypt*, **50**:25–40.

Hafez, M., and F. M. Gamal-Eddin. 1968b. The feeding habits of the horn-fly *Siphona irritans* L., with special reference to its biting cycle in nature in Egypt (Diptera: Muscidae). *Bull. Soc. Entomol. Egypt*, **50**:41–52.

Hagedorn, H. H. 1974. The control of vitellogenesis in the mosquito, *Aedes aegypti*. *Am. Zool.*, **14**:1207–17.

Hagen, D. H., and E. C. Wilmshurst. 1975. *The Control of Bovine Trypanosomiasis*. Agrochemicals Division, Boots Company, Nottingham, UK, 34 pp.

Haggard, H. W. 1929. *Devils, Drugs, and Doctors*. Harper and Brothers (reprint, paperback Pocket Books), 427 pp.

Hair, J. A.; J. R. Sauer; and K. A. Durham. 1975. Water balance and humidity preference in three species of ticks. *J. Med. Entomol.*, **12**:37–47.

Hall, D. G. 1948. *The Blowflies of North America*. Thomas Say Foundation, Entomological Society of America, 477 pp.

Hall, M. D., and J. T. Muir. 1913. A critical study of a case of myiasis due to *Eristalis*. *Arch. Intern. Med.*, **2**:193–203.

Hall, R. D., and W. B. Gross. 1975. Effect of social stress and inherited plasma corticosterone levels in chickens on populations of the northern fowl mite, *Ornithonyssus sylviarum*. *J. Parasitol.*, **61**:1096–1100.

Hall, R. E., and M. B. Madon. 1973. Envenomation by the green lynx spider *Peucetia viridans* (Hentz 1832), in Orange County, California. *Toxicon*, **11**:197–99.

Halstead, S. B. 1966. Mosquito-borne haemorrhagic fevers of south and southeast Asia. *Bull. WHO*, **35**:3–15.

Hammon, W. McD.; W. C. Reeves; B. Brookman; E. M. Izumi; and C. M. Gjullin. 1941. Isolation of the viruses of western equine and St. Louis encephalitis from *Culex tarsalis* mosquitoes. *Science*, **94**:328–30.

Hansens, E. J. 1956. Granulated insecticides against greenhead (*Tabanus*) larvae in the salt marsh. *J. Econ. Entomol.*, **49**:401–403.

Harley, K. L. S., and R. P. Wilkinson. 1971. A modification of pasture spelling to reduce acaricide treatments for cattle tick control. *Aust. Vet. J.*, **47**:108–1.

Harris, P.; D. F. Riordan; and D. Cooke. 1969. Mosquitoes feeding on insect larvae. *Science*, **164**:184–85.

Harrison, B. A. 1972. A new interpretation of affinities within the *Anopheles hyrcanus* complex of Southeast Asia. *Mosq. Syst.*, **4**:73–83.

Harrison, I. R.; B. H. Palmer; and E. C. Wilmshurst. 1973. Chemical control of cattle ticks—resistance problems. *Pesticide Sci.*, **4**:531–42.

Harrison, R. J.; E. Rossier; and B. Lemieux. 1975a. The first Canadian case of eastern equine encephalitis (EEE) observed in the Eastern Townships, Quebec. *Ann. Entomol. Soc. Québec*, **20**:27–32.

Harrison, R. J.; E. Rossier; and B. Lemieux.

1975b. (First case of Powassan encephalitis in Quebec.) *Ann. Entomol. Soc. Québec*, **20**:48–52.

Harwood, R. F. 1952. *The Function of the Pharynx in the Retention of Particulate Material by the Larvae of Culicine Mosquitoes*. Master's thesis, University of Illinois, 27 pp.

Harwood, R. F., and E. Halfhill. 1964. The effect of photoperiod on fat body and ovarian development of *Culex tarsalis* (Diptera: Culicidae). *Ann. Entomol. Soc. Am.*, **57**:596–600.

Hasslinger, M. A., and D. Jonas. 1975. Control of *Gasterophilus intestinalis* (De Geer, 1776) with dichlorvos. *Br. Vet. J.*, **131**:89–93.

Hatch, C. L.; J. W. Williams; P. A. Jarinko; and B. M. Tice. 1973. Guidelines for measuring proficiency as an aid in mosquito abatement program assessment. *Mosq. News*, **33**:228–33.

Hatziolos, B. C. 1967. *Cuterebra* larva causing paralysis in a dog. *Cornell Vet.*, **57**:129–45.

Haufe, W. O. 1957. Physical environment and behavior of immature stages of *Aedes communis* (Deg.) (Diptera: Culicidae) in subarctic Canada. *Can. Entomol.*, **89**:120–39.

Haufe, W. O., and L. Burgess. 1956. Development of *Aedes* at Fort Churchill, Manitoba, and prediction of dates of emergence. *Ecology*, **37**:500–519.

Haufe, W. O., and L. Burgess. 1960. Design and efficiency of mosquito traps based on visual response to patterns. *Can Entomol.*, **92**:124–40.

Hawking, F. 1974. Filariasis in India. *Prog. Drug Res.*, **18**:173–90.

Hawking, F., and M. Worms. 1961. Transmission of filarioid nematodes. *Annu. Rev. Entomol.*, **6**:413–32.

Hawking, F.; M. J. Worms; and K. Gammage. 1968. 23- and 48-hour cycles of malaria parasites in the blood; their purpose, production and control. *Trans. R. Soc. Trop. Med. Hyg.*, **62**:731–60.

Hayashi, A., and M. Hatsukade. 1970. (The toxic effects of various insecticides on adults of the moth fly.) *Botyu-Kagaku*, **35**:38–43.

Hayes, G. R.; P. P. Scheppf; and E. B. Johnson. 1971. An historical review of the last continental U.S. epidemic of dengue. *Mosq. News*, **31**:422–27.

Hayes, R. O.; L. D. Beadle; A. D. Hess; O. Sussman; and M. J. Bonese. 1962. Entomological aspects of the 1959 outbreak of eastern encephalitis in New Jersey. *Am. J. Trop. Med.*, **11**:115–21.

Hayes, R. O., and A. D. Hess. 1964. Climatological conditions associated with outbreaks of eastern encephalitis. *Am. J. Trop. Med.*, **13**:851–58.

Hayes, W. J. 1975. *Toxicology of Pesticides.* Williams & Wilkins, Baltimore, 580 pp.

Heath, A. C. G.; A. P. Millthorpe; and N. Eves. 1971. Pigeon mites and human infestation. *N. Z. Entomol.*, **5**:90–92.

Hendrickse, R. G. 1967. Interactions of nutrition and infection: experience in Nigeria. In Wolstenholme, G. E. W. and M. O'Connor (eds.). *Nutrition and Infection.* Churchill, London, 144 pp. (98–111).

Hennig, W. 1948–1952. *Die Larvenformen der Dipteren.* Part I, 185 pp.; Part II, 458 pp.; Part III, 628 pp. Akademie-Verlag, Berlin.

Henry, L. G., and S. McKeever. 1971. A modification of the washing technique for quantitative evaluation of the ectoparasite load of small mammals. *J. Med. Entomol.*, **8**:504–505.

Hermann, H. R.; R. Baer; and M. Barlin. 1975. Histology and function of the venom gland system in Formicine ants. *Psyche*, **82**:67–73.

Herms, W. B. 1925. Entomological observations on Fanning and Washington Islands. *Pan-Pac. Entomol.*, **2**:49–54.

Herms, W. B.; S. F. Bailey; and B. McIvor. 1935. The black widow spider. *Science*, **82**:395–96.

Herrer, A., and H. A. Christensen. 1975. Implication of *Phlebotomus* sand flies as vectors of bartonellosis and leishmaniasis as early as 1764. *Science*, **190**:154–55.

Hertig, M. 1927. A technique for artificial feeding of sandflies (*Phlebotomus*) and mosquitoes. *Science*, **65**:328–29.

Hertig, M. 1948. Sand flies of the genus *Phlebotomus*—a review of their habits, disease relationships, and control. *Proc. 4th Internat. Congr. Trop. Med. Malaria* (Abstracts). Washington, D.C.

Hess, A. D.; C. E. Cherubin; and L. C. Lamotte. 1963. Relation of temperature to activity of western and St. Louis encephalitis viruses. *Am. J. Trop. Med.*, **12**:657–67.

Hess, A. D., and R. O. Hayes. 1970. Relative potentials of domestic animals for zooprophylaxis against mosquito vectors of encephalitis. *Am. J. Trop. Med. Hyg.*, **19**:327–34.

Hewetson, R. W., and J. Nolan. 1968. Resistance of cattle to cattle ticks, *Boophilus microplus*. II. The inheritance of resistance to experimental infestations. *Aust. J. Ag. Res.*, **19**:497–505.

Hewitt, M., and S. M. Turk. 1974. *Cheyletiella*

sp. in the personal environment. With notes on the difference between *C. parasitivorax* Mégnin and *C. vasguri* Smiley. *Br. J. Dermatol.*, **90**:679–83.

Hibler, C. P., and J. L. Adcock. 1971. Elaeophorosis. In Davis, J. W., and R. C. Anderson (eds.), *Parastic Diseases of Wild Mammals.* Iowa State University Press, Ames, 364 pp (pp. 262–78).

Hickin, N. E. 1977. The cockroach. Vol. 2. *Insecticides and Cockroach Control.* St. Martin's Press, New York.

Hienton, T. E. 1974. *Summary of Investigations of Electric Insect Traps.* ARS, USDA Tech. Bull. No. 1498, 136 pp.

Hinton, H. E. 1958. The phylogeny of the panorpoid orders. *Annu. Rev. Entomol.*, **3**:181–206.

Hirst, L. F. 1953. *The Conquest of Plague.* Clarendon Press, Oxford, 478 pp.

Hite, J. M.; W. J. Gladney; J. L. Lancaster, Jr.; and W. H. Whitcomb. 1966. The biology of the brown recluse spider. University of Arkansas Ag. Exp. Sta. Bull., 711, 26 pp.

Ho, B. C., and M. M. J. Lavoipierre. 1975. Studies on filariasis. IV. The rate of escape of the third-stage larvae of *Brugia pahangi* from the mouthparts of *Aedes togoi* during the blood meal. J. Helminthol., **49**:65–72.

Hoare, C. A. 1972. *The Trypanosomes of Mammals: a Zoological Monograph.* Blackwell, Oxford, 749 pp.

Hoch, A. L.; R. W. Barker; and J. A. Hair. 1971. Further observations on the control of lone star ticks (Acarina: Ixodidae) through integrated control procedures. *J. Med. Entomol.*, **8**:731–34.

Hocking, B. 1960. Northern biting flies. *Annu. Rev. Entomol.*, **5**:135–52.

Hocking, B. 1971. Blood-sucking behavior of terrestrial arthropods. *Annu. Rev. Entomol.*, **16**:1–26.

Hockmeyer, W. T.; B. A. Schiefer; B. C. Redington; and B. F. Eldridge. 1975. *Brugia pahangi:* effects upon the flight capability of *Aedes aegypti. Exp. Parasitol.*, **38**:1–5.

Hoeden, J. van der 1968. Twenty years of epidemiological research on leptospirosis in Israel. *Refuah Vet.*, **24**:158–65.

Hoeppli, R. 1963. Early references to the occurrence of *Tunga penetrans* in tropical Africa. *Acta Trop.*, **20**:143–53.

Hoeppli, R. 1969. *Parasitic Diseases in Africa and the Western Hemisphere, Early Documentation and Transmission by the Slave Trade.*

Acta Trop. Supp. 10, Verlag für Recht und Gesellschaft, Basel, 240 pp.

Hoffmann, G. 1971. (The infectivity of different strains of *Boophilus* with *Babesia bigemina* as well as the influence on the ticks of host or parasite.) *Z. Tropenmed. Parasitol.,* **22:**270–84.

Hofstad, M. S. (ed.). 1972. *Diseases of Poultry,* 6th ed. Iowa State University Press, Ames, 1,176 pp.

Hoigné, R.; U. Klein; H. Fahrer; and U. Müller. 1974. (Acute allergic systemic reaction to bee and wasp stings: results and duration of specific hyposensitization with intracutaneous administration of allergen extracts.) *Schweiz. Med. Wochenschr.,* **104:** 221–28.

Holden, P.; R. O. Hayes; C. J. Mitchell; D. B. Francy; J. S. Lazuick; and T. B. Hughes. 1973. House sparrows, *Passer domesticus* (L.) as hosts of arboviruses in Hale County, Texas. I. Field studies, 1965–1969. *Am. J. Trop. Med. Hyg.,* **22:**244–53.

Holland, G. P. 1949. *The Siphonaptera of Canada.* Ottawa, Can. Dept. Ag. Tech. Bull., 70, 306 pp.

Holland, G. P. 1964. Evolution, classification, and host relationships of Siphonaptera. *Annu. Rev. Entomol.,* **9:**123–46.

Holway, R. T.; A. W. Morrill; and F. J. Santana. 1967. Mosquito control activities of the U.S. armed forces in the Republic of Vietnam. *Mosq. News,* **27:**297–307.

Homsher, P. J., and D. E. Sonenshine. 1973. The occurrence of unornamented dwarf American dog ticks, *Dermacentor variabilis;* is it genetically controlled? *Ann. Entomol. Soc. Am.,* **66:**689.

Honzáková, E. 1971. Survival of some Ixodid tick species submerged in water in laboratory experiments. *Folia Parasitol.,* **18:**155–59.

Hoogstraal, H. 1966. Ticks in relation to human diseases caused by viruses. *Annu. Rev. Entomol,* **11:**261–308.

Hoogstraal, H. 1967. Ticks in relation to human diseases caused by *Rickettsia* species. *Annu. Rev. Entomol.,* **12:**377–420.

Hoogstraal, H. 1970 and continuing. *Bibliography of Ticks and Tick-borne Diseases from Homer (about 800* B.C.*) to 31 December 1969.* 5 vols., with additional expected. Special Pub. U.S. Naval Medical Res. Unit No. 3 (NAMRU-3), Cairo, Egypt.

Hoogstraal, H. 1972. The influence of human activity on tick distribution, density, and diseases. *Wiad. Parazytol.,* **18 (4, 5, 6):**501–11.

Hoogstraal, H. (ed.). 1974. A translation of Crimean hemorrhagic fever. Papers from the Third Regional Workshop at Rostov-on-Don in May 1970. *Misc. Pub. Entomol. Soc. Am.,* **9:** 123–200.

Hoogstraal, H.; H. Y. Wassel; J. D. Converse; J. E. Keirans; C. M. Clifford; and C. J. Feare. 1976. *Amblyomma loculosum* (Ixodoidea: Ixodidae): identity, marine bird and human hosts, virus infection, and distribution in the southern oceans. *Ann. Entomol. Soc. Am.,* **69:**3–14.

Hooshmand-Rad, P., and N. J. Hawa. 1973. Transmission of *Theileria hirci* in sheep by *Hyalomma anatolicum anatolicum. Trop. Anim. Health Prod.,* **5:**103–109.

Hopkins, G. H. E. 1949. The host-associations of the lice of mammals. *Proc. Zool. Soc. London,* **119:**387–604.

Hopkins, G. H. E., and M. Rothschild, 1953, 1956, 1962, 1966, 1971. *An Illustrated Catalogue of the Rothschild Collection of Fleas (Siphonaptera) in the British Museum (Natural History).* London, British Museum (Natural History), Vol. I, 361 pp., Vol. II, 445 pp., Vol. III, 500 pp., Vol. IV, 549 pp., Vol. V, 530 pp.

Hopla, C. E. 1974. The ecology of tularemia. *Adv. Vet. Sci. Parasitol.,* **18:**25–53.

Horsfall, F. L., and I. Tamm (eds.). 1965. *Viral and Rickettsial Infections of Man.* Lippincott, Philadelphia, 1,282 pp.

Horsfall, W. R. 1955. *Mosquitoes: Their Behavior and Relation to Disease.* Ronald Press, New York, 723 pp.

Horsfall, W. R., and G. B. Craig, Jr. 1956. Eggs of floodwater mosquitoes IV. Species of *Aedes* common in Illinois (Diptera: Culicidae). *Ann. Entomol. Soc. Am.,* **49:**368–74.

Horsfall, W. R.; H. W. Fowler; L. J. Moretti; and J. R. Larsen. 1973. *Bionomics and Embryology of the Inland Floodwater* mosquito *Aedes vexans.* University of Illinois Press, Urbana, 211 pp.

Horsfall, W. R.; R. C. Miles; J. T. Sokatch. 1952. Eggs of floodwater mosquitoes. I. Species of *Psorophora* (Diptera: Culicidae). *Ann. Entomol. Soc. Am.,* **45:**618–24.

Horsfall, W. R.; R. J. Novak; and F. L. Johnson. 1975. *Aedes vexans* as a flood-plain mosquito. *Environ. Entomol.,* **4:**675–81.

Hörstrup, P., and R. Ackermann. 1973. (Tick-borne meningopolyneuritis [Garin-Bujadoux, Bannwarth].) *Fortsch. Neurol. Psychiatr.,* **41:**583–606.

Hoskins, W. M., and H. T. Gordon. 1956. Arthropod resistance to chemicals. *Annu. Rev. Entomol.,* **1:**89–122.

Houlihan, D. F. 1971. How mosquito pupae escape from the surface. *Nature,* **229:**489–90.

Hourrigan, J. L. 1968. Progress report on psoroptic sheep scabies. *J. Am. Vet. Med. Assoc.,* **152:**506–508.

Hourrigan, J. L., and A. L. Klingsporn. 1975. Epizootiology of bluetongue: the situation in the United States of America. *Aust. Vet. J.,* **51:**203–208.

Howard, L. O. 1921. Sketch history of medical entomology. In Revenel, M. P., *A Half Century of Public Health*. American Public Health Assoc., pp. 412–38.

Howe, G. M. (ed.). 1977. *A World Geography of Human Diseases*. Academic Press, London, 621 pp.

Howell, C. J. 1970. (*Leptoconops* mange.) *J. S. Afr. Vet. Assoc.,* **41:**71, 73.

Howell, C. J.; A. W. H. Neitz; and D. J. J. Potgieter. 1975. Some toxic, physical and chemical properties of the oral secretion of the sand tampan, *Ornithodoros savignyi* Audouin (1827). *Onderstepoort J. Vet. Res.,* **42:** 99–102.

Howell, F. G. 1975. The roles of host-related stimuli in the behavior of *Argas cooleyi* (Acarina: Argasidae). *J. Med. Entomol.,* **11:**715–23.

Hoy, J. B.; A. G. O'Berg; and E. E. Kauffman. 1971. The mosquitofish as a biological control agent against *Culex tarsalis* and *Anopheles freeborni* in Sacramento Valley rice fields. *Mosq. News,* **31:**146–52.

Hsu, S. H.; B. T. Wang; M. H. Huang; W. J. Wong; and J. H. Cross. 1975. Growth of Japanese encephalitis virus in *Culex tritaeniorhynchus* cell cultures. *Am. J. Trop. Med. Hyg.,* **24:**881–88.

Hubbard, C. A. 1947. *Fleas of Western North America*. Iowa State College Press, Ames, 533 pp.

Hudemann, H.; U. Mielke; and H. D. Schulze. 1972. (Experience in the control of Pharoah's ant in a large hospital.) *Z. Gesamte Hyg.,* **18:**593–96.

Hudson, A. 1970. Notes on piercing mouthparts of three species of mosquitoes [*Aedes stimulans* (Wlk.), *A. atropalpus* (Coq.) and *Wyeomyia smithii* (Coq.)] (Diptera: Culicidae) viewed with the scanning electron microscope. *Can. Entomol.,* **102:**501–509.

Hudson, A.; L. Bowman; and C. W. M. Orr. 1960. Effects of absence of saliva on blood feeding by mosquitoes. *Science,* **131:**1730–31.

Hudson, B. W., and T. J. Quan. 1975. Serologic observations during an outbreak of rat-borne plague in the San Francisco Bay area of California. *J. Wildl. Dis.,* **11:**431–36.

Huebner, R. J.; W. L. Jellison; and C. Pomerantz. 1946. Rickettsialpox—a newly recognized rickettsial disease. IV. Isolation of a rickettsia, apparently identical with the causative agent of rickettsialpox from *Allodermanyssus sanguineus,* a rodent mite. *Public Health Rep.,* **61:**1677–82.

Huff, C. 1931. A proposed classification of disease transmission by arthropods. *Science,* **74:**456–57.

Huffaker, C. B. (ed.) 1971. *Biological Control*. Plenum Press, New York, 511 pp.

Hughes, J. H. 1961. Mosquito interceptions and related problems in aerial traffic arriving in the United States. *Mosq. News,* **21:**93–100.

Hughes, L. E.; C. M. Clifford; R. Gresbrink; L. A. Thomas; and J. E. Keirans. 1976. Isolation of a spotted fever group rickettsia from the Pacific Coast tick, *Ixodes pacificus,* in Oregon. *Am. J. Trop. Med. Hyg.,* **25:**513–16.

Hughes, R. D. 1970. The seasonal distribution of bushfly (*Musca vetustissima* Walker) in southeast Australia. *J. Anim. Ecol.,* **39:**691–706.

Hughes, R. D., and W. L. Nicholas. 1974. The spring migration of the bushfly (*Musca vetustissima* Walk.): evidence of displacement provided by natural population markers including parasitism. *J. Anim. Ecol.,* **43:**411–28.

Hughes, R. D.; L. T. Woolcock; and P. Ferrar. 1974. The selection of natural enemies for the biological control of the Australian bushfly. J. Appl. Ecol., **11:**483–88.

Humphries, D. A. 1968. The host-finding behaviour of the hen flea, *Ceratophyllus gallinae* (Schrank) (Siphonaptera). *Parasitology,* **58:** 403–14.

Hunt, R. H. 1973. A cytological technique for the study of *Anopheles gambiae* complex. *Parassitologia,* **15:**137–39.

Hunt, W. B., and D. C. McLean. 1970. Fatal reactions to insect stings: their incidence in the State of Virginia (1954–1966); proposed methods of emergency and prophylactic therapy. *Ann. Allergy,* **28:**64–68.

Hunter, A. R. 1973. Head fly disease in Northumberland. *State Vet. J.,* **28:**239–42.

Hunter, A. R. 1975. Sheep head fly disease in Britain. *Vet. Rec.,* **97:**95–96.

Hunter, W. D., and F. C. Bishopp. 1911. *The Rocky Mountain Spotted Fever Tick*. USDA Bur. Ent. Bull. 105, 47 pp.

Hunter, W. D., and W. A. Hooker. 1907. *Information Concerning the North American Fever Tick*. USDA Bur. Ent. Bull. 72, 87 pp.

Hurlbut, H. W. 1966. Mosquito salivation and virus transmission. *Am. J. Trop. Med. Hyg.*, **15**:989–93.

Hurlbut, H. S. 1973. The effect of environmental temperature upon the transmission of St. Louis encephalitis virus by *Culex pipiens quinquefasciatus. J. Med. Entomol.*, **10**:1–12.

Husain, A., and W. E. Kershaw. 1971. The effect of filariasis on the ability of a vector mosquito to fly and feed and to transmit the infection. *Trans. R. Soc. Trop. Med. Hyg.*, **65**:617–19.

Hyland, K. E. 1963. Current trends in the systematics of acarines endoparastic in vertebrates. *Adv. Acarology*, **1**:365–73.

Hynes, H. B. N.; T. R. Williams; and W. E. Kershaw. 1961. Freshwater crabs and *Simulium neavei* in East Africa. I. Preliminary observations on the slopes of Mount Elgon in December, 1960, and January, 1961. *Ann. Trop. Med. Parasitol.*, **55**:197–201.

Iglisch, I. 1969. (An unusual occurrence of the human flea, *Pulex irritans* L. [Aphaniptera: Pulicidae], outdoors, and problems of its control.) *Anz. Schädlingsk. Pflanzens*. **42**:83–90.

Iglisch, I. 1970. (Experiences with the application of insecticides containing dichlorvos and propoxur for the control of swarms of wasps [Hymenoptera: Vespidae] in towns.) *Anz. Schädlingsk. Pflanzens*. **43**:21–26.

Ikeshoji, T. 1975. Chemical analysis of wood-creosote for species-specific attraction of mosquito oviposition. *Appl. Entomol. Zool.*, **10**:302–308.

Ikeshoji, T.; K. Saito; and A. Yano. 1975. Bacterial production of the ovipositional attractants for mosquitoes on fatty acid substrates. *Appl. Entomol. Zool.*, **10**:239–42.

Il'enko, V. I.; T. S. Gorozhankina; and A. A. Smorodintsev. 1970. (Main principles in the transovarian transmission of tick-borne encephalitis virus by tick vectors.) *Med. Parazitol. (Mosk.)*, **39**:263–69.

Imms, A. D. 1957. *A General Textbook of Entomology Including the Anatomy, Physiology, Development, and Classification of Insects*, 9th ed. (ed. by O. W. Richards and R. G. Davies.) Methuen, London.

Irving-Bell, R. J. 1974. Cytoplasmic factors in the gonads of *Culex pipiens* complex mosquitoes. *Life Sci.*, **14**:1149–51.

Ishijima, H. 1967. Revision of the third stage larvae of synanthropic flies of Japan (Diptera: Anthomyidae, Muscidae, Calliphoridae, and Sarcophagidae). *Jpn. J. Sanit. Zool.*, **18**: 47–100.

Ising, E. 1975. (Zoological aspects of the epidemiology of EEE in Europe.) *Angew. Zool.*, **62**:419–34.

Ismail, I. A. H., and E. I. Hammoud. 1968. The use of coeloconic sensillae on the female antenna in differentiating the members of the *Anopheles gambiae* Giles complex. *Bull. WHO*, **38**:814–21.

Ito, S.; J. W. Vinson; and T. J. McGuire, Jr. 1975. Murine typhus rickettsiae in the oriental rat flea. *Ann. N.Y. Acad. Sci.*, **266**:35–60.

Ivanov, V. I. 1975. (Anthropophily in *Lipoptena cervi* L. (Diptera: Hippoboscidae).) *Med. Parazitol. (Mosk.)*, **44**:491–95.

Ivey, M. C.; J. S. Palmer; and R. H. Washburn. 1976. Famphur and its oxygen analogue: residues in the body tissues of reindeer. *J. Econ. Entomol.*, **69**:260–62.

Jackson, G. J.; R. Herman; and I. Singer (eds.). 1969. *Immunity to Parasitic Animals*. Vol. 1. North-Holland Publ. Co., Amsterdam, 292 pp.

Jagannath, M. S.; K. S. Hegde; P. G. Mutthanna; S. A. Rahman; and G. R. Rajasekariah. 1972. A rare outbreak of *Ctenocephalides felis orientis* Jordan (1925) in poultry and its successful control with Sumithion. *Mysore J. Ag. Sci.*, **6**:515–16.

Jagannath, M. S.; K. S. Hegde; K. Shivaram; and K. V. Nagaraja. 1974. An outbreak of babesiosis in sheep and goats and its control. *Mysore J. Ag. Sci.*, **8**:441–43.

James. M. T. 1948. *The Flies that Cause Myiasis in Man*. USDA Publ. 631, 175 pp.

James, M. T. 1969. A study of the origin of parasitism. *Bull. Entomol. Soc. Am.*, **15**:251–53.

Jamnback, H. 1973. Recent developments in control of blackflies. *Annu. Rev. Entomol.*, **18**: 281–304.

Jamnback, H., and W. Wall. 1957. Control of salt marsh *Tabanus* larvae with granulated insecticides. *J. Econ. Entomol.*, **50**:379–82.

Janssen, J. A. H. A., and D. J. B. Wijers. 1974. *Trypanosoma simiae* at the Kenya coast. A correlation between virulence and the transmitting species of *Glossina. Ann. Trop. Med. Parasitol.*, **68**:5–19.

Janzen, H. G., and K. A. Wright. 1971. The salivary glands of *Aedes aegypti* (L.): an electron microscope study. *Can. J. Zool.*, **49**:1343–45.

Javadian, E., and W. W. MacDonald. 1974. The effect of infection with *Brugia pahangi* and *Dirofilaria repens* on the egg-production of *Aedes aegypti. Ann. Trop. Med. Parasitol.*, **68**:477–81.

Jayawardene, L. G. 1963. Larval development of *Brugia ceylonensis* Jayawardene, 1962, in *Aedes aegypti*, with a brief comparison of the infective larva with those of *Brugia* spp., *Dirofilaria repens* and *Artionema digitata*. *Ann. Trop. Med. Parasitol.*, **57:**359–70.

Jeffers, J. N. R. (ed.). 1972. *Mathematical Models in Ecology*. Blackwell, Oxford, 398 pp.

Jelliffe, E. F. P. 1968. Low birth-weight and malarial infection of the placenta. *Bull. WHO*, **38:**69–78.

Jellison, W. L.; B. Locker; and R. Bacon. 1953. *Index to the Literature of Siphonaptera of North America*. Supp. 1, 1939–1950. Mimeo. Manuscript, 246 pp. Hamilton, Mont., Rocky Mountain Laboratory, USPHS.

Jenkins, D. W. 1949. Trombiculid mites affecting man. IV. Revision of *Eutrombicula* in the American hemisphere. *Ann. Entomol. Soc., Am.*, **42:**289–318.

Jenkins, D. W. 1964. *Pathogens, Parasites, and Predators of Medically Important Arthropods. Annotated List and Bibliography*. Bull. WHO, Supplement to Vol. 30, 150 pp.

Jobsen, J. A. 1974. (Investigations on the application of the sterile male technique against *Dermatobia hominis* [L., Jr.]). *Entomol. Berichten*, **34:**1–3.

Johannsen, O. A. 1933–1937. *Aquatic Diptera*, Parts 1–4. Cornell University Ag. Exp. Sta. Mem. Nos. 164, 177, 205, and 210.

Johansen, C. A., and H. G. Davis. 1972. Toxicity of nine insecticides to the western yellowjacket. *J. Econ. Entomol.*, **65:**40–42.

Johnson, C. G. 1969. *Migration and Dispersal of Insects by Flight*. Methuen, London, 763 pp.

Johnston, D. E. 1964. *Psorergates bos, a New Mite Parasite of Domestic Cattle (Acari, Psorergatidae)*. Ohio Ag. Exp. Sta. Res. Circ. 129, 7 pp.

Johnston, D. E. 1968. *An Atlas of Acari. I. The Families of Parasitiformes and Opilioacariformes*. Publ. Acarology Publications, 1535 Trentwood Rd., Columbus, Ohio.

Johnston, J. H. 1975. Public policy on cattle tick control in New South Wales. *Rev. Market. Ag. Econ.*, **43:**3–39.

Jolivet, G. 1975. (A new sickness of aviary birds in France. Respiratory acariasis due to *Sternostoma tracheacolum* Lawrence 1948.) *Rec. Méd. Vét.*, **151:**273–77.

Jolivet, P. 1967. Les Alticides veneneux de l'Afrique du Sud. *L. Entomologiste*, **23**(4): 100–111.

Jones, B. M. 1950. The penetration of the host tissue by the harvest mite, *Trombicula autumnalis* Shaw. *Parasitology*, **40:**247–60.

Jones, C. M., and D. W. Anthony. 1964. *The Tabanidae (Diptera) of Florida*. USDA Tech. Bull. 1295, 85 pp.

Jones, R. H.; H. W. Potter, Jr.; and H. A. Rhodes. 1972. Ceratopogonidae attacking horses in south Texas during the 1971 VEE epidemic. *Mosq. News*, **32:**507–509.

Jordan, A. M. 1974. Recent developments in the ecology and methods of control of tsetse flies— a review. *Bull. Entomol. Res.*, **63:**361–99.

Jorge, R. 1928. *Les Faunes Regionales des Rongeurs et des Puces dans leur Rapports avec la Peste*. Masson et Cie., Paris, 306 pp.

Joseph, C.; M. A. U. Menom; K. R. Unnithan; and S. Raman. 1963. Studies on the comparative hospitability of *Salvinia auriculata* Aubet and *Pistia stratiotes* Linn., to *Mansoniodes annulifera* (Theobald) in Kerala. *Indian J. Malar.*, **17:**311–32.

Joubert, L.; E. Leftheriotis; and J. Mouchet. 1973. (Myxomatosis. II.) L'Expansion Scientifique Français, Paris, pp. 343–588.

Joyce, C. R., and P. Y. Nakagawa. 1964. New immigrant mosquito in Hawaii. *Public Health Rep.*, **79:**24.

Juricic, D.; B. E. Eno; and G. Parikh. 1974. Mathematical modeling of a virus vector: *Culex tarsalis*. *Biomed. Sci. Instrum.*, **10:**23–28.

Kaire, G. H. 1966. Isolation of tick paralysis toxin from *Ixodes holocyclus*. *Toxicon*, **4:**91–97.

Kaiser, M. N., and H. Hoogstraal. 1969. The subgenus *Persicargas* (Ixodoidea, Argasidae, *Argas*). 7. *A. (P.) walkerae*, new species, a parasite of domestic fowl in Southern Africa. *Ann. Entomol. Soc. Am.*, **62:**885–90.

Kaliner, G. 1970. (Histological studies on the lungs of hens infected with air sac mites [*Cytodites nudus*].) *Berl. Münch. Tieraerztl. Wochenschr.*, **83:**132–34.

Kalkofen, U. P. 1974. Public health implications of *Pulex irritans* infestation of dogs. *J. Am. Vet. Med. Assoc.*, **165:**903–905.

Kalkofen, U. P., and J. Greenberg. 1974. *Echidnophaga gallinacea* infestation in dogs. *J. Am. Vet. Med. Assoc.*, **165:**447–48.

Kappus, K. D., and C. E. Venard. 1967. The effects of photoperiod and temperature on the induction of diapause in *Aedes triseriatus* (Say). *J. Ins. Phys.*, **13:**1007–19.

Kartman, L. 1957. The concept of vector efficiency in experimental studies of plague. *Exp. Parasitol.*, **6:**599–609.

Kartman, L. 1970. Historical and oecological observations on plague in the United States. *Trop. Geogr. Med.*, **22:**257–75.

Kartman, L., and S. F. Quan. 1964. Notes on the fate of avirulent *Pasteurella pestis* in fleas. *Trans. R. Soc. Trop. Med. Hyg.*, **58:**363–65.

Kashin, P. 1966. Electronic recording of the mosquito bite. *J. Ins. Phys.*, **12:**281–86.

Kaufman, W. R., and J. E. Phillips. 1973. Ion and water balance in the Ixodid tick *Dermacentor andersoni*. I. Routes of ion and water excretion. *J. Exp. Biol.*, **58:**523–36.

Kavemba, L. 1975. Resettlement programme in Kigoma Region as a tsetse reclamation method. *E. Afr. J. Med. Res.*, **2:**135–38.

Kawashima. 1961. (Spiders, spider web toxin, and ophthalmitis.) *Ganka Rinsho I-Ho*, **55:**40–48.

Kean, B. H. 1974. Toxoplasmosis. In Jucker, E., *Progress in Drug Research*. Vol. 18, Tropical Diseases I. Birkhauser Verlag, Basel and Stuttgart. 498 pp. (pp. 205–10).

Kearby, W. H. 1975. Variable oakleaf caterpillar larvae secrete formic acid that causes skin lesions (Lepidoptera: Notodontidae). *J. Kans. Entomol. Soc.*, **48:**280–82.

Keegan, H. L. 1972. Venomous bites and stings in Mississippi. *J. Miss. State Med. Assoc.*, **13:**495–99.

Keegan, H. L., and W. R. Lockwood. 1971. Secretory epithelium in venom glands of two species of scorpion of the genus *Centruroides* Marx. *Am. J. Trop. Med. Hyg.*, **20:**770–85.

Keegan, H. L., and W. V. Macfarlane (eds.). 1963. *Venomous and Poisonous Animals and Plants of the Pacific Region*. Pergamon Press, Oxford, 456 pp.

Keh, B. 1967. A brief review of necrotic arachnidism or North American loxoscelism. *Calif. Vector Views*, **14:**48–50.

Keh, B., and J. H. Poorbaugh. 1971. Understanding and treating infestation of lice on humans. *Calif. Vector Views*, **18:**23–31.

Keirans, J. E. 1975. A review of the phoretic relationship between Mallophaga (Phthiraptera: Insecta) and Hippoboscidae (Diptera:Insecta). *J. Med. Entomol.*, **12:**71–76.

Keirans, J. E., and C. M. Clifford. 1978. The genus Ixodes in the United States: a scanning electron microscope study and key to the adults. *J. Med. Entomol.*, Suppl. No. 2, 149 pp.

Keller, H.; J. Eckert; and H. C. Trepp. 1972. (On the eradication of sarcoptic mange in pigs.) *Schweiz. Arch. Tierheilkd.*, **114:**573–82.

Kenaga, E., and C. S. End. 1974. *Commercial and Experimental Organic Insecticides (1974 Revision). Indexed as to Their Scientific, Common and Trade Names, Code Designations,*

Uses, Empirical Formulas, Manufacturers, Mammalian Toxicology and Chemical Structures. Special Pub. Entomol. Soc. Am., 74-1, 77 pp.

Kessel, J. F. 1967. Diethylcarbamazine in filariasis control. Proc. Papers 23rd Ann. Mtg. Am. Mosq. Contr. Assoc., pp. 17–22.

Khan, A. A.; H. I. Maibach; and D. L. Skidmore. 1973. A study of insect repellents. 2. Effect of temperature on protection time. *J. Econ. Entomol.*, **66:**437–38.

Khan, M. A. 1969. Systemic pesticides for use on animals. *Annu. Rev. Entomol.*, **14:**369–86.

Khan, M. A., and W. O. Haufe (eds.). 1972. *Toxicology, Biodegradation and Efficacy of Livestock Pesticides*. Swets and Zeitlinger, Amsterdam, 433 pp.

Kilpatrick, J. W., and C. T. Adams. 1967. Emergency measures employed in the control of St. Louis encephalitis epidemics in Dallas and Corpus Christi, Texas, 1966. Proc. Papers 23rd Ann. Mtg. Am. Mosq. Control Assoc., p. 53.

Kim, J. C. S. 1974. Distribution and life cycle stages of lung mites (*Pneumonyssus* spp.). *J. Med. Primatol.*, **3:**105–19.

Kim, J. C. S., and S. S. Kalter. 1975. Pathology of pulmonary acariasis in baboons (*Papio* sp.). *J. Med. Primatol.*, **4:**70–82.

Kim, K. C., and H. W. Ludwig. 1978. The family classification of the Anoplura. *System. Entomol.*, **3:**249–84.

Kimber, C. D.; R. E. Purnell; and S. A. Sellwood. 1973. The use of fluorescent antibody techniques to detect *Theileria parva* in the salivary glands of the tick *Rhipicephalus appendiculatus*. *Res. Vet. Sci.*, **14:**126–27.

King, R. C. (ed.). 1975. *Handbook of Genetics*. Vol. 3. Plenum, New York, 874 pp.

Kinghorn, A., and W. Yorke. 1912. On the transmission of human trypanosomes by *Glossina morsitans* Westw., and on the occurrence of human trypanosomes in game. *Ann. Trop. Med.*, **6:**1–23.

Kirk, R. W. (ed.). 1974. *Current Veterinary Therapy*. V. *Small Animal Practice*. Saunders, Philadelphia, 1,041 pp.

Kissam, J. B.; R. Noblet; and G. I. Garris. 1975. Large-scale aerial treatment of an endemic area with Abate granular larvicide to control black flies (Diptera: Simuliidae) and suppress *Leucocytozoon smithi* of turkeys. *J. Med. Entomol.*, **12:**359–62.

Knight, K. L. 1964. Qualitative methods for mosquito larval surveys. *J. Med. Entomol.*, **1:**109–15.

Knight, K. L. 1967. Distribution of *Aedes sollici-*

tans (Walker) and *Aedes taeniorhynchus* (Wiedemann) within the United States (Diptera: Culicidae). *J. Ga. Entomol. Soc.*, **2**:9–12.

Knight, K. L. 1978. *Supplement to a Catalogue of the Mosquitoes of the World* (Diptera: Culicidae). The Thomas Say Foundation, Entomological Soc. Am., College Park, Md., 107 pp.

Knight, K. L., and A. Stone. 1977. *A Catalog of the Mosquitoes of the World (Diptera: Culicidae)*, 2nd ed. Thomas Say Foundation (published by *Entomol. Soc. Am.*), 611 pp.

Knipling, E. F.; H. Laven; G. B. Craig; R. Pal; J. B. Kitzmiller; C. N. Smith; and A. W. A. Brown. 1968. Genetic control of insects of public health importance. *Bull. WHO*, **38**:421–38.

Knowles, R., and B. C. Basu. 1943. Laboratory studies on the infectivity of *Anopheles stephensi. J. Malar. Inst. India*, **5**:1–30.

Koeman, J. H., and J. H. Pennings. 1970. An orientational survey of the side-effects and environmental distribution of insecticides used in tsetse-control in Africa. *Bull. Environ. Contam. Toxicol.*, **5**:164–70.

Kohls, G. M.; H. Hoogstraal; C. M. Clifford; and M. N. Kaiser. 1970. The subgenus *Persicargas* (Ixodoidea, Argasidae, *Argas*). 9. Redescription and New World records of *Argas* (*P.*) *persicus* (Oken), and resurrection, redescription, and records of *A.* (*P.*) *radiatus* Railliet, *A.* (*P.*) *sanchezi* Dugès, and *A.* (*P.*) *miniatus* Koch, New World ticks misidentified as *A.* (*P.*) *persicus. Ann. Entomol. Soc. Am.*, **63**:590–606.

Kokorin, I. N.; V. E. Sidorov; and O. S. Gudima. 1969. (Some features of rickettsia development in the body of ticks.) *Parazitologiia*, **3**:193–95.

Königsmann, E. 1960. Zur Phylogenie der Parametabola unter besonderer Berücksichtigung der Phthiraptera. *Beitr. Entomol.*, **10**:705–44.

Korsh, P. V.; O. V. Ravdonikas; and N. B. Dunaev. 1975. (The distribution of epizootics of tularemia and Omsk haemorrhagic fever in populations of *Ondatra zibethica* in western Siberia.) *Zool. Zh.*, **54**:1697–1702.

Kovban, V. Z.; N. A. Voronkov; A. S. Makhovskiĭ; and V. E. Kononets. 1966. (A case of toxicosis of cattle caused by *Simulium.*) *Veterinariia*, **43**:88–90.

Krantz, G. W. 1970. *A Manual of Acarology.* O.S.U. Book Stores, Inc., Corvallis, Ore., 335 pp.

Krantz, G. W. 1978. *A Manual of Acarology,* 2nd ed. O.S.U. Book Stories, Inc., Corvallis, Ore., 509 pp.

Krinsky, W. L. 1976. Animal disease agents transmitted by horse flies and deer flies (Diptera: Tabanidae). *J. Med. Entomol.*, **13**:225–75.

Krinsky, W. L., and L. L. Pechuman. 1975. Trypanosomes in horse flies and deer flies in central New York State. *J. Parasitol.* **61**:12–16.

Kryuchechnikov, V. N., and V. E. Sidorov. 1969. (The distribution of the rickettsia *Dermacentroxenus sibiricus* in the organism of the tick *Dermacentor marginatus* Sulz. in different stages of its development). *Parazitologiia*, **3**:110–14.

Kuhlow, F. 1971. (On the epidemiology and spread of filariasis in South America.) In *Kongressbericht über die Tagung der Österreichischen Gesellschaft für Tropenmedizin (2nd) und der Tagung der Deutschen Tropenmedizinischen Gesellschaft (4th)*, Salzburg und Bad Reichenhall, pp. 205–10.

Kunz, S. E., and J. L. Eschle. 1971. Possible use of the sterile-male technique for eradication of the horn fly. In *Proceedings of a Symposium on the Sterility Principle for Insect Control or Eradication*. International Atomic Energy Agency, Vienna, 542 pp (pp. 145–56).

Kurihara, T.; M. Sasa; J. Miyamoto; and H. Sato. 1973. (Studies on the tolerance of the guppy, *Poecilia reticulata* [Peters], a natural enemy of mosquito larvae, to the septic pollution of water). *Jpn. J. Sanit. Zool.*, **24**:165–74.

Kurosa, K. 1958. (Studies on poisonous beetles. III. Studies on the life history of *Paederus fuscipes* Curtis [Staphylinidae].) *Jpn. J. Sanit. Zool.*, **9**:245–76.

Kusel'tan, I. V. 1967. (Nocardiosis of lambs in the Tadzhik SSR.) In Kovalenko, Y. R., and others (eds.). *Maloizuchennye zabolevaniya s.-kh zhivotnyka*, pp. 110–17. Izd. Kolos, Moscow.

Kusoz, V. N. 1955. (Ecological premises towards an understanding of the epizoology of the tick paralysis of sheep.) *Akad. Nauk. Kaz. SSR Inst. Zool. Trudy*, **3**:27–43.

Kutz, F. W. 1974. Evaluations of an electronic mosquito repelling device. *Mosq. News*, **34**:369–75.

Kutz, F. W., and R. C. Dobson. 1974. Effects of temperature on the development of *Dirofilaria immitis* (Leidy) in *Anopheles quadrimaculatus* Say and on vector mortality resulting from this development. *Ann. Entomol. Soc. Am.*, **67**:325–31.

Kutzer, E. 1966. (On the epidemiology of mange caused by *Sarcoptes.*) *Angew. Parasitol.*, **7**:241–48.

Kutzer, E., and W. Grünberg. 1969. (Transmission of sarcoptic mange from animals to man.) *Berl. Münch. Tieraerztl. Wochenschr.*, **82:**311–14.

Kwan, W. H., and F. O. Morrison. 1974. A summary of published information for field and laboratory studies of biting midges, *Culicoides* species (Diptera: Ceratopogonidae). *Ann. Soc. Entomol. Québec,* **19:**127–37.

Laake, E. W. 1936. Economic studies of screwworm flies, Cochliomyia species (Diptera, Calliphorinae), with special reference to the prevention of myiasis of domestic animals. *Iowa State Coll. J. Sci.,* **10:**345–59.

LaBrecque, G. C., and C. N. Smith. 1967. *Principles of Insect Chemosterilization.* Appleton-Century-Crofts, New York, 325 pp.

LaChance, L. E. 1974. *Status of the Sterile-Insect Release Method in the World. The Sterile-Insect Technique and Its Field Applications.* International Atomic Energy Agency, Vienna, pp. 55–62.

Lackman, D. B.; E. J. Bell; H. G. Stoenner; and E. G. Pickens. 1965. The Rocky Mountain spotted fever group of rickettsias. *Health Lab. Sci.,* **2:**135–41.

LaForce, F. M.; I. L. Acharya; G. Stott; P. S. Brachman; A. F. Kaufman; R. F. Clapp; and N. K. Shah. 1971. Clinical and epidemiological observations on an outbreak of plague in Nepal. *Bull. WHO,* **45:**693–706.

Lainson, R., and J. J. Shaw. 1974. (The Leishmanias and leishmaniasis of the new world with particular reference to Brazil.) *Bol. Of. Sanit. Panam.,* **76:**93–114.

Lancaster, J. L., Jr.; J. S. Simco; and R. Everett. 1969. *Pre-treated Rice Hull Litter for the Control of the Lesser Mealworm.* Rep. Ser. Ark. Ag. Exp. Sta. No. 174, 14 pp.

Lang, J. D.; L. D. Charlet; and M. S. Mulla. 1976. Bibliography (1864 to 1974) of house-dust mites *Dermatophagoides* spp. (Acarina: Pyroglyphidae), and human allergy. *Sci. Biol. J.,* **2:**62–83.

Larsen, B. 1973. Some observations on rat mite dermatitis. *Calif. Vector Views,* **20:**34–35.

Larsen, J. R., and R. F. Ashley. 1971. Demonstration of Venezuelan equine encephalomyelitis virus in tissues of *Aedes aegypti. Am. J. Trop. Med. Hyg.,* **20:**754–60.

LaSalle, R. N., and K. L. Knight. 1974. *Effects of Salt Marsh Impoundments on Mosquito Populations.* UNC-WRRI-74-92, 85 pp.

Lavoipierre, M. M. J. 1958. Studies on the host-parasite relations of filarial nematodes and their arthropod hosts. II. The arthropod as a host to the nematode; a brief appraisal of our present knowledge, based on a study of the more important literature from 1878 to 1957. *Ann. Trop. Med. Parasitol.,* **52:**326–45.

Lavoipierre, M. M. J. 1964. A new family of acarines belonging to the suborder Sarcoptiformes parasitic in the hair follicle of primates. *Ann. Natal Museum,* **16:**191–208.

Lavoipierre, M. M. J. 1965. Feeding mechanism of bloodsucking arthropods. *Nature,* **208:**302–303.

Lavoipierre, M. M. J. 1967. Feeding mechanism of *Haematopinus suis* on the transilluminated mouse ear. *Exp. Parasitol.,* **20:**303–11.

Lavoipierre, M. M. J., and A. J. Beck. 1967. Feeding mechanism of *Chiroptonyssus robustipes* on the transilluminated bat wing. *Exp. Parasitol.,* **20:**312–20.

Lavoipierre, M. M. J.; G. Dickerson; and R. M. Gordon. 1959. Studies on the methods of feeding of blood-sucking arthropods. I. The manner in which Triatomine bugs obtain their blood-meal, as observed in the tissues of the living rodent, with some remarks on the effects of the bite on human volunteers. *Ann. Trop. Med. Parasitol.,* **52:**235–50.

Lavoipierre, M. M. J., and M. Hamachi. 1961. An apparatus for observations on the feeding mechanism of the flea. *Nature,* **192:**998–99.

Lavoipierre, M. M. J., and M. Lavoipierre. 1966. An arthropod intermediate host of a pentastomid. *Nature,* **210:**845–46.

Lavoipierre, M. M. J., and C. Rajamanickam. 1973. Experimental studies on the life cycle of a lizard pentastomid. *J. Med. Entomol.,* **10:**301–302.

Lavoipierre, M. M. J., and R. F. Riek. 1955. Observations on the feeding habits of argasid ticks and on the effect of their bites on laboratory animals, together with a note on the production of coxal fluid by several of the species studied. *Ann. Trop. Med. Parasitol.,* **49:**96–113.

Lea, A. O. 1970. Endocrinology of egg maturation in autogenous and anautogenous *Aedes taeniorhynchus. J. Ins. Phys.,* **16:**1689–96.

Leach, T. M. 1973. African trypanosomiases. In Brandly, C. A., and C. E. Cornelius (eds.). 1973. *Adv. Vet. Sci. Comp. Med.,* Vol. 17, pp. 119–62.

Leahy, M. G.; S. Sternberg; C. Mango; and R. Galun. 1975. Lack of specificity in assembly pheromones of soft ticks (Acari: Argasidae). *J. Med. Entomol.,* **12:**413–14.

LeBerre, R. 1968. (Peritrophic membranes in arthropods. The part they play in digestion and their influence on the development of parasitic organisms [review of the literature]). *Cah. ORSTOM (Entomol. méd.),* **5:**143–204.

Lechleitner, R. R., and others. 1962. Die-off of a

Gunnison's prairie dog colony in central Colorado. I, II. *Zoonoses Res.*, **1:**185–224.

Lecks, H. I. 1973. The mite (*Dermatophagoides* spp.) and house dust allergy. A review of current knowledge and its clinical significance. *Clin. Pediatr.*, **12:**514–17.

Leclercq, M. 1960. *Revision Systematique et Biogeographique des Tabanidae (Diptera) Palearctiques. Vol. I. Pangoniinae et Chrysopinae.* Inst. Roy. Sci. Nat. Belgique, Mem., 2e Ser., fasc. 63, 77 pp.

Leclercq, M. 1969. *Entomological Parasitology.* English Edition. Pergamon Press, New York, 158 pp.

Leclercq, M. 1971. *Les Mouches Nuisables aux Animaux Domestiques.* Les Presses Agronomiques de Gembloux, Gembloux, Belgium, 199 pp.

Leclercq, M. 1974. (Myiases of the human digestive tract.) *Med. Chir. Dig.*, **3:**147–52.

Leclercq, M., and J. Lecomte. 1975. (Serious effects of the stings of aculeate Hymenoptera.) *Spectr. Internat.*, **18(2):**1–7, 10–14.

Leclerq, M., and J. Letawe-Genin. 1976. (Human hypodermosis. A recent case in Belgium.) *Spectr. Internat.*, **19(1):**2–7.

Leclercq, M., and J. Tinant-Dubois. 1973. (Entomology and forensic medicine. Unpublished observations.) *Bull. Méd. Lég. Toxicol.*, **16:**251–67.

Ledger, J. A. 1970. Ectoparasite load in a laughing dove with a deformed mandible. *Ostrich*, **41:**191–94.

LeDuc, J. W.; J. F. Burger; B. R. Eldridge; and P. K. Russell. 1975. Ecology of Keystone virus, a transovarially maintained virus. *Ann. N.Y. Acad. Sci.*, **266:**144–51.

Lee, C. W. 1973. Aerial spraying trials in West Africa for blackfly control. *PANS*, **19:**190–92.

Lee, D. J. 1968. Human myiasis in Australia. *Med. J. Aust.*, **1:**170–73.

Lee, R. 1974. Structure and function of the fascicular stylets and the labral and cibarial sense organs of male and female *Aedes aegypti* (L.) (Diptera: Culicidae). *Quaest. Entomol.* **10:** 187–215.

Lee, R. D. 1955. The biology of the Mexican chicken bug, *Haematosiphon inodorus* (Dugès). *Pan-Pac. Entomol.*, **31:**47–61.

Lees, A. D., and J. W. L. Beament. 1948. An egg-waxing organ in ticks. *Q. J. Micros. Sci.*, **89:**291–332.

Legner, E. F. 1970. Advances in the ecology of *Hippelates* eye gnats in California indicate means for effective integrated control. In *Proceedings and Papers, 38th Conf.* Calif. Mosq. Contr. Assoc., pp. 89–90.

Legner, E. F., and E. C. Bay. 1964. Natural exposure of *Hippelates* eye gnats to field parasitization and the discovery of one pupal and two larval parasites. *Ann. Entomol. Soc. Am.*, **57:**767–69.

Legner, E. F.; W. R. Bowen; W. D. McKeen; W. F. Rooney; and R. F. Hobza. 1973. Inverse relationships between mass of breeding habitat and synanthropic fly emergence and the measurement of population densities with sticky tapes in California inland valleys. *Environ. Entomol.*, **2:**199–205.

Legner, E. F.; W. R. Bowen; W. F. Rooney; W. D. McKeen; and G. W. Johnston. 1975. Integrated fly control on poultry ranches. *Calif. Ag.*, **29:**8–10.

Legner, E. F., and E. J. Dietrick. 1972. Inundation with parasitic insects to control filth breeding flies in California. In *Proceedings and Papers, 40th Ann. Conf.* Calif. Mosq. Contr. Assoc., pp. 129–30.

Legner, E. F., and R. A. Medved. 1973. Influence of *Tilapia mossambica* (Peters), *T. zillii* (Gervais) (Cichlidae) and *Mollienesia latipinna* Le Sueur (Poecoliidae) on pond populations of *Culex* mosquitoes and Chironomid midges. *Mosq. News*, **33:**354–64.

Legner, E. F., and R. A. Medved. 1974. The native desert pupfish, *Cyprinodon macularius* Baird and Girard, a substitute for *Gambusia* in malaria control. *Proceedings and Papers, 30th Ann. Mtg.* Am. Mosq. Contr. Assoc., 11. 58–59.

Legner, E. F.; I. Moore; and G. S. Olton. 1976. Tabular keys and biological notes to common parasitoids of synanthropic Diptera breeding in accumulated animal wastes. *Entomol. News*, **87:**113–24, 125–44.

Legner, E. F.; R. D. Sjogren; and I. M. Hall. 1974. The biological control of medically important arthropods. *CRC Crit. Rev. Environ. Cont.*, **4:**85–113.

Legner, E. F.; R. D. Sjogren; G. S. Olton; and L. Moore, 1976. Control of biting and annoying gnats with fertilizer. *Calif. Ag.*, **30:**14–17.

Leonard, G. J. 1972. The isolation of a toxic factor from sawfly (*Lophyrotoma interrupta* Klug) larvae. *Toxicon*, **10:**597–603.

Lepeš, T. 1974. Present status of the global malaria eradication programme and prospects for the future. *J. Trop. Med. Hyg.*, **77:**47–53.

Levi, H. W. 1958. Number of species of black-widow spiders (Theridiidae: Latrodectus). *Science*, **127:**1055.

Levi, H. W. 1959. The spider genus *Latrodectus* (Araneae, Theridiidae). *Trans. Am. Microsc. Soc.*, **78:**7–43.

Levi, I.; A. Bajric; and D. Hlubna. 1974. (The application of Alugan-concentrate in the therapy of *Cnemidocoptes* sp. of scabies in parrots.) *Veterinaria, (Yugoslavia)*, **23**:121–22.

Lewis, D. J. 1974. The biology of Phlebotomidae in relation to leishmaniasis. *Annu. Rev. Entomol.*, **19**:363–84.

Lewis, D. J. 1975. Functional morphology of the mouth parts in New World Phlebotomine sandflies (Diptera: Psychodidae). *Trans. R. Entomol. Soc. London*, **126**:493–532.

Lewis, D. J.; D. G. Young; G. B. Fairchild; and D. M. Minter. 1977. Proposals for a stable classification of the phlebotomine sandflies (Diptera: Psychodidae). *Syst. Entomol.*, **2**:319–32.

Lewis, I. J. 1970. Observations on the dispersal of larvae of the cattle tick *Boophilus microplus* (Can.). *Bull. Entomol. Res.*, **59**:595–604.

Lichtenstein, L. M.; M. D. Valentine; and A. K. Sobotka. 1974. A case for venom treatment in anaphylactic sensitivity to Hymenoptera sting. *N. Engl. J. Med.*, **290**:1223–27.

Lim, B. L., and C. E. Davie. 1970. The bite of a bird-eating spider *Lampropelma violaceopedes. Med. J. Malaya*, **24**:311–13.

Lindner, E. 1949. *Die Fliegen der Palaearktischen Region, Handbuch.* E. Schweizerbart'sche Verlagsbuchhandlung, Stuttgart, 422 pp.

Lindsay, D. R., and H. I. Scudder. 1956. Nonbiting flies and disease. *Annu. Rev. Entomol.*, **1**:323–46.

Liston, W. G. 1905. Plague, rats and fleas. *J. Bombay Nat. Hist. Soc.*, **16**:253–73.

Little, J. W.; J. Tay; and F. Biagi. 1966. A study of the susceptibility of triatomid bugs to some Mexican strains of *Trypanosoma cruzi. J. Med. Entomol.*, **3**:252–55.

Lofgren, C. S. 1970. Ultralow volume applications of concentrated insecticides in medical and veterinary entomology. *Annu. Rev. Entomol.*, **15**:321–42.

Lofgren, C. S.; W. A. Banks; and B. M. Glancey. 1975. Biology and control of imported fire ants. *Annu. Rev. Entomol.*, **20**:1–30.

Lomax, J. L. 1970. Native fish for mosquito control—a reassessment. *Proc. N.J. Mosq. Exterm. Assoc.*, **57**:185–90.

Londono, M. I. 1973. (Behavior and characteristics of the filaria *Dipetalonema viteae* in the tick *Ornithodoros tartakowskyi.*) *Antioquia Medica*, **23**(9/10):515–16.

Londt, J. G. H. 1975. A rapid spectrophotometric method for the monitoring of embryonic development in ticks (Acarina: Ixodoidea). *Onderstepoort J. Vet. Res.*, **42**:103–107.

Loomis, E. C.; E. T. Schmidtmann; and M. N. Oliver. 1974. A summary review on the distribution of *Ornithodoros coriaceus* Koch in California (Acarina: Argasidae). *Calif. Vector Views*, **21**:57–62.

Loomis, R. B. 1956. *The Chigger Mites of Kansas (Acarina, Trombiculidae).* U. Kans. Sci. Bull., **37** (part 2):1195–1443.

Ludlam, K. W.; L. A. Jachowski, Jr.; and G. F. Otto. 1970. Potential vectors of *Dirofilaria immitis. J. Am. Vet. Med. Assoc.*, **157**:1354–59.

Lumbreras, H.; W. Flores; and A. Escallón. 1959. Allergische Reaktionen auf Stiche von Reduviiden und ihre Bedeutung bei der Chagaskrankheit. *Z. Tropenmed. Parasitol.*, **10**:6–19.

Lumsden, L. L. 1958. St. Louis encephalitis in 1933: Observations on epidemiological features. *Public Health Rep.*, **73**:340–53.

L'vov, D. K.; G. N. Leonova; V. L. Gromashevskiĭ; N. P. Belikova; L. K. Berezina; A. V. Safronov; O. V. Veselovskaya; Y. P. Gofman; and S. M. Klimenko. 1974. (The isolation of Powassan virus from *Haemaphysalis neumanni* Dönitz, 1905, in the Maritime Province.) *Vopr. Virusol.*, No. 5, 538–41.

Lysenko, A. J. 1971. Distribution of leishmaniasis in the Old World. *Bull. WHO*, **44**:515–20.

MacDonald, G. 1957. *The Epidemiology and Control of Malaria.* Oxford University Press, London, 201 pp.

Macdonald, W. W. 1967. The influence of genetic and other factors on vector susceptibility to parasites. In Wright, J. W., and R. Pal, *Genetics of Insect Vectors of Disease.* Elsevier, New York, pp. 567–86.

Mackerras, I. M. 1954–1955. The classification and distribution of Tabanidae. *Aust. J. Zool.*, **2**:431–54; **3**:439–511, 583–633.

Mackerras, I. M. 1956, 1960, 1961. The Tabanidae of Australia. *Aust. J. Zool.*, **4**:376–443; **8**:1–152; **9**:827–905.

Mackerras, J. 1965–1966. Australian Blattidae (Blattodea). Parts I–VI. *Aust. J. Zool.*, **13**:841–82, 883–902, 903–27; **14**:305–34, 335–63, 593–618.

Mackie, F. P. 1907. The part played by *Pediculus corporis* in the transmission of relapsing fever. *Br. Med. J.*, **2**:1706–1709.

MacLaren, J. P.; J. L. Hawkins; W. P. Murdoch; and R. G. Bush. 1967. *Culicoides* control in the Canal Zone by water management. *Mosq. News*, **27**:513–19.

Maddock, D. R., and C. F. Fehn. 1958. Human ear invasions by adult Scarabaeid beetles. *J. Econ. Entomol.*, **51**:546–47.

Maier, W. A. 1976. (Arthropods as hosts and vectors of human parasites: pathology and defence reactions.) *Parasitenk.*, **48**:151–79.

Main, B. Y. 1976. *Spiders*. William Collins, Sydney, 296 pp.

Maĭorov, A. I. 1969. (The treatment of foxes and arctic foxes against sarcoptic mange.) *Veterinariia*, **46**:43–55.

Majeski, J. A., and G. G. Durst. 1975. Bite by the spider *Herpyllus ecclesiasticus* in South Carolina. *Toxicon*, **13**:377.

Majeski, J. A.; G. G. Durst; and K. T. McKee. 1974. Acute systemic anaphylaxis associated with an ant sting. *South. Med. J.*, **67**:365–66.

Makhan'ko, E. V. 1972a. (A study of haematophagy in flies of the genus Hydrotaea.) *Med. Parazitol. (Mosk.)*, **41**:53–55.

Makhan'ko, E. V. 1972b. (The pattern of feeding in flies of the genus *Hydrotaea*) *Med. Parazitol. (Mosk.)*, **41**:172–75.

Mallis. A. 1969. *Handbook of Pest Control. The Behavior, Life History, and Control of Household Pests*, 5th ed. MacNair-Dorland, New York, 1158 pp.

Manson, D. C. M. 1970. The spider mite family Tetranychidae in New Zealand. V. *Tetranychus (Tetranychus) moutensis* a new species of spider mite from flax (*Phormium tenax* Forst.). *N.Z. J. Sci.*, **13**:323–27.

Maramorosch, K., and R. E. Shope (eds.). 1975. *Invertebrate Immunity*. Academic Press, New York. 365 pp.

Maretić, Z. 1965. *Latrodectus* und Latrodectismus. *Natur. Mus.*, **95**:124–232.

Marín-Rojas, R. 1975. (Immunological control of the torsalo [*Dermatobia hominis* L. jr.]. Preliminary notes.) *Rev. Lat. Am. Microbiol.*, **17**:21–24.

Marks, G., and W. K. Beatty. 1976. *Epidemics*. Charles Scribner & Sons, New York, 323 pp.

Marshall, A. G. 1967. The cat flea, *Ctenocephalides felis felis* (Bouché, 1835) as an intermediate host for cestodes. *Parasitology*, **57**:419–30.

Marston, N. 1965. Recent modifications in the design of Malaise insect traps with a summary of the insects represented in collections. *J. Kans. Entomol. Soc.*, **38**:154–62.

Masawe, A. E. J., and H. Nsanzumuhire. 1975. Scabies and other skin diseases in pre-school children in Ujamaa villages in Tanzania. *Trop. Geogr. Med.*, **27**:288–94.

Mason, J., and P. Cavalie. 1965. Malaria epidemic in Haiti following a hurricane. *Am. J. Trop. Med.*, **14**:533–39.

Mathew, G., and K. S. Rai. 1975. Structure and formation of egg membranes in *Aedes aegypti* (L.) (Diptera: Culicidae). *Int. J. Ins. Morphol. Embryol.*, **4**:369–80.

Matsumura, F. 1975. *Toxicology of Insecticides*. Plenum Press, New York. 503 pp.

Matthysse, J. G.; C. J. Jones; and A. Purnasiri. 1974. *Development of Northern Fowl Mite (Ornithonyssus sylviarum [Canestrini and Fanzago]), (Acarina: Dermanyssidae). Populations on Chickens, Effects on the Host, and Immunology*. Search Agriculture 4 (9, Entomology XIII), 39 pp.

Mattingly, P. F. 1957. Genetical aspects of the *Aedes aegypti* problem. I. Taxonomy and bionomics. *Ann. Trop. Med. Parsitol*, **51**:392–408.

Mattingly, P. F. 1958. Genetical aspects of the *Aedes aegypti* problem II. Disease relationships, genetics and control. *Ann. Trop. Med. Parasitol.*, **52**:5–17.

Mattingly, P. F. 1964. The *Anopheles gambiae* complex. Some introductory notes. *Riv. Malar.*, **43**:165–66.

Mattingly, P. F. 1967. The systematics of the *Culex pipiens* complex. *Bull. WHO*, **37**:257–61.

Mattingly, P. F. 1969. *The Biology of Mosquito-borne Disease*. Allen & Unwin, London, 168 pp.

Mattingly, P. F. 1971. Contributions to the mosquito fauna of Southeast Asia. XII. Illustrated keys to the genera of mosquitoes (Diptera, Culicidae). *Contr. Am. Entomol. Inst.*, 7 (4):1–84.

Mattingly, P. F. 1973. Origins and evolution of the human malarias: the role of the vector. *Parassitologia*, **15**:169–72.

Mattingly, P. F. 1977. Names for the *Anopheles gambiae* complex. *Mosq. Systematics*, **9**:323–28.

Maugh, T. H. 1977. Malaria: resurgence in research brightens prospects. *Science*, **196**:413–16.

Maurand, J. 1975. (The microsporidia of Simuliid larvae: systematics, cytochemical, pathological and ecological data.) *Ann. Parasitol. Hum. Comp.*, **50**:371–96.

May, J. M. (ed.). 1961. *Studies in Disease Ecology*. Hafner, New York, 688 pp.

Mazurkiewicz, J. E., and E. M. Bertke. 1972. Ultrastructure of the venom gland of the scorpion, *Centruroides sculpturatus* (Ewing). *J. Morphol.*, **137**:365–84.

Mazzotti, L., and M. A. Bravo-Becherelle. 1963. Scorpionism in the Mexican Republic. In Keegan, H., and W. V. Macfarlane. *Venomous and Poisonous Animals and Noxious Plants of the Pacific Region.* Macmillan Publishing Co., Inc., New York, pp. 119–31.

McClelland, G. A. H. 1974. A worldwide survey of variation in scale pattern of the abdominal tergum of *Aedes aegypti* (L.) (Diptera: Culicidae). *Trans. R. Entomol. Soc. London,* **126:**239–59.

McConnell, E. E.; P. A. Basson; V. de Vos; B. J. Myers; and R. E. Kuntz. 1974. A survey of diseases among 100 free-ranging baboons (*Papio ursinus*) from the Kruger National Park. *Onderstepoort J. Vet. Res.,* **41:**97–167.

McCrae, A. W. R., and S. A. Visser. 1975. *Paederus* (Coleoptera: Staphylinidae) in Uganda. I: Outbreaks, clinical effects, extraction and bioassay of the vesicating toxin. *Ann. Trop. Med. Parasitol.,* **69:**109–20.

McCulloch, R. N., and I. J. Lewis. 1968. Ecological studies of the cattle tick, *Boophilus microplus,* in the North Coast District of New South Wales. *Aust. J. Ag. Res.,* **19:**689–710.

McCurnin, D. M. 1969. Malathion intoxication in military scout dogs. *J. Am. Vet. Med. Assoc.,* **155:**1359–63.

McEnroe, W. D. 1975. The effect of mean winter temperature around 0 C on the population size of the American dog tick, *Dermacentor variabilis* Say. (Acarina: Ixodidae). *Acarologia,* **17:**208–19.

McEnroe, W. D., and M. A. McEnroe. 1973. Questing behavior in the adult American dog tick *Dermacentor variabilis* Say. (Acarina: Ixodidae). *Acarologia,* **15:**38–42.

McGovern, T. P.; G. S. Burden; and M. Beroza. 1975. *n*-Alkanesulfonamides as repellents for the German cockroach (Orthoptera [Dictyoptera]: Blatellidae). *J. Med. Entomol.,* **12:** 387–89.

McGreevy, P. B.; G. A. H. McClelland; and M. M. J. Lavoipierre. 1974. Inheritance of susceptibility to *Dirofilaria immitis* infection in *Aedes aegypti. Ann. Trop. Med. Parasitol.,* **68:**97–109.

McIntosh, B. M. 1972. Rift Valley fever. 1. Vector studies in the field. *J. S. Afr. Vet. Assoc.,* **43:**391–95.

McKeever, S. 1977. Observations of *Corethrella* feeding on tree frogs (*Hyla*). *Mosq. News,* **37:**522–23.

McKelvey, J. J., Jr. 1973. *Man Against Tsetse: Struggle for Africa.* Cornell University Press, Ithaca, New York, 306 pp.

McKiel, J. A. 1959. Sensitization to mosquito bites. *Can. J. Zool.,* **37:**341–51.

McKiel, J. A.; E. J. Bell; and D. B. Lackman. 1967. *Rickettsia canada:* A new member of the typhus group of rickettsiae isolated from *Haemaphysalis leporispalustris* ticks in Canada. *Can. J. Microbiol.,* **13:**503–10.

McLean, D. M. 1975. *Arboviruses and Human Health in Canada.* Associate Committee on Scientific Criteria for Environmental Quality, National Research Council of Canada, Ottawa, 35 pp.

McLennan, H., and I. Oikawa. 1972. Changes in function of the neuromuscular junction occurring in tick paralysis (caused by *Dermacentor andersoni* Stiles). *Can. J. Physiol. Pharmacol.,* **50:**53–58.

McLintock, J.; A. N. Burton; J. A. McKiel; R. R. Hall; and J. G. Rempel. 1970. Known mosquito hosts of western encephalitis virus in Saskatchewan. *J. Med. Entomol.,* **7:**446–54.

McLintock, J., and K. R. Depner. 1954. A review of the life history and habits of the horn fly, *Siphona irritans* (L.) (Diptera, Muscidae). *Can. Entomol.,* **86:**20–33.

Mead-Briggs, A. R. 1964. Observations on the rabbit flea—a vector of myxomatosis. *Ann. Appl. Biol.,* **51:**338–42.

Means, R. G. 1973. Preliminary evaluation of the effectiveness of Mosquito Beater, a granular repellent, against mosquitoes and blackflies. *Mosq. News,* **33:**542–44.

Medley, J. G., and E. Ahrens. 1970. Life history and bionomics of two American species of fowl ticks (Ixodoidea, Argasidae, *Argas*) of the subgenus *Persicargas. Ann. Entomol. Soc. Am.,* **63:**1591–94.

Mehlhorn, H., and E. Schein. 1976. (Electron microscope studies on the development stages of *Theileria parva* [Theiler, 1904] in the gut of the tick vector *Hyalomma anatolicum excavatum* [Koch, 1844].) *Tropenmed. Parasitol.,* **27:**182–91.

Mehlhorn, H.; G. Weber; E. Schein; and G. Buscher. 1975. (Electron microscope studies on the developmental stages of *Theileria annulata* [Dschunkowsky, Luhs, 1904] in the intestine and haemolymph of *Hyalomma anatolicum excavatum* [Koch, 1844].) *Z. Parasitenkd.,* **48:**137–50.

Meifert, D. W.; G. C. LaBrecque; C. N. Smith; and P. B. Morgan. 1967. Control of house flies on some West Indies islands with metepa, apholate, and trichlorfon baits. *J. Econ. Entomol.,* **60:**480–85.

Meleney, W. P., and K. C. Kim. 1974. A comparative study of cattle-infesting *Haematopinus* with redescription of *H. quadripertusus* Fahrenholz, 1919 (*Anoplura: Haematopinidae*). *J. Parasitol.,* **60:**507–22.

Mellor, P. S. 1975. Studies on *Onchocerca cervicalis* Railliet and Henry 1910: V. The development of *Onchocerca cervicalis* larvae in the vectors. *J. Helminthol.,* **49:**33–42.

Mellor, P. S., and J. McCaig. 1974. The probable cause of "sweet itch" in England. *Vet. Rec.,* **95:**411–15.

Merinov, V. A. 1962. The significance of the immature stages of *Dermacentor nuttali* in the epizootology of north-Asian rickettsiosis (in Russian). *Med. Parazitol. (Mosk.),* **31:** 393–98.

Metcalf, C. L.; W. P. Flint; and R. L. Metcalf. 1962. *Destructive and Useful Insects.* McGraw-Hill, New York, 1,087 pp.

Metcalf, R. L. 1977. Model ecosystem approach to insecticide degradation: a critique. *Annu. Rev. Entomol.,* **22:**241–61.

Metcalf, R. L., and W. H. Luckmann (eds.). 1975. *Introduction to Insect Pest Management.* Wiley, New York, 587 pp.

Michaeli, D., and S. Goldfarb. 1968. Clinical studies on the hyposensitisation of dogs and cats to flea bites. *Aust. Vet. J.,* **44:**161–65.

Michener, C. D. 1973. The Brazilian honeybee— possible problem for the future. *Clin. Toxicol.,* **6:**125–27.

Miles, V. I. 1968. A carbon dioxide bait trap for collecting ticks and fleas from animal burrows. *J. Med. Entomol.,* **5:**491–95.

Miller, D. R. 1974. Sensitivity analysis and validation of simulation models. *J. Theor. Biol.,* **48:**345–60.

Miller, R. F., and E. C. Loomis. 1966. Control of hornflies on range cattle with systemic insecticides. *Calif. Ag.,* **20:**8–10.

Milner, K. C.; W. L. Jellison; and B. Smith. 1957. The role of lice in transmission of *Salmonella. J. Infect. Dis.,* **101:**181–92.

Minter, D. M., and E. Goedbloed. 1970. Recovery of viable trypanosomatid flagellates from naturally infected tsetse flies and Phlebotomine sandflies, previously preserved whole in liquid nitrogen. (Preliminary note.) *Trans. R. Soc. Trop. Med. Hyg.,* **64:**789–90.

Minton, S. A. 1974. *Venom Diseases.* C. C. Thomas, Springfield, 235 pp.

Mitchell, B. K. 1976. ATP reception by the tsetse fly, *Glossina morsitans* West. *Experientia,* **32:**192–93.

Miyaga, I. 1975. A new species of the genus *Corethrella* Coquillett from Japan (Diptera:

Chaoboridae). *Jpn. J. Sanit. Zool.,* **26:**25–29.

MMWR. See *Morbidity and Mortality Weekly Report.*

Molina, C.; J. M. Aiache; A. Tourreau; and A. Jeanneret. 1975. (An epidemiological and immunological investigation among cheese makers.) *Rev. Fr. Allerg. Immunol. Clin.,* **15:**89–91.

Molloy, D., and H. Jamnback. 1975. Laboratory transmission of Mermithids parasitic in blackflies. *Mosq. News,* **35:**337–42.

Moloo, S. K.; R. F. Steiger; and H. Hecker. 1970. Ultrastructure of the peritrophic membrane formation in *Glossina* Wiedemann. *Acta Trop.,* **27:**378–83.

Monro, H. A. U. 1969. *Manual of Fumigation for Insect Control.* FAO Ag. Studies No. 79, 381 pp.

Montabar, M. 1974. Malaria and the nomadic tribes of southern Iran. *Cah. ORSTOM, Sér. Entomol. Méd. Parisitol.,* **12:**175–78.

Monty, J. 1972. A review of the stable fly problem in Mauritius. *Rev. Ag. Sucriere de l'Île Maurice,* **51:**13–29.

Moor, P. P. de, and F. E. Steffens. 1970. A computer-simulated model of an arthropod-borne virus transmission cycle, with special reference to chikungunya virus. *Trans. R. Soc. Trop. Med. Hyg.,* **64:**927–34.

Moorhouse, D. E. 1972. Cutaneous lesions on cattle caused by stable fly. *Aust. Vet. J.,* **48:**643–44.

Moorhouse, D. E. 1973. On the morphogenesis of the attachment cement of some Ixodid ticks. *Proceedings 3rd International Congress of Acarology, pp. 527–29.*

Moorhouse, D. E., and A. C. G. Heath. 1975. Parasitism of female ticks by males of the genus *Ixodes. J. Med. Entomol.,* **12:**571–72.

Morbidity and Mortality Weekly Report (*MMWR*). Center for Disease Control, U.S. Department of Health, Education, and Welfare/Public Health Service.

Morel, P. C. 1974. (Methods of combating ticks related to their biology.) *Cah. Méd. Vét.,* **43:**3–23.

Morel, P. C., and G. Vassiliades. 1962. Les *Rhipicephalus* du groupe *sanguineus:* Espèces africaines (Acariens: Ixodoidea). *Rev. Élev. Med. Vet. Pays Trop.,* **15:**343–86.

Moreno, A. 1940. *Scorpiologia Cubana.* Rev. Univ. Habana, Nos. 23, 26, 27, 75 pp.

Morgan, C. E., and G. D. Thomas. 1974. *Annotated Bibliography of the Horn Fly, Haematobia irritans (L.), Including References on the Buffalo Fly, H. exigua (de Meijere), and other*

Species Belonging to the Genus Haematobia. Misc. Pub. Ag. Res. Serv., USDA, No. 1278, 134 pp.

Morgan, N. O.; L. G. Pickens; and R. W. Miller. 1972. Doorway curtains help exclude flies from dairy barns. *J. Econ. Entomol.,* **65:**1061–63.

Morgan, P. B.; R. S. Patterson; G. C. LaBrecque; D. E. Weidhaas; and A. Benton. 1975. Suppression of a field population of houseflies with *Spalangia endius. Science,* **189:**388–89.

Morikawa, T. 1958. Studies on myiasis. I. (A revision of human myiasis reported in Japan.) *Ochanomizu Egaku Zasshi,* **6:**1451–66 (in Japanese, with English summary).

Morini, E. G., and R. J. Roveda. 1974. (A contribution of acarology. 1. Mites considered non-parasitic, of importance in human and veterinary medicine.) *Rev. Med. Vet. Argentina,* **55:**111–15.

Morisod, A.; M. Brossard; C. Lambert; H. Suter; and A. Aeschlimann. 1972. (*Babesia bovis:* transmission by *Ixodes ricinus* [Ixodoidea] in the Rhone plain.) *Schweiz. Arch. Tierheilkd.,* **114:**387–91.

Morris, C. D.; R. H. Zimmerman; and L. Magnarelli. 1976. The bionomics of *Culiseta melanura* and *Culiseta morsitans dyari* in central New York State (Diptera: Culicidae). *Ann. Entomol. Soc. Am.,* **69:**101–105.

Morse, R. A.; D. A. Shearer; R. Boch; and A. W. Benton. 1967. Observations on alarm substances in the genus *Apis. J. Apicultural Res.,* **6:**113–18.

Moser, J. C. 1975. Biosystematics of the straw itch mite with special reference to nomenclature and dermatology. *Trans. R. Entomol. Soc.,* **127:**185–91.

Moss, W. W. 1968. An illustrated key to the (females of 13) species of the Acarine genus *Dermanyssus* (Mesostigmata: Laelapoidea: Dermanyssidae). *J. Med. Entomol.,* **5:**67–84.

Moss, W. W.; C. J. Mitchell; and D. E. Johnston. 1970. New North American host and distribution records for the mite genus *Dermanyssus. J. Med. Entomol.,* **7:**589–93.

Mougeot, G., and J. L. Poirot. 1975. (Larvae of *Cheyletiella* responsible for persistent anal pruritus.) *Nouv. Presse Med.,* **4:**1509

Mougey, Y., and O. Bain. 1976. (Passage of microfilariae into the haemocoel of the vector: stochastic models appropriate to various hypotheses on the mechanisms of limitation.) *Ann. Parasitol. Hum. Comp.,* **51:**95–110.

Mount, G. A.; R. H. Grothaus; K. F. Baldwin; and J. R. Haskins. 1975. ULV sprays of propoxur for control of *Trombicula alfreddugesi. J. Econ. Entomol.,* **68:**761–62.

Mount, G. A.; R. H. Grothaus; J. T. Reed; and K. F. Baldwin. 1976. *Amblyomma americanum:* area control with granules or concentrated sprays of diazinon, propoxur, and chlorpyrifos. *J. Econ. Entomol.,* **69:**257–59.

Muangyai, M. 1974. (*Quantitative Studies on Transovarial Infection of Boophilus microplus* [*Ixodoidea*] *with Babesia bigemina* [*Piroplasmea*].) Inaugural Dissertation, Tierärzt, Hochschule Hannover, German Federal Republic, 38 pp.

Muirhead-Thomson, R. C. 1968. *Ecology of Insect Vector Populations.* Academic Press, New York, 174 pp.

Mulhern, T. D. (ed.). 1973. *Manual for Mosquito Control Personnel.* California Mosquito Control Ass., 205 pp.

Mulla, M. S., and H. Axelrod. 1973. Organic wastes and soil additives as producers of *Hippelates* eye gnats (Diptera-Chloropidae). *Env. Entomol.,* **2:**409–13.

Mulla, M. S.; H. Axelrod; and T. Ikeshoji. 1974. Attractants for synanthropic flies: area-wide control of *Hippelates collusor* with attractive baits, *J. Econ. Entomol.,* **67:**631–38.

Mulla, M. S.; R. W. Dorner; G. P. Georghiou; and M. J. Garber. 1960. Olfacometer and procedure for testing baits and chemical attractants against *Hippelates* eye gnats. *Ann. Entomol. Soc. Am.,* **53:**529–37.

Mulla, M. S.; Y. S. Hwang; and H. Axelrod. 1973. Attractants for synanthropic flies: 3. Evaluation, development, and formulation of attractive baits against *Hippelates collusor. J. Econ. Entomol.,* **66:**1089–94.

Mulla, M. S., and R. B. March. 1959. Flight range, dispersal patterns and population density of the eye gnat *Hippelates collusor* (Townsend). *Ann. Entomol. Soc., Am.,* **52:**641–46.

Mulla, M. S.; R. L. Norland; T. Ikeshoji; and W. L. Kramer. 1974. Insect growth regulators for the control of aquatic midges. *J. Econ. Entomol.,* **67:**165–70.

Mulligan, H. W., and W. H. Potts. 1970. *The African Trypanosomiases.* Allen and Unwin, London, 950 pp.

Mulrennan, J. A.; L. A. Lewis; and R. H. Grothaus. 1975. Field tests with repellent treated wide-mesh netted jackets against the valley black gnat, *Leptoconops carteri. Mosq. News,* **35:**228–29.

Mulvey, P. M. 1972. Cot death survey. Anaphylaxis and the house dust mite. *Med. J. Aust.,* **2:**1240–44.

Muma, M. H. 1951. The arachnid order Solpugida in the United States. *Bull. Am. Mus. Nat. Hist.,* **97:**35–141.

Murphy, F. A. 1974. Arboviruses—ecologic and antigenic grouping. *Proc. Annu. Meet. U.S. Anim. Health Assoc.*, **78:**425–34.

Murray, E. S., and S. B. Torrey. 1975. Virulence of *Rickettsia prowazeki* for head lice. In Bulla, L. A., Jr., and T. C. Cheng (eds.). Pathology of invertebrate vectors of disease. *Ann. N.Y. Acad. Sci.*, **266:** 540 pp.

Nagatomi, A. 1977. Classification of lower Brachycera (Diptera). *J. Nat. Hist.*, **11:**321–35.

Nair, R. B., and P. A. Kurup. 1975. Investigations on the venom of the south Indian scorpion *Heterometrus scaber*. *Biochim. Biophys. Acta*, **381:**165–74.

Najera, J. A. 1974. A critical review of the field application of a mathematical model of malaria eradication. *Bull. WHO*, **50:**449–57.

Nappi, A. J., 1973. Effects of parasitization by the nematode, *Heterotylenchus autumnalis*, on mating and oviposition in the host, *Musca autumnalis*. *J. Parasitol.*, **59:**963–69.

Nash, T. A. M. 1969. *Africa's Bane: the Tsetse Fly*. Collins, London, 224 pp.

National Academy of Sciences. 1976. *Pest Control and Public Health, Pest Control: An Assessment of the Present and Alternative Technologies*, Vol. 5. National Academy of Sciences, Washington, D.C., 282 pp.

Nayar, J. K., and D. M. Sauerman. 1975a. Flight and feeding behavior of autogenous and anautogenous strains of the mosquito *Aedes taeniorhynchus*. *Ann. Entomol. Soc. Am.*, **68:**791–96.

Nayar, J. K., and D. M. Sauerman. 1975b. Physiological basis of host susceptibility of Florida mosquitoes to *Dirofilaria immitis*. *J. Ins. Phys.*, **21:**1965–75.

Nazarova, S. A. 1971. (Oribatid mites of the pasture-desert zone of Uzbekistan as the intermediate hosts of two species of *Moniezia*.) *Helminthol. Abstr.*, A. **40:**3327.

Neafie, R. C., and J. Piggott. 1971. Human pulmonary dirofilariasis. *Arch. Pathol.*, **92:** 342–49.

Nechaeva, L. K., and G. M. Panchenko. 1974. (The number of generations in rat fleas [in experimental conditions].) *Rev. Appl. Entomol.*, B, 64, abstract 2836.

Neitz, W. O. 1962. *The Different Forms of Tick Toxicosis: a Review*. 2nd meeting FAO/OIE expert panel on tick-borne diseases of livestock. Cairo, U.A.R., working paper No. 2, 22 pp.

Neitz, W. O. 1972. The experimental transmission of *Theileria ovis* by *Rhipicephalus evertsi mimeticus* and *R. bursa*. *Onderstepoort J. Vet. Res.*, **39:**83–86.

Nelson, B. C., and C. R. Smith. 1974. Intra- and

interspecific territoriality exhibited by *Spermophilus lateralis* at insecticide-bait stations during a plague epizootic. *Calif. Vector Views*, **21:**19–21.

Nelson, G. S. 1970. Onchocerciasis. *Adv. Parasitol.*, **8:**173–224.

Nelson, G. S.; R. B. Heisch; and M. Furlong. 1962. Studies in filariasis in East Africa II. Filarial infections in man, animals and mosquitoes on the Kenya coast. *Trans. R. Soc. Trop. Med. Hyg.*, **56:**202–17.

Nelson, V. A. 1969. Human parasitism by the brown dog tick. *J. Econ. Entomol.*, **62:** 710–12.

Nelson, W. A. 1962. Development in sheep of resistance to the ked *Melophagus ovinus* (L.). *Exp. Parasitol.*, **12:**41–51.

Nelson, W. A., and R. Hironaka. 1966. Effect of protein and vitamin A intake of sheep on numbers of the sheep ked, *Melophagus ovinus* (L.). *Exp. Parasitol.*, **18:**274–80.

Nelson, W. A., and D. M. Petrunia. 1969. *Melophagus ovinus*: feeding mechanism on transilluminated mouse ear. *Exp. Parasitol.*, **26:**308–13.

Nepoklonov, A. A.; P. I. Bryushinin; and E. N. Shal'kov. 1973. (An effective means against the reindeer warble-fly.) *Veterinariia*, No. 12, pp. 65–66.

Newell, I. M. 1963. Feeding habits in the genus *Balaustium* (Acarina, Erythraeidae), with special reference to attacks on man. *J. Parasitol.*, **49:**498–502.

Newlands, G. 1975. A revision of the spider genus *Loxosceles* Heinecken & Lowe, 1835 (Araneae: Scytodidae) in southern Africa with notes on the natural history and morphology. *J. Entomol. Soc. S. Afr.*, **38:**141–54.

Newson, H. D. 1977. Arthropod problems in recreational areas. *Annu. Rev. Entomol.*, **22:** 333–53.

Newton, W. H.; M. A. Price; O. H. Graham; and J. L. Trevino. 1972. Chromosomal and gonadal aberrations observed in hybrid offspring of Mexican *Boophilus annulatus* x *B. microplus*. *Ann. Entomol. Soc. Am.*, **65:**536–41.

Nicolet, J., and W. Büttiker. 1975. (Observations on infectious keratoconjunctivitis of cattle in Ivory Coast. 2. Study on the eye-frequenting Lepidoptera as vectors). *Rev. Élev. Méd. Vét. Pays Trop.*, **28:**125–32.

Nicoll, William. 1911. On the part played by flies in the dispersal of the eggs of parasitic worms. In *Reports to the Local Government Board on Public Health and Medical Subjects*, London n.s. No. 53, further reports (No. 4) on flies as carriers of infection.

Nicolle, C.; C. Comte; and E. Conseil. 1909. Transmission experimentals du typhus ex-anthimatique par le pou du corps. *C. R. Acad. Sci.*, **149**:486–89.

Nielsen, B. O., and O. Christensen. 1975. A mass attack by the biting midge *Culicoides nubeculosus* (Mg.) (Diptera, Ceratopogonidae) on grazing cattle in Denmark. A new aspect of sewage discharge. *Nord. Vet. Med.*, **27**: 365–72.

Nielsen, E. T., and J. S. Haeger. 1960. Swarming and mating in mosquitoes. *Misc. Pub. Entomol. Soc. Am.*, **1**:71–95.

Nikol'skiĭ, S. N., and S. A. Pozov. 1972. (The tick *Ixodes ricinus* as a vector of *Babesia capreoli* in roe deer.) *Veterinariia*, **4**:62.

Norris, K. R. 1966. Notes on the ecology of the bush-fly, *Musca vetustissima* Walk. (Diptera: Muscidae) in the Canberra district. *Aust. J. Zool.*, **14**:1139–56.

Norris, K. R., and M. D. Murray. 1964. *Notes on the Screw-worm fly, Chrysomya bezziana (Diptera: Calliphoridae) as a Pest of Cattle in New Guinea.* Commonwealth Sci. Ind. Res. Org. Tech. paper 6, 26 pp.

Nuttall, G. H. F. 1899. On the role of insects, arachnids, and myriapods as carriers in the spread of bacterial and parasitic disease of man and animals. A critical and historical study. *Johns Hopkins Hosp. Rep.*, **8**:1–154.

Nutting, W. B. 1976. Pathogenesis associated with hair follicle mites (Acari: Demodicidae). *Acarologia*, **17**:493–506.

Ogata, M. 1897. Ueber die Pestepidemie in Formosa. *Centrabl. Bakteriol.*, **21**:769–77.

Ogata, T.; H. Imai; and F. Coulston. 1971. Pulmonary acariasis in rhesus monkeys: electron microscopy study. *Exp. Mol. Pathol.*, **15**: 137–47.

Ogunba, E. O. 1972. The development of *Loa loa* (Guyot) in *Mansonia africana* (Theobald). *J. Med. Entomol.*, **9**:159–61.

Ohmori, M., and W. G. Banfield. 1974. *The Ultrastructure of the Mosquito, Aedes aegypti* (L.). Saikon Pub. Co., Tokyo, 169 pp.

Okulova, N. M., and V. A. Aristova. 1974. Influence of ectoparasites on a population of the northern red-backed vole of Siberia. *Sov. J. Ecol.*, **4**:522–27.

Older, J. J. 1970. The epidemiology of murine typhus in Texas, 1969. *J.A.M.A.*, **214**: 2011–17.

Oldroyd, H. 1952–1957. *The Horse Flies (Diptera) of the Ethiopian Region.* I (1952), 226 pp., II (1954), 341 pp., III (1957), 489 pp. Br. Mus. (Nat. Hist.), London.

Oldroyd, H. 1964. *The Natural History of Flies.* W. W. Norton & Co., Inc., New York, xiv + 324 pp.

Oldroyd, H. 1977. The Suborders of Diptera. *Proc. Entomol. Soc. Wash.*, **79**:3–10.

Oliver, J. H., Jr. 1971. Parthenogenesis in mites and ticks (Arachnida: Acari.) *Am. Zoologist,* **11**:283–99.

Oliver, J. H., Jr. 1977. Cytogenetics of mites and ticks. *Annu. Rev. Entomol.*, **22**:407–30.

Oliver, J. H.; P. R. Wilkinson; and G. M. Kohls. 1972. Observations on hybridization of three species of North American *Dermacentor* ticks. *J. Parasitol.*, **58**:380–84.

Olkowski, H., and W. Olkowski. 1976. Entomophobia in the urban ecosystem; some observations and suggestions. *Bull. Entomol. Soc. Am.*, **22**:313–17.

Olton, G. S., and E. F. Legner. 1975. Winter inoculative releases of parasitoids to reduce houseflies in poultry manure. *J. Econ. Entomol.*, **68**:35–38.

O'Meara, G. F.; J. W. Knight; and D. G. Evans. 1974. Experimental hybridization between *Aedes sollicitans* and *Aedes mitchellae. Mosq. News,* **34**:457–61.

Ori, M. 1975. (Studies on the poisonous spider, *Chiracanthium japonicum,* as a pest of medical importance. 1. Envenomation by the spider, *Chiracanthium japonicum.*) *Jpn. J. Sanit. Zool.*, **26**:225–29.

Orihel, T. C., and R. C. Lowrie. 1975. *Loa loa:* development to the infective stage in an American deerfly, *Chrysops atlanticus. Am. J. Trop. Med. Hyg.*, **24**:610–15.

Otsuka, H. 1974. (Studies on transmission of *Babesia gibsoni* Patton [1910] by *Haemaphysalis longicornis* Neumann [1901].) *Bull. Fac. Ag. Miyazaki U.*, **21**:359–67.

Oudemans, A. C. 1937. *Kritisch Historisch Overzicht der Akarologie. Tydschrift voor Entomologie intzegeven door de Nederlandsche Vereeniging,* Jan. 1926. (850 v.s. tot 1758), Vol. G. (1805–1850). E. J. Brill, Leiden, 3,379 pp.

Owen, D. 1972. *Common Parasites of Laboratory Rodents and Lagomorphs.* HMSO, London, 140 pp.

Owen, L. N. 1972. Demodectic mange in dogs immunosuppressed with antilymphocyte serum. *Transplantation,* **13**:616–17.

PAHO. 1972. *Venezuelan Encephalitis: Proceedings of the Workshop-Symposium on Venezuelan Encephalitis Virus, Washington, D.C.,* 14–17 Sept. 1971. Pan American Health Organization, Sci. Pub. 243, 416 pp.

PAHO, 1973. *Proceedings of the International Symposium on the Control of Lice and Louse-Borne Diseases, Washington, D.C.,* 4–6 December 1972. Pan American Health Organization, Sci. Pub. No. 263, 311 pp.

Paige, C. J., and G. B. Craig. 1975. Variation in filarial susceptibility among East African populations of *Aedes aegypti. J. Med. Entomol.,* **12:**485–93.

Pal, R. 1972. World Health Organization's programme on resistance of vectors to insecticides. *Proceedings XIIIth International Congress Entomology, Moscow,* **3:**216–17.

Pal, R., and L. E. LaChance. 1974. The operational feasibility of genetic methods for control of insects of medical and veterinary importance. *Annu. Rev. Entomol.,* **19:**269–92.

Pal, R., and R. H. Wharton (eds.). 1974. *Control of Arthropods of Medical and Veterinary Importance.* Plenum, New York, 138 pp.

Pampana, E. 1963. *A Textbook of Malaria Eradication.* Oxford, New York, 508 pp.

Pandit, C. J. 1971. India and the yellow fever problem. *Indian J. Med. Res.,* **59:**1523–47.

Panfilova, I. M. 1976. (The relation of ixodid ticks [*Ixodes persulcatus, Dermacentor silvarum* and *Haemaphysalis concinna*] to light.) *Zool. Zh.,* **55:**371–77.

Papasarathorn, T.; S. Areekul; S. Chermsirivatana; and S. Pinichpongse. 1961. (A study of the rove beetle [*Paederus fuscipes* Curt.] causing vesicular dermatitis in Thailand.) *J. Med. Assoc. Thail.,* **44:**60–81.

Papavero, N., *et al.* Various dates. A Catalogue of the Diptera of the Americas South of the United States. University of São Paulo, Brazil. 110 fascicles planned, most of which have been published, each bearing date of publication.

Paperna, I., and M. Giladi. 1974. Morphological variability, host range and distribution of ticks of the *Rhipicephalus sanguineus* complex in Israel. *Ann. Parasitol. Hum. Comp.,* **49:**357–67.

Parkin W. E.; W. McD. Hammon; and G. E. Sather. 1972. Review of current epidemiological literature on viruses of the California arbovirus group. *Am. J. Trop. Med. Hyg.,* **21:**964–78.

Parrish, H. M. 1963. Analysis of 460 fatalities from venomous animals in the United States. *Am. J. Med. Sci.,* **245:**129–41.

Pascoe, R. R. 1973. The nature and treatment of skin conditions observed in horses in Queensland. *Aust. Vet. J.,* **49:**35–40.

Passera, L. 1975. (Ants as provisional or intermediate hosts of helminths.) *Ann. Biol.,* **14:**227–59.

Paterson, H. E. 1975. The *Musca domestica* complex in Sri Lanka. *J. Entomol., B,* **43:**247–59.

Paterson, H. E., and K. R. Norris. 1970. The *Musca sorbens* complex: the relative status of the Australian and two African populations. *Aust. J. Zool.,* **18:**231–45.

Patnaik, B. 1973. Studies on stephanofilariasis in Orissa. III. Life cycle of *S. assamensis* Pande, 1936. *Z. Tropenmed. Parasitol.,* **24:**457–66.

Patterson, J. W. 1973. Prospects of using juvenile hormone mimics in the control of Triatomine bugs. *Trans. R. Soc. Trop. Med. Hyg.,* **67:**306.

Pavan, M. 1959. Biochemical aspects of insect poisons. *4th Intern. Congr. Biochem.,* **12:**15–36.

Peirce, M. A. 1974. Distribution and ecology of *Ornithodoros moubata porcinus* Walton (Acarina) in animal burrows in East Africa. *Bull. Entomol. Res.,* **64:**605–19.

Pennington, N. E., and C. A. Phelps. 1969. Canine filariasis on Okinawa, Ryukyu Islands. *J. Med. Entomol.,* **6:**59–67.

Perez-Inigo, C. 1974. (Dipterous and Coleopterous pseudoparasites of the human intestine.) *Graellsia,* **27:**161–76.

Perlman, F. 1967. Arthropod sensitivity. In Criep, L. H. (ed.), *Dermatologic Allergy: Immunology, Diagnosis, Management.* Saunders, Philadelphia, 605 pp.

Perlman, F.; E. Press; G. A. Googins; A. Malley; and H. Poarea. 1976. Tussockosis: reactions to Douglas fir tussock moth. *Ann. Allerg.,* **36:**302–307.

Petersen, J. J. 1973. Factors affecting mass production of *Reesimermis nielseni,* a nematode parasite of mosquitoes. *J. Med. Entomol.,* **10:**75–79.

Peterson, A. 1951. *Larvae of Insects: Coleoptera, Diptera, Neuroptera, Siphonoptera, Mecoptera, Trichoptera.* Part II. Edwards Brothers, Ann Arbor, Mich., 416 pp.

Peterson, B. V., et al. (Manual of the Nearctic Diptera.) In preparation.

Petrishcheva, P. A. 1961. (The duration of existence of natural foci of tick-borne spirochaetosis.) *Med. Parazitol. (Mosk.),* **30:**439–42.

Petrov, D. 1972. (Role of *Argas persicus* in the epidemiology of Newcastle disease.) *Vet. Med. Nauki,* **9**(7):13–17.

Petrova, B. K. 1971. (The larvae of the synanthropic flies of the genus *Morellia* R.-D [Diptera, Muscidae] of the south of the Maritime Province.) *Ent. Obozrenie,* **50:**227–35.

Peyton, E. L., and B. A. Harrison. 1979.

Anopheles (Cellia) dirus, a new species of the leucosphyrus group from Thailand (Diptera: Culicidae). *Mosq. Systematics,* **11:**40–52.

Phelps, R. J., and G. R. De Foliart. 1964. *Nematode Parasitism of Simuliidae.* University of Wisconsin Res. Bull. 245, 78 pp.

Philip, C. B. 1948. Tsutsugamushi disease (scrub typhus) in World War II. *J. Parasitol.,* **34:**169–91.

Philip, C. B. 1962. Transmission of yellow fever virus by aged *Aedes aegypti* and comments on some other mosquito-virus relationships. *Am. J. Trop. Med.,* **11:**697–701.

Philip, C. B., and W. Burgdorfer. 1961. Arthropod vectors as reservoirs of microbial disease agents. *Annu. Rev. Entomol.,* **6:**391–412.

Philip, C. B., and L. E. Rozeboom. 1973. Medico-veterinary entomology: A generation of progress. In Smith, Mittler, and Smith, 1973, *History of Entomology.* Annual Reviews, Inc., Palo Alto, pp. 330–60.

Pichon, G.; G. Perrault; and J. Laigret. 1974. (Parasite yield in filariasis vectors.) *Bull. WHO,* **51:**517–24.

Pinet, J. M.; J. Bernard; and J. Boistel. 1969. (Electrophysiological study of the receptors on the stylets of a blood-sucking bug: *Triatoma infestans.*) *C. R. Séances Soc. Biol.,* **163:**1939–46.

Pinheiro, F. P.; G. Bensabath; D. Costa, Jr.; O. M. Maroja; Z. C. Lins; and A. H. P. Andrade. 1974. Haemorrhagic syndrome of Altamira. *Lancet,* **1**(7859):639–42.

Plowright, W.; C. T. Perry; and A. Greig. 1974. Sexual transmission of African swine fever virus in the tick, *Ornithodoros moubata porcinus* Walton. *Res. Vet. Sci.,* **17:**106–13.

Plowright, W.; C. T. Perry; M. A. Peirce; and J. Parker, 1970. Experimental infection of the Argasid tick, *Ornithodoros moubata porcinus,* with African swine fever virus. *Arch. Gesamte Virusfors.,* **31:**33–50.

Poinar, G. O., Jr. 1972. Nematodes as facultative parasites of insects. *Annu. Rev. Entomol.,* **17:**103–22.

Poinar, G. O., Jr. 1975. *Entomogenous Nematodes. A Manual and Host List of Insect-Nematode Associations.* E. J. Brill, Leiden, 317 pp.

Poinar, G. O., and R. Hess. 1977. Virus-like particles in the nematode *Romanomermis culicivorax* (Mermithidae). *Nature,* **266:**256–57.

Pollitzer, R. 1954. *Plague.* WHO Mon. ser. No. 22, 698 pp.

Pollitzer, R. 1966. *Plague and Plague Control in the Soviet Union.* Bronx, N.Y., Fordham University, Institute of Contemporary Russian Studies, 478 pp.

Polls, I.; B. Greenberg; and C. Lue-Hing. 1975. Control of nuisance midges in a channel receiving treated municipal sewage. *Mosq. News,* **35:**533–37.

Pomerantz, C. 1959. Arthropods and psychic disturbances. *Bull. Entomol. Soc. Am.,* **5:**65–67.

Ponnampalam, J. T. 1975. The durian season in peninsular Malaysia as a factor in the epidemiology of malaria. *Trans. R. Soc. Trop. Med. Hyg.,* **69:**285.

Pospelova-Shtrom, M. V. 1969. On the system of classification of ticks of the family Argasidae Can., 1890. *Acarologia,* **11:**1–22.

Pratt, H. D., and R. F. Darsie, Jr. 1975. Highlights in medical entomology in 1974. *Bull. Entomol. Soc. Am.,* **21:**173–76.

Price, R. D. 1956. The multiplication of *Pasteurella tularensis* in human body lice. *Am. J. Hyg.,* **63:**186–97.

Princis, K. 1954. Wo ist die Urheimat von *Blatta orientalis* L. zu suchen? *Opusc. Entomol.,* **19:**202–204.

Probst, P. J. 1972. (On the biology of reproduction and the development of the poison glands in the scorpion *Isometrus maculatus* [De Geer, 1778] [Scorpiones: Buthidae].) *Acta Trop.,* **29:**1–87.

Prokopič, J., and S. Bilý. 1975. Beetles (Coleoptera) as new intermediate hosts of helminths. *Vestn. Ceskoslov. Spolecn. Zool.,* **39:**224–30.

Proverbs, M. D. 1969. Induced sterilization and control of insects. *Annu. Rev. Entomol.,* **14:**81–102.

Provost, M. W. 1972a. Environmental hazards in the control of disease vectors. *Environ. Entomol.,* **1:**333–39.

Provost, M. W. 1972b. The ecological relevance and migratorial significance of the rhythm of pupation in *Aedes taeniorhynchus* (Weidemann). *Proceedings XIIIth International Congress Entomology, Moscow,* **3:**228.

Pugh, A. O., and D. A. Parker. 1975. Plague: Rhodesia's first recorded outbreak. *Cent. Afr. J. Med.,* **21:**93–96.

Pull, J. H., and B. Grab. 1974. A simple epidemiological model for evaluating the malaria inoculation rate and the risk of infection in infants. *Bull. WHO,* **51:**507–16.

Purnell, R. E.; C. D. H. Boarer; and M. A. Peirce. 1971. *Theileria parva:* comparative infection rates of adult and nymphal *Rhipicephalus appendiculatus. Parasitology,* **62:**349–53.

Purnell, R. E.; D. Branagan; and C. G. D. Brown. 1970. Attempted transmission of some piroplasms by Rhipicephalid ticks. *Trop. Anim. Health Prod.,* **2:**146–50.

Pursley, R. E. 1973. Stinging Hymenoptera. *Am. Bee J.*, **113**:131, 132, 135.

Putnam, S. E. 1977. *Controlling Stinging and Biting Insects at Campsites.* ED&T 2689, USDA Forest Service, 22 pp.

Quélennec, G. 1976. Measurement of the susceptibility of blackfly [simuliid] larvae to insecticides. WHO(WHO/BC)76.622.

Quélennec, G.; E. Simonkovich; and M. Ovazza. 1968. (Research on a type of dam spillway unfavorable to the attachment of *S. damnosum.*) *Bull. WHO*, **38**:943–56.

Rabinovich, J. E. 1971. (Simulation with a digital computer of population regulation in Triatomine vectors of Chagas' disease by the parasite *Telenomus fariai* [Hymenoptera: Scelionidae], and of strategies for integrated control.) *Rev. Peruana Entomol.*, **14**:117–26.

Radda, A.; W. Schmidtke; and A. Wandeler. 1974. (Isolation of tick-borne encephalitis [TBE] virus from *Ixodes ricinus* collected in the Kanton Zurich, Switzerland.) *Zentralbl. Bakteriol. (Orig. A)*, **229**:268–72.

Radford, A. J. 1975. Millipede burns in man. *Trop. Geogr. Med.*, **27**:279–87.

Radford, C. D. 1943. Genera and species of parasitic mites (Acarina). *Parasitology*, **35**:58–81.

Radovsky, F. 1969. Adaptive radiation in the parasitic Mesostigmata. *Acarologia*, **15**:480–83.

Radovsky, F. J. 1967. *The Macronyssidae and Laelapidae (Acarina: Mesostigmata) Parasitic on Bats.* University of California Pub. Ent. 46, 288 pp.

Raghavan, R. S.; K. R. Reddy; and G. A. Khan. 1968. Dermatitis in elephants caused by the louse *Haematomyzus elephantis* (Piagot 1869). *Indian Vet. J.*, **45**:700–701.

Raimbert, A. 1869. Recherches experimentales sur la transmission du charbon par les mouches. *C. R. Acad. Sci.*, **69**:805–12. (Cited by Nuttall.)

Ranade, D. R., and D. V. Bhalchandra. 1976. A note on the natural infection in rat flea, *Xenopsylla cheopis* (Roths.) with *Trichinella spiralis* (Owen). *J. Communicable Dis.*, **8**:77–80.

Ranck, F. M., Jr.; J. H. Gainer; J. E. Hanley; and S. L. Nelson. 1965. Natural outbreak of eastern and western encephalitis in pen-raised chukars in Florida. *Avian Dis.*, **19**:8–20.

Rao, N. R.; O. P. Vig; and S. N. Agarwala. 1974. Transmission dynamics of malaria. Quantitative studies—part I; a stochastic model. *Bull. Haffkine Inst.*, **2**:71–78.

Rao, V. N. 1974. Problems of malaria eradication in India. *Prog. Drug Res.*, **18**:245–51.

Rapmund, G.; R. W. Upham; W. D. Kundin; C. Manikuraran; and T. C. Chan. 1969. Tran-

sovarial development of scrub typhus rickettsiae in a colony of vector mites. *Trans. R. Soc. Trop. Med. Hyg.*, **63**:251–58.

Rasnitsyn, S. P.; A. Alekseev; R. M. Gornostaeva; E. S. Kupriyanova; A. A. Potapov; and O. V. Razumova. 1974. (Negative results of a test of examples of sound generators intended to repel mosquitos.) *Med. Parazitol. (Mosk.)*, **43**:706–708.

Ravdonikas, O. V.; M. P. Chumakov, E. A. Solovey; D. Ivanov; and P. V. Korsh. 1971. (The importance of small mammals in transphase, interphase and interspecies transmission of OHF virus in Ixodid ticks in the course of maintenance in a natural focus.) *Trudy Inst. Poliomelita i Virusnykh Entsefalitov, Akad. Med. Nauk SSSR*, **19**:492–503.

Ravkin, Y. S.; V. F. Sapegina; and Y. I. Dokuchaeva. 1973. (On the nature of the role of birds in foci of tick-borne encephalitis in the southern taiga and subtaiga forests of western Siberia.) *Trudy Biol. Inst., Sibirsko Otdelenie, Akad. Nauk SSSR*, **14**:181–90.

Ray, A. C.; J. D. Norris; and J. C. Reagor. 1975. Benzene hexachloride poisoning in cattle. *J. Am. Vet. Med. Assoc.*, **166**:1180–82.

Rechav, Y.; G. B. Whitehead; and M. M. Knight. 1976. Aggregation response of nymphs to pheromone(s) produced by males of the tick *Amblyomma hebraeum* (Koch). *Nature*, **259**(5544):563–64.

Redington, B. C., and W. T. Hockmeyer. 1976. A method for estimating blood meal volume in *Aedes aegypti* using a radioisotope. *J. Ins. Phys.*, **22**:961–66.

Redman, J. F. 1974. Human envenomation by a lycosid. *Arch. Dermatol.*, **110**:111–12.

Reedy, L. M. 1975. Use of flea antigen in treatment of feline flea-allergy dermatitis. *Vet. Med. Small Anim. Clin.*, **70**:703–704.

Reeves, W. C.; R. E. Bellamy; A. F. Geib; and R. P. Scrivani. 1964. Analysis of the circumstances leading to abortion of a western equine encephalitis epidemic. *Am. J. Hyg.*, **80**:205–20.

Reeves, W. C.; R. E. Bellamy; and R. P. Scrivani. 1961. Differentiation of encephalitis virus infection rates from transmission rates in mosquito vector populations. *Am. J. Hyg.*, **73**:303–15.

Reeves, W. C.; W. M. Hammon; W. A. Longshore, Jr.; H. E. McClure; and A. F. Geib. 1962. *Epidemiology of the Arthropod-Borne Viral Encephalitides in Kern County California 1943–1952.* University of California Pub. Public Hlth. 4, 257 pp.

Řeháček, J.; D. Blaškovič; and W. F. Hink

(eds.). 1973. *Proceedings Third International Colloquium on Invertebrate Tissue Culture, Smolenice near Bratislava, June 22–25, 1971.* Slovak. Acad. Sci. Bratislava, 509 pp.

Reinert, J. F. 1975. Mosquito generic and subgeneric abbreviations (Diptera: Culicidae). *Mosq. Syst., 7:*105–10.

Reitblat, A. G., and A. M. Belokopytova. 1974. (Cannibalism and predation in flea larvae.) *Zool. Zh., 53:*135–37.

Reitblat, A. G.; N. P. Kalmykova; and P. F. Emel'yanov. 1974. (The eating of eggs and larvae of fleas by Gamasoid mites.) *Rev. Appl. Entomol.,* B, **64,** abstract 3076.

Remington, C. L. 1950. The bite and habits of a giant centipede (*Scolopendra subspinipes*) in the Philippine Islands. *Am. J. Trop. Med.,* **30:**453–55.

Reshetnikov, P. T. 1967. (The mite *Dermanyssus gallinae*—a vector of fowl spirochaetosis.) *Veterinariia,* **44:**48.

Rhoades, R. B.; W. L. Schafer; W. H. Schmid; P. F. Wubbena; R. M. Dozier; A. W. Townes; and H. J. Wittig. 1975. Hypersensitivity to the imported fire ant. A report of 49 cases. *J. Allergy Clin. Immunol.,* **56:**84–93.

Rhodes, A. P. 1975. Seminal degeneration associated with chorioptic mange of the scrotum of rams. *Aust. Vet. J.,* **51:**428–32.

Rice, P. L., and G. W. Douglas. 1972. Myiasis in man by *Cuterebra* (Diptera: Cuterebridae). *Ann. Entomol. Soc. Am.,* **65:**514–16.

Rice, P. L., and N. Gleason. 1972. Two cases of myasis in the United States by the African tumbu fly, *Cordylobia anthropophaga* (Diptera: Calliphoridae). *Am. J. Trop. Med. Hyg.,* **21:**62–65.

Rich, G. B. 1973. Grooming and yarding of spring-born calves prevent paralysis caused by the Rocky Mountain wood tick. *Can. J. Anim. Sci.,* **53:**377–78.

Rich, G. B., and J. D. Gregson. 1968. The first discovery of free-living larvae of the ear tick, *Otobius megnini* (Duges), in British Columbia. *J. Entomol. Soc. Br. Columb.,* **65:**22–23.

Richards, A. G. 1975. The ultrastructure of the midgut of hematophagous insects. *Acta Trop.,* **32:**83–95.

Richards, A. G., and P. A. Richards. 1977. The peritrophic membranes of insects. *Ann. Rev. Entomol.,* **22:**219–40.

Rickards, D. A. 1975. Cnemidocoptic mange in parakeets. *Vet. Med. Small Anim. Clin.,* **70:**729–31.

Ricketts, H. T. 1911. *Contribution to Medical Science.* University of Chicago Press, Chicago, 497 pp. (See pp. 278–450.)

Ricketts, H. T., and R. M. Wilder. 1910a. The transmission of the typhus fever of Mexico (tabardillo) by means of the louse (*Pediculus vestimenti*). *J.A.M.A.,* **54:**1304–1307.

Ricketts, H. T., and R. M. Wilder. 1910b. Further investigation regarding the etiology of tabardillo, Mexican typhus fever. *J.A.M.A.,* **55:**309–11.

Riek, R. F. 1966. The development of *Babesia* and *Theileria* spp. in ticks with special reference to those occurring in cattle. In Soulsby, E. J. L. *Biology of Parasites.* Academic Press, New York, pp. 15–32.

Riley, W. A. 1939. The possibility of intestinal myiasis in man. *J. Econ. Entomol.,* **32:**875–76.

Rioux, J. A.; H. Croset; J. Pech-Périères; E. Guilvard; and A. Belmonte. 1975. (Autogenesis in mosquitoes. Synoptic table of autogenous species.) *Ann. Parasitol. Hum. Comp.,* **50:**134–40.

Ristic, M.; J. Oppermann; S. Sibinovic; and T. N. Phillips. 1964. Equine piroplasmosis—a mixed strain of *Piroplasma caballi* and *Piroplasma equi* isolated in Florida and studied by the fluorescent-antibody technique. *Am. J. Vet. Res.,* **104:**15–23.

Rivosecchi, L., and L. Colombo. 1973. (Pollution of running waters in the Provence of Latina and changes in the Simuliid [Diptera, Nematocera] fauna.) *Parassitologia,* **15:**183–211.

Roberts, I. H., and W. P. Meleney. 1975. Variations among strains of *Psoroptes ovis* (Acarina: Psoroptidae) on sheep and cattle. *Ann. Entomol. Soc. Am.,* **64:**109–16.

Roberts, L. W.; D. M. Robinson; G. Rapmund; J. S. Walker; E. Gan; and S. Ram. 1975. Distribution of *Rickettsia tsutsugamushi* in organs of *Leptotrombidium* (*Leptotrombidium*) *fletcheri* (Prostigmata: Trombiculidae). *J. Med. Entomol.,* **12:**345–48.

Roberts, R. H. 1968. A feeding association between *Hippelates* (Diptera: Chloropidae) and Tabanidae on cattle: its possible role in transmission of anaplasmosis. *Mosq. News,* **28:**236–37.

Roberts, R. H. 1970. Tabanidae collected in a Malaise trap baited with CO_2. *Mosq. News,* **30:**52–53.

Rocha e Silva, E. O. da; J. C. R. de Andrade; and A. R. de Lima. 1975. (Importance of synanthropic animals in the control of endemic Chagas' disease.) *Saúde Púb.,* **9:**371–81.

Rodaniche, E. de; P. Galindo; and C. M. Johnson. 1959. Further studies on the experimental transmission of yellow fever by *Sabethes chloropterus. J. Trop. Med.,* **8:**190–94.

Rodeck, H. G. 1932. Arthropod designs on prehistoric Mimbres pottery. *Ann. Entomol. Soc. Am.*, **25**:688–93.

Rodhain, J. 1948. Susceptibility of the chimpanzee to *P. malariae* of human origin. *Am. J. Trop. Med.*, **28**:629–31.

Rodriguez, J. L., Jr.; and L. A. Riehl. 1959. Results with cockerels for house fly control in poultry droppings. *J. Econ. Entomol.*, **52**:542–43.

Rogers, A. J. 1962. Effects of impounding and filling on the production of sand flies (*Culicoides*) in Florida salt marshes. *J. Econ. Entomol.*, **55**:521–27.

Rogoff, W. M.; C. H. Gretz; T. B. Clark; H. A. McDaniel; and J. E. Pearson. 1977. Laboratory transmission of exotic Newcastle disease virus by *Fannia canicularis* (Diptera: Muscidae). *J. Med. Entomol.*, **13**:617–21.

Rohdenforf, B. B. (ed.). 1962. *Foundations of Paleontology. Arthropods, Tracheates and Chelicerates.* (In Russian) Pub. Acad. Sci., Moscow, 560 pp.

Romoser, W. S. 1973. *The Science of Entomology.* Macmillan Publishing Co., Inc., New York, 449 pp.

Ronald, N. C., and J. E. Wagner. 1973. Pediculosis of spider monkeys: a case report with zoonotic implications. *Lab. Anim. Sci.*, **23**:872–75.

Rosen, L.; R. B. Tesh; J. C. Lien; and J. H. Cross. 1978. Transovarial transmission of Japanese encephalitis virus by mosquitoes. *Science*, **199**:909–11.

Roth, L. M. 1948. A study of mosquito behavior. An experimental laboratory study of the sexual behavior of *Aedes aegypti* (Linnaeus). *Am. Midl. Naturalist*, **40**:265–352.

Roth, L. M. 1967. Water changes in cockroach öothecae in relation to the evolution of ovoviparity and viviparity. *Ann. Entomol. Soc. Am.* **60**:928–46.

Roth, L. M., and T. Eisner. 1962. Chemical defenses of arthropods. *Ann. Revu. Entomol.*, **7**:107–36.

Roth, L. M., and E. R. Willis. 1957. *The Medical and Veterinary Importance of Cockroaches.* Smithsonian Misc. Coll., Vol. 134, No. 10, 147 pp.

Rothenberg, R., and D. E. Sonenshine. 1970. Rocky Mountain spotted fever in Virginia: clinical and epidemiologic features. *J. Med. Entomol.*, **7**:663–69.

Rothschild, M. 1965. Fleas. *Sci. Am.*, **213**:44–53.

Rothschild, M. 1970. *Cheyletiella parasitivorax* (Megnin) (Acar., Cheyletidae) feeding upon the rabbit flea (*Spilopsyllus cunicula* [Dale]). *Entomol. Mon. Mag.*, **105**:216.

Rothschild, M. 1975. Recent advances in our knowledge of the order Siphonaptera. *Annu. Rev. Entomol.*, **20**:241–60.

Rothschild, M., and B. Ford. 1964. Maturation and egglaying of the rabbit flea (*Spilopsyllus cuniculi* Dale) induced by the external application of hydrocortisone. *Nature*, **203**:210–11.

Rothschild, M., and B. Ford. 1972. Breeding cycle of the flea *Cediopsylla simplex* is controlled by breeding cycle of host. *Science*, **178**:625–26.

Rothschild, M.; T. Reichstein; J. von Euw; R. Aplin; and R. R. M. Harman. 1970. Toxic Lepidoptera. *Toxicon*, **8**:293–99.

Rothschild, M., and R. Traub. 1971. A revised glossary of terms used in the taxonomy and morphology of fleas. In Hopkins, G. H. E., and M. Rothschild. *An Illustrated Catalogue of the Rothschild Collection of the Fleas in the British Museum* (Natural History)., **5**:8–85. Trustees of the British Museum (N.H.), London.

Rowley, W. A.; C. L. Graham; and R. E. Williams. 1968. A flight mill system for the laboratory study of mosquito flight. *Ann. Entomol. Soc. Am.*, **61**:1507–14.

Rozeboom, L. E. 1936. The life cycle of laboratory-bred *Anopheles albimanus* Wiedemann. *Ann. Entomol. Soc. Am.*, **29**:480–89.

Rozeboom, L. E., and R. W. Burgess. 1962. Dry-season survival of some plant-cavity breeding mosquitoes in Liberia. *Ann. Entomol. Soc. Am.*, **55**:521–24.

Rubtsov, I. A. 1964. Simuliidae (Melusinidae). In Lindner, *Die Fliegen der Palaearktischen Region*, Band III. **4**:1–689.

Rudnick, A.; N. J. Marchette; and R. Garcia. 1967. Possible jungle dengue—recent studies and hypotheses. *Jpn. J. Med. Sci. Biol.*, **20**, Suppl., pp. 69–74.

Rudolph, D., and W. Knulle. 1974. Site and mechanism of water vapour uptake from the atmosphere in Ixodid ticks. *Nature*, **249**(5452):84–85.

Rueger, M. E., and T. A. Olson. 1969. Cockroaches (Blattaria) as vectors of food poisoning and food infection organisms. *J. Med. Entomol.*, **6**:185–89.

Russell, F. E., and P. R. Saunders. 1967. *Animal Toxins.* Pergamon, Oxford, 428 pp.

Russell, F. E.; W. G. Waldron; and M. B. Madon. 1969. Bites by the brown spiders *Loxosceles unicolor* and *Loxosceles arizonica* in California and Arizona. *Toxicon*, **7**:109–17.

Russell, P. F. 1958. Malaria in the world today. *Am. J. Public Health,* **47:**414–20.

Russell, P. F. 1959. Insects and the epidemiology of malaria. *Ann. Revu. Entomol.,* **4:**415–34.

Russell, P. F.; L. E. Rozeboom; and A. Stone. 1943. *Keys to the Anopheline Mosquitoes of the World with Notes on Their Identification, Distribution, Biology, and Relation to Malaria.* American Entomological Society, 152 pp.

Russell, P. F.; L. S. West; R. D. Manwell; and G. MacDonald. 1963. *Practical Malariology.* Oxford, New York, 750 pp.

Ryckman, R. E. 1971. Plague vector studies. Part III. The rate deparasitized ground squirrels are reinfested with fleas under field conditions. *J. Med. Entomol.,* **8:**668–70.

Ryckman, R. E.; D. L. Folkes; L. E. Olsen; P. L. Robb; and A. E. Ryckman. 1965. Epizootology of *Trypanosoma cruzi* in southwestern North America. Parts I–VII. *J. Med. Entomol.,* **2:**87–108.

Sabrosky, C. W. 1941. The *Hippelates* flies or eye gnats: Preliminary notes. *Can. Entomol.,* **73:**23–27.

Saito, Y., and H. Hoogstraal. 1973. *Haemaphysalis (Kaiseriana) mageshimaensis* sp. n. (Ixodoidea: Ixodidae), a Japanese deer parasite with bisexual and parthenogenetic reproduction. *J. Parasitol.,* **59:**569–78.

Salimov, B. 1971. (The epizootological importance of ants in dicrocoeliosis.) *Veterinariia* No. 6, pp. 63–65.

Samšiňák, K.; E. Vobrázková; L. Malis; P. Palička; and K. Zitek. 1974. To the possible spread of scabies through bed linen. *Folia Parasitol.,* **21:**89–91.

Sanders, D. A. 1940. A *Musca domestica* and *Hippelates* flies, vectors of bovine mastitis. *Science,* **92:**286.

Sankaran, T., and H. Nagaraja. 1975. Observations on two sibling species of *Gryon* (Hymenoptera, Scelionidae) parasitic on Triatominae (Hemiptera) in India. *Bull. Entomol. Res.,* **65:**215–19.

Sardey, M. R., and S. R. Rao. 1973. Observations on the life-history and bionomics of *Rhipicephalus sanguineus* (Latreille, 1806) under different temperatures and humidities. *Indian Vet. J.,* **50:**863–67.

Sasa, M. 1961. Biology of chiggers. *Ann. Revu. Entomol.,* **6:**221–44.

Satchell, G. H., and R. A. Harrison. 1953. Experimental observations on the possibility of transmission of yaws by wound-feeding diptera, in western Samoa. *Trans. R. Soc. Trop. Med. Hyg.,* **47:**148–53.

Saugstad, E. S.; J. M. Dalrymple; and B. F. Eldridge. 1972. Ecology of arboviruses in a Maryland freshwater swamp. I. Population dynamics and habitat distribution of potential mosquito vectors. *Am. J. Epidemiol.,* **96:**114–22.

Savory, T. H., 1935. *The Arachnida.* Edward Arnold, London, 218 pp.

Scanlon, J. E.; J. A. Reid; and W. H. Cheong. 1968. Ecology of *Anopheles* vectors of malaria in the Oriental region. *Cah. ORSTOM, sér. Entomol. Méd.* **6:**237–46.

Schenone, H.; T. Letonja; and F. Knierim. 1975. (Some data on the venom apparatus of *Loxosceles laeta* and the toxicity of its venom for various animal species.) *Bol. Chil. Parasitol.,* **30:**37–42.

Schenone, H.; A. Rojas; H. Reyes; F. Villarroel; and G. Suarez. 1970. Prevalence of *Loxosceles laeta* in houses in central Chile. *Am. J. Trop. Med. Hyg.,* **19:**564–67.

Schlein, Y., and N. G. Gratz. 1973. Determination of the age of some Anopheline mosquitos by daily growth layers of skeletal apodemes. *Bull. WHO,* **49:**371–75.

Schmidt, G. D., and L. S. Roberts. 1977. *Foundations of Parasitology.* C. V. Mosby, St. Louis, 604 pp.

Schmidt, N. J., and E. H. Lennette. 1973. Advances in the serodiagnosis of viral infections. *Prog. Med. Virol.,* **15:**244–308.

Schnitzker, W. F. 1974. Grasshopper allergy. *Pediatrics,* **53:**280–81.

Schoenbaum, M., and K. Rauchbach. 1975. Morbidity and mortality of infant mice as a result of infestation by the mite *Ornithonyssus bacoti* (Hirst 1913). *Refuah Vet.,* **32:**24–6.

Schomberg, O., and D. E. Howell. 1955. Biological notes on *Tabanus abactor* Phil. and *aequalis* Hine. *J. Econ. Entomol.,* **48:**618–19.

Schreck, C. E. 1977. Techniques for the evaluation of insect repellents: a critical review. *Ann. Revu. Entomol.,* **22:**101–20.

Schreck, C. E.; I. H. Gilbert; D. E. Weidhaas; and K. H. Posey. 1970. Spatial action of mosquito repellents. *J. Econ. Entomol.,* **63:**1576–78.

Schreck, C. E.; K. Posey; and H. K. Gouck. 1975. Evaluation of the electrocutor grid trap baited with carbon dioxide against the stable fly, *Stomoxys calcitrans* (L.)(Diptera: Muscidae). *J. Med. Entomol.,* **12:**338–40.

Schreck, C. E.; N. Smith; and H. K. Gouck. 1976. Repellency of N,N-diethyl-M-toluamide (deet) and 2-hydroxymethyl cyclohexanecarboxylate against the deer flies *Chrysops atlan-*

ticus Pechuman and *Chrysops flavidus* Wiedemann. *J. Med. Entomol.*, **13**:115–18.

Schröder, P. 1975. (Pharoah's ant and its control with a special bait.) *Prakt. Schädlingsbekämpfer*, **27**:125–28.

Schultz, H. 1974. (A case of infection with snake mites and a brief review of other mites that occasionally infest man.) *Ugeskr. Laeger*, **136**:2752–53.

Schultz, M. G. 1968. A history of bartonellosis (Carrión's disease). *Am. J. Trop. Med. Hyg.*, **17**:503–15.

Schultz, M. G. 1974. Imported malaria. *Bull. WHO*, **50**:329–36.

Schulz-Key, H. 1975. (Investigations on the filariae of Cervids in southern Germany. 3. Filariae of roe deer [*Capreolus capreolus*] and fallow deer [*Dama dama*].) *Tropenmed. Parasitol.*, **26**:494–98.

Schumacher, H. H., and R. Hoeppli. 1963. Histochemical reactions to Trombiculid mites, with special reference to the structure and function of the "stylostome." *Tropenmed. Parasitol.*, **14**:192–208.

Scott, H. G., 1964. Human myiasis in North America (1952–1962 inclusive). *Fla. Entomol.*, **47**:255–61.

Scott, H. G., and R. M. Fine. 1967. A hazard for PCO's and their customers . . . straw itch mite dermatosis. *Pest Control*, **35**:19–20, 22–23.

Scott, H. G.; J. S. Wiseman; and C. J. Stojanovich. 1962. Collembola infesting man. *Ann. Entomol. Soc. Am.*, **55**:428–30.

Scudder, H. I. 1947. A new technique for sampling the density of housefly populations. *Public Health Rep.*, **62**:681–86.

Seal, S. C., and L. M. Bhattacharji. 1961. Epidemiological studies on plague in Calcutta. Part I. Bionomics of two species of rat fleas and distribution, densities and resistance of rodents in relation to the epidemiology of plague in Calcutta. *Indian J. Med. Res.*, **49**:974–007.

Seegar, W. S.; E. L. Schiller; W. J. L. Sladen; and M. Trpis. 1976. A Mallophaga, *Trinoton anserinum*, as a cyclodevelopmental vector for a heartworm parasite of waterfowl. *Science*, **194**:739–41.

Self, L. S.; R. J. Tonn; D. H. Bai; and H. K. Shin. 1973. Toxicity of agricultural pesticide applications to several mosquito species in South Korean rice fields. *Trop. Med.*, **15**:177–88.

Semtner, P. J., and J. A. Hair. 1973a. Distribution, seasonal abundance, and hosts of the Gulf Coast tick in Oklahoma. *Ann. Entomol. Soc. Am.*, **66**:1264–68.

Semtner, P. J., and J. A. Hair. 1973b. The ecology and behavior of the lone star tick (Acarina: Ixodidae). V. Abundance and seasonal distribution in different habitat types. *J. Med. Entomol.*, **10**:618–28.

Semtner, P. J., and J. A. Hair. 1975. Evaluation of CO_2-baited traps for survey of *Amblyomma maculatum* Koch and *Dermacentor variabilis* Say (Acarina: Ixodidae). *J. Med. Entomol.*, **12**:137–38.

Seneviratna, P.; N. Weerasinghe; and S. Ariyadasa. 1973. Transmission of *Haemobartonella canis* by the dog tick, *Rhipicephalus sanguineus*. *Res. Vet. Sci.*, **14**:112–14.

Sérié, C.; L. Andral; A. Poirier; A. Lindrec; and P. Neri. 1968. (Studies on yellow fever in Ethiopia. 6. Epidemiological study.) *Bull. WHO*, **38**:879–84.

Service, M. W. 1966. The replacement of *Culex nebulosus* Theo. by *Culex pipiens fatigans* Wied. (Diptera, Culicidae) in towns in Nigeria. *Bull. Entomol. Res.*, **56**:407–15.

Service, M. W. 1976. *Mosquito Ecology: Field Sampling Methods*. Applied Science Publishers, London, 583 pp.

Seventer, H. A. van. 1969. The disappearance of malaria in the Netherlands. Thesis. University of Amsterdam, Netherlands, 86 pp. (From *Rev. Appl. Entomol.*, B, **60**, Abst. 178.)

Sexton, D. J., and B. Haynes. 1975. Bird-mite infestation in a university hospital. *Lancet*, **1** (7904):445.

Seyedi-Rashti, M. A., and A. Nadim. 1975. Reestablishment of cutaneous leishmaniasis after cessation of anti-malaria spraying. *Trop. Geogr. Med.*, **27**:79–82.

Shanbaky, N. M., and G. M. Khalil. 1975. The subgenus *Persicargas* (Ixodoidea: Argasidae: *Argas*). 22. The effect of feeding on hormonal control of egg development in *Argas* (*Persicargas*) *arboreus*. *Exp. Parasitol.*, **37**:361–66.

Sharma, M. I. D.; J. C. Suri; N. L. Kalra, K. Mohan; and P. N. Swami. 1973. Epidemiological and entomological features of an outbreak of cutaneous leishmaniasis in Bikaner, Rajasthan, during 1971. *J. Communicable Dis.*, **5**:54–72.

Shaw, P. K.; T. Loren; and D. Juranek. 1976. Autochthonous dermal leishmaniasis in Texas. *Am. J. Trop. Med. Hyg.*, **25**:788–96.

Sheahan, B. J.; P. J. O'Connor; and E. P. Kelly. 1974. Improved weight gains in pigs following treatment for sarcoptic mange. *Vet. Rec.*, **95**:169–70.

Shelokov, A., and P. H. Peralta. 1967. Vesicular

stomatitis virus, Indiana type: an arbovirus infection of tropical sandflies and humans. *Am. J. Epidemiol.,* **86:**149–57.

Shemanchuk, J. A. 1972. Observations on the abundance and activity of three species of Ceratopogonidae (Diptera) in northeastern Alberta. *Can. Entomol.,* **104:**445–48.

Shemanchuk, J. A., and J. Weintraub. 1961. Observations on the biting and swarming of snipe flies (Diptera: *Symphoromyia*) in the foothills. of southern Alberta. *Mosq. News.* **21:**238–43.

Shephard, M. R. N. 1974. *Arthropods as Final Hosts of Nematodes and Nematomorphs. An Annotated Bibliography 1900–1972.* Tech. Comm. Commonwealth Inst. Helminth., No. 45, 248 pp.

Shim, J. C., and L. S. Self. 1973. Toxicity of agricultural chemicals to larvivorous fish in Korean rice fields. *Trop. Med.,* **15:**123–30.

Shinskii, G. E.; N. A. Sotnikova; Z. A. Rubtsova; T. S. Khamandritova; and A. G. Pashkina. 1973. (Comparative characteristics of some current methods of treating scabies.) *Vestn. Dermatol. Venerol.,* **8:**70–73.

Shirinov, F. B.; I. A. Farzaliev; Y. G. Alekperov; and A. A. Ibragimova. 1969. (The transmission of the virus of fowl-pox by *Argas persicus.*) *Veterinariia,* **46:**37–39.

Shirinov, F. B.; A. I. Ibragimova; and Z. G. Misirov. 1968. (The dissemination of the virus of fowl-pox by the mite *D. gallinae.*) *Veterinariia,* No. **4:**48–49.

Shishido, W. H., and D. E. Hardy. 1969. Myiasis of new-born calves in Hawaii. *Proc. Hawaiian Entomol. Soc.,* **20:**435–38.

Sholdt, L. L.; R. H. Grothaus; C. E. Schreck; and H. K. Gouck. 1975. Field studies using repellent-treated wide-mesh net jackets against *Glossina morsitans* in Ethiopia. *Afr. Med. J.,* **52:**277–83.

Shorey, H. H., and J. J. McKelvey, Jr. (eds.). 1977. *Chemical Control of Insect Behavior: Theory and Application.* Wiley, New York, 414 pp.

Shortt, H. E. 1973. *Babesia canis:* the life cycle and laboratory maintenance in its arthropod and mammalian hosts. *Int. J. Parasitol.,* **3:**119–48.

Shtakelberg, A. A. 1956. (*Synanthropic Flies of the Fauna of the U.S.S.R.*) Moscow, Acad. Sci. U.S.S.R., 164 pp.

Shulman, S. 1967. Allergic responses to insects. *Ann. Revu. Entomol.,* **12:**323–46.

Sikorowski, P. P., and C. H. Madison. 1967. Diseases of the Clear Lake Gnat (*Chaoborus astictopus*). *Mosq. News,* **28:**180–87.

Simmons, S. W. (ed.). 1959. *The Insecticide*

Dichlorodiphenyltrichloroethane and Its Significance. II. Human and Veterinary Medicine. Lehrb. Monogr. Geb. exakt. Wiss., chem. Reihe 10. Birkhäuser Verlag, Stuttgart, 570 pp.

Simond, P. L. 1898. La propagation de la peste. *Ann. Inst. Pasteur,* **12:**625.

Sinclair, A. N. 1975. The prevalence of fleece derangement in some Australian and New Zealand flocks infested with the sheep itch mite, *Psorergates ovis. N.Z. Vet. J.,* **23:**57–58.

Sinclair, A. N., and A. J. F. Gibson. 1975. Population changes of the itch mite, *Psorergates ovis,* after shearing. *N.Z. Vet. J.,* **23:**14.

Sinden, R. E., and P. C. C. Garnham. 1973. A comparative study on the ultrastructure of *Plasmodium* sporozoites within the oocyst and salivary glands, with particular reference to the incidence of the micropore. *Trans. R. Soc. Trop. Med. Hyg.,* **67:**631–37.

Singh, D. 1967. The *Culex pipiens fatigans* problem in South-East Asia: with special reference to urbanization. *Bull. WHO,* **37:**239–43.

Siverly, R. E. 1974. Observations on the biology of *Mansonia perturbans* (Walker) (Diptera, Culicidae) in Indiana. *Proc. Indiana Acad. Sci.,* **83:**216.

Sjogren, R. S. 1971. An effective repellent for *Leptoconops kerteszi* Kieffer (Diptera: Ceratopogonidae). *Mosq. News,* **31:**115–16.

Smith, A. 1961. Resting habits of *Anopheles gambiae* and *Anopheles pharoensis* in salt bush and in crevices in the ground. *Nature,* **190:**1220–21.

Smith, A. C. 1956. Fly prevention in dairy operations. *Calif. Vector News,* **3:**57, 59–60.

Smith, C. E. G. 1972. Human and animal ecological concepts behind the distribution, behaviour and control of yellow fever. *Bull. Soc. Pathol. Exot.,* **64:**683–89.

Smith, C. N., and others. 1963. *Factors Affecting the Protection Period of Mosquito Repellents.* Tech. Bull. USDA, No. 1285, 36 pp.

Smith, H. J. 1961. *Demodicidiosis in Large Domestic Animals. A Review.* Health Anim. Dir., Can. Ag., Ottawa, 56 pp.

Smith, J. J. B., and W. G. Friend. 1971: The application of split-screen television recording and electrical resistance measurement to the study of feeding in a blood-sucking insect (*Rhodnius prolixus*). *Can. Entomol.,* **103:** 167–72.

Smith, K. G. V. (ed.). 1973. *Insects and Other Arthropods of Medical Significance.* Pub. No. 720, British Museum (Natural History), London, 561 pp.

Smith, M. W. 1973. The effect of immersion in water on the immature stages of the Ixodid ticks—*Rhipicephalus appendiculatus* Neumann 1901 and *Amblyomma variegatum* Fabricius 1794. *Ann. Trop. Med. Parasitol.,* **67:**483–92.

Smith, R. D.; D. M. Sells; E. H. Stephenson; M. Ristic; and D. L. Huxsoll. 1976. Development of *Ehrlichia canis,* a causative agent of canine ehrlichiosis, in the tick *Rhipicephalus sanguineus* and its differentiation from a symbiotic rickettsia. *Am. J. Vet. Res.,* **37:**119–26.

Smith, R. F.; T. E. Mittler; and C. N. Smith (eds.). 1973. *History of Entomology.* Annual Reviews, Palo Alto, Calif., 517 pp.

Smith, T., and F. L. Kilbourne. 1893. *Investigations into the Nature, Causation, and Prevention of Texas or Southern Cattle Fever.* USDA Bur. An. Indust. Bull. 1, 301 pp.

Smith, T. A. 1969. The maturation of fly larvae following removal from the larval medium. *Calif. Vector Views,* **16:**73–78.

Smith, T. A., and D. D. Linsdale, 1967. First supplement to an annotated bibliography of the face fly, *Musca autumnalis* De Geer, in North America. *Calif. Vector Views,* **14:**74–76.

Smith, T. A.; D. D. Linsdale; and D. J. Burdick. 1966. An annotated bibliography of the face fly *Musca autumnalis* DeGeer, in North America. *Calif. Vector Views,* **13:**43–54.

Smith, W. W. 1957. Populations of the most abundant ectoparasites as related to prevalence of typhus antibodies of farm rats in an endemic murine typhus region. *Am. J. Trop. Med.,* **6:**581–89.

Snodgrass, R. E. 1935. *Principles of Insect Morphology.* McGraw-Hill, New York, 667 pp.

Snodgrass, R. E. 1944. *The Feeding Apparatus of Biting and Sucking Insects Affecting Man and Animals.* Smithsonian Misc. Coll., Vol. 102, No. 7 (Publ. No. 3773) Washington, D.C., 113 pp.

Snodgrass, R. E. 1946. *The Skeletal Anatomy of Fleas. (Siphonaptera).* Smithsonian Misc. Coll., Vol. 104, No. 18, Washington, D.C. (Publ. No. 3815), 89 pp.

Snodgrass, R. E. 1948. *The Feeding Organs of Arachnida, Including Mites and Ticks.* Smithsonian Misc. Coll., Vol. 110, No. 10. Washington, D.C. (Publ. No. 3944), 93 pp.

Snodgrass, R. E. 1952. *A Textbook of Arthropod Anatomy.* Comstock Publ. Co., Ithaca, N.Y., 363 pp.

Snodgrass, R. E. 1953. *The Metamorphosis of a Fly's Head.* Smithsonian Misc. Coll. 122 (3), 25 pp.

Snodgrass, R. E. 1959. *The Anatomical Life of the Mosquito.* Smithsonian Misc. Coll. 139 (8), 87 pp.

Sobey, W. R., and D. Conolly. 1971. Myxomatosis: the introduction of the European rabbit flea *Spilopsyllus cuniculi* (Dale) into wild rabbit populations in Australia. *J. Hyg.,* **69:**331–46.

Sobotka, A. K.; M. D. Valentine; A. W. Benton; and L. M. Lichtenstein. 1974. Allergy to insect stings. I. Diagnosis of IgE-mediated Hymenoptera sensitivity by venom-induced histamine release. *J. Allergy Clin. Immunol.,* **53:**170–84.

Soldatkin, I. S.; V. B. Rodnikovsky; and Y. V. Rudenchik. 1975. Development of plague epizootic as a statistical model. *Folia Parasitol.,* **22:**171–76.

Sonenshine, D. E. 1972. Ecology of the American dog tick, *Dermacentor variabilis,* in a study area in Virginia. 1. Studies on population dynamics using radioecological methods. *Ann. Entomol. Soc. Am.,* **65:**1164–75.

Sonenshine, D. E. 1974. Vector population dynamics in relation to tick-borne arboviruses: a review. *Phytopathology,* **64:**1060–71.

Sonenshine, D. E., and G. F. Levy. 1971. The ecology of the lone star tick, *Amblyomma americanum* (L.) in two contrasting habitats in Virginia (Acarina: Ixodidae). *J. Med. Entomol.,* **8:**623–35.

Sonenshine, D. E., and G. F. Levy. 1972. Ecology of the American dog tick, *Dermacentor variabilis,* in a study area in Virginia. 2. Distribution in relation to vegetative types. *Ann. Entomol. Soc. Am.,* **65:**1175–82.

Sonenshine, D. E.; A. H. Peters; and G. F. Levy. 1972. Rocky Mountain spotted fever in relation to vegetation in the eastern United States, 1951–1971. *Am. J. Epidemiol.,* **96:**59–69.

Sonenshine, D. E.; C. E. Yunker; C. M. Clifford; G. M. Clark; and J. A. Rudbach. 1976. Contributions to the ecology of Colorado tick fever virus. 2. Population dynamics and host utilization of immature stages of the Rocky Mountain wood tick, *Dermacentor andersoni. J. Med. Entomol.,* **12:**651–56.

Sönnichsen, N., and H. Barthelmes. 1976. (Epidemiological and immunological investigations on human scabies.) *Angew. Parasitol.,* **17:** 65–70.

Soper, F. L. 1936. Jungle fever: A new epidemiological entity in South America. *Rev. Hyg., Saude Pub.,* **10:**107–44.

Soper, F. L., and D. B. Wilson, 1943. *Anopheles gambiae in Brazil 1930–1940.* Rockefeller Foundation, New York, 262 pp.

Southcott, R. V. 1976. Arachnidism and allied syndromes in the Australian region. *Rec. Adelaide Childr. Hosp.*, **1**:97–186.

Southwood, T. R. E.; G. Murdie; M. Yasuno; R. J. Tonn; and P. M. Reader. 1972. Studies on the life budget of *Aedes aegypti* in Wat Samphaya, Bangkok, Thailand. *Bull. WHO*, **46**:211–16.

Spencer, R. W. 1972. A mechanical approach toward control of the greenhead fly. *Public Works*, **103**:90–92.

Spencer, T. S.; R. K. Shimmin; and R. F. Schoeppner. 1975. Field test of repellents against the valley black gnat, *Leptoconops carteri* Hoffman (Diptera: Ceratopogonidae). *Calif. Vector Views*, **22**:5–7.

Spieksma, F. T. M., and M. I. A. Spieksma-Boezeman. 1967. The mite fauna of house dust with particular reference to the house-dust mite *Dermatophagoides pteronyssinus* (Trouessart, 1897) (Psoroptidae: Sarcoptiformes). *Acarologia*, **9**:226–41.

Spielman, A. 1976. Human babesiosis on Nantucket Island: transmission by nymphal Ixodes ticks. *Am. J. Trop. Med. Hyg.*, **25**:784–87.

Spielman, A., and H. W. Levi. 1970. Probable envenomization by *Chiracanthium mildei*, a spider found in houses. *Am. J. Trop. Med. Hyg.*, **19**:729–32.

Spielman, A.; J. Piesman; and P. Etkind. 1977. Epizootology of human babesiosis. *J. N.Y. Entomol. Soc.*, **85**:214–16.

Spielman, A., and J. Wong. 1973. Studies on autogeny in natural populations of *Culex pipiens* III. Midsummer preparation for hibernation in anautogenous populations. *J. Med. Entomol.*, **10**:319–24.

Spielman, A., and J. Wong. 1974. Dietary factors stimulating oogenesis in *Aedes aegypti*. *Biol. Bull.*, **147**:433–42.

Srivastava, G. C. 1975. The intensity of infection in naturally infected *Formica pratensis* with the metacercariae of *Dicrocoelium dendriticum* in relation to their size. *J. Helmithol.*, **49**:57–64.

Stahnke, H. L. 1967. Scorpiology. *Turtox News*, **45**:218–23.

Stahnke, H. L. 1971. Some observations of the genus *Centruroides* (Buthidae, Scorpionida). *Entomol. News*, **82**:281–307.

Stampa, S. 1969. The control of *Ixodes rubicundus*, Neumann (1904) by alteration of its environment. *Rev. Appl. Entomol.*, Ser. B., **60** abst. No. 2300.

Stampa, S. 1971. The ecology of the Karoo paralysis tick (*Ixodes rubicundus* Neumann 1904):

microclimatological investigations. *Fort Hare Papers*, **5**:49–69.

Standfast, H. A., and A. L. Dyce. 1968. Attacks on cattle by mosquitoes and biting midges. *Aust. Vet. J.*, **44**:585–86.

Stanley, N. F., and M. P. Alpers (eds.). 1975. *Man-made Lakes and Human Health*. Academic Press, New York, 495 pp.

Stark, H. E., and V. I. Miles. 1962. Ecological studies of wild rodent plague in the San Francisco Bay area of California. VI. The relative abundance of certain flea species and their host relationships on co-existing wild and domestic rodents. *Am. J. Trop. Med.*, **11**:525–34.

Starratt, A. N., and C. E. Osgood. 1972. An oviposition pheromone of the mosquito *Culex tarsalis:* diglyceride composition of the active fraction. *Biochim. Biophys. Acta*, **280**:187–93.

Steelman, C. D. 1976. Effects of external and internal arthropod parasites on domestic livestock production. *Annu. Rev. Entomol.*, **21**:155–78.

Steelman, C. D., and A. R. Colmer. 1970. Some effects of organic wastes on aquatic insects in impounded habitats. *Ann. Entomol. Soc. Am.*, **63**:397–400.

Steelman, C. D.; J. E. Farlow; T. P. Breaud; and P. E. Schilling. 1975. Effects of growth regulators on *Psorophora columbiae* (Dyar and Knab) and non-target aquatic insect species in rice fields. *Mosq. News*, **35**:67–76.

Steinhaus, E. A. (ed.). 1963. *Insect Pathology, An Advanced Treatise*. Academic Press, New York, Vol. 1, 661 pp. Vol. 2, 689 pp.

Stelmaszyk, Z. J. 1975. (The technique of infecting ticks [Ixodidae] per anum). *Wiad. Parazytol.*, **21**:29–36.

Stephen, L. E. 1966. *Pig Trypanosomiasis in Tropical Africa*. Rev. Ser. Commonw. Bur. Anim. Hlth., No. 8, 65 pp.

Stephens, J. W. W., and H. B. Fantham. 1910. On the peculiar morphology of a trypanosome from a case of sleeping sickness and the possibility of its being a new species (*T. rhodesiense*). *Proc. R. Soc. London*, Ser. B, **83**:28–33.

Sterling, C. R.; M. Aikawa; and J. P. Vanderberg. 1973. The passage of *Plasmodium berghei* sporozoites through the salivary glands of *Anopheles stephensi:* an electron microscope study. *J. Parasitol.*, **59**:593–605.

Sterling, H. R.; R. G. Price; and K. O. Furr. 1972. Laboratory evaluation of insecticides on various surfaces and at various intervals for control of the brown recluse spider. *J. Econ. Entomol.*, **65**:1071–73.

Stewart, G. H. 1972. Dermatophilosis: a skin

disease of animals and man. Part I. *Vet. Rec.*, **91:**537–44.

Stiles, G. W. 1939. *Anaplasmosis in Cattle.* U.S. Dept. Agr. Circ. No. 154 (revised) 10 pp.

Stirrat, H. J.; J. McLintock; G. W. Schwindt; and K. R. Depner. 1955. Bacteria associated with wild and laboratory-reared horn flies, *Siphona irritans* (L.)(Diptera: Muscidae). *J. Parasitol.*, **41:**398–406.

Stoenescu, D.; A. Clipa; A. Gheorghiteanu; N. Curatu; P. Tanase; C. Toma; and S. Teodorescu. 1972. (On the clinical epizootiology and treatment of *Demodex canis* infestation.) *Arch. Vet.*, **9:**113–25.

Stoffolano, J. G., Jr. 1970. Nematodes associated with the genus *Musca* (Diptera: Muscidae). *Bull. Entomol. Soc. Am.*, **16:**194–203.

Stone, A. 1938. *The Horseflies of the Subfamily Tabanidae of the Nearctic Region.* USDA Misc. Pub. 305, 171 pp.

Stone, A. 1964. *Simuliidae and Thaumateidae. Insects of Connecticut. Part VI. The Diptera or True Flies of Connecticut.* State Geol. Nat. Hist. Survey Conn., Bull. 97, 126 pp.

Stone, A., *et al.* 1965. *A Catalog of the Diptera of America North of Mexico.* USDA Handbook 276. 1,696 pp.

Stone, B. F. 1963. Parthenogenesis in the cattle tick, *Boophilus microplus. Nature,* **200:**1233.

Stone, B. F.; P. Atkinson; and C. O. Knowles. 1974. Formamidine structure and detachment of the cattle tick *Boophilus microplus. Pestic. Biochem. Phys.,* **4:**407–16.

Stout, I. J.; C. M. Clifford; J. E. Keirans; and R. W. Portman. 1971. *Dermacentor variabilis* (Say) (Acarina: Ixodidae) established in southeastern Washington and northern Idaho. *J. Med. Entomol.,* **8:**132–47.

Strandtmann, R. W., and G. W. Wharton. 1958. *A Manual of Mesostigmatid Mites Parasitic on Vertebrates.* Institute of Acarology, College Park, Maryland. Contrib. No. 4, 330 pp.

Strenger, A. 1973. (On the biology of larval nourishment of *Ctenocephalides felis felis* B.) *Zool. Jahrbücher, Systematik, Ökol. Geog. Tiere,* **100:**64–80.

Strickland, R. K.; R. R. Gerrish; J. L. Hourrigan; and G. O. Schubert. 1976 (revision). *Ticks of Veterinary Importance.* Animal and Plant Health Inspection Service, USDA Ag. Handbook 485.

Strickland, R. K.; R. R. Gerrish; T. P. Kistner; and F. E. Kellogg. 1970. The white-tailed deer, *Odocoileus virginianus,* a new host for *Psoroptes* sp. *J. Parasitol.,* **56:**1038.

Stuckenberg, B. R. 1973. The Athericidae, a new family of lower Brachycera (Diptera). *Ann. Natal Mus.,* **21:**649–73.

Sudia, W. D., and R. W. Chamberlain. 1962. Battery-operated light trap, an improved model. *Mosq. News,* **22:**126–29.

Sudia, W. D., and R. W. Chamberlain. 1974. *Collection and Processing of Medically Important Arthropods for Arbovirus Isolation.* U.S. Dept. Health, Education, and Welfare, CDC, Atlanta, 29 pp.

Sudia, W. D.; V. F. Newhouse; C. H. Calisher; and R. W. Chamberlain. 1971. California group arboviruses: isolations from mosquitoes in North America. *Mosq. News,* **31:**576–600.

Sullivan, W. N.; R. Pal; J. W. Wright; J. C. Azurin; R. Okamoto; J. U. Mcuire; and R. M. Waters. 1972. Worldwide studies on aircraft disinsection at "blocks away." *Bull WHO,* **46:**485–91.

Sullivan, W. N.; M. S. Schechter; C. M. Amyx; and E. E. Crooks. 1972. Gas-propelled aerosols and micronized dusts for control of insects in aircraft. 1. Test protocol. *J. Econ. Entomol.,* **65:**1442–44.

Suter, P. R. 1964. Biologie von *Echidnophaga gallinacea* (Westw.) und Vergleich mit andern Verhaltenstypen bei Flöhen. *Acta Trop.,* **21:**193–238.

Sutherland, D. W. S. 1978. *Common Names of Insects and Related Organisms. (1978 Revision).* Special Pub. 78-1, Entomol. Soc. Am., 132 pp.

Sutherland, S. K. 1972a. The Sydney funnel-web spider (*Atrax robustus*). 1. A review of published studies on the crude venom. *Med. J. Aust.,* **2:**528–30.

Sutherland, S. K. 1972b. The Sydney funnel-web spider (*Atrax robustus*). 3. A review of some clinical records of human envenomation. *Med. J. Aust.,* **2:**643–47.

Sutherst, R. W., and R. H. Wharton. 1971. Preliminary considerations of a population model for *Boophilus microplus* in Australia. *Proceedings 3rd International Congress Acarology,* pp. 797–801.

Suzuki, T., and F. Sone. 1975. Filarial infection in vector mosquitos after mass drug administration in Western Samoa. *Trop. Med.,* **16:**147–56.

Swaminath, C. S.; H. E. Shortt; and L. A. R. Anderson. 1942. Transmission of Indian kala-azar to man by the bites of *Phlebotomus argentipes,* Ann. and Brun. *Indian J. Med.,* **30:**473–77.

Swaroop, S. 1959. Statistical considerations and

methodology in malaria eradication. WHO/MAL/240.

Sweatman, G. K. 1957. Life history, non specificity, and revision of the genus *Chorioptes*, a parasitic mite of herbivores. *Can. J. Zool.*, **35**:641–89.

Sweatman, G. K. 1958. On the life history and validity of the species in *Psoroptes*, a genus of mange mites. *Can. J. Zool.*, **36**:905–29.

Sweatman, G.; G. Tomey; and G. Katul. 1976. A technique for the continuous recording of tick feeding electrograms and temperature by telemetry from free-ranging cattle. *Int. J. Parasitol.*, **6**:299–305.

Swynnerton, C. F. M. 1936. The tsetse flies of East Africa. *Trans. R. Entomol. Soc.* (London), **84**:i–xxxvi, 1–579.

Syartman, M.; E. V. Potter; J. F. Finklea; T. Poon-King; and D. P. Earle. 1972. Epidemic scabies and acute glomerulonephritis in Trinidad. *Lancet*, **1**(7744):249–51.

Tager, A.; N. Lass; D. Gold; and J. Lengy. 1969. Studies on *Culex pipiens molestus* in Israel. IV. Desensitization attempts on children showing strophulus-like skin eruptions following bites of the mosquito. *Int. Arch. Allergy Appl. Immunol.*, **36**:408–14.

Talbot, N. 1968. Respiratory mites of poultry in Papua New Guinea. *Aust. Vet. J.*, **44**:530.

Talybov, A. N. 1975. (The life-span of *Ctenophthalmus wladimiri* Is.-Gurv., 1948 [Siphonaptera, Ctenophthalmidae] under laboratory conditions.) *Parazitologiia*, **9**:354–58.

Tarry, D. W., and A. C. Kirkwood. 1974. *Hydrotaea irritans:* the sheep headfly in Britain. *Br. Vet. J.*, **130**:180–88.

Tarry, D. W., and A. C. Kirkwood. 1976. Biology and development of the sheep headfly *Hydrotaea irritans* (Fall.) (Diptera, Muscidae). *Bull. Entomol. Res.*, **65**:587–94.

Tatchell, R. J. 1967. Salivary secretion in the cattle tick as a means of water elimination. *Nature*, **213**:940–41.

Tatchell, R. J., and D. E. Moorhouse. 1970. Neutrophils: their role in the formation of a tick feeding lesion. *Science*, **167**:1002–1003.

Tay, J., and F. Biagi. 1961. Accidentes por animales venenosos. *Rev. Fac. Med. U.N.A.M.*, **3**:811–19.

Teo, S. K., and J. S. Cheah. 1973. Severe reaction to the bite of the Triatomid bug (*Triatoma rubrofasciata*) in Singapore. *J. Trop. Med. Hyg.*, **76**:161–62.

Terent'ev, A. F. 1972. (The rôle of the blood-sucking insects in the ecology of the reindeer.) *Proceedings XIIIth. International Congress Entomology*, Moscow, **3**:260–61.

Terskikh, I. I.; D. B. Oboladze; and V. E. Sidorov. 1973. (Ticks as carriers of the causal agent of enzootic abortion of sheep.) *Veterinariia*, No. 2, 57–58.

Tesh, R. B.; B. N. Chaniotis; P. H. Peralta; and K. M. Johnson. 1974. Ecology of viruses isolated from Panamanian Phlebotomine sandflies. *Am. J. Trop. Med. Hyg.*, **23**:258–69.

Tesh, R. B., and D. J. Gubler. 1975. Laboratory studies of transovarial transmission of La Crosse and other arboviruses by *Aedes albopictus* and *Culex fatigans*. *Am. J. Trop. Med. Hyg.*, **24**:876–80.

Tesh, R. B.; P. H. Peralta; R. E. Shope; B. N. Chaniotis; and K. M. Johnson. 1975. Antigenic relationships among Phlebotomus fever group arboviruses and their implications for the epidemiology of sandfly fever. *Am. J. Trop. Med. Hyg.*, **24**:135–44.

Testi, B., and F. de Michelis: 1972. (An interesting parasitological discovery in monkeys imported into Italy for laboratory use.) *Zooprofilassi*, **27**:353–69.

Tewari, S. C.; I. P. Singh; and S. P. Kaduskar. 1974. A note on aetiological study of pox in swine. *Indian J. Anim. Sci.*, **44**:220–21.

Thatcher, V. E. 1968. Arboreal breeding sites of Phlebotomine sandflies in Panama. *Ann. Entomol. Soc. Am.*, **61**:1141–43.

Theiler, G. 1964. Ecogeographical aspects of tick distribution. In Davis, D. H. S. (ed.), *Ecological Studies in South Africa*, W. Junk, The Hague, pp. 284–300.

Theiler, G., and L. E. Salisbury. 1958. Zoological survey of the Union of South Africa. Tick survey X, XI. *Onderstepoort J. Vet. Res.*, **27**:599–610.

Theodor, O. 1958. Psychodidae—Phlebotominae. In Lindner, *Flieg. Palaeark. Reg.* **9c**: 1–55.

Theodor, O. 1965. On the classification of the American Phlebotominae. *J. Med. Entomol.*, **2**:171–97.

Théodoridès, J. 1950. The parasitological, medical and veterinary importance of Coleoptera. *Acta Trop.*, **7**:48–60.

Thiel, N. 1974. (*Q Fever and its Geographical Distribution with Special Reference to the Countries of Eastern and South-eastern Europe.*) Osteuropastudien der Hochschulen des Landes Hessen, 1, 65, 165 pp.

Thomas, L. A., 1963. Distribution of the virus of western equine encephalomyelitis in the mosquito vector, *Culex tarsalis*. *Am. J. Hyg.*, **78**:150–65.

Thompson, R. S.; W. Burdorfer; R. Russell; and B. J. Francis. 1969. Outbreak of tick-borne relapsing fever in Spokane County, Washington. *J.A.M.A.*, **210:**1045–50.

Thompson, W. H., and B. J. Beaty. 1977. Venereal transmission of La Crosse (California encephalitis) arbovirus in *Aedes triseriatus* mosquitoes. *Science*, **196:**530–31.

Thorp, R. W., and W. D. Woodson. 1945. *Black Widow, America's Most Poisonous Spider*. University of North Carolina Press, Chapel Hill, 222 pp.

Tod, M. E.; D. E. Jacobs; and A. M. Dunn. 1971. Mechanisms for the dispersal of parasitic nematode larvae. 1. Psychodid flies as transport hosts. *Helminthology*, **45:**133–37.

Trager, W. 1975. On the cultivation of *Trypanosoma vivax:* a tale of two visits in Nigeria. *J. Parasitol.*, **61:**3–11.

Traub, R. 1972a. Notes on fleas and the ecology of plague. *J. Med. Entomol.*, **9:**603.

Traub, R. 1972b. The relationship between the spines, combs and other skeletal features of fleas (Siphonaptera) and the vestiture, affinities and habits of their hosts. *J. Med. Entomol.*, **9:**601.

Traub, R. 1972c. The zoogeography of fleas (Siphonaptera) as supporting the theory of continental drift. *J. Med. Entomol.*, **9:**584–89.

Traub, R., and M. A. C. Dowling. 1961. The duration of efficacy of the insecticide dieldrin against the chigger vectors of scrub typhus in Malaya. *J. Econ. Entomol.*, **54:**654–59.

Traub, R., and C. L. Wisseman, 1968. Ecological considerations in scrub typhus. 3. Methods of control. *Bull. WHO*, **39:**231–37.

Traub, R., and C. L. Wisseman. 1974. The ecology of chigger-borne rickettsiosis (scrub typhus). *J. Med. Entomol.*, **11:**237–303.

Traub, R.; C. L. Wisseman; and N. Ahmad. 1967. The occurrence of scrub typhus infection in unusual habitats in West Pakistan. *Trans. R. Soc. Trop. Med. Hyg.*, **61:**23–53.

Traub, R.; C. L. Wisseman; and A. Farhang-Azad. 1978. The ecology of murine typhus—a critical review. *Trop. Dis. Bull.*, **75:**237–317.

Trejos, A.; R. Trejos; and R. Zeledón. 1971. (Arachnidism by *Phoneutria* in Costa Rica [Araneae: Ctenidae].) *Rev. Biol. Trop.*, **19:**241–48.

Trimble, R. M., and S. M. Smith. 1975. A bibliography of *Toxorhynchites rutilus* (Coquillett) (Diptera: Culicidae). *Mosq. Syst.*, **7:**115–25.

Triplett, R. F. 1973. Snesitivity to the imported fire ant: successful treatment with immunotherapy. *South. Med. J.*, **66:**477–480.

Trpiš, M. 1960. (*Mosquitoes of Rice Fields and Possibilities of their Control* [*Ecological Study*].) *Biol. Práce*, **6**(3), 137 pp.

Tsuji, H, and S. Ono. 1970. Laboratory evaluation of several bait factors against the German cockroach, *Blattella germanica* (L.). *Jpn. J. San. Zool.*, **20:**240–47 (1969).

Tu, A. T. 1977. *Venoms: Chemistry and Molecular Biology*. Wiley, New York, 560 pp.

Tugwell, P.; E. C. Burns; and J. W. Turner. 1969. Brahman breeding as a factor affecting the attractiveness or repellency of cattle to the horn fly. *J. Econ. Entomol.*, **62:**56–57.

Tully, J. G.; R. F. Whitcomb; H. F. Clark; and D. L. Williamson. 1977. Pathogenic mycoplasmas, cultivation and vertebrate pathogenicity of a new *Spiroplasma*. *Science*, **195:** 892–94.

Turner, K. J.; B. A. Baldo; R. F. Carter; and H. R. Kerr, 1975. Sudden infant death syndrome in South Australia. Measurement of serum IgE antibodies to three common allergens. *Med. J. Aust.*, **2:**855–59.

Turner, R. W.; M. Supalin; and A. P. Soeharto. 1974. Dynamics of the plague transmission cycle in Central Java (ecology of potential flea vectors). *Bull. Penelitian Kesehatan*, **2:**15–37.

Turner, W. J. 1976. *Fannia thelaziae*, a new species of eye-frequenting fly of the *benjamini* group from California and description of *F. conspicua* female (Diptera: Muscidae). *Pan-Pac. Entomol.*, **52:**234–41.

Turner, W. J. 1978. A case of severe human allergic reaction to bites of *Symphoromyia* (Diptera: Rhagionidae). *J. Med. Entomol.*, **15:**138–39.

Ueba, N., et al. 1972. Natural infection of swine by Japanese encephalitis virus and its modification by vaccination. Abst. Hyg. (1973)**48:**919.

Uetz, G. W. 1973. Envenomation by the spider *Trachelas tranquillus* (Hentz) (Araneae: Clubionidae). *J. Med. Entomol.*, **10:**227.

Ungureanu, E. M. 1972. Methods for dissecting dry insects and insects preserved in fixative solutions or by refrigeration. *Bull. WHO*, **47:**239–44.

Urlic, V.; D. Heneberg; N. Heneberg; A. Catipovic; R. Stajanovic; and J. Bakic. 1971. (Biological and epidemiological investigations of an endemic focus of murine typhus in Yugoslavia.) *Trop. Dis. Bull.*, **69**(11), 2140.

USDA. 1972. *Venezuelan Equine Encephalomyelitis, a National Emergency*. USDA, APHS-81-1, 24 pp.

USDA. 1973. *The Origin and Spread of Venezuelan Equine Encephalomyelitis*. USDA, APHIS 91–10, 51 pp.

USDA. 1976. *Controlling Chiggers*. USDA, ARS, *Home and Garden Bull*. No. 137, 10 pp.

USDHEW. 1966. *Pictorial Keys to Arthropods*,

Reptiles, Birds and Mammals. U.S. Department of Health, Education, and Welfare, PHS, CDC, Atlanta, Ga., 192 pp.

Usinger, R. L. 1934. Bloodsucking among phytophagous Hemiptera. *Can. Entomol.*, **66:** 97–100.

Usinger, R. L. 1944. *The Triatominae of North and Central America and the West Indies and Their Public Health Significance*. U.S. Public Health Bull. No. 288, 83 pp.

Usinger, R. L. 1966. *Monograph of Cimicidae (Hemiptera-Heteroptera)*. Thomas Say Foundation (Entomol. Soc. Am.), Vol. 7, 585 pp.

Uspenskaya, I. G. 1974. (Ecological differences in *Ixodes ricinus* L. in different parts of its distributional area.) Parazity Zhivonykh: Rastenii, No. 10, 123–33.

Vago, C. (ed.). 1972. *Invertebrate Tissue Culture*. Vol. II. Academic Press, London, 415 pp.

Vaivanijkul, P., and F. H. Haramoto. 1968. The biology of *Pyemotes boylei* Krczal (Acarina: Pyemotidae). *Proc. Hawaiian Entomol. Soc.*, **20:**443–54.

Van Emden, F. I. 1954. *Handbooks for the Identification of British Insects. Diptera Cyclorrhapha, Calyptrata* (1). *Section* (a). *Tachinidae and Calliphoridae*. Royal Entomological Society, London, 133 pp.

Vargas, L. 1972. (Key for the identification of the genera of mosquitoes of the Americas, using female characters.) *Bol Inform. Direccion Malariol. Saneamiento Ambiental*, **12:**204–206.

Vargas, L. 1974. Bilingual key to the New World genera of mosquitoes (Diptera: Culicidae) based upon fourth stage larvae. *Calif. Vector Views*, **21:**15–18.

Varma, M. G. R. 1962. Transmission of relapsing fever spirochetes by ticks. *Symp. Zool. Soc.* (London), **6:**61–82.

Varma, M. G. R. 1964. The acarology of louping ill. *Acarologia*, **6:**241–54.

Vasenin, A. A.; V. M. Kogan; G. A. Nekipelova; V. A. Leonov; L. F. Gerasimova; B. V. Shikharbeev; E. T. Shaparova; and V. T. Sherstov. 1975. (The epidemiological characteristics of foci of tick-borne encephalitis in transitional landscapes of Pre-Baikalia.) *Med. Parazitol. (Mosk.)*, **44:**521–24.

Vashkevich, R. B. 1972. (*Oedemagena tarandi* L.—vector of brucellosis.) *Veterinariia*, No. 4, 46–47.

Vater, G. 1973. (Virus transmission by mosquitoes in central Europe.) *Biol. Rundeschau*, **11:**234–38.

Velimirovic, B. 1974a. (Investigations on the epidemiology and control of plague in South Vietnam. Part I; Part II.) *Zentralbl. Bakteriol. (Orig. A.)*, **228:**482–508.

Velimirovic, B. 1974b. A review of the global epidemiological situation of plague (caused by *Yersinia pestis*) since the last Congress of Tropical Medicine and Malaria in 1968. *Zentralb. Bakteriol. (Orig. A)*, **229:**127–33.

Venters, D.; M. Spencer; and S. H. Christian. 1974. Field and laboratory methods in malaria entomology. *Papua New Guinea Med. J.*, **17:**36–41.

Ventura, A. K., and N. J. Ehrenkranz. 1975. Detection of Venezuelan equine encephalitis virus in rural communities of southern Florida by exposure of sentinel hamsters. *Am. J. Trop. Med. Hyg.*, **24:**715–17.

Vercammen-Grandjean, P. H. 1963. Valuable taxonomic characters of Trombiculidae, including correlations between larvae and nymphs. In Naegele, J. A. (ed.), *Advances in Acarology*, **1:**399–407.

Verhulst, A. 1976. *Tunga penetrans (Sarcopsylla penetrans)* as a cause of agalactia in sows in the Republic of Zaïre. *Vet. Rec.*, **98:**384.

Villalobos, D., and V. Neuman. 1975. (A study of the anti-scabies action of 5% thiabendazole used topically.) *Bol. Chil. Parasitol.*, **30:**2–5.

Vinograd, I. A.; I. A. Krasovskaya; G. A. Sidorova; A. A. Sazonov; R. Bosh; G. Robin; P. Ravis; J. Meno; D. K. L'vov; and L. Gonidek. 1975. (The isolation of Bhanja arbovirus from *Boophilus decoloratus* ticks in Cameroon.) Vopr. Virusol., No. 1, 63–67.

Vinogradskaya, O. N. (ed.). 1972. (*Guide to the Control of Insects and Acarina that are Carriers of the Agents of Diseases of Man.*) Meditsina, Moscow, 248 pp.

Vinson, S. B., and F. W. Platt. 1974. Third generation pesticides: the potential for the development of resistance by insects. *J. Agric. Food Chem.*, **22:**356–60.

Vitzthum, H. G. 1943. *Acarina.* 7 Lieferung. *Dr. H. G. Bronn's Klassen und Ordnungen des Thierreichs*, **5:**913–1011. Akademische Verlagsgesellschaft. Becker und Erler Kom. Ges.

Vockeroth, J. R. 1954. Notes on the identities and distribution of *Aedes* species of northern Canada, with a key to the females (Diptera: Culicidae). *Can. Entomol.*, **86:**241–55.

Volzhinskiĭ, D. V.; G. Y. Gromova; and A. Y. Alashkin. 1974. (Some data on the experimental study of tick-borne relapsing fever.) *Parazitol. Sbornik*, **26:**223–37.

Voorhorst, R. 1972. To what extent are house-dust mites (*Dermatophagoides*) responsible for complaints in asthma patients? *Allerg. Immunol.*, **18:**9–18.

Votava, C. L.; F. C. Rabalais; and D. C. Ashley. 1974. Transmission of *Dipetalonema viteae* by hyperparasitism in *Ornithodorus tartakovskyi*. *J. Parasitol.*, **60:**479.

Waddy, B. B. 1969. Prospects for the control of onchocerciasis in Africa: with special reference to the Volta River Basin. *Bull. WHO*, **40:**843–58.

Wagland, B. M. 1975. Host resistance to cattle tick (*Boophilus microplus*) in Brahman (*Bos indicus*) cattle. I. Responses of previously unexposed cattle to four infestations with 20,000 larvae. *Aust. J. Ag. Res.*, **26:**1073–80.

Wagner, R. E., and D. A. Reierson. 1975. Clothing for protection against venomous Hymenoptera. *J. Econ. Entomol.*, **68:**126–28.

Wagner-Jevseenko, O. 1958. Fortpflanzung bei *Ornithodorus moubata* und genitale Uebertragung von *Borrelia duttoni*. *Acta Trop.*, **15:**118–68.

Waldbauer, G. P. 1962. The mouth parts of female *Psorophora ciliata* (Diptera, Culicidae) with a new interpretation of the functions of the labial muscles. *J. Morphol.*, **111:**201–15.

Waldron, W. G. 1962. The role of the entomologist in delusory parasitosis (entomophobia). *Bull. Entomol. Soc. Am.*, **8:** 81–83.

Waldron, W. G. 1972. The entomologist and illusions of parasitosis. *Calif. Med.*, **117:** 76–78.

Waldron, W. G.; M. B. Madon; and T. Suddarth. 1975. Observations on the occurrence and ecology of *Loxosceles laeta* (Araneae: Scytodidae) in Los Angeles County, California. *Calif. Vector Views*, **22:**29–36.

Wall, W. J., Jr., and V. M. Marganian. 1973. Control of salt marsh *Culicoides* and *Tabanus* larvae in small plots with granular organophosphorus pesticides, and the direct effect on other fauna. *Mosq. News*, **33:**88–93.

Wallis, R. C.; J. J. Howard; A. J. Main; C. Frazier; and C. Hayes. 1974. An increase of *Culiseta melanura* coinciding with an epizootic of eastern equine encephalitis in Connecticut. *Mosq. News*, **34:**63–65.

Wallwork, J. A., and J. G. Rodriguez. 1961. Ecological studies on oribatid mites with particular reference to their role as intermediate hosts of Anaplocephalid cestodes. *J. Econ. Entomol.*, **54:**701–705.

Walton, G. A. 1962. The *Ornithodoros moubata* superspecies problem in relation to human relapsing fever epidemiology. *Symp. Zool. Soc. London*, No. 6, pp. 83–156.

Walton, G. A. 1964. The *Ornithodoros* "moubata" group of ticks in Africa. Control problems and implications. *J. Med. Entomol.*, **1:**53–64.

Walton, G. A. 1967. Relative status in Britain and Ireland of louping ill encephalitis virus. *J. Med. Entomol.*, **4:**161–67.

Wanstall, J. C., and I. S. de la Lande. 1974. Fractionation of bulldog ant venom. *Toxicon*, **12:**649–55.

Ward, R. A. 1962. Preservation of mosquitoes for malarial oocyst and sporozoite dissections. *Mosq. News*, **22:**306–307.

Ward, R. A., and J. E. Scanlon (eds.). 1970. Conference on anopheline biology and malaria eradication. *Misc. Pub. Entomol. Soc. Am.*, **7**(1):1–196.

Ward, R. D. 1977. New world leishmaniasis: A review of the epidemiological changes in the last three decades. *Proceedings XV International Congress Entomology*, pp. 505–22.

Warner, R. E. 1968. The role of introduced diseases in the extinction of the endemic Hawaiian avifauna. *Condor*, **70:**101–20.

Warren, M.; D. E. Eyles; and R. H. Wharton. 1962. Primate malaria infections in *Mansonia uniformis, Mosq. News*, **22:**303–304.

Warren, M., and R. H. Wharton. 1963. The vectors of simian malaria: Identity biology, and geographical distribution. *J. Parasitol.*, **49:** 829–904.

Washino, R. K., and J. G. Else. 1972. Identification of blood meals of hematophagous arthropods by the hemoglobin crystallization method. *Am. J. Trop. Med. Hyg.*, **21:**120–22.

Waterhouse, D. F. (ed.). 1970. *The Insects of Australia*. Melbourne University Press, Melbourne, 1029 pp.

Waterhouse, D. F. (ed.). 1974. *The Insects of Australia*. Supplement. Melbourne University Press, Melbourne, 146 pp.

Waters, K. S. 1972. Pasture spelling for tick control. *Queensland Ag. J.*, **98:**170–75.

Watt, D. D.; D. R. Babin; and R. V. Mlejnek. 1974. The protein neurotoxins in scorpion and Elapid snake venoms. *J. Agric. Food Chem.*, **22:**43–51.

Watt, J., and D. R. Lindsay. 1948. Diarrheal disease control studies. I. Effect of fly control in a high morbidity area. *Public Health Rep.*, **63:**1319–34.

Watt, J. C. 1971. The toxic effects of the bite of a Clubionid spider. *N.Z. Entomol.*, **5:**87–90.

Watt, K. E. F. 1962. Use of mathematics in population ecology. *Annu. Rev. Entomol.*, **7:** 243–60.

Watts, D. M., and B. F. Eldridge. 1975. Transovarial transmission of arboviruses by mosquitoes: a review. *Med. Biol.*, **53:**271–78.

Watts, D. M.; S. Pantuwatana; G. R. DeFoliart; T. M. Yuill; and W. H. Thompson. 1973. Transovarial transmission of LaCrosse virus (California encephalitis group) in the mosquito, *Aedes triseriatus. Science,* **182:**1140–41.

Watts, D. M.; W. H. Thompson; T. M. Yuill; G. R. DeFoliart; and R. P. Hanson. 1974. Overwintering of La Crosse virus in *Aedes triseriatus. Am. J. Trop. Med. Hyg.*, **23:** 694–700.

Webber, L. A., and J. D. Edman. 1972. Antimosquito behaviour of Ciconiiform birds. *Anim. Behav.*, **20:**228–32.

Webber, R. H. 1975. Theoretical considerations in the vector control of filariasis. *Southeast Asian J. Trop. Med. Public Health,* **6:**544–48.

Weidhaas, D. E. 1974. Simplified models of population dynamics of mosquitoes related to control technology. *J. Econ. Entomol.,* **67:**620–24.

Weidhaas, D. E.; S. G. Breeland; C. S. Lofgren; D. A. Dame; and R. Kaiser. 1974. Release of chemosterilized males for the control of *Anopheles albimanus* in El Salvador. IV. Dynamics of the test population. *Am. J. Trop. Med. Hyg.,* **23:**298–308.

Weidhaas, D. E.; R. S. Patterson; C. S. Lofgren; and H. Ford. 1971. Bionomics of a population of *Culex pipiens quinquefasciatus* Say. *Mosq. News,* **31:**177–82.

Weidhaas, D. E.; B. J. Smittle; R. S. Patterson; H. R. Ford; and C. S. Lofgren. 1973. Survival, reproductive capacity, and migration of adult *Culex pipiens quinquefasciatus* Say. *Mosq. News,* **33:**83–87.

Weinmann, C. J.; J. R. Anderson; P. Rubtzoff; G. Connolly; and W. M. Longhurst. 1974. Eyeworms and face flies in California. *Calif. Ag.,* **28**(11):4–5.

Weinmann, C. J., and R. Garcia. 1974. Canine heartworm in California, with observations on *Aedes sierrensis* as a potential vector. *Calif. Vector Views,* **21:**45–50.

Weisbroth, S. H.; S. Friedman; M. Powell; and S. Scher. 1974. The parasitic ecology of the rodent mite *Myobia musculi.* I. Grooming factors. *Lab. Anim. Sci.,* **24:**510–16.

Weisel, G. F. 1952. Animal names, anatomical terms and some ethnozoology of the Flathead Indians. *J. Wash. Acad. Sci.,* **42:**345–55.

Weiser, J. 1969. *An Atlas of Insect Diseases.* Irish University Press, Shannon, 43 pp.

Weiss, E. (ed.). 1971. *Arthropod Cell Cultures and their Application to the Study of Viruses.* Current Topics Microbiology and Immunology, Vol. 55, 288 pp.

Weitz, B. 1956. Identification of blood meals of blood-sucking arthropods. *Bull. WHO,* **15:** 473–90.

Weitz, B. 1963. The feeding habits of *Glossina. Bull. WHO,* **28:**711–29.

Welch, H. E., and I. A. Rubtsov. 1965. Mermithids (Nematoda: Mermithidae) parasitic in blackflies (Insecta: Simuliidae). Taxonomy and bionomics of *Gastromermis boophthorae. Can. Entomol.,* **97:**581–96.

Weller, B., and G. M. Graham. 1930. Relapsing fever in Central Texas. *J.A.M.A.,* **95:**1834–35.

West, L. C. 1951. *The House Fly. Its Natural History, Medical Importance and Control.* Cornell Pub. Co., Ithaca, N.Y., 584 pp.

West, L. S., and O. B. Peters. 1973. *An Annotated Bibliography of Musca domestica Linnaeus.* Dawsons, Folkestone, UK, 743 pp.

Wetzel, H. 1969. (Oculovascular myiasis of domestic animals in South and South-West Africa.) *Berl. Munch. Tieraerzt. Wochenschr.,* **82:**330–32.

Weyer, F. 1960. Biological relationships between lice and microbial agents. *Annu. Rev. Entomol.,* **5:**405–20.

Weyer, F. 1964. Experimentelle Übertragung von Rickettsien auf Artropoden. *Z. Tropenmed. Parasitol.,* **15:**131–38.

Weyer, F. 1968. (On the problem of the maintenance and conservation of strains of relapsing fever in the laboratory.) *Z. Tropenmed. Parasitol.,* **19:**344–51.

Wharton, G. W. 1976. House dust mites. *J. Med. Entomol.,* **12:**577–621.

Wharton, G. W., and H. S. Fuller. 1952. *A Manual of the Chiggers.* Entomological Society Washington, D.C., Mem. 4, 185 pp.

Wharton, R. H.; K. L. S. Harley; P. R. Wilkinson; K. B. Utech; and B. M. Kelley. 1969. A comparison of cattle tick control by pasture spelling, planned dipping, and tick-resistant cattle. *Aust. J. Agric. Res.,* **20:**783–97.

Whisler, H. C.; S. L. Zebold; and J. A. Shemanchuk. 1975. Life history of *Coelomomyces psorophorae. Proc. Natl. Acad. Sci. USA,* **72:**693–96.

Whitaker, J. O., Jr., and N. Wilson. 1974. Host and distribution lists of mites (Acari), parasitic and phoretic, in the hair of wild mammals of

North America, north of Mexico. *Am. Midl. Nat.*, **91**:1–67.

White, G. B. 1974. *Anopheles gambiae* complex and disease transmission in Africa. *Trans. R. Soc. Trop. Med. Hyg.*, **68**:278–98.

Whitfield, S. G.; F. A. Murphy; and W. D. Sudia. 1971. Eastern equine encephalitis virus: an electron microscopic study of *Aedes triseriatus* (Say) salivary gland infection. *Virology*, **43**:110–22.

Whitfield, S. G.; F. A. Murphy; and W. D. Sudia. 1973. St. Louis encephalitis virus: an ultrastructural study of infection in a mosquito vector. *Virology*, **56**:70–87.

Whitsel, R. H., and R. F. Schoeppner. 1966. Summary of a study of the biology and control of the valley black gnat, *Leptoconops torrens* Townsend (Diptera: Ceratopogonidae). *Calif. Vector Views*, **13**:17–25.

Whitten, M. J., and G. G. Foster. 1975. Genetical methods of pest control. *Annu. Rev. Entomol.*, **20**:461–76.

WHO. 1963. *Terminology of Malaria and of Malaria Eradication*. WHO, Geneva, Switzerland.

WHO. 1969. *African Trypanosomiasis. Report of a Joint FAO/WHO Expert Committee*. Tech. Rep. Ser. No. 434, 79 pp.

WHO. 1970. *Insecticide Resistance and Vector Control*. Tech. Rep. Ser. No. 443, 280 pp.

WHO. 1972a. *Manual Planning for Malaria Eradication and Malaria Control Programmes*. ME/72.10, 180 pp.

WHO. 1972b. *Vector Control in International Health*. WHO, Geneva, 144 pp.

WHO. 1972c. *Vector Ecology*. Tech. Rep. Ser. No. 501, 38 pp.

WHO. 1973a. *Manual on Larval Control Operations in Malaria Programmes*. WHO, Geneva, 199 pp.

WHO. 1973b, *Specifications for Pesticides Used in Public Health*, 4th ed. WHO, Geneva, 333 pp.

WHO. 1973c. Technical guide for a system of plague surveillance. *Weekly Epid. Rec.*, **48**:149–60.

WHO. 1973d. *The Use of Viruses for the Control of Insect Pests and Disease Vectors. Report of a Joint FAO/WHO Meeting on Insect Viruses, Geneva, 22–27 November 1972*. Tech. Rep. Ser., No. 531, 48 pp.

WHO. 1974a. *WHO Expert Committee on Filariasis, Third Report*. Tech. Rep. Ser. No. 542, 54 pp.

WHO. 1974b. *Ecology and Control of Rodents of Public Health Importance*. Tech. Rep. Ser. No. 553, 42 pp.

WHO. 1974c. *Equipment for Vector Control*. Geneva, ISBN 92-4-154035-4, 179 pp.

WHO. 1975a. *Tentative Instructions for Determining the Susceptibility or Resistance of Cockroaches to Insecticides*. (WHO/VBC/75.593), 4 pp.

WHO. 1975b. *Instructions for Determining the Susceptibility or Resistance of Reduviid Bugs to Organochlorine Insecticides*. (WHO/VBC/75.587), 5 pp.

WHO. 1975c. *Instructions for Determining the Susceptibility or Resistance of Body Lice and Head Lice to Insecticides*. (WHO/VBC/75.587), 5 pp.

WHO. 1975d. *Instructions for Determining the Susceptibility or Resistance of Adult Ticks to Insecticides*. (WHO/VBC/75.592), 5 pp.

WHO. 1975e. *Developments in Malaria Immunology. Report of a WHO Scientific Group*. Tech. Rep. Ser. No. 579, 68 pp.

WHO. 1976. *Resistance of Vectors and Reservoirs of Disease to Pesticides. Twenty-second Report of the WHO Expert Committee on Insecticides*. Tech. Rep. Ser., No. 585, 88 pp.

Wilcocks, C., and P. H. Manson-Bahr. 1972. *Manson's Tropical Diseases*, 17th ed., Williams & Wilkins, Baltimore, 1164 pp.

Wilkinson, P. R. 1953. Observations on the sensory physiology and behaviour of larvae of the cattle tick, *Boophilus microplus* (Can.)(Ixodidae). *Aust. J. Zool.*, **1**:345–56.

Wilkinson, P. R. 1957. The spelling of pasture in tick control. *Aust. J. Ag.*, **8**:414–23.

Wilkinson, P. R. 1967. The distribution of *Dermacentor* ticks in Canada in relation to bioclimatic zones. *Can. J. Zool.*, **45**:517–37.

Wilkinson, P. R. 1968. Phenology, behavior, and host-relations of *Dermacentor andersoni* Stiles in outdoor "rodentaria," and in nature. *Can. J. Zool.*, **46**:677–89.

Wilkinson, P. R. 1972. Sites of attachment of "prairie" and "montane" *Dermacentor andersoni* (Acarina: Ixodidae) on cattle. *J. Med. Entomol.*, **9**:133–37.

Wilkinson, P. R., and J. T. Wilson. 1959. Survival of cattle ticks in central Queensland pastures. *Aust. J. Ag. Res.*, **10**:129–43.

Willett, K. C. 1962. Recent advances in the study of tsetse-borne diseases. In Maramorosch, K., *Biological Transmission of Disease Agents*. Academic Press, New York, pp. 109–21.

Willett, K. C. 1963. Trypanosomiasis and the

tsetse fly problem in Africa. *Annu. Rev. Entomol.,* **8**:197–214.

Willett, K. C. 1967. *African Trypanosomiasis.* In Cockburn, A. (ed.), *Infectious Diseases, Their Evolution and Eradication.* Thomas, Springfield, Ill., 402 pp. (pp. 261–75.)

Williams, D. F. 1973. Sticky traps for sampling populations of *Stomoxys calcitrans. J. Econ. Entomol.,* **66**:1279–80.

Williams, J. A., and J. D. Edman. 1968. Occurrence of blood meals in two species of *Corethrella* in Florida. *Ann. Entomol. Soc. Am.,* **61**:1336.

Williams, J. E.; O. P. Young; and D. M. Watts. 1974. Relationship of density of *Culiseta melanura* mosquitoes to infection of wild birds with eastern and western equine encephalitis viruses. *J. Med. Entomol.,* **11**:352–54.

Williams, R. T.; P. J. Fullagar; C. Kogon; and C. Davey. 1973. Observations on a naturally occurring winter epizootic of myxomatosis at Canberra, Australia, in the presence of rabbit fleas (*Spilopsyllus cuniculi* Dale) and virulent myxoma virus. *J. Appl. Ecol.,* **10**:417–27.

Williams, R. W. 1946. A contribution to our knowledge of the bionomics of the common North American chigger, *Eutrombicula alfreddugesi* (Oudemans) with a description of a rapid collection method. *Am. J. Trop. Med.,* **26**:243–50.

Wilson, E. O. 1971. *The Insect Societies.* Belknap, Harvard University Press, Cambridge, Mass., 548 pp.

Wilson, V. J. 1972. Observations on the effect of dieldrin on wildlife during tsetse fly *Glossina morsitans* control operations in eastern Zambia. *Arnoldia,* **5**(34), 12 pp.

Wilson, V. J. 1975. Game and tsetse fly in eastern Zambia. *Occas. Papers Nat. Mus. Mon. Rhodesia, B,* **5**:339–404.

Wilton, D. P., and R. W. Fay. 1972. Air flow direction and velocity in light trap design. *Entomol. Exp. Appl.,* **15**:377–86.

Winget, R. N.; D. M. Rees; and G. C. Collett. 1969. Preliminary investigations of the brine flies in the Great Salt Lake, Utah. *Proceedings 22 Annu. Meet. Utah Mosq. Abatement Assoc.,* pp. 16–18.

Wingo, C. W. 1964. The status of *Loxosceles reclusa,* the brown recluse spider, as a public health problem. *Proc. N. Centr. Br. Entomol. Soc. Am.,* **19**:115–18.

Winney, R. 1975. Pyrethrins and pyrethroids in coils—a review. *Pyrethrum Post,* **13**:17–22.

Wirth, W. W., and W. R. Atchley. 1973. *A Review of the North American Leptoconops*

(Diptera: Ceratopogonidae). Graduate Studies, Texas Tech. University, No. 5, 57 pp.

Wolff, K. 1975. (Bird fleas as facultative ectoparasites of man.) *Schweiz. Rundschau Med.,* **64**:1173–75.

Wood, C. S. 1975. New evidence for a late introduction of malaria into the New World. *Curr. Anthropol.,* **16**:93–104.

Wood, D. M.; B. V. Peterson; D. M. Davies; and H. Gyorkos. 1963. The black flies (Diptera: Simuliidae) of Ontario. Part II. Larval identification, with descriptions and illustrations. *Proc. Entomol. Soc., Ontario,* **93**:99–129.

Woodard, D. B., and H. C. Chapman. 1970. Hatching of flood-water mosquitoes in screened and unscreened enclosures exposed to natural flooding of Louisiana salt marshes. *Mosq. News,* **30**:545–50.

Woodward, T. E., and W. F. Turner. 1915. *The Effect of the Cattle Tick upon the Milk Production of Dairy Cows.* USDA Bur. Anim. Ind., Bull. No. 147, 22 pp.

Wooff, W. R. 1967. Consolidation in tsetse reclamation. In International Science Council Trypanosomiasis Res., 11th Mtg., Org. Afr. Unity, No. 100, Publ. Sci. Tech. Res. Commn., pp. 141–48.

Woolley, T. A. 1961. A review of the phylogeny of mites. *Annu. Rev. Entomol.,* **6**:263–84.

Work, T. H. 1962. Kyasanur Forest disease—an infection of man by a virus of the RSS complex in India. *Proceedings 6th International Congress Tropical Medicine and Malaria,* **5**:180–96.

Wright, C. G.; H. C. McDaniel; H. E. Johnson; and C. E. Smith. 1973. American cockroach feeding in sewer access shafts on paraffin baits containing propoxur or Kepone plus a mold inhibitor. *J. Econ. Entomol.,* **66**:1277–78.

Wright, D. F. 1975. The systems approach as an aid to pest assessment. *N.Z. Entomol.,* **6**:24–28.

Wright, J. W. 1971. The WHO programme for the evaluation and testing of new insecticides. *Bull. WHO,* **44**:11–22.

Wright, J. W., and R. Pal (eds.). 1967. *Genetics of Insect Vectors of Disease.* Elsevier, Amsterdam, 794 pp.

Wu, L. T.; J. W. H. Chun; R. Pollitzer; and C. Y. Wu. 1946. *Plague, a Manual for Medical and Public Health Workers.* Shanghai, Weishengshu Nat. Quar. Service, 547 pp.

Xu, J. J., and L. C. Feng. 1975. (Studies on the *Anopheles hyrcanus* group of mosquitoes in China.) *Acta Entomol. Sinica,* **18**:77–104.

Yanes, de Ramirez, A. I. 1974. (Plan of a simulated model for the study of an Anopheline population.) *Rev. Fac. Agron. Univ. Cent. Venezuela,* **7:**75–88.

Yen, J. H., and A. R. Barr. 1973. The etiological agent of cytoplasmic incompatibility of *Culex pipiens. J. Invertebr. Pathol.,* **22:**242–50.

Yeo, R. R. 1972. The algae *Chara* and *Cladophora,* the problems they cause and their control. *Proc. Calif. Mosq. Cont. Assoc.,* **40:** 81–83.

Yinon, U.; A. Shulov; and J. Margalit. 1967. The hygroreaction of the larvae of the Oriental rat flea *Xenonpsylla cheopis* Rothsch. (Siphonaptera: Pulicidae). *Parasitology,* **57:**315–19.

Youdeowei, A. 1975. A simple technique for observing and collecting the saliva of tsetse flies (Diptera, Glossinidae). *Bull. Entomol. Res.,* **65:**65–67.

Young, E. 1975. Some important parasitic and other diseases of lion, *Panthera leo,* in the Kruger National Park. *J. S. Afr. Vet. Assoc.,* **46:**181–83.

Yu, H. S., and E. F. Legner. 1976. Regulation of aquatic Diptera by planaria. *Entomophaga,* **21:**3–12.

Yunker, C. E., and J. Cory. 1975. Plaque production by arboviruses in Singh's *Aedes albopictus* cells. *Appl. Microbiol.,* **29:**81–89.

Zahar, A. R. 1974. Review of the ecology of malaria vectors in the WHO Eastern Mediterranean Region. *Bull. WHO,* **50:**427–40.

Zavortink, T. J. 1964. *Mosquito Studies (Diptera, Culicidae) XXXVIII. The New World Species formerly Placed in Aedes (Finlaya).* Contrib. Am. Entomol. Inst., 8, No. 3, 206 pp.

Zeledón, R. 1974a. (The vectors of Chagas' disease in America.) *Simposio Internat. Sobre la Enfermedad de Chagas en America, Buenos Aires,* Dec. 1972; pp. 327–45.

Zeledón, R. 1974b. Epidemiology, modes of transmission and reservoir hosts of Chagas' disease. In Ciba Foundation, *Trypanosomiasis and Leishmaniasis with Special Reference to Chagas' Disease,* Symposium 20 (new series).

Elsevier, Amsterdam, 353 pp. (pp. 51–71.)

Zemskaya, A. A., and A. A. Pchelkina. 1967. (Gamasid mites and Q fever.) English Translation 296, Dep. Med. Zool., U.S. Nav. Med. Res. Unit no. 3, c/o Spanish Embassy, Cairo, U.A.R.

Zharov, A. A. 1973. (On the question of the geographical distribution of the mosquito *Aedes vexans* Meigen [Diptera; Culicidae]). *Med. Parazitol. (Mosk.), Bolezni,* **42:**11–16.

Zielke, E. 1973. (Investigations on the mechanism of transmission of filariae by mosquitoes.) *Tropenmed. Parasitol.,* **24:**32–35.

Zielke, E. 1975. (On the migration of the third stage larvae of *Wuchereria bancrofti* in *Anopheles gambiae.) Tropenmed. Parasitol.,* **26:**345–47.

Zielke, E. 1976. Studies on quantitative aspects of the transmission of *Wuchereria bancrofti. Tropenmed. Parasitol.,* **27:**160–64.

Zimmerman, E. C. 1948. *Insects of Hawaii.* Vol. 3, *Heteroptera.* University of Hawaii Press, Honolulu, 255 pp.

Zinsser, H. 1935. *Rats, Lice and History.* Little, Brown, Boston, 301 pp.

Zolotova, S. I., and B. M. Yakunin. 1973. (Development of the flea *Pluex irritans* L., 1758 in experimental conditions.) *Parazitologiia,* **7:**24–30.

Zulueta, J. de 1973. Malaria and Mediterranean history. *Parassitologia,* **15:**1–15.

Zumpt, F. 1965. *Myiasis in Man and Animals in the Old World.* Butterworths, London, 267 pp.

Zumpt, F. 1973. *The Stomoxyine Biting Flies of the World.* Gustav Fischer Verlag, Stuttgart, 175 pp.

Zumpt, F., and H. E. Paterson. 1953. Studies on the family Gasterophilidae with keys to the adults and maggots. *J. Entolmol. Soc. S. Afr.,* **16:**59–72.

Zurier, R. B.; H. Mitnick, D. Bloomgarden; and Weissmann. 1973. Effect of bee venom on experimental arthritis. *Ann. Rheum. Dis.,* **32:**466–70.

INDEX

Boldface numbers indicate primary discussion in text.
Italic numbers indicate pages on which illustrations appear.
Complete scientific names, including describers, are provided only for arthropods, and only in this index.

521